# Water Supply and Pollution Control

**Other HarperCollins books coauthored
by Warren Viessman, Jr.**

*Introduction to Hydrology,* third edition, by Viessman, Lewis, and
   Knapp
*Water Management, Technology and Institutions,* by Viessman
   and Welty

# Water Supply and Pollution Control

Warren Viessman, Jr.
University of Florida

Mark J. Hammer
Lincoln, Nebraska

## FIFTH EDITION

HarperCollins*CollegePublishers*

Sponsoring Editor: John Lencheck
Art Director: Julie Anderson
Project Coordination and Cover Design: Elm Street Publishing Services, Inc.
Cover Illustration/Photo: Comstock Inc./Comstock
Compositor: Bi-Comp, Inc.
Printer and Binder: R. R. Donnelley & Sons Company
Cover Printer: Lehigh Press Lithographers

**Water Supply and Pollution Control, Fifth Edition**

Library of Congress Cataloging-in-Publication Data

Viessman, Warren.
  Water supply and pollution control/Warren Viessman, Jr., Mark J.
  Hammer.—5th ed.
    p. cm.
  Includes bibliographical references and index.
  ISBN 0-06-500058-7
  1. Water-supply—Management. 2. Water—Purification. 3. Sewage—
  Purification. I. Hammer, Mark J., 1931–    . II. Title.
  TD353.V54   1992                                   92-17473
  628.1—dc20

92  93  94  95  9  8  7  6  5  4  3  2  1

To Bette and Audrey for their help and understanding

# Contents

# CHAPTER 10   Physical Treatment Processes   322

# CHAPTER 11   Chemical Treatment Processes   396

# Preface

This fifth edition of *Water Supply and Pollution Control* has been modernized in many ways. There is considerable new material on standards, treatment processes, water use and development, and sewer design. In particular, there are major revisions of the chapters, or sections, on water supply and use, water distribution, hydraulics and hydrology of sewer and storm drainage systems, water quality, physical treatment processes, chemical treatment processes, and biological treatment processes. New problems have been added, and greater clarity in presentation has been achieved. Consistent with the original intent of the book, the emphasis is on the application of scientific methods to problems associated with the development, movement, and treatment of water and wastewater. The book's tradition of presenting treatment processes in the context of what they can do, rather than in the context of water or wastewater treatment, is becoming more and more appropriate as we move toward the concept of total water management, recognizing that all waters are potential sources of supply. Water reuse is becoming an increasingly important consideration nationally, and additional material on this subject has been added. On the water supply side, more attention is paid to the sharing of water with natural systems and the impact of this sharing on the quantities of water available for traditional water-using sectors, including public water supply.

The authors wish to acknowledge the advice and assistance of students, professors, and practicing engineers who have reviewed and commented on previous editions. Particular recognition is given to those who helped prepare the manuscript for the fifth edition, namely Audrey Hammer and Bette Viessman. We are indebted to them for their perseverance and understanding and for the excellent quality of their work.

WARREN VIESSMAN, JR.
MARK J. HAMMER

# Chapter 1
# Introduction

The human search for pure water supplies must have begun in prehistoric times. Much of that earliest activity is subject to speculation. Some individuals may have led water where they wanted it through trenches dug in the earth. Later a hollow log was perhaps used as the first water pipe.

Thousands of years probably passed before our more recent ancestors learned to build cities and enjoy the convenience of water piped to the home and drains for water-carried wastes. Our earliest archeological records of central water supply and wastewater disposal date back about 5000 years, to Nippur of Sumeria. In the ruins of Nippur there is an arched drain with each stone being a wedge tapering downward into place [1].* Water was drawn from wells and cisterns. An extensive system of drainage conveyed the wastes from the palaces and residential districts of the city.

The earliest recorded knowledge of water treatment is in the Sanskrit medical lore and Egyptian wall inscriptions [2]. Sanskrit writings dating about 2000 B.C. tell how to purify foul water by boiling in copper vessels, exposure to sunlight, filtering through charcoal, and cooling in an earthen vessel.

There is nothing concerning water treatment in the biblical sanitary and hygienic code of the early Hebrews, although three incidents may be cited as examples of the importance of fresh water. At Morah, Moses is said to have sweetened bitter waters by casting into them a tree shown him by God [3]. During the wandering in the wilderness, the Lord commanded Moses to bring forth water by smiting a rock [4]. At a much later date, Elisha is said to have "healed unto this day" the spring water of Jericho by casting "salt" into it [5].

The earliest known apparatus for clarifying liquids was pictured on Egyptian walls in the fifteenth and thirteenth centuries B.C. The first picture, in a

---

* The numbers in square brackets refer to the references at the end of the chapter.

tomb of the reign of Amenhotep II (1447–1420 B.C.), represents the siphoning of either water or settled wine. A second picture, in the tomb of Rameses II (1300–1223 B.C.), shows the use of wick siphons in an Egyptian kitchen.

The first engineering report on water supply and treatment was made in A.D. 98 by Sextus Julius Frontinus, water commissioner of Rome. He produced two books on the water supply of Rome. In these he described a settling reservoir at the head of one of the aqueducts and pebble catchers built into most of the aqueducts. His writings were first translated into English by the noted hydraulic engineer Clemens Herschel, in 1899 [2].

In the eighth century A.D., an Arabian alchemist, Geber, wrote a rather specialized treatise on distillation that included various stills for water and other liquids.

The English philosopher Sir Francis Bacon wrote of his experiments on the purification of water by filtration, boiling, distillation, and clarification by coagulation. This work was published in 1627, one year after his death. Bacon also noted that clarifying water tends to improve health and increase the "pleasure of the eye."

The first known illustrated description of sand filters was published in 1685 by Luc Antonio Porzio, an Italian physician. He wrote a book on conserving the health of soldiers in camps, based on his experimence in the Austro-Turkish War. This was probably the earliest published work on mass sanitation. He described and illustrated the use of sand filters and sedimentation. Porzio also stated this his filtration method was the same as that of "those who built the Wells in the Palace of the Doges in Venice and in the Palace of Cardinal Sachett, at Rome" [2].

The oldest known archeological examples of water filtration are in Venice and the colonies it occupied. The ornate heads on the cisterns bear dates, but it is not known when the filters were placed. Venice, built on a series of islands, depended on catching and storing rainwater for its principal freshwater supply for over 1300 years. Cisterns were built, and many were connected with sand filters. The rainwater ran off the house tops to the streets, where it was collected in stone-grated catch basins and then filtered through sand into cisterns.

A comprehensive article on the water supply of Venice appeared in the *Practical Mechanics Journal* in 1863 [6]. The land area of Venice was 12.85 acres and the average yearly rainfall was 32 in. Nearly all of this rainfall was collected in 177 public and 1900 private cisterns. These cisterns provided a daily average supply of about 4.2 gallons per capita per day (gpcd). This low consumption was due in part to the absence of sewers, the practice of washing clothes in the lagoon, and the universal drinking of wine. The article explained in detail the construction of the cisterns. The cisterns were usually 10–12 ft deep. The earth was first excavated to the shape of a truncated inverted pyramid. Well-puddled clay was placed against the sides of the pit. A flat stone was placed in the bottom and a cylindrical wall was built from brick laid with open joints. The space between the clay walls and the central brick cylinder was filled with sand. The stone surfaces of the courtyards were sloped toward the cistern, where perforated stone blocks collected the water at the lowest point and discharged it to the filter sand. This water was always fresh and cool, with a temperature of about 52°F. These cisterns continued to be the principal water supply of Venice until about the sixteenth century.

Many experiments were conducted in the eighteenth and nineteenth centuries in England, France, Germany, and Russia. Henry Darcy patented filters in France and England in 1856 and anticipated all aspects of the American rapid-sand filter except coagulation. He appears to be the first to apply the laws of hydraulics to filter design [7]. The first filter to supply water to a whole town was completed at Paisley, Scotland, in 1804, but this water was carted to consumers [2]. In Glasgow, Scotland, filtered water was piped to consumers in 1807 [8].

In the United States little attention was given to water treatment until after the Civil War. Turbidity was not as urgent a problem as in Europe. The first filters were of the slow-sand type, similar to British design. About 1890 rapid-sand filters were developed in the United States, and coagulants were later introduced to increase their efficiency. These filters soon evolved to our present rapid-sand filters.

The drains and sewers of Nippur and Rome are among the great structures of antiquity. These drains were intended primarily to carry away runoff from storms and for the flushing of streets. There are specific instances where direct connections were made to private homes and palaces, but these were the exceptions, for most of the houses did not have such connections. The need for regular cleansing of the city and flushing of the sewers was well recognized by commissioner Frontinus of Rome, as indicated in his statement, "I desire that nobody shall conduct away any excess water without having received my permission or that of my representatives, for it is necessary that a part of the supply flowing from the water-castles shall be utilized not only for cleaning our city, but also for flushing the sewers."

It is astonishing to note that from the days of Frontinus to the middle of the nineteenth century there was no marked progress in sewerage. In 1842, after a fire destroyed the old section of the city of Hamburg, Germany, it was decided to rebuild this section of the city according to modern ideas of convenience. The work was entrusted to an English engineer, W. Lindley, who was far ahead of his time. He designed an excellent collection system that included many of the ideas now used. Unfortunately, the ideas of Lindley and their influence on public health were not recognized.

The history of the progress of sanitation in London probably affords a more typical picture of what took place in the middle of the nineteenth century. In 1847, following an outbreak of cholera in India that had begun to work westward, a royal commission was appointed to look into the sanitary conditions of London. This royal commission found that one of the major obstacles was the political structure, especially the lack of a central authority. The city of London was only a small part of the metropolitan area, comprising approximately 9.5% of the land area and less than 6% of the total population of approximately 2.5 million. This lack of a central authority made the execution of sewerage works all but impossible. The existing sewers were at different elevations, and in some instances the wastes would have had to flow uphill. In 1848 Parliament followed the advice of this commission and created the Metropolitan Commission of Sewers. That body and its successors produced reports that clearly showed the need for extensive sewerage works and other sanitary conditions [9]. Cholera appeared in London during the summer of 1848, and 14,600 deaths were recorded during 1849. In 1854 cholera claimed a mortality of 10,675 people in London. The connection was established between a contaminated water supply and spread of the disease, and it was determined that the absence of effective sewerage was a major hindrance in combatting the problem.

In 1855 Parliament passed an act "for the better local management of the metropolis," thereby providing the basis for the subsequent work of the Metropolitan Commission of Sewers, which soon after undertook the development of an adequate sewerage system. It will be noted that the sewerage system of London came as a result of the cholera epidemic, as was true of Paris.

The remedy for these foul conditions was to discharge human excrement into the existing storm sewers and install additional collection systems. This suggestion created the combined sewers of many older metropolitan areas. These storm drains had been constructed to discharge into the nearest watercourse. The addition of wastes to the small streams overtaxed the receiving capacities of the waters, and many of them were covered and converted into sewers. Much of the material was carried away from the point of entry into the drains, which in turn overtaxed the receiving waters. First the smaller and then the larger bodies of water began to ferment, creating a general health problem, especially during dry, hot weather. The solution has been the varying degrees of treatment currently practiced according to the capabilities of the receiving stream or lake to take the load.

The work on storm drainage in the United States closely paralleled that in Europe, especially England. Some difficulty was experienced because of the difference in the rainfall patterns in America from those of England. English rains are more frequent but less intense. In the United States storm drains must usually be larger for the same topographical conditions.

Today the enormous demands being placed on water supply and wastewater disposal facilities have necessitated the development and implementation of far broader concepts in environmental engineering than those envisioned only a few years ago. The standards for water quality have significantly increased concurrently with a marked decrease in raw-water quality. Evidence of water supply contamination by toxic and hazardous materials has become common, and concern about broad water-related environmental issues has heightened. As populations throughout the world multiply at an alarming rate, environmental control becomes a critical factor. Land and water management become increasingly important. Many European and Asian nations have reached the maximum populations that their land areas can bear comfortably. They are faced with the problem of providing for more people than their lands can conveniently support. The lesson is that populations increase, but water and land resources do not. Consequently, the use and control of these resources must be nearly perfect to maintain our way of life.

The charge to environmental engineers and scientists embraces scientific, technical, social, political, economic, and legal dimensions. The challenge is enormous. Few other scientific or engineering specialties face such an array of disparate, and at the same time, interrelated subject matter areas. The field is an exciting and demanding one. It requires the best human resources we have. The future of our world rests upon the decisions environmental scientists, engineers, and others will foster and on the actions that will flow from these decisions. Water supply and pollution control technology is a key foundation component of today's environmental decision-making processes, and its understanding and application can do much for the long-term benefit of society.

## REFERENCES

1. W. Durant, *Our Oriental Heritage* (New York: Simon and Schuster, 1954), p. 132.
2. M. N. Baker, *The Quest for Pure Water* (New York: American Water Works Association, 1949), pp. 1–3, 6–11.
3. Exodus 15:22–27.
4. Exodus 17:1–7.
5. 2 Kings 2:19–22 (King James version).
6. "The Water Cistern in Venice," *J. Franklin Inst*. 3rd Ser. 70 (1860): 372–373.
7. H. Darcy, *Les Fontaines Publiques de la Ville de Dijon; Distribution d'Eau et Filtrage des Eaux* (Paris: Victor Dalmont, 1856).
8. D. Mackain, "On the Supply of Water to the City of Glasgow," *Proc. Inst. Civil Engrs*. (*London*) 2 (1842–1843): 134–136.
9. First Report of the Metropolitan Sanitary Commission (London: 1848).

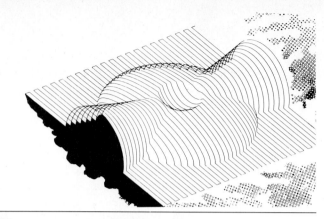

# Chapter 2
# Water Management: Institutions and Technology

Water management is multidimensional. It embraces planning, design, construction, operation, and maintenance. Its ingredients include technologic capability, societal attitudes, economic realities, political viewpoints, and environmental goals. Good water management will provide water of acceptable quality for purposes ranging from supplying municipal water needs to maintaining the life cycles of fish and wildlife. It includes both structural and nonstructural measures. It requires recognizing and taking advantage of interconnections between surface and groundwater bodies, exploiting the potential for coordinated use of existing facilities, acknowledging that water quantity and quality are a single issue, devising new ways to operate old systems, blending structural and nonstructural approaches, accepting that the expansiveness of water resources systems may require regional rather than local solutions to problems, and providing equity, insofar as possible, if not on a monetary basis, at least on a service basis, to all those affected. In concept, water management is simple; the trouble is that the boundaries of the physical systems that must be dealt with often differ markedly from the political boundaries that affect how water is used and developed. Furthermore, many historical, social, legal, and organizational factors have been narrowly focused and constrain, if not preclude, good water management.

## 2.1 FROM PROJECTS TO ISSUES

In the early twentieth century the construction of dams, waterways, water treatment plants, and wastewater treatment facilities was given priority. Irrigation works helped settle the West. Waterways improvements encouraged commerce and industry in populous areas of the East, South, and

Midwest. Municipal water and wastewater systems provided the basis for increasing urbanization and industrial growth in many localities. Now, however, most of the nation's rivers have been subjected to engineering controls, and many old water policies are no longer valid. Furthermore, numerous facilities constructed in the past are reaching their design lives, and the question of how to rehabilitate them is becoming important.

The maturity of our water infrastructure suggests that the exercise of good management practices be the basis for correcting deficiencies and making improvements. Broad issues, rather than the local interests that historically have been satisfied on a project-by-project basis, are moving to the forefront. A transition is underway. The question is whether it can hurdle the barriers created over the years.

Of interest is the fact that the National Water Commission's 1973 recommendations still stand as a model for modernizing water management policies [1]. The NWC's seven recurring themes are relevant to the subject of water supply and pollution control. They are:

Future water demands are not inevitable but are the result of policy decisions within the control of society. Good planning should be based on a range of plausible alternative futures.

National priorities are shifting from water resources development to restoring and enhancing water quality.

Water resources planning must be tied more closely to land use planning.

Water use efficiency should be emphasized, and policies to encourage wise water use and conservation practices should be promoted.

Sound economic principles should be incorporated in decisions on whether to build water projects. Beneficiaries should pay for the costs of the services they receive, and unjustified subsidies that distort allocation of scarce resources should be eliminated.

Laws and legal institutions should be reexamined in the light of contemporary water problems.

Development, management, and protection of water resources should be controlled at that level of government nearest the problem and most capable of effectively representing the vital interests involved.

## 2.2 INSTITUTIONS

The need for objective water management has long been recognized, but its implementation presents a problem. Successful application requires coordinating *human, agency, government,* and *special-group* interests and resolving conflicts among them. This institutional problem underlies most water issues. The influences of laws and regulations, political boundaries, agency missions, financing mechanisms, social customs, and the belief that water is free for the taking have all interacted to create a "water crisis" aura. The challenge is to bring about needed reforms in water-related institutions. If we are not able to do this, water shortages may be widespread, and correcting them could be costly and time consuming.

Our society is not an easy one to live in. The rate of technological change has been swift, and these changes are accelerating. Many institutions and policies, however, have not kept pace, leading to conflicts that must be resolved. Even when it appears that we know what to do, we often fail to do so owing to selfishness or an unwillingness to attack constraining influences. Entrenched institutions are hard to change, but they cannot be ignored. Like facilities, they must be kept current. The following are some of the most troublesome issues:

> The nonuniformity and sometimes conflicting aspects of state and federal water laws. At the federal level, Indian and federal reserved water rights issues are compelling. Uncertainties created by unquantified water rights constrain planning and inhibit investments in water management facilities.

> The failure of laws, agencies, and water users to recognize the interrelationships of surface waters and groundwaters. The use of water from one source often affects the availability of water from another.

> The general lack of groundwater and surface water use coordination.

> The separation by statutes and administrative processes of issues of water quality and water quantity. Water-quality decisions affect water quantity and vice versa, and yet these types of decisions are often made without regard for one another.

> The failure to recognize that water is not a free good. Users often pay far less than the cost of providing the water they use. Furthermore, fees collected for water services are sometimes channeled to purposes other than defraying new water project capital costs or meeting operation and maintenance expenses.

> The focus on individual water projects as opposed to comprehensive water resources goals. Historic approaches to water resources development have often been piecemeal and narrow in perspective. This has resulted in less than optimal regional systems of development and questionable payoffs from federal investments in facilities.

> The lack of effective national, state, and regional mechanisms for setting priorities for water resources investments. Good investment decisions require mechanisms to set water development priorities in the context of all options for water resources expenditures and within the broader framework of other needs for outlays of funds. The principal yardstick used at present is project-oriented benefit-cost analysis.

> The lack of mechanisms for ensuring that plans are implemented. A great deficiency in water resources planning systems has been the inability of planners to influence decisions through their plans. If plans are worth developing, they should be used in decision-making processes.

> The diffusion of legislative jurisdictions within a single level of government. Overlapping legislative committee jurisdictions, for example, foster inconsistencies in programs, duplication of efforts, and conflicts in management.

> The inability of federal, state, and local agencies to coordinate their programs.

The proliferation of regulations, some of which constrain rather than promote the effective use of the nation's waters.

Some institutional issues of particular relevance to the subject of this book are those related to land use, political boundaries, and appropriate payment for development of water sources and for transporting and treating water and wastes.

Land use and water use are inseparable. Almost everything done on the land surface affects the connected water resource in one way or another. Traditionally, land use and water management decisions have been made by different levels of government and/or agencies. The resulting conflicts and inconsistencies this creates are well known, but progress in overcoming them has been slow. Subjects of particular concern here include siting of landfills and industries, urbanization of rural areas, agricultural development and operations, and other land management practices.

Today, many locally rooted water resources issues have regional dimensions, and they must be dealt with in that context. Political subdivisions are often too small to deal effectively with resource management problems transcending their boundaries. The problem is that it is difficult, if not impossible, to manage a system properly if only some of its parts can be manipulated. Developing and implementing workable regional institutions is a challenge that water suppliers and wastewater managers must face.

There is a need for regional planning at all levels of government. Regional institutions can vary from authorities with broad powers to cooperative agreements among governmental units. There is no uniformly acceptable format, but given the right circumstances, regional arrangements for water and wastewater management are more likely to produce good results than the sum of independent actions by subregional entities. Regional organizations such as Florida's Water Management Districts, the Nebraska Natural Resource Districts, and the British Water Authorities are examples of formal regional institutions that work to address problems in a broad context [2, 3].

Finally, most Americans have been raised under the myth that water is a free good and that all are entitled to use it for any purpose, and in any manner, at little or no cost to themselves. This philosophy has resulted in subsidies for some water-using sectors and in a backlog of decaying infrastructures in others. It has also encouraged inefficient uses and generated excessive capital outlays for both water supply and wastewater works. It has fostered planning that generates, rather than manages, growth. Allocating and using water are costly processes. Both programs and facilities are involved. Questions must be raised relative to the costs to be incurred by various water management options and the benefits to be gained, and by whom, of implementing them. Sources of funds and economic incentives and disincentives must be considered. With federal funding cutbacks, state and local governments will have to bear an increasingly larger burden of costs. This could lead to greater efficiencies and promote more imaginative approaches to water management. At the same time, new and innovative financing approaches will have to be developed, and more attention will have to be paid to the worth of proposed projects and to nontraditional approaches to meeting water supply targets. Options for instituting efficient and equitable payments for water facilities and services include capital budgeting, establishment of user fees, creation of utilities, taxation, setting impact fees, and implementing pricing policies.

## 2.3 LEGAL CONSIDERATIONS

The competition for water, especially in regions of short supply, often gives rise to confrontations among users. Unfortunately, resolution of the conflicts created is often impeded by prevailing legal doctrines. Although water laws in some states have tried to integrate the actions of all interests, other states have not. Large-scale multiple-purpose projects have created novel problems of reshuffling rights between users and uses, between watersheds, between states, and between countries [4].

In many areas, opportunities for the development of new sources of water have been largely exhausted. It is becoming increasingly necessary for cities to acquire through legal action the rights to the additional water needed for growth. These additional waters may be available locally or conveyed to the places of use over considerable distances. For example, the Feather River project in California takes water from the northern Sierra Nevada and conveys it almost to the Mexican border. This project covers a distance of several hundred miles down the Sacramento Valley, up the San Joaquin Valley, over a mountain range, and then into southern California. Such transfers involve many legal and other institutional obstacles. The reader should understand that the solution of almost any water allocation problem, even if it is not on as grand a scale as the Feather River project, involves full understanding of the pertinent legal systems.

### State Water Rights Doctrines

Ownership is the right of one or more persons to possess and use property to the exclusion of others. Before rules can be prescribed governing the use or the rights to the use of water, its ownership must be determined.

**Riparian Rights**    The doctrine of riparian rights came to America from Roman law by several lines of descent. Under the philosophy of riparian rights the owner of land containing a natural stream or abutting a stream is entitled to receive the full natural flow of the stream without change in quality or quantity. The physical contact of land and water is an essential factor. The riparian owner is protected against the diversion of water except for domestic purposes upstream from his or her property and from the diversion of excess floodwaters toward his or her property. The riparian concept has been modified to provide for actual use of the water for beneficial purposes, such as for irrigation, industry, and the dilution of sewage effluent.

The riparian doctrine is generally followed in all states east of the Mississippi River. It is recognized to some degree in the six western states that extend from North Dakota to Texas on the 100th meridian and in the three states that border on the Pacific Ocean. In all other western states, it has been completely repudiated or has never been recognized. The supreme courts of New Mexico and Arizona have declared that the riparian doctrine never existed in those states.

Where the riparian doctrine is in effect, the owner of a tract of land contiguous to the channel of a surface stream, called riparian land, has certain rights to the water flowing in the stream. The owner of the land may divert what water he needs for domestic use. Water for irrigation and other commercial purposes must be reasonable with respect to the requirements of all others. This same principle applies to the ownership of land that overlies an underground stream or the underflow of a surface stream. In each case, the water right arises out of the physical contact of land and the water source. Limited use of the water on nonriparian land may be permitted in some states. This nonriparian use is considered to be without formal right rather than an exercise of the riparian right.


**The Appropriation Doctrine**    The appropriation doctrine has various antecedents, both ancient and modern. One view is that this principle had its roots in Roman civil law; another is that it developed from the local customs of the Southwest. The idea that it was brought to the New World by the Spaniards, who adopted it from Roman civil law, is supported by the fact that this doctrine materially influenced the early development of water law in several states in which the Spaniards and Mexicans settled.

The Mormons, in settling on public lands in Utah during the middle of the nineteenth century, developed their own system of appropriation. They diverted water from streams for irrigation. During this same period, gold was discovered in California on what were then Mexican public lands. The miners improvised local rules and regulations for taking and holding mining claims and claims for the use of water. Water was essential to hydraulic and placer mining, which was widely practiced in the area.

The principal feature of these appropriation doctrines is the concept of "first in time, first in line." The first to appropriate water for beneficial use has the right to as much water as is needed for that use, and this right must be completely satisfied prior to any other water use. The right to use water in excess of the amount required by the senior appropriator shifts to the appropriator next in line. This is true regardless of the appropriator's geographic position on the waterway. If water is flowing in a stream crossing a property, the property owner might not be allowed to divert any of the water for any purpose, even in periods of short water supply; only the priority of appropriation may be considered in establishing a water right.

In most western states water may be appropriated for use outside the watershed in which it naturally flows. The appropriation doctrine gives no preference to water uses on land contiguous to the water supply. Generally, the appropriators need not be the owners of the land in connection with which they propose to appropriate water. They must, however, be in lawful possession such as entrymen on public land or as lessees on private land to establish water rights.

The appropriative right is to a specific *amount* of water. The element of priority negates any obligation on the part of the appropriator to share the water with others. However, appropriators are not entitled to divert and use more water than can be put to beneficial use.

In the United States, the appropriation doctrine was first based on custom, which, in the absence of legislation, the courts recognized as controlling. Generally, throughout the West the early statutes codified prevailing local rules and regulations. Later, as development progressed and complications ensued, the appropriation doctrine became both statutory and judicial, with many high-court decisions and constitutional provisions.

**Groundwater Rights**   Groundwater poses an extremely difficult legal problem. Unlike the case of surface water, it is often difficult to determine the source and rates of recharge, the extent and variation of quality in storage, and the direction of water movement. Three basic rules cover the use of groundwater.

The first, or English, rule is one of absolute ownership. It allows the overlying landowner to take groundwater from the land at any time and in any quantity, regardless of the effect on the water table of a neighbor's land. Under this rule it would be possible for a landowner to exhaust the total groundwater supply of an area by heavy pumping. This rule has been qualified in some areas to limit the malicious and wasteful use of the water. The American rule, or rule of reasonable use, recognizes that the landowner has rights to the water under the land but that these rights may be limited. The rights to water are limited to its reasonable use in relationship to the overlying land. The third rule covering groundwater is the appropriation principle, whereby the water is allocated for specific uses.

The English doctrine is followed in some eastern and western states. The American principle of reasonable use is followed by many states. In several of the western states, the appropriative principle has been applied by statute, court decision, or both to the use of groundwater.

**Comparison of Doctrines**   Each of the doctrines has certain advantages and disadvantages. Furthermore, there has been a marked trend away from the point of view of the individual water user to that of large user groups encompassing whole river basins. In many instances, the present doctrines and projected uses are in direct conflict and are irreconcilable.

The appropriation doctrine provides for acquiring rights to water by putting it to beneficial use in accordance with procedures set forth in state statutes and judicial decisions. This assures the holders of the water the right to their full appropriative supply, in order of priority, whenever the water is available. Early priority holders thus have reasonable assurance of a water supply each year. This enables development to be predicated on a reasonably assured annual supply and allows major investments in water-dependent operations.

Water can be appropriated and stored for beneficial use on either a temporary or a seasonal basis. The operation of the doctrine allows a considerable measure of flexibility for changes in the exercise of water rights. The appropriative right is gained by use and lost by disuse. Therefore, theoretically, all water is available for beneficial use.

It is probable that the appropriation doctrine has allowed the multiple appropriation of the same water in some instances, and it is possible for a late priority to gain in time because of this condition. The problem arises where groundwaters and surface waters are appropriated and the excess has come about from groundwater storage. Late appropriators could use the percolating groundwater and materially diminish stream flow. An earlier (in point of time) water user is displaced therefore by a later appropriation. This problem is further complicated by state boundaries and the later developments of groundwater across a state line in areas where the groundwater is not subject to appropriation. Holders of later priorities have little assurance of water supply in most seasons, and the very late ones have no assurance except in very wet years.

Under the riparian doctrine, all landowners are assured of some water when it is available. This doctrine tends to freeze a large proportion of water to lands whether it is being used or not. This unused right can be held indefinitely without forfeiture and thus can materially affect the water-related development of an area. The riparian doctrine does not allow provision for storage of water that is essential for the full development of many water uses.

The unrestricted use of groundwater in many states is in direct conflict with both doctrines, as it is impossible to distinguish between groundwater and surface water under some conditions. Finally, there appears to be a need for some modification of existing doctrines to meet changing economic conditions in both the eastern and western portions of the United States.

## Acquiring Water Rights

The riparian right to water accrues when the land title passes. There is no formality of acquiring the water right other than that involved in acquiring the land. Rights to the use of percolating groundwater and vagrant surface water are secured in the same way.

Each of the 17 western states has a statutory procedure under which surface water may be appropriated, and most of the western states have some formal procedure for groundwater appropriation. Nearly all of the procedures contemplate applications to state officials for water. Permits or licenses are issued when the request is approved.

In some eastern states, legislation requires a permit from a state agency to take water for irrigation or other purposes from a watercourse or groundwater supply. It is doubtful in most states whether these permits establish any substantial water right. They have little effect on riparian rights, but they do provide some administrative restraint upon the exercise of water rights. They also provide a record of use that should be valuable in considering water rights legislation in the future.

Most problems associated with water use and development do not occur due to technologic inadequacies. Water can be transported almost anywhere if one has the ability to pay for it. More often, the problems we face arise from conflicts over allocation and the constraints on innovative solutions imposed by outdated, or opposing, laws and regulations. As a result, it seems inevitable that the interests of individual water users will come to be represented less and less by individual water rights acquired from the states and more and more by contracts with basin management districts that will hold mass water rights in trust for users.

## Federal Water Laws

The authority of the federal government over waters has been based on several sections of the U.S. Constitution. Article I, Section 8, Number 3, gives Congress authority over navigable waters that cross state boundaries or connect with the ocean. The term *navigation* has been rather loosely defined and could include most all waters in the broadest sense. Article IV, Section 3, Number 2, authorizes Congress to "dispose of and make all needful rules and regulations respecting the territory or other property belonging to the United States. . . ." The treaty powers of the president and Senate under Article VI, Number 2, have been used to designate and regulate water rights. Waters originating in the Rio Grande River in the United States are appropriated to Mexico according to a formula based on stream flow, as set up in a treaty between the United States and Mexico. Article I, Section 8, Number 1, states: "The Congress shall have power to lay and collect taxes, duties, imports, and excises, to pay the debts and provide for the common defense and general welfare of the United States. . . ." Basin development has contributed to the general welfare in the Tennessee Valley Authority and resulted in construction of the Wilson Dam on the Tennessee River during World War I to produce nitrate for ammunition.

A liberal interpretation of the Constitution allows for considerable federal regulation over the nation's waters. The U.S. Supreme Court has ruled that every state has the power, within its dominion, to change the rule of the common law referring to the rights of the riparian owner to the continuous natural flow of the stream and to permit appropriation of the water for such purposes as it deems wise [5].

**Federal Reserved Water Rights**  National parks, national forests, and other federal lands withdrawn from the public domain have rights to sufficient water to serve the purposes for which the reservations were intended. This is the federal reserved water right.

Originally, this notion applied only to Indian reservations. Later, the doctrine was extended to land reserved for national forests, parks, and recreational areas, national monuments, and wildlife refuges [2]. Reserved water rights may involve both surface water and groundwater. Only waters unappropriated at the time of the reservation are affected, however, and the federal government must compensate for damages to those holding rights predating those of the reservation.

Key issues are the quantity of water to be reserved and the use to which it is to be put. A 1978 court decision (*U.S.* v. *New Mexico*) held that reserved water may be used only for the original purpose of the federal reservation. This finding narrowed the scope of the reservation doctrine and in so doing removed some of the uncertainty associated with the doctrine. One other modifying influence on the doctrine is that the determination of federal reserved water rights may be made in state courts [2]. The fact that recent legislation and court decisions tend to restrict the reservation doctrine is of great importance to the states. A broad interpretation of the reservation doctrine could lead to confrontation with state constitutional or statutory provisions and generally interfere with the operation of state water rights systems. By integrating federal reserved water rights into existing state water rights systems, the limit of the rights can be determined, and planning uncertainties can be minimized.

**Tribal Water Rights**    The competition between Indian and non-Indian water rights poses some extraordinary problems. Most Indian reservations predate the large water development projects in the western United States, although the use of water in significant quantities by Indians has generally developed only in recent years.

The resource potentials of Indian reservations are enormous. In the northern Great Plains, for example, most Indian lands are underlain with large reserves of coal and other valuable minerals, many have outstanding recreational features, and many contain large areas suitable for agricultural development. Preliminary surveys indicate that Indian water requirements may constitute a significant portion of the annual flows in the Missouri River and its tributaries.

The tribes are concerned that water used for energy and other non-Indian development will adversely affect their water rights and lead to depletions of supplies critical for sustaining future economic developments on their reservations. They seek assurances that their water requirements will be properly considered in all planning scenarios.

Rational water planning is dependent on quantification of all existing and proposed water uses. Current studies of future water uses in the western United States have addressed the issue of Indian water requirements to greater or lesser degrees, but the fact remains that the quantities involved are often unknown or in dispute. Until this matter is resolved, estimates of future stream-flow depletions will be biased accordingly, and decisions on trade-offs with other users will be clouded.

## 2.4 TECHNOLOGY AND WATER MANAGEMENT

If the world's water problems could be solved without concern for political boundaries, social customs, agencies, laws, regulations, and politics, technical solutions would be the order of the day, but reality dictates that technology be exercised within institutional limits. Only those technical solutions that are politically feasible, socially acceptable, and legally permissible are likely to be implemented, short of lifting some or all of the constraining influences. To seek technical solutions without regard for their viability within institutional settings is to invite delay, added cost, or even total failure.

On the other hand, we do not have to accept that prevailing institutional systems are fixed forever. Factual analyses of the impact of existing constraints on addressing water issues must be provided. If benefits can be shown to increase as the result of modifying laws, changing regulations, developing new organizations, and so on, then prospects for reform will be enhanced. The approach taken should be to offer options that include both constrained and unconstrained solutions, documenting the benefits of each. In this way, options requiring institutional change can be weighed against those that do not, and choices can at least be informed. The tools of diagnosis and evaluation at our disposal permit prompt, in-depth evaluations of many courses of action. It is time to examine existing systems objectively to see if they can be operated more efficiently and, if so, what changes would be needed to bring this about.

A solution to the water supply problem of the Washington, D.C., metropolitan area (WMA) exemplifies what can be done. For years, political jurisdictions, special-interest groups, and agencies argued about plans for increasing the WMA water supply [2, 6]. Hundreds of proposals emerged, but none really gained support, mainly because the institutions that would have to implement them did not desire, or were unable to find, a common ground. Efforts to improve the region's water supply situation stagnated, but population growth did not. The result was that periodic low flows in the river became more and more troublesome.

Thoughts of an impending crisis spurred the desire to "do something," and in the late 1970s, a unique marriage of institutional cooperation and technological effort emerged. The Corps of Engineers, the states of Maryland and Virginia, the Interstate Commission on the Potomac River Basin, the Fairfax County Water Authority, the Washington Suburban Sanitary Commission, the Metropolitan Washington Area Council of Governments, and other key organizations worked together to provide a setting for coordination of river management policies that would benefit them all.

Analysts found a way to augment the region's water supply without the need for major new facilities. It involved dealing with a maze of engineering, social, environmental, economic, and political problems. The approach, adopted in 1982, was fresh and nontraditional. It focused on what could be done in the face of prevailing constraints. Modern techniques of systems analysis—linear programming, synthetic hydrology, statistical analysis, hydrologic modeling, and computer simulation—were merged to produce a predominantly nonstructural solution to the water supply problem. As a result, a problem dissolved that had existed for almost 30 years. What had appeared to be an impasse in regional cooperation was obliterated. Furthermore, between $200 million and $1 billion was potentially saved compared to previous alternatives. The environmental impact of the solution was minimal.

This example offers proof that nonstructural engineering alternatives, including better management of existing facilities, can be devised to solve difficult technical and institutional water resources problems at low economic and environmental costs. The coming of age of the use of modern systems analysis techniques for solving water resources problems is exemplified. Such approaches have great potential for solving problems nationwide, but to implement them we must consider the systems being dealt with in totality.

Finally, it must be recognized that, for management to be effective, sweeping institutional reforms may be required. Their nature will be determined by the scale of the problem being addressed, but it is clear that agency missions, federal and state laws, and citizen and special-interest group attitudes may all have to be tampered with. There will be winners and losers, but unless we can come to grips with those elements that influence our ability to provide more and better water, the actions taken will fall far short of those needed to do the job.

## 2.5 OUTLOOK FOR THE FUTURE

Most known water problems can be solved on the basis of our present capabilities. The problem is that constraining influences often force the wrong moves, ineffectual moves, or no moves at all. We have the analytical power to identify efficient and environmentally sound alternatives, but we have little ability to move special interests, whatever they may be, to compromise their kingdoms no matter how compelling the argument. In the past, water issues were often recognized in advance but were ignored for political and other reasons until they became crises. Unfortunately, we do not seem to be able to learn from these experiences. Problem solving under a crisis is destructive of both natural resources and the nation's economy. Fortunately, perhaps, we have overcome most historic calamities within reasonable periods of time through the massive application of resources. The problem is that as our resources are increasingly spread thinner, and as the number of people to be served and protected increases, problems that could have been accommodated 20 years ago with some ease could now result in a national catastrophe, one that could not be overcome quickly or without great human suffering. The changes needed to permit wise management of our water resources may be difficult to implement, but the alternative is forbidding. The severity of future water problems will depend heavily on our ability to change entrenched institutional elements and to introduce objectivity into water resources management. Unquestionably, management cannot be effective without some degree of control, and to permit that control, adjustments by governments and individuals will have to be made. On the other hand, reforms do not have to be destructive of all old beliefs and approaches; they can be built on them, as was done for the Potomac. What is needed is an approach with a wide focus, blending the physical and social elements already in place with the best prevailing technology.

On the bright side, societal attitudes toward water resources development and management have changed significantly in the last quarter-century. A new attitude on how water resources should be managed and what constitutes their beneficial use has emerged. Changes in viewpoint have infiltrated both federal and state laws and have played a major role in defining modern water management. Many states have taken strong environmental stands, and local regulations often impose more stringent conditions on water management practices than do federal laws. Significantly altered public perceptions have resulted, and extraordinary projects focused on environmental protection are now being actively promoted. If this momentum is sustained, a new era in water management will be the product.

## PROBLEMS

**2.1** Prepare a summary of the procedure for appropriating surface water in your state. How are these water rights administered?

**2.2** Prepare a summary of the procedure for appropriating groundwater in some state. Outline the most important rules for governing the use of this groundwater.

**2.3** Investigate a particular state's laws concerning stream pollution and outline the provisions that should be considered in the design of a municipal waste treatment plant.

**2.4** Prepare an outline of an interstate pollution control compact. Is your state a party to any interstate pollution control compact?

**2.5** Investigate the local ordinances concerning the disposal of industrial waste into the municipal system for a large city in your state. Prepare a summary of those provisions that should be considered in the design of an industrial waste treatment plant to be located in that city.

## REFERENCES

1. National Water Commission, "Water Policies for the Future," U.S. Gov't. Printing Office, Washington, D.C., June 1973.
2. W. Viessman, Jr. and C. Welty, *Water Management: Technology and Institutions* (Harper and Row, New York, 1985).
3. W. Viessman, Jr. and G. M. Biery-Hamilton, "An Analysis of State Water Resources Planning Processes in the United States," Volume III of *Comprehensive Review of Water Resources Policies, Planning and Programs in Florida*, Northwest Florida Water Management District, Havana, Florida, 1986.

4. F. J. Trelease, "Federal-State Relations in Water Law," National Technical Information Service, Springfield, VA, Accession No. PB 203 600 (1971).
5. *United States* v. *Rio Grande Dam and Irrigation Co.*, 174 U.S. 690, 43 L., ed. 1136, 19 Supp. 770.
6. D. P. Sheer, "Management of Water Resource Systems," *Natl. Forum* 69 (1) (1989): 9, 10.

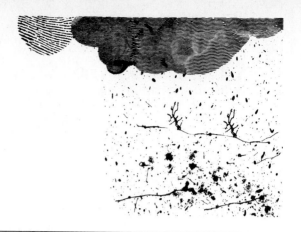

# Chapter 3
# Water Management: Environmental Considerations

A discovery some 8000 years ago eventually transformed the whole of human life. For perhaps one million years prior to that time, people had wandered individually or in small bands in search of whatever food nature might provide. The discovery that food could be produced, either by cultivating plants or by taming animals, changed entirely the way in which humankind lived. Once there was an assured control of the food supply, the great civilizations of antiquity could be built. But development in a purely agrarian society was slow and its benefits thinly distributed. The lot of a large segment of ancient society was hard, chiefly because all work was performed by muscle power.

In the second or first century B.C. came the second revolutionary discovery, that muscle power could be replaced by natural forces. The first machine to use these natural forces was water powered.

Around the tenth century, people began to use water power for other than agricultural purposes. Water-driven machines that could hammer metals, saw wood, and run oil presses were invented. As the supply of usable energy increased, there was rapid development of an industrial base.

Near the end of the eighteenth century, other great steps took place. A number of intricate waterpowered machines for such delicate processes as spinning and weaving were invented. At about the same time, fossil fuel (coal) was harnessed to heat water to steam and then force the steam under pressure to drive the pistons of work engines.

In the nineteenth century, oil and natural gas were added to the list of energy-producing fuels, and internal combustion engines were designed to use gasoline from the oil. By the end of the nineteenth century, electric energy could be developed from water power, transmitted great distances, and then converted to mechanical energy to do work.

Finally, in the twentieth century, humankind found a new and tremendous power source, nuclear energy. Fissionable minerals were added to the supply of energy-producing fuels, providing a vast new reservoir of potential energy. Fresh scientific knowledge, with its technological consequences and utilization by industry, led to discovery after discovery, invention after invention, at a consistent and ever-quickening pace. Such development has brought us to the situation as it exists today. There is no turning back. Progress under the influence of industrialization has produced ever-increasing demands for material goods and services.

In this rapidly accelerating process, nature has been shamelessly exploited. Environmental pollution has reached the point that we can no longer obtain all that is desired.

There are symptoms of impending shortages in many natural resources, shortages that are generally not caused by nature but by pollution and misuse. These shortages are a consequence of the human population explosion and the massive resultant effect our species is having on everything that occurs on this planet; that is, more people require more of the material things of life. Also, there has been a relaxation in the individual's struggle for simple survival along with a rapid drop in death rates all over the world. There have not been corresponding drops in birth rates.

Environmental engineers have been directly involved in the reduction of death rates, especially in the underdeveloped countries. One of the first programs available through the United Nations was in the area of preventive medicine, which includes a safe water supply and some sanitary means of waste disposal. The objectives of these programs have usually been achieved in a short period of time through outside effort with little participation by the country involved. The net result has been a sudden increase in population because of the reduced death rate, unaccompanied by the time-dependent problems of economic efficiency and social adjustment.

Although environmental pollution is related to the numbers of people, population control is not the entire answer. Population *concentration* is also a leading factor. More importantly, the degree of pollution depends on the standard of living and the state of technological development. To sustain this standard, we are organized to collect widely scattered resources, process them, change them in form, and distribute useful goods to the consumer. In so doing, contaminants are introduced by nearly all human activities.

Contaminants are also introduced by many natural phenomena—forest fires, collapsing river banks, and volcanic eruptions. A contaminant is considered a pollutant only if it can adversely affect something that man values and is present in high enough concentrations to do so. These pollutants are either unwanted byproducts of our activities or spent substances that have served their purposes. They tend to impair our economy and the quality of our life. Many pollutants can be carried long distances by air or water or on articles of commerce, threatening the health, longevity, livelihood, recreation, cleanliness, and happiness of citizens who have no direct involvement in their production but cannot escape their influence. The continued strength and welfare of a nation depends on the quantity and quality of its resources and on the quality of its environment.

Much of our technical capability has been devoted to producing commodities inevitably destined to contaminate the environment. One of the most spectacular examples is surely the automobile. Its exhaust and fumes pollute our atmosphere, it has brought about the need for ever-spreading highways, and it is ultimately discarded in junkyards that detract from the beauty of the landscape.

## 3.1 ENVIRONMENTAL REGULATION AND PROTECTION

Water pollution legislation originated in Congress with a bill passed in 1886. This bill forbade the dumping of impediments to navigation in New York Harbor. In 1899 Congress passed the Rivers and Harbors Act, which prohibited deposit of solid wastes into navigable waters. These early concerns with water pollution were strictly in the interests of navigation. The Public Health Service Act of 1912 had a section on waterborne diseases, and the Oil Pollution Act of 1924 was designed to prevent oil discharges from vessels into coastal waters, discharges that could damage aquatic life. This act gave pollution enforcement authority to the federal government if local efforts failed and included provision for matching grants for waste treatment facilities. Policy was strengthened with the Water Quality Act of 1965, which set water-quality standards for interstate waters.

In 1966 attention to water quality sharpened, owing to the efforts of President Johnson [1]. It was his position that entire river basins rather than localities should be considered in pollution control efforts. He proposed a "clean rivers demonstration program" in which the federal government would provide funds to interstate and/or regional water pollution control authorities on a first-ready, first-served basis. Those participating in the program would be required to have permanent water-quality planning organizations, water-quality standards, and implementation plans in effect for all waters of the basin designated.

The Clean Rivers Restoration Act of 1966 provided for a substantial increase in the level of funding appropriated for the construction of wastewater treatment facilities. Unfortunately, because of the Vietnam War, the construction grant program was not funded at the levels authorized.

After the Nixon administration took office, in 1969, Congress prodded it to take action in the areas of water pollution control and environmental policy. This prodding was supported by a strong environmental movement that had been swelling in the late 1960s. By 1970 the Nixon administration became convinced that there was need for a massive federal investment in sewage treatment plant construction. In his February 1970 message on environmental quality, President Nixon proposed a four-year, $10 billion program of state, federal, and local investment in wastewater treatment facilities. The federal share of this investment was to be $1 billion per year. While this amount lagged actual authorized funding levels and was less than many environmental advocates desired, it was much more than any previous presidential request [1].

In 1970 the National Environmental Policy Act was passed (NEPA, 1969). The act was praised by President Nixon, who proclaimed that the three-member Council on Environmental Quality (CEQ) would be a great asset in informing the president on important environmental issues. The Nixon administration promptly put the provisions of NEPA into effect. On March 5, 1970, the president issued an executive order, instructing all federal agencies to report on possible variances of their authorities, policies, and so on, with NEPA's purposes. Then, on April 30, 1970, the CEQ issued interim guidelines for the preparation of environmental impact statements.

In December 1970, as an outgrowth of the administration's environmental interests, a new independent body, the Environmental Protection Agency (EPA), was created. This organization assumed the functions of several existing agencies relative to matters of environmental management. It brought together under one roof all of the pollution control programs related to water, air, solid wastes, pesticides, and radiation. The EPA was seen by the administration as the most effective way of recognizing that the environment must be looked on as a single, interrelated system. It is noteworthy, however, that the creation of the EPA made even more pronounced the separation of water quality programs from other water programs.

Even with the enactment of NEPA, it was clear that a comprehensive response to water pollution issues was still lacking. It became evident during Congressional hearings in 1971 that, relative to the construction grants program, the program was underfunded. To rectify this situation, Congress passed the Water Pollution Control Act Amendments of 1972 (P.L. 92-500). Responding to public demand for cleaner water, the law culminated two years of intense debate, negotiation, and compromise and resulted in the most assertive step taken in the history of national water pollution control activities.

The act (P.L. 92-500) departed in several ways from previous water pollution control legislation. It expanded the federal role in water pollution control, increased the level of federal funding for construction of publicly owned waste treatment works, elevated planning to a new level of significance, opened new avenues for public participation, and created a regulatory mechanism requiring uniform technology-based effluent standards, together with a national permit system for all point-source dischargers as the means of enforcement.

In the strategy for implementation, Congress stated requirements for achievement of specific goals and objectives within specified time frames. The objective of the act is to restore and maintain the chemical, physical, and biological integrity of the nation's waters. Two goals and eight policies are articulated. The goals are

1. to reach, wherever attainable, a water quality that provides for the protection and propagation of fish, shellfish, and wildlife, and for recreation in and on the water; and
2. to eliminate the discharge of pollutants into navigable waters.

The policies are

1. to prohibit the discharge of toxic pollutants in toxic amounts;
2. to provide federal financial assistance for construction of publicly owned treatment works;
3. to develop and implement areawide waste treatment management planning;
4. to mount a major research and demonstration effort in wastewater treatment technology;
5. to recognize, preserve, and protect the primary responsibilities and roles of the states to prevent, reduce, and eliminate pollution;
6. to ensure, where possible, that foreign nations act to prevent, reduce, and eliminate pollution in international waters;
7. to provide for, encourage, and assist public participation in executing the act; and
8. to pursue procedures that drastically diminish paperwork and interagency decisions on procedures and prevent needless duplication and unnecessary delays at all levels of government.

The act provides for achievement of its goals and objectives in phases, with accompanying requirements and deadlines. It was intended to be more than a mandate for point-source discharge control. It embodied an entirely new approach to the traditional way Americans had used and abused their water resources. Construction grants for publicly owned treatment works were made available to encourage full waste treatment management, providing for

1. the recycling of potential sewage pollutants through the production of agriculture, silviculture, and aquaculture products, or any combination thereof;
2. the confined and contained disposal of pollutants not recycled;
3. the reclamation of wastewater; and
4. the ultimate disposal of sludge in a manner that will not result in environmental hazards.

These statutory provisions outline a long-term program to reduce water use, reduce the generation of wastes, and establish financially self-sustaining, public-owned pollution control facilities.

The 1972 amendments recognized the importance and urgency of the water-quality management problem. It was estimated by the National League of Cities and the U.S. Conference of Mayors, for example, that a financial commitment of from $33 billion to $37 billion would be needed for water pollution control programs during the remainder of the 1970s [2]. The 1972 act committed the federal government to 75% of the costs associated with the construction of wastewater treatment facilities and authorized $18 billion of contract authority.

After passage of Public Law 92-500, there was a transition from researching the water pollution problem to the implementation of solutions [2]. For example, Section 101 of the act states goals for fishable and swimmable waters and the prohibition of toxic discharges. These goals required that programs be implemented to reverse the threats that scientists had identified. The 1972 Clean Water Act provided the framework for a concerted effort on water pollution control. Contract authority to construct treatment facilities combined with meaningful enforcement procedures set in motion a policy to reverse the water-quality degrading practices of the past.

Not long after passage of the 1972 Clean Water Act, the Safe Drinking Water Act was passed (December 16, 1974). The purpose of that legislation was to ensure that water supply systems serving the public would meet minimum standards for the protection of public health. The act was designed to achieve uniform safety and quality of drinking water in the United States by identifying contaminants and establishing maximum acceptable levels. Prior to the Safe Drinking Water Act, it was possible to prescribe federal drinking-water standards only for water supplies used by interstate carriers. In contrast, after the act the EPA established federal standards to control the levels of harmful contaminants in drinking water supplied by all public water systems. It also established a joint federal-state system for ensuring compliance with these standards. The major provisions of the act are

1. the establishment of primary regulations for the protection of the public health;
2. the establishment of secondary regulations that are related to taste, odor, and appearance of drinking water;
3. the establishment of regulations to protect underground drinking-water sources by the control of surface injection;
4. the initiation of research on health, economic, and technological problems related to drinking-water supplies;
5. the initiation of a survey of rural water supplies; and
6. the allocation of funds to states for improving their drinking-water programs through technical assistance, training of personnel, and grant support.

In 1977, in response to an indicated need to address deficiencies in the 1972 act, the Clean Water Act was revised. The salient points of the 1977 act included the following:

1. States were specifically mandated primacy over water-quality and water-use issues.
2. Municipalities were given evidence of a federal commitment in the form of construction grants and training assistance.
3. The public received assurances of the priority of water quality in the form of effective enforcement and incentive provisions for governments and industries to achieve the goal of fishable and swimmable waters.
4. Industry received the necessary extensions of compliance deadlines under the effluent discharge limitations provision.
5. Environmental groups witnessed the incorporation of a Resource Defense Council/EPA consent decree into the law that established toxic effluent stan-

dards and set forth a comprehensible process to implement effluent limitations [2].

In 1986, the Safe Drinking Water Act of 1974 was amended [3]. The principal changes were focused on groundwater protection. A wellhead protection program was established. The program provides that states undertaking wellhead protection efforts are eligible to receive federal grants to aid them in these endeavors. The EPA guidelines for wellhead protection are somewhat unique in that they allow regional flexibility, rather than prescribe uniform national standards. The act also provides for sole-source aquifer protection. The objective is to protect from contamination recharge areas that are primary sources of drinking water. Drinking-water standards apply to these areas, and underground injection of effluent is regulated. Enforcement provisions of the act are strong, and in 1987 the first criminal conviction under the act was obtained [4].

The Clean Water Act was reauthorized in 1987 as the Water Quality Act of 1987 [5]. A major feature of the 1987 act was the addition of the goal of controlling nonpoint sources of pollution. It is the most pronounced federal excursion into this important water-quality management dimension. Agricultural fields, feedlots, and urban areas, including streets, are addressed. And while mandatory controls are not authorized, Congress did direct the states to conduct planning studies for the purpose of developing strategies for abating water pollution associated with nonpoint sources. A total of $400 million of federal funds was authorized to be used by the states to implement cleanup programs. Priority is to be given regulatory programs, innovative practices, and strategies that deal with groundwater contamination. The 1987 act provides for creation, by the states, of revolving funds to facilitate low-interest loans to local governments for sewage treatment improvements. It also provided more options for state-federal sharing of programs provided for by the National Pollution Discharge Elimination System (NPDES). The EPA and the states can now divide the categories of discharges regulated within each state.

As a result of the water pollution control efforts since 1972, the tide of pollution has been diminished. But there is still much to be done, particularly in the area of nonpoint pollution control. A summary of federal statutes governing, or affecting, water-quality protection is given in Table 3.1.

## 3.2  EFFECTS OF ENVIRONMENTAL REGULATIONS

The intended result of Congressional passage of pollution control laws was for the EPA and the states to issue enforceable regulations to improve the quality of the nation's waters. Pollution control programs were generated at every level of government to implement regulations, issue permits, inspect regulated facilities, and enforce established rules. In response, industries and

**TABLE 3.1**   A SUMMARY OF FEDERAL ENVIRONMENTAL
LEGISLATION: 1948–1987

| Year | Act |
|------|-----|
| 1948 | Federal Water Pollution Control Act |
| 1968 | Wild and Scenic Rivers Act |
| 1969 | National Environmental Policy Act |
| 1972 | Federal Water Pollution Control Act Amendments |
| 1972 | Marine Protection, Research and Sanctuaries Act |
| 1973 | Endangered Species Act |
| 1974 | Safe Drinking Water Act |
| 1976 | Resources Conservation and Recovery Act |
| 1976 | Toxic Substances Control Act |
| 1977 | Clean Water Act |
| 1980 | Comprehensive Environmental Response, Compensation, and Liability Act |
| 1984 | Resources Conservation and Recovery Act Amendments |
| 1986 | Superfund Amendment and Reauthorization Act |
| 1986 | Federal Safe Drinking Water Act Amendments |
| 1987 | Water Quality Act |

municipalities organized internal pollution control programs to stay abreast of regulatory requirements, work with plant personnel to attain compliance with regulations, learn about environmental monitoring and sampling techniques, and work with the regulatory agencies to obtain permits. In a sense, the 1970s was a period of institutionalization of the ideals of the environmental movement prevalent in the sixties.

The President's Council on Environmental Quality, in its 1981 report, stated that water pollution controls were showing positive results in the United States [6]. The EPA has also reported success stories of rivers and lakes slowly returning to their natural state. The point-source program is now well established and appears to be working well, at least for industrial sources. But there is still need for improvement in the operation of municipal waste treatment facilities.

By law, all waters must have designated "beneficial uses" that must be protected and met. These uses establish the water-quality criteria that must be considered in pollution control efforts. Using EPA guidelines, states apply a range of chemical, biological, habitat, and other parameters to establish criteria to protect specific designated uses. The EPA must approve the water-quality standards that result, and the states then apply them to determine the quality of their waters, consistent with supported uses. In 1990, it was reported that of 519,000 miles of streams assessed in 1988, 30% did not meet, or partially did not meet, the standards for their designated uses [7, 8]. These findings are summarized in Table 3.2. They give a good indication of where the nation stands in dealing with its water-quality problems.

**TABLE 3.2**   EXTENT TO WHICH THE NATION'S ASSESSED WATERS SUPPORT DESIGNATED USES

|  | River miles | Lake acres | Estuary square miles |
|---|---|---|---|
| Uses not supported | 10% (53,499) | 10% (1,591,391) | 6% (1,488) |
| Uses partially supported | 20% (104,632) | 17% (12,701,577) | 23% (6,078) |
| Uses fully supported | 70% (361,332) | 74% (12,021,044) | 72% (19,110) |
| Assessed miles | 519,412 | 16,313,962 | 26,628 |
| Total in U.S. | 1,800,000 | 39,400,000 | 36,000 |

*Source:* R. Savage, "The Clean Water Act: Accomplishments," Water Resources Update, Universities Council on Water Resources, Winter 1991, No. 84, Carbondale, IL, p. 31.

## 3.3 THE FUTURE OF POLLUTION CONTROL EFFORTS IN THE UNITED STATES

During the Reagan administration, a review of regulations and regulatory practices was initiated to determine whether costs of pollution control could be reduced by modifying the regulatory approach [9]. This led to a debate about the relative merits of water-quality versus technology-based standards, which had been the keystone of the Clean Water Act since 1972. *Water-quality standards* establish a designated use for a specified section of a water body, which is then balanced with the maximum amount of waste the water body can assimilate. *Technology-based standards* are effluent limitations based on the levels of pollutant removal that can be achieved by modern wastewater treatment technology.

Congress initiated the technology-based approach in 1972 because the water-quality-based approach of the 1960s had failed to work because of the difficulties of enforcement and the limited availability of data for use in water-quality models. The arguments in favor of a technology-based approach are as follows:

1. Technology-based standards are easy to enforce. This is important from an institutional perspective.
2. These standards are the first step toward the ultimate goal of zero discharge of pollutants to natural waters, as opposed to merely cleaning up waters to suit man's objectives (the basis for water-quality standards).
3. There is insufficient knowledge and resources to set water-quality standards for all pollutants and locations. Technology-based standards are an interim approach to avoid pollution.
4. Nationwide uniformity in treatment standards minimizes economic dislocations.
5. The approach promotes equity among dischargers. No one should have the right to discharge more into the environment simply because of geographic location.

The Reagan administration, however, contended that while technology-based standards were important in the past in providing impetus for local governments and industry to clean up pollution from their treatment facilities, the agency now had the ability and sophistication to regulate discharged pollutants under water-quality standards and that the Clean Water Act should be amended accordingly. The Reagan administration noted the following advantages of water-quality standards [9]:

1. Water-quality standards and the process by which they are adopted inherently encourage an assessment of costs and benefits, which is absent in the adoption and application of technology-based standards.
2. They foster scientific debate, which accelerates the advancement of the state of the art in predicting the fate and effect of pollutants.
3. The debate takes place in a local-state arena and heightens awareness of local government, policymakers, and the public of the importance of water pollution control in their communities.
4. The assertion of the primary right and responsibility of states to regulate pollutants is essential to establishing the appropriate balance of power between the federal establishment and state governments.
5. Water-quality-based decisions can avoid requirements of treatment for treatment's sake, which can result from application of technology-based standards [9].

For the present, it appears that technology-based effluent standards will continue to be the norm, even though they may be economically and socially inefficient [10]. But someday, a shift to water-quality standards may gain stronger support, particularly because holistic water management is beginning to approach reality.

Although the future is uncertain, it appears likely that water pollution control efforts during the remainder of this century will be characterized by

maintaining present levels of water quality where they are satisfactory, and improving water quality in locations exhibiting deficiencies [11];

intensifying efforts to bring municipal wastewater facilities into full compliance with standards;

accelerating efforts to deal with nonpoint pollution problems (about 65% of stream pollution can be attributed to nonpoint sources, agriculture being responsible for about 55% of all impaired stream mileage [7, 10]); and

developing and implementing new and innovative approaches to planning, financing, and enforcement.

The American public appears to be solidly committed to the goal of clean water. Billions of dollars have already been invested in water-quality control programs, and this trend is expected to continue.

## PROBLEMS

**3.1**  What are the main statutes under which the EPA controls water pollution? Which aspects of water pollution may be regulated under each act?

**3.2**  Define point and nonpoint water pollution sources. Give five examples of each.

**3.3**  List ten adverse health effects that can be caused by toxic chemical pollutants.

**3.4**  In your opinion, what is the most significant source of groundwater pollution in this country? Explain.

**3.5**  What is the most significant source of groundwater pollution in your state? Describe the sources and effects of contamination to the extent they are known.

**3.6**  Why is groundwater contamination so difficult to detect and clean up? Briefly describe some of the techniques used in aquifer restoration.

**3.7**  Identify the agencies in your state responsible for managing (a) water quality and (b) water quantity.

**3.8**  Do you believe the primary responsibility for pollution control should rest with the states, the federal government, or some mix? Explain your viewpoint.

**3.9**  Which agency, or agencies, regulate water quality in your state?

**3.10** Of the federal or state laws that you are familiar with, which have the greatest impact on water quality? Why?

## REFERENCES

1. B. H. Holmes, "History of Federal Water Resources Programs and Policies, 1961–1970," U.S. Department of Agriculture Misc. Publ. No. 1379 (Washington, DC, September 1979).
2. R. M. Linton, "The Politics of Clean Water," *Chemtech* (July 1982).
3. 42 U.S.C. 300(f) et seq.
4. *United States* v. *Jay Woods Oil Co. Inc.* (ED Mich No. 87 CR20012 BC), unreported opinion cited in the Environmental Reporter, vol. 18, no. 6, June 5, 1987, p. 502.
5. Public Law 100-4: Feb. 4, 1987.
6. Council on Environmental Quality, *Environmental Quality 1981: The Twelfth Annual Report of the Council on Environmental Quality* (Washington, D.C.: U.S. Government Printing Office, December 1981).
7. D. H. Moreau, "The Clean Water Act in Retrospect," Water Resources Update, Universities Council on Water Resources, Winter 1991, No. 84, Carbondale, IL, pp. 5–12.
8. U.S. Environmental Protection Agency, "National Water Quality Inventory: 1988 Report to Congress," Report No. EPA 440-4-90-003, Office of Water, Washington, D.C., April 1990.
9. W. Viessman, Jr. and C. Welty, *Water Management: Technology and Institutions* (New York: Harper and Row, 1985).
10. L. J. MacDonnell, "Restoring and Maintaining the Integrity of the Nation's Water: An Assessment of the Clean Water Act," Water Resources Update, Universities Council on Water Resources, Winter 1991, No. 84, Carbondale, IL, pp. 19–23.
11. R. Savage, "The Clean Water Act: Accomplishments," Water Resources Update, Universities Council on Water Resources, Winter 1991, No. 84, Carbondale, IL, pp. 30–34.

# Chapter 4
# Water Use and Wastewater Generation

The availability of water resources, geographically and temporally, the quality of these resources, the rates at which they are replenished and depleted, and the demands placed upon them by water users are determining factors in water management strategies. Estimates of future water uses, uncertain as they might be, are fundamental to efficient and/or equitable allocation of water resources. These estimates depend on an ability to forecast changes in population, agricultural and industrial activity, economic conditions, technology, and other related factors.

Relevant factors include water-use permitting; estimating water requirements; inventorying water supplies; defining new "reasonable beneficial uses"; infrastructure development, operation, maintenance, and replacement; policies for allocating water among competing users; strategic planning; interbasin transfers; saline water use; financing; education; conservation and reuse; research; drought management; coordinating water supply and water-quality management programs; and assessing risks tied to proposed courses of action.

Of similar concern is the subject of the quantities and timing of stormwater and sewage flows. These waters, although degraded in quality, are also part of the water supply potential, and they are the determining factors in decisions related to the extent and character of wastewater treatment required.

In this chapter, the major water-using sectors are defined, trends in water use and wastewater generation are reviewed, and techniques for forecasting water requirements and wastewater volumes are presented.

## 4.1 WATER-USING SECTORS

Decisions related to the allocation of water must be based on a knowledge of both the availability of water and the type and rate of use. Otherwise, inefficient allocations and undesirable levels of water-source depletions are likely to be the outcomes. This is particularly true when the resource is scarce and/or it is subject to a spectrum of competing demands. Accordingly, the major water-using sectors must be recognized, and the demands they place on their source waters quantified.

### Agriculture

Water is critical to agriculture. In arid and semiarid regions, water constitutes the difference between unproductive wastelands and highly productive food-generating units. In humid areas, rainfall is usually adequate to produce good crops, but even there supplemental irrigation is being relied on more and more to prevent crop failures and to improve the quality of the products produced. It is important to note that the use of supplemental irrigation in the eastern United States has been increasing rapidly [1].

Water requirements for irrigation are mostly seasonal, varying with climate and type of crops. In humid regions, water withdrawals for irrigation may range from about 10% of the total annual demand in May to 30% in September, while in arid and semiarid locations, rates of withdrawal are nearly uniform during the irrigation season. Irrigation water use conflicts with many other uses and thus creates special problems in the more arid regions. For example, irrigators attempt to store as much water as possible during the winter, when hydropower producers using the same reservoirs are eager to release the flows into their turbines to generate electricity. An important aspect of irrigation water use is that about 50% of the amount withdrawn is consumptively used.

### Steam Electric Generation

The principal use of water in steam electric generating facilities is for cooling to dissipate rejected heat, an inherent characteristic of heat engines. The amount of cooling water withdrawn depends on plant size, generator thermal efficiency, cooling heat transfer efficiency, and institutionally regulated limits on effluent temperatures. In 1985 generation of electricity ranked first in total water withdrawals (fresh plus saline water). About two thirds of the amount used was from freshwater sources [2]. If the demands for cooling water increase, limitations on freshwater resources will probably stimulate even greater interest in developing coastal sites with their potential for once-through cooling using saline water. *Once-through cooling* is the passage of water through cooling units followed by direct release to a receiving body of water without any recycling through water-cooling facilities. Withdrawals for once-through cooling are large, but little water is used consumptively.

## Cities and Other Communities

In 1985, central water supply systems furnished water to about 200 million people residing in municipal areas of the United States [1]. Approximately 43 million people living outside of these service areas had their own domestic systems, while more than 4 million people were reported to have no piped water supply. The principal domestic and commercial water uses are for drinking, cooking, meeting sanitary needs, lawn watering, swimming pool maintenance, street cleaning, fire fighting, and various aspects of city and park maintenance. While the public water-use sector is vital to our well-being, since it furnishes much of our drinking water, the total amount of water used by this sector is small when compared to water-using sectors such as irrigation and thermoelectric cooling. In 1985, the United States Geological Survey (USGS) reported it to represent only about 9% of the nation's freshwater withdrawals. In 1975, the national average daily per capita use from public supplies was about 170 gpcd; in 1985, it had increased to 183 gpcd [1, 2]. In 1975, strict household use from public water supply systems was about 90 gpcd for a family of four; in 1985, it was reported at 105 gpd for each person served [1, 2]. Note that the unit of measurement in the SI system is liters per capita per day (lpcd). Due to limited water supplies and increasing population, per capita demand may be expected to decline in the future.

Residential water-use rates are continually fluctuating, from hour to hour, from day to day, and from season to season. Average daily winter consumption is only about 80% of the annual daily average, whereas summer consumption averages are about 25% greater than the annual daily average. Figure 4.1 compares a typical winter day with a typical maximum summer day in Baltimore, Maryland. Note the hourly fluctuations and the tendency toward two peaks. Studies by Wolff indicate that hydrographs of systems serving predominantly residential communities generally show two peaks, the first between the hours of 7 A.M. and 1 P.M., the second in the evening between 5 and 9 P.M. [3]. During the summer, when lawn-sprinkling demands are high, the second peak is usually the greatest, while during the colder months or during periods of high rainfall, the morning peak is commonly the larger of the two.

In Figure 4.2 the effects of lawn-sprinkling demands are strikingly demonstrated. This figure shows the effects of rainfall on two different days—one a hot, dry day with no antecedent rainfall for 4 days and the other a day during which rainfall in excess of 1 in. was recorded. Lawn sprinkling has been found to represent as much as 75% of total daily volumes and as much as 95% of peak hourly demands where large residential lots are involved [4]. Peak hourly demands have been found to vary from average daily demands by as much as 1500% [3], but there is no rule that can be universally applied to determine what these might be. Variations are a function of the type of development, its age, geographic location, and extent of conservation practices. Figures 4.3 and 4.4 provide some guidance on peak demand values.

Fire-fighting demands must also be considered in municipal water system

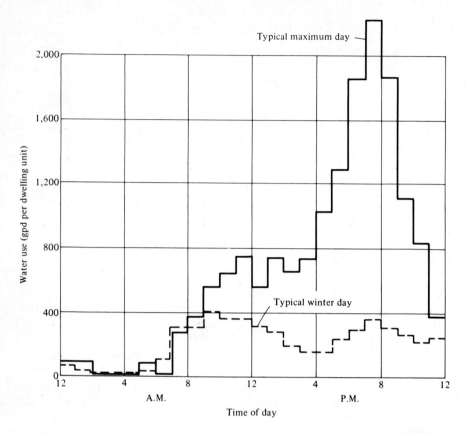

**Figure 4.1** Daily water-use patterns, maximum day and winter day. (From Residential Water-Use Research Project, Johns Hopkins University and Federal Housing Administration, 1963.)

design. The annual volumes required for fire-fighting purposes are small, but during periods of need the demand may be large and may govern the design of distribution systems, distribution storage, and pumping equipment. Recommendations as to the quantities of water to be used in fire fighting in high-valued community districts have been published by the National Board of Fire Underwriters (American Insurance Association) [5].

Fire-fighting requirements for residential areas vary from 500 to 3000 gpm, the required rate being a function of population density. Hydrant pressures should generally exceed 20 psi where motor pumpers are used; otherwise, pressures in excess of 100 psi might be required. If recommended fire flows cannot be maintained for the indicated time periods, community fire insurance rates may be adjusted upward.

The coincident draft (flow to be expected at the time the fire is being fought) during fire fighting is usually considered to be equal to the maximum daily demand since the probability of the maximum rate of water usage for

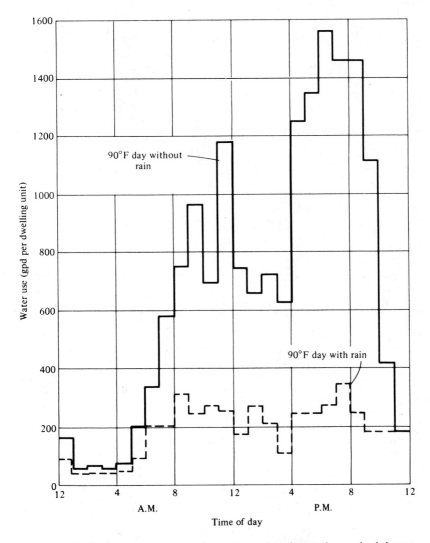

**Figure 4.2** Daily water-use patterns in R-6 area: Maximum day and minimum day. (From Residential Water-Use Research Project, Johns Hopkins University and Federal Housing Administration, 1963.)

community purposes occurring simultaneously with a major conflagration is slight.

The National Board of Fire Underwriters has proposed the following formula for computing fire flows of communities having less than 200,000 people:

$$Q = 1020 \sqrt{P}(1 - 0.01 \sqrt{P}) \tag{4.1}$$

where   $Q$ = demand in gpm
$\phantom{where}$   $P$ = population in thousands

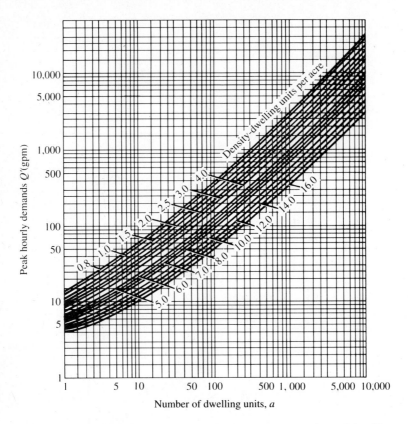

**Figure 4.3** Relation of total peak hourly demands to number of dwelling units in terms of housing density. (Courtesy of the Residential Water-Use Research Project of The Johns Hopkins University and the Office of Technical Studies of the Architectural Standards Division of the Federal Housing Administration, 1963.)

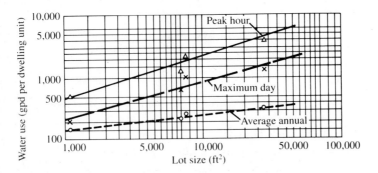

**Figure 4.4** Relationship of lot size to water use. (Courtesy of the Residential Water-Use Research Project of The Johns Hopkins University and the Office of Technical Studies of the Architectural Standards Division of the Federal Housing Administration, 1963.)

Such flows must often be provided for periods up to 10 hr or more. Gupta describes another equation for computing fire flows, based on construction type, floor area, and occupancy of a building, which was developed by the Insurance Services Office [6]. It states

$$Q = 18C\sqrt{A} \qquad (4.2)$$

where   $Q =$ required fire flow, gpm
$A =$ total floor area excluding the basement, ft$^2$
$C =$ a coefficient having values of 1.5 for wood frames, 1.0 for ordinary construction, 0.8 for noncombustible buildings, and 0.6 for fire-resistant construction

For this equation, the flow should be more than 500 gpm, but should not exceed 6000 gpm for a single-story structure, 8000 gpm for a single building, and 12,000 gpm for a single fire.

In addition to supplying water for homes, fire fighting, and perhaps some industrial purposes, most communities also must meet the needs of various commercial establishments. In considering commercial requirements, it is important to know both the magnitude and time of occurrence of peak flow. Table 4.1 indicates typical requirements and periods of maximum demand for apartments, motels, hotels, office buildings, shopping centers, laundries, washmobiles, and service stations [4]. Generally, it can be stated that commercial water users do not materially affect peak municipal demands. In fact, peak hours for many commercial establishments tend to coincide with the secondary residential peak period. The cessation of numerous commercial activities at about 6 P.M. precludes the imposition of large demands in the early evening, when sprinkling demands are often high.

The commercial peak-hour demands given in Table 4.1 are compared to peak-hour demands for a typical individual residence on a 1-acre lot. The data show that maximum commercial needs are considerably less important than peak sprinkling demands in determining peak loads on a distribution system subject to heavy sprinkling loads. An average figure of 20 gpcd is normally considered representative of commercial water consumption. The range is normally reported as 10–130 gpcd.

The many factors affecting municipal water use preclude any generalization that could apply to all areas. General trends and representative figures are useful, but it should be understood that local usage may vary considerably from reported averages. For design purposes, thorough studies must be made of past records of the type and pattern of community water use, the physical and climatic characteristics of the area, expected trends in development, projected population values, and other pertinent factors.

### ■ EXAMPLE 4.1
Given a residential area encompassing 500 acres with a housing density of six houses per acre. Assume a high-value residence with a fire flow requirement of 1000 gpm. Find (a) the combined draft and (b) the peak hourly demand.

**TABLE 4.1** COMMERCIAL WATER USE

| | Unit | Average annual demand (gpd) | Maximum hourly demand rate (gpd) | Hour of peak occurrence | Ratio, maximum hourly to average annual | Average annual demand per unit | Ratio, maximum hourly demand to R-40 demand[a] |
|---|---|---|---|---|---|---|---|
| Miscellaneous residential | | | | | | | |
| Apartment building | 22 units | 3,430 | 11,700 | 5–6 P.M. | 3.41 | 156 gpd/unit | 2.2:1 |
| Motel | 166 units | 11,400 | 21,600 | 7–8 A.M. | 1.89 | 69 gpd/unit | 4.0:1 |
| Hotels: | | | | | | | |
| Belvedere | 275 rooms | 112,000 | 156,000 | 9–10 A.M. | 1.39 | 407 gpd/room | 29:1 |
| Emerson | 410 rooms | 126,000 | | | | 307 gpd/room | |
| Office buildings | | | | | | | |
| Commercial Credit | 490,000 ft² | 41,400 | 206,000 | 10–11 A.M. | 4.89 | 0.084 gpd/ft² | 38:1 |
| Internal Revenue | 182,000 ft² | 14,900 | 74,700 | 11–12 A.M. | 5.01 | 0.082 gpd/ft² | 14:1 |
| State Office Building | 389,000 ft² | 27,000 | 71,800 | 10–11 A.M. | 2.58 | 0.070 gpd/ft²[b] | 13:1 |
| Shopping centers | | | | | | | |
| Towson Plaza | 240,000 ft² | 35,500 | 89,900 | 2–3 P.M. | 2.50 | 0.15 gpd/ft² | 17:1 |
| Hillendale | 145,000 ft² | 26,000 | | | | 0.18 gpd/ft² | |
| Miscellaneous commercial | | | | | | | |
| Laundries: | | | | | | | |
| Laundromat | Ten 8-lb washers | 1,840 | 12,600 | 11–12 A.M. | 6.85 | 184 gpd/washer | 2.3:1 |
| Commercial | Equivalent to ten 8-lb washers | 2,510 | 16,200 | 10–11 A.M. | 6.45 | 251 gpd/washer equivalent | 3.0:1 |
| Washmobile | Capacity of 24 cars per hour | 7,930 | 75,000 | 11–12 A.M. | 9.46 | 330 gpd per car per hour of capacity | 14:1 |
| Service station | 1 lift | 472 | 12,500 | 6–7 P.M. | 26.5 | 472 gpd/lift | 2.3:1 |

*Source*: Residential Water Use Research Project of The Johns Hopkins University and the Office of Technical Studies of the Architectural Standards Division of the Federal Housing Administration, 1963.

[a] Lot type R-40 (1 acre) peak hourly demand for single service is 5400 gpd.   [b] Exclusive of air conditioning.

*Solution.*   (a) Given that 1 acre contains 43,560 ft$^2$, each lot will be about 7000 ft$^2$ in size. Enter Fig. 4.4 with this value and find a maximum day value of 700 gpd per dwelling unit. For the 3000 dwelling units, this would be 3000 × 700 = 2,100,000 gpd or 1458 gpm. The combined draft is thus 1458 + 1000 = 2458 gpm. (b) From Fig. 4.3, for 3000 dwelling units and a density of 6.0, find a peak hourly demand of 2500 gpm. In this case the peak hourly flow would control the design since it exceeds the combined flow estimate.    ■

## Manufacturing

From 1970 to 1980 the manufacturing sector accounted for about 17% of total U.S. freshwater withdrawals. In 1985, industrial withdrawals represented only about 7% of the total withdrawals for all categories of water use [2]. This decline is attributed mainly to recycling and process changes. Manufacturing uses vary with the product produced, but they generally include both process waters and cooling waters. From about 1955 to 1975 the amount of fresh and saline water withdrawn for manufacturing purposes almost doubled. This water was recycled about twice before being returned to the source and diminished somewhat less than 10% by evaporation and incorporation into products. Although manufacturing water use is expected to increase in the future, recycling is also predicted to increase substantially, with the prospect that actual water withdrawals for this purpose will continue to show a decline. Consumptive use will increase, however.

## Natural Systems

Providing water for the preservation and benefit of fish and wildlife, protection of marshes and estuary areas, and for other environmentally oriented purposes is now considered a necessity. But such water uses are often in conflict with traditional uses, and resolving these conflicts is destined to become an increasingly common task. Furthermore, estimation of the quantities of water needed for these new uses is difficult. Scientific data needed to make good determinations are often lacking, and this presents some special problems since the quantities of water can be substantial. Particular topics of concern include instream flow requirements, maintenance of lake levels, freshwater releases to bays and estuaries, and water requirements for protecting fish and wildlife.

The water-related aspects of restoring, protecting, and managing natural systems are abundant. Dealing with them requires special policies, a good data base, and close coordination with a host of programs conducted under the auspices of various levels of government. Water management policies that encourage or result in excessive growth may trigger unwanted spillover effects on the environment. The drainage and/or reclamation of lands may provide additional opportunity for economic development, for example, but not without a price to be paid in disrupting prevailing ecosystems. Development and environmental protection can be partners, but only if care is exercised in modifying the landscape. Growth management policies that embrace the many dimensions that must be dealt with in managing natural systems are needed.

## Navigation

Water requirements for navigation on most river systems are seasonal. The greatest demands usually occur during the driest months of the year. Flows released for navigation limit the availability of water for irrigation and hydropower generation and for recreational uses at reservoir sites. They do, however, complement other instream uses.

Where navigation depths are maintained by low dams, there is usually little effect on other water uses in a river. These structures do not impound large volumes of water; rather they serve to provide greater uniformity of flows. Many advantages result from this type of operation, including benefits to fish and wildlife, recreation, pollution control, and aesthetics.

Large multipurpose reservoirs, such as those on the main stem of the Missouri River, also provide storage to meet periodic navigational flows. In such cases, reservoir operating policies must be designed to accommodate the conflicting requirements of other water uses for which storage is provided.

## Hydroelectric Power Generation

In the past, requirements for hydroelectric power were usually heaviest during the peak winter heating months, but with the increased use of air conditioning, demands for electricity are less seasonal, and in some cases the summer months are the most demanding. The use of hydroelectric facilities to provide peak power, as opposed to furnishing base load, is also becoming more common. Unfortunately, this type of operation increases conflicts with recreationists and others who favor little or no short-term fluctuations in reservoir levels. In general, conflicts between water use for electric power generation and use for other purposes stem from opposing seasonal requirements. For example, heavy summertime releases for navigation dictate maximizing storage during the winter, a situation in conflict with discharging from storage during the same period to produce electricity. Hydroelectric production does not adversely affect all water uses, however; for example, water passed through turbines can also be used downstream for navigation, flow augmentation, and other purposes.

## Recreation

About a fourth of the nation's outdoor recreation activity depends on water. In 1975 swimming, fishing, boating, water skiing, and ice skating accounted for about 3 billion activity days. By the year 2000, this figure is projected to increase to about 8 billion [1]. Water requirements for recreation are normally greatest in the summer. The sportsman and vacationer desire substantial stream flows and unvarying reservoir levels during this period. Such conditions are optimal for water-based recreation activities but conflict with many withdrawal uses.

### Energy Resource Development

Water can be used to produce energy via turbines driving electric generators. It can also be used to process energy-producing resources such as coal and oil shale and to aid in restoring lands despoiled during mining operations. The water requirements for extraction of coal, oil shale, uranium, and oil gas are not great, but secondary recovery operations for oil require large quantities of water. Substantial quantities of water may be used in coal slurry pipelines and for retorting oil shale. Synfuels conversion processes also require large quantities of water, and as stated earlier, withdrawal of water for cooling thermal electric plants is the largest category of total water use in the United States.

The availability of water is a factor in the location and design of energy conversion facilities, but these users can generally afford to pay high prices for water. Securing legal rights to water, rather than its availability, is often the critical issue in dry regions.

## 4.2 WATER-USE TRENDS

Analyses of water-use projections made since 1970 show that a rapidly increasing rate of per capita water use is less likely than estimators of the 1950s and 1960s would have believed. Another point is that national or regional trends are not always indicative of state and local trends. Thus, planners must be equipped to deal with development and management options at several geographic levels so that special local and regional influences can be accommodated. Although the overall tendency in water use to the year 2000 appears to be more conservative than that of the past, striking local variances can be expected.

Every five years the United States Geological Survey publishes a circular entitled "Estimated Water Use in the United States" [2]. This publication summarizes water use in each major water-using category and indicates trends over time. The data are available by state and by region. Table 4.2 gives 1990 water use estimates and Table 4.3 shows water-use trends from 1950 to 1985. Note that the total offstream withdrawal figure declined from 1980 to 1985. Figure 4.5 shows trends for each of the major sectors; only public supply and rural-use categories do not reflect a downward trend. That is not surprising, however, since these sectors are strongly associated with population, and as long as there is population growth, water use is likely to increase, although not necessarily at the same per capita rate. Figure 4.6a indicates the sources (surface and groundwater) from which fresh and saline water was withdrawn in 1985. Figure 4.6b indicates that of all of the water

**TABLE 4.2** TOTAL WATER WITHDRAWALS FOR OFFSTREAM WATER-USE CATEGORIES, BY STATE, 1990

| Fresh/Saline | Public supply Fresh | Domestic Fresh | Commercial Fresh | Irrigation Fresh | Livestock Fresh | Industrial Fresh | Industrial Saline | Mining Fresh | Mining Saline | Thermoelectric Fresh | Thermoelectric Saline | Total Fresh | Total Saline |
|---|---|---|---|---|---|---|---|---|---|---|---|---|---|
| Alabama | 707 | 28 | 3.5 | 94 | 141 | 784 | .0 | 11 | 9.1 | 6,310 | .0 | 8,080 | 9.1 |
| Alaska | 92 | 6.9 | 18 | .6 | .6 | 111 | .0 | 25 | 357 | 31 | .0 | 284 | 357 |
| Arizona | 691 | 128 | 46 | 5,450 | 73 | 167 | .3 | 118 | .7 | 99 | 6.1 | 6,770 | 7.1 |
| Arkansas | 309 | 51 | 222 | 5,250 | 189 | 177 | .0 | 2.5 | .0 | 1,640 | .0 | 7,940 | .0 |
| California | 5,830 | 321 | 234 | 27,900 | 398 | 43 | 33 | 12 | 310 | 246 | 12,000 | 34,900 | 12,300 |
| Colorado | 650 | 19 | 8.6 | 11,600 | 162 | 118 | .0 | 54 | 30 | 114 | .0 | 12,700 | 30 |
| Connecticut | 374 | 46 | 18 | 15 | 1.5 | 80 | 68 | 2.2 | .0 | 530 | 3,710 | 1,070 | 3,780 |
| Delaware | 85 | 11 | 1.8 | 32 | 2.4 | 65 | 6.0 | .0 | .0 | 831 | 333 | 1,030 | 339 |
| D.C. | .0 | .0 | .0 | .0 | .0 | .5 | 6.0 | .5 | .0 | 8.0 | .0 | 9.0 | .0 |
| Florida | 1,930 | 299 | 52 | 3,730 | 78 | 403 | 56 | 31.5 | .0 | 732 | 10,300 | 7,530 | 10,400 |
| Georgia | 937 | 100 | 43 | 441 | 46 | 657 | 33 | 12 | .0 | 3,030 | 33 | 5,260 | 65 |
| Hawaii | 238 | 9.9 | 40 | 755 | 7.2 | 43 | .6 | 1.4 | .0 | 95 | 1,550 | 1,190 | 1,150 |
| Idaho | 201 | 48 | 16 | 18,700 | 1,210 | 196 | .0 | 8.4 | .0 | 6.1 | .0 | 20,400 | .0 |
| Illinois | 1,860 | 115 | 173 | 78 | 63 | 464 | .0 | 69 | 25 | 15,200 | .0 | 18,000 | 25 |
| Indiana | 604 | 118 | 63 | 51 | 46 | 2,480 | .0 | 97 | .0 | 5,960 | .0 | 9,430 | .0 |
| Iowa | 322 | 45 | 27 | 23 | 121 | 219 | .0 | 34 | .0 | 2,070 | .0 | 2,560 | .0 |
| Kansas | 371 | 25 | 6.2 | 4,190 | 114 | 53 | .0 | 25 | .0 | 1,300 | .0 | 6,080 | .0 |
| Kentucky | 427 | 56 | 13 | 12 | 33 | 313 | .0 | 18 | .0 | 3,440 | .0 | 4,320 | .0 |
| Louisiana | 619 | 50 | 13 | 708 | 551 | 2,360 | 67 | 37 | .0 | 4,950 | .0 | 9,290 | 67 |
| Maine | 106 | 49 | 34 | 1.8 | 1.7 | 254 | .0 | 3.7 | .0 | 82 | 609 | 532 | 609 |
| Maryland | 798 | 67 | 26 | 29 | 27 | 70 | 379 | 28 | 21 | 421 | 4,550 | 1,470 | 4,950 |
| Massachusetts | 714 | 37 | 74 | 100 | 1.7 | 87 | .0 | 5.0 | .0 | 1,010 | 3,490 | 2,030 | 3,490 |
| Michigan | 1,400 | 123 | 35 | 240 | 29 | 1,680 | 3.7 | 55 | .9 | 8,060 | .0 | 11,600 | 4.6 |
| Minnesota | 479 | 167 | 78 | 179 | 67 | 133 | .0 | 205 | .0 | 1,780 | .0 | 3,080 | .0 |
| Mississippi | 320 | 33 | 16 | 1,880 | 411 | 269 | .0 | 3.4 | .0 | 386 | 316 | 3,320 | 316 |

| | | | | | | | | | | | | |
|---|---|---|---|---|---|---|---|---|---|---|---|---|---|
| Missouri | 677 | 62 | 22 | 371 | 55 | 85 | .0 | 25 | .1 | 4,580 | 1,060 | 5,870 | 1,060 |
| Montana | 135 | 16 | .0 | 9,000 | 52 | 57 | .0 | 6.2 | 13 | 33 | .0 | 9,300 | 13 |
| Nebraska | 301 | 47 | .2 | 5,100 | 139 | 41 | .0 | 131 | 4.7 | 2,180 | .0 | 6,940 | 4.7 |
| Nevada | 385 | 9.8 | 23 | 2,820 | 5.5 | 10 | .0 | 49 | 12 | 34 | .0 | 3,340 | 12 |
| New Hampshire | 95 | 27 | .6 | .9 | 1.0 | 37 | .0 | 2.8 | .0 | 255 | 894 | 420 | 894 |
| New Jersey | 1,040 | 68 | 17 | 58 | 2.1 | 326 | 1,020 | 110 | .0 | 599 | 9,550 | 2,220 | 10,600 |
| New Mexico | 273 | 24 | 17 | 3,020 | 20 | 6.3 | .0 | 80 | .0 | 50 | .0 | 3,490 | .0 |
| New York | 2,910 | 120 | 61 | 54 | 26 | 274 | .0 | 45 | 16 | 6,990 | 8,470 | 10,500 | 8,490 |
| North Carolina | 805 | 103 | 17 | 114 | 201 | 390 | 5.5 | 96 | .0 | 7,210 | .0 | 8,940 | 5.5 |
| North Dakota | 76 | 12 | .1 | 164 | 24 | 8.8 | .0 | 3.7 | .0 | 2,390 | .0 | 2,680 | .0 |
| Ohio | 1,300 | 134 | 36 | 18 | 34 | 354 | .0 | 243 | .0 | 9,550 | .0 | 11,700 | .0 |
| Oklahoma | 515 | 41 | 6.3 | 601 | 131 | 35 | .0 | 2.8 | 243 | 89 | .0 | 1,420 | 243 |
| Oregon | 470 | 64 | 711 | 6,500 | 21 | 284 | .0 | 1.5 | .0 | 15 | .0 | 8,070 | .0 |
| Pennsylvania | 1,730 | 141 | 24 | 14 | 53 | 1,870 | .0 | 252 | .0 | 5,750 | .0 | 9,830 | .0 |
| Rhode Island | 102 | 4.9 | 5.6 | 2.1 | .3 | 12 | .0 | 6.8 | .0 | .0 | 393 | 133 | 393 |
| South Carolina | 359 | 62 | 41 | 34 | 10 | 1,130 | .0 | 5.3 | .0 | 5,180 | 6.3 | 6,810 | 6.3 |
| South Dakota | 75 | 8.8 | 17 | 392 | 43 | 15 | .0 | 38 | .0 | 3.2 | .0 | 592 | .0 |
| Tennessee | 695 | 51 | 55 | 38 | 49 | 882 | 1,467 | 90 | 491 | 7,130 | .0 | 9,180 | .0 |
| Texas | 3,081 | 93 | 62 | 8,493 | 228 | 884 | 2.3 | 139 | 98 | 7,129 | 3,154 | 20,117 | 5,103 |
| Utah | 508 | 5.1 | 4.2 | 3,590 | 34 | 106 | .0 | 41 | 87 | 87 | .0 | 4,380 | .0 |
| Vermont | 39 | 17 | 3.8 | .5 | 6.1 | 44 | .0 | 3.7 | .0 | 519 | .0 | 632 | .0 |
| Virginia | 709 | 113 | 35 | 36 | 29 | 495 | 66 | 91 | .0 | 3,210 | 2,080 | 4,710 | 2,150 |
| Washington | 875 | 104 | 27 | 6,030 | 30 | 501 | 36 | 3.0 | .0 | 334 | .0 | 7,910 | 36 |
| West Virginia | 160 | 49 | 3.1 | .0 | 4.8 | 132 | .0 | 527 | .0 | 3,710 | .0 | 4,580 | .0 |
| Wisconsin | 595 | 90 | 11 | 151 | 99 | 468 | .0 | .2 | .0 | 5,100 | .0 | 6,510 | .0 |
| Wyoming | 88 | 8.5 | 1.6 | 7,160 | 27 | 16 | .0 | 101 | 19 | 184 | .0 | 7,580 | 19 |
| Puerto Rico | 404 | 6.7 | .1 | 140 | 8.3 | 11 | .0 | 2.6 | .0 | 2.6 | 2,470 | 576 | 2,470 |
| Virgin Islands | 4.5 | 1.6 | .9 | .0 | .0 | .0 | 14 | .0 | .0 | .0 | 103 | 7.1 | 117 |
| Total | 38,470 | 3,436 | 2,464 | 135,361 | 5,107 | 19,700 | 3,257 | 4,153 | 1,650 | 130,645 | 65,077 | 337,312 | 69,515 |

*Source*: USGS 1990, preliminary for 1990 report see [2]. Figures may not add to totals because of independent rounding.

**TABLE 4.3** SUMMARY OF ESTIMATED USE OF WATER IN THE UNITED STATES, IN THOUSANDS OF MILLION GALLONS PER DAY, AT FIVE-YEAR INTERVALS, 1950–1985

| | Year | | | | | | | | Percentage change 1980–85 |
|---|---|---|---|---|---|---|---|---|---|
| | [a]1950 | [a]1955 | [b]1960 | [b]1965 | [c]1970 | [c]1975 | [d]1980 | [d]1985 | |
| Population, in millions | 150.7 | 164.0 | 179.3 | 193.8 | 205.9 | 216.4 | 229.6 | 242.4 | +6 |
| Offstream use | | | | | | | | | |
| Total withdrawals | 180 | 240 | 270 | 310 | 370 | 420 | [e]440 | 400 | −10 |
| Public supply | 14 | 17 | 21 | 24 | 27 | 29 | 34 | 37 | +7 |
| Rural domestic and livestock | 3.6 | 3.6 | 3.6 | 4.0 | 4.5 | 4.9 | 5.6 | 7.8 | +39 |
| Irrigation | 89 | 110 | 110 | 120 | 130 | 140 | 150 | 140 | −6 |
| Industrial: | | | | | | | | | |
| Thermoelectric power use | 40 | 72 | 100 | 130 | 170 | 200 | 210 | 190 | −13 |
| Other industrial use | 37 | 39 | 38 | 46 | 47 | 45 | 45 | 31 | −33 |
| Source of water | | | | | | | | | |
| Ground: | | | | | | | | | |
| Fresh | 34 | 47 | 50 | 60 | 68 | 82 | [e]83 | 73 | −12 |
| Saline | [f] | .6 | .4 | .5 | 1 | 1 | .9 | .7 | −29 |
| Surface: | | | | | | | | | |
| Fresh | 140 | 180 | 190 | 210 | 250 | 260 | 290 | 260 | −8 |
| Saline | 10 | 18 | 31 | 43 | 53 | 69 | 71 | 60 | −16 |
| Reclaimed sewage | [f] | .2 | .6 | .7 | .5 | .5 | .5 | .6 | +22 |
| Consumptive use | [f] | [f] | 61 | 77 | [g]87 | [g]96 | [g]100 | [g]92 | −9 |
| Instream use | | | | | | | | | |
| Hydroelectric power | 1,100 | 1,500 | 2,000 | 2,300 | 2,800 | 3,300 | 3,300 | 3,100 | −7 |

*Source:* USGS 1985 [2].

[a] 48 States and District of Columbia.

[b] 50 States and District of Columbia.

[c] 50 States, District of Columbia, and Puerto Rico.

[d] 50 States, District of Columbia, Puerto Rico, and Virgin Islands.

[e] Revised.

[f] Data not available.

[g] Freshwater only.

[Data for 1950–1980 adapted from MacKichan (1951, 1957), MacKichan and Kammerer (1961), Murray (1968), Murray and Reeves (1972, 1977), and Solley and others (1983). The data generally are rounded to two significant figures; percentage changes are calculated from unrounded numbers.]

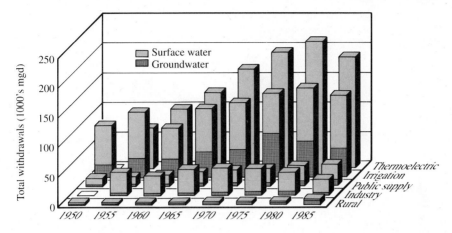

**Figure 4.5** Trends in water withdrawals by water-use category and source of supply for rural, industry, public supply, irrigation, and thermoelectric power uses, 1950–85 [2].

withdrawn in 1985, 71% was returned to the source after use, 6% was lost in irrigation conveyances, and 23% was consumptively used (evaporated or transpired). Of the amount of water consumptively used, irrigated agriculture was responsible for about 80%. U.S. Geological Survey staff indicate that 1990 data continue to support the notion of a more conservative water use trend, suggesting that there might have been a peaking-out of water use in the 1985–1990 period.

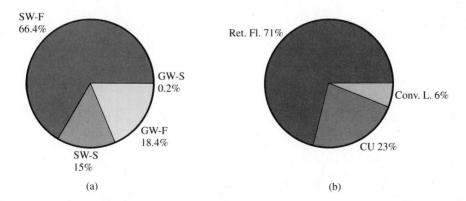

**Figure 4.6** Total offstream water withdrawals in the U.S. in 1985; (a) by source; (b) by disposition: SW-F = fresh surface water; SW-S = saline surface water; GW-F = fresh groundwater; GW-S = saline groundwater; Ret. Fl. = return flow; Conv. L. = irrigation conveyance loss; CU = consumptive use. (Data from USGS, [2].)

**TABLE 4.4**   ESTIMATED WATER USE IN THE YEAR 2000 FOR VARIOUS CATEGORIES FOR SEVERAL STATES

| State | Estimated total water withdrawal $W_T$ (bgd) | Municipal water withdrawals | | All withdrawals except steam electric, and agriculture | |
|---|---|---|---|---|---|
| | | (bgd) | (% of $W_T$) | (bgd) | (% of $W_T$) |
| Alabama | 24.5 | 1.5 | 6.12 | 4.5 | 18.36 |
| Alaska | 1.3 | 0.18 | 13.7 | 0.9 | 68.7 |
| Arizona | 10.3 | 1.3 | 12.62 | 2.3 | 22.33 |
| Pennsylvania | 16.5 | 1.5 | 9.1 | 6.5 | 39.39 |
| North Dakota | 8.1 | 0.7 | 8.64 | 0.7 | 8.64 |
| New Mexico | 4.6 | 0.4 | 8.69 | 0.4 | 8.69 |
| New York | 37.0 | 4.0 | 10.81 | 7.0 | 18.91 |
| Nebraska | 13.8 | 0.7 | 5.07 | 0.7 | 5.07 |
| Total | | | 74.75 | | 190.09 |

*Source:* "State and National Water Use Trends to the Year 2000," Committee on Environment and Public Works, U.S. Senate, Serial No. 96-12, May 1980.

■ **EXAMPLE 4.2**

Compare the estimated total water withdrawals for domestic and commercial use with total withdrawals for all purposes for the state water-use data given in columns 1–4 of Table 4.4. Then compare the sum of all water uses for these same states for purposes other than steam electric cooling and agriculture with the totals for the states in the year 2000.

*Solution.*

1. The comparisons are given in Table 4.4. The values in column 4 range from about 5% to almost 14% and are indicative of the relatively small fraction of municipal use to total water use. The figures in column 6 indicate that, in general, once agricultural and steam electric uses have been removed, the remaining water uses are normally well below 50% of the total.
2. The average of the values in column 5 is 9.34% (74.75/8). In like manner, the average of the values in column 6 is 23.76. A comparison of these values with the national average estimates made by the Water Resources Council in 1975 indicates that 12.1% for domestic and commercial use and 23.1% for all uses excepting agriculture and steam electric shows good agreement.

This example demonstrates the significance of irrigation water use and steam electric use contrasted with all other water-withdrawing categories. It can also be seen that the percentage of domestic and commercial use of the total is fairly uniform from state to state, at least for those states analyzed. The high percentages shown in column 6 for Pennsylvania and Alaska can be traced to the high percentage of industrial water use in these two states relative to that in the other states for which data were reported. ■

## 4.3  FACTORS AFFECTING WATER USE

To make reasonable forecasts of future water uses, we must understand the factors that affect them and the role these influences will play in the future. Important considerations are population size and character, water availability, economic conditions, attitudes toward conservation, the nature of regulatory programs, the types of management practices in force, statutory requirements, and social perspectives.

### Population

Population projections are important components of water-use estimating processes. The amount of water used in a locality is directly related to the size, disbursement, and composition of the local population. Consequently, errors in projecting population changes are carried over into water-use projections. Population projections used in water-use forecasting must be chosen carefully, and their probable variance recognized. In some cases, emphasis must be placed on estimating immigration rates and on dealing with the large transient tourist influx. For population as well as for other projections, the wise policy is to consider several alternative futures rather than a single futures forecast.

Population changes affect a locality's water requirements and the capability for providing the water needed to meet them. Water quality and thus the nature of water and waste treatment facilities are also influenced. Population is both a direct and an indirect measure of water use. Its influence, however, is felt most directly in the municipal water-use sector. Other subdivisions, such as thermoelectric cooling, are influenced mainly by such factors as technology and economics.

The level of a region's population depends on the combined effects of births, deaths, immigration, and emigration. Zero population growth is now a reality in some states and in many large metropolitan areas. If birth rates and death rates continue to converge, immigration could become the predominant factor in population change.

Forecasting future population is at best complex. There can be no exact determination, even though seemingly sophisticated mathematical equations are often used. War, technological developments, new scientific discoveries, government operations, and many other factors can drastically disrupt population trends. There is no way to predict many of these occurrences, and thus their impact cannot be estimated. Nevertheless, population forecasts are important and must be made. Both mathematical and graphical methods can be used in estimating future populations. Usually the computations are based on past census records for the area or on the records of what are considered to be similar communities. Most estimates are based on an extension of past trends and thus do not take into account such factors as the influx of workers when new industries settle in an area, the loss of residents due to curtailment of military activities, or changes in business or transportation facilities. To incorporate such happenings, information on anticipated industrial growth, local birth and death rates, government activities, and other related factors should be obtained and used. The local census bureau, the planning commission, the bureau of vital statistics, local utility companies, movers, and chambers of commerce are sources of information. Both short-term (up to ten-year) and long-term estimates are used in water supply planning.

The most widely employed mathematical or graphical methods for extending population data are (1) use of the arithmetical progression or uniform growth rate, (2) use of the constant percentage growth rate, (3) use of a decreasing rate of increase, (4) graphical extension, (5) graphical comparison with the growth rates of similar and larger cities, (6) the ratio method, based on a comparison of the local and national population figures for past census years, and (7) use of mathematical trends such as the Gompertz and logistic curves. Knowledge of the population of a region permits estimates to be made of the total quantities of water or wastewater generated in these areas. To design individual conveyance systems, however, additional information regarding the spatial distribution of the population to be served must also be obtained. Population densities may be estimated from data collected on existing areas. Table 4.5 can be used as a guide if local data are not available. Further information on population forecasting may be obtained from Viessman and Welty [7].

## Economic Conditions

Economic health is reflected in all aspects of resource management and development. Inflationary and other economic trends are influential in determining the nature and rate of growth in all sectors of the nation's economy. These trends also influence the availability of funds for water supply, wastewater treatment, and environmental and other programs; they also affect the attitudes of individuals. Uncertain economic conditions must be dealt with by water resources planners and managers as they attempt to estimate future needs and their timing. Adjustments in water-use forecasts must be provided

**TABLE 4.5**   GUIDE TO POPULATION DENSITY

| Area type | Number of persons per acre |
|---|---|
| Residential | |
|   Single-family units | 5–35 |
|   Multiple-family units | 30–100 |
| Apartments | 100–1000 |
| Commercial areas | 15–30 |
| Industrial areas | 5–15 |

*Source:* J. B. Wolff, "Peak Demands in Residential Areas," *J. Am. Water Works Assoc.* 87 (October 1961).

for as new information on changing economic circumstances provides better definition of unfolding trends.

A wide range in economic growth paths are possible for the United States and other nations between the present and the year 2000. According to the Edison Electric Institute, these alternatives are "fraught with uncertainty" [8]. This situation carries over into all areas of resource management and development and, accordingly, will influence projections of where and how much water is used. That water-use trends are so sharply influenced by economic and other conditions is good reason for the water resources planner to utilize all the information sources available in estimating the validity of projections.

## Desalination

Even though there are many technical and economic problems associated with the widespread use of saline-to-fresh water conversion, there appears to be an increased interest in this approach to augmenting water supplies. California and Florida already have a number of brackish-water conversion plants, and California, as a result of extended drought, is seriously considering seawater conversion. In February 1980 *Chemical and Engineering News* reported that one optimistic forecast sees a U.S. desalting capacity of about 30 bgd by the year 2000. Stricter water-quality regulations could also be partially met through the use of appropriately treated saline waters. In this case, the costs of conversion would not be a deterring factor since the costs would be small compared to other aspects of energy resources development. In addition, fairly large increases in saline-water use for power cooling are forecast. Any stepped-up tendency by industry to convert saline waters for various purposes would affect previously developed projections of freshwater use.

## Environmental Protection

Social attitudes toward environmental protection and enhancement strongly affect water resources management and/or development processes. This has been particularly true since the beginning of the environmental movement in the 1960s. Old conflicts among water users have been sharpened, and new conflicts among traditional water users, conservationists, and others have emerged. The issue of reserving instream flows and/or setting water levels for the enhancement or preservation of fish and wildlife, for example, is a case in point. Such nontraditional water allocations are often seen to be in conflict with those for customary uses, and heated political debates about them are not uncommon. Accordingly, water-use forecasts and/or allocations must be redesigned to take into account changing attitudes about what constitutes a beneficial use, and how society perceives the importance of allocating water for environmental and other purposes. Increased emphasis on the regulation of water quality also affects how water is used and the quantities used as well. Projections of future water needs will also have to be made respecting the constraining influences of these environmental actions.

## Conservation

Attractive alternatives to the development of new water supplies are conservation practices and the reuse of wastewaters. These approaches, although not a panacea, can at least delay the need for increasing water supplies and/or developing new facilities. Considering the large amount of freshwater used for irrigation, efforts to improve efficiency of water use in agriculture are expected to intensify. Furthermore, increases in population in coastal regions, with the associated problems of seawater intrusion, suggest that these areas are prime targets for accelerated programs in demand management and wastewater reuse and recycling. Unquestionably, wastewater reuse will become more popular as population pressures in urban areas increase, shortages of traditional water supplies become more pronounced, and those living in water-rich areas become more reluctant to transfer their water resources to other regions. The fact that about 50% to 75% of the freshwater withdrawn by most cities is discharged as waste indicates that the quantities of water can be substantial.

Estimates of water savings in agriculture range from 20% to 50%; for municipalities they may range up to 20% or more by changing water-use patterns and modifying plumbing codes. Obviously, the effects such practices have on future water uses must be taken into account. But assumptions made relative to the influence of conservation practices on water use will bear close scrutiny. Overly optimistic reductions could be projected, for example, if theoretical rather than practical considerations prevail.

## Management Practices

A host of management practices can significantly influence trends in water use, including interbasin transfers, weather modification, saline-water conversion, water reclamation and reuse, and genetic engineering. These and other factors must be taken into consideration by water planners as they formulate scenarios for the future, a task easier said than done. The opportunity for technologic change to affect water use in all sectors is significant, and it must be recognized in planning processes. New water-saving devices, process designs that reduce industrial water requirements, and modified varieties of crops are some examples.

## Social Perspectives

The attitudes of people shift with time regarding resource management issues. At the turn of the century, Floridians supported draining the Everglades to recover land for agricultural and other purposes. Today, there is a concern about preserving and/or restoring environmental values. In 20 years, it is hard to say what might take place. Nevertheless, these shifts in perspective and attitude have significant implications for water resources planners and managers regarding the amount of water used and its purpose. To the extent possible, water-use forecasts must incorporate trends in society's attitude toward the relationship between water use and development and environmental management.

## Tourism

Some states, such as Florida, have tourist populations that significantly exceed their resident populations. The Florida governor's office estimated, for example, that the annual number of tourists in 1980 was about 32 million, and that by 1990 this transient population would increase to about 55 million (1991 resident population was about 13 million). All of these nontaxpaying individuals expect adequate water supplies, and meeting them can be very stressful to local environmental systems as well as to water treatment and distribution infrastructures. Such impacts must be dealt with in forecasting future water supply needs.

## Technological Change

The opportunity for technological change to affect water use is significant. New water-saving devices, process designs that reduce industrial water requirements, and modified plant and crop varieties are among the many possibilities for reducing water use technologically. Such impacts will have to be assessed as water-use projections are made. As new technologies emerge, revisions in forecasts involving their influences will also be needed.

## Cost of Water

Changes in the cost of water have obvious implications for the quantities of water used in various sectors. It is well known that many water users pay only a small fraction of the cost of developing and furnishing their water supplies. This has led to inefficient water use and has fostered resistance to paying a fair price for the water used. A notable example is irrigated agriculture. In some parts of the western United States, farmers pay as little as $3.50/acre·ft for water whose replacement costs might be as high as $160/acre·ft. That subsidies of this magnitude discourage conservation is easy to see. It also follows that the implementation of pricing policies reflecting the true cost of water can have far-reaching effects on water-use trends [9].

## Interbasin Transfers

Regional water-use trends could be affected by interbasin transfers of water. In this regard, the potential of such undertakings should be evaluated when projections of water use are made. The degree of influence that shifts of water from one region to another might have on an area's total water use will depend on the scale of the transfer. Large reallocations such as those envisioned in the North American Water and Power Alliance plan could significantly alter certain water-use patterns, for example, in the southwestern United States [10].

## Weather Modification

Although climate is an important factor in a region's water-use practices, its effects are generally static over time. Now, however, weather—the day-to-day manifestation of climate—appears to be susceptible to change through human intervention. As a result, water-use trends may be affected in the future. The extent of these influences is difficult to predict at present because of the primitive state of weather modification technology and the many unanswered legal questions surrounding the inducement of weather changes. In the future, however, weather control could become an important factor in water resources planning in the United States and other parts of the world.

## 4.4 RECLAIMED WATER

Wastewater reuse offers attractive alternatives to developing new water supplies that might have to be transported from distant locations. Municipal waste flows are being widely used, especially in the more arid western and southwestern states. In California and Texas, for example, reuse projects are numerous. Wastewater reuse appears to be particularly attractive in areas where rainfall is low, evaporation is high, irrigation water use is intense, and interbasin transfers of water are being practiced or planned. Environmental regulations, water scarcity, and economic factors are expected to accelerate reuse of waste streams. This will undoubtedly affect water-use trends in the future, especially in the industrial and agricultural sectors. Figure 4.7 gives a good indication of the significance of wastewater reclamation to the water supply future of Orange County, California. By the year 2010, it is projected that about 20% of the Orange County Water District's water needs will be met by the use of reclaimed wastewater.

In the past the impetus for wastewater reuse was often the alleviation of water-quality problems caused by discharge of waste streams to surface waters or the desire to reduce costs of effluent treatment. Now, wastewater is being recognized more and more as a potentially valuable resource itself. In areas of the United States where the population is growing, particularly in the arid southwest, reuse of wastewater is seen as a means of easing the strains on tight water supplies. Other benefits of reuse include improved quality of surface waters, preservation of higher-quality water for potable consumption, and added recreational opportunities [11].

In direct reuse, wastewater is treated and piped directly into some type of water system without intervening travel dilution in natural surface water or

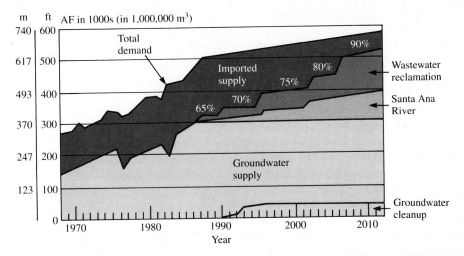

**Figure 4.7** Projected distribution of source waters for Orange County, California water supply to the year 2010. (From L. W. Owen and W. R. Mills, "California's Orange County Water District: A Model for Comprehensive Water Resources Management," in Water Resources: Planning and Management and Urban Water Resources, Proceedings of the 18th Annual Conference and Symposium of the Water Resources Planning and Management Division of ASCE, New York, NY, May 1991, p. 5.)

groundwater bodies [12]. For example, connection of a municipal wastewater plant to an irrigation site provides a direct water supply to that site. Indirect reuse involves a middle step between generation of reclaimed water and reuse. This commonly includes discharge, retention, and mixing with another water supply before reuse. Recycling refers to use by the same business, plant, or person two or more times in a coordinated, planned manner, sometimes with partial treatment between uses.

The most widely available and least variable source of wastewater for reuse is municipal wastewater. It can be relied on to provide a dependable continuous flow having fairly stable physical, chemical, and biological characteristics. Reuse of municipal wastewater may be accomplished in a number of ways with treatment ranging from none to the most advanced systems available, depending on the end use of the water. Municipal wastewater reuses are varied. For example, in California municipal wastewaters are reused for fiber and seed crop irrigation, landscape irrigation, orchard and vineyard irrigation, processed food crop irrigation, groundwater recharge, food crop irrigation (not processed), restricted recreational impoundments, pasture for dairy animals, and unrestricted recreational impoundments [13].

In industrialized sections of the United States, manufacturing operations can contribute heavily to the total wastewater generated in an area. The constituents of the wastewater are highly industry specific and may limit the reuse options available for that wastewater. On-site wastewater reuse by manufacturing plants has become more important as environmental standards are tightened by regulatory agencies.

In agriculture, wastewater from irrigation operations is a potential source of reusable water. However, the water is often highly contaminated with salts leached from the soil, and in many cases treatment is required if the water is to be reused on the fields. Desalination is technically feasible but often prohibitively expensive.

Wastewater reclamation processes must ensure removal of residual pollutants to such a degree as to make the water acceptable for the designated reuse. The pollutants that must be removed depend on the desired use of the water and its previous use. If municipal wastewater has undergone secondary treatment, the remaining pollutants that typically must be removed to make the wastewater suitable for reuse are nitrates, phosphates, total dissolved solids, microorganisms, and refractory organics such as trace levels of pesticides [12, 14]. All of these constituents need not be removed should their presence be advantageous in the product water. For example, phosphates and nitrates are desirable in reclaimed water for use in irrigation because the constituents act as nutrients for plant life. The state of California has established water-quality requirements for different levels of reuse [15].

For high-quality effluent, *tertiary* or advanced treatment processes are required to achieve the degree of contaminant removal necessary to make wastewater acceptable for reuse. These processes are generally proved technologies for wastewater reclamation and some are readily available as "off-the-shelf" package units, but economics is the driving factor in limiting the attractiveness of certain processes for widespread use.

**Municipal Reuse**    Wastewater can be reused for municipal purposes either directly or indirectly. Direct municipal reuse requires advanced treatment of wastewater to meet stringent water-quality requirements if the water is to be used for drinking-water purposes (not presently practiced in the United States). If a dual-distribution system is available, water of lesser quality may be directly used for nonpotable purposes, such as for landscape irrigation.

Indirect reuse is the reuse of treated municipal wastewater as a raw-water supply after the wastewater has entered, commingled, and become a part of a natural surface water or groundwater resource. Anyone located in an area where the water supply system uses surface water for intake at a point downstream from a discharging municipal wastewater treatment plant is indirectly reusing the wastewater. The natural assimilative capacity of the river acts to purify and dilute the discharged water to such a degree as to make it more acceptable as intake water for municipal water supplies. This practice has been accepted for years with little questioning of its public health significance.

Indirect municipal reuse of wastewater is also accomplished by discharging reclaimed water to groundwater aquifers (artificial recharge) and later withdrawing the water for potable or other municipal reuse. Discharge to aquifers can be carried out by either deep-well injection or by shallow surface spreading with subsequent percolation to groundwater. This latter practice benefits from the extra purification provided by percolation through the soil. Both practices benefit from dilution of reclaimed water with existing groundwater. The natural aquifer provides a convenient storage area that is beneficial in case of uneven demand.

There are still many unanswered questions regarding the health effects of indirect (via groundwater) and direct wastewater reuses for drinking-water purposes. There is a general fear on the part of the public that microorganisms and refractory organic chemicals in the wastewater designated for reuse may escape treatment, enter the water supply, and pose a threat to human health. Central unresolved issues involve deciding what chemical constituents are acceptable in potable water and at what concentrations. Water Factory 21 in California is injecting reclaimed water of high quality into selected aquifers with the goal of reusing it for potable purposes once health officials can agree on water-quality standards that are acceptable to the public.

Thus far, nowhere in the United States has wastewater been directly reused as drinking water. The goal of direct reuse is, however, being considered for Denver, Colorado. In 1970, when the Denver Water Board predicted future water shortages, a series of pilot studies was initiated to determine the feasibility of wastewater reclamation for direct potable consumption. With current usage of 100,000 acre·ft/yr from surface supplies and 250,000 acre·ft/yr from interbasin water transfers, Denver perceived that an additional source of water was needed to allow for future growth. Pilot studies were completed in 1979. Although the water reclamation technology has proven effective on a pilot-scale basis, the question of water quality remains. To provide data for a complete assessment of health-related issues, Denver built a 1-mgd demonstration plant consisting of lime clarification, recarbonation, filtration, selective ion exchange, first-stage carbon adsorption, ozonation, second-stage carbon adsorption, reverse osmosis, air stripping, and chlorine dioxide disinfection. The plant is designed to provide the water needed for complete analytical quality testing and health effects research, as well as the design criteria for scale-up to an operating facility. The testing program is being used to compare the quality of direct potable reuse water with Denver's current drinking water quality. This program, coupled with an extensive public education campaign, may pave the way for direct potable reuse in Denver by the turn of the century [16, 17].

**Industrial Reuse**  Scarcity of resources, environmental constraints, and economics are major factors that have caused industry to turn increasingly to recycling and reusing its wastewater, and to take in less raw water [18]. These practices include (1) treating some or all process wastewaters to make them suitable for plant process makeup water, e.g., cooling tower water; (2) recirculating the same water within a unit process or group of processes before discharge; and (3) sequential use of effluent from one process as input into another with optional treatment between processes [19–21]. These practices also provide a savings in energy (for pumping, heating, and treating) as well as in water costs.

Reclaimed municipal wastewater is also a suitable source for certain industrial processes, depending on the economics of treatment and distribution and the quality of the water. The prime applications have been for use as cooling water and boiler feedwater. Other uses will become feasible as the quality of the reclaimed water increases [18].

**Irrigation Reuse** Direct irrigation with wastewater has become more widespread since land treatment of municipal wastewater has been encouraged by the EPA as an "innovative and alternative" wastewater treatment technology. Both can be accomplished at the same time, yielding a savings in energy and in cost of treatment. Additional benefits include augmenting groundwater supplies, utilization of the nutrients in the wastewater as fertilizer, and preservation of the quality of surface waters.

Land treatment involves the spreading of wastewater on land and making use of the natural filtration and adsorption properties of the soil to remove contaminants as wastewater percolates downward to the groundwater zone. Low-rate land application systems are suitable for irrigation. The detention times allow plants to take up nutrients from the wastewater. There are public health concerns, however, regarding the levels of nitrates, pathogens, and refractory organics that may be present in the wastewater.

Indirect reuse for irrigation is quite common, either from aquifers that have been recharged with reclaimed water or from rivers downstream of a wastewater discharge where the river water is withdrawn for irrigation purposes.

**Storm Water Reuse** In urban areas where storm water is collected by a system separate from that which collects the municipal wastewater, a source of reusable water is provided to the city that, with a certain amount of treatment, is acceptable for a number of uses [22]. Storm water is typically low in biochemical oxygen demand (BOD; see Chapter 8) (comparable to the BOD level of effluent from a secondary municipal wastewater treatment plant) but high in suspended solids (often at a level comparable to raw municipal wastewater). Since the water is characteristically high in solids and has a highly variable flowrate, physical treatment processes are most suitable. By employing various combinations of physical unit processes, effluents of various qualities can be produced. This flexibility is desirable because the end use of the water dictates the minimum degree of treatment needed to make it suitable for reuse.

**Recreational Uses** The project at Santee, California, is a demonstration of the reuse of wastewaters for recreational purposes [23, 24]. The Santee County Water District in San Diego County, California, has undertaken a reclamation project that has provided for the development of a series of artificial lakes fed by the effluent of a standard activated-sludge treatment plant. The effluent from this plant is held in a 16-acre lagoon for 30 days (lake 1). It is then chlorinated and spread onto a percolation area of sand and gravel through which it travels for about 0.5 mi. The percolated water is collected in an interceptor trench and then flows into lake 4 (11 acres), then into lake 3 (7.5 acres), and ultimately into lake 2 (6.8 acres), after which the overflow passes on to Sycamore Canyon Creek. Lakes 3 and 4 are used for fishing and boating. Playgrounds and picnic areas line the shores. In 1965 lake 5 was approved for swimming by the California Department of Public Health. This approval followed three years of study by an Epidemiology Advisory Committee of the California Department of Public Health. The state health director indicated that biologic, hydrologic, bacteriologic, and virologic information supported approval by the committee. This community recreational area is provided through the utilization of wastewaters and permits an aquatic environment in an area that affords little attraction of this type. The public acceptance of these facilities has been exceptional. Considering the important emphasis being placed on the recreational use of water resources, there is little doubt that expanded development of facilities of this type is forthcoming.

Reclaimed wastewaters are also being used successfully for the irrigation of golf courses in many areas of the southwest. Examples are the El Toro Marine Air Base near Santa Ana, California; Camp Pendleton Marine Base near San Diego; the Naval Ordnance Test Station near China Lake in California; Eagle Pass, Texas; Prescott, Arizona; Santa Fe, New Mexico; and Los Alamos, New Mexico.

**Economics of Reuse**    If the no-discharge goal of the 1972 Federal Water Pollution Control Act is achieved, the qua ty of effluent discharge would exceed that of most natural water supplies. Thus, the reuse of all effluent waters would be achieved without additional cost.

The true direct net cost of treatment for reuse can be expressed as follows [25]:

1. the cost of advanced treatment to make wastewater suitable for reuse;
2. minus the cost of pollution control treatment measures otherwise necessary to achieve water-quality standards;
3. minus the cost of water treatment of the supply being considered as an alternative to reuse;
4. plus or minus the difference in conveyance costs between the reusable supply and its alternative, including allowance for the cost of separate supply lines if reuse is contemplated for the industrial water supply only.

The costs of secondary and advanced waste treatment of wastewater for reuse will vary from one locality and situation to another, and reuse may not be the best water management alternative in all cases.

## 4.5  WATER AVAILABILITY

The availability of water is as important a consideration as is an understanding of how much water is needed for anticipated uses. Several supply-side aspects deserve comment. They include the location, extent, and quality of (1) natural surface-water and groundwater sources and (2) of sources resulting from water-use practices. Traditionally, natural sources have been considered almost exclusively by those desiring to expand water supply capability. These sources are the principal water stocks, but other, nontraditional sources of supply must be given more attention in the future. Discharges from treatment plants, urban stormwater flows, and irrigation return flows all may be looked upon as sources of water supply. Often, these flows are very large relative to some water supply needs. It is not uncommon for about 75% of treated potable water to be returned to a discharge point after its use. Along Florida's lower east coast, for example, the quantities of water generated for potable water use and then discharged as treated waste flows constitute a significant fraction of the daily water supplied to that area. The infrastructure needed to efficiently reuse all of these waters is not yet in place, and the economic and public health aspects of reuse are constraining factors, but these discharges have never been cast prominently in water-source assessments or considered appropriately in water supply planning strategies. It is time that such sources be given full consideration as alternative soruces of water supply. Furthermore, many states have access to brackish and/or seawater sources, and while the costs of their conversion and their associated brine waste problems have suppressed most large-scale saline-water conversion programs in the United States, to date, there is increasing interest in using them, particularly as water shortages become more frequent and conversion costs decrease. Finally, importation of water across political and natural boundaries may be a viable water supply option, providing that political, economic, and environmental circumstances are favorable.

## 4.6  WATER USE

The water supply problem is one of balancing supply and demand. Basically, demand-side considerations include identification of the water-using sectors (traditional, such as industry, and nontraditional, such as fish and wildlife), assessment of the quantities of water being used by these sectors, identification of trends in water use, forecasts of future needs, analyses of water allocation strategies and regulatory practices, studies of the feasibility of reducing demands (during critical periods and in a sustained mode), research on water-saving methods, and analyses of existing infrastructures to ascertain the feasibility of combining and/or operating them differently to achieve water supply targets. Many of the demand-side factors are more political and sociological in character than they are technical. Regulatory and pricing practices are often the key features.

# 4.7 WATER-USE FORECASTING

Forecasting is the art and science of looking ahead; it is the core of planning processes. Options for selecting a forecasting procedure range from the exercise of judgment to the use of complex mathematical models. Forecasts are required for periods varying from less than a day to more than 50 years. Unfortunately, the more distant the planning horizon, the more questionable is the forecast. It is therefore important that a forecasting technique be selected carefully and its limitations understood. Furthermore, it is recommended that an array of alternative futures, rather than a single projection, be used to guide decision-making processes. Forecasts should be recognized for what they are—approximations of what might occur, not accurate portrayals of what will be. Commonly used forecasting methods include projections based on historic data, the use of models and simulation, and qualitative and holistic techniques [7].

The Delphi technique, an expert polling process, and scenario writing, the design of plausible futures, are examples of qualitative/holistic approaches, while trend extrapolation and trend impact analysis are history-based procedures [7]. Simulation is the process of mimicking the dynamic behavior of a system over time. A simulation model is a surrogate of the real-world system it is designed to represent. The results of simulation computer runs are widely used to describe (forecast) the future states of water or other systems being studied.

Many forecasting models have been designed to address water supply issues. In general, they range from rough trend extrapolations and crude correlations of variables (mainly population and water use) to complex mathematical representations of the dynamics of land and water use. A mathematical modeling approach and a geographic information systems (GIS) approach are presented to illustrate the variety of techniques available.

## IWR-MAIN Model

IWR-MAIN is a computerized water-use forecasting system that includes a range of forecasting models, social and economic parameter-generating procedures, and data management techniques [26–30]. The model features a high level of disaggregation of water-use categories and user flexibility in selecting forecasting methods and assumptions. It is designed to deal principally with urban water uses: residential, commercial/institutional, manufacturing, and public/unaccounted. The model also provides for forecasts of water use that recognize the influences of conservation policies. Figures 4.8 and 4.9 show the relationships among the principal elements of the model. Figure 4.10 indicates the data requirements and the general procedure for deriving disaggregate water-use estimates in response to data input by the user.

The IWR-MAIN model differs from most conventional water demand models in that its forecast is sectorially (industrial use, for example), spatially,

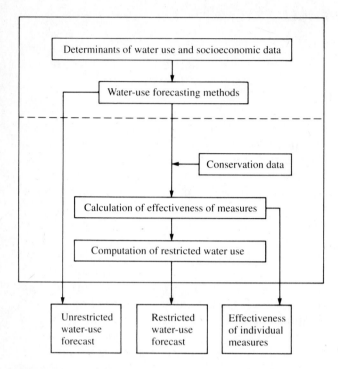

Figure 4.8 The IWR-MAIN water-use forecasting system. (Courtesy of Planning and Management Consultants Ltd., Carbondale, IL, [29].)

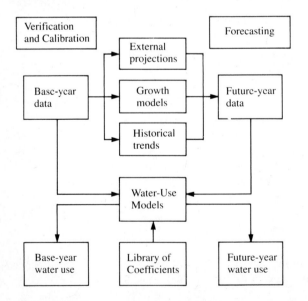

Figure 4.9 Major components of the IWR-MAIN system (after Dziegielewski, 1987 [27]).

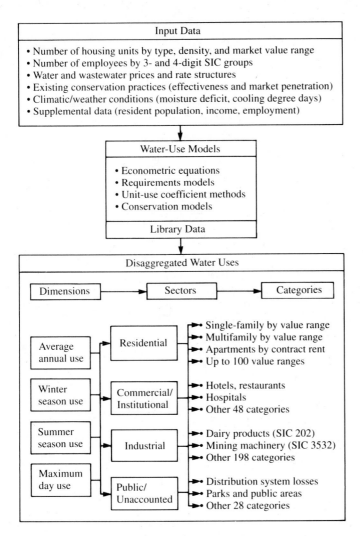

| Input Data |
|---|
| • Number of housing units by type, density, and market value range |
| • Number of employees by 3- and 4-digit SIC groups |
| • Water and wastewater prices and rate structures |
| • Existing conservation practices (effectiveness and market penetration) |
| • Climatic/weather conditions (moisture deficit, cooling degree days) |
| • Supplemental data (resident population, income, employment) |

| Water-Use Models |
|---|
| • Econometric equations |
| • Requirements models |
| • Unit-use coefficient methods |
| • Conservation models |
| Library Data |

Disaggregated Water Uses

| Dimensions | → | Sectors | → | Categories |

Average annual use → Residential →
→ Single-family by value range
→ Multifamily by value range
→ Apartments by contract rent
→ Up to 100 value ranges

Winter season use → Commercial/Institutional →
→ Hotels, restaurants
→ Hospitals
→ Other 48 categories

Summer season use → Industrial →
→ Dairy products (SIC 202)
→ Mining machinery (SIC 3532)
→ Other 198 categories

Maximum day use → Public/Unaccounted →
→ Distribution system losses
→ Parks and public areas
→ Other 28 categories

**Figure 4.10** IWR-MAIN procedure for estimating water use (after Dziegielewski, 1987 [27]).

and seasonally disaggregate (i.e., broken down into basic components); and it is a function of all likely determinants of water demand. Separate forecasts can be made for each identified sector, for specific study areas such as cities, counties, and load zones, for summer and winter daily water use, and for average daily, annual, and peak-day water use. In addition, the model provides for inclusion of up to 18 different demand management or conservation practices.

Output from the IWR-MAIN model can be used to aid in the following tasks:

planning to meet future water supply needs;

sizing and expanding distribution systems;

preparing contingency plans for water shortages;

evaluating the effectiveness of conservation practices;

performing sensitivity analyses using varying assumptions about water prices, climatic factors, and other determinants of water use; and

assessing utility revenues with improved precision.

Various water-use estimating techniques and equations are employed in the computational subroutines of the model (Tables 4.6 and 4.7 are examples).

To use the IWR-MAIN model, one needs a complete set of data for a base year. Verification runs are made to determine whether the internal water-use models need to be adjusted (calibrated) so that the model's output conforms to the base-year and other historic-year observations. The base-year data can then be projected to any forecast year by using a set of parameters or explanatory variables. Each parameter must be projected to the forecast year for which water use is to be estimated. In the IWR-MAIN model there are three options for doing this: projection by internal growth models, projection by extrapolation of local historic data; and the use of projections made externally. A water conservation effectiveness algorithm permits estimating the impacts of existing and/or proposed demand reduction measures if the user desires to employ it.

The IWR-MAIN model has been applied to water-use forecasting in the United States, including Phoenix, Arizona; Springfield, Illinois; El Paso,

**TABLE 4.6**  ORGANIZATION OF THE IWR-MAIN SYSTEM

| Sector | Water-use category | Forecast method |
|---|---|---|
| Residential | Metered, sewered residences<br>Flat-rate, sewered residences<br>Flat-rate, septic tank residences<br>Master-metered apartments | Econometric demand model<br>Mult. coef. rqrmts. model<br>Mult. coef. rqrmts. model<br>Mult. coef. rqrmts. model |
| Commercial/<br>  Institutional | Up to 50 user categories, defined<br>  as groups of 4-digit SIC* codes | Unit use coefficients<br>  (per employee) |
| Industrial | Up to 200 user categories,<br>  presently including 198 3-digit<br>  and 4-digit SIC-coded<br>  manufacturing categories | Unit use coefficients<br>  (per employee) |
| Public/<br>  Unaccounted | Up to 30 user categories, such<br>  as distribution system losses,<br>  free service | Unit use coefficients or<br>  per capita requirements |

* SIC is an abbreviation for Standard Industrial Classification.

*Source:* Courtesy of Planning and Management Consultants Ltd., Carbondale, IL [29].

**TABLE 4.7** RESIDENTIAL WATER-USE MODELS

| Dependent variable | Empirical equation[a] | Significance | | |
|---|---|---|---|---|
| | | $F$ | $F_{0.05}$ | $R^2$ |
| 1. Winter use (indoor) in metered and sewered areas (single-family) | $Q_1 = 234 + 1.451V - 45.9P_a - 2.59I_a$ | — | — | — |
| 2. Winter use (indoor) in flat rate areas with sewers (multifamily, apartments) | $Q_2 = 28.9 + 1.576V + 33.6D_p$ | 42.1 | 4.1 | 0.89 |
| 3. Winter use (indoor) in flat-rate areas with septic tanks (multifamily) | $Q_3 = 30.2 + 39.5D_p$ | 72.0 | 10.1 | 0.96 |
| 4. Sprinkling use (outdoor) in metered and sewered areas (western U.S.) | $Q_4 = 28.213P_s^{-0.703}V^{0.429}$ | 7.2 | 4.7 | 0.67 |
| 5. Sprinkling use (outdoor) in flat-rate areas (East and West) | $Q_5 = 18.275V^{0.783}$ | 10.5 | 6.0 | 0.64 |
| 6. Summer use (indoor and outdoor) in metered and sewered areas (eastern U.S.) | $Q_6 = 385.0 + 2.876V - 285.8P_s - 4.35I_s + 157.77(B \cdot MD)$ | — | — | — |
| 7. Maximum-day sprinkling use (outdoor) in metered and sewered areas (western U.S.) | $Q_7 = 2227.34E_m^{2.06}V^{0.413}$ | 10.0 | 4.7 | 0.74 |
| 8. Maximum-day sprinkling use (outdoor) in metered and sewered areas (eastern U.S.) | $Q_8 = 0.000046B^{0.118}E_m^{-10.4}P_s^{-1.25}V^{0.931}$ | 4.6 | 4.5 | 0.75 |
| 9. Maximum-day sprinkling use (outdoor) in flat-rate areas (East and West) | $Q_9 = 1609.59B^{0.943}V^{0.523}$ | 15.7 | 5.8 | 0.86 |

*Source:* Howe and Linaweaver [38], Howe [37], Boland et al. [50]. Courtesy of Planning and Management Consultants Ltd., Carbondale, IL [29].

[a] Regression coefficients of price and home value adjusted to constant 1980 dollars.

$V$ = market value of the residence in $1000s
$P_a$ = sum of water and sewer charges that vary with indoor water use in $/1000 gals
$I_a$ = effective Nordin/Taylor bill difference evaluated at average indoor use in $/billing period
$D_p$ = number of persons per dwelling unit
$P_s$ = sum of water and sewer charges that vary with outdoor use in $/1000 gals
$I_s$ = effective Nordin/Taylor bill difference applicable to average outdoor use in $/billing period

$B$ = irrigable landscape per dwelling unit in acres
$MD$ = summer season moisture deficit in inches
$E_m$ = maximum-day evapotranspiration in inches

Texas; Oklahoma; and Southern California. Results of the model's application in six areas in which water use exceeded 3 mgd revealed errors of estimation ranging from 0.3% to 3.1%. For smaller areas, errors have been recorded up to 8.4%. No water-use data are required to use the IWR-MAIN system, and the backcasts that have been made were based entirely on demographic and socioeconomic data.

IWR-MAIN is a mathematical model for urban water use that incorporates built-in data libraries, forecasting options, and procedures for incorporating demand reduction policies. It is a "data-intensive" model that allows a high degree of disaggregation. As of 1988, IWR-MAIN represented the state-of-the-art in its class, consolidating most of the elements of other contemporary approaches to water-use estimating.

The IWR-MAIN model has been expanded to incorporate benefit-cost measures for determining the economic feasibility of conservation measures, and to include long-term drought mitigation plans and emergency drought conservation programs. Figure 4.11 illustrates the interactions of the IWR-MAIN forecasting system, the benefit-cost procedure, and the drought optimization (DROPS) method.

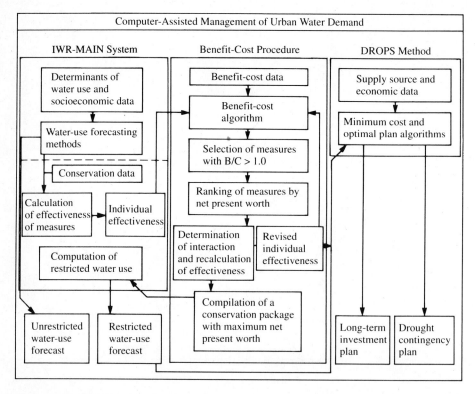

**Figure 4.11** Computer-assisted management of urban water demand. (Courtesy of Planning and Management Consultants Ltd., Carbondale, IL, [29].)

## Alafia River Basin Model

The Alafia River Basin model is an interesting departure from other approaches for projecting water use. It incorporates a GIS data base and display capability as well as a forecasting component to simulate future land-use demand scenarios. The model is a spatially distributed and time-incremented land- and water-use model of the Alafia River Basin in Florida [31]. Total water use is estimated for three major water-use categories: urban, agricultural, and phosphate mining per square mile section per month. Most of the baseline data on land use were obtained from county tax appraiser tax tapes by grouping various detailed land-use codes. Baseline data on water use were estimated from metered systems records, agricultural benchmark farm operations, water utilities records, and other sources of information.

While dynamic simulation models have been widely used for many years, there are few examples of models that spatially distribute results. Accordingly, a series of computer programs fitted to simulating change and spatial location of land use on a regional basis had to be designed. Five elements emerged as necessary for carrying out this task: (1) design of a baseline data base of existing basin features; (2) dynamic simulation of basin growth; (3) spatial distribution of projected growth within the basin; (4) projection of water use to accommodate estimated growth; and (5) program documentation for technology transfer. Graphical map representations of land and water use in the basin were developed as output, but the authors cautioned that these should not be interpreted as accurate reflections of land or water use at specific locations. They were intended to indicate the regional character of the data.

The simulation program employed was seeded by data collected to emulate the environmental, cultural, and developmental trends experienced within the basin. The simulation model projected past trends through the present into the future. A three-step procedure typified the simulation of future land-use and water-use patterns: (1) simulation of future land-use demand scenarios using an energetics circuit model; (2) allocation of future land-use demand to specific geographic areas within the basin; and (3) calculation of future water use and pumpage based on projected land use. Figure 4.12 illustrates the modeling concept employed to project future land and water use in the basin.

The forecasting model employed was designed to predict alternative futures for the basin based on current conditions, history, and future scenarios. It includes natural, agricultural, phosphate mining, and urban subsystems. As the model operates, the various subsystems compete for available land and resources within the basin and for supporting goods and services outside the basin boundary. Model outputs are future land acreage and land-use type. The model forecasts the total acreages of natural land, agricultural crop land, phosphate mining land, and urban land for a specified forecast period. The acreage estimated for each land-use type is then applied to the land distribution model by year.

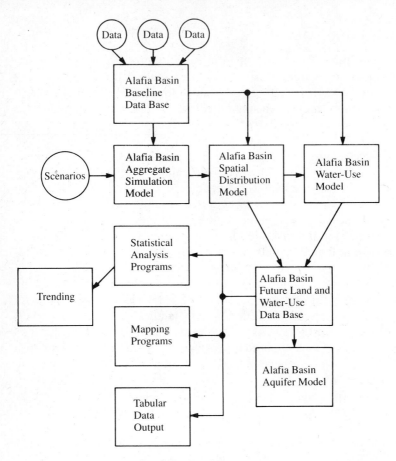

**Figure 4.12** Modeling concept employed to project future land and water use within the Alafia River Basin, Florida (after Alexander, 1983 [31]).

Once land-use patterns are projected, water-use coefficients are applied to them to yield estimates of total water use by land section and water-using sector. The model estimates both water use and water pumpage. Pumpage is located at the point of withdrawal, while usage is determined at the point of consumption.

There are many other models for forecasting water use, but most of them are narrow in scope, and almost all have an urban focus. Prediction equations for many models are empirical and have been derived from regression analysis and/or other related techniques [32–36]. Most models developed since 1970 include the cost of supplying water as a parameter [29, 31, 33, 36–39]. The IWR-MAIN model, which is widely used, and the Alafia River Basin model, which is more experimental, represent the types of approaches that can be taken in water-use forecasting. Further information on forecasting models may be found in the references at the end of the chapter [40–50].

## 4.8  QUANTITIES OF WASTES

To develop treatment strategies for wastes generated by various sectors, we must determine the amount, timing, and characteristics of these wastes. In this section, quantities and timing are discussed; wastewater characteristics are addressed later in the chapters on water quality and treatment.

The principal wastewater producers are cities, industries, and agricultural operations. Generally, these waste streams are handled independently, although some industrial wastes are transported and treated in municipal systems. The time variation of flows is an especially important consideration since most conveyances are gravity-flow systems and they must be able to accommodate minimum flows at sufficient velocities for self-cleansing, as well as the peak flows. The flow in municipal sewers consists mainly of community wastes plus infiltration, although if there are illicit connections or if the sewer is a combined one, stormwater flows must also be considered.

## 4.9  WASTE FLOWS FROM URBAN AREAS

The quantity of sewage generated by a community depends on its population and thus the per capita contribution of sewage. Therefore, for accurate estimates of sewage flows, reliable population studies are required. Figure 4.13 gives a comparison of water use and wastewater flow on days without lawn sprinkling. The data are from the Pine Valley Subdivision in Baltimore

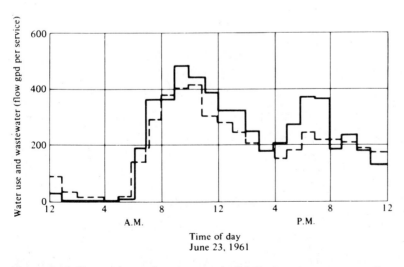

**Figure 4.13** Comparison of water use (solid line) and wastewater flow (dashed lines) on days when little sprinkling occurred. (From Residential Water-Use Research Project, Johns Hopkins University and Federal Housing Administration, 1963.)

County, Maryland [51]. Domestic sewage flows are highly variable throughout the day and, as in the case of the hydrograph of water use, two distinct
peaks have been observed. The primary peak takes place in the morning
hours, and the secondary peak occurs about dinner time and maintains itself
during the evening hours. Extraneous flows resulting from infiltration or
storm runoff tend to distort the basic hydrograph shape. Infiltration rates
generally tend to gradually increase the total daily volume but do not ordinarily alter the twin-peaked character of the hydrograph. Storm runoff that
enters the system may impose almost instantaneous changes; if the quantity
is large, the entire characteristic of the hydrograph may be changed. Estimation of the various components of the flow is essential for design purposes.
A 1963 study by Lentz of wastewater flows in communities in California,
Florida, Missouri, and Maryland provides considerable useful information
regarding residential flows and the components of these flows [51].

## Average Rates of Flow

Lentz and Linaweaver have shown that when residential water is not being
used for consumptive purposes (principally lawn sprinkling) and when infiltration and exfiltration do not produce large flow components, the wastewater
flow is essentially equal to the water use [4, 51]. Thus, average daily water-
use rates that do not reflect sprinkling demands can be used to estimate
annual average domestic wastewater flows. It is generally reported that about
60%–80% of the total water supplied to a community becomes wastewater.
Low ratios generally apply to semiarid regions. Ordinarily, the annual variation in the ratio of sewage to water supply in a city is not great, and thus the
amount of water used by a city is a good indicator of the amount of sewage
that will be generated. Sometimes, however, illicit drains and water use from
privately owned sources produce quantities of sewage larger than public
water withdrawals. When such external factors are not present, sewage flows
from residential areas are often less than about 100 gpcd.

## Variability in Sewage Flows

Sewage flowrates vary by source and with time. In most municipalities
the sources may be residences, institutions such as hospitals and schools,
commercial establishments, and industries. It is thus necessary for the designer and manager to determine the mix of these elements and to estimate
their contributions. Guideline figures are reported in the literature, but these
may not be reliable, especially where industrial wastes are involved [52]. A
case-by-case assessment should always be made before sewers are sized or
treatment plant capacities set. For institutions, flows may vary from as low
as 10 gpcd for schools to 175 gpcd for hospitals. Hotels may produce flows
of about 100 gpcd, while small business may generate only about 20–60 gal
per day per employee.

**TABLE 4.8**   RESIDENTIAL WASTEWATER
FLOWS AS RATIOS TO THE
AVERAGE

| Description of flow | Ratio to the average |
| --- | --- |
| Maximum daily | 2.25 : 1 |
| Maximum hourly | 3 : 1 |
| Minimum daily | 0.67 : 1 |
| Minimum hourly | 0.33 : 1 |

Sewage flow patterns in residential areas closely resemble water-use patterns for those areas, with the exception of a time lag as shown on Fig. 4.13. The magnitude of this lag varies with the situation, but it is usually on the order of a few hours. Where infiltration from rainfall events or other water uses enters sewers, the hydrographs produced in the sewers may vary considerably from those generated during dry periods. Otherwise, the variation in daily flow patterns for most residential areas is quite small. Where sewers receive significant quantities of wastes from industrial operations, the amounts and timing of flows are affected by prevailing industrial practices. Table 4.8 gives an indication of the variation in residential wastewater flows as ratios to the average. In the absence of site-specific data, such figures may be used to estimate high and low flows. Data on the ratio of peak flows and minimum flows to average daily flow have been summarized by Gupta [6]. They show that the peak to average daily flow ratio ranges from about 3.0 for cities of about 10,000 to about 1.5 for cities of one million. For the same size cities, the minimum flow to average daily flow ranges from about 0.3 to about 0.7, respectively.

Ordinarily, residential, industrial, and commercial sewage flows are estimated for an area and then combined to obtain the composite sewage hydrograph. In this way, detailed information regarding the nature of each flow component can be introduced into calculations.

## Infiltration and Exfiltration

Infiltration and exfiltration are both functions of height of the groundwater table in the vicinity of the sewer, type and tightness of sewer joints, and soil type. Exfiltration is undesirable since it may tend to pollute local groundwaters, while infiltration has the effect of reducing the capacity of the sewer for conveying the waste flows for which it was designed. If the sewer is well above the groundwater table, infiltration will occur only during or after periods of precipitation, when water is percolating downward through the soil. Where groundwater tables are high and sewers are not tight, infiltration rates in excess of 60,000 gpd/mi of sewer might be experienced. Rates of 3500–5000 gpd/mi/24 hr for an 8-in. pipe, 4500–6000 for a 12-in. pipe, and

10,000–12,000 for a 24-in. pipe represent the range in which the greater number of specifications fall [53]. Common practice is to design for the peak-design rate of wastewater flow plus 30,000 gpd infiltration per mile of sewer and house connections [6, 54]. This allowance represents average conditions and should be revised by the designer as required on the basis of the physical characteristics of the area and the type of pipe joint to be used.

## Stormwater Runoff

Except for large combined sewers, storm runoff should be excluded from the sewerage system. Storm runoff may enter at manholes or through illicit roof drains connected to the sanitary system. Quantities of flow that enter in this manner vary with degree of enforcement of regulations and types of preventive measures taken. The American Society of Civil Engineers reports that tests on leakage through manhole covers show that 20–70 gpm may enter a manhole cover submerged by 1 in. of water [54]. Rates of this magnitude may be considerably in excess of average wastewater flows. Small sewers can be surcharged easily by very few roof-drain connections. For example, a rainfall of 1 in./hr on a 1000-ft roof area will contribute flows in excess of 10 gpm. The average domestic sewage flow from a dwelling with this approximate roof area (consider four persons) would equal only about 1.5% of this.

### ■ EXAMPLE 4.3

Estimate the maximum hourly, average daily, and minimum hourly residential sewage flows from an area occupied by 700 people. Consider the length of sewer and house drains equal to 1.3 mi. Give results in gpd and lpd.

*Solution.*   Based on 1985 USGS water-use data (Section 4.2) and assuming a return flow of 75%, the average daily per capita flow would be 105 gpcd × 0.75 = 78.8 gpcd or 298.3 lpcd. Thus, the total average daily domestic flow would be 78.8 × 700 = 55,160 gpd or 208,810 lpd.

   Infiltration = 30,000 gpd/mi × 1.3 = 39,000 gpd or 147,615 lpd

Using data from Table 4.8, we have

   Maximum hourly flow   = 39,000 + 55,160 × 3 = 204,480 gpd
                                      or
                                      774,045 lpd

   Total average daily flow = 39,000 + 55,160 = 94,160 gpd or
                                      356,396 lpd

   Minimum hourly flow   = 39,000 + 55,160/3 = 57,387 gpd
                                      or
                                      217,218 lpd                                   ■

## 4.10  INDUSTRIAL WASTE VOLUMES

Industrial waste volumes are highly variable in both quantity and quality, depending principally on the product produced. Since very little water is consumed in industrial processing, large volumes are often returned as waste. These wastes may include toxic metals, chemicals, organic materials, biologic contaminants, and radioactive materials. The design of treatment processes for these wastes is a highly specialized operation. Where industrial wastes must be processed in municipal sewage treatment works, accurate estimates of the time distribution and total volume of the load are necessary, together with a complete analysis of the characteristics of the waste. Under these circumstances, metering and analyzing the industrial waste is normally required. For more complete information on volumes as well as characteristics of industrial wastes, the reader should consult appropriate references [55–75].

## 4.11  AGRICULTURAL WASTES

The quantities and character of wastes from agricultural lands are highly variable. The most important pollutants found in runoff from agricultural areas are sediment, animal wastes, wastes from industrial processing of raw agricultural products, plant nutrients, forest and crop residues, inorganic salts and minerals, and pesticides [76–78]. Because waste volumes are determined by numerous factors for any given area, it is impossible to indicate any general rules. Nevertheless, the importance of these wastes to any region should not be overlooked. It is clear that large quantities of agricultural drainage are discharged into streams, rivers, and lakes. Regional water-quality control will not be a reality unless these wastes are considered along with those of the municipalities and industries. A comprehensive treatment of this topic is beyond the scope of this book, but the student should not minimize the importance of the subject. More detailed information can be found in the references cited.

## PROBLEMS

**4.1**  The Elephant Butte reservoir has a capacity of 2.64 million acre·ft. How many years would this supply the city of Las Cruces (population 60,000) if evaporation losses are neglected? Assume a use rate of 166 gpcd.

**4.2**  If the minimum flow of a stream having a 200-mi$^2$ watershed is 0.10 cfs/mi$^2$, what population could be supplied continuously from the stream? Assume that there is only distribution storage (maximum withdrawal = stream flow) and that the water-use rate is 175 gpcd.

**4.3**  Estimate the 1990, 1995, and 2000 population of your community by plotting the historic data on arithmetic coordinate paper and extending the curve by eye.

**4.4**  Compare the maximum hourly domestic sewage flow from ten houses (four persons per house) with the roof drainage from these houses if the roof dimensions are 60 × 35 ft and the rainfall intensity is 2.0 in./hr. Approximately what size of sewer would be required to handle (a) the domestic flow alone and (b) the combined flow if the pipe is laid on a 1.1% grade?

**4.5**  Make a comparison between the annual water requirements of an 1800-acre irrigation farm and a city of 60,000 population. Assume an irrigation requirement of 3 acre·ft/yr/acre.

**4.6**  Obtain historic and projected water-use figures for your state (consult your state water planning agency for data). Analyze the historic trends and try to explain the major influences on water use in each of the principal water-using sectors. Do you think the trends projected for the future are reasonable? Why? If not, why not? What could be done to modify these trends? How much reduction could be expected in each sector?

**4.7**  For the region in which you reside, make a determination of which water-using sectors are most dominant. How did you arrive at this determination? Do you think the past trends will continue into the future? If so, why? If not, why not? Are there water supply problems in the region? If so, could these be alleviated by modifying water-use rates in one or more of the water-using sectors? How much of a reduction below current rates of use do you think could be achieved? What revision in facilities or systems operation would be required to bring about this reduction?

**4.8**  Obtain historic and future population data on the municipality in which you reside. Consult your local water department and obtain historic and projected water-use trends for your city. Estimate the current per capita water-use rate. Use this figure to project water requirements to the year 2000. How does your projection compare with that of the water department? By how much could the per capita water-use rate be reduced, if any, by 2000? If this level of reduction were achieved, by how much would the city's water requirement be reduced by 2000?

**4.9**  Given a residential area encompassing 600 acres with a housing density of four houses per acre. Assume a high-value residence with a fire flow requirement of 1000 gpm. Find (a) the combined draft and (b) the peak hourly demand.

**4.10**  Estimate the maximum hourly, average daily, and minimum hourly residential sewage flows from an area occupied by 650 people. Consider the length of sewer and house drains equal to 1.2 mi. Give results in gpd and lpd.

**4.11**  Given that a residential community has an area of 10 mi², assume a population density and calculate the required fire flow. Give results in gpm and lpm.

**4.12**  Consider a 1000-acre residential area with a housing density of four dwellings per acre. Estimate the peak hourly water-use requirement.

**4.13**  If 100 acres of farmland were developed for urban housing (four houses per acre), what would be the difference in average annual water requirements after the changeover? Assume that the irrigation requirement is 2.5 acre·ft of water for a growing season of six months.

**4.14**  Estimate the average hourly, average daily, and minimum hourly residential sewage flows from an area serving a population of 1200. Assume that the length of sewer and house drains equals 4.0 km and that infiltration occurs.

**4.15** The population of a state was 7 million in 1980. Consider that by the year 2010, it is expected to increase to 10 million. Consider that the amount of fresh water withdrawn in 1980 was 2.5 bgd. Estimate the amount of fresh water that might be withdrawn in 2010. State your assumptions.

**4.16** A community had a population of 200,000 in 1980 and it is expected that this will increase to 260,000 by 2000. The water treatment capacity in 1980 was 43 mgd. A survey showed that the average per capita water-use rate was 150 gpcd. Estimate the community's water requirements in 2000 assuming (a) no change in use rate and (b) a reduced rate of 135 gpcd. Will expanded treatment facilities be needed by 2000 for condition a? for condition b?

**4.17** For the community described in Prob. 4.16, assume that the treatment plant capabilities in 1980 were 35 mgd. If the water-use rate in 2000 is 140 gpcd, will expanded treatment facilities be needed by 2000? In about what year would they be needed if the answer above is yes? What reduction in water-use rate would be needed to eliminate the need for new facilities until 2000? Would a reduction of this magnitude appear to be attainable?

**4.18** Consider that a state had a total water withdrawal of 4.5 bgd in 1980. This was distributed as follows: municipal use 1.0 bgd; industrial use 1.5 bgd; and steam electric cooling 2 bgd. Assume a 1980 population of 10 million and a projection of 13 million for the year 2000. It is estimated that an additional 3000 MW of electric generating capacity will be required by 2000 and that the cooling water requirements will be 50 gal/kW·h. An industrial expansion of 10% is also expected. Estimate the total water that will be withdrawn in 2000 for each sector and for all sectors combined. State your assumptions. Assume a plant capacity factor of 0.6 (water use will relate to 3000 MW × 0.6).

**4.19** Given that a residential community has an area of 26 km², assume a population density and calculate the required fire flow. Give results in lpm.

**4.20** Consider a 450-acre residential area with a housing density of four dwellings per acre. Estimate the peak hourly water-use requirement and the peak hourly sewage flow.

**4.21** Estimate the maximum hourly, average daily, and minimum hourly residential sewage flows from an area occupied by 700 people. Consider the length of sewer and house drains equal to 2.0 km. Give results in lpd and gpd.

# REFERENCES

1. U.S. Water Resources Council, *The Nation's Water Resources 1975–2000* (Washington, DC: U.S. Government Printing Office, December 1978).
2. "Estimated Water Use in the United States, 1985," U.S. Geol. Survey Circular 1004, 1985.
3. J. B. Wolff, "Peak Demands in Residential Areas," *J. Am. Water Works Assoc.* 87 (October 1961).
4. F. P. Linaweaver, Jr., "Report on Phase One, Residential Water Use Research Project," The Johns Hopkins University, Department of Sanitary Engineering, Baltimore, October, 1963.
5. *Standard Schedule for Grading Cities and Towns of the United States with*

*Reference to Their Fire Defenses and Physical Conditions* (New York: National Board Fire Underwriters, 1956).

6. R. S. Gupta, *Hydrology and Hydraulic Systems* (Englewood Cliffs, NJ: Prentice-Hall, 1989).

7. W. Viessman, Jr., and C. Welty, *Water Management: Technology and Institutions* (New York: Harper & Row, 1985).

8. Edison Electrical Institute, "Economic Growth in the Future-II," unpublished draft report (Washington, DC: 1980).

9. J. Vitullo-Martin, "Ending the Southwest's Water Binge," *Fortune,* February 23, 1981.

10. C. W. Howe and K. W. Easter, *Interbasin Transfers of Water* (Baltimore: The Johns Hopkins Press, 1971).

11. R. Von Dohren, "Consider the Many Reasons for Reuse," *Water and Wastes Eng.* 17 (9) (1980): 7478.

12. SCS Engineers, Inc., "Contaminants Associated with Direct and Indirect Reuse of Municipal Wastewater," U.S. EPA Report No. EPA600/178019, NTIS No. PB 280 482 (March 1978).

13. J. Crook, "Reliability of Wastewater Reclamation Facilities," California Department of Health, Water Sanitation Section (Berkeley, CA: 1976).

14. W. E. Garrison, and R. P. Miele, "Current Trends in Water Reclamation Technology," *J. Am. Water Works Assoc.* 69 (7) (1977): 364–369.

15. California Department of Health, "Wastewater Reclamation Criteria," California Administrative Code, Title 22, Division 4, Water Sanitation Section (Berkeley, CA: 1975).

16. S. W. Work, M. R. Rothberg, and K. J. Miller, "Denver's Potable Reuse Project: Pathway to Public Acceptance," *J. Am. Water Works Assoc.* 72 (8) (1980): 435–440.

17. U.S. Congress, "Water Reuse Project Underway in Denver," *Congressional Record,* pp. E5191, E5192, October 22, 1979.

18. J. E. Matthews, "Industrial Reuse and Recycle of Wastewaters: Literature Review," EPA Report No. EPA/600/280183, U.S. EPA, Ada, Oklahoma (1980).

19. B. A. Carnes, J. M. Eller, and J. C. Martin, "Reuse of Refinery and Petrochemical Wastewaters," *Indus. Water Eng.* 9 (4) (1979): 25.

20. L. E. Streebin, L. W. Canter, and J. R. Palafox, "Water Conservation and Reuse in the Canning Industry," Proceedings of the 26th Industrial Waste Conference, Purdue University, Lafayette, Indiana, Part II (1971), p. 766.

21. C. A. Caswell, "Water Reuse in the Steel Industry," *Complete Water Reuse: Industry's Opportunity* (New York: American Institute of Chemical Engineers, 1973), p. 384.

22. R. Field and C. Fan, "Recycling Urban Stormwater for Profit," *Water/Eng. Management* 128 (4) (1981).

23. J. Merrell and R. Stoyer, "Reclaimed Sewage Becomes a Community," *American City* (April 1964): 97.

24. J. C. Merrell, Jr., W. F. Jopling, R. F. Bott, A. Katko, and H. E. Pintler, "Santee Recreational Project, Santee, California, Final Report," FWPCA Report WP-20-7 (Cincinnati: Publication Office, Ohio Basin Region, Federal Water Pollution Control Administration, 1967).

25. National Water Commission, *Water Policies for the Future* (Washington, DC: U.S. Government Printing Office, 1973).

26. J. J. Boland, IWR-MAIN System Improvements: Phase III—Growth Models,

Consultant Report, Contract DACW72-84-C-0004 (U.S. Army Engineer Institute for Water Resources, Fort Belvoir, VA, 1987).

27. B. Dziegielewski, "The IWR-MAIN Disaggregate Water Use Model," Proceedings of the 1987 Annual UCOWR Meeting, UCOWR, Southern Illinois University, Carbondale, Illinois (1987).

28. Planning and Management Consultants, Ltd., "A Disaggregate Water Use Forecast for the Phoenix Water Service Area," Report presented to City of Phoenix, Water and Wastewater Department (March 1986).

29. Planning and Management Consultants, Ltd., "IWR-Main Water Use Forecasting System, System Description", Carbondale, Illinois (January 1987).

30. Planning and Management Consultants, Ltd., "Municipal and Industrial Water Use in the Metropolitan Water District Service Area," Carbondale, Illinois (November, 1987).

31. J. F. Alexander, Jr., P. D. Zwick, J. J. Miller, and M. H. Hoover, "Alafia River Basin, Land and Water Use Projection Model," Vol. I, Series No. 12 (Department of Urban and Regional Planning, University of Florida, 1983).

32. D. W. Berry and G. W. Bonem, "Predicting the Municipal Demand for Water," *Water Resources Res.* 10(6) (1974): 1239–1242.

33. H. S. Foster and B. R. Beattie, "Urban Residential Demand for Water in the United States," *Land Economics* 55(1) (1979): 43.

34. S. L. Franklin and D. R. Maidment, "An Evaluation of Weekly and Monthly Time Series Forecasts of Municipal Water Use," *Water Resources Bull.*, Paper No. 86038, Am. Water Resources Assoc., Vol. 2, No. 4, August 1986.

35. D. T. Lauria and C. H. Chiang, *Models For Municipal And Industrial Water Demand Forecasting in North Carolina* (Department of Environmental Sciences and Engineering, University of North Carolina, Chapel Hill, NC, 1975).

36. S. T. Wong, "A Model of Municipal Water Demand: A Case Study on Northeastern Illinois," *Land Economics* 48(1) (1972): 34.

37. C. W. Howe, "The Impact of Price on Residential Water Demand: Some New Insights," *Water Resources Res.* 18(4) (1982): 713–716.

38. C. W. Howe and F. P. Linaweaver, "The Impact of Price on Residential Water Demand and Its Relation to System Design and Price Structure," *Water Resources Res.* 3(1) (1967): 13–32.

39. D. E. Agthe and R. B. Billings, "Dynamic Models of Water Demand," *Water Resources Res.* 16(3) (1980): 476–480.

40. AWWA Committee on Water Use, "Review of The Johns Hopkins University Research Project Method for Estimating Residential Water Use," *J. Am. Water Works Assoc.* 65(5) (1973): 300–301.

41. J. P. Heaney, G. D. Lynne, N. Khanal, W. C. Martin, C. L. Sova, and R. Dickinson, "Agricultural and Municipal Water Demand Projection Models," University of Florida, Gainesville, Florida, March 1981.

42. B. Dziegielewski, J. J. Boland, and D. D. Baumann, "An Annotated Bibliography on Techniques of Forecasting Demand for Water," IWR Contract Report 81-C03, U.S. Army Corps of Engineers, Fort Belvoir, VA, 1981.

43. J. J. Boland, D. D. Baumann, and B. D. Ziegielewski, "An Assessment of Municipal and Industrial Water Use Forecasting Approaches," IWR Contract Report 81-C05, Army Corps of Engineers, Fort Belvoir, VA, 1981.

44. B. T. Bower, I. Gouevsky, D. R. Maidment, and W. R. D. Sewell, *Modeling Water Demands* (Academic Press, New York, 1984).

45. Comptroller General, "Water Supply for Urban Areas: Problems in Meeting

Future Demand," Report to The Congress of The United States, CED-79-56, June 15, 1979.

46. Congressional Research Service, Library of Congress, *State and National Water Use Trends to the Year 2000,"* Serial No. 96-12 (Gov't. Printing Office, Washington, DC, May 1980).

47. J. R. Kim and R. H. McCuen, "Factors for Predicting Commercial Water Use," *Water Resources Bull.* 15(4) (1979): 1073–1080.

48. D. R. Maidment and S. Miaou, "Daily Water Use in Nine Cities," *Water Resources Res.* 22(6) (1986): 845–851.

49. C. T. Osborn, J. E. Schefter, and L. Shabman, "The Accuracy of Water Use Forecasts: Evaluation and Implications," *Water Resources Bull.,* Paper No. 84245, Am. Water Resources Assoc., Vol. 22, No. 1, February 1986.

50. J. J. Boland, E. M. Opitz, B. Dziegielewski, and D. D. Baumann, IWR MAIN Modifications, Technical Consultant Report, Contract DACW-72-84-C-0004, Task 2, U.S. Army Engineer Institute for Water Resources, Fort Belvoir, VA, 1985.

51. J. J. Lentz, "Special Report No. 4 of the Residential Sewage Research Project to the Federal Housing Administration," The Johns Hopkins University, Department of Sanitary Engineering, Baltimore, May, 1963.

52. Metcalf and Eddy, Inc., *Wastewater Engineering* (New York: McGraw-Hill, 1972).

53. C. R. Vlezy and J. M. Sprague, *Sewage Ind. Wastes* 27(3) (1955).

54. "Design and Construction of Sanitary and Storm Sewers," Manual of Engineering Practice No. 37 (New York: American Society of Civil Engineers, 1960).

55. "Development Document for Proposed Effluent Limitations Guidelines and New Source Performance Standards for the Iron and Steel Industry," EPA 440/1-82/024 (1982).

56. "Development Document for Proposed Effluent Limitations Guidelines and New Source Standards of Performance for the Nonferrous Metals Industry," EPA 440/1-83/019-b (1983).

57. "Development Document for Proposed Effluent Limitations Guidelines and New Source Performance Standards for the Bauxite Refining Subcategory of the Aluminum Segment of the Nonferrous Metals Manufacturing Point Source Category," EPA 440/1-74/091-c (1974).

58. "Development Document for Proposed Effluent Limitations Guidelines and New Source Performance Standards for the Copper, Nickel, Chromium, and Zinc Segments of the Electroplating Point Source Category," EPA 440/1-74/003-a (1974).

59. "Development Document for Proposed Effluent Limitations Guidelines and New Source Performance Standards for the Petroleum Refining Industry," EPA 440/1-82/014 (1982).

60. "Development Document for Proposed Effluent Limitations Guidelines and New Source Performance Products for Inorganic Chemicals, Alkali and Chlorine Industry," EPA Contract No. 68-01-1513 (draft) (June 1973).

61. "Development Document for Proposed Effluent Limitations Guidelines and New Source Performance Standards for the Organic Chemicals Industry," EPA 440/1-74/009-a (1974).

62. "Development Document for Proposed Effluent Limitations Guidelines and New Source Performance Standards for the Fertilizer and Phosphate Manufacturing Industry," EPA 440/1-74/011-a and 74/006-a (1974).

63. "Development Document for Proposed Effluent Limitations Guidelines and

New Source Performance Standards for the Plastic and Synthetics Industry,'' EPA 440/1-83/009-b (1983).

64. ''Development Document for Proposed Effluent Limitations Guidelines and New Source Performance Standards for the Major Inorganic Products Segment of the Inorganic Chemicals Manufacturing Point Source Category,'' EPA 440/1-74/007-a (1974).

65. ''Development Document for Proposed Effluent Limitations Guidelines and New Source Performance Standards for the Phosphorus-Derived Chemicals Segment of the Phosphate Manufacturing Point Source Category,'' EPA 440/1-74/006-a (1974).

66. ''Development Document for Proposed Effluent Limitations Guidelines and New Source Performance Standards for the Synthetic Resins Segment of the Plastics and Synthetic Materials Manufacturing Point Source Category,'' EPA 440/1-74/010-a (1974).

67. ''Development Document for Proposed Effluent Limitations Guidelines and New Source Performance Standards for the Canned and Preserved Fish and Seafoods Processing Industry,'' EPA 440/1-80/020 (1980).

68. ''Development Document for Proposed Effluent Limitations Guidelines and New Source Performance Standards for the Meat Packing Industry,'' EPA 440/1-74/012-a (1974).

69. ''Development Document for Proposed Effluent Limitations Guidelines and New Source Performance Standards for the Dairy Products Processing,'' EPA 440/1-74/021-a (1974).

70. ''Development Document for Proposed Effluent Limitations Guidelines and New Source Performance Standards for the Citrus, Apple and Potato Segment of the Canned and Preserved Fruits and Vegetables Processing Point Source Category,'' EPA 440/1-74/027-a (1974).

71. ''Development Document for Proposed Effluent Limitations Guidelines and New Source Performance Standards for the Unbleached Kraft and Semichemical Pulp Segment of the Pulp, Paper, and Paperboard Mills Point Source Category,'' EPA 440/1-74/025-a (1974).

72. ''Development Document for Proposed Effluent Limitations Guidelines and New Source Performance Standards for the Builders Paper and Roofing Felt Segment of the Builders Paper and Board Mills Point Source Category,'' EPA 440/1-80/025-b (1980).

73. ''Development Document for Proposed Effluent Limitations Guidelines and New Source Performance Standards for the Textile Mills,'' EPA 440/1-82/022 (1982).

74. ''Development Document for Proposed Effluent Limitations Guidelines and New Source Performance Standards for the Leather Tanning and Finishing Industry,'' EPA 440/1-82/016 (1982).

75. ''Development Document for Proposed Effluent Limitations Guidelines and New Source Performance Standards for Steam Electric Power Plants,'' EPA 440/1-82/029 (1982).

76. R. C. Loehr, ''Animal Wastes—A National Problem,'' *Proc. Am. Soc. Civil Engrs., J. San. Eng. Div.* 95 (SA2) (1969): 189–221.

77. ''Control of Agriculture-Related Pollution,'' A report to the President submitted by the Secretary of Agriculture and the Director of the Office of Science and Technology, Washington, DC (January 1969).

78. ''Agricultural Waste Waters,'' Proceedings Symposium of Agricultural Waste Waters, Report No. 10, Water Resources Center, University of California, Davis, California (April 1966).

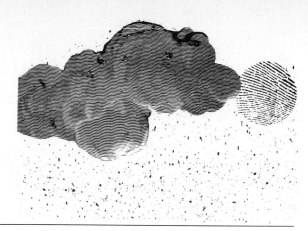

# Chapter 5
# The Availability of Water

Not many years ago, water resources management was focused almost exclusively on water supply, flood control, and navigation. Today, protecting the environment, ensuring safe drinking water, and providing aesthetic and recreational experiences compete equally for the allocation of water resources and for funds for water management and/or development. An environmentally conscious public is pressing for greater reliance on improved management practices, having fewer structural components, to solve the nation's water problems. The notion of continually striving to provide more water has been replaced by one of husbanding this precious resource.

Water is located in all regions of the earth. The problem is that the distribution, quality, quantity, and mode of occurrence are highly variable from one locale to another.

The most voluminous water source is the oceans. It is estimated that they contain about 1060 trillion acre·ft of water [1]. The most valuable water supply (in terms of quality or freshness) is contained within the atmosphere, on the earth's surface, or underground. This supply, however, amounts to only about 3% of that contained in the oceans.

## Water Quantity

Water resources vary widely in regional and local patterns of availability. The supply is dependent on topographic and meteorological conditions as they influence precipitation and evapotranspiration. Quantities of water stored are dependent to a large extent on the physical features of the earth and on the earth's geological structure. Table 5.1 shows the major components of the water resources of the continental United States.

### 5.1 SOIL MOISTURE

Soil moisture is the most broadly used water source on the earth's surface. Agriculture and natural plant life are dependent on it for sustenance. The quantity of water stored as soil moisture at any specified time is small,

**TABLE 5.1**    SUMMARY DATA CONCERNING WATER RESOURCES OF THE
CONTINENTAL UNITED STATES

|  | Square miles | Acre-feet $(\times 10^6)$ |
|---|---|---|
| Gross area of continental United States | 3,080,809 | — |
| Land area, excluding inland water | 2,974,726 | — |
| Volume of average annual precipitation | — | 4,750 |
| Volume of average annual runoff (discharge to sea) | — | 1,372 |
| Estimated total usable groundwater | — | 47,500 |
| Average amount of soil moisture | — | 635 |
| Estimated total lake storage | — | 13,000 |
| Total reservoir storage (capacity of 5000 acre·ft or more) | — | 365 |

*Source:* E. A. Ackerman and G. O. Löf, *Technology in American Water Development* (Baltimore, MD: The Johns Hopkins Press, 1959).

however. Estimates indicate an equivalent layer about 4.6 in. in thickness distributed over 57 million $mi^2$ of land surface. This in itself would be insufficient to support adequate plant growth without renewal. It is therefore important that the frequency with which this supply is renewed and the length of time it remains available for use be known. The supply of soil moisture is dependent on geographical location, climatic conditions, geologic structure, and soil type. Variations may be experienced on a seasonal, weekly, or even daily basis.

It is considered that the natural supply of soil moisture in most of the agricultural areas of this country is less than optimum for crop growth during an average year. It is evident, then, that a greater understanding of optimum water requirements for crops is essential if we are economically and efficiently to supply water artificially to overcome natural soil moisture deficiencies.

## 5.2  SURFACE WATERS AND GROUNDWATER

Surface waters are nonuniformly distributed over the earth's surface. Of the U.S. land mass, only about 4% is covered by rivers, lakes, and streams. The volumes of these freshwater sources depend on geographic, landscape, and temporal variations and on the impact of human activities.

For surface waters, historic records of streamflows, lake levels, and climatic data are used to identify trends and to indicate deficiencies in data bases. Since surface-water supplies are always in a state of transition, hydrologic models become valuable tools for estimating future water supply scenarios based on assumed sequences of hydrologic variables, such as precipitation, temperature, and evaporation and for projected physical manipulations of the surface-water containment system. The verification of hydrologic models depends heavily on adequate historic data for testing, and where data voids exist every effort should be made to fill them.

Groundwater supplies are much more widely distributed than surface waters, but local variations are found as a result of the variety of soils, rocks, and geologic structures located beneath the land surface.

The importance of groundwater to the health and well-being of humans is well documented. Groundwater is a major source of fresh water for public consumption, industrial uses, and crop irrigation. More than half of the fresh water used in Florida for all purposes, for example, comes from groundwater sources, and about 90% of that state's population depends on groundwater for its potable water supply. The need to husband this resource is clear. Quantity and quality dimensions are both very important.

The need for an adequate data base to determine the stock of groundwater and to estimate its change over time is very great because groundwater systems are not as easily defined as those for surface water. Groundwater storage volumes and transmission rates are affected by soil properties and geologic conditions, and these are often highly variable in space and not amenable to simple quantification.

## 5.3 RUNOFF DISTRIBUTION

Approximately 30% of the average annual rainfall in the United States is estimated to appear as surface runoff. The allocation of this water is directly related to precipitation patterns and thus to meteorologic, geographic, topographic, and geologic conditions. In the West, large regions are devoid of permanent runoff, and some localities, such as Death Valley, California, receive no runoff for years at a time. In contrast, some areas in the Pacific Northwest average about 6 ft of runoff annually. Mountain regions are usually the most productive of runoff, whereas flat areas, especially those experiencing lower precipitation rates, are generally poor runoff producers.

Runoff is distributed in a far-from-uniform pattern over the continental United States. In addition, it is subject to seasonal variations and definite annual periods during which the concentration is significantly greater. For example, about 75% of the runoff in the semiarid and arid regions of the country occurs during a period of a few weeks following the snowmelt on the upper watershed. Even in the well-watered east, an uneven distribution prevails, which has been one of the factors in the expansion of supplemental irrigation to be found there. Figure 5.1 indicates the four major runoff regions in the United States.

## 5.4 GROUNDWATER DISTRIBUTION

The usable groundwater storage in the United States is estimated to be about 48 billion acre·ft. This vast reservoir is distributed across the nation in quantities determined primarily by precipitation, evapotranspiration, and geologic structure. There are two components to this supply: one, a part of

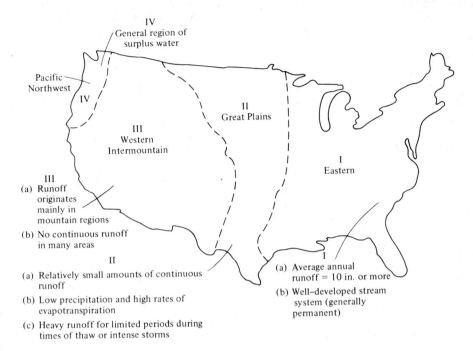

**Figure 5.1** Major runoff regions of the United States.

the hydrologic cycle; the other, water trapped underground in past ages and no longer naturally circulated in the cycle.

Figure 5.2 shows the principal groundwater areas of the United States as depicted by Thomas [2]. Generally, it is evident that the mountain regions in the east and west, the northern Great Plains, and the granitic and metamorphic rock areas of New England and the southern Piedmont do not contain important groundwater supplies.

Aquifers may be generally classified into four categories:

1. Those directly connected to surface supplies which are replenished by gravitational water and which release water to surface flow. Gravels found in floodplains or river valleys are examples.
2. "Regional" aquifers occurring east of the 100th meridian. These aquifers produce some of the largest permanent groundwater yields and have moderate to high rates of recharge. Good examples are found in the Atlantic and Gulf coastal plain areas.
3. Low recharge aquifers between the 100th and 120th meridians. These aquifers have relatively little inflow compared to potential or actual drafts. Although storage volumes are often large, the low rate of replenishment indicates that the water must be considered more of a minable material than a renewable resource. This possible "limited-life" category poses particular problems of development and management.
4. Aquifers subject to saline-water intrusion. These are usually found in coastal regions, but inland saline waters also exist, principally in the western states.

Patterns show areas underlain by
aquifers generally capable of yielding to individual wells 50 gpm or
more of water containing not more
than 2000 ppm of dissolved solids
(includes some areas where more
highly mineralized water is actually
used)

Watercourses in which groundwater can be replenished
by perennial streams

Buried valleys not now occupied by perennial streams

Unconsolidated and semiconsolidated aquifers

Consolidated rock aquifers

Both unconsolidated and consolidated rock aquifers

Not known to be underlain by aquifers that will
generally yield as much as 50 gpm to wells

**Figure 5.2** Groundwater areas in the United States. [From "Water," in *The Yearbook of Agriculture* (Washington, DC: U.S. Government Printing Office, 1955).]

# Water Quality

Although water quality and water quantity are inextricably linked, water quality deserves special attention because of its implications for affecting the public health and the quality of life. Even with the large federal investments in pollution control since 1972, the President's Council on Environmental Quality reports that the nation's waters continue to be damaged by pollution and misuse. Pollutants reach water bodies from both point and nonpoint sources. Municipal wastes, urban and agricultural runoff, and industrial wastes are principal offenders. Of special importance are the vestiges of past toxic and hazardous materials disposal that are now being transported by surface-water and groundwater systems. The impacts of polluting activities are widespread, and they affect the public health, the economy, and the environment [3, 4].

## 5.5 GROUNDWATER

Groundwater quality is influenced by the quality of its source. Changes in source waters or degraded quality of source supplies may seriously impair the quality of the groundwater supply. Municipal and industrial wastes entering an aquifer are major soruces of organic and inorganic pollution. Large-scale organic pollution of groundwaters is infrequent, however, since significant quantities of organic wastes usually cannot be easily introduced underground. The problem is quite different with inorganic solutions, since these move easily through the soil and once introduced are removed only with great difficulty. In addition, the effects of such pollution may continue for indefinite periods since natural dilution is slow and artificial flushing or treatment is generally impractical or too expensive. The number of harmful enteric organisms is generally reduced to tolerable levels by the percolation of water through 6 or 7 ft of fine-grained soil [5]. However, as the water passes through the soil, a significant increase in the amounts of dissolved salts may occur. These salts are added by soluble products of soil weathering and of erosion by rainfall and flowing water. Locations downstream from heavily irrigated areas may find that the water they are receiving is too saline for satisfactory crop production. These saline contaminants are difficult to control because removal methods are exceedingly expensive. A possible solution is to dilute with water of lower salt concentration (wastewater treatment plant effluent, for example) so that the average water produced by mixing will be suitable for use.

## 5.6 SURFACE WATER

The primary causes of deterioration of surface-water quality are municipal and domestic wastewater, industrial and agricultural wastes (organic, inorganic, heat), and solid and semisolid refuse. A municipality obtaining its water supply from a surface body may find its source so fouled by wastes and toxic chemicals that it is unsuitable or too costly to treat for use as a water supply. Fortunately, waste products discharged by cities and industry are being controlled at the point of initiation. This has been borne out by successes in cleaning up such watersheds as the Delaware and Susquehanna in the eastern United States.

# Hydrology and Water Management

In the past, water resources engineers and planners focused mainly on the developmental aspects of water resources: building dams, canals, distribution networks, and other infrastructure elements. Today, management is the byword, and it centers on applying both structural and nonstructural measures to problems related to water supply, water allocation, water quality, floods, droughts, and environmental protection. The emphasis is on innovative technical approaches to problem solving, a more holistic view of water management, and improved operating efficiency of the water and wastewater systems already in place. But regardless of whether the focus is conservation, protection, or development, good water management requires an understanding of the hydrology and quality of the relevant water sources.

## 5.7  THE WATER BUDGET

In theory, accounting for the water resources of an area is relatively simple. The basic procedure involves the separate evaluation of each factor in the water budget so that a quantitative comparison of the available water resources with the known or anticipated water requirements of the area can be made. In practice, however, the evaluation of the water budget is often quite complex, and extensive and time-consuming investigations are generally required [6].

Both natural and artificial gains and losses in the water supply must be considered. The primary natural gains to surface bodies are those that result from direct runoff caused by precipitation and effluent seepage of groundwater. Evapotranspiration and unrecovered infiltration are the major natural losses. Dependable dry-season supplies can be increased through diversion from other areas, through low-flow augmentation, through saline-water conversion, and perhaps in the future through induced precipitation. On the other hand, diversions out of the basin will decrease the quantity of water available.

After the gross dependable water supply has been estimated, the net dependable supply may be determined at any point of interest by subtracting the quantity used, detained, or lost as a result of man's activities from the gross supply. When water is withdrawn from a river, a decrease in flow between the point of withdrawal and the point of return is experienced. As the water is used, part of it will be lost to the atmosphere through various forms of consumption. These consumptive losses are accumulative downstream and effect a permanent decrease in the dependable water supply. Temporary decreases in dependable supply occur along the watercourse between intake and discharge points. The dependable supply in the reach of watercourse between the two points is thus diminished by the amount of the withdrawal.

A water supply may be considered adequate for present needs but inadequate to provide for future requirements. In many cases, it is therefore necessary to predict future water needs based on estimated changes in population, industrial development, and agricultural practices and on changes in water policy and technology that will affect the supply and use of water.

In a gross sense, the water budget is given by

$$\text{inflows} - \text{outflows} = \text{change in storage} \tag{5.1}$$

where the inflows are all sources of water, natural and man made, entering the region; the outflows are all the movements of water out of the region, including evaporation, streamflows, etc.; and the change in storage is the increase or decrease in storage over time for all natural (surface and underground) and all artificial reservoirs.

*Water consumption* and *water use* are terms that have frequently been employed indiscriminately and interchangeably. Several definitions of *consumption* or *consumptive use* may be found in the literature. A special committee of the American Water Works Association defines consumptive use as water used in connection with vegetative growth, food processing, or incidental to an industrial process, which is discharged to the atmosphere or incorporated in the products of the process [7]. A slight modification of this definition that includes air-conditioning losses will be adopted here.

*Withdrawal use* is the use of water for any purpose which requires that it be physically removed from the source. Depending on the use to which the water is put, a significant percentage may be returned to the original source and then be available for reuse.

*Nonwithdrawal use* is the use of water for any purpose which does not require that it be removed from the original source, such as water used for navigation.

Certain water losses, although not "consumptive" in the sense of the definition, may have the effect of reducing the available water supply of an area. For example, dead storage (storage below outlet elevations) in impoundments is unavailable for downstream use. Diversion of water from one drainage basin to another represents an additional form of *nonconsumptive loss*. An example of this is the use of the Delaware River basin for the municipal supply of New York City. This decreases the total flow in the Delaware River below the point of diversion. New York is required, however, to augment low flows through compensating the downstream interests for diversion losses. Water contaminated or polluted during use that cannot be economically treated for reuse also constitutes a real loss from the total supply.

# Surface-Water Supplies

Surface-water supplies may be generally categorized as perennial or continuous unregulated rivers, rivers or streams containing impoundments, or natural lakes. Evaluation of the surface-water resources of an area requires a determination of the general characteristics of the region. The area should be located geographically on the basis of established references. All available data on the climate, hydrology, geology, and topography of the area should be accumulated. The type of industrial, agricultural, and residential development and the predicted growth rates of these factors are necessary information. A description of the principal metropolitan areas and a knowledge of the economy of the region is also important. An evaluation of the natural resources of the area and the impact of their development on the hydrology and economy of the area is an additional requirement.

## 5.8  BASIN CHARACTERISTICS AFFECTING RUNOFF

Important natural characteristics of the basin affecting streamflow are the basin's topographic and geologic features.

Topography determines the slopes and location of drainage channels and the storage capacity of the basin. Channel slope and configuration are directly related to the rate of flow in a basin and the magnitude of the peak flows. A steep watershed will generally indicate a rapid rate of runoff with little storage, whereas relatively flat areas are subject to considerable storage and lower rates of flow.

The soils in a basin affect infiltration capacity and the ability of underground strata to transmit or hold groundwater. A thorough understanding of the characteristics of the underground formations is necessary to properly evaluate these factors.

## 5.9  NATURAL AND REGULATED RUNOFF

*Natural runoff* is defined as runoff that is unaffected by any other than natural influences. Runoff subject to withdrawals by man or artificial storage is defined as *regulated runoff*. If an unregulated stream is to be developed as a primary water source, the safe yield will be the lowest dry-weather flow of the stream. Under this condition, the user will always have an adequate supply, provided that the maximum requirements do not exceed this minimum flow. If during any time interval the expected demands exceed the lowest dry-weather flow, periods of water shortage can be anticipated unless supplementary supplies are provided.

Regulated runoff is normally the type of runoff for which information is available, although many streams for which runoff records are at hand were unregulated at the time they were first gaged. Most records are seriously affected by artificial regulation from upstream storage works or by the diversion of flows into or out of the stream at points above the gaging station. These withdrawals from the basin or diversions into the basin from outside sources affect the hydrology of the basin. Fortunately, diversions and withdrawals usually lend themselves to accurate estimates, since they are for specific purposes and are gaged or can be measured with little difficulty.

The safe yield of a stream that is regulated approaches the average annual flow as storage approaches full development. Economical yields are between the safe yields for unregulated and fully regulated flows. Usually safe yields of 75%–90% of the mean annual flow can be realistically developed. Through regulation, the greatest benefits are normally derived from a stream and, by means of diversified usage, a method of achieving the maximum economy in the utilization of the water is provided.

## 5.10  STORAGE

Water may be stored for single or multiple uses, such as navigation, flood control, hydroelectric power, agriculture, water supply, pollution abatement, recreation, and flow augmentation. Either surface or subsurface storage can be utilized, but both necessitate a reservoir, which affects the hydrology of a basin.

Reservoirs regulate stream flow for beneficial use by storing water for later release. The term *regulation* can be defined as the amount of water stored or released from storage in a period of time, usually 1 yr. The ability of a reservoir to regulate river flow depends on the ratio of its capacity to the volume of river flow. Evaluation of the regulation provided by existing storage facilities can be determined by studying the records of typical reservoirs. A representative group of reservoirs having detention periods from 0.01 to 20 yr is given by Langbein, with the usable capacity, detention period, and annual regulation of each [8].

About 190 million acre·ft of water, representing approximately 13% of the total river flow, has been made available through reservoir storage development in the United States [8]. The degree of storage development is exceedingly variable but is generally greatest in the Colorado River Basin and least in the Ohio River Basin. Substantial increases in water supply can be attained through the development of additional storage, but water regulation of this type follows a law of diminishing returns. There are limitations on the amount of storage that can be used. The storage development of the Colorado River Basin, for example, may be approaching (if not already in excess of) the maximum useful limit.

Reservoir construction schedules suggest that high development levels may be widespread in the not-too-distant future. Water supply, pollution control, and recreation appear to be the future primary objectives of water storage.

# Reservoirs

Where natural storage in the form of ponds or lakes is not available, artificial impoundments or reservoirs must often be constructed if optimum development of the surface supply is to be obtained.

## 5.11  DETERMINATION OF REQUIRED RESERVOIR CAPACITY

The amount of storage that must be provided is a function of the expected demands and the inflow to the impoundment. Mathematically this may be stated as follows:

$$\Delta S = I - O \tag{5.2}$$

where    $\Delta S$ = change in storage volume during a specified time interval
$I$ = total inflow volume during this period
$O$ = total outflow volume during this period

Normally, $O$ will be the draft requirement imposed by the various types of use, but it may also include evaporation and transpiration and flood discharges during periods of high runoff when inflow may greatly exceed draft plus available storage and outflow seepage from the bottom or sides of the reservoir.

Because the natural inflow to any impoundment area is often highly variable from year to year, season to season, or even day to day, it is obvious that the reservoir function must be that of redistributing this inflow with respect to time so that the projected demands are satisfied.

## 5.12 CLASSICAL METHOD OF COMPUTATION

Several approaches may be taken in the selection of reservoir capacities. Actual or synthetic records of streamflow and a knowledge of the proposed operating rules of the reservoir are fundamental to all solutions. Determination of storage may be accomplished by either graphical or analytical techniques.

Historically, one of the most used methods of determining storage has been the selection of some low-flow period considered to be critical. The most severe drought of record might be selected, for example. Once the critical period is chosen, storage is usually calculated by the mass-curve analysis introduced in 1883 by Rippl [9]. This method evaluates the cumulative deficiency between outflow and inflow $(O - I)$ and selects the maximum cumulative value as the required storage. Examples 5.1 and 5.2 illustrate the procedure.

■ **EXAMPLE 5.1**
Find the storage capacity required to provide a safe yield of 67,000 acre·ft/yr for the data given in Fig. 5.3.

*Solution.*    Construct tangents at *A, B,* and *C* having slopes equal to 67,000 acre·ft/yr. Find the maximum vertical ordinate between the inflow mass curve and the constructed draft rates. From Fig. 5.3 the maximum ordinate is found to be 38,000 acre·ft, which is the required capacity.    ■

This example shows that the magnitude of the required storage capacity depends entirely on the time period chosen. Since the period of record given covers only 5.5 yr, it is clear that a design storage of 38,000 acre·ft might be totally inadequate for the next 3 yr, for example. Unless the frequency of the flow conditions used in the design is known, little can be said regarding the long- or short-term adequacy of the design.

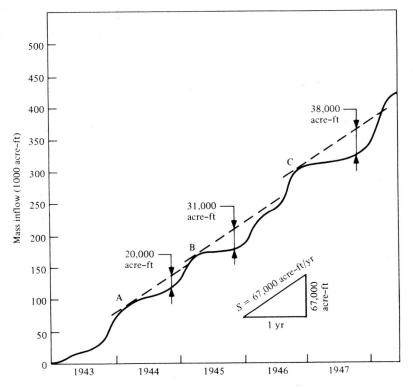

**Figure 5.3** Reservoir capacity for a specified yield as determined by use of a mass curve.

Example 5.1 also illustrates the fact that the period during which storage must be provided is dependent on hydrologic conditions. Since reservoir yield is defined as the amount of water that can be supplied during a specific time interval, choice of the interval is critical. For distribution reservoirs, a period of 1 day is often sufficient. For large impounding reservoirs, periods of several months, a year, or several years storage may be required.

■ **EXAMPLE 5.2**

Consider an impounding reservoir that is expected to provide for a constant draft of 448 million gallons (mil gal)/$mi^2$/yr. The following record of monthly mean inflow values is representative of the critical or design period. Find the storage requirement.

| Month | F | M | A | M | J | J | A | S | O | N | D | J | F |
|---|---|---|---|---|---|---|---|---|---|---|---|---|---|
| Observed inflow (mil gal/$mi^2$/month) | 31 | 54 | 90 | 10 | 7 | 8 | 2 | 28 | 42 | 108 | 92 | 22 | 50 |

**Solution.** Set up the tabulation of values shown in Table 5.2. It can be seen that the maximum cumulative deficiency is 131.5 mil gal/$mi^2$, which occurs in September. The

**TABLE 5.2**    STORAGE REQUIREMENT COMPUTATIONS

| Month | Inflow $I$ (mil gal/mi$^2$/ month) | Draft $O$ (mil gal/mi$^2$/ month) | Cumulative inflow $\Sigma I$ | Deficiency, $O - I$ | Cumulative deficiency,[a] $\Sigma(I - O)$ |
|---|---|---|---|---|---|
| F | 31 | 37.3 | 31 | 6.3 | 6.3 |
| M | 54 | 37.3 | 85 | −16.7 | 0 |
| A | 90 | 37.3 | 175 | −52.7 | 0 |
| M | 10 | 37.3 | 185 | 27.3 | 27.3 |
| J | 7 | 37.3 | 192 | 30.3 | 57.6 |
| J | 8 | 37.3 | 200 | 29.3 | 86.9 |
| A | 2 | 37.3 | 202 | 35.3 | 122.2 |
| S | 28 | 37.3 | 230 | 9.3 | 131.5 |
| O | 42 | 37.3 | 272 | −4.7 | 126.8 |
| N | 108 | 37.3 | 380 | −70.7 | 56.1 |
| D | 98 | 37.3 | 478 | −60.7 | 0 |
| J | 22 | 37.3 | 500 | 15.3 | 15.3 |
| F | 50 | 37.3 | 550 | −12.7 | 2.6 |

[a] Only positive values of cumulative deficiency are tabulated.

number of months of draft is 131.5/37.3 = 3.53, or stated differently, enough water must be stored to supply the region for about 3.5 months.    ■

This example gives a numerical answer to the question posed in determining a design storage. It does not, however, give an expression for the probabilities of the shortages or excesses that may result from this design. Past practice has been to use the lowest recorded flow of the stream as the critical period. Obviously, this approach overlooks the possibility that a more serious drought might occur, with a resultant yield less than the anticipated safe yield.

## 5.13  FREQUENCY OF EXTREME EVENTS

If a hydrologic event has a true recurrence interval of $T_R$ years, the probability that this magnitude will be equaled or exceeded in any particular year is

$$P = \frac{1}{T_R} \tag{5.3}$$

where $T_R$ is the recurrence interval of the event. *Recurrence interval* is defined as the average interval in years between the occurrence of an event of stated magnitude and an equal or more serious event.

Both annual series and partial duration series are used in estimating the recurrence intervals of extreme events [6]. An annual series is composed of one significant event for each year of record. The nature of the event depends on the object of the study. Usually the event will be a maximum or minimum

**TABLE 5.3** PROBABILITY THAT AN EVENT HAVING A PRESCRIBED RECURRENCE INTERVAL WILL BE EQUALED OR EXCEEDED DURING A SPECIFIED DESIGN PERIOD

| $T_R$ (yr) | Design period (yr) | | | | | |
|---|---|---|---|---|---|---|
|  | 1 | 5 | 10 | 25 | 50 | 100 |
| 1 | 1.0 | 1.0 | 1.0 | 1.0 | 1.0 | 1.0 |
| 2 | 0.5 | 0.97 | 0.999 | $1.0^a$ | $1.0^a$ | $1.0^a$ |
| 5 | 0.2 | 0.67 | 0.89 | 0.996 | $1.0^a$ | $1.0^a$ |
| 10 | 0.1 | 0.41 | 0.65 | 0.93 | 0.995 | $1.0^a$ |
| 50 | 0.02 | 0.10 | 0.18 | 0.40 | 0.64 | 0.87 |
| 100 | 0.01 | 0.05 | 0.10 | 0.22 | 0.40 | 0.63 |
| 200 | 0.005 | 0.02 | 0.05 | 0.12 | 0.22 | 0.39 |

[a] Values are approximate.

flow. A partial duration series consists of all events exceeding, in significance, a base value. The two series compare favorably at the larger recurrence intervals, but for the smaller recurrence intervals, the partial duration series will normally indicate events of greater magnitude.

There are two possibilities regarding an event; it either will or will not occur in a specified year. The probability that at least one event of equal or greater significance than the $T_R$-yr event will occur in any series of $N$ years is shown in Table 5.3. For example, there exists a probability of 0.22 that the 100-yr event will occur in a design period of 25 yr.

Table 5.3 was derived by means of the binomial distribution, which gives the probability $p(X; N)$ that a particular event will occur $X$ times out of $N$ trials as

$$p(X; N) = \binom{N}{X} P^X (1 - P)^{N-X}$$

(5.4)

$$\binom{N}{X} = \frac{N!}{X!(N - X)!}$$

where $P$ is the probability that an event will occur in each individual trial ($P = 1/T_R$ in this case). Now, if we let the number of occurrences equal zero ($X = 0$) in a given period of years $N$ (number of trials) and substitute this value in Eq. 5.4, the result is

$$p(0; N) = (1 - P)^N$$

(5.5)

This is the probability of zero events equal to or greater than the $T_R$-year event. Then the probability $Z$ that at least one event equal to or greater than the $T_R$-year event will occur in a sequence of $N$ years is given by

$$Z = 1 - \left(1 - \frac{1}{T_R}\right)^N$$

(5.6)

Solution of Eq. 5.6 for various values of $N$ and $T_R$ provided the data for Table 5.3.

## 5.14 PROBABILISTIC MASS TYPE OF ANALYSIS

Low-flow information is the basis for reservoir design and for studies of the waste-assimilative capacity of streams. For example, a customary critical period used by water-quality personnel is the average low flow for 7 consecutive days occurring on the average of once in 10 yr. For reservoir design, the critical period will usually be measured in months or even years and return periods are normally on the order of 20, 50, or 100 yr.

Information on the probability of occurrence of droughts of various severities during any single year or during any specified period of years can be developed from climatological and hydrologic records. Using such information, the engineer can gain considerable insight into the risks associated with reservoir development. A knowledge of these risks permits an estimate of the numerical odds for any specified yield. The designer may thus evaluate the performance of the reservoir under any set of risks he chooses.

In designing a reservoir to meet a specific draft, certain basic information must be at hand. First, the duration of the low-flow or critical period of design must be known. Second, the magnitude of the critical low flow must be determined. Third, the frequency of occurrence of the critical event must be known.

The question of the critical period can be answered by experimenting with a number of low-flow durations and then selecting, by judgment or by policy, some duration with which to work. From existing or synthetic streamflow data, a series of magnitudes of critical flows for the specified duration can be obtained. This information answers the question of magnitude. Finally, by assigning recurrence intervals to critical events, the frequency of events can be established.

## 5.15 LOSSES FROM STORAGE

The design of an impounding reservoir must include an evaluation of storage losses that may result from natural or artificial phenomena. Natural losses occur through evaporation, seepage, and siltation, while artificial losses are usually the product of withdrawals made to satisfy prior water rights.

Once a dam has been built and the impoundment filled, the exposed water-surface area is increased significantly over that of the natural stream. The resultant effect is that losses are incurred through increased evaporation. In addition, the opportunity for the derivation of runoff from the flooded land is eliminated. On the credit side, gains are made through the catchment of direct precipitation. These phenomena are commonly known as the *water-surface effect*. Net gains, then, usually result in well-watered regions. In arid lands, losses are the typical outcome, since evaporation generally exceeds precipitation.

The magnitude of seepage losses depends mainly on the geology of the region. If porous strata underlie the reservoir valley, considerable losses can occur. On the other hand, where permeability is low, seepage may be negligible. A thorough subsurface exploration is a prerequisite to the adequate evaluation of such losses.

Sedimentation studies are valuable in planning for the development of surface water supplies by impoundment [3, 8, 10]. Since the useful life of a reservoir will be materially affected by the deposition of sediment, a knowledge of sedimentation rates is important in reaching a decision regarding the feasibility of its construction.

The rate and characteristic of the sediment inflow can be controlled by using sedimentation basins, providing vegetative screens, and by employing various erosion control techniques [3]. Dams can be designed so that part of the sediment load can be passed through or over them. A last resort is the physical removal of sediment deposits. Normally this is not economically feasible.

### ■ EXAMPLE 5.3

Determine the expected life of the Lost Valley Reservoir. The initial capacity of the reservoir is 45,000 acre·ft, and the average annual inflow is 76,000 acre·ft. A sediment

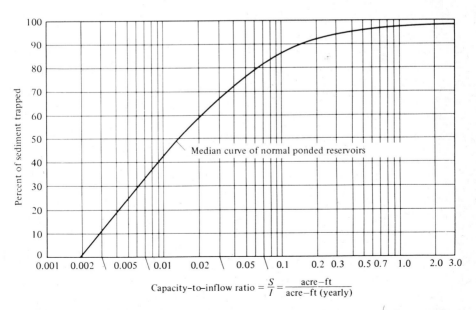

**Figure 5.4** Relationship between reservoir sediment trap efficiency and the capacity/inflow ratio. [Developed from data by G. M. Bruce, "Trap Efficiency of Reservoirs," *Trans. Am. Geophys. Union* 34 (1953): 407–418.]

**TABLE 5.4    DETERMINATION OF PROBABLE LIFE OF THE LOST VALLEY RESERVOIR**

| (1) | (2) | (3) | (4) | (5) | (6) | (7) |
|---|---|---|---|---|---|---|
| Reservoir capacity (acre·ft) | Volume increment (acre·ft) | Capacity/ inflow ratio: (1) ÷ 76,000 | Percent sediment trapped, from Fig. 5.4 | Average percent sediment trapped per volume increment | Acre-feet sediment trapped annually: (5) × 176 | Number of years required to fill the volume increment: (2) ÷ (6) |
| 45,000 | 5000 | 0.59 | 96.5 | | | |
| 40,000 | 5000 | 0.52 | 96.1 | 96.3 | 169 | 30 |
| 35,000 | 5000 | 0.46 | 95.8 | 95.9 | 169 | 30 |
| 30,000 | 5000 | 0.39 | 95.0 | 95.4 | 168 | 30 |
| 25,000 | 5000 | 0.33 | 94.5 | 94.7 | 167 | 30 |
| 20,000 | 5000 | 0.26 | 93.0 | 93.8 | 165 | 30 |
| 15,000 | 5000 | 0.20 | 92.0 | 92.5 | 163 | 31 |
| 10,000 | 5000 | 0.13 | 88.0 | 90.0 | 158 | 32 |
| Total number of years of useful life | | | | | | 213 |

inflow of 176 acre·ft/yr is reported. Assume that the useful life of the reservoir is exceeded when 77.8% of the original capacity is lost.

*Solution.*    The solution is obtained through application of the data given in Fig. 5.4. The results are tabulated in Table 5.4.    ∎

Problems of the type illustrated in Examples 5.2 and 5.3 are especially suited to the use of spreadsheet analyses such as those facilitated by Lotus 1-2-3 and Quattro Pro. These and other powerful analytic tools offer the opportunity of quick adjustments to parameters and speedy recalculation of values. More will be said about this in Chapter 7, in which the use of spreadsheets for designing sewers is illustrated.

# Groundwater

Groundwater storage is considerably in excess of all artificial and natural surface storage in the United States, including the Great Lakes [3]. This enormous groundwater reserve sustains the continuing outflow of streams and lakes during prolonged periods, which often follow the relatively few runoff-producing rains each year. However, the relation between groundwater and surface storage is one of mutual interdependence. For example, groundwater intercepted by wells as it moves toward a stream represents a real diversion just as if the water had actually been taken from the stream.

Increased usage and development of groundwater supplies have stimulated the search for knowledge regarding the occurrence, origin, and movement of these supplies. The study of groundwater is vital. It is also exceedingly complex, because groundwater location and movement are determined primarily by the geology of the area. This geologic role cannot be overemphasized. Groundwater distribution in the United States was discussed in Section 5.4.

Practically all groundwater can be considered to be part of the hydrologic cycle, even though small amounts may enter the cycle from other origins. *Connate water,* water entrapped in the pores of sedimentary rock at the time of deposition, is an example.

## 5.16  THE SUBSURFACE DISTRIBUTION OF WATER

Groundwater distribution may be generally categorized into zones of aeration and saturation. The *saturation zone* is one in which all the voids are filled with water under hydrostatic pressure. The *aeration zone,* in which the interstices are filled partly with air and partly with water, may be subdivided into several subzones. Todd classifies these as follows [10, 11]:

1. The *soil-water zone* begins at the ground surface and extends downward through the major root zone. Its total depth is variable and dependent on soil type and vegetation. The zone is unsaturated except during periods of heavy infiltration. Three categories of water classification may be encountered in this region: *hygroscopic water,* which is adsorbed from the air; *capillary water,* which is held by surface tension; and *gravitational water,* which is excess soil water draining through the soil.
2. The *intermediate zone* extends from the bottom of the soil-water zone to the top of the capillary fringe and may vary from nonexistence to several hundred feet in thickness. The zone is essentially a connecting link between the near-ground surface region and the near-water table region through which infiltrating waters must pass.
3. The *capillary zone* extends from the water table to a height determined by the capillary rise which can be generated in the soil. The capillary zone thickness is a function of soil texture and may vary not only from region to region but also within a local area.
4. At the *groundwater zone,* groundwater fills the pore spaces completely, and porosity is therefore a direct measure of storage volume. Part of this water (specific retention) cannot be removed by pumping or drainage because of molecular and surface-tension forces. The specific retention is the ratio of the volume of water retained against gravity drainage to the gross volume of the soil.

The water that can be drained from a soil by gravity is known as the *specific yield.* It is expressed as the ratio of the volume of water that can be drained by gravity to the gross volume of the soil. Values of specific yield are dependent on soil particle size, shape and distribution of pores, and degree of compaction of the soil. Average values of specific yield for alluvial aquifers range from 10% to 20%. Meinzer and others have proposed numerous procedures for determining specific yield [12].

## 5.17  AQUIFERS

An *aquifer* is a water-bearing stratum or formation capable of transmitting water in quantities sufficient to permit development. Aquifers may be considered as falling into two categories, confined and unconfined, depending on whether or not a water table or free surface exists under atmospheric pressure. The storage volume within an aquifer is changed whenever water is recharged to or discharged from an aquifer. In the case of an unconfined aquifer, this may be easily determined as

$$\Delta S = S_y \Delta V \tag{5.7}$$

where   $\Delta S$ = change in storage volume
   $S_y$ = average specific yield of the aquifer
   $\Delta V$ = volume of the aquifer lying between the original water table and the water table at a later, specified time

For saturated, confined aquifers, pressure changes produce only slight changes in storage volume. In this case, the weight of the overburden is supported partly by hydrostatic pressure and partly by the solid material in the aquifer. When the hydrostatic pressure in a confined aquifer is reduced by pumping or other means, the load on the aquifer increases, causing its compression, with the result that some water is forced from it. Decreasing the hydrostatic pressure also causes a small expansion, which in turn produces an additional release of water. For confined aquifers, the water yield is expressed in terms of a storage coefficient $S_c$. This storage coefficient may be defined as the volume of water that an aquifer takes in or releases per unit surface area of aquifer per unit change in head normal to the surface. Figure 5.5 illustrates the classifications of aquifers.

In addition to water-bearing strata exhibiting satisfactory rates of yield, there are also non-water-bearing and impermeable strata. An *aquiclude* is an impermeable stratum that may contain large quantities of water but whose transmission rates are not high enough to permit effective development. An *aquifuge* is a formation that is impermeable and devoid of water.

## 5.18  FLUCTUATIONS IN GROUNDWATER LEVEL

Any circumstance that alters the pressure imposed on underground water will also cause a variation in the groundwater level. Seasonal factors, change in stream and river stages, evapotranspiration, atmospheric pressure changes, winds, tides, external loads, various forms of withdrawal and re-

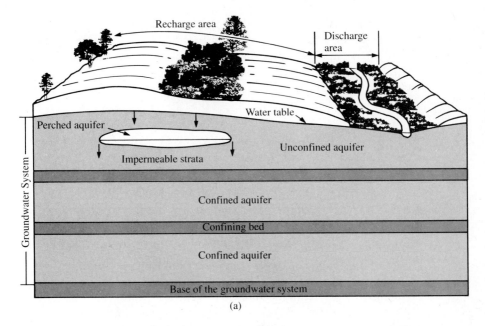

(a)

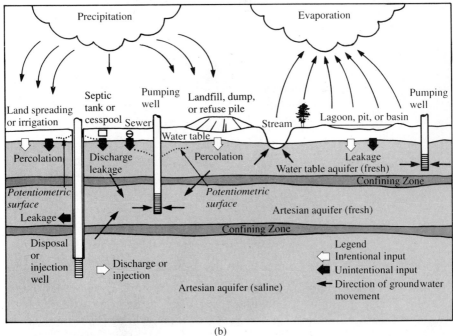

(b)

**Figure 5.5** Definition sketches of groundwater systems and mechanisms for recharge and withdrawal: (a) aquifer notation [13] and (b) components of the hydrologic cycle affecting groundwater [14].

charge, and earthquakes all may produce fluctuations in the level of the water table or the piezometric surface, depending on whether the aquifer is free or confined [10]. It is important that the engineer concerned with the development and utilization of groundwater supplies be aware of these factors. He should also be able to evaluate their importance relative to the operation of a specific groundwater basin.

## 5.19 SAFE YIELD OF AN AQUIFER

The development engineer must be able to determine the quantity of water that can be produced from a groundwater basin in a given period of time. He must also be able to evaluate the consequences that will result from the imposition of various drafts on the underground supply. A knowledge of the safe yield of an aquifer is therefore exceedingly important. Following are pertinent definitions:

1. The *safe yield* is the quantity of water that can be withdrawn annually without the ultimate depletion of the aquifer.
2. The *maximum sustained yield* is the maximum rate at which water can be withdrawn on a continuing basis from a given source.
3. The *permissive sustained yield* is the maximum rate at which withdrawals can be made legally and economically on a continuing basis for beneficial use without the development of undesired results.
4. The *maximum mining yield* is the total storage volume in a given source that can be withdrawn and used.
5. The *permissive mining yield* is the maximum volume of water that can be withdrawn legally and economically, to be used for beneficial purposes, without causing an undesired result.

After studying these definitions, the reader should note that extreme caution must be exercised in the development and utilization of groundwater resources. These resources are finite and not inexhaustible. If drafts are imposed such that the various recharge mechanisms will balance them over a period of time, no difficulty will result. On the other hand, excessive drafts may reduce underground storage volumes to a point at which economic development is no longer feasible. The mining of water, as with any other natural resource, will ultimately result in exhaustion of the supply. Figure 5.6 indicates those areas in the United States where perennial overdrafts occur [3, 10].

Methods for determining the safe yield of an aquifer have been proposed by Hill, Harding, Simpson, and others [10]. The Hill method based on groundwater studies in Southern California and Arizona will be presented here. In this method, the annual change in groundwater table elevation or piezometric surface elevation is plotted against the annual draft. The points can be fitted by a straight line, provided that the water supply to the basin is fairly uniform.

That draft which corresponds to zero change in elevation is considered to be the safe yield. The period of record should be such that the supply during this period approximates the long-time average supply. Even though the draft during the period of record may be an overdraft, the safe yield can be deter-

**Figure 5.6** Groundwater reservoirs with perennial overdraft. [From H. E. Thomas, "Water," in *The Yearbook of Agriculture* (Washington, DC: U.S. Government Printing Office, 1955).]

mined by extending the line of best fit to an intersection with the zero change in elevation line. An example of this procedure is given in Fig. 5.7.

It should be noted that the safe yield of a groundwater basin may be variable with time. This is because the groundwater basin conditions under which the safe yield was determined are subject to change, and these changes will be reflected in the modified value of the safe yield.

## 5.20 GROUNDWATER FLOW

The rate of movement of water through the ground is of an entirely different magnitude than that through natural or artificial channels or conduits. Typical values range from 5 ft/day to a few feet per year. Methods for determining these transmission rates are primarily based on the principles of fluid flow represented by *Darcy's law*. Mathematically, this law may be stated as

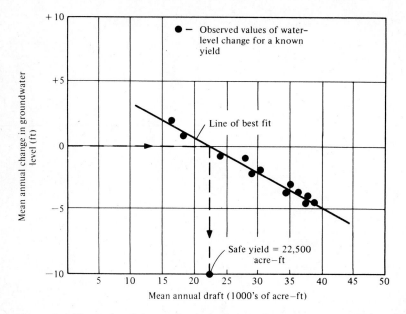

**Figure 5.7** Example of determination of the safe yield by the Hill method.

$$V = kS \tag{5.8}$$

where  $V$ = velocity of flow
$\phantom{where}$ $k$ = coefficient having same units as velocity
$\phantom{where}$ $S$ = slope of hydraulic gradient

Darcy's law is limited in its applicability to flows in the laminar region. The controlling criterion is the *Reynolds number*.

$$N_R = \frac{Vd}{\nu} \tag{5.9}$$

where  $V$ = flow velocity
$\phantom{where}$ $d$ = mean grain diameter
$\phantom{where}$ $\nu$ = kinematic viscosity

For Reynolds numbers of less than unity, groundwater flow may be considered laminar. Departure from laminar conditions develops normally in the range of Reynolds numbers from 1 to 10, depending on grain size and shape. Under most conditions, with the exception of regions in close proximity to collecting devices, the flow of groundwater is laminar and Darcy's law applies.

■ **EXAMPLE 5.4**
$\phantom{xx}$ **a.** Find the Reynolds number for the portion of an aquifer distant from any collection device where water temperature is 50°F ($\nu = 1.41 \times 10^{-5}$ ft²/sec), flow velocity is 1.0 ft/day, and mean grain diameter is 0.09 in.

**b.** Find the Reynolds number for a flow 4 ft from the centerline of a well being pumped at a rate of 3800 gpm if the well completely penetrates a confined aquifer 28 ft thick. Assume a mean grain diameter of 0.10 in., a porosity of 35%, and $\nu = 1.41 \times 10^{-5}$ ft²/sec.

*Solution.*

**a.** Using Eq. 5.9, we obtain

$$N_R = \frac{Vd}{\nu}$$

where

$$V = \frac{1.0}{86,400} \text{ fps} \quad d = \frac{0.09}{12} \text{ ft}$$

$$N_R = \frac{1.0}{86,400} \times \frac{0.09}{12} \times \frac{1}{1.41 \times 10^{-5}}$$

$$= 0.0062 \quad \text{(including laminar flow)}$$

**b.** Using Eq. 5.9 yields

$$N_R = \frac{Vd}{\nu} = \frac{Q}{A}\frac{d}{\nu}$$

$$Q = 3800 \times 2.23 \times 10^{-3} = 8.46 \text{ cfs}$$

$$V = 8.46/2\pi \, rh \times \text{porosity}$$

$$= 8.46/8\pi \times 28 \times 0.35$$

$$= 0.0344 \text{ fps}$$

$$N_R = \frac{0.0344 \times 0.10}{1.41 \times 10^{-5} \times 12} = 20.3 \quad \text{(beyond Darcy's law range)} \qquad ■$$

To compute discharge, it is necessary to multiply Eq. 5.8 by the effective cross-sectional area. The equation then becomes

$$Q = pAkS \qquad (5.10)$$

where    $p$ = porosity or ratio of void volume to total volume of the mass
   $A$ = gross cross-sectional area

and the other terms are as previously defined. By combining $k$ and $p$ into a single term, Eq. 5.10 may be written in its most common form,

$$Q = KAS \qquad (5.11)$$

where $K$ is known as the *coefficient of permeability*. A number of ways of expressing $K$ may be found in the literature. The U.S. Geological Survey defines the standard coefficient of permeability $K_s$ as the number of gallons of water per day that will flow through a medium of 1-ft² cross-sectional area under a hydraulic gradient of unity at 60°F. The field coefficient of

**TABLE 5.5**   THE STANDARD COEFFICIENT OF
PERMEABILITY FOR VARIOUS
MATERIALS
Typical Values

| Material | Approximate range in $K_s$ (gpd/ft²) |
|---|---|
| I    Clean gravel | $10^6$–$10^4$ |
| II   Clean sands; mixtures of clean sands and gravels | $10^4$–10 |
| III  Very fine sands; silts; mixtures of sand, silt, and clay; stratified clays, etc. | 10–$10^{-3}$ |
| IV   Unweathered clays | $10^{-3}$–$10^{-4}$ |

permeability is obtained directly from the standard coefficient by correcting
for temperature.

$$K_f = K_s \frac{\mu_{60}}{\mu_f} \tag{5.12}$$

where   $K_f$ = field coefficient
$\mu_{60}$ = dynamic viscosity at 60°F
$\mu_f$ = dynamic viscosity at field temperature

An additional term that is much used in groundwater computations is the
coefficient of transmissibility $T$. It is equal to the field coefficient of permeability multiplied by the saturated thickness of the aquifer in feet. Using this
terminology, Eq. 5.10 may also be written

$$Q = T \times \text{section width} \times S \tag{5.13}$$

Table 5.5 gives typical values of the standard coefficient of permeability
for a range of sedimentary materials. It should be noted that the permeabilities
for specific materials vary widely. Traces of silt and clay can significantly
decrease the permeability of an aquifer. Differences in particle orientation
and shape can cause striking changes in permeability within aquifers composed of the same geologic material. Careful evaluation of geologic information is absolutely essential if realistic values of permeability are to be used
in groundwater-flow computations.

It is of interest to note that Darcy's equation is analogous to the electrical
equation known as *Ohm's law*,

$$i = \frac{1}{R} E \tag{5.14}$$

where    $i$ = current
    $R$ = resistance
    $E$ = voltage

The quantities $i$ and $Q$, $K$ and $1/R$, and $E$ and $S$ are comparable. This equivalency permits the use of electrical models in solving many ground-water-flow problems [3, 10, 11].

## 5.21 HYDRAULICS OF WELLS

The collection of groundwater is accomplished primarily through the construction of wells or infiltration galleries. Numerous factors are involved in the numerical estimation of the performance of these collection works. Some cases are amenable to solution through the utilization of relatively simple mathematical expressions. Other cases can be solved only through graphical analysis or the use of various kinds of models. Several of the less difficult cases will be discussed here. The reader is cautioned not to be misled by the simplicity of some of the solutions presented and should observe that many of these are special-case solutions and are not indiscriminately applicable to all groundwater-flow situations. A more thorough treatment of groundwater and seepage problems may be found in numerous references [10, 15, 16].

### Flow to Wells

A well system may be considered to be composed of three elements—the well structure, the pump, and the discharge piping. The well itself contains an open section through which flow enters and a casing through which the flow is transported to the ground surface. The open section is usually a perforated casing or a slotted metal screen that permits the flow to enter and at the same time prevents collapse of the hole. Occasionally gravel is placed at the bottom of the well casing around the screen.

When a well is pumped, water is removed from the aquifer immediately adjacent to the screen. Flow then becomes established at locations some distance from the well in order to replenish this withdrawal. Owing to the resistance to flow offered by the soil, a head loss is encountered and the piezometric surface adjacent to the well is depressed. This is known as the *cone of depression* (Fig. 5.8). The cone of depression spreads until a condition of equilibrium is reached and steady-state conditions are established.

The hydraulic characteristics of an aquifer (which are described by the storage coefficient and the aquifer permeability) may be determined by laboratory or field tests. The three most commonly used field methods are the application of tracers, use of field permeameters, and aquifer performance tests [3, 10]. A discussion of aquifer performance tests will be given here along with the development of flow equations for wells.

Aquifer performance tests may be classified as either equilibrium or

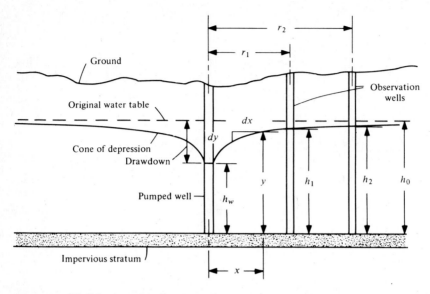

**Figure 5.8** Well in an unconfined aquifer.

nonequilibrium tests. In the former, the cone of depression must be stabilized for the flow equation to be derived. In the latter, the derivation includes the condition that steady-state conditions have not been reached. Thiem published the first performance tests based on equilibrium conditions in 1906 [17].

The basic equilibrium equation for an unconfined aquifer can be derived using the notation of Fig. 5.8. In this case, the flow is assumed to be radial, the original water table is considered to be horizontal, the well is considered to fully penetrate the aquifer of infinite areal extent, and steady-state conditions must prevail. Then the flow toward the well at any location $x$ from the well must equal the product of the cylindrical element of area at that section and the flow velocity. Using Darcy's law, this becomes

$$Q = 2\pi xy K_f \frac{dy}{dx} \tag{5.15}$$

where
$$2\pi xy = \text{area at any section}$$
$$K_f \, dy/dx = \text{flow velocity}$$
$$Q = \text{discharge, cfs}$$

Integrating over the limits specified below yields

$$\int_{r_1}^{r_2} Q \frac{dx}{x} = 2\pi K_f \int_{h_1}^{h_2} y \, dy \tag{5.16}$$

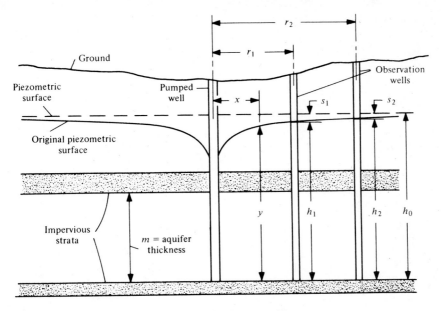

**Figure 5.9** Radial flow to a well in a confined aquifer.

$$Q \ln\left(\frac{r_2}{r_1}\right) = \frac{2\pi K_f (h_2^2 - h_1^2)}{2} \tag{5.17}$$

and

$$Q = \frac{\pi K_f (h_2^2 - h_1^2)}{\ln(r_2/r_1)} \tag{5.18}$$

This equation may then be solved for $K_f$ to yield

$$K_f = \frac{1055 Q \log(r_2/r_1)}{h_2^2 - h_1^2} \tag{5.19}$$

where ln has been converted to log, $K_f$ is in gallons per day per square foot, $Q$ is in gallons per minute, and $r$ and $h$ are measured in feet. If the drawdown is small compared with the total aquifer thickness, an approximate formula for the discharge of the pumped well can be obtained by inserting $h_w$ for $h_1$ and the height of the aquifer for $h_2$ in Eq. 5.18.

The basic equilibrium equation for a confined aquifer can be obtained in a similar manner, using the notation of Fig. 5.9. The same assumptions apply. Mathematically the flow in cubic feet per second may be determined as follows:

$$Q = 2\pi x m K_f \frac{dy}{dx} \tag{5.20}$$

Integrating, we obtain

$$Q = 2\pi K_f m \frac{h_2 - h_1}{\ln(r_2/r_1)} \qquad (5.21)$$

The coefficient of permeability may be determined by rearranging Eq. 5.21 to the form

$$K_f = \frac{528Q \log(r_2/r_1)}{m(h_2 - h_1)} \qquad (5.22)$$

where $Q$ is in gallons per minute, $K_f$ is the permeability in gallons per day per square foot, and $r$ and $h$ are measured in feet.

■ **EXAMPLE 5.5**
Determine the permeability of an artesian aquifer being pumped by a fully penetrating well. The aquifer is composed of medium sand and is 90 ft thick. The steady-state pumping rate is 850 gpm. The drawdown of an observation well 50 ft away is 10 ft, and the drawdown in a second observation well 500 ft away is 1 ft.

*Solution.*

$$K_f = \frac{528Q \log(r_2/r_1)}{m(h_2 - h_1)}$$

$$= \frac{528 \times 850 \times \log(10)}{90 \times (10 - 1)}$$

$$= 554 \text{ gpd/ft}^2 \qquad ■$$

For a steady-state well in a uniform flow field where the original piezometric surface is not horizontal, a somewhat different situation than that previously assumed prevails. Consider the artesian aquifer shown in Fig. 5.10. The heretofore assumed circular area of influence becomes distorted in this case. This problem may be solved by application of potential theory or by graphical means; or, if the slope of the piezometric surface is very slight, Eq. 5.21 may be applied without serious error.

Referring to the definition sketch of Fig. 5.10, a graphical solution to this type of problem will be discussed. First, an orthogonal flow net consisting of flow lines and equipotential lines must be constructed. The construction should be performed so that the completed flow net will be composed of a number of elements that approach little squares in shape. Harr [15] is a good source of information on this subject for the interested reader. A comprehensive discussion cannot be provided here.

Once the net is complete, it may be analyzed by considering the net geometry and using Darcy's law in the manner of Todd [10]. In the definition sketch of Fig. 5.10, the hydraulic gradient is

$$h_g = \frac{\Delta h}{\Delta s} \qquad (5.23)$$

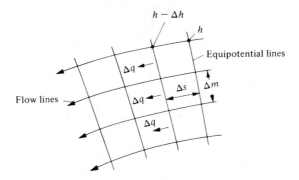

Definition Sketch of a Segment of a Flow Net

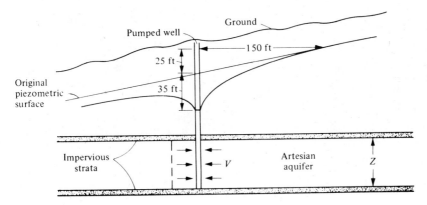

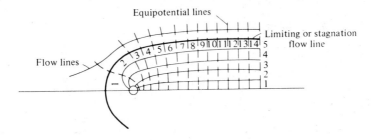

**Figure 5.10** Well in a uniform flow field and flow-net definition.

and the flow increment between adjacent flow lines is

$$\Delta q = K \frac{\Delta h}{\Delta s} \Delta m \tag{5.24}$$

where, for a unit thickness, $\Delta m$ represents the cross-sectional area. If the flow net is properly constructed so that it is orthogonal and composed of little square elements,

$$\Delta m \approx \Delta s \qquad\qquad (5.25)$$

and

$$\Delta q = K\Delta h \qquad\qquad (5.26)$$

Now considering the entire flow net,

$$\Delta h = \frac{h}{n} \qquad\qquad (5.27)$$

where $n$ is the number of subdivisions between equipotential lines. If the flow is divided into $m$ sections by the flow lines, the discharge per unit width of the aquifer will be

$$q = \frac{Kmh}{n} \qquad\qquad (5.28)$$

Knowledge of the aquifer permeability and the flow-net geometry permits solution of Eq. 5.28.

### ■ EXAMPLE 5.6

Find the discharge to the well of Fig. 5.10 by using the applicable flow net. Consider the aquifer to be 35 ft thick, $K_f$ to be $3.65 \times 10^{-4}$ fps, and the other dimensions as shown.

*Solution.* Using Eq. 5.28,

$$q = \frac{Kmh}{n}$$

where  $h = (35 + 25) = 60$ ft

$m = 2 \times 5 = 10$

$n = 14$

$q = \dfrac{3.65 \times 10^{-4} \times 60 \times 10}{14}$

$= 0.0156$ cfs per unit thickness of the aquifer

The total discharge $Q$ is thus

$Q = 0.0156 \times 35 = 0.55$ cfs  or  245 gpm    ■

When a new well is first pumped, a large portion of the discharge is produced directly from the storage volume released as the cone of depression develops. Under these circumstances, the equilibrium equations overestimate permeability and therefore the yield of the well. Where steady-state conditions are not encountered—as is usually the case in practice—a nonequilibrium equation must be used. Two approaches can be taken, the rather rigorous method of Theis or a simplified procedure such as that proposed by Cooper and Jacob [18, 19].

In 1935, Theis published a nonequilibrium approach which takes into consideration time and the storage characteristic of the aquifer [18]. His method utilizes an analogy between heat transfer described by the Biot-Fourier law and groundwater flow to a well. Theis states that the drawdown $s$ in an observation well located at a distance $r$ from the pumped well is given by

$$s = \frac{114.6Q}{T} \int_u^\infty \frac{e^{-u}}{u} \, du \qquad (5.29)$$

where  $T$ = transmissibility
         $Q$ = discharge, gpm

and $u$ is given by

$$u = \frac{1.87r^2 S_c}{Tt} \qquad (5.30)$$

where  $S_c$ = storage coefficient
         $T$ = transmissibility
         $t$ = time since the start of pumping, days

The integral in Eq. 5.29 is usually known as the well function of $u$ and is commonly written $W(u)$. It may be evaluated from the infinite series

$$W(u) = -0.577216 - \log_e u + u - \frac{u^2}{2 \times 2!} + \frac{u^3}{3 \times 3!} - \cdots \qquad (5.31)$$

The basic assumptions employed in the Theis equation are essentially the same as those for the Thiem equation except for the non-steady-state condition.

Equations 5.29 and 5.30 can be solved by comparing a log-log plot of $u$ versus $W(u)$, known as a "type curve," with a log-log plot of the observed data $r^2/t$ versus $s$. In plotting the type curve, $W(u)$ is the ordinate and $u$ is the abscissa. The two curves are superimposed and moved about until some of their segments coincide. In doing this, the axes must be maintained parallel. A coincident point is then selected on the matched curves and both plots are marked. The type curve then yields values of $u$ and $W(u)$ for the selected point. Corresponding values of $s$ and $r^2/t$ are determined from the plot of the observed data. Inserting these values in Eqs. 5.29 and 5.30 and rearranging, values for the transmissibility $T$ and the storage coefficient $S_c$ may be found.

Often, this procedure can be shortened and simplified. When $r$ is small and $t$ large, Jacob found that values of $u$ are generally small [19]. Thus, the terms in the series of Eq. 5.31 beyond the second term become negligible and the expression for $T$ becomes

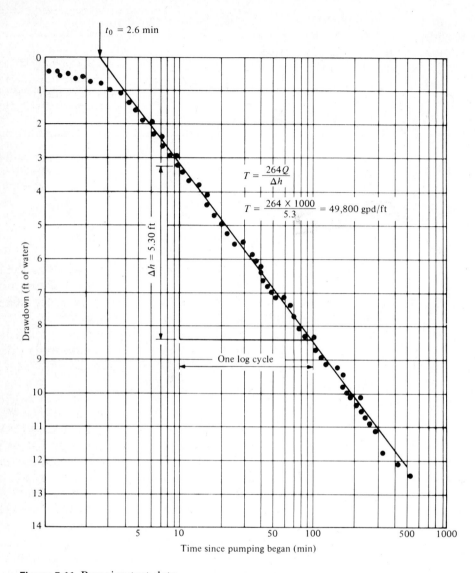

**Figure 5.11** Pumping test data.

$$T = \frac{264Q(\log t_2 - \log t_1)}{h_0 - h} \qquad (5.32)$$

which can be further reduced to

$$T = \frac{264Q}{\Delta h} \qquad (5.33)$$

where    $\Delta h$ = drawdown per log cycle of time, $(h_0 - h)/(\log t_2 - \log t_1)$
         $Q$ = well discharge, gpm

$h_0$ and $h$ are defined as shown on Fig. 5.9, and $T$ is the transmissibility in gallons per day per foot. Field data on drawdown $h_0 - h$ versus $t$ are plotted on semilogarithmic paper. The drawdown is plotted on the arithmetic scale as shown in Fig. 5.11. This plot forms a straight line, the slope of which permits the determination of the formation constants using Eq. 5.33 and

$$S_c = \frac{0.3 T t_0}{r^2} \qquad (5.34)$$

where $t_0$ is the time that corresponds to zero drawdown.

### ■ EXAMPLE 5.7

Using the data given in Fig. 5.11, find the coefficient of transmissibility $T$ and the storage coefficient $S_c$ for the aquifer given $Q$ = 1000 gpm and $r$ = 300 ft.

*Solution.* Find the value of $\Delta h$ from the graph. This is 5.3 ft. Then, using Eq. 5.33, we obtain

$$T = \frac{264Q}{\Delta h} = \frac{264 \times 1000}{5.3}$$

$$= 49,800 \text{ gpd/ft}$$

Using Eq. 5.34 yields

$$S_c = \frac{0.3 T t_0}{r^2}$$

Note from Fig. 5.11 that $t_0$ = 2.6 min. When we convert to days, this becomes

$$t_0 = 1.81 \times 10^{-3} \text{ day}$$

and

$$S_c = \frac{0.3 \times 49,800 \times 1.81 \times 10^{-3}}{(300)^2}$$

$$= 0.0003 \qquad ■$$

### ■ EXAMPLE 5.8

Find the drawdown at an observation point 200 ft away from a pumping well. Given that $T = 3.0 \times 10^4$ gpd/ft, the pumping time is 12 days, $S_c = 3 \times 10^{-4}$, and $Q$ = 300 gpm.

*Solution.* From Eq. 5.29, $u$ can be computed,

$$u = \frac{1.87 \times (200)^2 \times 3 \times 10^{-4}}{3 \times 10^4 \times 12} = 6.23 \times 10^{-5}$$

Referring to Table 5.6 and interpolating, we estimate $W(u)$ to be 9.1. Then, by Eq. 5.30, the drawdown is found to be

$$s = \frac{114.6 \times 9.1 \times 300}{3 \times 10^4} = 10.41 \text{ ft} \qquad ■$$

## 5.22 BOUNDARY EFFECTS

Only the effect of pumping a single well has been previously considered. If more than one well is pumped in an area, however, a composite effect or interference due to the overlap of the cones of depression will result. In this case, the drawdown at any location is obtained by summing the individual drawdowns of the various wells involved. An additional problem is that of boundary conditions. The previous derivations have been based on the supposition of a homogeneous aquifer of infinite areal extent. A situation such as this is rarely encountered in practice. Computations based on this assumption are often sufficiently accurate, however, provided that field conditions closely approximate the basic hypotheses. Boundary effects may be evaluated by using the theory of image wells proposed by Lord Kelvin, through the use of electrical and membrane analogies and through the use of relaxation procedures. For a detailed discussion of these topics, the reader is referred to the many references on groundwater flow [10, 11, 15].

## 5.23 REGIONAL GROUNDWATER SYSTEMS

The methods discussed so far have related mostly to the flow of water to individual wells. The analysis of regional groundwater systems is an important aspect of water resources planning, however, and it is appropriate to introduce some methodology for analyzing these large systems [20–24].

Aquifers are important not only because they are the vehicles for providing water supplies, but also from the water-quality standpoint. As in the case of surface waters, if groundwater sources are so contaminated that their use for any purpose or purposes is impaired or precluded, then regardless of the quantity of water available, the source may have to be abandoned or untapped. The methods of regional groundwater systems analysis deal with both the quality and quantity aspects of groundwaters and are thus applicable to a wide variety of water management problems.

The approach to regional analysis of groundwater systems is modeling. Models may be physical, electric analog, or mathematical. The classification that will be discussed here is the mathematical model—a set of equations representative of the physical processes that occur in an aquifer, subject, of course, to the simplifying assumptions that must be made. The models may be deterministic, probabilistic, or a combination of the two. In the discussions that follow, we shall limit ourselves to deterministic mathematical models. These models describe cause-effect relationships stemming from known features of the physical system under study. Figure 5.12 characterizes the procedure for developing a deterministic mathematical model. Based on a study of the region of interest, and armed with an understanding of the mechanics of groundwater flow, a conceptual model is formulated. This is translated into a mathematical model of the system, which is usually a partial differential equation accompanied by appropriate boundary and initial condi-

**TABLE 5.6** VALUES OF $W(u)$ FOR VARIOUS VALUES OF $u$

| $u$ | 1.0 | 2.0 | 3.0 | 4.0 | 5.0 | 6.0 | 7.0 | 8.0 | 9.0 |
|---|---|---|---|---|---|---|---|---|---|
| $\times 1$ | 0.219 | 0.049 | 0.013 | 0.0038 | 0.0011 | 0.00036 | 0.00012 | 0.000038 | 0.000012 |
| $\times 10^{-1}$ | 1.82 | 1.22 | 0.91 | 0.70 | 0.56 | 0.45 | 0.37 | 0.31 | 0.26 |
| $\times 10^{-2}$ | 4.04 | 3.35 | 2.96 | 2.68 | 2.47 | 2.30 | 2.15 | 2.03 | 1.92 |
| $\times 10^{-3}$ | 6.33 | 5.64 | 5.23 | 4.95 | 4.73 | 4.54 | 4.39 | 4.26 | 4.14 |
| $\times 10^{-4}$ | 8.63 | 7.94 | 7.53 | 7.25 | 7.02 | 6.84 | 6.69 | 6.55 | 6.44 |
| $\times 10^{-5}$ | 10.94 | 10.24 | 9.84 | 9.55 | 9.33 | 9.14 | 8.99 | 8.86 | 8.74 |
| $\times 10^{-6}$ | 13.24 | 12.55 | 12.14 | 11.85 | 11.63 | 11.45 | 11.29 | 11.16 | 11.04 |
| $\times 10^{-7}$ | 15.54 | 14.85 | 14.44 | 14.15 | 13.93 | 13.75 | 13.60 | 13.46 | 13.34 |
| $\times 10^{-8}$ | 17.84 | 17.15 | 16.74 | 16.46 | 16.23 | 16.05 | 15.90 | 15.76 | 15.65 |
| $\times 10^{-9}$ | 20.15 | 19.45 | 19.05 | 18.76 | 18.54 | 18.35 | 18.20 | 18.07 | 17.95 |
| $\times 10^{-10}$ | 22.45 | 21.76 | 21.35 | 21.06 | 20.84 | 20.66 | 20.50 | 20.37 | 20.25 |
| $\times 10^{-11}$ | 24.75 | 24.06 | 23.65 | 23.36 | 23.14 | 22.96 | 22.81 | 22.67 | 22.55 |
| $\times 10^{-12}$ | 27.05 | 26.36 | 25.96 | 25.67 | 25.44 | 25.26 | 25.11 | 24.97 | 24.86 |
| $\times 10^{-13}$ | 29.36 | 28.66 | 28.26 | 27.97 | 27.75 | 27.56 | 27.41 | 27.28 | 27.16 |
| $\times 10^{-14}$ | 31.66 | 30.97 | 30.56 | 30.27 | 30.05 | 29.87 | 29.71 | 29.58 | 29.46 |
| $\times 10^{-15}$ | 33.96 | 33.27 | 32.86 | 32.58 | 32.35 | 32.17 | 32.02 | 31.88 | 31.76 |

*Source*: After L. K. Wenzel, "Methods for Determining Permeability of Water Bearing Materials with Special Reference to Discharging Well Methods," U.S. Geological Survey, Water-Supply Paper 887, Washington, DC, 1942.

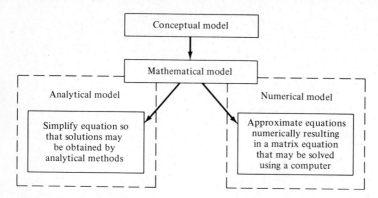

**Figure 5.12** Logic diagram for developing a mathematical model. (Courtesy of the National Water Well Association, Worthington, OH.)

tions. Continuity and conservation of momentum considerations are featured in the model and are represented over the extent of the region of concern. Darcy's law, discussed earlier, is widely used to describe conservation of momentum. Other model features include artesian or water table condition designation and dimensionality, i.e., one, two, or three dimensional. Where the modeling objective includes water-quality or heat transport considerations, other equations describing conservation of mass for the chemical species involved and conservation of energy are required. Typically used relationships are Fick's law for chemical diffusion and Fourier's law for heat transport.

After a conceptual model has been selected, the next step is to obtain a solution to the mathematical formulation by simplifying the groundwater equation through assumptions such as radial flow or, alternatively, by using numerical methods such as finite difference or finite elements to represent the governing partial differential equations. Using a finite difference approach, for example, the region of interest is divided into grid elements and the continuous variables of concern are represented as discrete variables at nodal points in the grid. In this way, the continuous differential equation defining head or other features is replaced by a finite number of algebraic equations that define the head or other variables at nodal points. Such models find wide application in prediction of site-specific aquifer behavior. Although a choice of model should be based on the study involved, numerical models have proven to be effective where irregular boundaries, heterogeneities, or highly variable pumping or recharge rates are expected [20]. Figure 5.13 indicates four types of groundwater models and their application.

If it is decided that a groundwater model is needed for a proposed study, the next step is to consider the requirements for carrying out a successful modeling effort. Normally, several steps are involved. These include data collection and preparation, matching of observed histories,

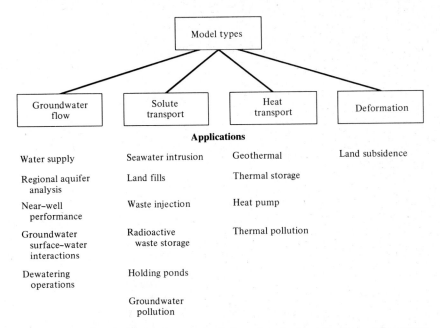

**Figure 5.13** Types of groundwater models and typical applications. (Courtesy of the National Water Well Association, Worthington, OH.)

and predictive simulation. Figure 5.14 indicates the various categories of model use.

The first step in model data preparation involves specifying the region's boundaries. These boundaries may be physical, such as an impervious layer, or arbitrary, such as the choice of some small subregion. After the overall boundaries are defined, the region is then divided into discrete elements by superimposing a rectangular or polygonal grid (see Fig. 5.15).

Having defined an appropriate grid, the modeler must specify the controlling aquifer parameters (such as storage coefficients, transmissivities, etc.) and set initial conditions. This must be accomplished for each grid element. Where solute transport models are required, additional parameters such as hydrodynamic dispersion properties must also be specified. Results of model runs include determination of hydraulic heads at node points for each time step during the period of interest. Where solute and heat transport are involved, concentrations and temperatures may also be determined at nodes for each time interval.

After the aquifer parameters have been set, the model is operated using these initial values and the output is checked with recorded history. This process is known as history matching. This matching procedure is used to refine parameter values and to determine boundaries and flow conditions at the boundaries. Comparisons between historic conditions and modeled conditions are made and parameters adjusted until satisfactory fits are ob-

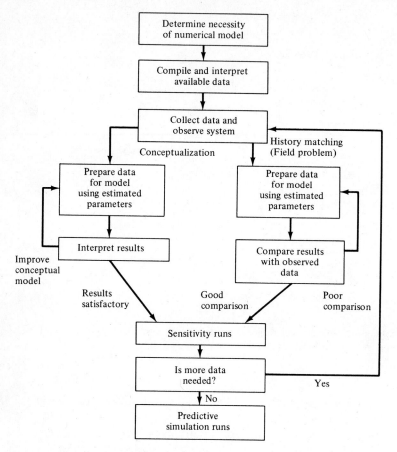

**Figure 5.14** Diagram showing model use. (Courtesy of the National Water Well Association, Worthington, OH.)

tained. This is known as the *calibration process*. There is no rule that specifies when adequate matching is achieved, however. This determination must be made by the modeler based on an understanding of the problem and the use to which the model's results are to be put.

Once calibration is completed and it is determined that the model can be expected to yield dependable results, the model can be directed toward analyzing many different types of management and development options so that the outcomes of different courses of action that might be followed in the future can be assessed. Observations of model performance under varying conditions is an asset in determining courses of action to be prescribed for future aquifer operation or development. Groundwater models may be used to estimate natural and artificial recharge, effects of boundaries, effects of well location and spacing, effects of varying rates of drawdown and recharge, rates of movement of hazardous wastes, saltwater intrusion, and other factors [20].

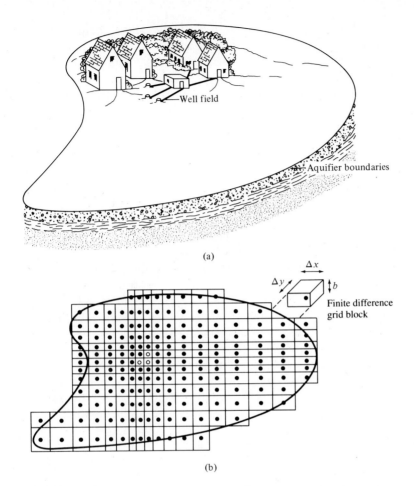

(a)

(b)

**Figure 5.15** (a) Map view of aquifer showing well field and boundaries. (b) Finite difference grid for aquifer study, where $\Delta x$ is the spacing in the $x$ direction, $\Delta y$ is the spacing in the $y$ direction, and $b$ is the aquifer thickness. Solid dots: block-center nodes; open circles: source-sink nodes. (Courtesy of the National Water Well Association, Worthington, OH.)

Finally, while numerical and other groundwater models have much to recommend them, caution must be exercised if they are not to be misused. According to Prickett, there are three common ways to misuse models [21]: overkill, inappropriate prediction, and misinterpretation. To avoid such pitfalls, the modeler and model user must be fully aware of the limitations and sources of errors in the model used. Underlying assumptions must be carefully understood.

Numerical models are important tools for analyzing groundwater problems and making recommendations for their solution. To be effective, however, the groundwater hydrologist must have a full understanding of the aquifer being studied, must be aware of alternative modeling approaches, and must be cognizant of model limitations and restrictions imposed by simplifying assumptions. Used appropriately, models can be powerful decision-making aids. Used inappropriately, they can lead to erroneous and sometimes damaging proposals.

## 5.24 SALTWATER INTRUSION

Saltwater contamination of freshwater aquifers presents a serious water-quality problem in island locations; in coastal areas; and occasionally inland, as in Arizona, where some aquifers contain highly saline waters. Because fresh water is lighter than saltwater (specific gravity of seawater is about 1.025), it will usually float above a layer of saltwater. When an aquifer is pumped, the original equilibrium is disturbed and saltwater replaces the fresh water. Under equilibrium conditions, a drawdown of 1 ft in the freshwater table will result in a rise by the saltwater of approximately 40 ft. Pumping rates of wells subject to saltwater intrusion are therefore seriously limited. In coastal areas, recharge wells are sometimes used in an attempt to maintain a sufficient head to prevent seawater intrusion. Injection wells have been used effectively in this manner in southern California.

A prime example of freshwater contamination by seawater is noted in Long Island, New York [25]. During the first part of this century, the rate of pumping far exceeded the natural recharge rate. The problem was further complicated because stormwater runoff from the highly developed land areas was transported directly to the sea. This precluded the opportunity for this water to return to the ground. As pumping continued, the water table dropped well below sea level and saline water entered the aquifer. The result was such a serious impairment of local water quality that Long Island was forced to transport its water supply from upper New York State.

## 5.25 GROUNDWATER RECHARGE

The volumes of groundwater replaced annually through natural mechanisms are relatively small because of the slow rates of movement of groundwaters and the limited opportunity for surface waters to penetrate the earth's surface. To supplement this natural recharge process, a recent trend toward artificial recharge has been developing. In 1955, over 700 million gallons of water per day were artificially recharged in the United States [26]. This water was derived from natural surface sources and returns from air conditioning, industrial wastes, and municipal water supplies. The total recharge volume was equal, however, to only about 1.5% of the groundwater withdrawn that year. In California, for example, artificial recharge is presently a primary method of water conservation. During the period 1957–1958 a daily recharge volume of about 560 mil gal was reported for 63 projects in that state alone [26].

Numerous methods are employed in artificial recharge operations. One of the most common plans is the utilization of holding basins. The usual practice is to impound the water in a series of reservoirs arranged so that the overflow of one will enter the next, and so on. These artificial storage works are generally formed by the construction of dikes or levees. A second method is the modified streambed, which makes use of the natural water supply. The stream channel is widened, leveled, scarified, or treated by a combination of methods to increase its recharge capabilities. Ditches and furrows are also used. The basic types of arrangement are the contour type, in which the ditch follows the contour of the ground; the lateral type, in which water is diverted into a number of small furrows from the main canal or channel; and the tree-shaped or branching type, where water is diverted from the primary channel into successively smaller canals and ditches. Where slopes are relatively flat and uniform, flooding provides an economical means of recharge. Normal practice is to spread the recharge water over the ground at relatively small depths so as not to disturb the soil or native vegetation. An additional method is the use of injection wells. Recharge rates are normally less than pumping rates for the same head conditions, however, because of the clogging that is often encountered in the area adjacent to the well casing. Clogging may result from the entrapment of fine aquifer particles, from suspended material in the recharge water which is subsequently strained out and deposited in the vicinity of the well screen, from air binding, from chemical reactions between recharge and natural waters, and from bacteria. For best results the recharge water should be clear, contain little or no sodium, and be chlorinated.

Richter and Chun have outlined several factors worthy of consideration relative to the selection of artificial recharge project sites [26]. Essentially they are the availability and character of local and imported water supply, factors relating to infiltration rates of the native soils, operation and maintenance problems, net benefits, water quality, and legal considerations.

## 5.26 CONCURRENT DEVELOPMENT OF GROUNDWATER AND SURFACE-WATER SOURCES

The maximum practical conservation of our water resources is based on the coordinated development of groundwater and surface-water supplies. Geologic, hydrologic, economic, and legal factors must be carefully considered.

Concurrent utilization is primarily founded on the premise of transference of impounded surface water to groundwater storage at optimum rates [27, 28]. Annual water requirements are generally met by surface storage while groundwater storage is used to meet cyclic requirements covering periods of dry years. The operational procedure involves a lowering of groundwater levels during periods of below-average precipitation and a subsequent raising of levels during wet years. Transfer rates of surface waters to underground storage must be large enough to ensure that surface-water reservoirs will be drawn down sufficiently to permit impounding significant volumes during periods of high runoff. To provide the required maximum transfer capacity, methods of artificial recharge such as spreading, ponding, injecting, or returning flows from irrigation must be used.

The coordinated use of groundwater and surface-water sources will result in the provision of larger quantities of water at lower costs. As an example, it has been found that the conjunctive operation of the Folsom Reservoir (California) and its groundwater basin yields a conservation and utilization efficiency of approximately 82% as compared with about 51% efficiency for the operation of the surface reservoir alone [29]. There is little doubt that the inclusion of groundwater resources should be given very careful consideration in future planning for large-scale water development projects.

In general, the analysis of a conjunctive system consisting of a dam and an aquifer requires the solution of three fundamental problems. The first is to establish the design criteria for the dam and the recharge facilities. The second is to determine the service area for the combined system. Finally, a set of operating rules that define the reservoir drafts and pumpages to be taken from the aquifer is required. A mathematical model for an analysis such as this has been proposed by Buras [27].

## PROBLEMS

**5.1**  Compare the amounts of water required by the various users in your state. What is the relative worth of water in its various uses?

**5.2**  A flow of 120 mgd is to be developed from a 200-mi$^2$ watershed. At the flow line, the area's reservoir is estimated to cover 4200 acres. The annual rainfall is 38 in., the annual runoff is 13 in., and the annual evaporation is 47 in. Find the net gain or loss in storage this represents. Calculate the volume of water evaporated in acre-feet and cubic meters.

**5.3**  A flow of 4.8 m$^3$/sec is to be developed from a 500-km$^2$ watershed. At the flow line, the area's reservoir is estimated to cover 1700 hectares. The annual rainfall is 97 cm, the annual runoff is 30 cm, and the annual evaporation is 120 cm. Find the net gain or loss in storage this represents. Calculate the volume of water evaporated in cubic meters.

**5.4**  Discuss how you would go about collecting data for an analysis of the water budget of a region. What agencies would you contact? What other sources of information would you seek out?

**5.5**  For some area of your choice, make a plot of mean monthly precipitation versus time. Explain how this fits the pattern of seasonal water uses for the area. Will the form of precipitation be an important consideration?

**5.6**  Given the following 10-yr record of annual precipitation, plot a rough precipitation frequency curve. Tabulate the data to be plotted and show the method of computation. The data are annual precipitation in inches: 28, 21, 33, 26, 29, 27, 19, 28, 18, 22. (*Note:* The frequency in percent of years is $1/T_R \times 100$.)

**5.7**  Given the 10-yr record of annual precipitation that follows, develop and plot a precipitation frequency curve. The precipitation values in cm are 70, 54, 89, 66, 75, 69, 48, 72, 46, and 56.

**5.8**  Find a maximum reservoir storage requirement if a uniform draft of 726,000 gpd/mi² from a specific stream is to be maintained. The following record of average monthly runoff values is given:

| Month | A | M | J | J | A | S | O | N | D | J | F | M | A | M | J |
|-------|---|---|---|---|---|---|---|---|---|---|---|---|---|---|---|
| Runoff (mgal/mi²) | 97 | 136 | 59 | 14 | 6 | 5 | 3 | 7 | 19 | 13 | 74 | 96 | 37 | 63 | 49 |

**5.9**  Using the information given in Table 5.3, plot recurrence interval in years as the ordinate, design period in years as the abscissa, and construct a series of recurrence interval-design period probabilities that an event will not be exceeded during the design period. Use arithmetic coordinate paper. [*Note:* To conform to this, probabilities in the table must be subtracted from 1.0. Where sufficient information is not provided by the table, probabilities may be computed using $P_n = (1 - 1/T_R)^n$, where $n$ is the design period in years.]

**5.10**  Given the following 50-month record of mean monthly discharge, find the magnitude of the 20-month low flow. The consecutive average monthly flows (cfs) were 14, 17, 19, 21, 18, 16, 18, 25, 29, 32, 34, 33, 30, 28, 20, 23, 16, 14, 12, 13, 16, 13, 12, 12, 13, 14, 16, 13, 12, 11, 10, 12, 10, 9, 8, 7, 6, 4, 6, 7, 8, 9, 11, 9, 8, 6, 7, 9, 13, 17.

**5.11**  Find the expected life of a reservoir having an initial capacity of 45,000 acre·ft. The average annual inflow is 63,000 acre·ft, and a sediment inflow of 180 acre·ft/yr is reported. Consider the useful life of the reservoir to be exceeded when 80% of the original capacity is lost. Use 5,000-acre·ft volume increments. From Fig. 5.4, obtain values of percent sediment trapped.

**5.12**  Given the following data relating mean annual change in groundwater level to mean annual draft, find the safe yield.

| Mean annual change in groundwater level (ft) | +1 | +2 | −1 | −3 | −4 | +1.5 | +1.2 | −2.6 |
|---|---|---|---|---|---|---|---|---|
| Mean annual draft (thousands of acre·ft) | 23 | 19 | 31 | 42 | 44 | 21 | 19 | 33 |

**5.13**  A 12-in. well fully penetrates a confined aquifer 100 ft. thick. The coefficient of permeability is 600 gpd/ft². Two test wells located 45 and 120 ft away show a difference in drawdown between them of 8 ft. Find the rate of flow delivered by the well.

**5.14** Determine the permeability of an artesian aquifer being pumped by a fully penetrating well. The aquifer is composed of medium sand and is 100 ft thick. The steady-state pumping rate is 1200 gpm. The drawdown in an observation well 75 ft away is 14 ft, and the drawdown in a second observation well 500 ft away is 1.2 ft. Find $K_f$ in gallons per day per square foot.

**5.15** Consider a confined aquifer with a coefficient of transmissibility $T = 680$ ft$^3$/day/ft. At $t = 5$ min, the drawdown $s = 5.6$ ft; at 50 min, $s = 23.1$ ft; and at 100 min, $s = 28.2$ ft. The observation well is 75 ft away from the pumping well. Find the discharge of the well.

**5.16** Assume that an aquifer is being pumped at a rate of 300 gpm. The aquifer is confined and the pumping test data are given below. Find the coefficient of transmissibility $T$ and the storage coefficients $S$ for $r = 60$ ft.

| Time since pumping started (min) | 1.3 | 2.5 | 4.2 | 8.0 | 11.0 | 100.0 |
|---|---|---|---|---|---|---|
| Drawdown $s$ (ft) | 4.6 | 8.1 | 9.3 | 12.0 | 15.1 | 29.0 |

**5.17** You are given the following data: $Q = 59,000$ cfd, $T = 630$ cfd, $t = 30$ days, $r = 1$ ft, and $S_c = 6.4 \times 10^{-4}$. Consider this to be a nonequilibrium problem. Find the drawdown $s$. Note that for

$u = 8.0 \times 10^{-9}$     $W(u) = 18.06$

$u = 8.2 \times 10^{-9}$     $W(u) = 18.04$

$u = 8.6 \times 10^{-9}$     $W(u) = 17.99$

**5.18** Find the drawdown at an observation point 250 ft away from a pumping well. Given that $T = 3.1 \times 10^4$ gpd/ft, the pumping time is 10 days, $S_c = 3 \times 10^{-4}$, and $Q = 280$ gpm.

**5.19** A 12-in. well fully penetrates a confined aquifer 100 ft thick. The coefficient of permeability is 600 gpd/ft$^2$. Two test wells located 40 and 120 ft away show a difference in drawdown between them of 9 ft. Find the rate of flow delivered by the well.

**5.20** Find the permeability of an artesian aquifer being pumped by a fully penetrating well. The aquifer is 130 ft thick and is composed of medium sand. The steady-state pumping rate is 1300 gpm. The drawdown in an observation well 65 ft away is 12 ft, and in a second well 500 ft away it is 1.2 ft. Find $K_f$ in gpd/ft$^2$.

**5.21** Consider a confined aquifer with a coefficient of transmissibility of 700 ft$^3$/day-ft. At $t = 5$ min, the drawdown is 5.1 ft; at 50 min, $s = 20.0$ ft; at 100 min, $s = 26.2$ ft. The observation well is 60 ft from the pumping well. Find the discharge from the well.

**5.22** An 18-in. well fully penetrates an unconfined aquifer 100 ft deep. Two observation wells located 90 and 235 ft from the pumped well are known to have drawdowns of 22.5 ft and 20.6 ft, respectively. If the flow is steady and $K_f = 1300$ gpd/ft$^2$, what is the discharge?

**5.23** A well is pumped at the rate of 500 gpm under nonequilibrium conditions. For the data given below, find the formation constants $S$ and $T$. Use the Theis method.

| $r^2/t$ | Average drawdown, $h$ (ft) |
|---|---|
| 1,250 | 3.24 |
| 5,000 | 2.18 |
| 11,250 | 1.93 |
| 20,000 | 1.28 |
| 45,000 | 0.80 |
| 80,000 | 0.56 |
| 125,000 | 0.38 |
| 180,000 | 0.22 |
| 245,000 | 0.15 |
| 320,000 | 0.10 |

# REFERENCES

1. E. A. Ackerman and G. O. Löf, *Technology in American Water Development* (Baltimore: The Johns Hopkins Press, 1959).
2. H. E. Thomas, "Underground Sources of Water," *The Yearbook of Agriculture, 1955* (Washington, DC: U.S. Government Printing Office, 1956).
3. W. Viessman, Jr. and C. Welty, *Water Management: Technology and Institutions* (New York: Harper and Row, 1985), pp. 44–45.
4. W. Viessman, Jr., "United States Water Resources Development," *Natl. Forum* 69(1) (Winter 1989): 6, 7.
5. J. Hirshleifer, J. DeHaven, and J. Milliman, *Water Supply—Economics, Technology, and Policy* (Chicago: University of Chicago Press, 1960).
6. W. Viessman, Jr., G. L. Lewis, and J. W. Knapp, *Introduction to Hydrology* (New York: Harper and Row, 1989).
7. Task Group A4, D1, "Water Conservation in Industry," *J. Am. Water Works Assoc.* 45 (December 1958).
8. W. B. Langbein, "Water Yield and Reservoir Storage in the United States," *U.S. Geol. Survey Circular* (1959).
9. W. Rippl, "The Capacity of Storage Reservoirs for Water Supply," *Proc. Inst. Civil Engrs. (London) 71* (1883): 270.
10. D. K. Todd, *Ground Water Hydrology* (New York: Wiley, 1960).
11. R. A. Freeze and J. A. Cherry, *Groundwater* (Englewood Cliffs, NJ: Prentice-Hall, 1979).
12. O. E. Meinzer, "Outline of Methods of Estimating Ground Water Supplies," *U.S. Geol. Survey Water Supply Paper No. 638C* (1932).
13. R. C. Heath, "Groundwater Regions of the United States," Geological Survey Water Supply Paper No. 2242. Washington, DC: U.S. Government Printing Office, 1984.
14. U.S. Environmental Protection Agency, "The Report to Congress: Waste Disposal Practices and Their Effects on Groundwater," Executive Summary, U.S. EPA, PB 265–364, 1977.
15. M. E. Harr, *Groundwater and Seepage* (New York: McGraw-Hill, 1962).
16. M. Muskat, *The Flow of Homogeneous Fluids Through Porous Media* (Ann Arbor, MI: J. W. Edwards, 1946).

17. G. Thiem, *Hydrologische Methodern* (Leipzig: Gebhart, 1906), p. 56.
18. C. V. Theis, "The Relation Between the Lowering of the Piezometric Surface and the Rate and Duration of Discharge of a Well Using Ground Water Storage," *Trans. Am. Geophys. Union* 16 (1935): 519–524.
19. H. H. Cooper, Jr., and C. E. Jacob, "A Generalized Graphical Method for Evaluating Formation Constants and Summarizing Well-Field History," *Trans. Am. Geophys. Union* 27 (1946): 526–534.
20. J. W. Mercer and C. R. Faust, *Ground-Water Modeling* (Worthington, OH: National Water Well Association, 1981).
21. T. A. Prickett, "Ground-water Computer Models—State of the Art," *Ground Water* 17 (2) (1979): 121–128.
22. C. A. Appel and J. D. Bredehoeft, "Status of Groundwater Modeling in the U.S. Geological Survey," *U.S. Geol. Survey Circular* 737 (1976).
23. Y. Bachmat, B. Andres, D. Holta, and S. Sebastian, "Utilization of Numerical Groundwater Models for Water Resource Management," U.S. Environmental Protection Agency Report EPA-600/8-78-012.
24. J. E. Moore, "Contribution of Ground-water Modeling to Planning," *J. Hydrol.* 43 (October 1979).
25. J. F. Hoffman, "How Underground Reservoirs Provide Cool Water for Industrial Uses," *Heating, Piping, Air Conditioning* (October 1960).
26. R. C. Richter and R. Y. D. Chun, "Artificial Recharge of Ground Water Reservoirs in California," *Proc. Am. Soc. Civil Engrs., J. Irrigation Drainage Div.* 85, (IR4) (December 1959).
27. N. Buras, "Conjunctive Operation of Dams and Aquifers," *Proc. Am. Soc. Civil Engrs., J. Hydraulics Div.* 89 (HY6) (November 1963).
28. F. B. Clendenen, "A Comprehensive Plan for the Conjunctive Utilization of a Surface Reservoir with Underground Storage for Basin-wide Water Supply Development: Solano Project, California." D. Eng. thesis, University of California, Berkeley, CA (1959), p. 160.
29. "Ground Water Basin Management," *Manual of Engineering Practice No. 40* (New York: American Society of Civil Engineers, 1961).

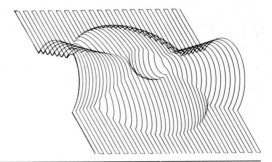

# Chapter 6
# Conveying and Distributing Water

Water is transported to locations where it is to be used and/or treated by aqueducts of various types and materials. It is then conveyed to individual users or use points through distribution networks.

## 6.1 TYPES OF AQUEDUCTS

Selection of an aqueduct type depends on topography, head availability, construction practices, economic factors, and water quality. Since the transported water must be safeguarded against contamination, the use of open channels or pipelines operating at low pressure must be considered in that context.

### Open Channels

Open channels are designed to convey water under conditions of atmospheric pressure. By this definition the hydraulic gradient and free-water surface are coincident. If the channel is supported on or above the ground, it is classified as a flume. Open channels may be covered or open and may take on a variety of shapes.

The choice of an open channel as the means of conveyance will usually be predicated on suitable topographic conditions that permit gravity flow with minimal excavation or fill. If the channel is unlined, the perviousness of the soil must be considered relative to seepage losses. Other considerations of importance are the potential pollution hazard and evaporative losses.

Open channels may be lined with concrete, bituminous materials, butyl rubber, vinyl, synthetic fabrics, or other products to reduce the resistance to flow, minimize seepage, and lower maintenance costs. Flumes are usually constructed of concrete, steel, or timber.

## Pipelines

Pipelines are usually built where topographic conditions preclude the use of canals. Pipelines may be laid above or below ground or may be partly buried. Most modern pressure conduits are built of concrete, steel, cast iron, asbestos cement, or plastic (polyvinyl chloride, pvc).

Pipelines used in important transportation systems may require gate valves, check valves, air-release valves, drains, surge control equipment, expansion joints, insulation joints, manholes, and pumping stations. The appurtenances are provided to ensure safe and efficient operation, provide for easy inspection, and facilitate maintenance. Check valves are normally located on the upstream side of pumping equipment and at the beginning of each rise in the pipeline to prevent backflow. Gate valves are often spaced about 1200 ft apart so that the intervening section of line can be drained for inspection or repair and on either side of a check valve to permit its removal for inspection or repair. Air-release valves are needed at the high points in the line to release trapped gases and to vent the line to prevent vacuum formation. Drains are located at low points to permit removal of sediment and allow the conduit to be emptied. Surge tanks or quick-opening valves provide relief from problems of hydraulic surge.

## Tunnels

Where it is not practical or economical to lay a pipeline on the surface or provide an open trench for underground installation, a tunnel is selected. Tunnels are well suited to mountain or river crossings. They may be operated under pressure or act as open channels.

## 6.2 HYDRAULIC CONSIDERATIONS

The analysis of flows in a water conveyance system is carried out through application of basic principles of open-channel and closed-conduit hydraulics. It is assumed that the student already has been exposed to these concepts in courses in hydraulics or fluid mechanics.

Except for sludges, most flows may be treated hydraulically in the same manner as clean water even though considerable quantities of suspended material are being carried. The Hazen–Williams and Manning formulas are used extensively in water transportation problems. The Hazen–Williams formula is used primarily for pressure conduits, while the Manning equation has found its major application in open-channel problems. Both equations are applicable when normal temperatures prevail, a relatively high degree of

**TABLE 6.1**   SOME VALUES OF THE
              HAZEN–WILLIAMS COEFFICIENT

| Pipe material | $C$ | Pipe material | $C$ |
|---|---|---|---|
| New cast iron | 130 | New welded steel | 120 |
| 5-yr-old cast iron | 120 | Asbestos cement | 140 |
| 20-yr-old cast iron | 100 | Plastic | 150 |
| Average concrete | 130 | | |

turbulence is developed, and ordinary commercial materials are used [1–4]. The Hazen–Williams equation is

$$V = 1.318CR^{0.63}S^{0.54} \tag{6.1}$$

where  $V$ = velocity, fps
       $C$ = a coefficient, which is a function of the material and age of the conduit
       $R$ = hydraulic radius (flow area divided by wetted perimeter), ft
       $S$ = slope of energy grade line, ft/ft

For circular conduits flowing full, the equation may be restated

$$Q = 0.279CD^{2.63}S^{0.54} \tag{6.2}$$

where  $Q$ = flow, mgd
       $D$ = pipe diameter, ft

and

$$Q = 0.278CD^{2.63}S^{0.54}$$

where  $Q$ = flow, m³/s
       $D$ = pipe diameter, m

Some values of $C$ for use in the Hazen–Williams formula are given in Table 6.1. A nomograph that facilitates the solution of this equation is given in Fig. 6.1.

The Manning equation is stated in the form

$$V = 1.49R^{0.66}S^{0.5}/n \tag{6.3}$$

where  $V$ = velocity of flow, fps
       $n$ = coefficient of roughness
       $R$ = hydraulic radius, ft
       $S$ = slope of energy grade line

In metric units

$$V = R^{0.66}S^{0.5}/n$$

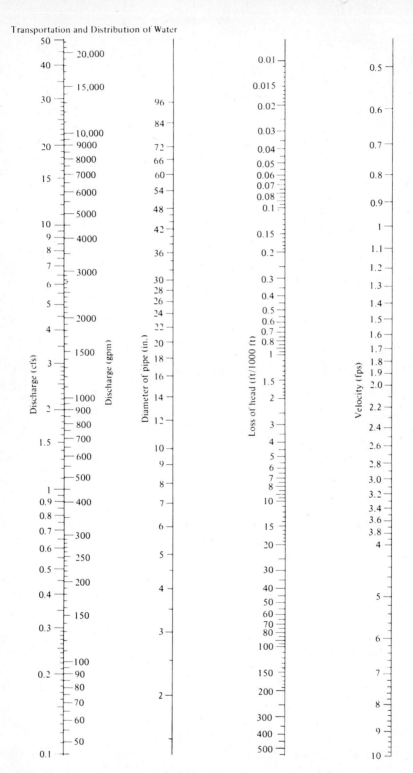

**Figure 6.1** Nomograph for the Hazen–Williams formula with $C = 100$.

**TABLE 6.2**  VALUES OF MANNING'S ROUGHNESS COEFFICIENT

| Material | $n$ | Material | $n$ |
|---|---|---|---|
| Concrete | 0.013 | Corrugated metal pipe | 0.022 |
| Cast-iron pipe | 0.015 | Bituminous concrete | 0.015 |
| Vitrified clay | 0.014 | Uniform, firm sodded earth | 0.025 |
| Brick | 0.016 | | |

where  $V$ = velocity of flow, m/s
$n$ = coefficient of roughness
$R$ = hydraulic radius, m
$S$ = slope of energy grade line

The equation is applicable as long as $S$ does not materially exceed 0.10. In channels having no uniform roughness, an average value of $n$ is selected. Where the cross-sectional roughness changes considerably, as in a channel with a paved center section and grassed outer sections, common practice is to compute the flow for each section independently and sum these flows to obtain the total. For most purposes, $n$ is considered to be a constant. It is actually a function of pipe diameter, however, and should be adjusted for pipe diameters exceeding several feet. As in the case of the Hazen–Williams equation, nomographs are available to permit rapid computations (Fig. 7.1). Values of $n$ for use in Manning's equation are indicated in Table 6.2.

Head lost as a result of pipe friction can be computed by solving Eq. 6.1 or 6.3 for $S$ and multiplying by the length of the pipeline. A slightly more direct method is to use the Darcy–Weisbach equation,

$$h_L = fLV^2/2Dg \qquad (6.4)$$

where  $h_L$ = head loss, ft
$L$ = pipe length, ft
$D$ = pipe diameter, ft
$f$ = friction factor
$V$ = flow velocity, fps

The friction factor is related to the Reynolds number and the relative roughness of the pipe. For conditions of complete turbulence, Fig. 6.2 relates the friction factor to pipe geometry and characteristics.

In transportation systems, the pipe friction head loss is usually predominant and other losses can ordinarily be neglected without serious error. In short pipelines, such as found in water treatment plants, the minor losses may be exceedingly important. If there is doubt, it is always best to include them. Minor losses result from valves, fittings, bends, and changes in flow characteristics at inlets and outlets. For turbulent-flow conditions, minor

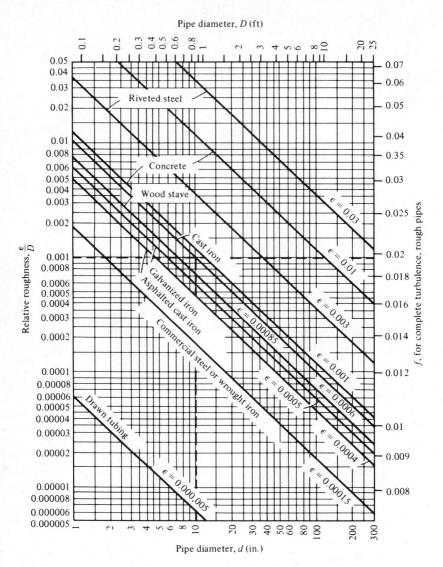

**Figure 6.2** Relative roughness of pipe materials and friction factors for complete turbulence. (Courtesy of the Crane Company, Chicago, IL.)

losses are customarily expressed as a function of velocity head. Adequate information is available from manufacturers and other sources on losses in various types of valves and fittings [1–5].

## ■ EXAMPLE 6.1

Consider that water is pumped 10 mi from a reservoir at elevation 100 ft to a second reservoir at elevation 230 ft. The pipeline connecting the reservoirs is 48 in. in diameter and is constructed of concrete with an absolute roughness of 0.003. If the flow is 25 mgd and the efficiency of the pumping station is 80%, what will be the monthly power bill if electricity costs 8 cents/kWh?

**Solution.**

1. Writing the energy equation between a point on the water surface of reservoir $A$ and a point on the water surface of reservoir $B$, one obtains

$$Z_A + \frac{P_A}{W} + \frac{V_A^2}{2g} + H_p = Z_B + \frac{P_B}{W} + \frac{V_B^2}{2g} + H_L$$

2. Letting $Z_A = 0$, and noting that $P_A = P_B$ is equal to the atmospheric pressure and that $V_A = V_B = 0$ for a large reservoir, one finds that the equation reduces to

$$H_P = Z_B + H_L$$

   where    $H_P$ = head developed by the pump
   $H_L$ = total head lost between $A$ and $B$, including pipe friction and all minor losses

3. Using Fig. 6.2, determine the value of $f$ as 0.0182.
4. Using Eq. 6.4, find the pipe friction head loss. Assuming that the minor losses are negligible in this problem, this is equal to $H_L$:

$$H_L = f\frac{L}{D}\frac{V^2}{2g}$$

   $V$ must be determined before Eq. 6.4 can be solved:

$$V = \frac{Q}{A} = \frac{25 \times 10^6 \times 1.55}{\pi \times 4 \times 10^6} = 3.09 \text{ fps}$$

$$H_L = 0.0182 \times \frac{5280 \times 10}{4} \times \frac{(3.09)^2}{64.4}$$

$$= 35.6 \text{ ft}$$

5. $H_p = (230 - 100) + 35.6$
   $= 130 + 35.6 = 165.6 \text{ ft·lb/lb}$

   the energy imparted by the pump to the water.
6. The power requirement may be computed as

$$P = Q\gamma H_p$$

$$= 25 \times 1.55 \times 62.4 \times 165.6 = 400,000 \text{ ft·lb/sec}$$

7. For 80% efficiency, the power requirement is

$$\frac{400,000}{0.80} = 500,000 \text{ ft · lb/sec}$$

8. $5.00 \times 10^5 \times 3.766 \times 10^{-7} = 18.8 \times 10^{-2} \text{ kWh/sec}$

The number of kilowatt-hours per 30-day month is then

$$18.8 \times 10^{-2} \times 30 \times 864 \times 10^2 = 485,000 \text{ kWh/month}$$

**9.** The monthly power cost is therefore $485,000 \times 0.08 = \$38,800$.    ■

## 6.3  DESIGN CONSIDERATIONS

The design of transportation systems involves a determination of hydraulic adequacy, structural adequacy, and economic efficiency. The required water-way area is a function of the flow to be carried, the head available, the character of the conduit material, and limiting velocities.

### Location

The location of an aqueduct is based on engineering and economic considera-tions. Since the terminal locations are dictated by the source of supply and the region to be served, the problem becomes one of finding the most practical and economic route between them. The choice of location has an obvious bearing on the type of aqueduct that can be built. Aqueducts built to grade require topography such that cut-and-cover operations can be closely bal-anced. Pressure aqueducts, on the other hand, can follow the topography. Pumping and material and construction costs must be given full consider-ation.

### Sizing

The size and configuration of the aqueduct finally selected will in all probabil-ity be variable along the route.

For a given type of aqueduct (pipe, tunnel, flume, canal), the size will usually be determined on the basis of hydraulic, economic, and construction considerations. Occasionally, construction practices dictate a minimum size in excess of that required to handle the flow under the prevailing hydraulic conditions (available head). This condition would most likely be encountered where a tunnel is involved. Hydraulic factors that control the design are the head available and permissible velocities. Available heads are affected by reservoir drawdown and local pressure requirements. Limiting velocities are based on the character of the water to be transported and the need to protect transmission lines against excessive pressures which might be developed through hydraulic surge. Where silt is transported with the water, minimum velocities of about 2.5 fps should be maintained. Maximum velocities must preclude pipe erosion or hydraulic surge problems and are ordinarily between 10 and 20 fps [6]. The usual range in velocities is from about 4 to 6 fps.

Where power generation is involved, pumping costs and/or the worth of power and conduit costs jointly determine the conduit size. For single gravity-flow pipelines, the size should be determined such that all the head available is consumed by friction.

## Economics

Hydraulic head has economic value. It costs money to produce the head at the upstream end of a system, but this head can then be used for increased flow, for power production, or a combination of these factors. A definite relationship always exists among aqueduct size, hydraulic gradient, and the value of head. In some cases, construction costs are related to the elevation of the hydraulic gradient. The elevation of the gradient also affects pumping costs and power production values, as does the slope of the hydraulic gradient. In long lines composed of different types of conduits passing through varied topography, a means of coordinating conduit types, choosing dam elevations, and selecting pump lifts or power drops is important. The problem can be approached through a joint application of hydraulic and economic principles [7, 8].

In any conduit, sufficient hydraulic slope must exist to obtain the required flow. Steep slopes generate high velocities with smaller conduit requirements. When sufficient fall is available, steep slopes are often economical. On the other hand, if head can be generated only by pumping or through construction of a dam, flatter slopes calling for larger conduits are probably necessary to reduce the cost of the lift. Apparently, then, some combination of lift and slope will yield the optimum economy.

Usually in designing water transmission lines some controlling feature establishes the elevation of the line at a specified point. Examples of possible governing features are dam heights, tunnel locations, terminal reservoirs, and hilltops. These controls are valuable aids in carrying out the overall system design.

## Strength

Water conveyance structures are required to resist numerous forces such as those resulting from water pressure within the conduit, hydraulic surge (transient internal pressure generated when the velocity of flow is rapidly reduced), external loads, forces at bends or changes in cross section, expansion and contraction, and flexural stresses. The student will find these topics adequately covered in the references [4, 5, 9].

# Distribution Systems

Water distribution systems are ordinarily designed to adequately satisfy the water requirements for a combination of domestic, commercial, industrial, and fire-fighting purposes. The system should be capable of meeting the demands placed on it at all times and at satisfactory pressures. Pipe systems, pumping stations, storage facilities, fire hydrants, house service connections, meters, and other appurtenances are the main elements of the system [3].

## 6.4 SYSTEM CONFIGURATIONS

Distribution systems may be generally classified as grid systems, branching systems, or a combination of these. The configuration of the system is dictated primarily by street patterns, topography, degree and type of development of the area, and location of treatment and storage works. Figure 6.3 illustrates the nature of several basic types of systems. A grid system is usually preferred to a branching system, since it can furnish a supply to any point from at least two directions. The branching system does not permit this type of circulation, since it has numerous terminals or dead ends. A grid or combination system can also incorporate loop feeders, which act to distribute the flow to an area from several directions.

In locations where sharp changes in topography occur (hilly or mountainous regions) it is common practice to divide the distribution system into two or more service areas or zones. This precludes the difficulty of extremely high pressure in low-lying areas in order to maintain reasonable pressures at higher elevations. Usual practice is to interconnect the various systems, with the interconnections closed off by valves during normal operations.

## 6.5 BASIC SYSTEM REQUIREMENTS

The performance of a distribution system can be judged on the basis of the pressures available in the system for a specific rate of flow [10, 11]. Pressures should be great enough to adequately meet consumer and fire-fighting needs. At the same time, they should not be excessive, since the development of pressure head is an important cost consideration. In addition, as pressures increase, leakage increases, and money is then spent to transport and process a product that is wasted. Because the investment in a distribution system is exceedingly large, it is important that the design be optimized economically.

For commercial areas, pressures in excess of 60 pounds per square inch, gage (psig), are usually required. Adequate pressures for residential areas usually range from 40 to 50 psig. In tower buildings it is often necessary to provide booster pumps to elevate the water to upper floors. Storage tanks are usually provided at the highest level and distribution is made directly from them.

The capacity of the distribution system is determined on the basis of local water needs plus fire demands as outlined in Chapter 4. Pipe sizes should be selected so that high velocities are avoided. Once the flow has been determined, pipe sizes can be selected by assuming velocities of from 3 to 5 fps. Where fire-fighting requirements are to be met, a minimum diameter of 6 in. is recommended. The National Board of Fire Underwriters recom-

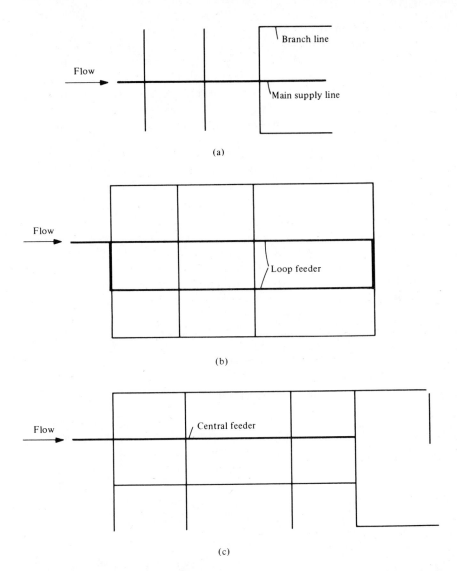

**Figure 6.3** Types of water distribution systems: (a) branching, (b) grid, and (c) combination.

mends 8 in. as a minimum but permits 6-in. pipes in grid systems provided the length between connections does not exceed 600 ft.

## 6.6 HYDRAULIC DESIGN

To effect the hydraulic design of a water distribution system, information must be available on the anticipated local rates of water consumption, the manner in which these design flows are distributed geographically, and the required pressure gradients for the system. The first is obtained in the manner indicated in Chapter 4. It should be reemphasized that the designer ought to investigate both the maximum day rate plus fire protection and the maximum hourly rate to determine which will govern the design.

The spatial distribution of consumption can be estimated by studying population densities and commercial and industrial use patterns that are known or predicted for the region. Consider the design of a feeder to an area composed of residential, commercial, and industrial users. In investigating the peak hour, for example, it will be important to have the predicted hydrograph for each type of user so that the specific hour in which the summation of the three component flows is greatest will be used for the design. Students are cautioned that the regional maximum hour might well coincide with the residential maximum within the region but not coincide with the commercial or industrial peaks, or vice versa. For this reason, information on the hourly variation of water use for all users is extremely valuable. Once the water consumption has been estimated, it is usual practice to consider it to be concentrated at specified points on the feeder–main system. Computations based on the concentration of consumption in this manner normally appear to correlate well with field observations.

Distribution systems are usually designed so that reasonably uniform pressures prevail. A pressure of 30 psi is normally considered to be the minimum desirable in any area, with the exception that during a serious fire it may be permissible to allow the pressure to drop to about 20 psi. Main feeders should be designed for pressures between 40 and 75 psi whenever possible [1, 10].

The analysis of a distribution system is often simplified by first skeletonizing the system. This might involve the replacement of a series of pipes of varying diameter with one equivalent pipe or replacing a system of pipes with an equivalent pipe. An equivalent pipe is one in which the loss of head for a specified flow is the same as the loss in head of the system it replaces. An example will illustrate the method of analysis.

■ **EXAMPLE 6.2**
Considering the pipe system shown in Fig. 6.4, replace (a) pipes BC and CD with an equivalent 12-in. pipe and (b) the system from B to D with an equivalent 20-in. pipe.

*Solution.*

    **a.** Assume a discharge through BCD of 8 cfs. Using the Hazen–Williams nomograph (Fig. 6.1), find the head loss for BC = 6.1 ft/1000 ft and for CD = 11 ft/1000 ft.

        The total head loss between B and D is therefore

$$6.1 \times \frac{200}{1000} + 11.0 \times \frac{500}{1000} = 6.72 \text{ ft}$$

    Using a discharge of 8 cfs, Fig. 6.1 indicates a head loss of 45 ft/1000 ft for a 12-in. pipe. The equivalent length of 12-in. pipe is therefore

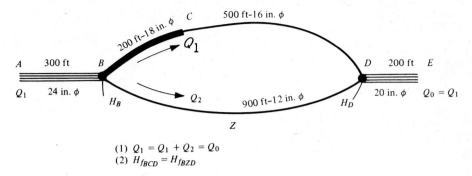

Figure 6.4 Example 6.2.

$$L_{12} = \frac{6.72 \times 1000}{45} = 149 \text{ ft}$$

**b.** Assume a total head loss between $B$ and $D$ of 5.0 ft. For the 12-in. equivalent pipe this is 33.5 ft/1000 ft, and for the 900 ft of 12-in. pipe it is 5.5 ft/1000 ft. Using these values and Fig. 6.1, the discharges for the two pipes are found to be 6.8 and 2.6 cfs, respectively. The total flow is thus 9.4 cfs at a head loss of 5 ft. For this discharge, a 20-in. pipe will have a head loss of 4.8 ft/1000 ft. The equivalent 20-in. pipe to replace the whole system will be

$$\frac{5}{4.8} \times 1000 = 1042 \text{ ft long} \qquad \blacksquare$$

The analysis of a simple hydraulic system such as that shown in Fig. 6.4 presents little difficulty. A slightly more complex system is shown in Fig. 6.5. The method of equivalent pipes will fail to yield a solution in this case

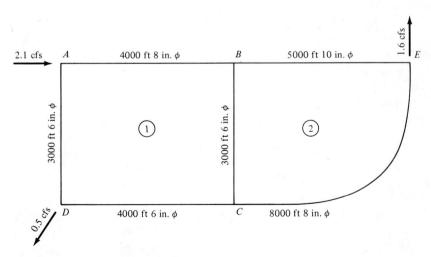

**Figure 6.5** Simple pipe network.

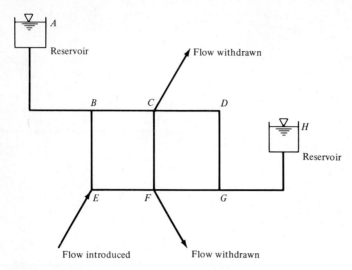

**Figure 6.6** Pipe network showing two fixed-grade nodes.

because there are crossover pipes (pipes that operate in more than one circuit and a number of withdrawal points throughout the system). This type of system is susceptible to network analysis.

## Pipe Networks

Most municipal water distribution systems are complex mazes containing pumps, storage elements, and pipelines of a variety of sizes. As pointed out in the preceding section, the ordinary methods of hydraulic analysis must be extended to take into account the looping characteristics of networks, changing reservoir levels, pumping, etc. Special techniques of network analysis come into play in this case. In these methods, iterative solutions based on initial assumptions lead to either balancing flows in a system or balancing heads in a system. The underlying principles are those of preserving mass continuity and ensuring energy conservation.

Pipe networks may be described as being composed of a number of constant-diameter pipe sections containing pumps and fittings. The ends of each pipe section are called nodes. In Fig. 6.5, points A and B are nodes, for example. Nodes may be either fixed grade or junction. Junction nodes are points where pipes meet and where flow may be introduced or withdrawn. Fixed-grade nodes are points where a constant grade is maintained. Connections to storage tanks or reservoirs or constant-pressure regions are examples (Fig. 6.6). The nodes shown in Fig. 6.5 are all of the junction variety, while nodes A and H in Fig. 6.6 are of the fixed-grade type. Networks are divided into loops for computational purposes. Primary loops, loops 1 and 2 in Fig. 6.5, are defined as those closed pipe circuits in the network which have no other closed pipe circuits within them. Using the foregoing terminology, we may write

$$P = J + L + F - 1 \tag{6.5}$$

where   $P$ = number of pipes
   $J$ = number of junction nodes
   $L$ = number of loops
   $F$ = number of fixed-grade nodes

This identity is directly related to the fundamental hydraulics equations that describe steady-state flow in a network. For Fig. 6.5, $J = 5, L = 2,$ $F = 0$, and thus the sum minus 1 is 6, the number of pipes in the complete network. Equations used to analyze steady-state pipe networks fall into two main categories—loop equations and node equations. The loop equations express mass conservation and energy conservation in terms of the discharge in a pipe section, while the node equations express mass continuity in terms of elevations or grades at junction nodes.

There are numerous commercially available computer programs that can handle the analysis of flows in pipe systems with any configuration and with a variety of components, such as storage tanks, pumps, check valves, pressure regulating valves, and variable-pressure supplies [1, 2, 12–18]. One such program, KYPIPE, was developed by Wood at the University of Kentucky [18].

**Loop Equations**   Equation 6.5 is the basis for formulation of a set of equations describing the hydraulic performance of a pipe network. In terms of unknown flows in the pipes, mass continuity and energy conservation equations can be written for the pipes and nodes. For each loop, an energy conservation equation can be written. A mass continuity equation can also be written for each node. For junction nodes, the inflow to the junction must be balanced by the outflow. This can be written

$$\sum Q_{in} - \sum Q_{out} = Q_e \quad (J \text{ equations}) \tag{6.6}$$

where $Q_{in}$ is the inflow, $Q_{out}$ is the outflow, and $Q_e$ is the external flow into the system or the withdrawal from the system at the node. For the primary loops, energy conservation can be described by

$$\sum h_L = \sum E_p \quad (L \text{ equations}) \tag{6.7}$$

where $h_L$ is the pipe energy loss (minor losses included) and $E_p$ is the energy introduced into the system by pumps. For loops having no pumps, the sum of the energy losses around the loop is zero. Note that a sign convention is used for loops. Clockwise flows might be considered positive and counterclockwise flows negative, for example.

Where there are $F$ fixed-grade nodes, $F - 1$ independent energy conservation equations can be written for pipe paths between any two fixed-grade nodes. These equations take the form

$$\Delta E = \sum h_L - \sum E_p \quad (F - 1 \text{ equations}) \tag{6.8}$$

where $\Delta E$ is the difference in elevation (grade) between the two fixed-grade nodes. Any connected path between the two fixed-grade nodes can be selected. This can be done by selecting a series of pipes so that the ending node of one path is the starting node for the next, etc. This procedure will produce the needed $F - 1$ equations with no redundancy [12].

Equation 6.7 can be considered as a special case of Eq. 6.8 where the difference in elevation ($\Delta E$) is zero for a closed-loop path. It follows that the energy conservation equations for a pipe network can be expressed by $L + F - 1$ energy equations described by Eq. 6.8. The continuity and energy equations that describe the pipe network are $P$ in number. They form a set of simultaneous nonlinear algebraic equations (loop equations) that describe steady-state flow conditions in a pipe network. To analyze an existing or proposed pipe network, the loop equations are solved to determine the flow in each pipe. In order to effect a solution, the terms in the energy equations must be expressed as functions of flow. Expressions for frictional losses in pipes, minor losses in fittings, and pump energy are needed.

Frictional losses in pipes are expressed as

$$h_{\text{LP}} = K_P Q^n \tag{6.9}$$

where $K_P$ is a constant incorporating pipe length, diameter, and roughness and $n$ is an exponent. The values of $K_P$ and $n$ are generally determined by the selection of the Darcy–Weisbach, Hazen–Williams, or Manning equations for the expression of energy losses.

The minor losses in a section of pipe result from fittings, valves, meters, or other insertions that affect the flow. They are expressed

$$h_{\text{LM}} = K_M Q^2 \tag{6.10}$$

where $K_M$ is the minor loss constant, a function of the sum of the minor loss coefficients for all fittings in the length of pipe ($\sum M$) and the pipe diameter. It is given by

$$K_M = \sum M / 2gA^2 \tag{6.11}$$

where $A$ is the cross-sectional area of the pipe.

The term in the energy equations representing pumping energy can be expressed in several ways. A constant power input can be specified or a curve can be fitted to data obtained from pump operations. In any event, the relationship between pump energy $E_P$ (head developed by the pump) and the flow $Q$ can be represented by

$$E_P = P(Q) \tag{6.12}$$

If the pump operates at constant power, then using the relationship for horsepower (hp = $Qwh/550$, where $w$ is the specific weight of water in pounds per ft$^3$, $h$ is the head in feet, $Q$ is the flow in cfs, and 550 is a conversion

**Figure 6.7** Pump notation for the node equation.

factor), $P(Q)$ is given by $550 \text{ hp}/62.4Q$. Letting $Z = 550 \text{ hp}/62.4$, $P(Q)$ can be written, for constant power, as $Z/Q$.

Combining Eqs. 6.9–6.12, the energy relationships in terms of discharge become

$$\Delta E = \sum (K_P Q^n + K_M Q^2) - P(Q) \tag{6.13}$$

Equation 6.13 and the continuity equations (Eq. 6.6) comprise the set of $P$ simultaneous equations that must be solved in a loop analysis. There is no direct solution to these nonlinear algebraic relationships, but several algorithms for determining an answer are available. They will be discussed in subsequent sections.

**Node Equations**   Solution of the loop equations begins generally with an assumption of flowrates in the pipe network. Computations proceed until adjustment in flows are considered to be within tolerable levels. When using node equations, adjustments are made in initial assumptions of head.

When considering nodes, the principal relationship used is the continuity equation (Eq. 6.6). The discharge in a section of pipe connecting nodes such as $A$ and $B$ (Fig. 6.5) is expressed in terms of the grade (head) at junction node $A$ ($H_a$), the grade at junction node $B$ ($H_b$), and the resistance offered by the pipeline ($K_{ab}$). This can be expressed

$$Q_{ab} = [(H_a - H_b)/K_{ab}]^{1/n} \tag{6.14}$$

where it is assumed that the pipe section is free of pumps and the head loss is calculated as

$$h_L = KQ^n \tag{6.15}$$

and $K$ is determined as indicated for Eq. 6.9. Combine Eqs. 6.6 and 6.14:

$$\sum_{b=1}^{N} \left[ \pm \left( \frac{H_a - H_b}{K_{ab}} \right)^{1/n} \right] = Q_e \tag{6.16}$$

which expresses continuity at a given junction node where $N$ pipes join. The sign of the term in the summation depends on the direction of flow into or out of the junction. A total of $J$ junction node equations result. This basic set can be expanded to include pumps where they exist. For each pump encountered, junction nodes are specified at the pump inlet and outlet, locations $b$ and $c$ in Fig. 6.7. Two additional equations are thus generated, one at the suction side and the other at the discharge side of the pump [12]. These equations involve the unknown heads (grades) on either side of the pump.

Following the notation of Fig. 6.7, an equation utilizing flow continuity in the suction and discharge lines can be written

$$H_a - H_b = \frac{K_{ab}}{K_{cd}}(H_c - H_d) \tag{6.17}$$

Another equation can be developed that relates the head change across the pump to the discharge in either the inlet or outlet pipe. Where the pump being considered is being operated at constant power, the relationship in terms of the outlet line discharge, according to Eq. 6.12, is

$$H_c - H_b = P\left[\left(\frac{H_c - H_d}{K_{cd}}\right)^{1/n}\right] \tag{6.18}$$

Equations 6.16 to 6.18 constitute the complete set of pipe network node equations. All of them are expressed in terms of the unknown grades at junction nodes and in terms of the suction and discharge grades at pumps within the system. This set of equations is also nonlinear, and thus direct solution is impossible. Commonly used algorithms involving the node equations are discussed in later sections.

## Algorithms for Solving Loop Equations

Several methods are widely used to solve the loop equations [12]. All use gradient methods to accommodate the nonlinear flow terms in Eq. 6.13. The gradient method is derived from the first two terms of the Taylor series expansion. Any function $f(x)$ that is continuous (differentiable) can be approximated as follows:

$$f(x) \approx f(x_0) + f'(x_0)(x - x_0) \tag{6.19}$$

Examination of the right-hand side of Eq. 6.19 reveals that the approximation has reduced $f(x)$ to a linear form. However, if $f$ is a function of more than one variable, Eq. 6.19 can be generalized as follows:

$$f[x(1), x(2), \ldots] = f[x(1)_0, x(2)_0, \ldots] + \frac{\partial f}{\partial x(1)}[x(1) - x(1)_0] \tag{6.20}$$

$$+ \frac{\partial f}{\partial x(2)}[x(2) - x(2)_0] + \cdots$$

in which the partial derivatives are evaluated at some $x(1) = x(1)_0$, $x(2) = x(2)_0$, etc.

The right-hand side of Eq. 6.13 represents the grade difference across a pipe carrying a discharge of $Q$. This can be stated as

$$f(Q) = K_P Q^n + K_M Q^2 - P(Q) \tag{6.21}$$

Substituting an estimated $Q_i$ for $Q$ and denoting $f(Q_i)$ by $H_i$, Eq. 6.21 becomes

$$H_i = f(Q_i) = K_P Q_i^n + K_M Q_i^2 - P(Q_i) \tag{6.22}$$

Differentiating Eq. 6.21 and setting $Q = Q_i$ gives the gradient of the function at $Q = Q_i$. Thus,

$$f'(Q_i) = nK_P Q_i^{n-1} + 2K_M Q_i - P'(Q_i)$$

Denoting $f'(Q_i)$ by $G_i$,

$$G_i = nK_P Q_i^{n-1} + 2K_M Q_i - P'(Q_i) \tag{6.23}$$

Both the function $H_i$ and its gradient $G_i$ evaluated at $Q = Q_i$ are used in algorithms for solving the loop equations.

**Single-Path Adjustment (P) Method**    This solution technique is the oldest and most well known of all the loop methods. It was first described by Hardy Cross [19]. Originally, the method was restricted to closed-loop networks and provided only for line losses. A generalization of the procedure is described below [12, 20].

1. An initial set of flowrates that satisfy continuity at each junction node is selected.
2. A flow adjustment factor is computed for each path $(L + F - 1)$ to satisfy the energy equation for that path. Continuity is maintained in this process.
3. Step 2 is repeated, building on improved solutions until the average correction factor is within an acceptable limit.

Equation 6.13 is used to compute the adjustment factor for a path using the gradient method to linearize the energy equations. Thus,

$$f(Q) = f(Q_i) + f'(Q_i)\,\Delta Q \tag{6.24}$$

in which $\Delta Q = Q - Q_i$, where $Q_i$ is the estimated discharge. Applying Eq. 6.24 to Eq. 6.13 and solving for $\Delta Q$ gives

$$\Delta Q = \frac{\Delta E - \sum H_i}{\sum G_i} \tag{6.25}$$

which is the flow adjustment factor to be applied to each pipe in the path. The numerator represents the imbalance in the energy relationship due to incorrect flowrates. The procedure reduces this to a negligible quantity. Flow adjustment is carried out for all $L$ fundamental (closed) loops and $F - 1$ pseudoloops in the network.

The Hardy Cross method of network analysis permits the computation of rates of flow through a network and the resulting head losses in the system [19]. It is a relaxation method by which corrections are applied to assumed flows or assumed heads until an acceptable hydraulic balance of the system is achieved.

The Hardy Cross analysis is based on the principles that (1) in any system continuity must be preserved and (2) the pressure at any junction of pipes is

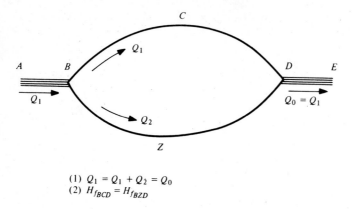

(1) $Q_1 = Q_1 + Q_2 = Q_0$
(2) $H_{f_{BCD}} = H_{f_{BZD}}$

**Figure 6.8** Derivation of the Hardy Cross method.

single valued. Referring to the simple network of Fig. 6.8, the elements of the procedure can be explained. First, the system must be defined in terms of pipe size, length, and roughness. Then, for any inflow $Q_I$, the system can be balanced hydraulically only if $H_{f_{BCD}} = H_{f_{BZD}}$. This restriction limits the possibilities to only one value of $Q_1$ and $Q_2$ which will satisfy the conditions.

Derivation of the basic equation for balancing heads by correcting assumed flows will now be given for the loop of Fig. 6.8 [19]. First, find the required inflow $Q_I$. Then arbitrarily divide this flow into components $Q_1$ and $Q_2$. The only restriction on the selection of these values is that $Q_1 + Q_2 = Q_I$. An attempt should be made, however, to select realistic values. Since the procedure involves a number of trials, the amount of work involved will be dependent on the accuracy of the value originally selected. For example, in the network shown, $BCD$ is considerably larger in diameter than $BZD$. A logical choice would therefore assume that $Q_1$ will be larger than $Q_2$. The final solution to the problem will be the same regardless of the original choice, but much more rapid progress results from reasonable initial assumptions.

After $Q_1$ and $Q_2$ have been chosen, $H_{f_{BCD}}$ and $H_{f_{BZD}}$ can be computed using the Hazen–Williams or some other pipe-flow formula. Remembering that the Hazen–Williams equation is of the form of Eq. (6.2):

$$Q = 0.279 \, CD^{2.63} S^{0.54}$$

the equation may be rewritten

$$Q = K_a S^{0.54} \tag{6.26}$$

where $K_a$ is a constant when we are dealing only with a single pipe of specified size and material. Rearranging this equation and substituting $H_f/L$ for $S$ yields Eq. (6.9):

$$H_f = KQ^n$$

where $n = 1.85$ in the Hazen–Williams equation. Equation 6.9 is convenient for expressing head loss as a function of flow in network analyses.

If the computed values of $H_{f_{BCD}}$ and $H_{f_{BZD}}$ are not equal (which is usually the case on the first trial), a correction must be applied to the initial values. Call this correction $\Delta Q$. If, for example, $H_{f_{BCD}} > H_{f_{BZD}}$, then the new value for $Q_1$ will be $Q_1 - \Delta Q = Q_1'$ and the new value for $Q_2$ must be $Q_2 + \Delta Q = Q_2'$. The corresponding values of head loss will be $H_{f_{BCD}}'$ and $H_{f_{BZD}}'$. If $\Delta Q$ is the true correction, then

$$H_{f_{BCD}}' - H_{f_{BZD}}' = 0 = K_1(Q_1 - \Delta Q)^n - K_2(Q_2 + \Delta Q)^n$$

The binomials may be expanded as follows:

$$K_1(Q_1^n - n\,\Delta Q Q_1^{n-1} + \cdots) - K_2(Q_2^n + n\,\Delta Q Q_2^{n-1} + \cdots) = 0$$

If $\Delta Q$ is small, the terms in the expansion involving $\Delta Q$ to powers greater than unity can be neglected. Therefore,

$$K_1 Q_1^n - nK_1\,\Delta Q Q_1^{n-1} - K_2 Q_2^n - nK_2\,\Delta Q Q_2^{n-1} = 0$$

Substituting $H_{f_{BCD}}$ for $K_1 Q_1^n$, $H_{f_{BZD}}$ for $K_2 Q_2^n$, and rewriting the terms $KQ^{n-1}$ as $K(Q^n/Q)$, one has

$$H_{f_{BCD}} - \Delta Q n K_1 \frac{Q_1^n}{Q_1} - H_{f_{BZD}} - \Delta Q n K_2 \frac{Q_2^n}{Q_2} = 0$$

$$H_{f_{BCD}} - H_{f_{BZD}} = \Delta Q n \left( \frac{H_{f_{BCD}}}{Q_1} + \frac{H_{f_{BZD}}}{Q_2} \right)$$

and

$$\Delta Q = \frac{H_{f_{BCD}} - H_{f_{BZD}}}{n(H_{f_{BCD}}/Q_1 + H_{f_{BZD}}/Q_2)} \tag{6.27}$$

Expanding this expression to the more general case gives the following equation for the flow correction $\Delta Q$:

$$\Delta Q = -\sum H \bigg/ n \sum \left( \frac{H}{Q} \right) \tag{6.28}$$

Application of this equation involves an initial assumption of discharge and a sign convention for the flow. Either clockwise or counterclockwise flows may be considered positive, and the terms in the numerator will bear the appropriate sign. For example, if the counterclockwise direction is considered positive, all $H$ values for counterclockwise flows will be positive and all $H$ values for clockwise flows will be negative. The denominator, however, is the absolute sum without regard to sign convention. The correction $\Delta Q$ has a single direction for all pipes in the loop, and thus the sign convention must also be considered in applying the correction. Example 6.3 illustrates

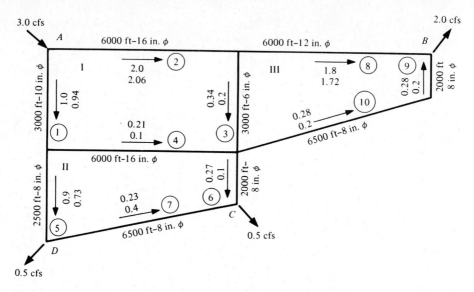

**Figure 6.9** Pipe network analyzed by the Hardy Cross method. (The clockwise direction is considered positive. The flows are the initial assumed and final corrected values.)

the application of the procedure to a network problem. Example 6.4 illustrates the solution of a similar problem by digital computer.

### ■ EXAMPLE 6.3

Given the network, the inflow at $A$, and outflows at $B$, $C$, and $D$ in Fig. 6.9, find the flows in the individual pipes comprising the network.

*Solution.*   The computational procedure is given in Table 6.3. The final flows are shown on Fig. 6.9.    ■

It should be noted that since pipes 3 and 4 appear in more than one loop, they are subject to the combined correction for loops I and III and I and II, respectively.

A similar procedure to that just discussed is to assume values of $H$ and then balance the flows by correcting the assumed heads. The mechanics of the two methods are the same and the applicable relationship,

$$\Delta H = -n \sum Q \Big/ \sum \left(\frac{Q}{H}\right)$$

(6.29)

can be derived in a manner similar to that for Eq. 6.28. The number of trials required for the satisfactory solution of any problem using Eq. 6.28 or 6.29 depends to a large extent on the accuracy of the initial set of assumed values and on the desired degree of accuracy of the results.

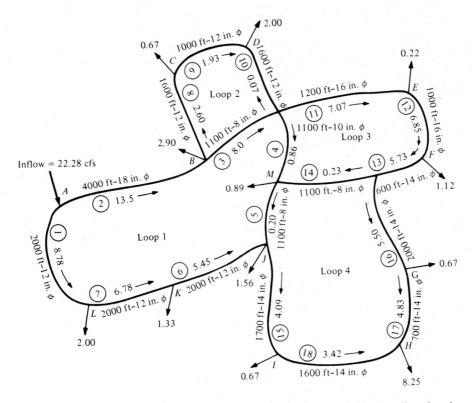

**Figure 6.10** Pipe network analyzed in Example 6.4. (The clockwise direction is considered positive. The flows shown are the initial assumptions, in cfs. Pipe numbers are circled.)

## ■ EXAMPLE 6.4

Given the network, inflow at $A$ and outflows at $B$ through $M$ on Fig. 6.10, find the flows in the individual pipes comprising the network.

**Solution.** The procedure is outlined in the following steps. Note that the definitions of variables used in the flow diagram and computer program are given in the comment statements at the beginning of the computer program. The printout of input data immediately follows the program, which in turn is followed by a table of final results.

1. Prepare a flow diagram for the computations as shown in Fig. 6.11.
2. Using the flow diagram as a guide, write a program in FORTRAN IV to solve Eq. 6.28 repeatedly until the flow correction is less than $\varepsilon$, a specified small value.
3. Run the program and obtain the required results.    ■

### FORTRAN IV *Program for Hardy Cross Pipe Network Analysis*
```
C  HARDY CROSS ANALYSIS II
C  THIS PROGRAM ANALYZES A PIPE NETWORK BY THE HARDY CROSS METHOD
```

**TABLE 6.3** COMPUTATIONS FOR EXAMPLE 6.3

| | | | | Trial I | | | | |
|---|---|---|---|---|---|---|---|---|
| 1 | 2 | 3 | 4 | 5 | 6 | 7 | 8 | 9 |
| Loop no. | Pipe no. | Pipe diameter (in.) | Length (ft) | $Q$ (cfs) | $H_L$ (ft) | $\dfrac{H_L}{Q}$ | $n\Sigma\left(\dfrac{H_L}{Q}\right)$ | $\Sigma H_L$ |
| I | 1 | 10 | 3000 | −1.0 | − 6.9 | 6.9 | | |
| | 2 | 16 | 6000 | +2.0 | + 4.92 | 2.46 | | |
| | 3 | 6 | 3000 | +0.2 | + 4.05 | 20.30 | | |
| | 4 | 6 | 6000 | −0.1 | − 1.92 | 19.2 | 90.3 | + 0.15 |
| II | 4 | 6 | 6000 | +0.1 | + 1.92 | 19.2 | | |
| | 6 | 8 | 2000 | +0.1 | + 0.16 | 1.6 | | |
| | 7 | 8 | 6500 | −0.4 | − 7.8 | 19.5 | | |
| | 5 | 8 | 2500 | −0.9 | −13.7 | 15.2 | 102.5 | −19.42 |
| III | 3 | 6 | 3000 | −0.2 | − 4.05 | 20.3 | | |
| | 8 | 12 | 6000 | +1.8 | +16.2 | 9.0 | | |
| | 9 | 8 | 2000 | −0.2 | − 0.7 | 3.5 | | |
| | 10 | 8 | 6500 | −0.2 | − 2.28 | 11.4 | 81.7 | + 9.8 |
| | | | | Trial III | | | | |
| I | 1 | | | −0.95 | − 6.3 | 6.64 | | |
| | 2 | | | +2.05 | + 5.1 | 2.48 | | |
| | 3 | | | +0.35 | +11.85 | 33.90 | | |
| | 4 | | | −0.20 | − 8.1 | 40.50 | 154.8 | + 2.55 |
| II | 4 | | | +0.20 | + 8.1 | 40.50 | | |
| | 6 | | | +0.25 | + 1.08 | 4.32 | | |
| | 7 | | | −0.25 | − 3.51 | 14.1 | | |
| | 5 | | | −0.75 | −10.00 | 13.33 | 133.8 | − 4.31 |
| III | 3 | | | −0.35 | −11.85 | 33.90 | | |
| | 8 | | | +1.70 | +14.38 | 8.46 | | |
| | 9 | | | −0.30 | − 1.44 | 4.80 | | |
| | 10 | | | −0.30 | − 4.68 | 15.61 | 116.3 | − 3.57 |

```
C  NUML = NUMBER OF LOOPS
C  NUMPI = TOTAL NUMBER OF PIPES
C  EPS = EPSILON THE ACCEPTABLE ERROR
C  L IS THE LOOP INDEX
C  K IS THE NUMBER OF PIPES IN A LOOP. THE VALUE IS OBTAINED
C  FROM J(L).
C  ID(L,K) IS AN INDEX WHICH ENABLES NUMBERING THE PIPES IN EACH LOOP
C  FROM ONE TO THE NUMBER OF PIPES IN THE LOOP WHILE STILL RETAINING
C  THE ORIGINAL PIPE NUMBERING SCHEME.
C  ID(L,K) IS AN INDEX WHICH INCLUDES THE PIPE NUMBER AND A SIGN H
C  THE SIGN IS POSITIVE FOR ALL PIPES NOT COMMON TO MORE THAN ONE
```

| | | | Trial II | | | | |
|---|---|---|---|---|---|---|---|
| 10 | 5 | 6 | 7 | 8 | 9 | 10 | 11 |
| $\Delta Q$ | $Q_1$ | $H_L$ | $\dfrac{H_L}{Q}$ | $n\Sigma\left(\dfrac{H_L}{Q}\right)$ | $\Sigma H_L$ | $\Delta Q$ | |
| −0.002 | −1.0 | − 6.9 | 6.9 | | | +0.05 | |
| −0.002 | +2.0 | + 4.92 | 2.46 | | | +0.05 | |
| +0.12 | +0.32 | + 9.45 | 29.5 | | | +0.03 | |
| −0.19 | −0.29 | −15.6 | 53.8 | 171.5 | −8.13 | +0.09 | |
| +0.19 | +0.29 | +15.6 | 53.8 | | | −0.09 | |
| +0.19 | +0.29 | + 1.4 | 4.83 | | | −0.04 | |
| +0.19 | −0.21 | − 2.4 | 11.43 | | | −0.04 | |
| +0.19 | −0.71 | − 9.0 | 12.69 | 153.1 | +5.6 | −0.04 | |
| −0.12 | −0.32 | − 9.45 | 29.5 | | | −0.03 | |
| −0.12 | +1.68 | +14.4 | 8.58 | | | +0.02 | |
| −0.12 | −0.32 | − 1.6 | 5.0 | | | +0.02 | |
| −0.12 | −0.32 | − 5.2 | 16.25 | 109.9 | −1.85 | +0.02 | |
| | | | Trial IV | | | | |
| −0.016 | −0.966 | − 6.59 | 6.83 | | | +0.027 | −0.94 |
| −0.016 | +2.034 | + 5.09 | 2.51 | | | +0.027 | +2.06 |
| −0.047 | +0.303 | + 9.0 | 29.70 | | | +0.038 | +0.34 |
| −0.048 | −0.248 | −12.0 | 48.30 | 163.5 | −4.50 | +0.035 | −0.21 |
| +0.048 | +0.248 | +12.0 | 48.30 | | | −0.035 | +0.21 |
| +0.032 | +0.282 | + 1.30 | 4.61 | | | −0.008 | +0.27 |
| +0.032 | −0.218 | − 2.92 | 13.4 | | | −0.008 | −0.23 |
| +0.032 | −0.718 | − 9.24 | 12.85 | 146.3 | +1.14 | −0.008 | −0.73 |
| +0.047 | −0.303 | − 9.0 | 29.70 | | | −0.038 | −0.34 |
| +0.031 | +1.731 | +15.25 | 8.80 | | | −0.011 | +1.72 |
| +0.031 | −0.269 | − 1.2 | 4.46 | | | −0.011 | −0.28 |
| +0.031 | −0.269 | − 3.89 | 14.48 | 106.2 | +1.16 | −0.011 | −0.28 |

```
C  LOOP. FOR PIPES FALLING IN TWO LOOPS, THE SIGN IS POSITIVE FOR THE
C  FIRST LOOP IN WHICH THE PIPE APPEARS AND NEGATIVE FOR THE SECOND
C  LOOP. FOR EXAMPLE, IF PIPE 3 IS COMMON TO LOOPS 1 AND 2, ID(1,3)
C  = +3 AND ID (2,1) = −3 WHERE 3 IS THE THIRD PIPE IN LOOP 1 AND
C  THE FIRST PIPE IN LOOP 2.
C  D IS THE PIPE DIAMETER IN FEET
C  Q(I) IS THE FLOW IN PIPE I (CFS)
C  THE OVERALL PIPE NETWORK IS BROKEN UP INTO L LOOPS
C  PL IS THE PIPE LENGTH IN THOUSANDS OF FEET. FOR A PIPE LENGTH OF
C  6,000 FEET, PL WOULD BE 6.
C  S IS THE SLOPE IN FEET PER THOUSAND FEET. IT IS COMPUTED USING THE
```

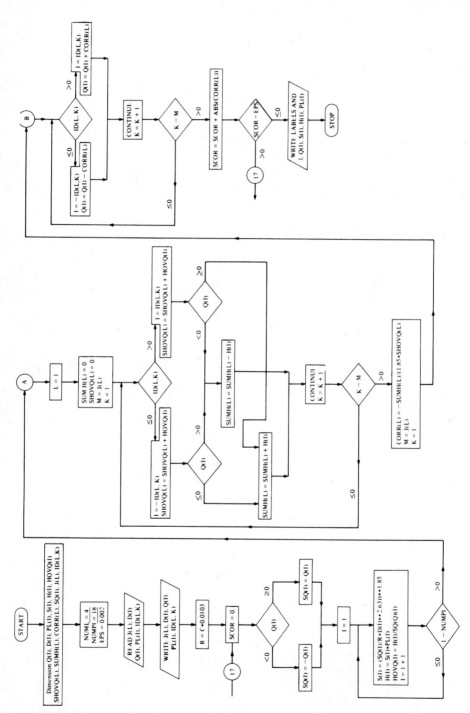

**Figure 6.11** Flow diagram for Hardy Cross pipe network analysis.

```
C  HAZEN WILLIAMS PIPE FORMULA.  S = (Q/(0.0103*C*D**2.63))**1.85
C  WHERE C IS THE HAZEN WILLIAMS C.
C  FOR VALUES OF C WHICH DIFFER FOR INDIVIDUAL PIPES A SIMPLE
C  PROGRAMMING CHANGE TO READ THE C VALUES IN AS DATA WOULD BE
C  SUFFICIENT.
C  H IS THE HEAD LOSS IN EACH PIPE IN FEET.  H = S*PL
C  HOVQ IS THE RATIO H/Q.
C  SHOVQ IS THE SUM OF THE RATIOS H/Q.
C  SUMH IS THE SUM OF THE HEAD LOSSES H.
C  CORR IS THE CORRECTION TO BE APPLIED TO THE LOOPS.
C  CORR = -SUMH/(1.85*SHOVQ).
C  SCOR IS THE SUM OF THE ABSOLUTE VALUES OF THE CORR(L)'S FOR EACH
C  LOOP. THIS VALUE IS COMPARED TO AN EPSILON. IF THE VALUE OF SCOR
C  IS LESS THAN EPS THE COMPUTATION STOPS. IF SCOR IS GREATER THAN
C  EPS, ANOTHER ITERATION IS MADE.
      DIMENSION Q(18),D(18),PL(18),S(18),H(18),HOVQ(18),SHOVQ(4),SUMH(4)
     1,CORR(4),SQ(18),J(4),ID(4,7)
      C = 110.
      NUML = 4
      NUMPI = 18
      EPS = 0.002
      READ(5,20)(J(L),  L = 1,NUML)
   20 FORMAT(4I5)
      WRITE(6,36)  (J(L),  L = 1,NUML)
   36 FORMAT(1X,4I5)
      READ(5,21)  (D(I),Q(I),PL(I),I=1,NUMPI)
   21 FORMAT(9F6.3)
      WRITE(6,29)  (D(I),Q(I),PL(I),I=1,NUMPI)
   29 FORMAT(1X,9F6.3)
      DO 23 L = 1,NUML
      M = J(L)
      DO 23 K = 1,M
      READ(5,24) ID(L,K)
   23 WRITE(6,35)ID(L,K)
   24 FORMAT(I5)
   35 FORMAT(1X,I5)
      R = C*0.0103
   17 SCOR = 0.
      DO 1 I = 1,NUMPI
      IF(Q(I).LT.0.) SQ(I) = -Q(I)
      IF(Q(I).GE.0.) SQ(I) = Q(I)
      S(I) = (SQ(I)/(R*D(I)**2.63))**1.85
      H(I) = S(I)*PL(I)
    1 HOVQ(I) = H(I)/SQ(I)
      DO 16 L = 1,NUML
      SUMH(L) = 0.
      SHOVQ(L) = 0.
      M = J(L)
      DO 7 K = 1,M
      IF(ID(L,K))3,3,4
```

```
  3 I = -ID(L,K)
    SHOVQ(L) = SHOVQ(L) + HOVQ(I)
    IF(Q(I))5,5,6
  4 I = ID(L,K)
    SHOVQ(L) = SHOVQ(L) + HOVQ(I)
    IF(Q(I))6,5,5
  6 SUMH(L) = SUMH(L) - H(I)
    GO TO 7
  5 SUMH(L) = SUMH(L) + H(I)
  7 CONTINUE
    CORR(L) = -SUMH(L)/(1.85*SHOVQ(L))
    M = J(L)
    DO 8 K = 1,M
    IF(ID(L,K))10,10,11
 10 I = -ID(L,K)
    Q(I) = Q(I) - CORR(L)
    GO TO 8
 11 I = ID(L,K)
    Q(I) = Q(I) + CORR(L)
  8 CONTINUE
 16 SCOR = SCOR + ABS(CORR(L))
    IF(SCOR - EPS.GT.0.) GO TO 17
    WRITE(6,25)
 25 FORMAT(1H1,20X,15HTABLE OF VALUES,///)
    WRITE(6,26)
 26 FORMAT(7X,1HI,11X,4HQ(I),13X,4HS(I),8X,7HH(I)ABS,8X,5HPL(I),///)
    DO 27 I = 1,NUMPI
 27 WRITE(6,28) I,Q(I),S(I),H(I),PL(I)
 28 FORMAT(6X,I2,3X,E14.4,3X,E14.4,3X,E14.4,3X,F6.3)
    STOP
    END
```

*Input Data*

```
  7     4     5     6
1.000-8.780 2.000 1.500  13.500 4.000 1.500 8.000 1.000
0.833 0.860 1.100 0.666   0.200 1.100 1.000-5.450 2.000
1.000-6.780 2.000 1.000   2.600 1.600 1.000 1.930 1.000
1.000-0.070 1.600 1.333   7.070 1.200 1.333 6.850 1.000
1.167 5.730 0.600 0.666   0.230 1.100 1.167-4.090 1.700
1.167 5.500 2.000 1.167   4.830 0.700 1.167-3.420 1.600
  1
  2
  3
  4
  5
  6
  7
  8
  9
 10
```

```
-3
-4
11
12
13
14
15
-5
-14
16
17
18
```

*Computer Output for Solution of Example 6.4*

| I | Q(I) | S(I) | H(I)ABS | PL(I) |
|---|------|------|---------|-------|
| 1 | −0.7030E 01 | 0.2928E 02 | 0.5856E 02 | 2.000 |
| 2 | 0.1525E 02 | 0.1706E 02 | 0.6823E 02 | 4.000 |
| 3 | 0.1003E 02 | 0.7856E 01 | 0.7856E 01 | 1.000 |
| 4 | 0.3048E 01 | 0.1518E 02 | 0.1669E 02 | 1.100 |
| 5 | 0.1605E 01 | 0.1376E 02 | 0.1514E 02 | 1.100 |
| 6 | −0.3700E 01 | 0.8932E 01 | 0.1786E 02 | 2.000 |
| 7 | −0.5030E 01 | 0.1576E 02 | 0.3153E 02 | 2.000 |
| 8 | 0.2321E 01 | 0.3768E 01 | 0.6029E 01 | 1.600 |
| 9 | 0.1651E 01 | 0.2007E 01 | 0.2007E 01 | 1.000 |
| 10 | −0.3489E 00 | 0.1133E 00 | 0.1812E 00 | 1.600 |
| 11 | 0.6632E 01 | 0.6491E 01 | 0.7789E 01 | 1.200 |
| 12 | 0.6412E 01 | 0.6098E 01 | 0.6098E 01 | 1.000 |
| 13 | 0.5292E 01 | 0.8166E 01 | 0.4899E 01 | 0.600 |
| 14 | −0.5530E 00 | 0.1917E 01 | 0.2109E 01 | 1.100 |
| 15 | −0.3745E 01 | 0.4308E 01 | 0.7324E 01 | 1.700 |
| 16 | 0.5845E 01 | 0.9814E 01 | 0.1963E 02 | 2.000 |
| 17 | 0.5175E 01 | 0.7835E 01 | 0.5485E 01 | 0.700 |
| 18 | −0.3075E 01 | 0.2992E 01 | 0.4787E 01 | 1.600 |

■

In using the Hardy Cross method to analyze large distribution systems, it is often useful to reduce the system to a skeleton network of main feeders [10]. Where the main feeder system has a very large capacity relative to that of the smaller mains, field observations indicate that this type of skeletonizing yields reasonable results. Where no well-defined feeder system is apparent, serious errors may result from skeletonizing. Figure 6.12 illustrates a skeletonized distribution network consisting of arterial mains only. Figure 6.13 shows how a portion of the distribution system of Fig. 6.12 (that part lying within the dashed rectangle) looked before skeletonizing. A more complete discussion of such procedures is given by Reh [10]. The analysis of a large network may also be expedited by balancing portions of the system successively instead of analyzing the whole network simultaneously.

Normally, minor losses are neglected in network studies, but they can

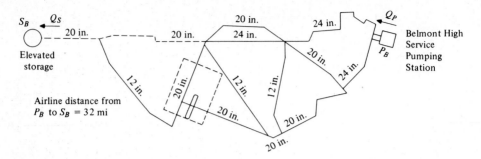

**Figure 6.12** Arterial pipe network of the Belmont High Service District, Philadelphia. (Courtesy of the Civil Engineering Department, University of Illinois, Urbana.)

easily be introduced as equivalent lengths of pipe when it is felt that they should be included. Where $C$ values are determined from field measurements, they invariably include a component due to the various minor losses encountered. McPherson gives a good discussion of local losses in water distribution networks [21].

The construction of pressure contours helps to isolate shortcomings in the hydraulic performance of distribution systems. Contours are often drawn with intervals of 1–5 ft of head loss but may have other intervals depending on local circumstances. For a given set of operating rules applicable to a particular network, the pressure contours indicate the distribution of head loss and are helpful in showing regions where head losses are excessive. Figure 6.14 illustrates contours constructed for a distribution network.

Computations for a given network may be reduced considerably by making use of the proportional flow method outlined by McPherson when studies of different inputs and outflows are needed [22]. In applying the technique, demands are first concentrated at key points in the network and a solution effected in the normal manner. Then, if the inputs and demands on the system are changed in a truly proportional manner, a new set of pipe flows and head losses may be readily obtained. McPherson states that *proportional load* may be defined as the design assumption that each consolidated demand on the system fluctuates about its mean value in direct proportion to the manner in which the total system load fluctuates. Note that the assumption of proportional loading is not always realistic, and in such cases the procedure outlined here does not apply. The significance of this condition can best be illustrated by referring to Fig. 6.8. In the simple network pictured, we may state that

(1) $Q_I = Q_0 = Q_1 + Q_2$

(2) $H_1 - H_2 = 0$   or   $K_1 Q_1^n - K_2 Q_2^n = 0$

Solving these equations simultaneously yields

$$Q_1 = AQ_I \quad \text{and} \quad Q_2 = BQ_I \tag{6.30}$$

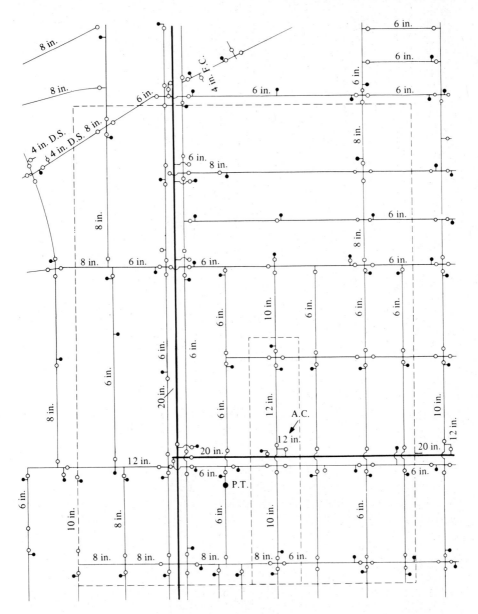

**Figure 6.13** Intermediate grid sector, Belmont High Service District, Philadelphia. (Courtesy of the Civil Engineering Department, University of Illinois, Urbana.)

where $A$ and $B$ are constants for the system. Thus, it can be seen that if $Q_I$ is doubled, $Q_1$ and $Q_2$ are also doubled. Extending this thinking to a slightly more complex system having several outflows in place of the single one, $Q_L$, as in the network of Fig. 6.8, the same reasoning holds, provided the drafts and inflow are all increased or decreased by the same percentage. New values

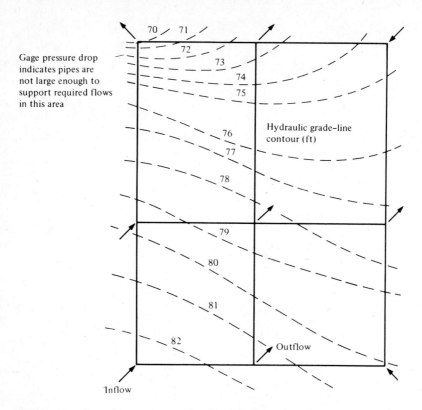

**Figure 6.14** Pressure contours of a distribution network.

for $Q_1$, $Q_2$, . . . may then be found by using a set of equations such as Eq. 6.30 without the need for making a complete network analysis.

McPherson has shown that under proportional loading conditions the head loss of a complex network can be reduced to the following equation [22]:

$$\sum \frac{h}{Q_d^m} = \Phi \left(\frac{Q_p}{Q_d}\right)^n \tag{6.31}$$

where     $h$ = loss in head between an input and a point in the district (such as an elevated storage location)
  $Q_d$ = demand placed on the network
  $Q_p$ = input rate at the point from which $h$ is referenced

To determine values of $\Phi$ and $n$, only two network balances (computations) are required. If $Q_p = Q_d$, which is the case when there is no equalizing storage, $\Phi$ can be determined by a single network balance.

**Simultaneous Path Adjustment (SP) Method**  To improve convergence over the P methods, a method of solution has been developed that simultaneously adjusts the flowrate in each path of pipes represented by an energy equation. The method is as follows [12, 20]:

1. An initial set of flowrates satisfying continuity at each junction node is determined.
2. A flow adjustment factor is simultaneously computed for each loop to satisfy the energy equations and avoid disturbance of the continuity balance.
3. Step 2 is repeated using improved solutions until the flow adjustment factor is within a specified limit.

The simultaneous solution of $L + F - 1$ equations is required to determine the loop flow adjustment factors. Each equation includes the contribution for a particular loop as well as contributions from all other loops that have pipes in common.

For loop $j$ the head change required to balance the energy equation is expressed in terms of the flow change in loop $j$ $(\Delta Q_j)$ and the flow changes in adjacent loops $(\Delta Q_k)$ as follows:

$$f(Q) = f(Q_i) + \frac{\partial f}{\partial Q}\Delta Q_j + \frac{\partial f}{\partial Q}\Delta Q_k$$

or

$$f(Q) = f(Q_i) + f'(Q_i)\,\Delta Q_j + f'(Q_i)\,\Delta Q_k \tag{6.32}$$

With the substitution $f(Q) = \Delta E$, $f(Q_i) = \Sigma\,H_i$, and $f'(Q) = \Sigma\,G_i$, Eq. 6.32 becomes

$$\Delta E - \sum H_i = \left(\sum G_i\right)\Delta Q_j + \sum (G_i\,\Delta Q_k) \tag{6.33}$$

in which $\Sigma\,H_i$ is the sum of the head changes for all pipes in loop $j$, $(\Sigma\,G_i) \times \Delta Q_j$ is the sum of all gradients for the same pipes times the flow change for loop $j$, and $\Sigma\,(G_i\,\Delta Q_k)$ is the sum of the gradients for pipes common to loops $j$ and $k$ multiplied by the flow change for loop $k$.

A set of simultaneous linear equations is thus formed in terms of flow adjustment factors for each loop representing an energy equation. The solution of these linear equations provides an improved solution for another trial until a specified convergence criterion is met.

**Linear (L) Method**   This procedure involves the solution of the basic hydraulic equations for a pipe network [12, 20]. In the method, the energy equations are linearized using a gradient approximation. This is accomplished in terms of an approximate discharge $Q_i$ as follows:

$$f(Q) = f(Q_i) + f'(Q_i)(Q - Q_i)$$

Introducing the expressions $H_i$ and $G_i$, defined as before, the foregoing equation becomes

$$\sum G_i Q = \sum (G_i Q_i - H_i) + \Delta E \tag{6.34}$$

This relationship is employed to formulate $L + F - 1$ energy equations, which, together with the $J$ continuity equations, form a set of $P$ simultaneous linear equations in terms of the flowrate in each pipe. A significant advantage of this scheme is that an arbitrary set of initial flowrates, which need not satisfy continuity, can be used to start the iteration. A flowrate based on a mean flow velocity of 4 ft/sec has been used by Wood [12]. Successive trials are carried out until the change in flowrate between successive trials becomes insignificant.

The use of the linear (L) method of analysis is illustrated by an example developed by Wood [12]. The calculations are given for one trial. The system analyzed is shown in Fig. 6.15, which includes the necessary data and shows the numbers assigned to the pipe sections and junction nodes. The system includes a globe valve in pipe 7 that imposes a significant minor loss. Other minor losses are neglected. A pump with constant-power input is included in pipe 2. The useful horsepower (hp) for this pump is given as 5. Thus, Eq. 6.12 becomes

$$E_P = P(Q) = Z/Q$$

and the pump constant $Z = 550 \, hp_u/62.4 = 44.07$ in the example. The pump terms used in the equations for $H_i$ and $G_i$ (Eqs. 6.22 and 6.23) are

$$P(Q_i) = Z/Q_i \quad \text{and} \quad P'(Q_i) = -Z/Q_i^2$$

The Hazen–Williams equation is employed in the example for head loss calculations. Using that expression, one obtains the pipeline constant:

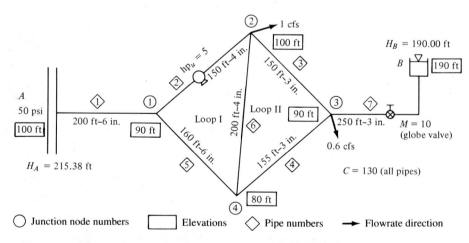

**Figure 6.15** Wood's sample pipe system: ○, junction node numbers; □, elevations; ◇, pipe numbers; →, flowrate direction. (After D. T. Wood, "Algorithms for Pipe Network Analysis and Their Reliability." Res. Rep. No. 127, University of Kentucky, Water Resources Research Institute, Lexington, KY, 1981.)

**TABLE 6.4**  FORMULAS FOR DETERMINING THE
PIPE CONSTANT $K_P$

| Equation type | $K_P$ | Units for | | | |
|---|---|---|---|---|---|
| | | $Q$ | $L$ | $d$ | $h_f$ |
| Hazen–Williams | $\dfrac{4.73L}{C^{1.85}D^{4.87}}$ | cfs | ft | ft | ft |
| | $\dfrac{10.44L}{C^{1.85}D^{4.87}}$ | gpm | ft | in | ft |
| | $\dfrac{10.70L}{C^{1.85}D^{4.87}}$ | m³/s | m | m | m |
| Darcy–Weisbach | $\dfrac{fL}{39.70D^5}$ | cfs | ft | ft | ft |
| | $\dfrac{fL}{32.15D^5}$ | gpm | ft | in | ft |
| | $\dfrac{fL}{12.10D^5}$ | m³/s | m | m | m |

$$K_P = \frac{4.73L}{C^{1.852}D^{4.87}}$$

where $n = 1.852$, $L$ is the pipe length, and $D$ is the pipe diameter. Formulas for computing the pipeline constant in English and metric units for both the Hazen–Williams and Darcy–Weisbach equations are given in Table 6.4.

Table 6.5 summarizes values of pipeline, minor loss, and pump constants for each pipe in the example system for a $C$ value of 130.

In the example, four mass continuity equations for the four junction nodes and three energy equations are required. The energy equations include one for each of the two loops noted ($\Delta E = 0$) and one additional energy equation for pipes connecting the two fixed-grade nodes ($\Delta E = 25.38$ ft).

An arbitrary set of initial flowrates is defined to start the procedure. A flowrate based on a mean flow velocity of 4 ft/sec is used for this purpose. The initial flowrates and corresponding values for $G$ and $H$ are shown in

**TABLE 6.5**  PIPE SYSTEM
CONSTANTS

| Pipe no. | $K_P$ | $K_M$ | $Z$ |
|---|---|---|---|
| 1 | 3.36 | 0 | 0 |
| 2 | 18.18 | 0 | 44.1 |
| 3 | 73.78 | 0 | 0 |
| 4 | 76.24 | 0 | 0 |
| 5 | 2.69 | 0 | 0 |
| 6 | 24.23 | 0 | 0 |
| 7 | 122.97 | 64.44 | 0 |

**TABLE 6.6**  VALUES OF $Q_i$, $G_i$, AND $H_i$ FOR
            THE LINEAR METHOD

| Pipe no. | $Q_i$ | $G_i$ | $H_i$ |
|----------|-------|-------|-------|
| 1 | 0.7854 | 5.072 | 2.151 |
| 2 | 0.3491 | 375.42 | − 123.66 |
| 3 | 0.1963 | 34.14 | 3.619 |
| 4 | 0.1963 | 35.278 | 3.740 |
| 5 | 0.7854 | 4.057 | 1.721 |
| 6 | 0.3491 | 18.308 | 3.451 |
| 7 | 0.1963 | 82.204 | 8.517 |

Table 6.6. A total of four continuity equations and three energy equations are to be solved simultaneously. The continuity equations (Eq. 6.6) are

$$-Q_1 + Q_2 + Q_5 = 0 \qquad \text{(junction 1)}$$

$$-Q_2 + Q_3 - Q_6 = -1.0 \quad \text{(junction 2)}$$

$$-Q_3 - Q_4 + Q_7 = -0.6 \quad \text{(junction 3)}$$

$$Q_4 - Q_5 + Q_6 = 0 \qquad \text{(junction 4)}$$

The energy equations (Eq. 6.34) are derived using the data in Table 6.6. Calculations on the right-hand side of these equations for the two loops and for the path $AB$ between the two fixed-grade nodes are given in Table 6.7. The left-hand side of the equations is the sum of the products of the $G_i$ and

**TABLE 6.7**  CALCULATIONS FOR ENERGY EQUATIONS FOR THE
            LINEAR (L) METHOD

| | Pipe no. and sign | $G_i \times Q_i$ | $H_i$ | $\Delta E$ for loop |
|---|---|---|---|---|
| Loop I | 2+ | 375.42 × 0.3491 = +131.059 | +123.66 | |
| | 6− | 18.308 × 0.3491 = −6.391 | +3.451 | 0 |
| | 5− | 4.057 × 0.7854 = −3.186 | +1.721 | |
| | $\Sigma G_i Q = \Sigma (G_i Q_i - H_i) + \Delta E = 121.482 + 128.832 + 0.0 = 250.31$ | | | |
| Loop II | 6+ | 18.308 × 0.3491 = +6.391 | −3.451 | |
| | 3+ | 34.14 × 0.1963 = +6.702 | −3.169 | 0 |
| | 4− | 35.278 × 0.1963 = −6.925 | +3.740 | |
| | $\Sigma G_i Q = \Sigma (G_i Q_i - H_i) + \Delta E = 6.168 - 3.330 + 0.0 = 2.838$ | | | |
| Path $AB$ | 1+ | 5.072 × 0.7854 = +3.984 | −2.151 | |
| | 2+ | 375.42 × 0.3491 = +131.059 | +123.66 | 25.38 |
| | 3+ | 34.14 × 0.1963 = +6.702 | −3.619 | |
| | 7+ | 82.204 × 0.1963 = +16.137 | −8.517 | |
| | $\Sigma G_i Q = \Sigma (G_i Q_i - H_i) + \Delta E = 157.882 + 109.373 + 25.38 = 292.635$ | | | |

$Q$ for each pipe in the loop or path. Notice that a sign convention must be used in accounting. The resulting three energy equations follow.

$$5.072Q_1 + 375.42Q_2 + 34.14Q_3 + 82.204Q_7 = 292.63 \quad \text{(path } AB\text{)}$$

$$375.42Q_2 - 4.057Q_5 - 18.308Q_6 = 250.304 \quad \text{(loop I)}$$

$$34.14Q_3 - 35.278Q_4 + 18.308Q_6 = 2.837 \quad \text{(loop II)}$$

The solution of these equations (in cfs) is $Q_1 = 1.725$, $Q_2 = 0.705$, $Q_3 = 0.262$, $Q_4 = 0.463$, $Q_5 = 1.020$, $Q_6 = 0.557$, and $Q_7 = 0.125$. These are used to formulate a second set of equations (only the energy equations change) and a second solution is obtained. The procedure continues until a specified convergence criterion is met. After five iterations the final flows were found to be $Q_1 = 1.73$, $Q_2 = 1.37$, $Q_3 = 0.37$, $Q_4 = 0.36$, $Q_5 = 0.36$, $Q_6 = 0.001$, and $Q_7 = 0.131$.

## Algorithms for Solving Node Equations

The two most widely used node methods are the single-node adjustment (N) method and the simultaneous node adjustment (SN) method [12, 20]. The N method was originally described by Hardy Cross [19]. The procedure is as follows:

1. A reasonable grade is assumed for each junction node in the system. The better the initial assumptions, the fewer the required trials.
2. A grade adjustment factor for each junction node that tends to satisfy continuity is computed.
3. Step 2 is repeated using improved solutions until a specified convergence criterion is met.

The grade adjustment factor is the change in grade at a particular node ($\Delta H$) that will result in satisfying continuity and considering the grades at adjacent nodes fixed. For convenience the required grade correction is expressed in terms of $Q_i$, the flow based on the grades at adjacent nodes before adjustment. With the gradient approximation,

$$f(Q) = f(Q_i) + f'(Q_i)\,\Delta Q$$

and substituting terms defined previously, one derives the flow correction

$$\Delta Q = \sum \left(\frac{1}{G_i}\right) \Delta H \tag{6.35}$$

where $\Delta H = H - H_i$, and the grade adjustment factor and $\Delta Q$ represent the flow corrections required to satisfy continuity at the nodes. From Eq. 6.6,

$$\Delta Q = \sum Q_i - Q_e \tag{6.36}$$

Thus, from Eqs. 6.35 and 6.36,

$$\Delta H = \left( \sum Q_i - Q_e \right) \Big/ \sum \frac{1}{G_i} \tag{6.37}$$

In Eq. 6.37, inflow is assumed positive. The numerator represents the unbalanced flowrate at the junction node (see also Eq. 6.29).

The flowrate $Q_i$ in a pipe section before adjustment is computed from

$$Q_i = (\Delta H_i / K)^{1/n}$$

in which $\Delta H_i$ is the grade change based on assumed initial values of grade.

When pumps are located in a pipeline, the following expression can be used to determine $Q_i$:

$$\Delta H_i = K Q_i^n - P(Q_i) \tag{6.38}$$

Equation 6.38 is solved using an approximation procedure. Adjustment of the grade for each junction node is made following each trial until a selected convergence criterion is satisfied [12, 20]. The SN method is based on simultaneous solution of the basic network node equations. These equations must be linearized in terms of approximate values of grade (head). Details of the procedure are reported in the references given [12, 20].

**Newton–Raphson Method**    The Newton–Raphson Method is a widely used numerical method for solving systems of nonlinear equations [1]. The method is applicable to problems that can be expressed in the form $F(H) = 0$, where the solution is that value of $H$ that causes $F$ to become zero. Application of the technique to a simple system where there is only one equation with one unknown illustrates the principle involved. Here, the derivative of $F$ can be approximated as

$$\frac{dF}{dH} = \frac{F(H + \Delta H) - F(H)}{\Delta H} \tag{6.39}$$

With an initial assumption of $H$, the solution is obtained by determining the value of $H + \Delta H$ that forces $F$ to zero. By setting $F(H + \Delta H)$ to zero, the solution for $\Delta H$ becomes

$$\Delta H = \frac{-F(H)}{dF/dH} \tag{6.40}$$

The value of $H$ used in the next step of the iterative process than becomes $H + \Delta H$. Iterations continue until $F$ closely approaches zero.

Analysis of the types of pipe networks encountered in practice usually means dealing with large numbers of equations and unknowns. The Newton–Raphson method can be applied to either the $N - 1$, $\Delta H$ equations (Eq. 6.8) or the $\Delta Q$ equations exemplified by (Eq. 6.24). For each node, a head equation of the form of Eq. 6.16 is written

$$F(H_b) = \sum_{b=1}^{N} \left[ \pm \left( \frac{H_a - H_b}{K_{ab}} \right)^{1/n} \right] - Q = 0 \tag{6.41}$$

where   $N$ = is the number of pipes that join at a node
        $K$ = is the pipeline constant
        $n$ = is 1.85 for the Hazen–Williams equation
        $Q$ = is the flow withdrawn at the node

The sign of the term in the summation depends on the direction of flow into or out of the junction. If $F(i)$ is the value of $F$ at iteration $i$, then it follows that

$$dF = F(i + 1) - F(i) \tag{6.42}$$

The same change can also be approximated as

$$dF = \frac{\partial F}{\partial H_1} \Delta H_1 + \frac{\partial F}{\partial H_2} \Delta H_2 + \cdots + \frac{\partial F}{\partial H_k} \Delta H_k \tag{6.43}$$

where $\Delta H$ is the change in $H$ in the iteration from $i$ to $i + 1$. The problem is one of iteratively determining the values of $\Delta H$ so that the end result is that $F(i + 1)$ becomes zero. The process involves setting Eq. 6.42 equal to Eq. 6.43. A system of $k$ linear equations with $k$ unknowns of the form $\Delta H$ results. These equations can be solved by various linear methods [1].

The solution is obtained by selecting initial values of $H$, calculating the partial derivatives of each $F$ with respect to $H$, and solving the set of linear equations to find a new value of $H$. The process is repeated until all of the calculated $F$ values are sufficiently close to zero. Note that the derivative of the terms in Eq. 6.41 is of the form

$$\frac{dF}{dH} = \pm \frac{(H_a - H_b)^{(1/n - 1)}}{n(K_{ab})^{1/n}} \tag{6.44}$$

The following example illustrates the Newton–Raphson procedure.

■ **EXAMPLE 6.5**
Given the simple pipe network of Fig. 6.16, find the value of $H_1$ if $C = 100$ and $n = 1.85$.

*Solution.*   Equations 6.41 and 6.44 are applied along with

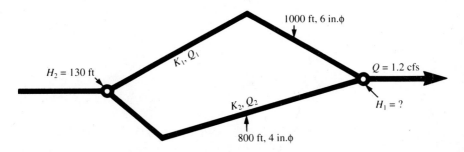

**Figure 6.16** Network diagram for Example 6.5.

**TABLE 6.8** CALCULATED VALUES OF $H$, $F$,
AND $dF/dH$ FOR EXAMPLE 6.5

| Iteration no. | $H$ | $F$ | $dF/dH$ |
|:---:|:---:|:---:|:---:|
| 1. | 100.00 | −0.504 | −0.0125 |
| 2. | 59.68 | −0.097 | −0.0084 |
| 3. | 48.13 | −0.002 | −0.0078 |
| 4. | 47.87 | | |

$$H(i + 1) = H(i) - \frac{F}{dF/dH} \tag{6.45}$$

In this case there are only two pipes, and only one equation must be solved for $F$, $dF/dH$, and $H(i + 1)$ at each iteration. Based on an initial assumption of 100 for $H_1$, the calculated values of $F$, $dF/dH$, and $H(i + 1)$ are displayed in Table 6.8. The first set of calculations follows.

1. Values of $K_1$ and $K_2$ are calculated by using the formula in Table 6.4 for the Hazen–Williams equation, where $Q$ is in cfs and $L$ and $d$ are in ft.

   $K_1 = (4.73 \times 1000)/[(100)^{1.85}(0.5)^{4.87}]$

   $= 275.95$

   In like manner, by substituting the appropriate values and solving, $K_2$ is found to be 167.78.

2. Assuming that $H_1 = 100$ and using Eq. 6.41,

   $$F(H)_1 = \left(\frac{130 - 100}{275.95}\right)^{0.54} + \left(\frac{130 - 100}{167.78}\right)^{0.54} - 1.2$$

   $$= -0.504$$

3. Using Eq. 6.44,

   $$\left(\frac{dF}{dH}\right)_1 = \frac{-1}{1.85(275.95)^{0.54}(130 - 100)^{0.46}}$$

   $$\frac{-1}{1.85(167.78)^{0.54}(130 - 100)^{0.46}}$$

   $$= -0.0125$$

4. The new value of $H$ is found from Eq. 6.45:

   $H_{i+1} = 100 - (-0.504/-0.0125)$

   $= 59.68$

5. The procedure is repeated, and the solutions are recorded in Table 6.5.
6. Noting that the sum of the $Q$'s must equal 1.2 cfs, the equations for $Q$ in pipes 1 and 2 are solved by using the head loss calculated to see if their sum is

correct. From Eq. 6.14,

$$Q_1 = [(130 - 47.87)/275.95]^{0.54} = 0.52 \text{ cfs}$$

$$Q_2 = [(130 - 47.87)/167.78]^{0.54} = 0.68 \text{ cfs}$$

$$Q_1 + Q_2 = 0.52 + 0.68 = 1.2 \text{ cfs}$$

which checks.
7. If the check had shown that the sum of the $Q$'s did not equal 1.2 very closely, additional iterations would be required.     ∎

## Selection of a Method of Network Analysis

In all decisions regarding the selection of an appropriate analytical tool, the designer or analyst must first answer the question of what is expected of the design or analysis. The network methods described herein are all equipped to accommodate many features of pipe systems. The methods are not all equal in their breadth, however, and they are not all equal in their ability to converge on a solution. Thus, the analyst should know something about the virtues and deficiencies of these tools [13]. The discussion that follows points out some features that are worth understanding.

Node equations are easy to formulate because they include only contributions from adjacent nodes. Loop equations require the identification of an appropriate set of energy equations, including terms for all pipes in the primary loops and for paths between fixed-grade nodes. Formulating this set of equations is considerably more difficult than formulating node equations.

The procedures described herein are all iterative. Computations continue until a specified convergence criterion is met. The solutions are therefore approximate, although they can be very accurate. The ability of an algorithm to produce an acceptable solution is important, and thus knowledge of the convergence problems associated with a technique proposed for use is basic to achieving good results.

A solution is considered satisfactory when all the basic equations are satisfied to a high degree of accuracy. When using loop equations, continuity is always exactly satisfied. The loop algorithms proceed to satisfy the energy equations iteratively, and the degree to which heads are unbalanced for the energy equations is evidence of solution accuracy. For methods based on node equations, iterations are carried out to satisfy continuity at junction nodes and the unbalance in continuity is the indicator of solution accuracy.

Studies by Wood, using an extensive data base, have shown that the P, N, and SN methods may exhibit significant convergence problems [12]. Thus, great care should be exercised when using these commonly employed tools.

Failures of the SN method are characterized by difficulty in meeting a reasonable convergence criterion. When this occurs in a limited number of trials, further trials are usually of no benefit. The failure rate for this method is quite high, and the use of results obtained employing this method are not recommended unless accuracy is obtained in a small number of trials [12]. It has also been found that algorithms based on the node equations (N and SN methods) fail to provide reliable results where low-resistance lines are encountered. This is attributable to the fact that solution algorithms for these equations do not incorporate an exact continuity balance.

For most network methods, failure rates are reduced when assumed initial values are close to the final values. However, there is no way to assume that this condition will be met, and even when it is, an excellent set of initial conditions does not guarantee convergence.

Both the SP and L methods have been found to provide excellent convergence, and the attainment of a reasonable convergence criterion is sufficient to assure great accuracy. Convergence failure for these methods is very rare, but since a gradient method is used to handle the nonlinear terms, there is always the possibility of convergence problems. Poor pump descriptions pose special problems, for example. The L method appears to have some advantages over the SP method. Assumed arbitrary flowrates need not satisfy continuity as the continuity conditions are already incorporated into the basic set of equations. This method also allows a more straightforward and reliable inclusion of hydraulic components such as check valves, closed lines, and pressure-regulating valves. Although the SP method has significantly fewer equations to solve, the use of sparse matrix techniques to handle the larger matrix generated by the L method has somewhat negated this advantage.

Wood has concluded that, if possible, either the SP or L method should be employed for pipe network analysis and that convergence is virtually assured if reasonable data are employed. Of the two methods the L method appears to have slightly better convergence characteristics [12].

## Use of Optimization Techniques in Pipe Network Analysis

Since the late 1960s, mathematical programming techniques such as linear programming, dynamic programming, and other approaches have been adopted for use in analyzing a wide variety of water supply and distribution problems. A detailed discussion of these techniques is beyond the scope of this book, but a few words about how optimization methods might be used in network analysis seem in order. The interested reader should consult appropriate references [20, 23].

One problem of pipe network analysis is the identification of measures by which the goodness of solutions can be gaged. Traditional approaches often give little good insight into this issue, particularly for large systems. Optimization models might prove to be useful in eliminating the vagueness surrounding definition of "close." But the use of optimization techniques is hampered by problems of convexity and nonlinearity inherent in water distribution systems. Research is needed on topics such as incorporating time variable storage in network elements to permit solving transient network problems and developing economic models to minimize construction, operation, and maintenance costs for water distribution systems.

**Computer Solutions**    Both the analog computer and the digital computer are used in analyzing distribution networks. The first deals with continuous physical variables, while the second is concerned only with numerical values. The analog computer acts as a physical model of the system to be studied and produces results that are limited in accuracy only by the physical elements of the model. The digital computer is theoretically limited in accuracy only by the reliability of the original input data.

The pioneering analog computer was the McIlroy Fluid Network Analyzer. The McIlroy analyzer uses fluistors, which are nonlinear resistances. The voltage drop $E$ for a fluistor with a direct current $I$ is given by

$$E = RI^m \tag{6.46}$$

where $R$ is the resistance. Any value of $m$ between 1.85 and 2.00 can be approximated by the judicious selection of fluistors. It is evident that Eq. 6.46 is in a form identical to that of Eq. 6.9, $H = KQ^n$. As a result of this analogy, a dc network can be developed to simulate any desired hydraulic network. Once the system is energized, measured voltages and currents, converted into units of discharge and head loss, can be read directly. A detailed discussion of the McIlroy analyzer can be found in several of the references at the end of the chapter [11, 21, 24, 25].

Solutions of the Hardy Cross method by digital computer involve a computation of the flow correction from Eq. 6.28:

$$\Delta Q = - \sum H \bigg/ n \sum \left( \frac{H}{Q} \right)$$

for each loop of the network after an initial assumption of flows has been made. Repeated sets of corrections are computed and applied until a satisfactorily balanced system is attained. The process of repeated trials is known as the iterative convergence technique. *Feedback* is the term applied to the modification of an initial value by some fraction of the corresponding output. The Hardy Cross procedure can be said to apply a feedback operation to sequential calculations, a procedure extremely well suited to the digital computer.

Using the iteration process combined with a convergence technique, the digital computer adjusts an initial set of flowrates to any desired degree of balance. The information obtained from the computer is usually in the form of tabulated head loss and directional flowrate for individual pipe branches. Example 6.4 illustrates the solution of a network problem by digital computer.

Numerous computer programs for the solution of pipe networks are available. Many of these are very sophisticated and are designed to handle large systems with pumping stations, storage facilities, and other features. Such programs are constantly being improved and the interested reader should consult the literature for detail [12, 14, 20, 23, 26–28]. The elements of network theory presented herein are basic to all of these, however, and should serve as adequate background [15–17, 29].

## 6.7 DISTRIBUTION SYSTEM DESIGN AND ANALYSIS

The design of a water distribution network involves the selection of a system of pipes so that the predicted design flows can be carried with head losses that do not exceed those deemed necessary for adequate operation of the system. Normally, the design flows should be based on estimated future requirements, since a distribution system is expected to provide effective service for many years (often as long as 100 yr). A sequence of evaluation, design, and layout operations is as follows [1–3, 30, 31]:

1. Review available data on the distribution system. These include maps, construction plans, billing records, planning studies, zoning regulations, population figures, water-use studies, and any other data relevant to the system being analyzed.
2. For existing water systems, determine pipe ages, the roughness of pipe interiors, pipe lengths and diameters, and the locations of pipes and appurtenances. Supply sources, pumping station locations and characteristics, and storage tank locations and volumes must also be determined.
3. Prepare a detailed map of the existing system or the proposed system.
4. Forecast population growth and distribution to the end of the design period (10–50 yr). Project water-use patterns (spatial and temporal) for domestic, industrial, and commercial uses to be served by the system. These estimates must also extend over the design life of the network. Water-use estimates are based on population projections and anticipated trends in commercial and industrial activity.
5. Develop a computer model of the existing or proposed system. In analyzing a network it is common to simplify the system by eliminating nonessential small lines, combining pipes using the equivalent pipe method, and assuming water use to be concentrated at takeoff points (nodes). Ordinarily, the model selected will be one of those described under network analysis.
6. Use the computer model of the existing system to evaluate historical conditions. If observed data and model runs compare favorably, the model may be considered to be "validated." The model can then be presumed to be adequate for evaluating proposed modifications to, or extensions of, the existing system. Applying the model to the existing system will quickly identify areas of low pressure, pipelines having high head loss characteristics, and overloaded parts of the network. Proposed replacements and extensions of the system can then be evaluated to see if they will resolve problems or meet new requirements.
7. Design new or extended water distribution systems on the basis of population and other data discussed previously. As in the case of analyzing existing systems, a suitable network algorithm must be chosen. In general, these are the steps followed in distribution system design:

**a.** On a development plan of the area to be serviced, sketch the tentative location of all water mains that will be needed to supply the area. The completed drawing should differentiate between proposed feeder mains and smaller service mains. The various pipelines comprising the system should be interconnected at intervals of 1200 ft or less. Looped feeder systems are desirable and should be used wherever possible. Two small feeder mains running parallel several blocks apart are preferable to a single large main with an equal or slightly larger capacity than the two mains combined.

**b.** Using estimated values of the anticipated design flows, select appropriate pipe sizes by assuming velocities of from 3 to 5 fps.

**c.** Mark the position of building service connections, fire hydrants, and valves. Service connections form the link between the distribution system and the individual consumer. Normally, the practice is one customer per service pipe. Figure 6.17 illustrates the details of a typical service connection for a private residence. Fire hydrants are located to provide complete protection to the area covered by the distribution system. Recommendations regarding average area per hydrant for various populations and required fire flow are given by the National Board of Fire Underwriters. Hydrants generally should not be farther than about 500 ft apart. Figure 6.18 illustrates a typical fire hydrant setting.

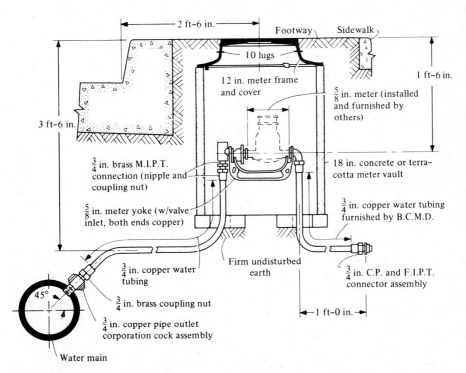

**Figure 6.17** Typical installation of $\frac{3}{4}$-in. metered domestic service. (Courtesy of Baltimore County, MD, Department of Public Works.)

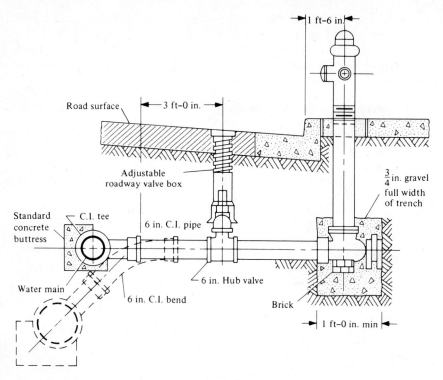

**Figure 6.18** Typical fire-hydrant setting. (Courtesy of Baltimore County, MD, Department of Public Works.)

    **d.** Apply projected water demands (including fire flow) to the network and calculate residual pressures.

    **e.** Compare calculated pressures to standards. Identify areas of projected less-than-adequate service.

    **f.** Check head losses in individual pipes to find excesses. These will usually occur in pipes where flow velocities are greater than 1.5 m/s (5 fps).

    **g.** Add pipes or replace high-head-loss pipes with larger pipes. Run the model again and see if residual pressures under maximum future loads are adequate. If not, try other additions or changes until the system is adequate for anticipated future loads. The system must be able to handle maximum daily loads plus fire loads without decreasing residual pressures below minimum standards. The overall objective is to design a system that will meet projected water demands at the least cost while incorporating appropriate safety measures for looping or duplicate lines so that line breaks or other disturbances will not isolate users from a water supply. Finally, the network analysis and resulting design will be no better than the assumptions made about water demands, pipe roughnesses, and so on.

**8.** Estimate construction costs for the proposed improvements.

**9.** Prepare a construction schedule for the identified improvements or new system that is consistent with the financial capabilities of the city.

# 6.8 DISTRIBUTION RESERVOIRS AND SERVICE STORAGE

Distribution reservoirs provide service storage to meet the widely fluctuating demands often imposed on a distribution system, to accommodate fire fighting and emergency requirements, and to equalize operating pressures. They are elevated at or just below ground level.

The main categories of storage reservoirs are surface reservoirs, stand-pipes, and elevated tanks. Common practice is to line surface reservoirs with concrete, gunite, asphalt, or an asphaltic membrane. Surface reservoirs may be covered or uncovered. Whenever possible a cover should be considered for the prevention of contamination of the water supply by animals or humans and to prevent the formation of algae.

Standpipes or elevated tanks are normally employed where the construction of a surface reservoir would not provide sufficient head. A standpipe is essentially a tall cylindrical tank whose storage volume includes an upper portion (the useful storage), which is above the entrance to the discharge pipe, and a lower portion (supporting storage), which acts only to support the useful storage and provide the required head. For this reason, standpipes over 50 ft high are usually uneconomical. Steel, concrete, and wood are used in the construction of standpipes and elevated tanks. When it becomes more economical to build the supporting structure for an elevated tank than to provide for the supporting storage in a standpipe, the elevated tank is used. A good discussion of the construction details and characteristics of various kinds of distribution reservoirs is given by Babbitt, Doland, and Cleasby [32].

Distribution reservoirs should be located strategically for maximum benefits. Normally the reservoir should be near the center of use, but in large metropolitan areas a number of distribution reservoirs may be located at key points. Reservoirs providing service storage must be high enough to develop adequate pressures in the system they are to serve. A central location decreases friction losses by reducing the distance from supply point to the area served. Positioning the reservoir so that pressures may be approximately equalized is an additional consideration of importance. Figure 6.19 will illustrate this point. The location of the tank as shown in part (a) results in a very large loss of head by the time the far end of the municipality is reached. Thus, pressures too low will prevail at the far end or excessive pressures will be in evidence at the near end. In part (b) it is seen that pressures over the whole municipal area are much more uniform for periods of both high and low demand. Note that during periods of high demand the tank is supplying flow in both directions (being emptied), whereas during periods of low demand the pump is supplying the tank and the municipality.

The amount of storage to be provided is a function of the capacity of the distribution network, the location of the service storage, and the use to which it is to be put. When water treatment facilities are required, it is preferable

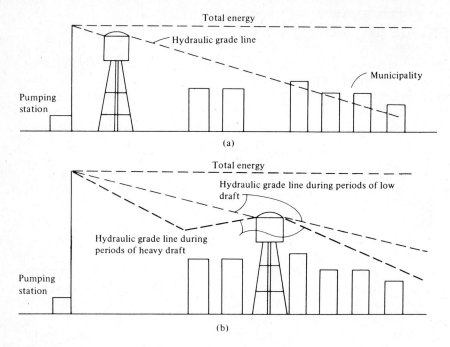

**Figure 6.19** Pressure distribution as influenced by the location of a distribution reservoir.

to operate them at a uniform rate such as the maximum daily rate. It is also usually desirable to operate pumping units at constant rates. Demands on the system in excess of these rates must therefore be met by storage, previously defined as operating storage. Requirements for fire-fighting purposes should be sufficient to provide fire flows for 10–12 hr in large communities and for 2 hr or longer in smaller ones. Emergency storage is provided to sustain the community's needs during periods when the inflow to the reservoir is shut off, for example, through a failure of the supply work, failure of pumping equipment, or need to take a supply line out of service for maintenance or repair. The length of time the supply system is expected to be out of service dictates the amount of emergency storage to be provided. Emergency storage volumes sufficient to last for several days are common.

The amount of storage required for emergency and fire-fighting purposes is readily computed once the time period over which these flows are to be provided has been selected [33]. An emergency storage of 3 days for a community of 8000 having an average use rate of 150 gpcd is $3 \times 150 \times 8000 = 3.6$ mil gal. Given that a fire flow of 2750 gpm must be provided for a duration of 10 hr, this means a total fire-fighting storage of 1.65 mil gal. To the sum of these values, an additional equalizing or operating storage requirement would be added. The determination of this volume is slightly more complex and needs further explanation.

**TABLE 6.9**  HOURLY DEMAND FOR THE MAXIMUM DAY

| (1) | (2) | (3) | (4) | (5) | (6) | |
|---|---|---|---|---|---|---|
| Time | Average hourly demand rate (gpm) | Hourly demand (gal) | Cumulative demand (gal) | Hourly demand as a percent of average | Average hourly demand minus hourly demand: 286,250 − (3) | |
| | | | | | − | + |
| 12 A.M. | 0 | 0 | 0 | 0 | — | — |
| 1 A.M. | 2,170 | 130,000 | 130,000 | 45.4 | | 156,250 |
| 2 | 2,100 | 126,000 | 256,000 | 44.1 | | 160,250 |
| 3 | 2,020 | 121,000 | 377,000 | 42.3 | | 165,250 |
| 4 | 1,970 | 118,000 | 495,000 | 41.3 | | 168,250 |
| 5 | 1,980 | 119,000 | 614,000 | 41.6 | | 167,250 |
| 6 | 2,080 | 125,000 | 739,000 | 43.2 | | 161,250 |
| 7 | 3,630 | 218,000 | 957,000 | 76.2 | | 68,250 |
| 8 | 5,190 | 312,000 | 1,269,000 | 108.9 | 25,750 | |
| 9 | 5,620 | 337,000 | 1,606,000 | 117.8 | 50,750 | |
| 10 | 5,900 | 354,000 | 1,960,000 | 123.6 | 67,750 | |
| 11 | 6,040 | 363,000 | 2,323,000 | 126.7 | 76,750 | |
| 12 | 6,320 | 379,000 | 2,702,000 | 132.4 | 92,750 | |
| 1 P.M. | 6,440 | 387,000 | 3,089,000 | 135.2 | 100,750 | |
| 2 | 6,370 | 382,000 | 3,471,000 | 133.4 | 95,750 | |
| 3 | 6,320 | 379,000 | 3,850,000 | 132.4 | 92,750 | |
| 4 | 6,340 | 381,000 | 4,231,000 | 133.0 | 94,750 | |
| 5 | 6,640 | 399,000 | 4,630,000 | 139.5 | 112,750 | |
| 6 | 7,320 | 439,000 | 5,069,000 | 153.3 | 152,750 | |
| 7 | 9,333 | 560,000 | 5,629,000 | 195.5 | 273,750 | |
| 8 | 8,320 | 499,000 | 6,128,000 | 174.4 | 212,750 | |
| 9 | 5,050 | 303,000 | 6,431,000 | 105.8 | 16,750 | |
| 10 | 2,570 | 154,000 | 6,585,000 | 53.8 | | 132,250 |
| 11 | 2,470 | 148,000 | 6,733,000 | 51.7 | | 138,250 |
| 12 | 2,290 | 137,000 | 6,870,000 | 47.9 | | 149,250 |
| Total | | 6,870,000 | | | 1,466,500 | 1,466,500 |

Average hourly demand $= \dfrac{6,870,000}{24} = 286,250$ gal

To compute the required equalizing or operating storage, a mass diagram or hydrograph indicating the hourly rate of consumption is required. The procedure to be used in determining the needed storage volume follows:

1. Obtain a hydrograph of hourly demands for the maximum day. This may be obtained through a study of available records, by gaging the existing system during dry periods when lawn-sprinkling demands are high, or by using available design criteria such as presented in Chapter 4 to predict a hydrograph for some future condition of development.
2. Tabulate the hourly demand data for the maximum day as shown in Table 6.9.
3. Find the required operating storage by using mass diagrams such as in Figs. 6.20 and 6.21, the hydrograph of Fig. 6.22, or the values tabulated in column 6 of Table 6.9.

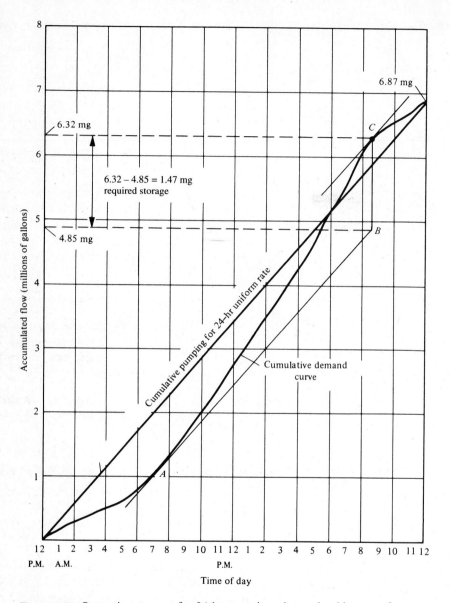

**Figure 6.20**  Operating storage for 24-hr pumping, determined by use of a mass diagram.

The required operating storage is found by using a mass diagram with the cumulative pumping curve plotted on it. Figure 6.20 illustrates this diagram for a uniform 24-hr pumping rate. Note that the total volume pumped in 24 hr must equal the total 24-hr demand, and thus the mass curve and cumulative pumping curve must be coincident at the origin and at the end of the day. Next, construct a tangent to the mass curve parallel to the pumping

.curve at point *A* in the figure. Then draw a second parallel tangent to the mass curve at point *C* and drop a vertical from *C* to an intersection with tangent *AB* at *B*. The required storage is equal to the magnitude of the ordinate *CB* measured on the vertical scale. In the example shown, the necessary storage volume is found to be 1.47 mil gal for a 24-hr pumping period.

Readers should recognize that the reservoir is full at *A*, is empty at *C*, is filling whenever the slope of the pump curve exceeds that of the cumulative demand curve, and is being drawn down when the rate of demand exceeds the rate of pumping.

It is often desirable to operate an equalizing reservoir so that pumping will take place at a uniform rate but for a period less than 24 hr. In small communities, for example, it is often advantageous to pump only during the normal working day. It may also be more economical to operate the pumping station at off-peak periods when electric power rates are low [32].

Figure 6.21 illustrates the operation of a storage reservoir where pumping occurs during the period between 6 A.M. and 6 P.M. only. To find the required storage in this case, construct the cumulative pumping curve *ED* so that the total volume of 6.87 mil gal is pumped uniformly from 6 A.M. to 6 P.M. Then project point *E* vertically upward to an intersection with the cumulative demand curve at *A*. Construct line *AC* parallel to *ED*. Point *C* will be at the intersection of line *AC* with the vertical extended upward from 6 P.M. on the abscissa. The required storage equals the value of the ordinate *CBD*. Numerically it is 2.55 mil gal and exceeds the storage requirement for 24-hr pumping.

Another graphical solution to the storage problem may be obtained as outlined in Fig. 6.22. The figure is a plot of the demand hydrograph for the maximum day. For uniform 24-hr pumping, the pumping rate will be equal to the mean hourly demand. This is shown as line *PQ*. The required storage is then obtained by planimetering or determining in some other manner the area between curve *BEC* and line *PQ*. Conversion of this area to units of volume yields the required storage of 1.47 mil gal. The required storage for 24-hr pumping may also be determined by summing either the plus or minus values of column 6 in Table 6.9.

Unless pumping follows the demand curve or demand hydrograph, storage will be required. Figure 6.22 shows that a maximum pumping rate of about 9400 gpm will be required with no storage, whereas if storage is provided, a maximum pumping rate of 4775 gpm (about 50% of that required with no storage) will suffice. This example tends to illustrate the economics of providing operating storage.

Variable-rate pumping is normally not economical. In practice, it is common to provide storage and pumping facilities so that pumping at the average rate for the maximum day can be maintained. On days of lesser demand, some pumping units will stand idle. Another operational procedure is to provide enough storage for pumping at the average rate for the average day, with idle reserve capacity, and then to overload all available units on the maximum day. Provision of pumping and storage capacity to meet peak

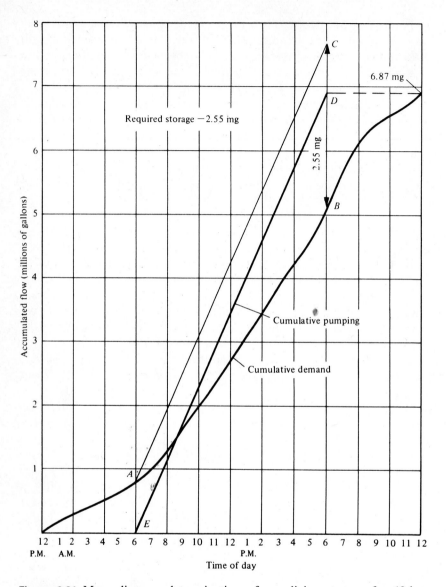

**Figure 6.21** Mass diagram determination of equalizing storage for 12-hr pumping.

demands experienced for a few hours only every few years has been found to be economically impractical.

Analyses of distribution systems are commonly concerned with the pipe network, topographic conditions, pumping station performance, and the operating characteristics of the distribution storage. One, all, or any combination of these features may be the object of study. Where multiple sources of

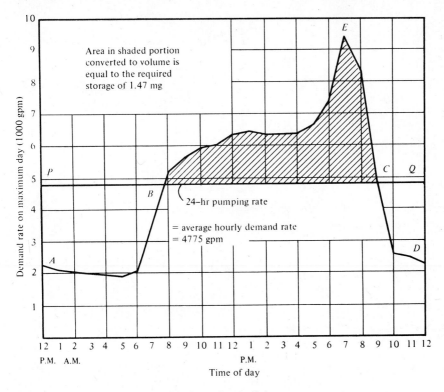

**Figure 6.22** Graphical determination of equalizing storage.

supply operate under variable-head conditions, the hydraulic balancing of the system becomes more complex. The simple system of Fig. 6.23 illustrates this point.

Considering that the demand for water by the municipal load center fluctuates hourly, it is evident there are essentially two modes of operation of the given distribution system. When municipal requirements are light, such as in the early morning hours, the pumping station will meet these demands and in addition supply the reservoir. The solution of the problem may then be had by writing the equations

(1)     $Q_1 - Q_D = Q_2$

(2)     $Z_P + E_P = Z_{LC} + E_{R_2} + H_{f_1}$

(3)     $Z_{LC} + E_{R_2} = Z_T + H_{f_2}$

(2 + 3)   $H_{f_1} + H_{f_2} = E_P + Z_P - Z_T$

where     $Q_1$ = flow from the pump
          $Q_D$ = municipal demand
          $Q_2$ = flow to the tank

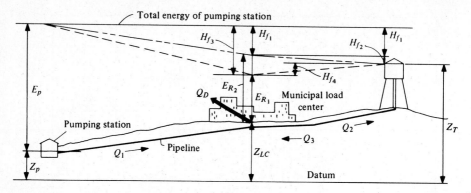

**Figure 6.23** Modes of operation of a distribution system.

$$
\begin{aligned}
Z_P, Z_{LC}, Z_T &= \text{elevation above the arbitrary datum } (Z_T = \text{elevation of} \\
&\quad\ \text{water surface in tank)} \\
E_P &= \text{energy produced by the pump} \\
E_{R_2} &= \text{residual energy of the load center (pressure head plus} \\
&\quad\ \text{velocity head)} \\
H_{f_1}, \text{ etc.} &= \text{friction-head losses}
\end{aligned}
$$

If $Q_D$, $Z_P$, $Z_{LC}$, and $Z_T$ are specified, the equations may be solved by selecting values of $E_{R_2}$ and then solving for $H_{f_1}$ and $H_{f_2}$. When a solution is reached such that Eq. $(2 + 3)$ is satisfied, $Q_1$ and $Q_2$ may be computed.

When demands are high, both the tank and the pump will supply the community. The direction of flow will then be reversed in the line from the tank to the pump and the applicable equation will be

(1) $\qquad Q_1 + Q_3 = Q_D$

(2) $\qquad Z_P + E_P = Z_{LC} + E_{R_1} + H_{f_3}$

(3) $\quad Z_{LC} + E_{R_1} + H_{f_4} = Z_T$

Again, an assumed value for $E_R$ will be taken and trial solutions carried out until equation (1) is satisfied. Note that the foregoing illustration is a simple case, since $Z_T$ has been specified. Actually, since $Z_T$ fluctuates with time, it is necessary to have information on storage volume available versus water elevation in the tank so that at any specified condition of draft, the actual value for $Z_T$ can be determined and used in the computations.

Water distribution systems generally are considered to be a composite of four basic constituents: the pipe network, the storage, the pump performance, and the pumping station and its suction source. These components must be integrated into a functioning system for various schedules of demand. A thorough analysis of each system must be made to ensure that it will operate satisfactorily under all anticipated combinations of demand and hydraulic component characteristics. The system may work well under one set of conditions but will not necessarily be operable under some other set. A comprehensive system balance requires an hourly simulation of performance for the expected operating schedule [21, 22].

There is an infinite number of possible arrangements of the basic components in a distribution system. The hydraulic analyses applicable to each system are, however, common to all and have been illustrated in detail in this chapter.

# Pumping

Pumping equipment forms an important part of the transportation and distribution facilities for water and wastes [32]. Requirements vary from small units used to pump only a few gallons per minute to large units capable of handling several hundred cubic feet per second. The two primary types of pumping equipment of interest here are the centrifugal pump and the displacement pump. Airlift pumps, jet pumps, and hydraulic rams are also used in special applications. In water and sewage works, the centrifugal pump finds the widest use. *Centrifugal pumps* have a rotating element (impeller) that imparts energy to the water. *Displacement pumps* are often of the reciprocating type, in which a piston draws water into a closed chamber and then expels it under pressure. Reciprocating pumps are widely used to handle sludge in sewage treatment works.

Electric power is the primary source of energy for driving pumping equipment, but gasoline, steam, and diesel power are also used. Often a standby engine powered by one of these forms is included in primary pumping stations to operate in emergency situations when electric power fails.

## 6.9 PUMPING HEAD

Anticipated operating conditions of the system must be known to effectively design a pumping station [34]. A knowledge of the *total dynamic head* (TDH) against which the pump must operate is needed. The TDH is composed of the difference in elevation between the pump centerline and the elevation to which the water is to be raised, the difference in elevation between the level of the suction pool and the pump centerline, the frictional losses encountered in the pump, pipe, valves and fittings, and the velocity head. Expressed in equation form, this becomes

$$\text{TDH} = H_L + H_F + H_V \tag{6.47}$$

where $H_L$ is the total static head or elevation difference between the pumping source and the point of delivery, $H_F$ is the total friction head loss, and $H_V$ is the velocity head $V^2/2g$. Figure 6.24 illustrates the total static head.

## 6.10 POWER

For a known discharge and total pump lift, the theoretical horsepower required may be found by using

$$hp = Q\gamma H/550 \tag{6.48}$$

where
$Q$ = discharge, cfs
$\gamma$ = specific weight of water
$H$ = total dynamic head
$550$ = conversion from foot-pounds per second to horsepower

The actual horsepower required is obtained by dividing the theoretical horsepower by the efficiency of the pump and driving unit.

## 6.11 CAVITATION

Cavitation is the phenomenon of cavity formation or the formation and collapse of cavities [35]. Cavities are considered to develop when the absolute pressure in a liquid reaches the vapor pressure related to the liquid temperature. Under severe conditions, cavitation can result in breakdown of pumping equipment. As the net positive suction head (NPSH) for a pump is reduced,

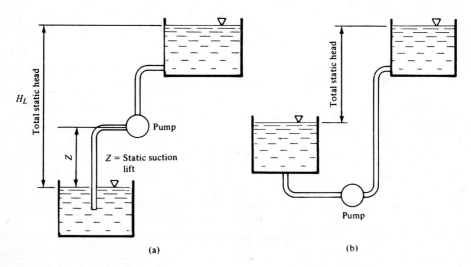

**Figure 6.24** Total static head: (a) Intake below the pump centerline and (b) intake above the pump centerline.

a point is reached where cavitation becomes detrimental. This point is usually referred to as the minimum net positive suction head ($\text{NPSH}_{\min}$) and is a function of the type of pump and the discharge through the pump. NPSH is calculated as

$$\text{NPSH} = \frac{V_1^2}{2g} + \frac{p_1}{\gamma} - \frac{p_v}{\gamma} \tag{6.49}$$

where $V_1$ is the velocity of flow at the centerline of the inlet to the pump, $p_1$ is the pressure at the centerline of the pump inlet, and $p_v$ is the vapor pressure of the fluid. Referring to Fig. 6.24a and writing the energy equation between the intake pool and the inlet to the pump, we have

$$\frac{p_a}{\gamma} = \frac{V^2}{2g} + \frac{p_1}{\gamma} + Z + h_L$$

or

$$\frac{p_a}{\gamma} - Z - h_L = \frac{V^2}{2g} + \frac{p_1}{\gamma}$$

where $p_a$ is atmospheric pressure and $h_L$ is the head loss in the intake.
Subtracting $p_v/\gamma$ from both sides, we have

$$\frac{p_a}{\gamma} - Z - h_L - \frac{p_v}{\gamma} = \frac{V^2}{2g} + \frac{p_1}{\gamma} - \frac{p_v}{\gamma}$$

This may be written

$$\frac{p_a}{\gamma} - Z - h_L - \frac{p_v}{\gamma} = \text{NPSH}$$

The minimum value of the static lift is then determined as

$$Z_{\min} = \frac{p_a - p_v}{\gamma} - \text{NPSH}_{\min} - h_L \tag{6.50}$$

The required NPSH for any pump must be obtained from the manufacturer. This value can then be checked against the proposed installation using Eqs. 6.49 and 6.50 to ensure that the available NPSH is greater than the manufacturer's requirement.

## 6.12 SYSTEM HEAD

The system head is represented by a plot of TDH versus discharge for the system being studied. Such plots are very useful in selecting pumping units [36]. It should be clear that the system head curve will vary with flow since $H_F$ and $H_V$ are both a function of discharge. In addition, the static head $H_L$ may vary as a result of fluctuating water levels and similar factors, and it is often necessary to plot system-head curves covering the range of variations in static head. Figure 6.25 illustrates typical system-head curves for a fluctuating static water level.

## 6.13 PUMP CHARACTERISTICS

Each pump has its own characteristics relative to power requirements, efficiency, and head developed as a function of rate of flow. These relationships are usually given as a set of pump characteristic curves for a specified speed. They are used in conjunction with system-head curves to select correct pumping equipment for a particular installation. A set of characteristic curves are shown in Fig. 6.26.

At no flow, the head is known as the *shutoff head*. The pump head may rise slightly or fall from the shutoff value as discharge increases. Ultimately, however, the head for any centrifugal pump will fall with increase in flow. At maximum efficiency, the discharge is known as the *normal* or *rated discharge* of the pump. Varying the pump discharge by throttling will lower the efficiency of the unit. By changing the speed of the pump, discharge can be varied within a certain range without a loss of efficiency. The most practical and efficient approach to a variable-flow problem is to provide two or more pumps in parallel so that the flow may be carried at close to the peak efficiency of the units which are operating.

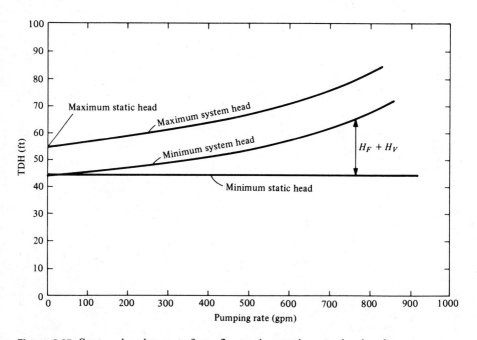

**Figure 6.25** System head curves for a fluctuating static pumping head.

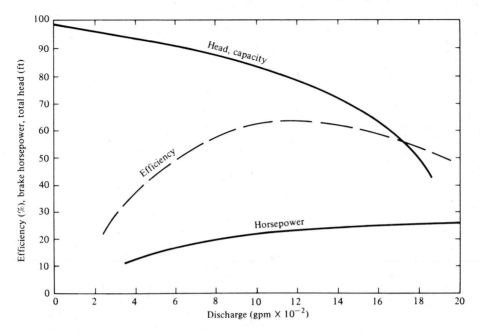

**Figure 6.26** Typical pump characteristic curves.

The normal range of efficiencies for centrifugal pumps is between 50% and 85%, although efficiencies in excess of 90% have been reported. Pump efficiency usually increases with the size and capacity of the pump [35].

## 6.14  SELECTION OF PUMPING UNITS

Normally the engineer is given the system-head characteristics and is required to find a pump or pumps to deliver the anticipated flows. This is done by plotting the system-head curve on a sheet with the pump characteristic curves. The operating point is at the intersection of the system-head curve and the pump-head capacity curve. This gives the head and flow at which the pump will be operating. A pump should be selected so that the operating point is also as close as possible to peak efficiency. This procedure is shown in Fig. 6.27.

Pumps may be connected in series or in parallel. For series operation at a given capacity, the total head equals the sum of the heads added by each pump. For parallel operation, the total discharge is multiplied by the number of pumps for a given head. It should be noted, however, that when two pumps are used in series or parallel, neither the head nor capacity for a given system head curve is doubled (Fig. 6.27).

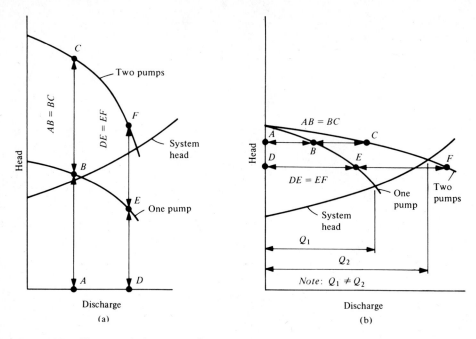

**Figure 6.27** Characteristic curves for (a) series and (b) parallel pump operations of equal pumps.

### ■ EXAMPLE 6.6

A pumping station is to be designed for an ultimate capacity of 1200 gpm at a total head of 80 ft. The present requirements are that the station deliver 750 gpm at a total head of 60 ft. One pump will be required as a standby.

***Solution.***    The total curve for dynamic head versus discharge is plotted as shown in Fig. 6.28. Values for the curve are obtained as indicated in Section 6.9.

Consider that three pumps will ultimately be needed (one as a standby). Determine the design flows as follows:

1. two pumps at 1200 gpm at 80 ft of TDH;
2. one pump at 1200/2 = 600 gpm at 80 ft of TDH;
3. one pump must also be able to meet the present requirements of 750 gpm at 60 ft of TDH.

From manufacturers' catalogs, two pumps, *A* and *B*, are found that will meet the specifications. The characteristic curves for each pump are shown in Fig. 6.28. The intersection of the characteristic curves with the system-head curve indicates that pump *A* can deliver 750 gpm at a TDH of 60 ft while pump *B* can deliver 790 gpm at a TDH of 62 ft. A check of the efficiency curves for each pump indicates that pump *B* will deliver the present flow at a much greater efficiency than pump *A*. Therefore, select pump *B*.

For the present, select two pumps of type *B* and use one as a standby. For the future, add one more pump of type *B*.    ■

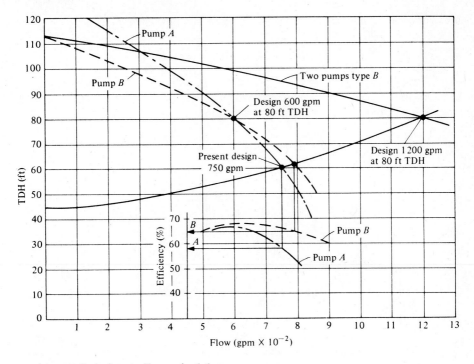

**Figure 6.28** Solution to Example 6.6.

Operating characteristics for a wide range of pump sizes and speeds are available from pump manufacturers. Special equipment manufactured to satisfy the prescribed requirements of the customer must customarily satisfy an acceptance test after it has been installed.

## PROBLEMS

**6.1** Given a V-shaped channel with a bottom slope of 0.001, a top width of 10 ft, and a depth of 5 ft, determine the velocity of flow. Find the discharge in cfs and $m^3/s$.

**6.2** A trapezoidal channel measures 3 m across the top and 1 m across the bottom. The depth of flow is 1.5 m. For $s = 0.005$ and $n = 0.012$, determine the velocity and rate of flow.

**6.3** Given an 18-in. concrete conduit with a roughness coefficient of $n = 0.013$, $s = 0.02$, and a discharge capacity of 15 cfs, what diameter pipe is required to quadruple the capacity?

**6.4** Find the dimensions of a rectangular concrete channel to carry a flow of 150 $m^3/s$, with a bottom slope of 0.015 and a mean velocity of 10.2 m/s.

**6.5** Determine the head loss in a 46-cm concrete pipe with an average velocity of flow of 3.2 m/s and a length of 30 m.

**6.6** Find the discharge from a full-flowing cast-iron pipe 24 in. in diameter having a slope of 0.004.

**6.7** Refer to the figure and assume that reservoirs $A$, $B$, and $C$ have water surface elevations of 150 ft, 90 ft, and 40 ft, respectively, and are connected by a system of concrete pipes of length $L_1 = 2400$ ft, $L_2 = 1500$ ft, and $L_3 = 5500$ ft, and where their respective diameters are 8, 12, and 21 in. Find the discharge in each pipe and the elevation of the hydraulic gradient at $P$.

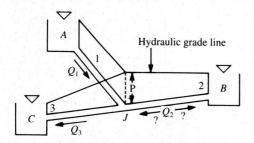

**6.8** Three reservoirs $A$, $B$, and $C$ are connected by a branching cast-iron pipe system. If the pipe lengths are $L_1 = 3000$ ft, $L_2 = 2200$ ft, and $L_3 = 1600$ ft, the respective pipe diameters are 15, 10, and 18 in., and the surface-water elevations of two of the three reservoirs are $A = 125$ ft and $B = 55$ ft, find the surface-water elevation of reservoir $C$. Assume that the flow to reservoir $C$ is 15 cfs.

**6.9** Solve Prob. 6.8 by assuming that the flow in pipe 3 is from reservoir $C$ rather than to reservoir $C$.

**6.10** Three riveted steel pipes are connected in series, with the flow through the system being 10 m³/s. Find the total head loss if the pipe diameters and lengths are $D_1 = 60$ cm, $D_2 = 40$ cm, and $D_3 = 54$ cm, $L_1 = 400$ m, $L_2 = 450$ m, and $L_3 = 750$ m. Assume the friction factor $f = 0.0125$.

**6.11** Given the same lengths and diameters of the pipes in series as in Prob. 6.10, determine the total flow if the system head loss is 50 m.

**6.12** Consider the pipe system in the figure. If the flow in $BCD$ is 6 cfs, find   (a) the flow in $BED$,   (b) the total flow, and   (c) a length of 16-in. pipe equivalent to the two parallel pipes.

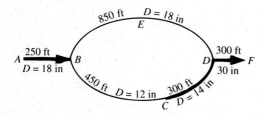

**6.13** A flow of 15 m³/s is divided into three parallel pipes of diameters 30, 20, and 45 cm and lengths of 30, 40, and 25 m respectively. Find the head loss and distribution of flow. Assume $f = 0.015$.

**6.14** If a system of parallel pipes has diameters of 18, 8, and 21 in. and lengths of 50, 95, and 60 ft, respectively, find the total flow in the system. Assume $f = 0.024$ and the total head loss is 100 ft.

**6.15** From the given layout, determine the length of an equivalent 8-cm pipe.

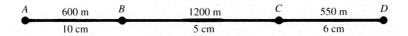

**6.16** For the pipe layout in Prob. 6.15, find the diameter of an equivalent 1000-m pipe.

**6.17** For the pipe network shown, determine the direction and magnitude of flow in each pipe, using the Hardy Cross method. Assume $C = 100$.

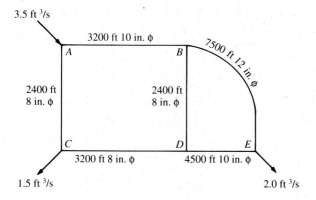

**6.18** Water is pumped 9 mi from a reservoir at elevation 100 ft to a second reservoir at elevation 210 ft. The pipeline connecting the reservoirs is 54 in. in diameter. It is concrete and has an absolute roughness of 0.003. If the flow is 25 mgd and pumping-station efficiency is 80%, what will be the monthly power bill if electricity costs 3 cents per kilowatt-hour?

**6.19** A reservoir at elevation 700 ft is to supply a second reservoir at elevation 460 ft. The reservoirs are connected by 1300 ft of 24-in. cast-iron pipe and 2000 ft of 20-in. cast-iron pipe in series. What will be the discharge delivered from the upper reservoir to the lower one?

**6.20** Consider that the water needs of a city for the next 40 yr may be met by constructing a single 10-in. main for the first 15 yr followed by a second 10-in. main for the next 25 yr. At this time both mains will be in service. An alternative solution is to build one 18-in. line which will serve the entire 40-yr period. Consider that the cost per foot installed is $20 for the first 10-in. line, $26 for the next 10-in. line, and $31 for the 18-in. line. The combined interest and amortization rate is 7.0% per year. Find the most economical method of meeting the community's water requirements.

**6.21** It is necessary to pump 6000 gpm of water from a reservoir at an elevation of 900 ft to a tank whose bottom is at an elevation of 1100 ft. The pumping unit is located at elevation 900. The suction pipe is 24 in. in diameter and very short, so head losses may be neglected. The pipeline from the pump to the upper tank is 410 ft long and is 20 in. in diameter. Consider the minor losses in the line to equal 2.5 ft of water. The maximum depth of water in the tank is 38 ft, and the supply lines are cast iron. Find the maximum lift of the pump and the horsepower required for pumping if the pump efficiency is 76%.

**6.22** If a flow of 5.0 cfs is to be carried by an 11,000-ft cast-iron pipeline without exceeding a head loss of 137 ft, what must the pipe diameter be?

**6.23** A 48-in. water main carries 79 cfs and branches into two pipes at point $A$. The branching pipes are 36 and 20 in. in diameter and 2800 and 5000 ft long, respectively. These pipes rejoin at point $B$ and again form a single 48-in. pipe. If the friction factor is 0.022 for the 36-in. pipe and 0.024 for the 20-in. pipe, what will the discharge be in each branch?

**6.24** Water is pumped from a reservoir whose surface elevation is 1390 ft to a second reservoir whose surface elevation is 1475 ft. The connecting pipeline is 4500 ft long and 12 in. in diameter. If the pressure during pumping is 80 psi at a point midway on the pipe at elevation 1320 ft, find the rate of flow and the power exerted by the pumps. Also, plot the hydraulic grade line. Assume that $f = 0.022$.

**6.25** A concrete channel 18 ft wide at the bottom is constructed with side slopes of 2.2 horizontal to 1 vertical. The slope of the energy gradient is 1 in 1400 and the depth of flow is 4.0 ft. Find the velocity and the discharge.

**6.26** A rectangular channel is to carry 200 cfs. The mean velocity must be greater than 2.5 fps. The channel bottom width should be about twice the channel depth. Find the channel cross section and the required channel slope.

**6.27** A rectangular channel carries a flow of 10 cfs/ft of width. Plot a curve of specific energy versus depth. Compute the minimum value of specific energy and the critical depth. What are the alternative depths for $Es = 5.0$?

**6.28** Determine an equivalent pipe for the system shown.

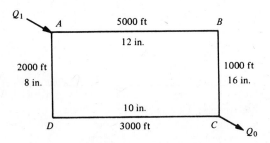

**6.29** From the diagram, compute (a) the total head loss from $A$ to $C$, (b) $P_a$ if $P_c$ = 25 psi, and (c) the flow in each line.

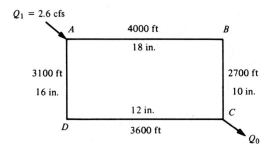

$Q_1 = 2.6$ cfs

**6.30** In the system shown, determine the flow distribution by a Hardy Cross analysis. Find the sizes of pipes $EF$ and $GF$. Use a Hazen—Williams $C$ of 100.

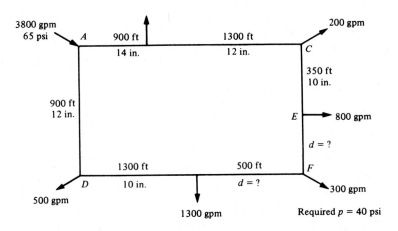

**6.31** Compute the flow in the lines of the diagram shown. Assume that the elevation at $A$ is 1500 ft and the elevation at $D$ is 1420 ft. The pressure at $D$ is 38 psi. Find the pressure at $A$.

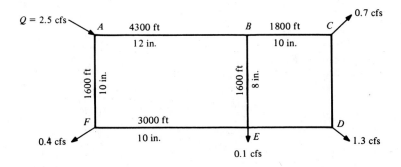

**6.32** In the diagram shown, $P_A$ = 70 psi and $P_C$ = 50 psi. Find the discharge in all lines and determine the size of lines *BC* and *CD*.

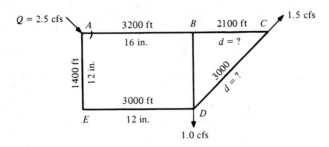

**6.33** Given the pipe layout shown, determine the length of an equivalent 18-in. pipe.

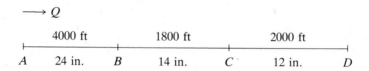

**6.34** Given the pipe layout shown, find the diameter of an equivalent 3000-ft pipe.

```
         1000 ft              1600 ft              2000 ft
Q ──→ ├────────────┼──────────────────┼────────────────────┤
      A   18 in.   B      14 in.       C      12 in.        D
```

**6.35** Given the following average hourly demand rates in gallons per minute, find the uniform 24-hr pumping rate and the required storage.

| 12 P.M. | 0 | 12 A.M. | 6300 |
|---|---|---|---|
| 1 A.M. | 1900 | 1 P.M. | 6500 |
| 2 | 1800 | 2 | 6460 |
| 3 | 1795 | 3 | 6430 |
| 4 | 1700 | 4 | 6500 |
| 5 | 1800 | 5 | 6700 |
| 6 | 1910 | 6 | 7119 |
| 7 | 3200 | 7 | 9000 |
| 8 | 5000 | 8 | 8690 |
| 9 | 5650 | 9 | 5220 |
| 10 | 6000 | 10 | 2200 |
| 11 | 6210 | 11 | 2100 |
| | | 12 | 2000 |

**6.36** Solve Prob. 6.35 if the period of pumping is from 6 A.M. to 6 P.M. only.

**6.37** Solve Prob. 6.35 by the method outlined in Fig. 6.24, assuming 24-hr pumping.

**6.38** Use the data given in Prob. 6.20 and add a fire-flow requirement of 4.7 mgd. Refer to Fig. 6.23 and assume that this flow requirement can be delivered to the city by the pumping station–tank system shown. The pipeline from the pumping station to the town is 30 in. in diameter and 5.6 mi long, and the pipeline from the town to the tank is 28 in. in diameter and 3.8 mi long. Elevations are 415 ft at the pump centerline, 570 ft at the ground surface in the town, 637 ft at the ground surface at the tank, and 700 ft at the bottom of the tank. The tank is 30 ft high and contains a total of 1.5 mil gal storage. Find the pressure in town during the peak average hour plus fire flow if the pumping station can maintain a maximum pressure of 130 psi and the tank is half full.

**6.39** Determine the total dynamic head of a pumping system where the total static head is 50 ft, the total friction head loss is 5 ft, and the velocity head is 10 ft.

**6.40** Calculate the horsepower requirements for the system described in Prob. 6.39 if the flow is 25 cfs.

**6.41** Plot the system head and the pump characteristic data given. At what point should the pump operate?

| Total dynamic head (ft) | 55 | 60 | 65 | 70 | 75 | 80 | 85 | 90 |
|---|---|---|---|---|---|---|---|---|
| Flow (gal/min) | | 200 | 500 | 700 | 850 | 975 | 1075 | 1200 | 1300 |

| Efficiency (%) | Horsepower (hp) | Total head (ft) | Flow (gal/min) |
|---|---|---|---|
| 22 | 8 | 95 | 225 |
| 37 | 12 | 93 | 400 |
| 49 | 17 | 90 | 600 |
| 57 | 20 | 87 | 800 |
| 63 | 22 | 84 | 1000 |
| 64 | 23 | 78 | 1200 |
| 62 | 24 | 72 | 1400 |
| 59 | 25 | 64 | 1600 |
| 54 | 26 | 49 | 1800 |

# REFERENCES

1. T. M. Walski, *Analysis of Water Distribution Systems* (New York: Van Nostrand Reinhold, 1984).
2. R. S. Gupta, *Hydrology and Hydraulic Systems* (Englewood Cliffs, NJ: Prentice-Hall, 1989).

3. H. W. Gehm and J. I. Bregman, editors, *Handbook of Water Resources and Pollution Control*, (New York: Van Nostrand Reinhold, 1976).

4. H. M. Morris, *Applied Hydraulics in Engineering* (New York: The Ronald Press, 1963).

5. "Flow of Fluids Through Valves, Fittings, and Pipe," *Tech. Paper No. 410*, Chicago, Crane Co., 1957.

6. N. Joukowsky, "Water Hammer" (translated by O. Simin), *Proc. Am. Water Works Assoc.* 24 (1904): 341–424.

7. J. M. Edmonston and E. E. Jackson, "The Feather River Project and the Establishment of the Optimum Hydraulic Grade Line for the Project Aqueduct," Metropolitan Water District of Southern California, Los Angeles (1963).

8. J. Hinds, "Economic Water Conduit Size," *Eng. News-Record* (January 1937).

9. G. E. Russell, *Hydraulics* (New York: Holt, Rinehart and Winston, 1957).

10. C. W. Reh, "Hydraulics of Water Distribution Systems," *University of Illinois Engineering Experiment Station Circular No. 75* (February 1962).

11. M. B. McPherson and J. V. Radziul, "Water Distribution Design and the McIlroy Network Analyzer," *Proc. Am. Soc. Civil Engrs.* 84, Paper 1588 (April 1958).

12. D. J. Wood, "Algorithms for Pipe Network Analysis and Their Reliability," Research Report No. 127, University of Kentucky, Water Resources Research Institute, Lexington, KY (1981).

13. T. M. Walski et al., "Battle of the Network Models: Epilogue," *J. Water Resources Planning Management,* 113(2) (March 1987): 191–203.

14. D. J. Wood and C. O. A. Charles, "Hydraulic Network Analysis Using Linear Theory," *Proc. Am. Soc. Civil Engrs., Hydraulics Div.* 98 (HY7) (July 1972).

15. R. Epp and A. G. Fowler, "Efficient Code for Steady-State Flows in Networks," *J. Hydraulics Division,* American Society of Civil Engineers, New York (January 1970).

16. T. M. Walski, "Sherlock Holmes Meets Hardy–Cross or Model Calibration in Austin, Texas," *J. Am. Water Works Assoc.* (March 1990): 34–38.

17. T. M. Walski, "Case Study: Pipe Network Calibration Issues," *J. Water Resources Planning Management* 112(2) (April 1986): 238–249.

18. D. J. Wood, "Pipe Network Analysis Programs: KYPIPE," Civil Engineering Software Center, University of Kentucky, Lexington, Kentucky (1987).

19. H. Cross, "Analysis of Flow in Networks of Conduits or Conductors," *University of Illinois Bull. No. 286* (November 1936).

20. Mun-Fong Lee, "Pipe Network Analysis," Non-Thesis Project, Department of Environmental Engineering Sciences, University of Florida, Gainesville, (November 1983).

21. M. B. McPherson, "Applications of System Analyzers: A Summary," *University of Illinois Engineering Experiment Station Circular No. 75* (February 1962).

22. M. B. McPherson, "Generalized Distribution Network Head Loss Characteristics," *Proc. Am. Soc. Civil Engrs., J. Hydraulics Div.* 86 (HY1) (January 1960).

23. W. Viessman, Jr., and C. Welty, *Water Management: Technology and Institutions,* (New York: Harper & Row, 1985).

24. M. S. McIlroy, "Direct Reading Electric Analyzer for Pipeline Networks," *J. Am. Water Works Assoc.* 42(4) (April 1950).

25. V. A. Appleyard and F. P. Linaweaver, Jr., "The McIlroy Fluid Analyzer in Water Works Practice," *J. Am. Water Works Assoc.* 49(1) (January 1957).

26. R. W. Jeppson, "Steady Flow Analysis of Pipe Networks," Utah State University, Department of Civil Engineering, Logan, UT (September 1974).

27. J. L. Gerlt and G. F. Haddix, "Distribution System Operation Analysis Model," *J. Am. Water Works Assoc.* (July 1975).

28. R. DeMoyer, Jr., and L. B. Horwitz, "Macroscopic Distribution System Modeling," *J. Am. Water Works Assoc.* (July 1975).

29. R. W. Jepson, *Analysis of Flow in Pipe Networks* (Ann Arbor, MI: Ann Arbor Science, 1977).

30. Freese and Nichols, Inc., "Water Distribution System Studies and Energy Evaluations," Fort Worth, TX (1983).

31. E. L. Thackston, Notes on Network Analysis, unpublished, Vanderbilt University (1982).

32. H. E. Babbitt, J. J. Doland, and J. L. Cleasby, *Water Supply Engineering* (New York: McGraw-Hill, 1962).

33. J. E. Kiker, "Design Criteria for Water Distribution Storage," *Public Works* (March 1964).

34. L. E. Ormsbee and T. M. Walski, "Identifying Efficient Pump Combinations," *J. Am. Water Works Assoc.* (January 1989): 30–34.

35. R. M. Olson, *Engineering Fluid Mechanics,* 3rd ed. (New York: IEP, 1973).

36. T. M. Walski, and L. E. Ormsbee, "Developing System Head Curves for Water Distribution Planning," *J. Am. Water Works Assoc.* (July 1989): 63–66.

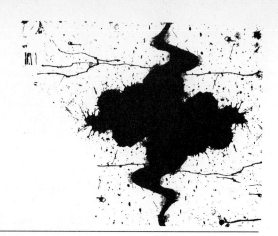

# Chapter 7
# Wastewater and Stormwater Systems

The collection and transportation of surface drainage, sewage, and other waste flows from residential, commercial, industrial, and rural sites pose problems different from those encountered in water supply. For example, sewage must be transported to treatment locations as quickly as possible to prevent the development of septic conditions. Furthermore, many waste flows contain solids that must be kept in suspension during transport periods. Most waste discharges exhibit fluctuations in time, but stormwater flows are especially subject to rapid rises and falls. Such temporal variations affect conveyance designs and must be carefully considered in hydraulic calculations. The varying quality of waste flows also requires that the types of conveyance used and the materials of which they are constructed be given special attention.

## Hydraulic Considerations

Wastewater systems are usually designed as open channels except where lift stations are required to overcome topographic barriers. The hydraulic problems associated with these flows are complicated in some cases by the quality of the fluid, the highly variable nature of the flows, and the fact that an unconfined or free surface exists. The driving force for open-channel flow is gravity. The forces retarding flow are derived from viscous shear along the channel bed.

### 7.1 UNIFORM FLOW

For the condition of uniform flow, the velocity in an open channel is usually determined by Manning's equation,

$$V = \frac{1.49}{n} R^{2/3} S^{1/2} \qquad (7.1)$$

which has been discussed in Chapter 6 (Eq. 6.3). Table 6.2 gives values of the roughness coefficient $n$ for various materials used in open channels. Figure 7.1 is a nomograph which facilitates the solution of the equation for various values of $n$.

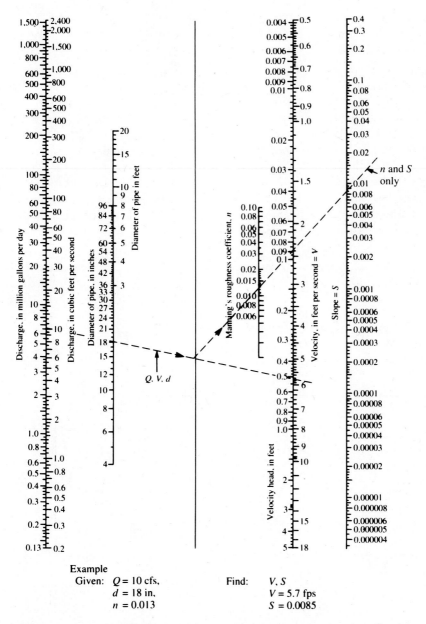

Example

Given: $Q = 10$ cfs,    Find:    $V, S$
$d = 18$ in,    $V = 5.7$ fps
$n = 0.013$    $S = 0.0085$

**Figure 7.1** Nomograph based on Manning's formula for circular pipes flowing full.

Many sanitary sewers and storm drains are circular in cross section, and thus it is cumbersome to compute values for the hydraulic radius and the cross-sectional area for conditions when the pipe is flowing partially full. Figure 7.2 is useful in computing partial-flow values from full-flow conditions. Example 7.1 illustrates this technique.

■ **EXAMPLE 7.1**

Given a discharge flowing full of 16 cfs and a velocity of 8 fps, find the velocity and depth of flow when $Q = 10$ cfs.

*Solution.* Enter the chart at 10/16 = 62.5% of value for full section. Obtain a depth of flow of 57.5% of full-flow depth and a velocity of $1.05 \times 8 = 8.4$ fps. ■

The depth at which uniform flow occurs in an open channel is termed the *normal depth $d_n$*. This depth can be computed by using Manning's equation for discharge after the cross-sectional area $A$ and the hydraulic radius $R$ have been translated into functions of depth. The solution for $d_n$ is often obtained through trial and error.

For a specified channel cross section and discharge, there are three possible values for the normal depth, all dependent on the channel slope.

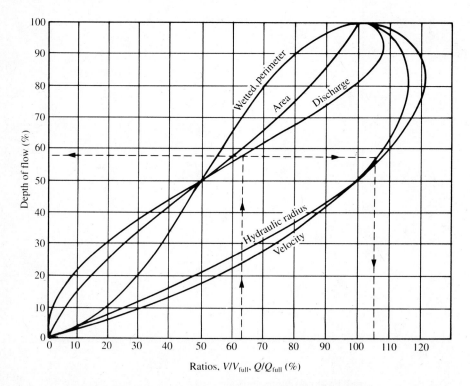

**Figure 7.2** Hydraulic elements of a circular section for constant $n$.

Uniform flow for a given discharge may occur at critical depth, at less than critical depth, or at greater than critical depth. Critical depth occurs when the specific energy is a minimum. Specific energy is defined as

$$E_s = d + \frac{V^2}{2g} \tag{7.2}$$

where  $d$ = depth of flow
$V$ = mean velocity

Flow at critical depth is highly unstable, and designs indicating flow at or near critical depth are to be avoided. For any value of $E_s$ above the minimum, two alternative depths of flow are possible. One is greater than critical depth, the other less than critical depth. The former case indicates subcritical flow, the latter supercritical flow. The critical depth for a channel can be found by taking the derivative of Eq. 7.2 with respect to depth, setting this equal to zero, and solving for $d_c$. For mild slopes, the normal depth is greater than $d_c$ and subcritical flow prevails. On steep slopes, the normal depth is less than the critical depth and flow is supercritical. Once the critical depth has been computed, critical velocity is easily obtained. The critical velocity for a channel of any cross section can be shown to be

$$V_c = \sqrt{g \frac{A}{B}} \tag{7.3}$$

where  $V_c$ = critical velocity
$A$ = cross-sectional area of the channel
$B$ = width of the channel at the water surface

The critical slope can then be found by using Manning's equation.

In practice, uniform flows are encountered only in long channels after a transition from nonuniform conditions. Nevertheless, a knowledge of uniform-flow hydraulics is extremely important, as numerous varied flow problems are solved through partial applications of uniform-flow theory. A basic assumption in gradually varied flow analyses, for example, is that energy losses are considered to be the same as for uniform flow at the average depth between two sections along the channel that are closely spaced.

## 7.2 GRADUALLY VARIED FLOW AND SURFACE PROFILES

Gradually varied flow results from gradual changes in depth which take place over relatively long reaches of a channel. Abrupt changes in the flow regime are classified as rapidly varied flow. Problems in gradually varied flow are widespread and represent the majority of flows in natural open channels and many of the flows in man-made channels. They can be caused by such factors as change in channel slope, cross-sectional area, or roughness or by obstructions to flow, such as dams, gates, culverts, and bridges. The pressure distribution in gradually varied flow is hydrostatic, and the streamlines are considered to be approximately parallel.

The significance of gradually varied flow problems may be illustrated by considering the following case. Assume that the maximum design flow for a uniform rectangular canal will occur at a depth of 8 ft under uniform-flow conditions. For these circumstances the requisite canal depth, including 1 ft of freeboard, will be 9 ft. Now consider that a gate is placed at the lower end of this canal. Assume that at maximum flow the depth immediately upstream from the gate will have to be 12 ft in order to produce the required flow through the gate. The depth of flow will then begin decreasing gradually in an upstream direction and approach the uniform depth of 8 ft. Obviously, unless the depth of flow is known all along the channel, the channel depth cannot be designed.

Determination of the water-surface profile for a given discharge in an open channel is of importance in solving many engineering problems. There are 12 classifications of water-surface profiles, or *backwater curves* [1]. Figure 7.3 illustrates several of these and also a typical change which might take place in the transition from one type of flow regime to another. For any channel, the applicable backwater curve will be a function of the relationship between the actual depth of flow and the normal and critical depths for the

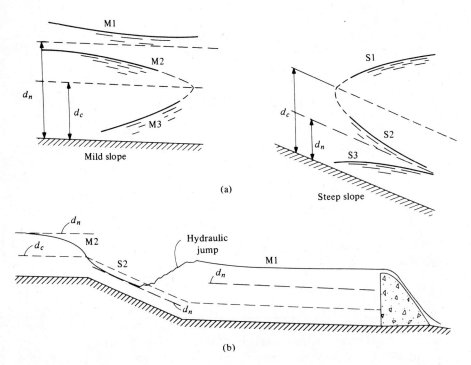

**Figure 7.3** Gradually varied flow profiles.

channel. Often it is helpful to sketch the type curves for a given problem before attempting to effect an actual solution. In doing this, the type curves given in Fig. 7.3 are useful. Additional type curves are presented by Woodward and Posey [1].

Numerous procedures have been proposed for computing backwater curves. The one to be discussed here is known as the direct-step method. Referring to Fig. 7.4, the energy equation may be written

$$Z_1' + d_1 + \frac{V_1^2}{2g} = d_2 + \frac{V_2^2}{2g} + H_f$$

A rearrangement of this equation yields

$$\left(\frac{V_2^2}{2g} + d_2\right) - \left(\frac{V_1^2}{2g} + d_1\right) = Z_1' - H_f$$

or

$$E_2 - E_1 = S_c L - \overline{S}_e L$$

so

$$L = \frac{E_2 - E_1}{S_c - \overline{S}_e} \tag{7.4}$$

where   $E_2, E_1$ = values of specific energy at sections 1 and 2
$S_c$ = slope of the channel bottom
$\overline{S}_e$ = slope of the energy gradient

The value of $S_e$ is obtained by assuming that (1) the actual energy gradient is

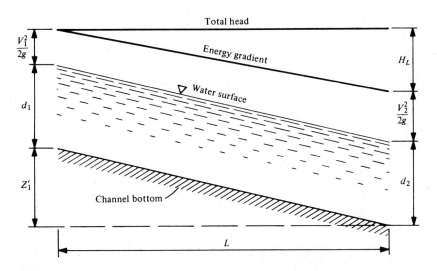

**Figure 7.4** Definition sketch for the direct step method of computing backwater curves.

the same as that obtained for uniform flow at a velocity equal to the average of the velocities at sections 1 and 2 (this can be found using Manning's equation) or (2) the slope of the energy gradient is equal to the mean of the slopes of the energy gradients for uniform flow at the two sections. The procedure in using Eq. 7.4 is to select some starting point where the depth of flow is known. A second depth is then selected in an upstream or downstream direction, and the distance from the known depth to this point is computed. Then, using the second depth as a reference, a third is selected and another length increment calculated and so on. The results obtained will be reasonably accurate provided the selected depth increments are small, since the energy-loss assumptions are fairly accurate under these conditions.

### ■ EXAMPLE 7.2

Water flows in a rectangular concrete channel 10 ft wide, 8 ft deep, and inclined at a grade of 0.10%. The channel carries a flow of 245 cfs and has a roughness coefficient $n$ of 0.013. At the intersection of this channel with a canal, the depth of flow is 7.5 ft. Find the distance upstream to a point where normal depth prevails, and determine the surface profile.

*Solution.*   Using Manning's equation and the given value of discharge, the normal depth is found to be 4 ft. Critical depth $y_c$ is determined by solving Eq. 7.3 after rearranging and substituting $by_c$ for $A$ and $b$ for $B$, where $b$ equals the width of the rectangular channel. Then

$$V_c = \sqrt{g\left(\frac{by_c}{b}\right)}$$

$$= \sqrt{gy_c}$$

$$y_c = \frac{V_c^2}{g} = \frac{Q^2}{A^2 g}$$

$$y_c^3 = \frac{Q^2}{gb^2} \tag{7.5}$$

$$y_c = \sqrt[3]{\frac{Q^2}{gb^2}}$$

Substituting the given values for $Q$ and $b$ yields

$$y_c = \sqrt[3]{\frac{(245)^2}{32.2 \times (10)^2}}$$

and

$$y_c = 2.65 \text{ ft}$$

Since $y > y_n > y_c$, an M1 profile will depict the water surface. The calculations then follow the procedure indicated in Table 7.1.

The distance upstream from the junction to the point of normal depth is 6788 ft.

**TABLE 7.1  CALCULATIONS FOR THE SURFACE PROFILE OF EXAMPLE 7.2**

| $y$ | $A$ | $V$ | $\dfrac{V^2}{2g}$ | $E = d + \dfrac{V^2}{2g}$ | $P$ | $R$ | $R^{4/3}$ | $S_e = \dfrac{V^2}{(1.49)^2 R^{4/3}/n^2}$ | $\bar{S}_e$ | $\Delta E$ | $S_c - \bar{S}_e$ | $\Delta L = \dfrac{\Delta E}{S_c - \bar{S}_e}$ |
|---|---|---|---|---|---|---|---|---|---|---|---|---|
| 7.5 | 75 | 3.27 | 0.166 | 7.666 | 25 | 3.0 | 4.33 | 0.000188 | 0.000207 | 0.475 | 0.000793 | 598 |
| 7.0 | 70 | 3.50 | 0.191 | 7.191 | 24 | 2.91 | 4.15 | 0.000225 | 0.000279 | 0.932 | 0.000721 | 1292 |
| 6.0 | 60 | 4.08 | 0.259 | 6.259 | 22 | 2.73 | 3.82 | 0.000333 | 0.000436 | 0.885 | 0.000564 | 1570 |
| 5.0 | 50 | 4.90 | 0.374 | 5.374 | 20 | 2.50 | 3.40 | 0.000539 | 0.000762 | 0.792 | 0.000238 | 3328 |
| 4.0 | 40 | 6.12 | 0.582 | 4.582 | 18 | 2.22 | 2.90 | 0.000984 | | | | |

$$L = \Sigma\,\Delta L = 6788$$

The surface profile can be plotted by using the values in Table 7.1 for $y$ and $\Delta L$. In practice, greater accuracy could be obtained by using smaller depth increments, such as 0.25 ft, but the computational procedure would be the same.    ∎

## 7.3  VELOCITY CONSIDERATIONS

Both maximum and minimum velocities are prescribed for waste transportation systems. Minimum velocities are set to ensure that suspended matter does not settle out in the conduit, while maximum velocities are set to prevent erosion of the channel material.

Normally, velocities in excess of 20 fps are to be avoided in concrete or tile sewers, and whenever possible velocities of 10 fps or less should be used. Specially lined inverts (pipe bottoms) are sometimes employed to combat channel erosion. The average value of the channel shearing stress is related to both erosion and sediment deposition.

Using the Chézy equation for velocity, Fair and Geyer [2] have shown that the minimum velocity for self-cleansing $V_m$ can be determined according to

$$V_m = C \sqrt{\frac{K(\gamma_s - \gamma)}{\gamma} d} \tag{7.6}$$

where $\gamma_s$ is the specific weight of the particles, $\gamma$ is the specific weight of water, $d$ is the particle diameter, and $C$ is the Chézy coefficient (equal to $1.49R^{1/6}/n$ as evaluated by Manning). The value of $C$ is selected with consideration given to the containment of solids in the flow. The value of $K$ must be found experimentally and appears to range from 0.04 for the initiation of scour to more than 0.8 for effective cleansing [2].

If nonflocculating particles have mean diameters between 0.05 and 0.5 mm, the minimum velocity may be determined [3] using

$$V_m = \left[ 25.3 \times 10^{-3} g \frac{d(\rho_s - \rho)}{\rho} \right]^{0.816} \left( \frac{D\rho_m}{\mu_m} \right)^{0.633} \tag{7.7}$$

where  $V_m$ = minimum velocity, fps
$\quad\quad d$ = mean particle diameter, ft
$\quad\quad D$ = pipe diameter, ft
$\quad\quad \rho_m$ = mean mass density of the suspension, pcf
$\quad\quad \rho_s$ = particle density, pcf
$\quad\quad \mu_m$ = viscosity of the suspension, lbm/ft·sec
$\quad\quad \rho$ = liquid mass density, pcf

The value of $\rho_m$ may be determined using

$$\rho_m = X_v \rho_s + (1 - X_v)\rho \tag{7.8}$$

where $X_v$ is the volumetric fraction of the suspended solids. Ordinarily, minimum velocities are 2 and 3 fps for sanitary sewers and storm drains, respectively.

# Design of Sanitary Sewers

The design of sanitary sewers involves (1) the application of the principles given in Chapter 4 to the estimation of design flows and (2) the application of hydraulic engineering principles to devise appropriate conveyances for transporting these flows [2, 4, 5]. Most sewers are designed for a capacity life of 25–50 yr. These conveyances must be capable of handling peak and minimum flows without the deposition of suspended solids. In general, circular sanitary sewers are designed to flow at between one-half and full depth when carrying their design discharges. Ordinarily they should not be designed to run full since some airspace is desirable for ventilation and suppression of sulfide generation. Sewers at system extremities should be designed to flow about half-full at peak flow. Large interceptor and trunk sewers are less subject to flow variations than are small collecting sewers, and greater depths of flow are justified for them. Gupta notes that pipes up to 16 in. (400 mm) in diameter are usually designed to flow half-full; those between 16 and 35 in. (900 mm) to flow two-thirds full; and larger pipes are usually designed to flow at three-quarters to full depth [4]. Once the flow the sewer must carry has been determined, a pipe size and slope must be established. Since sewers are commonly located beneath roads and streets, their slopes are generally made to conform to street slopes. This practice also tends to minimize excavation costs. Where street slopes do not permit generating minimum carrying velocities (2 ft/sec), sewer grades are set accordingly. Table 7.2

**TABLE 7.2**   SLOPES REQUIRED TO MAINTAIN VELOCITIES OF 2 ft/sec[a]

| Q (cfs) | Slope (ft/1000) |
|---------|-----------------|
| 0.1 | 9.2 |
| 0.2 | 6.1 |
| 0.3 | 4.8 |
| 0.4 | 4.1 |
| 0.6 | 3.22 |
| 0.8 | 2.73 |
| 1.0 | 2.39 |
| 1.5 | 1.89 |
| 2.0 | 1.59 |
| 3.0 | 1.26 |
| 4.0 | 1.06 |

[a] For circular pipes where the flow is not less than 0.1% and not more than 95% of capacity [6].

summarizes minimum slopes for a range of flow values. In cul-de-sacs and other short dead-end street sections, the designer will often find it impossible to meet minimum velocity requirements. Such pipes may require periodic flushing to scour accumulated sediments.

## 7.4  HOUSE AND BUILDING CONNECTIONS

Connections from the main sewer to houses or other buildings are commonly constructed of vitrified clay, concrete, or asbestos cement pipe. Building connections are usually made on about a 2% grade with 6-in. or larger pipe. Grades less than 1% are to be avoided, as are extremely sharp grades.

## 7.5  COLLECTING SEWERS

Collecting sewers gather flows from individual buildings and transport the material to an interceptor or main sewer. Local standards usually dictate the location of these sewers, but they are commonly located under the street paving on one side of the storm drain, which is usually centered. The collecting sewer should be capable of conveying the flow of the present and anticipated population of the area it is to serve. Design flows are the sum of the peak domestic, commercial, and industrial flows and infiltration. The collecting sewer must transport this design flow when flowing full. Grades requiring minimum excavation while maximum and minimum restrictions on velocity are satisfied are best. Manholes are normally located at changes in direction, grade, pipe size, or at intersections of collecting sewers. For 8-, 10-, or 12-in. sewers, manholes spaced no farther than about 400 ft apart permit inspection and cleaning when necessary. For larger sizes, the maximum spacing can be increased. The minimum size pipe allowed by most codes is 8 in. Note that where changes in sewer direction occur and/or where flows combine, usually at manholes, head losses occur. When flow velocities are in the usual range for sewers, a drop of about 0.1 ft or 30 mm across a manhole is usually sufficient to account for this (Fig. 7.5). For large sewers such as interceptors and trunks, head losses may require greater adjustments (see Section 7.18 and Fig. 7.19). At junctions where sewer size increases, common practice is to keep the crowns of the incoming and outgoing pipes, or the 0.8 depth points of these pipes, at the same elevation. A typical plan and profile of a residential sewer are given in Fig. 7.5.

## 7.6  INTERCEPTING SEWERS

Intercepting sewers are expected to carry flows from the collector sewers in the drainage basin to the point of treatment or disposal of the wastewater. These sewers normally follow valleys or natural streambeds of the drainage area. When the sewer is built through underdeveloped areas, precaution

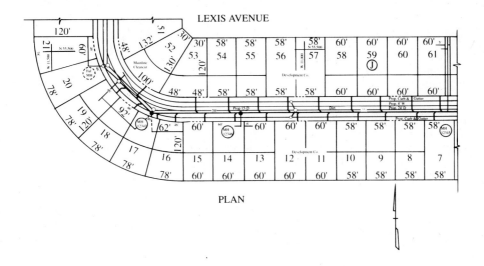

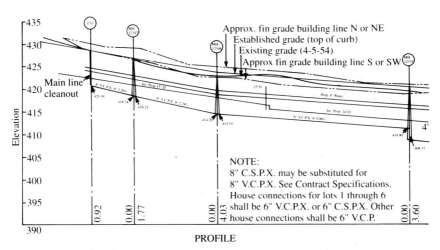

**Figure 7.5** Typical sewer plan and profile.

should be taken to ensure the proper location of manholes for future connections. For 15- to 27-in. sewers, manholes are constructed at least every 600 ft; for larger sewers, increased spacing is common. Manholes or other transition structures are usually built at every change in pipe size, grade, or alignment. For large sewers, horizontal or vertical curves are sometimes constructed. Grades should be designed so that the criteria regarding maximum and minimum velocities are satisfied.

## 7.7 MATERIALS

Collecting and intercepting sewers are constructed of asbestos cement sewer pipe, cast-iron pipe, concrete pipe, vitrified clay pipe, brick, plastic or PVC, and bituminized fiber pipe. Care should be exercised in designing the system so that permissible structural loadings are not exceeded for the material selected. Information on pipe loading is readily available from the Clay Products Association, the Portland Cement Association, the Cast Iron Pipe Manufacturers, and others.

## 7.8  SYSTEM LAYOUT

The first step in designing a sewerage system is to establish an overall system layout that includes a plan of the area to be sewered, showing roads, streets, buildings, other utilities, topography, soil type, and the cellar or lowest floor elevation of all buildings to be drained. Where part of the drainage area to be served is undeveloped and proposed development plans are not yet available, care must be taken to provide adequate terminal manholes that can later be connected to the constructed system serving the area. If the proposed sewer connects to an existing one, an accurate location of the existing terminal manhole, giving invert elevation, size, and slope, is essential.

On the drainage area plan just described, a tentative layout of collecting sewers and the intercepting sewer or sewers should be made. When feasible, the sewer location should minimize the length required while providing service to the entire area. Length may be sacrificed if the shorter runs would require costlier excavation. Normally, the sewer slope should follow the ground surface so that waste flows can follow the approximate path of the area's surface drainage. In some instances, it may be necessary to lay the sewer slope in opposition to the surface slope, or to pump wastes across a drainage divide. This situation can occur when a developer buys land lying in two adjacent drainage basins and, for economic or other reasons, must sewer the whole tract through only one basin.

Intercepting sewers or trunk lines are located to pass through the lowest point in the drainage area (the outlet) and to extend through the entire area to the drainage divide. Normally, they follow major natural drainage ways and are located in a designated right-of-way or street. Land slopes on both sides should be toward the intercepting sewer. Lateral or collecting sewers are connected to the intercepting sewer and proceed upslope to the drainage divide. Collecting sewers should be located in all streets or rights-of-way within the area to provide service throughout the drainage basin. Figure 7.6 illustrates a typical sewer layout. Note that the interceptor sewer is located in the stream valley and collector sewers transport flows from the various tributary areas to the interceptor.

A general guideline for sewer system layout and design has been proposed by Thackston. Basic steps include the following [7]:

Obtain or develop a topographic map of the area to be served.

Locate the drainage outlet. This is usually near the lowest point in the area and is often along a stream or drainage way.

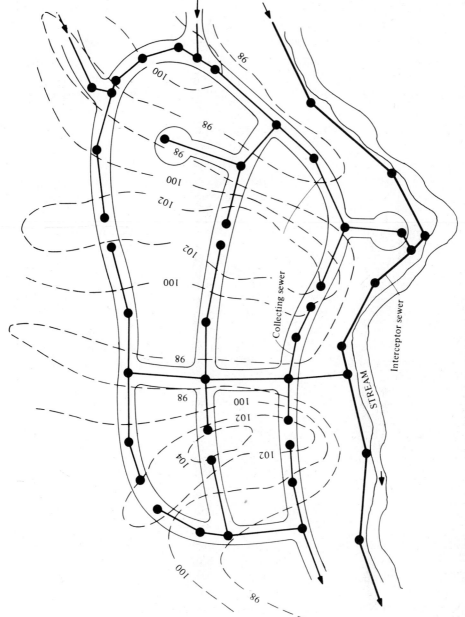

**Figure 7.6** Typical layout for the design of a sewerage system.

Sketch in a preliminary pipe system to serve all of the contributors (Fig. 7.6).

Pipes must be located so that all users or future users can readily tap on. They must also be located so as to provide access for maintenance and thus are ordinarily placed in streets or other rights-of-way.

Insofar as practical, sewers should follow natural drainage ways so as to minimize excavation and pumping requirements. Large trunk sewers are usually constructed in low-lying areas closely paralleling streams or channels. In general, pipes should cross contours at right angles.

Establish preliminary pipe sizes. Eight inches (usually the minimum allowable) can serve several hundred residences even at minimal grades.

Revise the layout so as to optimize flow-carrying capacity at minimum cost. Pipe lengths and sizes should be kept as small as possible, pipe slopes should be maximized within tolerances for velocity, excavation depth should be minimized, and the number of appurtenances should be kept as small as possible. Traditional hydraulic design procedures combined with the use of optimization techniques can be utilized in this process [8].

Try to avoid pumping across drainage boundaries. Pumping stations are costly and add maintenance problems. Furthermore, energy costs for pumping can be very significant. A balance must be made between excavation and pumping. While no hard-and-fast rule can be given, pumping should be considered only when excavations exceeding about 20–40 ft are contemplated.

## 7.9 HYDRAULIC DESIGN

### Open-Channel Flow

The hydraulic design of a sanitary sewer can be carried out in a systematic manner. Figure 7.7 is typical of the way in which data can be organized to facilitate computations. The essential elements of hydraulic design require that peak design flows be carried by the pipe at velocities great enough to prevent sedimentation, yet small enough to prevent erosion. To minimize head losses at transitions and eliminate backwater effects, the hydraulic gradient must not change abruptly or slope in a direction adverse to the flow at changes in horizontal direction, pipe size, or quantity of flow. Sewers are usually designed to follow closely the grade of the ground surface or street paving under which they are laid. In addition, the depth of cover should be kept as close to the minimum (below the frost line) as effective hydraulic design will permit to minimize excavation costs. The sewer cannot conflict

JOB NAME ______________________ J. O. ________ DISTRICT ________ DATE ________

PREPARED BY ______________________ OF ______________ (FIRM) SHEET ________ OF ________

CHECKED BY ______________________

| Area Symbol | Area Acres | Population Density | | Population | | Nondomestic Average Flow (mgd) | | Average Sewage Flow (mgd) | | Peak Flow (mgd) | | Infiltration (mgd) | Peak Industrial Waste Flow (mgd) | | Design Flow (mgd) | | Approx. Grade | Tentative Size | Velocity | |
|---|---|---|---|---|---|---|---|---|---|---|---|---|---|---|---|---|---|---|---|---|
| | | Pres. | Ult. | Pres. | Ult. | Pres. | Ult. | Pres. | Ult. | Pres. | Ult. | | Pres. | Ult. | Pres. | Ult. | | | + | ++ |
| (1) | (2) | (3) | (4) | (5) | (6) | (7)* | (8)* | (9) | (10) | (11) | (12) | (13) | (14) | (15) | (16) | (17) | (18) | (19) | (20) | (21) |
| | | | | | | | | | | | | | | | | | | | | |

*Notes*:
Where columns 7, 8, 14, and 15 are used, show calculations on back of form.
* Includes average sewage flow from institutional, industrial, and commercial establishment.
\+ Velocity based upon present average sewage flow.
++ Velocity based upon ultimate peak design flow.

| Revised | |
|---|---|
| Date | By |
| | |
| | |
| | |

| BALTIMORE COUNTY DEPARTMENT OF PUBLIC WORKS | APPROVED<br>DIRECTOR<br>DIVISION OF ENGINEERING<br>DATE | REVIEWED | STANDARD SANITARY SEWER DESIGN<br>FLOW TABULATION FORM | |
|---|---|---|---|---|

**Figure 7.7** Typical computation sheet for sanitary sewer design. (Courtesy of the Baltimore County, Maryland, Department of Public Works.)

with other subsurface utilities or structures and cannot usually be laid in the same trench as a water main.

Pipe sizes are determined in the following manner. A profile of the proposed sewer route is drawn and the hydraulic gradient at the downstream end of the sewer noted (normally the elevation of the hydraulic gradient of the sewer being met). Where discharge is to a treatment plant or open body of water, the hydraulic gradient is the elevation of the free water surface at this point. At the beginning elevation of the hydraulic gradient, a tentative gradient approximately following the ground surface is drawn upstream to the next point of control (usually a manhole). Exceptions to this occur when the surface slope is less than adequate to provide cleansing velocities, where obstructions preclude using this slope, or where adequate cover cannot be maintained. Using the tentative gradient slope, a pipe size is then selected that comes closest to carrying the design flow at the desired depth. Usually, a standard size pipe will not be found that will carry the flow at the exact depth and gradient investigated. It is common practice to then select the next largest pipe size, modify the slope, or both. The choice depends on a comparison of pipe cost savings versus excavation costs and on local conditions, such as the placement of other utilities in the right-of-way. But regardless of velocity and pipe size calculations, pipe diameter is never decreased on downstream reaches because this can lead to sediment accumulation and blockage at the point of reduction.

Hydraulic computations using Manning's equation (Eq. 7.1, Fig. 7.1) for full pipe flow are straightforward. Commonly, however, flow at less than full is encountered or desired in practice. In such cases, relationships involving geometric properties (Table 7.3) and hydraulic elements (Fig. 7.2) of circular sections facilitate calculations. Their use is illustrated in the following examples.

■ **EXAMPLE 7.3**

Find the peak hourly flow in mgd and $m^3/s$ for an 800-acre urban area having the following features: domestic flows 90 gpcd, commercial flows 15 gpcd, infiltration 600 gpd/acre, and population density 20 persons per acre.

*Solution.*  Calculate the average daily flows for each contributor.

    Domestic:    90 × 20 × 800  = 1.44 mgd
    Commercial: 15 × 20 × 800  = 0.24 mgd
    Infiltration:      600 × 800  = 0.48 mgd

    Total = 1.44 + 0.24 + 0.48 = 2.16 mgd

Referring to Table 4.8, select a peak hour to average day ratio of 3.0.

    Peak hourly flow = 3.0 × 2.16    = 6.48 mgd

                = 6.48 × 0.044 = 0.29 $m^3/s$    ■

**TABLE 7.3**   GEOMETRIC RELATIONSHIPS
FOR CIRCULAR PIPES

| $d/d_m$ | $R/d_m$ | $AR^{2/3}/d_m^{8/3}$ |
|---|---|---|
| 0.01 | 0.0066 | 0.0000 |
| 0.05 | 0.0326 | 0.0015 |
| 0.10 | 0.0635 | 0.0065 |
| 0.15 | 0.0929 | 0.0152 |
| 0.20 | 0.1206 | 0.0273 |
| 0.25 | 0.1466 | 0.0427 |
| 0.30 | 0.1709 | 0.0610 |
| 0.35 | 0.1935 | 0.0820 |
| 0.40 | 0.2142 | 0.1050 |
| 0.45 | 0.2331 | 0.1298 |
| 0.50 | 0.2500 | 0.1558 |
| 0.55 | 0.2649 | 0.1825 |
| 0.60 | 0.2776 | 0.2092 |
| 0.65 | 0.2881 | 0.2358 |
| 0.70 | 0.2962 | 0.2608 |
| 0.75 | 0.3017 | 0.2840 |
| 0.80 | 0.3042 | 0.3045 |
| 0.85 | 0.3033 | 0.3212 |
| 0.90 | 0.2890 | 0.3324 |
| 0.95 | 0.2864 | 0.3349 |
| 1.00 | 0.2500 | 0.3117 |

$d_m$ = full-flow depth
$d$ = actual depth of flow
$R$ = hydraulic radius
$A$ = area of flow

### ■ EXAMPLE 7.4

A sewer with a flow of 10 cfs enters manhole 6. The distance downstream to manhole 7 is 400 ft. The finished street surface elevation at manhole 6 is 166.3 ft, and at manhole 7 it is 164.3 ft. Note that the crown of the pipe must be at least 6 ft below the street surface at each manhole. For a Manning $n = 0.013$, find the required standard size pipe to carry the flow under (a) full flow and (b) one-half of full flow. The flow velocity must be at least 2 ft/sec and must not exceed 10 ft/sec.

*Solution.*   The fall of the sewer in 400 ft is 2 ft if the street grade is used as the slope. The initial assumption for $S$ is 0.005 ft/1000. If this slope is not sufficient to sustain a velocity of 2 fps, it will be modified later. In like manner, if the calculated velocity exceeds 10 fps the slope will have to be reduced.

a. Using the nomograph in Figure 7.1 and the values $Q = 10$ and $S = 0.005$, we determine that a 21-in.-diameter pipe is required. This pipe will support a discharge somewhat higher than the design value of 10, but it is the closest standard size. The velocity associated with a 21-in. pipe flowing full at a slope of 0.005 is between 4 and 5 fps and thus meets the design criteria. For full flow, the solution is to choose a 21-in. pipe laid at a slope of 0.5 ft/100 ft.

b. For flow at $d = 0.5d_m$, Manning's equation, Fig. 7.2, and Table 7.3 are used. From Fig. 7.2, it can be seen that for half-full depth, the velocity is the same as for full depth, and geometrically $A$ and $R$ are one-half their full-depth values.

The pipe size for one-half flow is calculated from the equation

$$Q = 1.49/nAR^{2/3}S^{1/2}$$

by inserting known values for $Q$, $n$, and $S$. From Table 7.3 note that $AR^{2/3}/d_m^{8/3} = 0.156$ for $d/d_m = 0.5$, so the equation can be solved for $d_m$:

$$10 = (1.49/0.013)(0.156d_m^{8/3})(0.005)^{0.5}$$

$$d_m^{8/3} = 7.91 \quad \text{and} \quad d_m = 2.17 \text{ ft}$$

Select the closest standard pipe size of 27 in. diameter.

From Fig. 7.1, it can be seen that for a 27-in. pipe laid at a slope of 0.005, the velocity at full flow (also at halfflow) is 5.5 fps. This satisfies the specifications.

The flow in the 27-in. pipe at half-full conditions is calculated to determine how close it comes to the design flow of 10 cfs:

$$Q = (1.49/0.013)(0.156)(0.005)^{0.5}(d_m)^{8/3}$$

$$= 10.9 \text{ cfs}$$

Thus, the 27-in. pipe is flowing at slightly less than half-depth under design conditions of $Q = 10$ cfs. For this pipe and slope, the full-flow is $10.9 \times 2 = 21.8$ cfs. From Fig. 7.2 with $Q/Q_{full} = 10/21.8$, it can be seen that the actual flow depth under design conditions is about $0.48d_m$. ∎

### ■ EXAMPLE 7.5

A sewer system is to be designed for the urban area of Fig. 7.8. Street elevations at manhole locations are shown on the figure. It has been determined that the population density is 40 persons per acre and that the sewage contribution per capita is 100 gpd. In addition, there is an infiltration component of 600 gallons per acre per day (gpad). Local regulations require that no sewers be less than 8 in. in diameter. Assume an $n$ value of 0.013.

*Solution.* The design procedure follows a series of steps and the format of Tables 7.4 and 7.5.

1. Determine manhole locations. These are already indicated for the example on Fig. 7.8, but in practice they are determined on the basis of changes in direction, junctions of pipes, and maximum lengths for various pipe sizes, as discussed earlier. Pipe lengths are tabulated in column 3 of Table 7.4.
2. Determine street elevations at manhole locations. These are given on Fig. 7.8, but they would be obtained from street profiles of the type shown on Fig. 7.5.
3. Using the plan showing the location of manholes, determine the distance between them. For the example, the lengths are given in column 3 of Table 7.4.
4. Calculate the street slopes between manholes. This is done by taking the differences in elevations between manholes and dividing them by the distances between the manholes. For example, the slope between M-6 and M-5 is

$$S = (112.19 - 109.23)/470 = 0.0063 \text{ ft/ft}$$

Calculated slopes are shown in column 5 of Table 7.5.
5. Determine the size of contributing areas to the sewer segments. This is done

| Manhole | Street El. |
|---------|-----------|
| 1 | 101.3 |
| 2 | 104.18 |
| 3 | 105.33 |
| 4 | 107.25 |
| 5 | 109.23 |
| 6 | 112.19 |
| 7 | 116.6 |
| 8 | 112.04 |
| 9 | 115.04 |
| 10 | 117.46 |
| 11 | 113.77 |
| 12 | 110.29 |
| 13 | 115.8 |
| 14 | 111.92 |
| 15 | 108.58 |
| 16 | 116.37 |
| 17 | 112.57 |
| 18 | 108.89 |

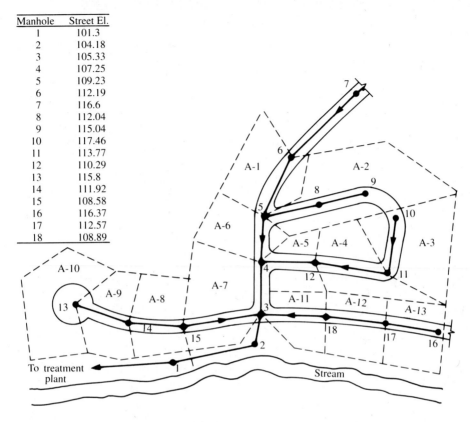

**Figure 7.8** Sewer system for Example 7.5.

by using finished-grade topography and knowledge of the building connection schemes for contributing elements, such as houses, to the sewer segments. For the example, the contributing area boundaries are shown on Fig. 7.8, and the area sizes in acres, both incremental and cumulative, are tabulated in Table 7.4, columns 4–7.

6. Calculate the flow per contributing acre for each of the 13 areas. In this example, the per-acre flows are all the same, but they can be highly variable. They are calculated by using procedures described in Sections 4.9 to 4.11 and in Example 7.3. For this problem, the per-acre flows are calculated as follows:

   **a.** residential flow = (40 per/acre) × (100 gpcd) = 4000 gpad

   **b.** infiltration = 600 gpad

   avg. daily flow = 4600 gpad

   **c.** assuming the design flow = 3 × avg. daily flow, peak flow = 4600 × 3 = 13,800 gpad

   **d.** converting to mgd, the design flow per acre is 0.0138 mgd/a. This value is entered in column 8 of Table 7.4.

7. Calculate the sewage flows in pipes. This is done by multiplying the number of contributing acres by 0.0138 (the per-acre contribution). The

**TABLE 7.4** CALCULATION OF SEWAGE FLOWS FOR EXAMPLE 7.5

| 1 | 2 | | 3 | 4 | 5 | 6 | 7 | 8 | 9 | 10 |
|---|---|---|---|---|---|---|---|---|---|---|
| | Manhole | | | | | | | | | |
| Pipe no. | From | To | Pipe lgth. (ft) | Incr. area (acres) | Cum. for br. (acres) | Incr. Mn. (acres) | Cum. Mn. (acres) | Sewage (mgd/a) | Flow (mgd) | Flow (cfs) |
| 1 | 7 | 6 | 630 | — | — | — | 87 | 0.0138 | 1.20 | 1.86 |
| 2 | 6 | 5 | 470 | — | — | 5.1 | — | 0.0138 | — | — |
| — | 6 | 5 | — | — | — | — | 92.1 | 0.0138 | 1.27 | 1.97 |
| 3 | 9 | 8 | 390 | 12.1 | 12.1 | — | — | 0.0138 | 0.17 | 0.26 |
| 4 | 8 | 5 | 385 | — | — | 4.8 | — | 0.0138 | 0.17 | 0.26 |
| 5 | 5 | 4 | 330 | — | — | — | — | 0.0138 | — | — |
| — | 5 | 4 | — | — | — | — | 109 | 0.0138 | 1.50 | 2.33 |
| 6 | 10 | 11 | 410 | 8.7 | 8.7 | — | — | 0.0138 | 0.12 | 0.19 |
| 7 | 11 | 12 | 400 | 6.3 | 15 | — | — | 0.0138 | 0.21 | 0.32 |
| 8 | 12 | 4 | 380 | 4.7 | 19.7 | — | — | 0.0138 | 0.27 | 0.42 |
| 9 | 4 | 3 | 370 | — | — | — | 128.7 | 0.0138 | 1.78 | 2.75 |
| 10 | 16 | 17 | 380 | 5 | 5 | — | — | 0.0138 | 0.07 | 0.11 |
| 11 | 17 | 18 | 400 | 4.9 | 9.9 | — | — | 0.0138 | 0.14 | 0.21 |
| 12 | 18 | 3 | 405 | 4.3 | 14.2 | — | — | 0.0138 | 0.20 | 0.30 |
| 13 | 13 | 14 | 400 | 13.1 | 13.1 | — | — | 0.0138 | 0.18 | 0.28 |
| 14 | 14 | 15 | 380 | 5.3 | 18.1 | — | — | 0.0138 | 0.25 | 0.39 |
| 15 | 15 | 3 | 411 | 9.7 | 28.1 | — | — | 0.0138 | 0.39 | 0.60 |
| 16 | 3 | 2 | 230 | — | — | — | 171 | 0.0138 | 2.36 | 3.66 |
| 17 | 2 | 1 | 600 | — | — | — | 171 | 0.0138 | 2.36 | 3.66 |

Column 4 = incremental area contributing to pipe.
Column 5 = cumulative area for branches entering collecting sewer.
Column 6 = incremental area contributing directly to collecting sewer.
Column 7 = cumulative area contributing to a sewer.
Column 8 = calculated contributing sewage flow in mgd/acre.
Column 9 = pipe flow in mgd [0.0138 × (area)].
Column 10 = values in column 9 converted to cfs [(col. 9) × (1.55)].

**TABLE 7.5  CALCULATION OF PIPE SIZES AND VELOCITIES FOR EXAMPLE 7.5**

| 1 | 2 | | 3 | 4 | 5 | 6 | 7 | 8 | 9 | 10 | 11 | 12 | 13 | 14 | 15 | 16 |
|---|---|---|---|---|---|---|---|---|---|---|---|---|---|---|---|---|
| | Manhole | | Pipe lgth. | Des. flow | Str. sl. | Min. sl. | Des. sl. | Pipe size | Pipe size | Pipe flow | $Q/Q$-full | $d/d$-max | Flow-$d$ | V-full | $V/V$-full | $V$ |
| Pipe no. | From | To | (ft) | (cfs) | (ft/ft) | (ft/ft) | (ft/ft) | (cal.-in) | (sel.-in) | full | (cfs) | | (in) | (fps) | — | (fps) |
| 1 | 7 | 6 | 630 | 1.86 | 0.0070 | 0.0027 | 0.007 | 10.04 | 12 | 2.98 | 0.62 | 0.58 | 6.96 | 3.80 | 1.05 | 3.99 |
| 2 | 6 | 5 | 470 | — | — | — | — | — | 12 | 2.83 | 0.70 | 0.62 | 7.44 | 3.60 | 1.08 | 3.89 |
| — | 6 | 8 | — | 1.97 | 0.0063 | 0.0016 | 0.0063 | 10.26 | — | — | — | — | — | — | — | — |
| 3 | 9 | 8 | 390 | 0.26 | 0.0077 | 0.0054 | 0.0077 | — | 8 | 1.06 | 0.25 | 0.35 | 2.8 | 3.04 | 0.82 | 2.49 |
| 4 | 8 | 5 | 385 | 0.26 | 0.0073 | 0.0054 | 0.0073 | — | 8 | 1.03 | 0.25 | 0.35 | 2.8 | 2.96 | 0.82 | 2.42 |
| 5 | 5 | 4 | 330 | — | — | — | — | — | 12 | 2.76 | 0.84 | 0.7 | 8.4 | 3.52 | 1.12 | 3.94 |
| — | 5 | 4 | — | 2.33 | 0.006 | 0.0015 | 0.006 | 10.93 | — | — | — | — | — | — | — | — |
| 6 | 10 | 11 | 410 | 0.19 | 0.009 | 0.0064 | 0.009 | — | 8 | 1.14 | 0.17 | 0.29 | 2.32 | 3.28 | 0.73 | 2.40 |
| 7 | 11 | 12 | 400 | 0.32 | 0.0087 | 0.0047 | 0.0087 | — | 8 | 1.13 | 0.28 | 0.36 | 2.88 | 3.23 | 0.83 | 2.68 |
| 8 | 12 | 4 | 380 | 0.42 | 0.008 | 0.0037 | 0.008 | — | 8 | 1.08 | 0.39 | 0.44 | 3.52 | 3.09 | 0.92 | 2.85 |
| 9 | 4 | 3 | 370 | 2.75 | 0.0052 | 0.0013 | 0.0052 | 11.63 | 15 | 4.66 | 0.59 | 0.55 | 8.25 | 3.80 | 1.04 | 3.95 |
| 10 | 16 | 17 | 380 | 0.11 | 0.01 | 0.0089 | 0.01 | — | 8 | 1.21 | 0.09 | 0.2 | 1.6 | 3.46 | 0.55 | 1.90 |
| 11 | 17 | 18 | 400 | 0.21 | 0.0092 | 0.006 | 0.0092 | — | 8 | 1.16 | 0.18 | 0.3 | 2.4 | 3.32 | 0.74 | 2.46 |
| 12 | 18 | 3 | 405 | 0.3 | 0.0088 | 0.006 | 0.0088 | — | 8 | 1.13 | 0.26 | 0.35 | 2.8 | 3.24 | 0.82 | 2.66 |
| 13 | 13 | 14 | 400 | 0.28 | 0.0097 | 0.005 | 0.0097 | — | 8 | 1.19 | 0.24 | 0.34 | 2.7 | 3.41 | 0.8 | 2.73 |
| 14 | 14 | 15 | 380 | 0.39 | 0.0088 | 0.0042 | 0.0088 | — | 8 | 1.13 | 0.34 | 0.41 | 3.28 | 3.24 | 0.89 | 2.89 |
| 15 | 15 | 3 | 411 | 0.6 | 0.0079 | 0.0032 | 0.0079 | — | 8 | 1.07 | 0.56 | 0.53 | 4.24 | 3.07 | 1.03 | 3.17 |
| 16 | 3 | 2 | 230 | 3.66 | 0.005 | 0.0013 | 0.005 | 12.94 | 15 | 4.57 | 0.80 | 0.68 | 10.2 | 3.73 | 1.11 | 4.14 |
| 17 | 2 | 1 | 600 | 3.66 | 0.0048 | 0.0013 | 0.0048 | 12.94 | 15 | 4.48 | 0.82 | 0.69 | 10.35 | 3.65 | 1.12 | 4.09 |

Column 5 = street slope; column 6 = minimum slope from Table 7.2; column 7 = design slope, larger of values in columns 5 and 6.

Column 8 = full flow $Q$ determined using Manning's equation; column 9 = closest standard pipe size.

Column 10 = full flow in selected pipe size (Manning's equation); column 11 = ratio of values in columns 4 and 10.

Column 12 = value from Fig. 7.2 for ratio of column 11; column 13 = value in column 9 × value in column 12.

Column 14 = $Q/A$ for full pipe flow; column 15 = ratio determined from Fig. 7.2 for the $d/d$-max value of column 12.

Column 16 = value in column 14 × value in column 15.

flows in mgd and cfs are shown in columns 9 and 10 of Table 7.4. Note that these calculations are performed by the spreadsheet used in the example. Spreadsheets are excellent tools for solving problems of this type. In this case QUATTRO PRO 3.0 was used, but many other spreadsheet programs are available.

8. Enter street slopes and minimum slopes for pipes in Table 7.5 (column 7). Note that the minimum slopes are estimated from values in Table 7.2. The initial design slope, adjusted later if need be, is the street slope unless that slope is less than the minimum slope required or other conditions make it necessary to assign a different grade.

9. Calculate the pipe size required to handle the design flow under full-flow conditions. Calculations made in the spreadsheet use Manning's equation with $n = 0.013$. Computed diameters are given in column 8 of Table 7.5. Note that a pipe size of 8 in. (the minimum allowable) is assigned to a number of pipes without calculating a diameter. The reasoning for that assignment is as follows. A look at the design flows of column 4, Table 7.5 shows that many of them are small, less than 0.6 cfs. For these flows, the minimum design slope is 0.0073. From Manning's nomograph, Fig. 7.1, it can be seen that an 8-in. pipe at a slope of 0.0073 can carry a flow of about 1 cfs at a velocity of about 3 fps. Accordingly, it is clear that an 8-in. pipe will handle all flows up to 0.6 cfs for the slopes given. An 8-in. pipe is thus selected for these flows and tabulated in column 9 of Table 7.5.

   The calculated pipe sizes (column 8, Table 7.5) are modified to the next largest standard pipe size or to a size that accommodates the design flow at a desired depth. These selected sizes are entered in column 9 of Table 7.5. Note that for the range of pipe sizes encountered in this problem, a desirable depth of flow is normally from about half to three-fourths full. Many of the 8-in. pipes, because of the low flows carried, have design flow depths that are less than half the maximum depth. This cannot be avoided and does not create a problem as long as cleansing velocities are maintained.

10. Calculate full flows for selected pipe sizes (column 10, Table 7.5). The design slope and a Manning's $n$ of 0.013 are used.

11. Compute ratios of $Q/Q_{full}$ (column 11, Table 7.5). Using these values and entering Fig. 7.2, we determine $d/d_{max}$ values and $V/V_{full}$ values and enter them in columns 12 and 15 of Table 7.5. Multiplying the depth ratios by the full-flow depths for each pipe gives the flow depths for each pipe (column 13, Table 7.5).

12. The full-flow velocity is calculated as $Q/A$ (column 14, Table 7.5). Multiplying the full-flow velocities by the velocity ratio gives the flow velocity for each pipe (column 16, Table 7.5).

13. The calculated values of $V$ are checked to see that they are at least 2 fps. All velocities agree except the velocity for pipe 16–17, 1.9 fps, but this value is considered close enough. Had the calculated value been much less, the design slope would have to be modified. ∎

## ■ EXAMPLE 7.6

Given an invert elevation of 105.19 for the pipe leaving M-6 in Fig. 7.8, calculate the invert elevations in and out of M-5 and M-4. Use the slopes and pipe lengths given in Example 7.5.

*Solution.*    Calculations are as follows:

105.19 − (0.0063)(470) = 102.23      invert at entrance to M-5

102.23 − 0.1 = 102.13      drop across M-5

102.13 − (0.006)(330) = 100.15      invert at entrance to M-4

100.15 − 0.25 = 99.9      invert out of M-4

A drop across a manhole of 0.1 was applied where no change in pipe size occurred. At M-4, the upstream pipe is 12 in., and the downstream pipe is 15 in. In this case a drop of 15 − 12 = 3 in. or 0.25 ft was used. The profile of Fig. 7.5 illustrates this process. ■

## Pressure Flow

Where pumping of sewage is required, force mains (pressure conduits) must be designed to carry the flows. The costs of pumping and associated equipment are important considerations. The hydraulics of these systems follows the principles already discussed. As in the case of open-channel sewers, pressure sewers must be able to transport sewage at velocities sufficient to avoid deposition and yet not so high as to create pipe erosion problems. Force mains are generally 8 in. in diameter or greater, but for small pumping stations, smaller pipes may sometimes be acceptable. According to Metcalf and Eddy, the following $C$ values for use in the Hazen–Williams equation for calculating friction losses in force mains are valid [9]:

100 for unlined cast-iron pipe;

120 for cement-lined cast-iron pipe, reinforced concrete pipe, asbestos-cement pipe, and plastic pipe;

110 for steel pipe having bituminous or cement-mortar lining and exceeding 20 in. in diameter.

Velocities encountered in force main operations are often in the range of 3–5 fps. In designing force mains, high points in lines should be avoided if possible. This eliminates the need for air-relief valves. Good design practice dictates that the hydraulic gradient should lie above the line at all points during periods of minimum-flow pumping.

## 7.10 PROTECTION AGAINST FLOODWATERS

Because the volume of sanitary wastewater is extremely small compared with flood flows, it is important that sewers be constructed to prevent admittance of large surface-runoff volumes. This will preclude overloads on treatment plants with their resultant reduction in degree of treatment or complete elimination of treatment in some instances.

Where interceptor sewers are built along streambeds, manhole stacks frequently are raised above a design flood level, such as the 50-yr level. In addition, the manhole structures are waterproofed. Where stacks cannot be raised, watertight manhole covers are employed. Such measures as these keep surface drainage into the sewer at a minimum.

## 7.11 INVERTED SIPHONS

An inverted siphon is a section of sewer constructed below the hydraulic gradient due to some obstruction and operates under pressure. The term *depressed sewer* is actually more appropriate, since no real siphon action is involved. Usually, two or more pipes are needed for a siphon in order to handle flow variability. Normally the water-surface elevation at the entrance and exit to a siphon is fixed. Under these conditions, the hydraulic design consists of selecting a pipe or pipes that will carry the design flow with a head loss equivalent to the difference in entrance and exit water-surface elevations. A transition structure is generally required at the entrance and exit of the depressed sewer to properly proportion or combine the flows.

The minimum flow in a siphon must be great enough to prevent deposition of suspended solids. Normally, velocities of less than 3 fps are unsatisfactory. When the siphon is required to handle flows that vary considerably during any 24-hr period, it is customary to provide two or more pipes. A small pipe is provided to handle the low flows; intermediate flows may be carried by the smallest pipe and a larger pipe. Maximum flows may require the use of three or more pipes. By subdividing the flow in this manner, adequate cleansing velocities are ensured for all flow magnitudes. The entrance transition structure is designed to channel the flow properly into the various pipes. Figure 7.9 illustrates a typical transition structure.

## 7.12 WASTEWATER PUMPING STATIONS

It is often necessary to accumulate wastewater at a low point in the collection system and pump it to treatment works or to a continuation of the system at a higher elevation. Pumping stations consist primarily of a wet well, which intercepts incoming flows and permits equalization of pump loadings, and a bank of pumps, which lift the wastewater from the wet well. In most cases, centrifugal pumps are used and standby equipment is required for emergency purposes. Selection of centrifugal pumps is made according to the principles outlined in Chapter 6.

Pumping of sewage may be necessitated by a number of circumstances. The area of concern may be lower in elevation than the nearest trunk sewer. It may also be that the area to be served lies outside of the drainage area in which the sewage treatment plant to be used is located. Such circumstances require pumping sewage flows across drainage divides. Excavation costs to construct gravity-flow systems may also be such that pumping is more

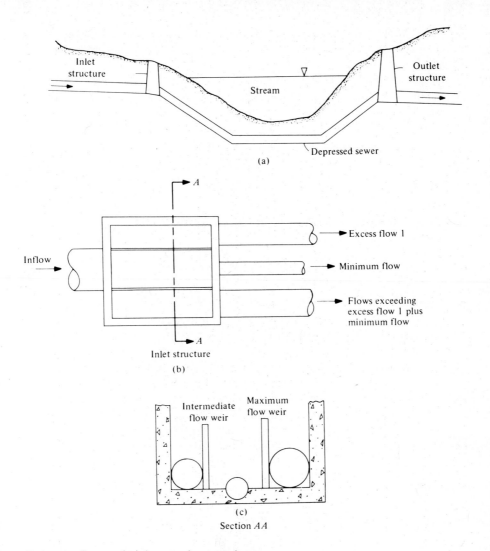

**Figure 7.9** Inverted siphon or depressed sewer.

economical than constructing a gravity sewer. In any event, circumstances and trade-offs will dictate the course of action to be followed. Additional details of pumping station designs may be found in *Wastewater Engineering* [9].

# Design of Storm Drainage Systems

Rapid and effective removal of storm runoff was a luxury not found in many cities in the early nineteenth century. Today, however, the modern city dweller has come to think of this as an essential service. Urban drainage facilities have progressed from crude ditches and stepping stones to the present intricate coordinated systems of curbs, gutters, inlets, and underground conveyances.

Handling surface runoff in urban drainage areas is a complex and costly undertaking with several primary difficulties—notably quantity and variability. Volumes of surface runoff can be exceedingly large during intense storms, yet such storms may occur only on a very infrequent basis. This poses the problem of building drainage works that perhaps are used for only a short time.

## 7.13 HYDROLOGIC CONSIDERATIONS

The hydrologic phase of urban drainage design is concerned with determining the magnitude, distribution, and timing of the various runoff events. Maximum events are of utmost importance since they are the basis for the design of major structures. In some cases, a knowledge of peak flow will suffice, but if storage or routing considerations are important, the volume and time distribution must also be known.

Runoff that occurs on any drainage area is a function of climate and the physical characteristics of the area. Factors that may be pertinent in precipitation–runoff relationships include precipitation type, rainfall intensity, duration, and distribution; storm direction; antecedent precipitation; initial soil moisture conditions; soil type; evaporation; transpiration; and the size, shape, slope, elevation, directional orientation, and land-use characteristics of the drainage area. Indirect and artificial drainage must also be considered when applicable.

A brief study of these items should be sufficient proof that the hydrologic problem is complex. If urban drainage works are effectively and economically to serve the areas for which they are designed, considerable emphasis must be placed on the determination of accurate and reliable estimates of flow. Unfortunately, it is all too common that months are spent on the structural design of hydraulic conveyances, while only meager computations are made of the flow magnitudes on which these designs are based. Except in the case of very important structures, computations are often founded on formulas of questionable reliability or on the assumption of a past maximum flow plus a factor of safety. The greatest number of hydraulic failures have been caused by faulty determinations of runoff magnitude, not by structural inadequacies.

Peak flows result from excess surface-runoff volumes. The conditions that may generate these excesses are intense storms, snowmelt, and snowmelt combined with rainfall. Maximum flows on urban areas in the United States usually result from high-intensity short-duration rainfall of the thunderstorm type, whereas floods on large drainage basins are habitually caused by a combination of rainfall and snowmelt. The particular factors that produce maximum flow on a specific drainage area must be determined if reasonable reliability is to be accorded the estimated quantities of discharge.

## 7.14 DESIGN FLOW

*Design flow* is the maximum flow that a specified structure can pass safely. It is significant that the first question to be answered in relation to the design flow is its probability of occurrence. Should a drainage system be designed to carry the maximum possible flow, the 5%, the 1%, or some other chance discharge? Once this question has been answered, the magnitude of flow having the selected "design" frequency must be determined. Note particularly that there is little relationship between the normal life of the drainage works and the frequency of the design flow.

Table 7.6 shows, for example, that there is a 22% chance that the 100-yr storm will occur in any 25-yr period. This is good reason for questioning the design of a structure for a life expectancy of 25 yr, then using design flows expected on the average of once in 25 yr or less. If such a design is used, the chances are good that the structure will be damaged, destroyed, or at least overloaded before it has served its useful life.

Selection of a design frequency must be based on consideration of potential damage to human life, property damage, and inconvenience. Human life cannot be judged in terms of monetary values. If it is apparent that failure of a proposed drainage system or structure will imperil human lives, the design should be appraised accordingly. Property damage is a purely economic consideration, and design flows can be determined on the basis of the limiting size flow against which it is economically practical to protect. Inconvenience is an intangible quantity for the most part but can be related to economic

**TABLE 7.6**   PROBABILITY THAT AN EVENT HAVING A PRESCRIBED RECURRENCE INTERVAL WILL BE EQUALED OR EXCEEDED DURING A SPECIFIED PERIOD

| $T_R$ (yr) | Period (yr) | | | | | |
|---|---|---|---|---|---|---|
|  | 1 | 5 | 10 | 25 | 50 | 100 |
| 1 | 1.0 | 1.0 | 1.0 | 1.0 | 1.0 | 1.0 |
| 2 | 0.5 | 0.97 | 0.999 | $1.0^a$ | $1.0^a$ | $1.0^a$ |
| 5 | 0.2 | 0.67 | 0.89 | 0.996 | $1.0^a$ | $1.0^a$ |
| 10 | 0.1 | 0.41 | 0.65 | 0.93 | 0.995 | $1.0^a$ |
| 50 | 0.02 | 0.10 | 0.18 | 0.40 | 0.64 | 0.87 |
| 100 | 0.01 | 0.05 | 0.10 | 0.22 | 0.40 | 0.63 |

[a] Values are approximate.

values under many circumstances and should also be heeded in selecting a design frequency.

## 7.15 PROCEDURES FOR ESTIMATING RUNOFF

Procedures used in estimating runoff magnitude and frequency can be generally categorized as (1) empirical approaches, (2) statistical or probability methods, and (3) methods relating rainfall to runoff [10]. Historically, numerous empirical equations have been developed for use in the prediction of runoff. Most of them have been based on the correlation of only two or three variables and, at best, have given only rough approximations. Many are applicable only to specific localities—a fact that should be carefully considered before they are used. In most cases, the frequency of the computed flow is unknown. Formulas of this type are useful only where a more reliable means is not available—and then only with a complete understanding of the relationship and the exercise of due caution.

Statistical analyses provide good results if sufficient records are available and if no significant changes in stream regimen are experienced or expected in the future. Estimates generally are based on the use of duration or probability curves. In applying these curves, the larger the sample population, the greater the validity of the estimate. For example, a determination of the 10-yr peak flow based on records of only 10 yr might be seriously in error, whereas the same answer computed from a 100-yr record would yield good results. About ten independent samples normally can be expected to provide satisfactory estimates of maximum-flow magnitudes of any frequency. The need for long-term records emphasizes the limitation inherent in the use of probability methods, since extreme values expected to occur only on the average of once in 50–100 yr or more are not often obtainable within the available record. They must therefore be determined by extrapolation of the probability curve, an extremely dangerous undertaking. Most streams in the United States do not have reliable gaging records older than 50 yr, and only a few records exist for urban areas.

Of the methods relating rainfall to runoff, the unit hydrograph method, the rational method, and various simulation models are widely used. A discussion of these follows.

### Unit Hydrograph Method

The *unit hydrograph method* is a valuable tool for estimating runoff magnitudes of various frequencies that may occur on a specific stream [11]. To use this approach it is necessary to have continuous records of runoff and precipitation for the particular drainage area. Determinations must be made of infiltration capacity variations throughout the year. The method is limited to areas for which precipitation patterns do not vary markedly. On large drainage basins, hydrographs must be developed for the various reaches and then synthesized into a single design flow at the critical location.

A unit hydrograph represents the runoff volume of 1 in. from a specified drainage area for a particular rainfall duration. A separate unit hydrograph is theoretically required for every possible rainfall length of interest. Ordinarily, however, variations of $\pm 25\%$ from any duration are considered acceptable. In addition, unit hydrographs for short periods can be synthesized into hydrographs for longer durations.

Once a unit hydrograph has been derived for a particular drainage area and rainfall duration, the hydrograph for any other storm of equal duration can be obtained. This new hydrograph is developed by applying the unit hydrograph theorem, which states that the ordinates of all hydrographs resulting from equal unit time rains are proportional to the total direct runoff from that rain. The condition may be stated mathematically as

$$\frac{Q_s}{V_s} = \frac{Q_u}{1} \tag{7.9}$$

where  $Q_s$ = magnitude of a hydrograph ordinate of direct runoff having a volume equal to $V_s$ (in inches) at some instant of time after the start of runoff

  $Q_u$ = ordinate of the unit hydrograph having a volume of 1 in. at the same instant of time

Storms of reasonably uniform rainfall intensity, having a duration of about 25% of the drainage area lag time (the difference in time between the center of mass of the rainfall and the center of mass of the resulting runoff), and producing a total of 1 in. or more are most suitable in deriving a unit hydrograph.

Essential steps in the development of a unit hydrograph are as follows:

1. Analyze the streamflow hydrograph to permit separation of the surface runoff from the groundwater flow. It can be accomplished by any one of several methods [10, 12].
2. Measure the total volume of surface runoff (direct runoff) from the storm producing the original hydrograph. The volume is equal to the area under the hydrograph after the groundwater flow has been removed.
3. Divide the ordinates of the direct runoff by the total direct runoff volume in inches. The resulting plot of the answers versus time is a unit graph for the basin.
4. Finally, the effective duration of the runoff-producing rain for this unit graph must be found. The answer can be obtained by a study of the hydrograph of the storm used.

The steps listed are used in deriving the unit hydrograph from an isolated storm. Other procedures are required for complex storms or in developing synthetic unit graphs when few data are available. Unit hydrographs may also be transposed from one basin to another under certain circumstances.

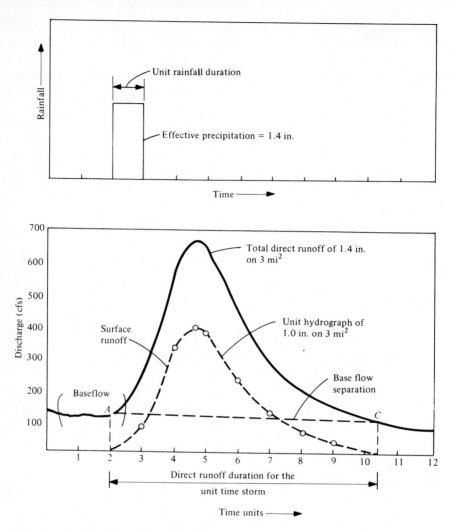

**Figure 7.10** Derivation of a unit hydrograph from an isolated storm.

An example will serve to illustrate the derivation and application of the unit hydrograph.

### ■ EXAMPLE 7.7
Using the hydrograph given in Fig. 7.10, derive a unit hydrograph for the 3-mi² drainage area. From this unit hydrograph, derive a hydrograph of direct runoff for the rainfall sequence given in Table 7.8.

*Solution.*

1. Separate the base or groundwater flow so that the total direct runoff hydrograph may be obtained. As stated previously, a number of procedures are reported in the literature. A common method is to draw a straight line *AC*

**TABLE 7.7**    DETERMINATION OF A UNIT HYDROGRAPH FROM AN
ISOLATED STORM

| (1) | (2) | (3) | (4) Total direct runoff: (2) − (3) | Unit hydrograph ordinate: (4) ÷ 1.4 |
|---|---|---|---|---|
| Time unit | Total runoff (cfs) | Base flow (cfs) | (cfs) | (cfs) |
| 1 | 110 | 110 | 0 | 0 |
| 2 | 122 | 122 | 0 | 0 |
| 3 | 230 | 120 | 110 | 78.7 |
| 4 | 578 | 118 | 460 | 328 |
| 4.7 | 666 | 116 | 550 | 393 |
| 5 | 645 | 115 | 530 | 379 |
| 6 | 434 | 114 | 320 | 229 |
| 7 | 293 | 113 | 180 | 129 |
| 8 | 202 | 112 | 90 | 64.2 |
| 9 | 160 | 110 | 50 | 35.7 |
| 10 | 117 | 105 | 12 | 8.6 |
| 10.5 | 105 | 105 | 0 | 0 |
| 11 | 90 | 90 | 0 | 0 |
| 12 | 80 | 80 | 0 | 0 |

beginning when the hydrograph starts an appreciable rise and ending where the recession curve intersects the base-flow curve. The most important point here is to be consistent in methodology from storm to storm.

2. Determine the duration of the effective rainfall (rainfall that actually produces surface runoff). The effective rainfall volume must be equivalent to the volume of direct surface runoff. Usually the unit time of the effective rainfall will be 1 day, 1 hr, 12 hr, or some other interval appropriate for the size of drainage area being studied. As stated before, the unit storm duration should not exceed about 25% of the drainage area lag time. The effective portion of the rainstorm for this example is given in Fig. 7.10 together with its duration. The effective volume is 1.4 in.

3. Project the base length of the unit hydrograph down to the abscissa, giving the horizontal projection of the base-flow separation line $AC$. It should be noted that the unit hydrograph theory assumes that for all storms of equal duration, regardless of intensity, the period of surface runoff is the same.

4. Tabulate the ordinates of direct runoff at the peak rate of flow and at sufficient other positions to determine the hydrograph shape. Note that the direct runoff ordinate is the ordinate above the base-flow separation line. See Table 7.7.

5. Compute the ordinates of the unit hydrograph by using Eq. 7.9. In this example, the values are obtained by dividing the direct runoff ordinates by 1.4. Table 7.7 outlines the computation of the unit hydrograph ordinates.

6. Using the values from Table 7.7, plot the unit hydrograph as shown in Fig. 7.10.

7. Using the derived ordinates of the unit hydrograph, determine the ordinates of the hydrographs for each consecutive rainfall period as given in Table 7.8.

**TABLE 7.8**  UNIT HYDROGRAPH APPLICATION

| Time unit sequence | Rain unit number | Effective rainfall (in.) | Hydrograph ordinates[a] for rainfall unit. . . | | |
|---|---|---|---|---|---|
| | | | 1 | 2 | 3 |
| 1 | 1 | 0.7 | 55.1 | — | — |
| 2 | 2 | 1.7 | 229 | 134 | — |
| 2.7 | 3 | 1.2 | 275 | — | — |
| 3 | — | — | 265 | 558 | 94.3 |
| 3.7 | — | — | — | 668 | — |
| 4 | — | — | 161 | 664 | 393 |
| 4.7 | — | — | — | — | 472 |
| 5 | — | — | 90.5 | 389 | 455 |
| 6 | — | — | 44.9 | 219 | 275 |
| 7 | — | — | 25.0 | 109 | 155 |
| 8 | — | — | 6.0 | 60.7 | 77 |
| 9 | — | — | — | 14.6 | 42.8 |
| 10 | — | — | — | — | 10.3 |

[a] Values are obtained by multiplying effective rainfall values by unit hydrograph ordinates.

**8.** Determine the synthesized hydrograph for unit storms 1–3 by plotting the three hydrographs and summing the ordinates. The procedure is indicated in Fig. 7.11.  ■

The unit hydrograph method provides the entire hydrograph resulting from a particular storm and offers certain definite advantages over procedures that produce peak flows alone. It has the disadvantage of requiring both rainfall and runoff data for its derivation and of being limited in its application to a particular drainage basin. A number of refinements to the procedure may be found in the literature [12, 13]. Procedures for producing synthetic unit hydrographs covering areas where adequate records are not available have also been derived [13].

## Rational Method

A second rainfall–runoff relationship that will be discussed here is known as the *rational method*. It was first proposed in 1889 and is currently a commonly used method in this country for computing quantities of stormwater runoff from small areas [6]. The rational formula relates runoff to rainfall in the following manner:

$$Q = cia \qquad (7.10)$$

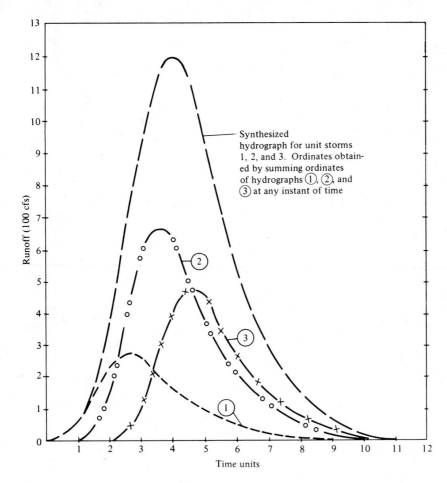

**Figure 7.11** Synthesized hydrograph derived by the unit hydrograph method.

where   $Q$ = peak runoff rate, cfs
    $c$ = runoff coefficient, which is actually the ratio of the peak runoff
        rate to the average rainfall rate for a period known as the time
        of concentration
    $i$ = average rainfall intensity, in./hr, for a period equal to the time
        of concentration
    $a$ = drainage area, acres

It should be noted that the assignment of cfs to $Q$ is satisfactory for all
practical purposes, since 1.008 cfs equals 1 in. of rainfall in 1 hr on an area
of 1 acre.

    Assumptions basic to the rational method are that (1) the maximum
runoff rate to any design location is a function of the average rate of rainfall
during the time of concentration and (2) the maximum rate of rainfall occurs

during the time of concentration. The variability of the storm pattern is not taken into consideration. The time of concentration $t_c$ is defined as the flow time from the most remote point in the drainage area to the point in question. Usually it is considered to be composed of an overland flow time or, in most urban areas, an inlet time plus a channel flow time.

The channel flow time can be estimated with reasonable accuracy from the hydraulic characteristics of the sewer. Normally, the average full-flow velocity of the conveyance for the existing or proposed hydraulic gradient is used. The channel flow time is then determined as the flow length divided by the average velocity.

The inlet time consists of the time required for water to reach a defined channel such as a street gutter, plus the gutter flow time to the inlet. Numerous factors, such as rainfall intensity, surface slope, surface roughness, flow distance, infiltration capacity, and depression storage, affect inlet time. Because of this, accurate values are difficult to obtain. Design inlet flow times of from 5 to 30 min are used in practice. In highly developed areas with closely spaced inlets, inlet times of 5–15 min are common; for similar areas with flat slopes, periods of 10–15 min are common; and for very level areas with widely spaced inlets, inlet times of 20–30 min are frequently used [6].

Inlet times are also estimated by breaking the flow path into various components, such as grass, asphalt, and so on, then computing individual times for each surface and adding them. In theory this would seem desirable, but in practice so many variables affect the flow that the reliability of computations of this type is questionable. The standard procedure is to use inlet times that have been found through experience to be applicable to the various types of urban areas.

Izzard found the concentration time for small experimental plots without developed channels to be [14]

$$t_c = \frac{41bL_o^{1/3}}{(ki)^{2/3}} \tag{7.11}$$

where  $t_c$ = time of concentration, min
$b$ = coefficient
$L_o$ = overland flow length, ft
$k$ = rational runoff coefficient (see Table 7.9)
$i$ = rainfall intensity, in./hr, during time $t_c$

The equation is valid only for laminar flow conditions where the product $iL_o$ is less than 500. The coefficient $b$ is found using

$$b = \frac{0.0007i + C_r}{S_o^{1/3}} \tag{7.12}$$

where  $S_o$ = surface slope
$C_r$ = coefficient of retardance

Values of $C_r$ are given in Table 7.10. Note that reliability of this equation is also strongly influenced by the selection of parameters.

**TABLE 7.9**  SOME VALUES OF THE
RATIONAL RUNOFF
COEFFICIENT $c$

| Surface type | $c$ Value[a] |
|---|---|
| Bituminous streets | 0.70–0.95 |
| Concrete streets | 0.80–0.95 |
| Driveways, walks | 0.75–0.85 |
| Roofs | 0.75–0.95 |
| Lawns; sandy soil | |
|   Flat, 2% | 0.05–0.10 |
|   Average, 2%–7% | 0.10–0.15 |
|   Steep, 7% | 0.15–0.20 |
| Lawns, heavy soil | |
|   Flat, 2% | 0.13–0.17 |
|   Average, 2%–7% | 0.18–0.22 |
|   Steep, 7% | 0.25–0.35 |

[a] See Ref. 6.

The runoff coefficient $c$ is the component of the rational formula that requires the greatest exercise of judgment by the engineer. It is not amenable to exact determination, since it includes the influence of a number of variables, such as infiltration capacity, interception by vegetation, depression storage, and antecedent conditions. As used in the rational equation, the coefficient $c$ represents a fixed ratio of runoff to rainfall, while in actuality it is not fixed and may vary for a specific drainage basin with time during a particular storm, from storm to storm, and with change in season. Fortunately, the closer the area comes to being impervious, the more reasonable the selection of $c$ becomes. This is true since for highly impervious areas $c$ approaches unity, and for these areas the nature of the surface is much less variable for changing seasonal, meteorological, or antecedent conditions. The rational method therefore is best suited for use on urban areas, where a high percentage of imperviousness is common.

At present there is no precise method for evaluating the runoff coefficient $c$, although some research has been directed toward that end [15]. Common

**TABLE 7.10**  IZZARD'S RETARDANCE
COEFFICIENT $C_r$

| Surface | $C_r$ |
|---|---|
| Smooth asphalt | 0.007 |
| Concrete paving | 0.012 |
| Tar and gravel paving | 0.017 |
| Closely clipped sod | 0.046 |
| Dense bluegrass turf | 0.060 |

engineering practice is to make use of average values of the coefficient for various surface types normally found in urban regions. Table 7.9 lists some values of the runoff coefficient as reported in the American Society of Civil Engineers' *Manual on the Design and Construction of Sanitary and Storm Sewers* [6].

Figure 7.12 relates the rational $c$ to imperviousness, soil type, and lawn

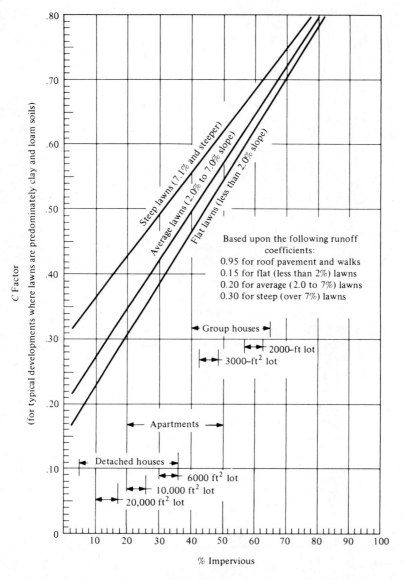

**Figure 7.12** $c$ factors for typical developments in clay soils. (Courtesy of the Baltimore County, Maryland, Department of Public Works.)

slope. This graph is used in designing storm drains in Baltimore County, Maryland. Most engineering designers make use of information reported in similar tabular or graphical form, inserting local conditions following their experience and practice.

In applying the rational method, a rainfall intensity $i$ must be used that represents the average intensity of a storm of given frequency for the time of concentration. The frequency chosen is largely a matter of economics. Factors related to the choice of a design frequency have already been discussed. Frequencies of 1–10 yr are commonly used where residential areas are to be protected. For higher-value districts, 10–20 yr or higher return periods often are selected. Local conditions and practice normally dictate the selection of these design criteria.

After $t_c$ and the rainfall frequency have been ascertained, the rainfall intensity $i$ is usually obtained by making use of a set of rainfall intensity–duration–frequency curves such as those shown in Fig. 7.13. Entering the curves on the abscissa with the appropriate value of $t_c$ and then projecting upward to an intersection with the desired frequency curve allows $i$ to be found by projecting this intersection point horizontally to an intersection with the ordinate. If an adequate number of years of local rainfall records is available, curves similar to Fig. 7.13 may be developed. Otherwise, data compiled

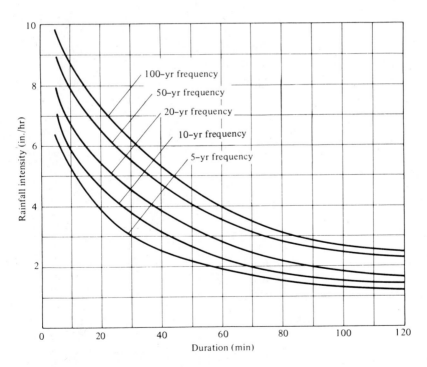

**Figure 7.13** Typical intensity–duration–frequency curves.

by the Weather Bureau, the Department of Commerce, the Department of Agriculture, and other government agencies, data that are available for numerous localities and regions, can be used.

Generally, the rational method should be used only on areas that are smaller than about 2 mi² (approximately 1280 acres) in size. For areas larger than 100 acres due caution should be exercised. Most urban drainage areas served by storm drains become tributary to natural drainage channels or large conveyances before they reach 100 acres or more in size, however, and for these tributary areas the rational method can be put to reasonable use.

## Simulation Models

Simulation models find widespread use in estimating stormwater flows generated on urban areas. Where sufficient data are available, simulation models can be used to test proposed designs and/or systems' operating procedures [8, 10]. One of the earliest simulation models to find wide use is the Stanford Watershed Model developed by Crawford and Linsley in 1966 [16]. Since that time the model has undergone many revisions by its authors and others [8, 10]. The version discussed here is the *hydrocomp simulation program* (HSP) [17]. This and other versions of the model have been used extensively for synthesizing continuous hydrographs of hourly or daily stream flows for watersheds of varying sizes and character.

The HSP is actually a sequence of computational routines representing the major hydrologic processes. Input to the model includes precipitation, potential evapotranspiration, temperature, radiation, dew point, and wind. The last four of these are needed only if snowmelt is involved.

What the model does is to account for the initial moisture stored in the watershed and then determine the fate of a precipitation input. The hydrologic budget is balanced for selected time steps so that the precipitation input can be converted to stream flow after the appropriate allocations to interception, evapotranspiration, and upper, lower, and deep groundwater storages have been made. In simple terms, the following equation is solved at each time-step for stream flow:

$$R = P - \text{ET} - \Delta S \tag{7.13}$$

where   $P$ = precipitation
        $R$ = streamflow
       $\text{ET}$ = evapotranspiration
       $\Delta S$ = the total change in storage

The equation is balanced at the end of each period, and the calculations proceed from interval to interval until there is no additional input of data [17].

Another widely used simulation model is the EPA's Storm Water Management Model (SWMM) [10].

# 7.16 STORMWATER INLETS

Stormwater inlet capacity has received little emphasis in the design of storm drainage systems for highways and streets. It is nevertheless of great importance, because regardless of the adequacy of the underground drainage system, proper drainage cannot result unless stormwater is quickly and efficiently collected and introduced into the system.

A reliable knowledge of the behavior of stormwater inlets has broad implications: overdesign and inefficient use of the drainage system due to inadequate inlets can be avoided, debris and leaf stoppage of inlets can be reduced, and traffic interference on streets and highways can be minimized. No specific inlet type can be considered best for all conditions of use. Street grade, cross slope, and depression geometry affect the hydraulic efficiency. Eliminating stoppages or minimizing traffic interference often take precedence over hydraulic considerations in design.

Ideally, a simple opening across the flow path is the most effective type of inlet structure. Construction of this type would be impractical and unsafe, however, and the opening must therefore be covered with a grate or located in the curb. Unfortunately, grates obstruct the fall of water and often serve to collect debris, while curb openings are not in the direct path of flow. Nevertheless, a compromise of this type must be made.

Increasing the street cross slope will increase the depth of flow of the gutter, gutter depressions will concentrate flows at the inlet, and curb and gutter openings can be combined. These and other modifications provide increased inlet capacities, although some of them are not compatible with high-volume traffic.

Numerous inlet designs are seen across the United States, most of which have been developed from the practical experience of engineers or by rule-of-thumb procedures. The hydraulic capacity of many of these designs is totally unknown, and estimates frequently are considerably in error. Only in recent years have laboratory studies been conducted to produce efficient designs and develop a better understanding of inlet behavior [18].

Four major types of inlets are being built: curb inlets, gutter inlets, combination inlets, and multiple inlets. A multitude of varieties is possible in each classification. A brief general description of the basic types follows:

a. *curb inlet*—a vertical opening in the curb through which gutter flow passes.
b. *gutter inlet*—a depressed or undepressed opening in the gutter section through which the surface drainage falls, covered by one or more grates.
c. *combination inlet*—an inlet composed of both curb and gutter openings, which acts as an integrated unit. Gutter openings may be placed directly in front of the curb opening (contiguous combination inlet) or upstream or downstream of the gutter opening (offset inlet). Combination inlets may be depressed or undepressed. Figure 7.14 depicts a typical combination inlet. Figure 7.15 relates the capacity of the inlet to the percent gutter slope.
d. *multiple inlets*—closely spaced interconnected inlets acting as a unit. Identical inlets end to end are called "double inlets."

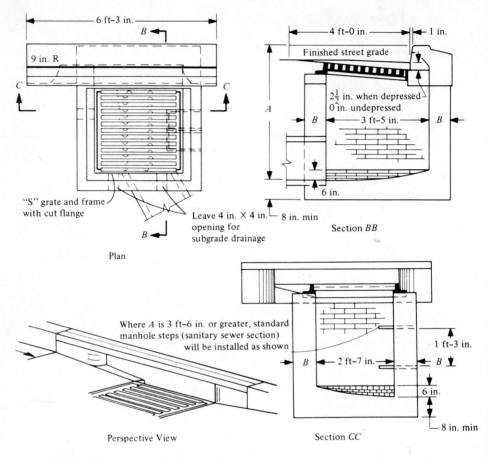

**Figure 7.14** Baltimore type *S* combination inlet. (Courtesy of the Baltimore County, Maryland, Department of Public Works.)

Several general design recommendations regarding stormwater inlets may be made. The final selection of the optimum inlet type for a specific location will necessarily have to be based on the exercise of engineering judgment relative to the importance of clogging, traffic hazard, safety, and cost. In general, use cross slopes that are as steep as possible considering traffic safety and comfort. Locate and design the inlets so there will be a 5%–10% bypass in gutter flow. This greatly increases the inlet capacity. The bypass flow should not be great enough to inconvenience pedestrians or vehicular traffic, however. On streets where parking is permitted or where vehicles are not expected to travel near the curb, use contiguous combination curb-and-gutter inlets with longitudinal grate bars, or build depressed gutter inlets if clogging is not a problem. Where clogging is important, use depressed curb inlets if the gutter flow is small. For large flows, use depressed combination inlets with the curb opening upstream. On streets having slopes in excess

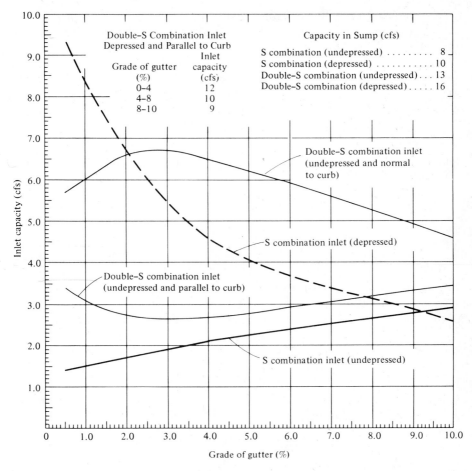

**Figure 7.15** Inlet capacity curves for type *S* combination inlets. The inlet capacities have been reduced by 20% for the estimated effects of debris partially clogging the inlet (on slopes). The capacities are based on standard $7\frac{3}{16}$-in. curb and gutter and 1–18 average crown slope for a distance of about 5 ft from the face of the curve. (Courtesy of the Baltimore County, Maryland, Department of Public Works.)

of 5%, where traffic passes close to the curb, use deflector inlets if road dirt will not pack in the grooves. For flat slopes or where dirt is a problem, use undepressed gutter inlets or combination inlets with longitudinal bars only. For streets having flat grades or sumps (lows), pitch the grade toward the inlet on both ends. This will have the effect of providing a sump at the inlet. For true sump locations, use curb openings or combination inlets. The total open area, not the size and arrangement of bars, is important because the inlet will act as an orifice. Normally, sump inlets should be overdesigned because of the unique clogging problems that develop in depressed areas.

Inlets should be constructed in all sumps and at all street intersections where the quantity of flow is significant or where nuisance conditions warrant such construction. Inlets are required at intermediate points along streets where the curb and gutter capacity would be exceeded without them. Inlet capacities should be equal to or greater than the design flows. As shown in Fig. 7.15, inlet capacity is related to gutter slope and must be taken into account when selecting an inlet for a specific location.

Rapid and efficient removal of surface runoff from streets and highways is exceedingly important for maximum safety and minimum nuisance. Street grade, crown slope, inlet type, grade design, and tolerable bypass flow are all important factors in the selection of inlet structures. Reliable information on the hydraulic performance of various inlet works is essential if effective designs are to be made.

## 7.17 SYSTEM LAYOUT

Storm drainage systems may be closed conduit, open conduit, or some combination of the two. In most urban areas, the smaller drains frequently are closed conduits, and as the system moves downstream, open channels are often employed. Because quantities of stormwater runoff are usually quite large when contrasted with flows in most water mains and sanitary sewers, the drainage works needed to carry these flows are also quite large and occupy an important position from the standpoint of utilities placement. Common practice is to build the storm drains along the centerline of the street, then offset the water main to one side and the sanitary sewer to the other.

In order to produce a workable system layout, a map of the area is needed showing contours, streets, buildings, other existing utilities, natural drainage ways, and areas for future development. As in the case of the sanitary sewer system, the storm drains will lie principally under the streets or in designated drainage rights of way. The entire area to the physical divide should be served by the drain or by an extension of it.

In laying out the draining system, the first step is to tentatively locate all the necessary inlet structures. Once this has been done, a skeleton pipe system connecting all the inlets within a particular drainage area is drawn along with the location of all manholes, wye branches, special bend or junction structures, and the outfall.

Customarily, a manhole or junction structure will be required at all changes in grade, pipe size, direction of flow, and quantity of flow. Figure 7.16 illustrates the details of a typical manhole. Note that today most of these structures are precast concrete rather than brick. Storm drains are usually not built smaller than 12 in. in diameter because pipes of lesser size tend to clog readily with debris and therefore present serious malfunction problems. Maximum manhole spacing for pipes 27 in. and under should not exceed about 600 ft. For larger pipes, no maximum is prescribed, and the normal

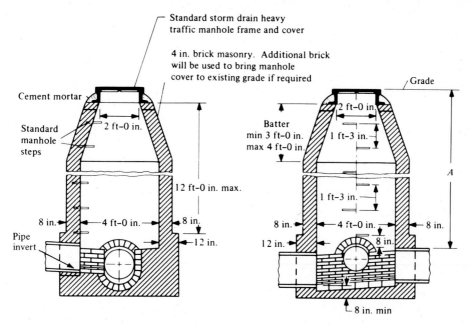

**Figure 7.16** Typical storm drainage manhole details. (Courtesy of the Baltimore County, Maryland, Department of Public Works.)

requirements for structures should provide access to the drain for inspection, cleaning, or maintenance.

## 7.18 HYDRAULIC DESIGN OF URBAN STORM DRAINAGE SYSTEMS

The basic hydraulic principles outlined at the beginning of the chapter are applicable to the design of an urban drainage system. The calculations for the drainage area shown in Fig. 7.17 are presented in detail to illustrate the mechanical procedure and the rational method as applied to an actual design problem. The overall Mextex area is an urban residential area made up of single-family dwellings and is divided into eight subareas which are tributary to individual stormwater inlets.

■ **EXAMPLE 7.8**
Design a storm drainage system to carry the flows from the eight inlet areas given in Fig. 7.17. It will be assumed that a 10-yr frequency rainfall satisfies the local design requirements. Assume clay soil to be predominant in the area, with average lawn slopes.

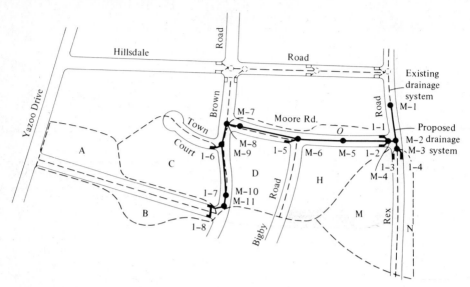

**Figure 7.17** Plan view of the Mextex storm drainage system. Drainage areas: A, 1.93; B, 1.34; C, 3.19; D, 2.66; H, 2.25; M, 2.75; N, 1.27; O, 0.40.

### Solution.

1. Prepare a drainage area map showing drainage limits, streets, impervious areas, and direction of surface flow.
2. Divide the drainage area into subareas tributary to the proposed stormwater inlets (Fig. 7.17).
3. Compute the acreage and imperviousness of each area.
4. Calculate the required capacity of each inlet, using the rational method. Assume a 5-min inlet time to be appropriate and compute inlet flows for a rainfall intensity of 7.0 in./hr. This is obtained by using the 10-yr frequency curve in Fig. 7.13 with a 5-min concentration time. Appropriate $c$ values are obtained from Fig. 7.12 by entering the graph with the calculated percentage imperviousness (the percent of the inlet area covered by streets, sidewalks, drives, roofs, etc.), projecting up to the average lawn-slope curve, and reading $c$ on the ordinate. Computations for the inlet flows are tabulated in Table 7.11.
5. Select the type inlets required to adequately drain the flows in Table 7.11. The choice will be based on a knowledge of the street slopes and their relation to various inlet capacities. Inlet capacity curves such as those given in Fig. 7.15 would be used. For the purposes of this example, no actual selections will be made, but the reader should recognize that this is the next logical step and an exceedingly important one.
6. Beginning at the upstream end of the system, compute the discharge to be carried by each successive length of pipe, moving downstream. These calculations are summarized in Table 7.12. Note that at each point downstream where a new flow is introduced, a new time of concentration must be determined as well as new values of $c$ and drainage area size. As the upstream inlet areas are combined to produce a larger tributary area at some design point, a revised

**TABLE 7.11**   REQUIRED STORMWATER INLET CAPACITIES FOR EXAMPLE 7.8

| (1) Inlet | (2) Area designation | (3) Area (acres) | (4) Percent impervious | (5) $C$ | (6) Rainfall intensity | $Q$: (3) × (5) × (6) (cfs) |
|---|---|---|---|---|---|---|
| I-1 | O | 0.40 | 49 | 0.57 | 7.0 | 1.59 |
| I-2 | H | 2.25 | 20 | 0.35 | 7.0 | 5.52 |
| I-3 | M | 2.75 | 26 | 0.40 | 7.0 | 7.70 |
| I-4 | N | 1.27 | 26 | 0.40 | 7.0 | 3.55 |
| I-5 | D | 2.66 | 26 | 0.40 | 7.0 | 7.43 |
| I-6 | C | 3.19 | 24 | 0.38 | 7.0 | 8.48 |
| I-7 | A | 1.93 | 23 | 0.37 | 7.0 | 4.99 |
| I-8 | B | 1.34 | 29 | 0.42 | 7.0 | 3.93 |

$c$ value representing these combined areas must be obtained. Usually the procedure is to take a weighted average of all the individual $c$ values of which the larger area is composed. For example, in the computation of the flow to be carried by the pipe from M-9 to M-8, the tributary area is A + B + C = 6.46 acres, and the composite value of $c$ will be

$$c = \frac{\Sigma c_i a_i}{\Sigma a_i} = \frac{0.37 \times 1.93 + 0.42 \times 1.34 + 0.38 \times 3.19}{6.46} = 0.38$$

At the design location the value of $t_c$ will be equal to the inlet time at I-8 plus the pipe flow time from I-8 to M-9 (see Table 7.12), which must be known to permit solving the rainfall intensity to be used in computing the runoff from composite area A + B + C.

7. Using the computed discharge values, select tentative pipe sizes for the approximate slopes given in column 8 of Table 7.12. Once the pipe sizes are known, flow velocities between input locations can be determined. Normally, these velocities are approximated by computing the full-flow velocities for maximum discharge at the specified grade. These velocities are then used to compute channel flow time for estimating the time of concentration. If upon completing the hydraulic design, enough change has been made in any concentration time to alter the design discharge, new values of flow should be computed. Generally, this will not be the case.

8. Using the pipe sizes selected in step 7, draw a profile of the proposed drainage system. Begin the profile at the point farthest downstream, which can be an outfall into a natural channel, an artificial channel, or an existing drain, as in the case of the example. In constructing the profile, be certain that the pipes have at least the minimum required cover. Normally 1.5–2 ft is sufficient. Pipe slopes should conform to the surface slope wherever possible. At all manholes indicate the necessary change in invert elevation. In this example, where there is no change in pipe size through the manhole, a drop of 0.2 ft will be used. Where the size decreases upstream through a manhole, the upstream invert will be set above the downstream invert a distance equal to the difference in the two diameters. In this way the crowns are kept at the same elevation. A part of the profile of the drainage system in the example is given in Fig. 7.18.

**TABLE 7.12** COMPUTATION OF DESIGN PIPE FLOWS FOR THE STORM DRAINAGE SYSTEM OF EXAMPLE 7.8

| Pipe section | Tributary area | Area (acres) | Flow time (min) Inlet | Flow time (min) Pipe | Flow time (min) Total | Rainfall intensity $i$ | $c$ | $Q$ (cfs) | Pipe Slope (%) | Pipe Size (in.) | Pipe Full-flow velocity (fps) | Length (ft) |
|---|---|---|---|---|---|---|---|---|---|---|---|---|
| I-8–I-7 | B | 1.34 | 5 | 0.10 | 5 | 7.0 | 0.42 | 3.93 | 1.0 | 15 | 5.2 | 30 |
| I-7–M-11 | A + B | 3.27 | 5 | 0.13 | 5.10 | 7.0 | 0.39 | 8.93 | 1.0 | 18 | 5.9 | 46 |
| M-11–M-10 | A + B | 3.27 | — | 0.24 | 5.23 | — | 0.39 | 8.93 | 1.0 | 18 | 5.9 | 85 |
| M-10–M-9 | A + B | 3.27 | — | 0.37 | 5.47 | — | 0.39 | 8.93 | 2.0 | 18 | 8.1 | 178 |
| I-6–M-9 | C | 3.19 | 5 | — | 5 | 7.0 | 0.38 | 8.48 | 1.0 | 18 | 5.9 | 40 |
| M-9–M-8 | A + B + C | 6.46 | — | 0.21 | 5.80 | 6.9 | 0.38 | 16.90 | 1.8 | 21 | 8.5 | 110 |
| M-8–M-7 | A + B + C | 6.46 | — | 0.11 | 6.01 | — | 0.38 | 16.90 | 1.8 | 21 | 8.5 | 57 |
| M-7–M-6 | A + B + C | 6.46 | — | 0.47 | 6.12 | — | 0.38 | 16.90 | 1.6 | 21 | 8.1 | 230 |
| I-5–M-6 | D | 2.66 | 5 | — | 5 | 7.0 | 0.40 | 7.43 | 2.0 | 15 | 7.4 | 19 |
| M-6–M-5 | A + B + C + D | 9.12 | — | 0.38 | 6.59 | 6.8 | 0.39 | 24.20 | 2.0 | 24 | 10.0 | 230 |
| M-5–M-4 | A + B + C + D | 9.12 | — | 0.42 | 6.97 | — | 0.39 | 24.20 | 1.9 | 24 | 9.8 | 247 |
| I-1–M-4 | O | 0.40 | 5 | — | 5 | 7.0 | 0.57 | 1.59 | 3.0 | 15 | 9.0 | 19 |
| I-2–M-4 | H | 2.25 | 5 | — | 5 | 7.0 | 0.35 | 5.52 | 3.0 | 15 | 9.0 | 17 |
| M-4–M-2 | A + B + C + D + O + H | 11.77 | — | 0.05 | 7.39 | 6.6 | 0.39 | 30.3 | 1.5 | 27 | 9.4 | 29 |
| I-3–M-3 | M | 2.75 | 5 | — | 5 | 7.0 | 0.40 | 7.70 | 2.0 | 15 | 7.4 | 15 |
| I-4–M-3 | N | 1.27 | 5 | — | 5 | 7.0 | 0.40 | 3.55 | 2.0 | 15 | 7.4 | 20 |
| M-3–M-2 | M + N | 4.02 | — | — | — | 7.0 | 0.40 | 11.30 | 1.8 | 18 | 7.8 | 37 |
| M-2–M-1 | A + B + C + D + O + H + M + N | 15.79 | — | — | 7.44 | 6.6 | 0.39 | 40.6 | 1.4 | 30 | 9.8 | 176 |

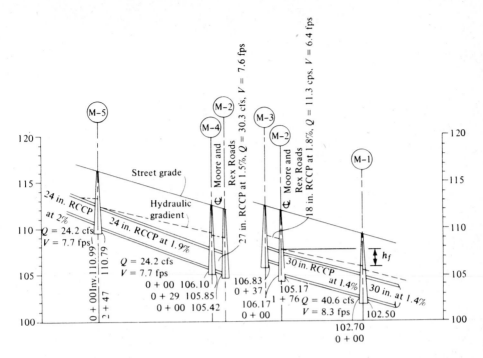

**Figure 7.18** Profile of part of the Mextex storm drain, showing the hydraulic gradient.

9. Compute the position of the hydraulic gradient along the profile of the pipe. If this gradient lies less than 1.5 ft below the ground surface, it must be lowered to preclude the possibility of surcharge during the design flow. Note that the value of 1.5 ft is arbitrarily chosen here. In practice, local standards indicate the limiting value. Hydraulic gradients may be lowered by increasing pipe sizes, decreasing head losses at structures, designing special transitions, lowering the system below ground, or a combination of these means.

Computations for a portion of the hydraulic gradient of the example will now be given. Head losses in the pipes are determined by applying Manning's equation, assuming $n = 0.013$ in this example. Head losses in the structures will be determined by using the relationships defined in Figs. 7.19 and 7.20. These curves were developed for surcharged pipes entering rectangular structures but may be applied to wye branches, manholes, and junction chambers as pictured on the curves [19]. The $A$ curve is used to find entrance and exit losses, the $B$ curve to evaluate the head loss due to an increased velocity in the downstream direction. The loss is designated as the difference between the head losses found for the downstream and upstream pipe ($V_{h-2} - V_{h-1}$). In cases where the greatest velocity occurs upstream, the difference will be negative and may be applied to offset other losses in the structure. The $C$ loss results from a change in direction in a manhole, wye branch, or bend structure. The $D$ loss is related to the effects produced by the entrance of secondary flows into the structure. Several examples of the use of these curves are shown in Figs. 7.19 and 7.20.

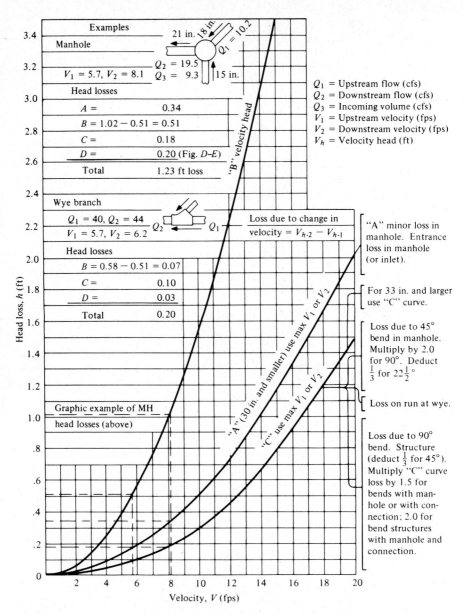

**Figure 7.19** Types $A$, $B$, and $C$ head losses in structures. (Courtesy of the Baltimore County, Maryland, Department of Public Works.)

Computations for the hydraulic gradient shown in Fig. 7.18 are as follows:

1. Begin at the elevation of the hydraulic gradient at the upstream end of the existing 30-in. reinforced-concrete culvert pipe (RCCP). This elevation is 105.50. The existing hydraulic gradient is shown in Fig. 7.18.
2. Compute the head losses in manhole M-1 using Figs. 7.19 and 7.20.

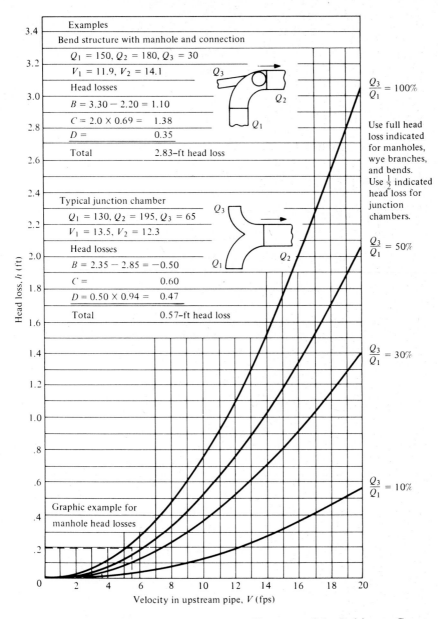

**Figure 7.20** Type $D$ head loss in structures. (Courtesy of the Baltimore County, Maryland, Department of Public Works.)

$A$ loss $= 0.36$ ft ($V = 8.3$ fps $= Q/A$)
$B$ loss $= 0$; no change in velocity, $V_1 = V_2$
$C$ loss $= 0$; no change in direction
$D$ loss $= 0$; no secondary flow
   total $= 0.36$ ft

The hydraulic gradient therefore rises in the manhole to an elevation of $105.50 + 0.36 = 105.86$ ft, plotted in M-1 in Fig. 7.18.

3. Compute the head loss due to friction in the 30-in. drain from M-1 to M-2. Assume that $n = 0.013$. Using Manning's equation, the head loss per linear foot of drain is

$$S = \frac{(nV)^2}{2.21R^{4/3}}$$

and from M-1 to M-2,

$$S = \frac{(0.013 \times 8.3)^2}{2.21 \times 0.534} = 0.00986$$

The total frictional head loss is therefore

$$hf = S \times L = 0.00986 \times 176 = 1.73 \text{ ft}$$

Elevation of the hydraulic gradient at the downstream end of M-2 is thus $105.86 + 1.73 = 107.59$ ft. This elevation is plotted in Fig. 7.18, where the hydraulic gradient in this reach is drawn in.

4. Compute the head losses in M-2.

$A = 0.36 \ (V = 8.3 \text{ fps})$

$B = 1.07 - 0.90 = 0.17 \ (V_2 = 8.3, \ V_1 = 7.6 = Q/A \text{ for 27-in. drain})$

$C = 0.20 \times 2.0$ (multiply by 2 for 90° bend in manhole—see Fig. 7.19)

$\quad = 0.40$

$D = 0.22$ for $Q_3/Q_1 = 11.3/30.3 = 37\%$

Total head loss in M-2 equals 1.15 ft and the elevation of the hydraulic gradient in M-2 is therefore $107.59 + 1.15 = 108.74$ ft.

5. Compute the friction head loss in the section of pipe from M-2 to M-3.

$$S = \frac{(0.013 \times 6.4)^2}{2.21 \times 0.272} = 0.0113$$

$$hf = 0.0113 \times 37 = 0.42 \text{ ft}$$

The elevation of the hydraulic gradient at the downstream end of M-3 is therefore $108.74 + 0.42 = 109.16$ ft. Plot this point on the profile and draw the gradient from M-2 to M-3. ∎

The hydraulic gradient in this example was computed under the assumption of uniform flow. In closed-conduit systems, if the pipes are flowing full or the system is surcharged (the usual design flow conditions), the method will produce good results. Where open conduits are used, or in partial-flow systems, the hydraulic gradient can be determined by computing surface profiles in the manner described in Section 7.2.

Computations for the remainder of the hydraulic gradient are identical to those just given and will not be presented. It should be noted that none of the gradients shown in Fig. 7.18 come within 1.5 ft of the surface, so no revisions are needed. If too close to the surface, it would be necessary to modify all or a portion of the system to lower the gradient. This could be accomplished by reducing head losses, by increasing the depth of the system, or both. The choice would depend primarily on cost.

## 7.19 STORMWATER DETENTION/RETENTION

Historically, stormwater was considered a resource to be disposed of quickly. The notion was to remove it from the contributing area as rapidly as possible. More recently, however, stormwater flows have been looked upon as a resource to be captured and utilized for groundwater recharge, recreation, and other purposes and/or detained as a water-quality control or peak-flow reduction measure. The principal mechanism employed is storage. This approach is also consistent with the regulatory policies in many states that require storage for water-quality control in urban areas [20].

The idea behind detention storage is to reduce the rate of delivery of stormwater at a discharge point. Retention is employed where no downstream releases are to be made. In this case, the volumes of water generated are ultimately infiltrated and/or evaporated. Storage requirements for detention and retention basins are estimated by analyzing anticipated storm events. Flows from detention basins are calculated by using standard routing techniques [10, 20]. Most detention basins have a low-flow outlet and an emergency spillway or other type of control structure for dealing with high flows. Retention and detention basins may comprise parking lot storage, earth ponds, concrete tanks, depressed park lands, roofs, and capacity within the drains themselves.

Figure 7.21 illustrates a typical detention pond configuration. Note that

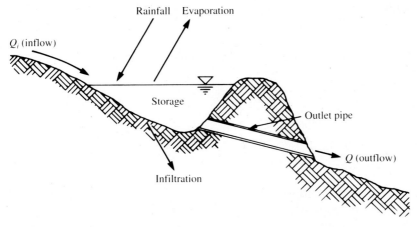

**Figure 7.21** A detention basin configuration.

detention basin design and analysis is founded on storage–discharge relationships. If in Fig. 7.21, $Q_i$ represents the inflow to the pond, $Q$ represents the
outflow, and $S$ represents storage in the pond, then a continuity equation of
the form

$$Q_i - Q = dS/dt \qquad (7.14)$$

can be written, where $dS/dt$ is the rate of change of storage with respect to
time. If one considers that storage increments can be calculated as the
product of mean surface area of the pond, at a specified depth, and the depth
increment, it can be stated that

$$\frac{dS}{dt} = A(h)\frac{dh}{dt} \qquad (7.15)$$

where $A(h)$ is a relationship equating pond surface area $A$ to a reference depth
(or elevation), and $dh/dt$ is the rate of change of depth with respect to
time. Depth–area relationships for reservoirs can be obtained by developing
contours for the reservoir and then relating the surface areas contained within
these contours to the elevations associated with them. Incremental volumes
can be determined by using average end-area or other formulas. Often, a
relationship of the form

$$A = Kh^m \qquad (7.16)$$

can be derived relating surface area to depth. In Eq. 7.16, $A$ is the surface
area, $K$ and $m$ are constants for the site, and $h$ is the depth of water in the
reservoir above a reference point such as the elevation of the centerline of
the outlet [2, 20]. Where the outlet is a pipe as shown in Fig. 7.21, the outflow
can be calculated from the orifice equation

$$Q = C_d A_o (2gh)^{0.5} \qquad (7.17)$$

where $Q$ is the outflow, $C_d$ is a discharge coefficient, $A_o$ is the orifice area, $h$ is
the elevation above the orifice, and $g$ is the acceleration of gravity. Discharge
coefficients for submerged pipes behaving as orifices range from about 0.62
to 1.0, depending on whether they are sharp-edged or well rounded [21].

If we insert Eqs. 7.17 and 7.15 in Eq. 7.14, the result is

$$Q_i - C_d A_o (2gh)^{0.5} = A(h)\frac{dh}{dt} \qquad (7.18)$$

Solving for $dh/dt$, we get

$$\frac{dh}{dt} = \frac{Q_i - C_d A_o (2gh)^{0.5}}{A(h)} \qquad (7.19)$$

which is the governing differential equation for depth as a function of time.
Specifying the right side of Eq. 7.19 as $f(h, t)$, then

$$dh/dt = f(h, t) \qquad (7.20)$$

This equation can be solved numerically in several ways. The Euler method is described here and is applied to an example which follows. This procedure is simple but coarse [20, 22].

If a time step of $\Delta t$ is assumed, then the derivative in Eq. 7.20 can be approximated as

$$\frac{dh}{dt} = \frac{h(t + \Delta t) - h(t)}{\Delta t} \tag{7.21}$$

If the right side of Eq. 7.20 is substituted into this relationship for $dh/dt$, the equation can be solved for $h(t + \Delta t)$ as follows:

$$h(t + \Delta t) = h(t) + \Delta t f(h, t) \tag{7.22}$$

Note that the function $h(t)$ is always evaluated at time step $t$ rather than at step $t + \Delta t$. Since $f(h, t)$ is quantified by Eq. 7.19, the calculation of successive values of $h$ can be accomplished in an organized and straightforward manner. The procedure is illustrated in Example 7.9.

### ■ EXAMPLE 7.9
Consider the sketch of Fig. 7.21 and the input hydrograph given in column 1 of Table 7.13. In addition, consider that the discharge pipe is 8 in. in diameter (0.203 m), that $C_d$ is 0.9, that the pond surface area is related to elevation by the expression $A = 400h^{0.7}$ (where $h$ is the depth above the culvert centerline and $A$ is the surface area in m²), and the initial depth in the pond above the outlet centerline is 0.5 m.

*Solution.*   A spreadsheet analysis is suggested. For this example, QUATTRO PRO was used. For ease in calculations, we calculate the product of $C_d A_o$ to be 0.02918. Substituting this value, the numerical value of the product $2g = (9.8 \text{ m/sec}) \times 2 = 19.6$, and the relationship for $h$ in Eq. 7.19, we get

$$\frac{dh}{dt} = \frac{Q_i(t) - 0.02918(19.6h)^{0.5}}{400h^{0.7}}$$

With this relationship, a solution for successive values of $h$, at quarter-hour (900-sec) time steps, can be carried out as illustrated in Table 7.13. In the table, the values in columns 1 and 2 are given, as is the initial value of depth (0.5 m) in column 3. Column headings are described in the table footnotes and follow the terminology of Eqs. 7.14 to 7.22. The calculated outflow hydrograph is plotted on Fig. 7.22 along with the inflow hydrograph. It is easy to see from the plot that the inflow rates are significantly attenuated by the detention structure. Note that the drop in outflow at time step 1 occurs because there was no inflow at $t = 0$ and there was already a positive head on the outlet of 0.5 m at the start of inflow. Accordingly, the system was draining at that time.                                                                            ■

In the design mode, it is easy to see that for selected inflow sequences and reservoir and outlet dimensions, a determination can be made of flow attenuation, and the aforementioned parameters can be modified if the amount of attenuation is unacceptable.

**TABLE 7.13**   SPREADSHEET DETENTION POND ANALYSIS

| (1) Time (hr) | (2) $Q_{in}$ (cms) | (3) $h(t)$ (m) | (4) $Q_{out}$ (cms) | (5) A-surf m$^2$ | (6) $dh/dt$ m/sec | (7) $h(t + dt)$ (m) |
|---|---|---|---|---|---|---|
| 0 | 0 | 0.5 | 0.09141 | 246.2289 | −0.00037 | 0.165882 |
| 0.25 | 0.18 | 0.165882 | 0.064552 | 113.7418 | 0.001015 | 1.079381 |
| 0.5 | 0.36 | 1.079381 | 0.164664 | 421.9708 | 0.000463 | 1.496004 |
| 0.75 | 0.54 | 1.496004 | 0.193855 | 530.2892 | 0.000653 | 2.083476 |
| 1 | 0.72 | 2.083476 | 0.228773 | 668.6702 | 0.000735 | 2.744645 |
| 1.25 | 0.9 | 2.744645 | 0.262575 | 810.9617 | 0.000786 | 3.452055 |
| 1.5 | 1.1 | 3.452055 | 0.294476 | 952.1697 | 0.000846 | 4.213444 |
| 1.75 | 0.99 | 4.213444 | 0.325334 | 1094.728 | 0.000607 | 4.75988 |
| 2 | 0.88 | 4.75988 | 0.345787 | 1192.277 | 0.000448 | 5.163135 |
| 2.25 | 0.77 | 5.163135 | 0.360137 | 1262.116 | 0.000325 | 5.455403 |
| 2.5 | 0.66 | 5.455403 | 0.37019 | 1311.713 | 0.000221 | 5.65425 |
| 2.75 | 0.55 | 5.65425 | 0.376876 | 1345.001 | 0.000129 | 5.770095 |
| 3 | 0.44 | 5.770095 | 0.380717 | 1364.231 | 4.35E-05 | 5.809204 |
| 3.25 | 0.33 | 5.809204 | 0.382005 | 1370.698 | −3.8E-05 | 5.775058 |
| 3.5 | 0.22 | 5.775058 | 0.380881 | 1365.053 | −0.00012 | 5.668986 |
| 3.75 | 0.11 | 5.668986 | 0.377367 | 1347.453 | −0.0002 | 5.490405 |
| 4 | 0 | 5.490405 | 0.371375 | 1317.598 | −0.00028 | 5.236733 |
| 4.25 | 0 | 5.236733 | 0.362695 | 1274.683 | −0.00028 | 4.980649 |
| 4.5 | 0 | 4.980649 | 0.353715 | 1230.723 | −0.00029 | 4.721985 |
| 4.75 | 0 | 4.721985 | 0.344408 | 1185.625 | −0.00029 | 4.460547 |
| 5 | 0 | 4.460547 | 0.334738 | 1139.283 | −0.00029 | 4.196114 |
| 5.25 | 0 | 4.196114 | 0.324664 | 1091.574 | −0.0003 | 3.928429 |
| 5.5 | 0 | 3.928429 | 0.314138 | 1042.349 | −0.0003 | 3.657191 |
| 5.75 | 0 | 3.657191 | 0.303099 | 991.433 | −0.00031 | 3.382045 |
| 6 | 0 | 3.382045 | 0.291475 | 938.6108 | −0.00031 | 3.10256 |
| 6.25 | 0 | 3.10256 | 0.279172 | 883.6172 | −0.00032 | 2.818213 |
| 6.5 | 0 | 2.818213 | 0.266071 | 826.1171 | −0.00032 | 2.528346 |
| 6.75 | 0 | 2.528346 | 0.252017 | 765.6768 | −0.00033 | 2.232118 |
| 7 | 0 | 2.232118 | 0.236793 | 701.7169 | −0.00034 | 1.928414 |
| 7.25 | 0 | 1.928414 | 0.220096 | 633.4322 | −0.00035 | 1.615695 |
| 7.5 | 0 | 1.615695 | 0.201461 | 559.6437 | −0.00036 | 1.291713 |
| 7.75 | 0 | 1.291713 | 0.180133 | 478.4936 | −0.00038 | 0.952899 |
| 8 | 0 | 0.952899 | 0.154716 | 386.7167 | −0.0004 | 0.592832 |
| 8.25 | 0 | 0.592832 | 0.122033 | 277.4036 | −0.00044 | 0.196912 |
| 8.5 | 0 | 0.196912 | 0.070331 | 128.2479 | −0.00055 | −0.29665 |

(1) Time sequence

(2) Inflow data

(3) Head above outlet centerline, m

(4) Outflow calculated using orifice equation

(5) Surface area of pond calculated using power relationship

(6) $dh/dt$ = (inflow − outflow)/(surface area)

(7) $h(t + dt) = h(t) + dt[f(h, t)]$

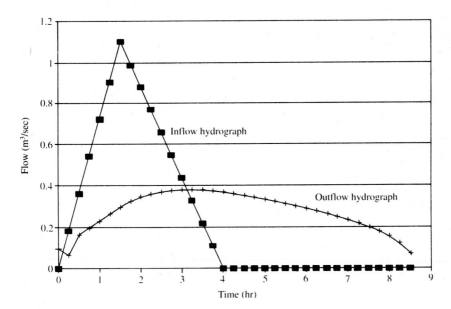

**Figure 7.22** Detention pond hydrographs.

# PROBLEMS

**7.1** Wastewater flows in a rectangular concrete channel 6.0 ft wide and 3.0 ft deep. The design flow is 30 cfs. Find the critical velocity. Also find the slope of the channel if the flow velocity is to be 2.5 fps.

**7.2** A 16-in. sewer pipe flowing full is expected to carry 5.8 cfs. The $n$ value is 0.011. The minimum flow is $\frac{1}{12}$ that of the maximum. Find the depth and velocity at minimum flow.

**7.3** The sewage from an area of 400 acres is to be carried by a circular sewer at a velocity not less than 2.5 fps. Manning's $n = 0.013$. The population density is 16 persons per acre. Find the maximum hourly and minimum hourly flows. Determine the pipe size required to handle these flows and the required slope.

**7.4** Water flows in a rectangular concrete channel 9.5 ft wide and 8 ft deep. The channel invert has a slope of 0.10% and the applicable $n$ value is 0.013. The flow carried by the channel is 189 cfs. At an intersection of this channel with a canal, the depth of flow is 7.1 ft. Find the distance upstream to a point where normal depth prevails. Plot the surface profile.

**7.5** A 54-in. sewer ($n = 0.013$) laid on a 0.17% grade carries a flow of 19 mgd. At a junction with a second sewer, the sewage depth is 36 in. above the invert. Plot the surface profile back to the point of uniform depth.

**7.6** Determine the minimum velocity and gradient required to transport $\frac{3}{16}$-in. gravel through a 42-in.-diameter pipe, given $n = 0.013$ and $K = 0.05$.

**7.7** An inverted siphon is to carry a minimum dry-weather flow of 1.5 cfs, a maximum dry-weather flow of 3.7 cfs, and a storm flow of 48.0 cfs in three pipes. Select the proper diameters to ensure velocities of 3.0 fps in all pipes. Make a detailed sketch of your design. Assume that the siphon goes under a highway with a 3.2-ft drop and is 73 ft long.

**7.8** Given the following 25-yr record of 24-hr maximum annual stream flows (cfs), plot on log-probability paper these values versus the percent of years during which runoff was equal to or less than the indicated value. Find the peak flow expected on the average of  (a) once every 5 yr;  (b) once every 15 yr.

| | | | | |
|---|---|---|---|---|
| 220 | 196 | 89 | 53 | 47 |
| 200 | 129 | 76 | 50 | 38 |
| 218 | 142 | 62 | 52 | 36 |
| 199 | 127 | 67 | 49 | 32 |
| 180 | 118 | 54 | 47 | 28 |

**7.9** Given the following unit storm, storm pattern, and unit hydrograph, determine the composite hydrograph.

Unit storm = 1 unit of rainfall for 1 unit of time

| Actual storm | (time units) | 1 | 2 | 3 | 4 |
|---|---|---|---|---|---|
| Pattern | (rainfall units) | 1 | 2 | 4 | 3 |

Unit hydrograph: triangular with base length of 4 time units, time of rise of 1 time unit, and maximum ordinate of $\frac{1}{2}$ rainfall unit height.

**7.10** Given a unit rainfall duration of 1 time unit, an effective precipitation of 1.5 in., and the following hydrograph and storm sequence, determine  (a) the unit hydrograph and  (b) the hydrograph for the given storm sequence.

*Storm Sequence*

| Time units | 1 | 2 | 3 | 4 |
|---|---|---|---|---|
| Precipitation (in.) | 0.4 | 1.1 | 1.8 | 0.9 |

*Hydrograph*

| Time units | 1 | 2 | 3 | 4 | 4.5 | 5 | 6 |
|---|---|---|---|---|---|---|---|
| Flow (cfs) | 101 | 96 | 218 | 512 | 610 | 580 | 460 |

| Time units | 7 | 8 | 9 | 10 | 11 | 12 | 13 |
|---|---|---|---|---|---|---|---|
| Flow (cfs) | 320 | 200 | 180 | 100 | 86 | 60 | 50 |

**7.11** Do Prob. 7.10 if the storm sequence is as follows:

| Time units | 1 | 2 | 3 | 4 |
|---|---|---|---|---|
| Precipitation (in.) | 0.6 | 1.3 | 1.5 | 0.8 |

**7.12** Compute, by the rational method, the design flows for the pipes shown in the accompanying sketch: I-1 to M-4, M-4 to M-3, M-3 to M-2, M-2 to M-1, and M-1 to outfall. A-1 = 2.1 acres, $c_1$ = 0.5; A-2 = 3.0 acres, $c_2$ = 0.4; A-3 = 4.1 acres, $c_3$ = 0.7. Given pipe flow times are I-1 to M-3, 1 min, and M-3 to M-1, 1.5 min.

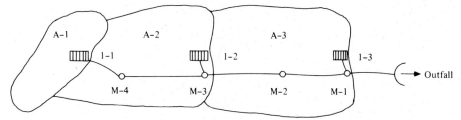

**7.13** Given the data below, design a section of sanitary sewer to carry these flows at a minimum velocity of 2.5 fps. Assume that the minimum depth of the sewer below the street must be 8.5 ft and that an $n$ value of 0.013 is applicable. Make a plan and profile drawing. (*Note:* The invert into manhole F must be 937 ft. Assume an invert drop of 0.2 ft across each manhole.)

| Manholes | Distance between manholes (ft) | Flow (gpm) | Street elevations at manholes (ft) |
|---|---|---|---|
| A to B | 180 | 1200 | A—978 |
| B to C | 300 | 1450 | B—972 |
| C to D | 450 | 2200 | C—964 |
| D to E | 300 | 2500 | D—958 |
| E to F | 400 | 3000 | E—954 |
|  |  |  | F—949 |

**7.14** For an inlet area of 2.0 acres having an imperviousness of 0.50 and a clay soil, find the peak rate of runoff for the 5-, 10-, and 20-yr storms.

**7.15** Using the manhole spacing-elevation data given in Prob. 7.13 and the data below, design a storm drainage system such that manhole A is at the upper end of the drainage area and manhole F is replaced by an outfall to a stream. The outfall invert elevation is equal to 940.3 ft. Design for the 5-yr storm.

| Inflow point | Incremental area contributing to inflow point (acres) | Imperviousness of areas (%) |
|---|---|---|
| A to B | 3.0 | 41 |
| B to C | 2.7 | 50 |
| C to D | 5.3 | 55 |
| D to E | 3.6 | 45 |
| Total area = 14.6 acres |  |  |

**7.16** An impounding reservoir is to provide for constant withdrawal of 430 million gallons per square mile per year. The following record of monthly inflow values was selected from a critical period and was chosen as a basis for design. Find the amount of storage required in $Mg/mi^2$.

| Month | J | F | M | A | M | J | J | A | S | O | N | D |
|---|---|---|---|---|---|---|---|---|---|---|---|---|
| Gauged inflow ($Mg/mi^2/mo$) | 23 | 48 | 53 | 87 | 12 | 9 | 7 | 1 | 22 | 34 | 98 | 28 |

**7.17** Given the following storm pattern, unit storm, and unit hydrograph, determine the composite hydrograph.

Unit storm = 1 in. effective rainfall in 1 hr

| Actual storm pattern    Time | Volume of effective rainfall (in.) |
|---|---|
| 1:00 | 2 |
| 2:00 | 2.5 |
| 3:00 | 7 |
| 4:00 | 3 |

Unit hydrograph (triangular) with base length = 5 hr, time to peak flow = 1.5 hr, and maximum ordinate = 0.4 in./hr.

**7.18** Using the derived unit hydrograph from Example 7.4, calculate the storm hydrograph for the following rainstorm pattern: time unit 1, rainfall = 1 in.; time unit 2, rainfall = 3 in.; time unit 3, rainfall = 2 in. Assume the storm data are effective rainfall.

**7.19** The following hydrograph was measured on a creek for a storm that produced a total of 1.9 in. of rainfall in 2 hr. Losses due to depression storage, infiltration, and interception amounted to 0.4 in.

| | Time from beginning of storm (hr) | Discharge ($ft^3/sec$) |
|---|---|---|
| | 0 | 73 |
| Storm begins | 6 | 75 |
| | 12 | 900 |
| | 18 | 3400 |
| | 24 | 4350 |
| | 30 | 2900 |
| | 36 | 1450 |
| | 42 | 600 |
| | 48 | 450 |
| | 54 | 375 |
| | 60 | 250 |
| | 66 | 180 |
| | 72 | 135 |

A spillway design storm is 10 in. of effective rainfall in 8 hr. Estimate the inflow hydrograph for the design storm.

**7.20** The ordinates of a 2-hr unit hydrograph are given in the following table. Derive a 4-hr unit hydrograph from the 2-hr unit hydrograph by    (a) the lagging method and    (b) the S-hydrograph method.

| Time unit | 2-hr Unit hydrograph ordinates |
|:---------:|:------------------------------:|
| 0 | 0 |
| 1 | 200 |
| 2 | 400 |
| 3 | 300 |
| 4 | 200 |
| 5 | 100 |
| 6 | 0 |

**7.21** Find the peak hourly flow in mgd and m³/s for a 1000-acre urban area having the following features: domestic flows 100 gpcd, commercial flows 15 gpcd, infiltration 650 gpd/acre, and population density 20 persons per acre.

**7.22** A sewer with a flow of 11 cfs enters manhole 2. The distance downstream to manhole 3 is 450 ft. The finished street surface elevation at manhole 2 is 165.0 ft; at manhole 3 it is 162.8 ft. Note that the crown of the pipe must be at least 6 ft below the street surface at each manhole. For a Manning's $n$ = 0.013, find the required standard size pipe to carry the flow under   (a) full flow and   (b) one-half full flow. The flow velocity must be at least 2 ft/sec and must not exceed 10 ft/sec.

**7.23** A sewer system is to be designed for the urban area of Fig. 7.8. Street elevations at manhole locations are shown on the figure. It has been determined that the population density is 50 persons per acre and that the sewage contribution per capita is 100 gpd. In addition, there is an infiltration component of 650 gpad. Local regulations require that no sewers be less than 8 in. in diameter. Assume an $n$ value of 0.013.

**7.24** Given an invert elevation of 105.1 for the pipe leaving M-6 in Fig. 7.8, and using the pipe sizes determined in Prob. 7.23, calculate the invert elevations in and out of M-5 and M-4 if the street elevation at M-5 is 108.95 ft and at M-4 is 106.82 ft.

**7.25** Solve Prob. 7.23 if the population density is 45 persons per acre and the infiltration component is 625 gpad.

**7.26** Find the peak hourly flow in mgd and m³/s for a 700-acre urban area having the following features: domestic flows 100 gpcd, commercial flows 18 gpcd, infiltration 650 gpd/acre, and population density 25 persons per acre.

**7.27** A sewer with a flow of 8 cfs enters manhole 2. The distance downstream to manhole 3 is 380 ft. The finished street surface elevation at manhole 2 is 135.0 ft; at manhole 3 it is 133.9 ft. Note that the crown of the pipe must be at least 6 ft below the street surface at each manhole. For a Manning's $n$ = 0.013, find the required standard size pipe to carry the flow under   (a) full flow and   (b) one-half full flow. The flow velocity must be at least 2 ft/sec and must not exceed 10 ft/sec.

**7.28** Modify the data given in Example 7.9 so that the initial depth is 0.6 m and the outlet pipe is 10 in. in diameter. Use a spreadsheet approach to solve the problem.

**7.29** Solve Prob. 7.28 if the inflows given are all increased by 10%.


# REFERENCES

1. S. M. Woodward and C. J. Posey, *Hydraulics of Steady Flow in Open Channels* (New York: Wiley, 1955).
2. G. M. Fair and J. C. Geyer, *Water Supply and Waste-Water Disposal* (New York: Wiley, 1954).
3. K. E. Spells, *Trans. Inst. Chem. Engrs.* (London) 33 (1955).
4. R. S. Gupta, *Hydrology and Hydraulic Systems* (Englewood Cliffs, NJ: Prentice-Hall, 1989).
5. H. W. Gehm and J. I. Bregman, editors, *Handbook of Water Resources and Pollution Control* (New York: Van Nostrand Reinhold, 1976).
6. "Design and Construction of Sanitary and Storm Sewers," *Manual of Engineering Practice No. 37* (New York: American Society of Civil Engineers, 1960).
7. E. L. Thackston, "Notes on Sewer System Design," unpublished, Vanderbilt University, 1982.
8. W. Viessman, Jr., and C. Welty, *Water Management: Technology and Institutions* (New York: Harper & Row, 1985).
9. Metcalf and Eddy, Inc., *Wastewater Engineering* (New York: McGraw-Hill, 1972).
10. W. Viessman, Jr., G. L. Lewis, and J. W. Knapp, *Introduction to Hydrology* (New York: Harper & Row, 1989).
11. L. K. Sherman, "Streamflow from Rainfall by the Unit Hydrograph Method," *Eng. News-Record* 108 (1932): 501–505.
12. R. K. Linsley, Jr., M. A. Kohler, and J. L. H. Paulhus, *Applied Hydrology* (New York: McGraw-Hill, 1949), pp. 387–411, 444–464.
13. B. S. Barnes, "Consistency in Unitgraphs," *Proc. Am. Soc. Civil Engrs., J. Hydraulics Div.* 85 (August 1959): 39–61.
14. C. F. Izzard, "Hydraulics of Runoff from Developed Surfaces, *Proc. Highway Res. Bd.* 26 (1946): 129–150.
15. J. C. Schaake, "Report No. XI of the Storm Drainage Research Project," The Johns Hopkins University, Department of Sanitary Engineering, Baltimore, (September 1963).
16. N. H. Crawford and R. K. Linsley, "Digital Simulation in Hydrology: Stanford Watershed Model IV," Department of Civil Engineering, Stanford University, Stanford, California, Technical Report 39 (July 1966).
17. Hydrocomp International, Inc., "Hydrocomp Simulation Programming Operations Manual," 4th ed., Palo Alto, California (January 1976).
18. "The Design of Storm Water Inlets," The Johns Hopkins University, Department of Sanitary Engineering and Water Resources, Baltimore (June 1956).
19. "Baltimore County Design Manual," Baltimore County Department of Public Works, Towson, MD (1955).
20. P. B. Bedient and W. C. Huber, *Hydrology and Flood Plain Analysis* (New York: Addison-Wesley, 1988).
21. R. L. Daugherty, J. G. Franzini, and E. J. Finnemore, *Fluid Mechanics with Engineering Applications* (New York: McGraw Hill, 1985).
22. R. W. Hornbeck, *Numerical Methods,* (Englewood Cliffs, NJ: Prentice-Hall, 1982).

# Chapter 8
# Water Quality

The common uses of water most dependent on quality are domestic and industrial water supplies, propagation of fish and wildlife, water-based recreation, shellfish culturing, and esthetic enjoyment. Drinking water, considered the highest beneficial use, must be free of health hazards, that is, pathogens, toxins, and carcinogens. Esthetic factors, such as temperature, clarity, taste and odor, and chemical balance, are also important.

## Microbiological Quality

A variety of pathogens are present in domestic wastewater with the kinds and concentrations relating to the health of the contributing community. Although a reduction of pathogens occurs in conventional wastewater treatment, even after chlorination the effluent still contains persistent pathogens. (For unrestricted water reuse, filtration after chemical coagulation and chlorination with an extended contact time are necessary to remove all pathogens.) Thus, surface waters downstream from wastewater discharges may contain pathogens, and their removal is required in water supply treatment. Since testing water samples for pathogens is not practical, nonpathogenic fecal coliform bacteria are used as indicators for the potential presence of pathogens. The following sections discuss waterborne diseases and coliform bacteria as indicator organisms.

### 8.1 WATERBORNE DISEASES

Many human diseases are transmitted by the feces of an infected person getting into the mouth of another person. This travel from anus to mouth, referred to as the fecal-oral route, may be direct from person to person by contaminated fingers or indirect through food or water. In addition, some pathogens may reinfect by inhalation of dust or aerosol droplets, and a few (notably hookworm) can penetrate through the skin. Some of these communicable diseases are endemic in form; that is, they are native to a particular region or country. The four major groups of pathogens are viruses, bacteria, protozoans, and helminths (worms).

## Pathogens Excreted in Human Feces

Viruses are obligate, intracellular parasites that replicate only in living hosts' cells. Being composed of only complex organic compounds, they lack the metabolic systems for self-reproduction. The size range of enteric viruses is 20–100 nm, about 1/50 the size of bacterial cells, and require electron microscopy for viewing. Human feces contain over 100 serotypes of enteric viruses. Of those, the groups listed in Table 8.1 are most likely to be transmitted by water. Persons infected by ingesting these viruses do not always become ill, but disease is a possibility in persons infected with any of the enteric viruses, particularly the hepatitis A virus. Several diseases involving the central nervous system, and more rarely the skin and heart, are caused by enteroviruses. Waterborne outbreaks of infectious hepatitis have occurred, but the most common route of transmission is by person-to-person contact. Infectious hepatitis may cause diarrhea and jaundice, and result in liver damage. Human beings are the reservoir for all of the enteric viruses.

Bacteria are microscopic single-celled plants that use soluble food and are capable of self-reproduction without sunlight. Their approximate range in size is 0.5–5 $\mu$m (500–5000 nm). The feces of healthy persons contain 1 to 1000 million of each of the following groups of bacteria per gram: enterobacteria, enterococci, lactobacilli, clostridia, bacteroides, bifidobacteria, and eubacteria. *Escherichia coli,* the common fecal coliform, is in the enterobacteria group. For many bacterial infections of the intestines, the major symptom is diarrhea. The most serious waterborne diseases are typhoid fever, paratyphoid fever, dysentery, and cholera. Typhoid and paratyphoid result in high fever and infection of the spleen, gastrointestinal tract, and blood. Dysentery causes diarrhea, bloody stools, and sometimes fever. Cholera symptoms are diarrhea, vomiting, and dehydration. While all of these diseases are debilitating and can cause death if untreated, their transmission can be controlled by pasteurization of milk, sanitary disposal of wastewater, and disinfection of water supplies.

Protozoans infecting humans are intestinal parasites that replicate in the host and exist in two forms. Trophozoites live attached to the intestinal wall where they actively feed and reproduce. At some time during the life of a trophozoite, it releases and floats through the intestines while making a morphologic transformation into a cyst for protection against the harsh environment outside of the host. This cyst form is infectious for other persons by the fecal-oral route of transmission. The cysts are 10–15 $\mu$m in length, significantly larger than intestinal bacteria. The two most common protozoal diseases are diarrhea and dysentery (Table 8.1). *Entamoeba histolytica*

**TABLE 8.1**  TYPICAL PATHOGENS EXCRETED IN HUMAN FECES

| Pathogen group and name | Associated diseases | Category for transmissibility[a] |
|---|---|---|
| Virus | | |
|   Adenoviruses | Respiratory, eye infections | I |
|   Enteroviruses | | |
|     Polioviruses | Aseptic meningitis, poliomyelitis | I |
|     Echoviruses | Aseptic meningitis, diarrhea, respiratory infections | I |
|     Coxsackie viruses | Aseptic meningitis, herpangina, myocarditis | I |
|   Hepatitis A virus | Infectious hepatitis | I |
|   Reoviruses | Not well known | I |
|   Other viruses | Gastroenteritis, diarrhea | I |
| Bacterium | | |
|   *Salmonella typhi* | Typhoid fever | II |
|   *Salmonella paratyphi* | Paratyphoid fever | II |
|   Other salmonellae | Gastroenteritis | II |
|   *Shigella* spp. | Bacillary dysentery | II |
|   *Vibrio cholerae* | Cholera | II |
|     Other vibrios | Diarrhea | II |
|     *Yersinia enterocolitica* | Gastroenteritis | II |
| Protozoan | | |
|   *Entamoeba histolytica* | Amoebic dysentery | I |
|   *Giardia lamblia* | Diarrhea | I |
| Helminth | | |
|   *Ancylostoma duodenale* (Hookworm) | Hookworm | III |
|   *Ascaris lumbricoides* (Roundworm) | Ascariasis | III |
|   *Hymenolepis nana* (Dwarf tapeworm) | Hymenolepiasis | I |
|   *Necator americanus* (Hookworm) | Hookworm | III |
|   *Strongyloides stercoralis* (Threadworm) | Strongyloidiasis | III |
|   *Trichuris trichiura* (Whipworm) | Trichuriasis | III |

*Source:* Adapted from R. G. Feachem, D. J. Bradley, H. Garelick, and D. D. Mara, *Sanitation and Disease, Health Aspects of Excreta and Wastewater Management; World Bank Studies in Water Supply and Sanitation 3.* (Chichester: Wiley, 1983).

[a]  I = Nonlatent, low infective dose
  II = Nonlatent, medium to high infective dose, moderately persistent
 III = Latent and persistent

causes amoebic dysentery that is severely debilitating to the human host. (Amoebic dysentery, while common in tropical climates, is nontransmittable in temperate climates.) *Giardia lamblia* causes the less severe gastrointestinal infection of giardiasis, resulting in diarrhea, nausea, vomiting, and fatigue. Human beings are the reservoir for both of these infectious protozoans.

Helminths are intestinal worms that (except for *Strongyloides*) do not multiply in the human host. Therefore, the worm burden in an infected person is directly related to the number of infective eggs ingested. The worm burden is also related to the severity of the infected person's disease symptoms. Eggs are excreted in the host's feces. Of the helminths listed in Table 8.1, most can be transmitted by ingestion of contaminated water or food after a latent period of several days. Hookworms live in the soil and, after molting, can infect humans by penetrating their skin. With a heavy worm infection, the symptoms can be anemia, digestive disorder, abdominal pain, and debility. Helminth eggs are commonly 40–60 $\mu$m in length and denser than water.

Human carriers exist for all enteric diseases. Thus, in communities where a disease is endemic, a proportion of the healthy persons excrete pathogens in feces. In some infections, the carrier condition may cease along with symptoms of the illness, but in others it may persist for months, years, or a lifetime. The carrier condition exists for most bacterial and viral infections, including the dreaded diseases of cholera and infectious hepatitis. The asymptomatic carriers of *Entamoeba histolytica* and *Giardia lamblia* are primarily responsible for continued transmission of these intestinal protozoans. In light helminthic infections, the human host may have only minor symptoms of illness while passing eggs in feces for more than a year.

## Factors Affecting Transmission of Diseases

The transmission of waterborne diseases is influenced by latency, persistence, and infective dose of the pathogens. Latency is the period of time between excretion of a pathogen and its becoming infective to a new host. No excreted viruses, bacteria, and protozoans have a latent period. Among the helminths, only a few have eggs or larvae passed in feces that are immediately infectious to humans. The majority of helminths require a distinct latent period either for eggs to develop the infectious stage or to pass through an intermediate to complete their life cycles. For example, *Ascaris lumbricoides* has a latency of 10 days and hookworms about 7 days. Persistence is measured by the length of time that a pathogen remains viable in the environment outside a human host. The transmission of persistent microorganisms can follow a long route, for example, through a wastewater treatment system, and still infect persons located remotely from the original host. In general, persistence increases from viruses, the least persistent, to protozoans, to bacteria, to helminths having persistence measured in months. Infective dose is the number of organisms that must be ingested to result in disease. Usually, the minimum infective dose for viruses and protozoans are low and less than for bacteria, while a single helminth egg or larva can infect. Median infective dose is that dose required to infect half of those persons exposed [1].

The transmission characteristics of pathogens can be categorized based on latency, persistence, and infective dose, as shown in the right column of Table 8.1. Category I comprises infections that have a low median infective dose (less than 100) and are infective immediately upon excretion. These infections are transmitted person-to-person where personal and domestic hygiene are poor. Therefore, control of these diseases requires improvements in personal cleanliness and environmental sanitation, including food preparation, water supply, and wastewater disposal. Category II comprises all bacterial diseases having a medium to high median infective dose (greater than 10,000) and are less likely to be transmitted by person-to-person contact than category I infections. In addition to the control measures given for category I, wastewater collection, treatment, and reuse are of greater importance, particularly if personal hygiene and living standards are high enough to reduce person-to-person transmission. Category III contains soil-transmitted helminths that are both latent and persistent. Their transmission is less related to personal cleanliness since these helminth eggs are not immediately infective to human beings. Most relevant is the cleanliness of vegetables grown in fields exposed to human excreta by reuse of wastewater for irrigation and sludge for fertilization. Effective wastewater treatment is necessary to remove helminth eggs, and sludge stabilization is necessary to inactivate the removed eggs.

## Waterborne Diseases in the United States

Waterborne disease outbreaks are not a major cause of illness in the United States. The number of outbreaks reported during the 35-year period of 1946–1980 was 672, which affected more than 150,000 persons [2]. Based on a served population of 150,000,000, this is only an average illness rate of 4400 persons per year throughout the country or about 1 per 34,000 persons per year. During the 10-year period of 1971–1980, 121 outbreaks occurred in 65,000 community systems for an average of 1.9 per 10,000 community systems per year, and 157 outbreaks occurred in 150,000 noncommunity systems for an average of 1.0 per 10,000 systems per year.

The causative agent was not determined in half of the waterborne disease outbreaks reported [2]. The most common identified bacterial diseases were gastroenteritis (salmonellosis) and dysentery (shigellosis). Gastroenteritis is an inflammation of the lining membrane of the stomach and intestines, and dysentery is diarrhea with bloody stools and sometimes fever. The most serious viral disease identified as waterborne is infectious hepatitis, resulting in loss of appetite, fatigue, nausea, and pain. The most characteristic feature of the disease is a yellow color that appears in the white of the eyes and skin, hence the common name yellow jaundice. None of the outbreaks of waterborne infectious hepatitis have occurred in municipal water systems.

Giardiasis is the most common waterborne protozoal disease and is characterized by diarrhea that usually lasts one week or more and may be accompanied by abdominal cramps, bloating, flatulence, fatigue, and weight loss. A unique feature of giardiasis is transmission to humans through beavers serving as amplifying hosts. In mountainous regions, beavers have been infected by upstream contamination with human excreta containing *G. lamblia*. After being infected, they return millions of cysts to the water for every one ingested, amplifying the number of *Giardia* cysts in clear mountain streams.

Outbreaks of disease have occurred in downstream resorts and towns where the water supplies withdrawn from these streams were not properly treated to remove the cysts. Although all of these waterborne gastrointestinal diseases can stress sensitive individuals, they are not like the dreaded bacterial diseases of cholera and typhoid that resulted in hundreds of deaths in the United States during the early years of this century. "During the 35-year period since 1946 the rate (of deaths resulting from waterborne disease outbreaks) has been reduced to one death per year." [2]

## 8.2 COLIFORM BACTERIA AS INDICATOR ORGANISMS

Public water supplies, reclaimed waters, and wastewater effluents are not tested for pathogens to determine microbiological quality. Laboratory analyses for pathogens are difficult to perform, quantitatively unreliable, and, for some pathogenic microorganisms, impossible to perform. Therefore, the microbial quality is based on testing for an indicator organism, i.e., a microorganism whose presence is evidence that the water has been polluted with feces of humans or warm-blooded animals.

Nonpathogenic fecal coliform bacteria, as typified by *Escherichia coli,* that reside in the human intestinal tract are excreted in large numbers in feces, averaging about 50 million coliforms per gram. Untreated domestic wastewater generally contains more than 3 million coliforms per 100 ml. Pathogenic microorganisms causing enteric diseases in humans originate from fecal discharges of diseased persons. Consequently, water contaminated by fecal pollution is identified as being potentially dangerous by the presence of coliform bacteria. Hence, coliform bacteria are indicator organisms of fecal contamination and possible presence of pathogens. Nevertheless, some genera of the coliform group of bacteria found in water and soil are not of fecal origin but grow and reproduce on organic matter outside the intestines of humans and animals. These coliforms indicate neither fecal contamination nor the possible presence of pathogens. In laboratory testing, the term *total coliforms* refers to coliform bacteria from feces, soil, or other origin, and *fecal coliforms* refers to coliform bacteria from human or warm-blooded animal feces.

The reliability of coliform bacteria to indicate the presence of pathogens in water depends on the persistence of the pathogens relative to coliforms. For pathogenic bacteria, the die-off rate is greater than coliforms outside the intestinal tract of humans. Thus, exposure in the water environment reduces the number of pathogenic bacteria relative to coliform bacteria. Viruses, protozoal cysts, and helminth eggs are more persistent than coliform bacteria. For example, the threshold chlorine residual effective as a bactericide may not inactivate enteric viruses, is ineffective in killing protozoal cysts, and cannot harm helminth eggs. In contrast, filtration through natural sand aquifers for a sufficient distance, or granular media in a treatment plant after chemical coagulation, can entrap cysts and eggs because of their relatively large size while allowing viruses to be carried through suspended in the water. In surface-water treatment, coliforms are a reliable indicator of the safety of the processed water for human consumption, since the treatment included chemical coagulation and filtration to remove cysts, eggs, and suspended matter for effective chlorination of the clear water to inactivate viruses and kill bacteria. In a similar manner, coliforms can be used as an indicator of water quality for reuse of reclaimed wastewater provided the treatment processes physically remove the persistent protozoal cysts and helminth eggs.

## Drinking Water Standards

The maximum contaminant level goal (MCLG) in the Safe Drinking Water Act is zero coliforms. The recommended testing procedure is the presence–absence test, which does not provide a coliform count.

The maximum contaminant level (MCL) allows for a limited number of positive samples because of inadvertent contamination. Coliform bacteria are common in the natural environment, for example, on dirty water faucets, on the hands of the person collecting the water sample, and in dust and soil. Even with careful sampling procedures, an occasional water sample is likely to test positive for coliform bacteria of nonfecal origin. When a positive test occurs, multiple repeat samples are required to identify whether the contamination is actual or inadvertent, and the positive sample is tested further to determine if the coliforms are fecal coliforms.

The number of water samples tested for monitoring a public water supply is based on the population served. For a population under 1000, one sample per month is required; from 1000 to 100,000, one sample per 1000 population is required per month; above 100,000 the number of samples required is less than one per 1000 population. For a population under 33,000, only one sample may test positive per month for total coliforms for no violation and, above 33,000, no more than 5.0% may test positive. Violation of this MCL requires public notification and an evaluation to determine the source of contamination and risk of contamination with pathogens.

The Safe Drinking Water Act rules also specify the disinfection treatment of public water supplies in addition to the coliform MCL of the water in the distribution system. The regulations on treatment technique are to ensure removal of pathogens more persistent than coliform bacteria during treatment. Strictly speaking, the coliform standard for drinking water applies only after proper treatment and disinfection of water supplies. (Sections 11.23–11.25 present the treatment techniques specified by the Environmental Protection Agency for disinfection of public water supplies.)

## Wastewater Effluent Standards

Conventional treatment removes 99%–99.9% of pathogenic microorganisms in the raw wastewater; however, the effluent still contains significant concentrations of excreted viruses, bacteria, protozoal cysts, and helminth eggs. The kinds of pathogens in the wastewater depend on the health of the contributing human population. If discharged to recreational waters, effluents are generally chlorinated at dosages in the range of 8–15 mg/l with a minimum contact time of 30 min at peak hourly flow. Satisfactory effluent disinfection is normally defined by an average fecal coliform count of 200 per 100 ml or less. This is a reduction from about 1,000,000 per 100 ml in the biologically treated effluent. The safety of disposing of chlorinated effluent by dilution in surface water is based on the argument that this reduction eliminates the great majority of bacterial pathogens and inactivates large numbers of viruses. In fact, viruses and bacteria may be harbored and protected in suspended organic matter (allowable suspended solids in a secondary effluent is 30 mg/l), and protozoal cysts and helminth eggs are resistant to chlorination. (For further discussion, refer to Section 11.26.)

Adequate disinfection for unrestricted reuse of wastewater effluents requires removal or inactivation of all pathogens. The processes following biological treatment are tertiary filtration to physically remove helminth eggs, protozoal cysts, and suspended solids and long-term chlorination to inactivate viruses and bacteria. The tertiary scheme may be conventional coagulation–sedimentation–filtration or direct filtration without sedimentation. The subsequent disinfection requires rapid mixing of the chlorine followed by a long contact time (usually about 2 hr) in a tank simulating plug flow. The common effluent standard for unrestricted reuse for irrigation is 2.2 fecal coliforms per 100 ml and a turbidity of less than 2 NTU (nephelometric turbidity unit). (For further discussion, refer to Section 11.27.)

# Chemical Quality of Drinking Water

Drinking-water standards of the U.S. Environmental Protection Agency (EPA) apply to all public water supply systems that have at least 15 service connections or that regularly serve an average of 25 or more individuals daily at least 60 days a year. Risk assessment is the process applied in developing standards for toxic chemicals. The primary standards, which are approval limits for health, are specified for inorganic chemicals, organic chemicals, radionuclides, and turbidity. Secondary standards that recommend limits for aesthetics include inorganic chemicals, dissolved salts, corrosivity, color, and odor.

## 8.3  ASSESSMENT OF CHEMICAL QUALITY

Treated drinking water can contain trace amounts of toxic chemicals often so low in concentration that predicting an observable effect on human health is difficult. Reliable information on the toxicity to humans of most chemicals is difficult to obtain, since it must usually be based on uncontrolled accidental or occupational exposures. Therefore, data from long-term animal studies are applied to evaluate chronic exposure and carcinogenic risk to humans.

## Risk Assessment

Risk assessment is the scientific evaluation of toxic chemicals, human exposure, and adverse health effects. Identification of a hazard results from exploratory studies on laboratory animals or case reports of actual human exposure. Based on the inference of toxicity, dose–response relationships are determined on laboratory animals between specific quantities of the substance and associated physical responses, such as growth of tumors, birth defects, or neurologic deficits. Then follows an evaluation of exposure and assessment. The purpose is to describe the magnitude and duration of exposure to human populations both in the past and anticipated in the future. Finally, a quantitative estimate of risk to humans is predicted from the expected exposure levels by applying a dose–response model. If human exposure data are not available, which is often the case, the quantitative estimate is a hypothetical risk based on dose–response data from studies of laboratory animals.

Risk assessment has been used extensively for estimating the risk of developing cancer [3] and is now being applied to assess risks to development, reproduction, and neurologic functioning [4]. Development effects include embryo and fetal death, growth retardation, and malfunctions. Developmental toxicity studies are very complex, and applying results to risk assessment remains problematic. Also highly complex is reproductive toxicity affecting any event from germ cell formation and sexual functioning in the parents through sexual maturation of the offspring. Neurotoxicity in humans ranges from cognitive, sensory, and motor impairments to immune system deficits. Numerous chemicals, including many pesticides, are known neurotoxins.

## Chronic Noncarcinogenic Toxicity

For noncarcinogens and nonmutagens, human exposure should be less than the threshold level causing chronic disease. The acceptable daily intake (ADI) of a noncarcinogenic chemical is defined as the dose anticipated to be without lifetime risk to humans when taken daily, expressed in milligrams of chemical per kilogram of body weight per day. It is an empirically derived value arrived at by combining exposure knowledge and uncertainty concerning the relative risk of a chemical.

The ADI is based on toxicity data from chronic (long-term) feeding studies of laboratory animals to identify the highest no-observed-adverse-effect level (NOAEL). The dosage of a toxin given to laboratory animals is assumed to be physiologically equivalent to humans on the basis of body weight. (Typical weights are 0.3 kg for a rat, 10 kg for a dog, and 70 kg for an adult human.) In other words, the intake per unit body weight of a rat or dog is equated to the intake per unit body weight of a human. After determining the NOAEL in animals, an uncertainty factor (safety factor) is applied to reduce the allowable human intake to account for uncertainties involved in extrapolating from animals to humans. The uncertainty factors used in calculating acceptable daily intakes for establishing drinking-water standards are a factor of 10 when chronic human exposure data are available and are supported by chronic oral toxicity data in animal species, by a factor of 100 when good chronic oral toxicity data are available in some animal species but not in humans, and by a factor of 1000 with limited chronic animal toxicity data [5]. Since calculated intake is based on the total amount of chemical ingested, the safe level for drinking water must account for intake from other sources, i.e., food and air. Finally, the ADI allowed in drinking water is converted from body weight to concentration in water by assuming the average human adult weighs 70 kg (154 lb) and consumes 2 liters of water per day.

The following equation summarizes the concept of acceptable daily intake.

$$ADI = \frac{(NOAEL, mg/kg \cdot d)(70 \text{ kg})}{(UF)(2 \text{ l/d})} \tag{8.1}$$

where     ADI = acceptable daily intake in drinking water, mg/l
      NOAEL = no-observed-adverse-effect level based on laboratory
              animal toxicity data, mg/kg·d of animal body weight
      70 kg = body weight of an adult human
         UF = uncertainty factor in the range of 10–1000
       2 l/d = daily adult water consumption

In drinking-water regulations, the ADI is expressed as the MCL. For some chemicals, the MCL may be further reduced by allocating only a portion of the ADI to water with other common sources being food and air.

For example, consider the chemical aldicarb, which is a systemic insecticide with high mammalian toxicity. This chemical is used on cotton with only a few registered uses on food crops, including potatoes, peanuts, and sugar beets. By oral dose, it inhibits blood cholinesterases that are enzymes responsible for hydrolyzing esters of choline. The NOAEL on both rats and dogs was established as 0.1 mg/kg·d. From Eq. 8.1 with an uncertainty factor of 100, the ADI equals 0.035 mg/l. Then, assuming 20% of the intake apportioned to water, the allowable concentration becomes 0.007 mg/l or 7 $\mu$g/l [4].

The ADI is a judgment regarding acceptable levels of chronic exposure to a chemical and is neither an estimate of risk nor a guarantee of absolute safety [5]. The ADI concept is not recommended for evaluating the intake of organic chemicals that are suspected carcinogens.

## Carcinogenic Toxicity

The development of cancer appears to have three distinct stages: initiation, promotion, and progression. Each stage is influenced by such factors as age, heredity, diet, metabolic activity, and exposure to carcinogenic chemicals. The most widely used tests for carcinogen evaluation are long-term animal bioassays.

The hazard of ingesting a chemical assessed as a confirmed or suspected carcinogen can be evaluated in terms of dose-related risk. The two major problems in assessing risk are (1) extrapolation from observed risks in relatively high exposure levels used in laboratory animal studies to the low levels of exposure to humans and (2) extrapolation of the estimated risk from laboratory animals to humans. The assessment based on high-dose animal bioassay to low-dose human exposure further suffers from a lack of basic knowledge concerning the disease process in animals and humans and the total lack of data regarding potential synergistic and antagonistic reactions of chemicals.

The estimate of risk from results of animal bioassays is made by first converting the laboratory animal dose to the physiologically equivalent human dose. One method of conversion is on the basis of relative skin areas of test animals and the human body. This is justified from the observation that effects of acute toxicity in humans on a dose per unit of body surface area are in the same range as those in experimental animals [3]. The linearized multistage mathematical model, based on the one-hit theory of cancer initiation, is generally used for low-dose risk estimation [3]. Although its application cannot be proved or disproved by current scientific data, it is considered the best available model that yields an estimate of risk representing a plausible upper limit. The actual risk is not likely to be higher than the risk predicted by this model.

For example, consider the insecticide lindane, which is related to tumorigenic effects observed in laboratory mice and neurologic impairment in humans. The risk estimate in humans at a concentration in drinking water of 1.0 $\mu$g/l and daily consumption of 1 l/d is between $3.3 \times 10^{-6}$ and $8.1 \times 10^{-6}$ [3]. (Most experts agree that current technologies for assessing cancer and neurotoxicity risk cannot generate a single precise estimate of human risk, and risks are best expressed in terms of ranges or confidence intervals.) The upper 95% confidence estimate of risk at the same chemical dose is from $5.6 \times 10^{-6}$ to $13 \times 10^{-6}$. These risk estimates are expressed as the probability of cancer after a lifetime consumption of 1 liter of water per day containing 1.0 $\mu$g/l of chemical. The MCL for lindane is 0.0002 mg/l. With 2 l/d water consumption at an MCL of 0.2 $\mu$g/l, the numerical risks decrease to 0.4 of the values for 1.0 l/d and 1.0 $\mu$g/l. They become $1.3 \times 10^{-6}$ to $3.2 \times 10^{-6}$ and $2.2 \times 10^{-6}$ to $5.2 \times 10^{-6}$. The average value of $3.0 \times 10^{-6}$ means that 70-yr lifetime exposure to lindane would be expected to produce one excess case of cancer for every 340,000 persons exposed.

For an example of carcinogenic risk assessment, refer to the discussion of chloroform in Section 11.21.

## 8.4 CHEMICAL CONTAMINANTS

The chemicals listed in Table 8.2 are currently regulated in drinking water by the EPA. Retaining the chemicals on this list and/or the associated maximum contaminant levels depends on the results of ongoing risk assessments. A chemical may be deleted when future studies lack convincing evidence of risk to human health, or the use of the chemical may be banned, resulting in significantly reduced presence in water. On the other hand, new chemicals may be added, resulting from additional risk assessments. In the future, drinking-water standards are likely to be flexible, with continuous adjustments to keep up with the changing chemical environment. (A reader interested in a specific contaminant is advised to contact the EPA, or state regulatory agency, to confirm the most recent regulatory status and MCL.)

The MCL is an enforceable standard for protection of human health. Monitoring requirements are specific, with a prescribed schedule of routine sampling and check sampling to confirm the results if a MCL is exceeded. The frequency of sampling varies from every 3 years for chemicals in ground-water to every 3 months for selected chemicals in treated surface waters. Testing is performed in an approved laboratory by specified methods. The MCLG is a nonenforceable health goal. MCLs are set as close to MCLGs as feasible based on best control management, treatment technology, and other means while considering cost.

Generally, the MCLG for a carcinogenic chemical is zero. Treatment technique rather than an MCL is specified for selected chemicals. Acrylamide and epichlorohydrin are used during water treatment in flocculants to decrease turbidity. The treatment technique requirements limit the concentration of these chemicals in polymers and their application.

The regulation of specific MCLs depends on the kind of water system. All of the standards are applicable to community systems and nontransient systems that supply water to the same people for a long period of time, e.g., schools and factories. Transient systems that serve different people for a short time (e.g., campgrounds, parks, and highway rest stops) are only required to meet the MCLs of those contaminants with health effects caused by short-term exposure.

### Inorganic Chemicals

The sources of trace metals are associated with the natural processes of chemical weathering and soil leaching and with human activities, such as mining and manufacturing. Corrosion in distribution piping and customers' plumbing can also add trace metals to tap water.

Arsenic, barium, cadmium, chromium, mercury, and selenium are toxic

**TABLE 8.2** CHEMICAL DRINKING-WATER STANDARDS, MAXIMUM CONTAMINANT LEVELS IN MILLIGRAMS PER LITER

### Inorganic chemicals

| | | | |
|---|---|---|---|
| Arsenic | 0.05 | Lead | $TT^a$ |
| Barium | 2. | Mercury | 0.002 |
| Cadmium | 0.005 | Nitrate (as N) | 10. |
| Chromium | 0.1 | Nitrite (as N) | 1. |
| Copper | $TT^a$ | Nitrate + Nitrite | 10. |
| Fluoride$^b$ | 4.0 | Selenium | 0.05 |

Asbestos 7 million fibers/liter (longer than 10 $\mu$m)

### Volatile organic chemicals

| | | | |
|---|---|---|---|
| Benzene | 0.005 | 1,2-Dichloropropane | 0.005 |
| Carbon tetrachloride | 0.005 | Ethylbenzene | 0.7 |
| p-Dichlorobenzene | 0.075 | Monochlorobenzene | 0.1 |
| o-Dichlorobenzene | 0.6 | Tetrachloroethylene | 0.005 |
| 1,2-Dichloroethane | 0.005 | 1,1,1-Trichloroethane | 0.2 |
| 1,1-Dichloroethylene | 0.007 | Trichloroethylene | 0.005 |
| cis-1,2-Dichloroethylene | 0.07 | Vinyl chloride | 0.002 |
| trans-1,2-Dichloroethylene | 0.1 | | |

### Organic chemicals

| | | | |
|---|---|---|---|
| Acrylamide | $TT^a$ | Heptachlor | 0.0004 |
| Alachlor | 0.002 | Heptachlor epoxide | 0.0002 |
| Aldicarb | 0.003 | Lindane | 0.0002 |
| Aldicarb sulfone | 0.003 | Methoxychlor | 0.04 |
| Aldicarb sulfoxide | 0.004 | Pentachlorophenol | 0.001 |
| Atrazine | 0.003 | Polychlorinated biphenyls | 0.0005 |
| Carbofuran | 0.04 | Styrene | 0.1 |
| Chlordane | 0.002 | Toluene | 1. |
| Dibromochloropropane | 0.0002 | Toxaphene | 0.003 |
| Endrin | 0.0002 | Xylenes (total) | 10. |
| Epichlorohydrin | $TT^a$ | 2,4-D | 0.07 |
| Ethylene dibromide | 0.00005 | 2,4,5-TP (Silvex) | 0.05 |

### Trihalomethanes

| | |
|---|---|
| Total trihalomethanes | 0.10 |

### Radionuclides

| | |
|---|---|
| Radium-226 | 20 pCi/l |
| Radium-228 | 20 pCi/l |
| Gross alpha particle activity | 15 pCi/l |
| Beta particle and photon radioactivity | 4 mrem/yr |
| Radon | 300 pCi/l |
| Uranium | 20 $\mu$g/l |

### Turbidity

Turbidity$^c$ 0.5 NTU

$^a$ Treatment technique (TT) requires modification or improvement of water processing to reduce the contaminant concentration.

$^b$ Many states require public notification at least annually of fluoride in excess of 2.0 mg/l to warn consumers of potential dental fluorosis.

$^c$ Turbidity is a performance standard in filtration treatment of surface waters to ensure removal of *Giardia lamblia* cysts.

metals affecting the internal organs of the human body. *Arsenic* is widely distributed in waters at low concentrations, with isolated instances of higher concentrations in well waters. It is also found in trace amounts in food [3]. *Barium,* one of the alkaline earth metals, occurs naturally in low concentrations in most surface waters and in many treated waters [3]. *Cadmium* can be introduced into surface waters in amounts significant to human health by improper disposal of industrial wastewaters. Nevertheless, the major sources are food, cigarette smoke, and air pollution; hence, the MCL is set so that less than 10% of the total intake is expected to be from water consumption. The health effects of cadmium can be acute, resulting from overexposure at a high concentration, or chronic, caused by accumulation in the liver and renal cortex.

*Chromium* is amphoteric and can exist in water in several valence states. The content in natural waters is extremely low because it is held in rocks in essentially insoluble trivalent forms. Acute systemic poisoning can result from high exposures to hexavalent chromium; trivalent is relatively innocuous. *Mercury* is a scarce element in nature and has been banned for most applications with environmental exposure, such as mercurial fungicides. The biological magnification of mercury in freshwater food fish is the most significant hazard to human health. The mercury becomes available in the food chain through the transformation of inorganic mercury to organic methylmercury by microorganisms present in the sediments of lakes and rivers. Toxicity via the oral route is related mainly to methylmercury compounds rather than to inorganic mercury salts or metallic mercury. Symptoms of methylmercury poisoning include mental disturbance, ataxia, and impairment of speech, hearing, vision, and movement. *Selenium* is a trace metal naturally occurring in soils derived from some sedimentary rocks. Surface streams and groundwater in seleniferous regions contain variable concentrations. Cattle grazing on seleniferous vegetation suffer from "blind staggers." Effects on human health have not been clearly established—a low-selenium diet is beneficial, whereas high doses can produce undesirable physical manifestations.

*Lead* exposure occurs through air, soil, dust, paint, food, and drinking water. Lead toxicity effects the red blood cells, nervous system, and kidneys, with young children, infants, and fetuses being most vulnerable. Depending on local conditions, the contribution of lead from drinking water can be a minor or major exposure for children. Lead is not a natural contaminant in surface waters or groundwaters and is rarely in source water. It is a corrosion by-product from high-lead solder joints in copper piping, old lead-pipe goosenecks connecting the service lines to the water main, and old brass fixtures. Lead pipe and brass fixtures with high-lead content are not installed today, and lead-free solder has replaced the old 50% lead–50% tin solder in water piping. Since dissolution of lead requires an extended contact time, lead is most likely to be present in tap water after being in the service connection piping and plumbing overnight. Therefore, the first-flush sample concentration is the highest expected, and if less than 0.015 mg/l no corrective action is required.

*Copper* as an indicator of corrosivity, has a no-action level of 1.3 mg/l in first-flush samples. (Refer to Section 11.33 for a discussion of corrosion of lead pipe and solder.)

*Fluoride,* the naturally occurring form of fluorine, is commonly found in trace amounts in most soil and rock. Groundwaters usually contain fluoride ion dissolved from geologic formations. Surface waters generally contain lesser amounts, under 0.3 mg/l, except when contaminated by industrial wastes. Excessive concentration of fluoride in drinking water causes the dental disease of fluorosis, also referred to as mottling. This disease in mildest form results in very slight, opaque whitish areas on some of the posterior teeth. With greater severity, the fluorosis is widespread and the color of the teeth is gray to black. Absence or low concentration of fluoride in drinking water results in a high incidence of dental caries in children's teeth. The optimum fluoride concentration in drinking water protects teeth from decay without causing noticeable fluorosis. Food is another source of fluoride in the diet; however, as a contributor of fluorides, many studies have shown that the average dietary intake with food is a constant amount throughout the United States. Hence, the dental effect of fluoride results primarily from the concentration in the public water supply. Since water consumption is influenced by climate, the recommended optimum concentrations listed in Table 8.3 are based on the annual average of the maximum daily air temperatures. Cities with water supplies deficient in natural fluoride have successfully provided supplemental fluoridation to optimum levels to reduce the rate of dental decay in children [6].

*Nitrate* is the common form of inorganic nitrogen found in water solution. In agricultural regions, heavy fertilizer application results in unused nitrate migrating down into the groundwater. As a result, groundwater withdrawn

**TABLE 8.3**  RECOMMENDED OPTIMUM CONCENTRATIONS OF FLUORIDE BASED ON THE ANNUAL AVERAGE OF THE MAXIMUM DAILY AIR TEMPERATURES

| Temperature range (°F) | Recommended optimum (mg/l) |
|---|---|
| 53.7 and below | 1.2 |
| 53.8 to 58.3 | 1.1 |
| 58.4 to 63.8 | 1.0 |
| 63.9 to 70.6 | 0.9 |
| 70.7 to 79.2 | 0.8 |
| 79.3 to 90.5 | 0.7 |

by private and public wells is likely to have measurable concentrations of nitrate, and in the same regions well waters in many rural communities can exceed the recommended limit of 10 mg/l of nitrate nitrogen. Surface waters can be contaminated by nitrogen from both discharge of municipal waste-water and drainage from agricultural lands. The health hazard of ingesting excessive nitrate in water is infant methemoglobinemia. In the intestine of an infant, nitrate can be reduced to nitrite that is absorbed into the blood oxidizing the iron of hemoglobin. This interferes with oxygen transfer, resulting in cyanosis and giving the baby a blue color. During the first 3 months of age, infants are particularly susceptible. Incidents of infant methemoglobinemia are extremely rare since most mothers in regions of known high-nitrate drinking water use either bottled water or a liquid formula requiring no dilution. Methemoglobinemia is readily diagnosed and rapidly reversed by injecting methylene blue into the infant's blood. Healthy adults are able to consume large quantities of nitrate in drinking water without adverse effects. The principal sources of nitrate in the average adult diet are saliva and vegetables amounting to about 130 mg/day [3]. Two liters per day at 10 mg/l equals only 20 mg/day. Justified by epidemiological evidence on the occurrence of methemoglobinemia in infants, the standard of 10 mg/l is the maximum contaminant level for water with no observed adverse health effects.

## Organic Chemicals

In the evaluation of organic chemicals, EPA groups them into three categories, depending on the evidence of carcinogenicity. Category I chemicals are probable human carcinogens, based on human or animal risk assessment, and are assigned MCLGs of zero. Category II chemicals are not regulated as human carcinogens, but the MCLGs are lower than MCLs based on inconclusive evidence of carcinogenicity. Category III chemicals causing chronic disease without evidence of carcinogenicity are given MCLs based on the acceptable daily intake.

The MCLs of chemicals are set as close to the MCLGs as feasible, including the feasibility of laboratory testing. Gas chromatography has high sensitivity and reliability in detecting concentrations down to a few micrograms per liter; therefore, for many carcinogenic and highly toxic chemicals the MCLs are set at the lowest level of quantitative measurement achievable in a good laboratory. For example, several of the volatile organic chemicals considered carcinogens have MCLGs of zero and MCLs of 0.005 mg/l, which is the practical quantitative level of measurement. This quantity of chemical in drinking water is extremely small. The amount of water theoretically consumed by a person in a lifetime of 70 years at 2 l per day is 51,100 l (13,500 gal). For a concentration of 0.005 mg/l, the amount of organic chemical added to this quantity of water would be 3 to 7 drops, depending on the specific gravity of the chemical. Thus, the MCL allows the ingestion of only about 5 drops of a carcinogenic chemical in a person's lifetime.

*Volatile organic chemicals* (VOCs) are produced in large quantities for use in industrial, commercial, agricultural, and household activities. The adverse health effects of VOCs include cancer and chronic effects on liver, kidney, and nervous system. Because of their volatility, air stripping is a proposed method of removal from water (Section 11.41). Volatility also reduces their concentrations in surface waters. Groundwater contamination is more common because VOCs have little affinity for soils and are diminished only by dispersion and diffusion, which is often limited. Those most frequently detected in contaminated groundwaters are trichloroethylene, a degreasing solvent in metal industries and a common ingredient in household cleaning products; tetrachloroethylene, a dry-cleaning solvent and chemical intermediate in producing other compounds; carbon tetrachloride, used in the manufacture of fluorocarbons for refrigerants and solvents; 1,1,1-trichloroethane, a metal cleaner; 1,2-dichloroethane, an intermediate in manufacture of vinyl chloride monomers; and vinyl chloride, used in the manufacture of plastics and polyvinyl chloride resins.

Most of the organic chemicals listed in Table 8.2 are *insecticides* and *herbicides*. Pesticides may be present in surface waters receiving runoff from either agricultural or urban areas where these chemicals are applied. Groundwaters can be contaminated by pesticides, manufacturing wastewaters, spillage, or infiltration of rainfall and irrigation water. Alachlor, aldicarb, atrazine, carbofuran, ethylene dibromide, and dibromochloropropane have been detected in drinking waters. Most pesticides can be absorbed into the human body through the lungs, skin, and gastrointestinal tract. From acute exposure, the symptoms in humans are dizziness, blurred vision, nausea, and abdominal pain. Chronic exposure of laboratory animals indicates possible neurologic and kidney effects and, for some pesticides, cancer.

## Trihalomethanes

Trihalomethanes (THMs) are produced by chlorination of surface waters and groundwaters containing natural organic substances from decaying vegetation, such as humic and fulvic acids. THMs are derivatives of methane where three of the four hydrogen atoms have been replaced by three atoms of chlorine, bromine, or iodine. Chloroform is the THM most commonly found in drinking water and is usually present in the highest concentration. In some cases, brominated THMs dominate as a result of naturally occurring bromide in the water. Bromide ions are oxidized by aqueous chlorine to bromine, and, since it is more reactive, bromine substitutions can dramatically increase the total THM level. The general reaction producing THMs is

$$\text{Chlorine} + \text{(bromide ion or iodide ion)} + \text{precursors} \qquad (8.2)$$
$$= \text{trihalomethanes and other halogenated compounds}$$

The reactions are not instantaneous but continue for an extended period of time following chlorination of the water. (For further discussion, refer to Section 11.21.)

The MCL for total THMs is based on a 12-month running average value rather than a single test. A minimum of four water samples is collected quarterly from the distribution system, 25% from extremities and 75% based on population distribution. The total THM concentration in each sample is the sum of the concentrations of $CHCl_3$, $CHBrCl_2$, $CHBr_2Cl$, and $CHBr_3$. The quarterly reported value is the arithmetic average of the total THM content of all samples. Finally, the contaminant level is calculated by averaging these quarterly values with the measured levels from the previous three quarters.

## Radionuclides

Radioactive elements decay by emitting alpha, beta, or gamma radiations caused by transformation of the nuclei to lower energy states. An alpha particle is the helium nucleus (2 protons + 2 neutrons); e.g., radon-222 decays to polonium-218 and emits helium-4. A beta particle is an electron emitted from the nucleus as a result of neutron decay; e.g., radium-228 decays to actinium-228 and emits $\beta^-$. In these processes, the helium nucleus emitted as an alpha particle or the electron ejected as a beta particle changes the parent atom into a different element. A gamma ray is a form of electromagnetic radiation; other forms are light, infrared and ultraviolet radiations, and x rays. Gamma decay involves only energy loss and does not create a different element. Alpha, beta, and gamma radiations have different energies and masses, thus producing different effects on matter. Each is capable of knocking an electron from its orbit around the nucleus and away from the atom, which is a process referred to as ionization. Radiation is detected by ionization, and high-reactive ions taken into the human body can lead to deleterious health effects, such as cancer.

The ability to penetrate matter varies among nuclear radiations. Most alpha particles are stopped by a single thickness of paper, while most gamma rays pass through the human body, as do x rays. Since alpha particles are stopped by short penetrations, more energy is deposited and does more damage per unit volume of matter receiving radiation.

Radioactivity in drinking water can be from natural or artificial radionuclides. Radium-226 is usually found in groundwater as a result of geologic conditions. Radioactivity from radium is widespread in surface waters because of fallout from testing of nuclear weapons. In some localities this radioactivity could be increased by small releases from nuclear power plants and industrial users of radioactive materials. The average amount of background radiation from cosmic rays and terrestrial sources is about 100 mrem/yr. [7]. Only a small portion of this unavoidable background radiation comes from drinking water containing radionuclides.

The recommended allowable dose from radioisotopes in drinking water supplies is very low. If a water supply were constituted in such a way as to contain either average or likely amounts of radioactivity, a total-body dose of

0.244 mrem/yr would be accumulated [3]. This is less than 1% of background. Although the dose to bone would be considerably higher, because strontium and radium are bone seekers, even this dose constitutes less than 10% of the total average natural background [3]. Estimates were made of three possible types of risks that could be induced by this magnitude of radiation: developmental and teratogenic, genetic, and somatic. Although a developing fetus is sensitive to radiation, this low dosage delivered from drinking water during the sensitive periods of gestation is so small that no measurable effects of the radation will be found [3]. The lowest dose level at which any effect has been reported is 3 mrem/day or 1100 mrem/yr, in contrast to 0.244 mrem/yr. For genetic risk to the general population, the maximum permissible dose of artificial radiation is 170 mrem/yr, excluding medical uses of radiation. This amounts to a 5-rem genetic dose in each 30-yr generation, which is insignificant in increasing the current incidence of genetic diseases [3]. The natural background of radiation can be estimated to cause 4.5–45 fatal cases of cancer per year per million people, depending on the risk model used to make the calculation. Less than 1% of this is contributed to radionuclides in drinking water [3].

The National Academy of Sciences concluded that the radiation associated with most water supplies is such a small proportion of the normal background that it is difficult, if not impossible, to measure any adverse health effects with certainty. In a few water supplies, however, radium can reach concentrations that pose a higher risk of bone cancer for the people exposed [3]. Because of the many uncertainties of the environmental effects, the EPA has adopted the MCLs given in Table 8.2.

The measurement unit of pCi/l is $10^{-12}$ curie per liter, with a curie being the activity of 1 g of radium. The rem (radiation equivalent man) is a unit of radiation dose equivalence that is numerically equal to the absorbed dose in rad multiplied by a quality factor, to describe the actual damage to tissue from the ionizing radiation. The mrem is 1/1000 of a rem. The rad is the unit of dose or radiation adsorbed. One rad deposits 100 ergs of energy in 1 g of matter.

In testing, gross alpha activity that could include radium radiation is determined first. If the value is greater than 20 pCi/l, separate tests for Rd-226 and Rd-228 are conducted. Beta activity is primarily from artificial radionuclides, e.g., contamination from nuclear weapons testing. The screening test is gross beta particle activity, since the decay products of fission are beta and gamma emitters. If the measurement is less than 50 pCi/l, tritium and strontium-90 activities are determined and converted to mrem/yr units. If the gross beta activity is greater than 50 pCi/l, analyses are performed to identify the radionuclides present to determine the mrem/yr.

## Turbidity

Insoluble particulates impede the passage of light through water by scattering and absorbing the rays. This interference of light passage is referred to as turbidity. The standard is a suspension of silica of specified particle size selected so that a 1.0-mg/l suspension measures as 1.0 NTU. The common method of measurement uses a photoelectric detector and nephelometry to measure the intensity of scattered light. In surface-water filtration for treatment of drinking water, turbidity in the filtered water must be equal to or less than 0.5 NTU in at least 95% of the measurements taken each month.

### Secondary Standards

Standards for aesthetics are recommended for characteristics that render the water less desirable for use; they are not related to health hazards and nonenforceable by the EPA.

Excessive color, foaming, or odor cause customers to question the safety of drinking water and result in complaints from users. Chloride or sulfate ion concentrations greater than 250 mg/l, or dissolved solid concentrations greater than 500 mg/l, can have taste and laxative properties. Sodium sulfate and magnesium sulfate are laxatives, with the common names of Glauber salt and Epsom salt, respectively. The laxative effect may be noticed by travelers or new consumers drinking waters high in sulfates; however, most persons become acclimated in a relatively short time. Excessive dissolved salts can also affect the taste of coffee and tea. For persons on a sodium-restricted diet, which is usually 2000 mg of sodium per day, the recommended maximum concentration in their drinking water is 100 mg/l [3]. For a severely restricted diet of 500 mg of sodium per day, the recommended maximum concentration is 20 mg/l of sodium ion.

Iron and manganese, above 0.3 mg/l and 0.05 mg/l, respectively, are objectionable because of brown stains imparted to laundry and porcelain and the bittersweet taste of iron. A noncorrosive water with an alkaline pH is desirable to reduce the probability of pipe corrosion contributing iron and other trace metals to the water by dissolution from water mains and plumbing.

# Quality Criteria for Surface Waters

All surface waters should be of adequate quality to support aquatic life and be aesthetically pleasing. Additionally, if needed as a source of supply, the water should be treatable by conventional processes to provide a potable supply meeting the drinking-water standards. Many lakes, reservoirs, and rivers are also maintained at a quality suitable for swimming, water skiing, and boating.

## 8.5 POLLUTION EFFECTS ON AQUATIC LIFE

A normal, healthy stream or lake has a balance of plant and animal life represented by great species diversity. Pollution disrupts this balance, resulting in a reduction in the variety of individuals and dominance of the surviving organisms. Complete absence of species normally associated with a particular habitat reveals extreme degradation. Of course, biological diversity and population counts are meaningful only if existing communities in a polluted environment are compared to those normally present in that particular habitat. Fish are good indicators of water quality, and no perennial river can be considered in satisfactory condition unless a variety of fish can live and survive in it.

Being an end product of the aquatic food chain, fish reflect both satisfactory water quality and a suitable habitat for food supply, shelter, and breeding sites. Even though depletion of dissolved oxygen is commonly blamed, poisons appear to cause the most damage to plant and animal life in surface waters [8]. The effects of toxins are frequently magnified by environmental conditions; for example, temperature has a direct influence on morbidity. At a given concentration of toxin, a rise of 10°C generally halves the survival time of fish; poisons therefore become more lethal in rivers during the summer. Many toxic substances become more lethal with decreasing dissolved oxygen content. Also, the rate of oxygen consumption of fish is altered by the presence of toxins, and their resistance to low oxygen levels can be impaired. The pH of a water within the allowable range of 6.5–9 can influence some poisons. The dissolved salt content can also influence toxicity, particularly the presence of calcium, which reduces the adverse effect of some heavy metals. For example, cadmium, copper, lead, nickel, and zinc decrease in toxicity with increased hardness in the water.

Poisonous effects on fish life also relate to the character of the watercourse, species of fish, and season of the year. During the winter fish are much more resistant because of the cold water. The rapid rise of temperature in spring and hot periods in summer create critical times when fish are susceptible to unfavorable conditions and likely to die. During spawning even slight pollution can cause damage to salmon and trout [8].

Acute toxic effects (24–16-hr exposure) have been studied for most toxic pollutants. In contrast, few data are available on chronic toxicity. Applying a safety factor to the median lethal concentration can result in a criterion that is either too conservative or unsafe for long-term exposure. Chronic effects often occur in the species population rather than in the individual. If eggs fail to develop or the sperm does not remain viable, the species would be eliminated from an ecosystem because of reproductive failure. Physiologic stress can make a species less competitive, resulting in a gradual population decline or absence from an area. The same phenomenon could occur if a crustacean that serves as a vital food during the larval period of a fish's life is eliminated. Finally, biological accumulation of certain toxins can result in acute effects in fish that are the ultimate consumers in the aquatic food chain.

## 8.6 CONVENTIONAL WATER POLLUTANTS

The common pollutants are biochemical oxygen demand (BOD), suspended solids, fecal coliforms, pH, ammonia nitrogen, phosphorus, oil and grease, and chlorine residual. For these pollutants the EPA has developed water-quality criteria consisting of numerical limits; their rationale is based on bioassays of aquatic organisms [9]. Bioassays are the best method for determining safe concentrations for conventional pollutants to aquatic organisms. Test species, usually fish, are exposed to various concentrations of a pollutant in water for a specified time span of 96 hr or less in laboratory tanks. The median lethal concentration ($LC_{50}$) is the level that kills 50% of the test organisms. The maximum allowable toxin concentration in surface waters is usually between 0.1 and 0.01 of the $LC_{50}$ value. The uncertainty factor of 10–100 is to account for long-term exposure and other constituents already present in the river water creating additional physiologic stresses. These criteria may be modified to take into account the variability of local waters in establishing state standards.

Water-quality standards associate particular numerical limits with the designated beneficial uses for specific surface waters, thus recognizing that use and criteria are interdependent. Local conditions commonly considered are natural background levels of pollutants and other constituents such as hardness, the presence or absence of sensitive aquatic species, characteristics of the biological community, temperature and weather, flow characteristics, and synergistic or antagonistic effects of combinations of pollutants. In general, EPA criteria are considered to be conservative estimates of pollutant concentrations that can be safely tolerated by a general ecosystem, whereas state standards address site-specific pollution problems.

State water quality standards also provide the basis for effluent standards through the National Pollutant Discharge Elimination System (NPDES) permit program. The EPA requires each state to establish effluent limitations and performance standards for sources of water pollution, including wastewater treatment plants, industries, power plants, and confined agricultural operations. Effluent limits allow discharge of a specified amount of conventional pollutants and either limit or prohibit emission of toxic pollutants. Secondary treatment—the national goal for all municipal wastewaters—is defined as producing an effluent with an average BOD of less than 30 mg/l and suspended solids of less than 30 mg/l. Effluent pH is limited to the range 6.0–9.0. Depending on the uses of dilutional capacity of the receiving watercourse, other conventional pollutants commonly limited by an NPDES permit are fecal coliforms, residual chlorine, ammonia nitrogen, and oil and grease. Phosphorus discharge is controlled where the receiving water is a lake or estuary subject to eutrophication.

The *dissolved oxygen* standard establishes lower limits to protect propagation of fish and other aquatic life, enhance recreation and reduce the possibility of odors resulting from decomposition of organic matter, and maintain a suitable quality for water treatment. The primary pollutant associated with depletion of dissolved oxygen is carbonaceous BOD. In addition, sedimentation of suspended solids can cause a buildup of decomposing organic matter in sediments, and dissolved ammonia can contribute to oxygen depletion by nitrification. Fish vary in their oxygen requirements according to species, age, activity, temperature, and nutritional state. In general, the minimum dissolved-oxygen level needed to support a diverse population of fish is 5 mg/l. Cold-water fish require stringent limitations, 6 mg/l with a minimum of 7 mg/l at spawning times, and warm-water species, being more tolerant, need 4–5 mg/l. A typical minimum standard for water supply, recreation, and shellfish harvesting is 4 mg/l. The EPA criteria are a minimum 5.0 mg/l for freshwater aquatic life and maintenance of aerobic conditions throughout a body of water for esthetic considerations [9].

*Suspended solids* interfere with the transmission of light and can settle out of suspension covering a streambed or lake bottom. Turbid water interferes with recreational use and aesthetic enjoyment. Excess suspended solids adversely affect fish by reducing their growth rate and resistance to disease, preventing the successful development of fish eggs and larvae, and reducing the amount of food available. Settleable solids covering the bottom damage invertebrate populations and fill gravel spawning beds. The EPA criterion states that solids should not reduce the depth of the compensation point (penetration of sunlight) for photosynthetic activity by more than 10% from the seasonally established norm for aquatic life [9].

*Oil and grease* contaminants include a wide variety of organic compounds having different physical, chemical, and toxicological properties. Common sources are petroleum derivatives and fats from vegetable oil and meat processing. Domestic water supplies need to be virtually free from oil and grease, particularly from the tastes and odors that emanate from petroleum products. Surface waters should be free of floating fats and oils. Based on EPA criteria, individual petrochemicals should not exceed 0.01 of the median lethal concentration that kills 50% ($LC_{50}$) of a specific aquatic species during an exposure period of 96 hr [9].

*Fecal coliform bacteria* indicate the possible presence of pathogenic organisms. The correlation between coliforms and human pathogens in natural waters is not, however, absolute since these bacteria can originate from both the feces of humans and other warm-blooded animals. Coliforms from the intestinal tract of a human cannot be distinguished from those of animals. Therefore, the significance of testing in pollution surveys depends on a knowledge of the river basin and probable source of the observed fecal coliforms. The EPA criterion for fecal coliform bacteria in bathing waters is a logarithmic mean of 200 per 100 ml, based on a minimum of five samples taken over a 30-day period, with not more than 10% of the total samples exceeding 400 per 100 ml. Since shellfish may be eaten without being cooked, the strictest coliform criterion applies to shellfish cultivation and harvesting. The EPA criterion states that the mean fecal coliform concentration should not exceed 14 per 100 ml with not more than 10% of the samples exceeding 43 per 100 ml.

*Residual chlorine* resulting from disinfection of wastewater effluents is very toxic to fish. When chlorine is added to wastewater, chloramines are formed by reacting with ammonia. These can be eliminated in wastewater effluents by the addition of a reducing agent such as sulfur dioxide [9]. The EPA criteria for total chlorine residual are 2.0 $\mu$g/l for salmonid fish and 10.0 $\mu$g/l for other freshwater and marine organisms.

*Un-ionized ammonia* is toxic to fish and other aquatic animals. When ammonia dissolves in water, a portion reacts with the water to form ammonium ions ($NH_4^+$) with the balance remaining as un-ionized ammonia ($NH_3$). The concentration of un-ionized ammonia increases with increasing pH, increases with increasing temperature, and decreases with decreasing ionic strength. The EPA criterion for freshwater aquatic life is 0.02 mg/l of un-ionized ammonia, which is based on salmonid fish [9]. This value was calculated by applying a safety factor of 10 to the lowest reported lethal concentration ($LC_{50}$) of 0.2 mg/l of un-ionized ammonia nitrogen for rainbow trout fry. Since un-ionized ammonia cannot be measured, its concentration in water is based on the measured concentration of total ammonia ($NH_3 + NH_4^+$). Values of total ammonia nitrogen that result in concentrations of 0.02 mg/l of un-ionized ammonia for 20°C are 5.1 mg/l at pH 7.0, 1.6 at pH 7.5, 0.52 at pH 8.0, and 0.18 at pH 8.5 [9]. For more tolerant fish species these maximum allowable concentrations are very conservative. Interpretation of bioassay data on warm-water species suggests that the acute toxic concentration of ammonia at the gill surface is 0.4 mg/l [10]. Using a safety factor of 5 yields a no-effect un-ionized ammonia concentration of 0.08 mg/l. Values of total ammonia nitrogen that result in concentrations of 0.08 mg/l of un-ionized ammonia for 20°C and an alkalinity of 200 mg/l are 23 mg/l at pH 7.0, 9.0 at pH 7.5, 5.0 at pH 8.0, and 2.3 at pH 8.5 [10].

Controversy over ammonia criteria results from the high concentration of ammonia nitrogen in treated effluents from municipal wastewater plants. After biological treatment without nitrification an average domestic wastewater contains 24 mg/l of ammonia nitrogen (Table 12.1). The dilution ratio of ammonia-free water in a receiving watercourse at pH 8.0 to wastewater effluent containing 24 mg/l to protect warm-water fish is 3.8; for cold-water salmonid fish the required dilution ratio is 23.5. Ammonia can be converted to nitrate to reduce the effluent concentration by additional aeration in biological treatment. This nitrification process, usually a second stage of aeration, adds significant cost to wastewater treatment.

The *pH of surface waters* is specified for protection of fish life and to control undesirable chemical reactions, such as the dissolution of metal ions in acidic waters. Many substances increase in toxicity with changes in pH. For example, the ammonium ion is shifted to the much more poisonous form of un-ionized ammonia as the pH of water rises above neutrality. The EPA criteria for pH are 6.5–9.0 for freshwater aquatic life, 6.5–8.5 for marine aquatic life, and 5–9 for domestic water supplies [9].

*Phosphate phosphorus* is a key nutrient stimulating excessive plant growth—both weeds and algae—in lakes, estuaries, and slow-moving rivers. Cultural eutrophication is the accelerated fertilization of surface waters arising from phosphate pollution associated with discharge of wastewaters and agricultural drainage. Since phosphate removal is feasible by chemical precipitation in wastewater treatment, effluent permits for municipal and industrial discharges to lakes, or streams that flow into lakes, commonly limit the concentration to 1.0 mg/l of phosphate phosphorus; this is equivalent to about 90% removal from domestic wastewater. For lakes in northern United

States to be free of algal nuisances, the generally accepted upper concentration limit in impounded water when completely mixed in the spring of the year is 0.01 mg/l of orthophosphate. Hammer and Mac Kichan present discussions of cultural eutrophication, allowable nutrient loadings, and several ecological studies of eutrophic lakes and reservoirs [8].

## 8.7 TOXIC WATER POLLUTANTS

Numerous organic chemicals and several inorganic ions, mostly heavy metals, are classified as toxic water pollutants. To qualify as a priority toxin, a substance must be an environmental hazard and known to be present in polluted waters. Toxicity to fish and wildlife may be related to either acute or chronic effects on the organisms themselves or to humans by bioaccumulation in food fish. Persistence in the environment (including mobility and degradability) and treatability are also important factors. Currently, the EPA list of priority toxic pollutants consists of over 100 substances. While some are well-documented toxins, others have limited data supporting their hazard in the environment.

The majority of the toxic pollutants can be categorized into ten groups. *Halogenated aliphatics* are used in fire extinguishers, refrigerants, propellants, pesticides, and solvents. Health effects include damage to the central nervous system and liver. *Phenols* are industrial compounds used primarily in production of synthetic polymers, pigments, and pesticides and occur naturally in fossil fuels. They impart objectionable taste and odor at very low concentrations, taint fish flesh, and vary in toxicity depending on chlorination of the phenolic molecule. *Monocyclic aromatics* (excluding phenols and phthalates) are used in the manufacture of chemicals, explosives, dyes, fungicides, and herbicides. These compounds are central nervous system depressants and can cause damage to liver and kidneys. *Ethers* are solvents for polymer plastics. They are suspected carcinogens and aquatic toxins. *Nitrosamines,* used in production of organic chemicals and rubber, are suggested carcinogens. *Phthalate esters* are used in production of polyvinylchloride and thermoplastics. They are an aquatic toxin and can be biomagnified. *Polycyclic aromatic hydrocarbons* are in pesticides, herbicides, and petroleum products. *Pesticides* of concern are those that biomagnify in the food chain and are persistent in nature; common in this group of pesticides are chlorinated hydrocarbons. Aldrin, dieldrin, and chlordane applications are already restricted by the EPA. *Polychlorinated biphenyls* (PCBs) were banned from production in 1979. They are readily assimilated by the aquatic environment and still persist in sediments and fish; PCBs were used in electric capacitors and transformers, paints, plastics, insecticides, and other industrial products. *Heavy metals* vary in toxicity, and some are subject to biomagnification.

The general goal of wastewater treatment in the NPDES is to reduce the discharge of toxic pollutants to an insignificant level. Toxicity reduction evaluation of a treatment plant is the first step. For municipal plants, the objectives are to evaluate operation and performance to identify and correct treatment deficiencies causing effluent toxicity; identify the toxic substances in the effluent; trace the toxins to their sources in the wastewater collection system; and implement appropriate remedial measures to reduce effluent toxicity.

## Chemical Evaluation of Effluent

Gross toxicity of a wastewater effluent is evidenced in the receiving watercourse by a reduction in species diversity. Presence of a poison disrupts the normal balance of plant and animal life, which is represented by a great variety of individuals. Complete absence of species normally associated with a particular aquatic habitat is evidence of severe degradation. Obviously, a fish kill is an example associated with extreme acute toxicity from the discharge of a lethal quantity of a poison. Periodic upset of the biological process in wastewater treatment is an indication of industrial toxins being discharged to the sewer collection system and passing through the treatment plant. If the evidence of toxicity is less dramatic, the presence of toxins in a wastewater effluent may be unnoticed.

The conventional approach, if toxicity is indicated, is to test effluent samples for specific substances. This can be very costly unless the toxins likely to be present can be reduced to a reasonable number. Initial tests should be for those substances related to wastes from industries served by the sewer system. After the toxic pollutants have been quantitatively identified, the NPDES permit is modified to ensure protection of aquatic life in accordance with appropriate criteria [11]. Remedial measures must be taken by industrial pretreatment and improved municipal treatment to meet the revised effluent standards. Mandatory monitoring for toxins in the effluent is established by the NPDES discharge permit. Potential deficiencies of this analytical approach through chemical analyses are the possible presence of undetected toxins and combinations of synergistic toxins.

## Biomonitoring of Effluent

A bioassay to determine toxicity is performed by exposing selected aquatic organisms to wastewater effluent in a controlled laboratory environment [12]. The bioassay screening test for acute toxicity is a compliance investigation within the NPDES program. In the static test, effluent is placed in laboratory containers, the test organisms added, and the units held in a controlled environment with termination after 24 hr. A control test in prepared water is conducted in parallel for quality assurance of the test organisms and test procedures. An enhanced bioassay method is the flow-through test where fresh effluent is continuously supplied to the laboratory containers with continuous outflow. The effluent is usually a 24-hr composite sample that has been filtered, aerated, and stabilized at a temperature of 20°C for testing warm-water species (12°C for cold-water species).

From the several recommended species of vertebrates and invertebrates, the common warm-water test organisms are fathead minnow (*Pimephales promelas*) and cladocera (*Daphina*). The fathead minnow, a popular bait fish, feeds primarily on algae and grows to an average length of 50 mm with a normal life span less than 3 years. *Daphina* are invertebrates that feed on algae, grow to a maximum length of 4–6 mm, and have a life span of 40–60 days. Ten to 20 of each species are tested in several containers under controlled temperature, light intensity, dissolved oxygen, pH, and food supply. The results of a screening test are expressed as percentages of surviving minnows and cladocera.

Definitive bioassays are conducted for longer test periods at effluent dilutions in a geometric series, such as 100%, 50%, 25%, 12.5%, and 6.25%. For chronic toxicity, the recommended test durations are extended to 7 days by renewing the diluted effluent waters every 24 hr in static tests or continuously in the flow-through tests [13]. The dilution water is a synthetic moderately hard water or filtered water from the receiving watercourse upstream from the wastewater discharge. Based on the results, the no-observed-effect concentration (NOEC) is determined for effluent in the diluted test samples. This is compared against the instream (receiving water) waste concentration (IWC) calculated for the quantity of wastewater discharge diluted with the 7-consecutive-day one-in-ten-year low flow. The IWC should be less than the NOEC with a margin of safety [12].

# Selected Pollution Parameters

Total solids, suspended solids, BOD, chemical oxygen demand (COD), and coliform bacteria are common parameters used in water and wastewater engineering. Knowledge of testing procedures is essential to understand the meaning of these terms. For a detailed description of testing procedures for these and other substances, refer to *Standard Methods for the Examination of Water and Wastewater* [14].

## 8.8  TOTAL AND SUSPENDED SOLIDS

The term *total solids* refers to the residue left in a drying dish after evaporation of a sample of water or wastewater and subsequent drying in an oven (Fig. 8.1). After a measured volume is placed in a porcelain dish, the water is evaporated from the dish on a steam bath. The dish is then transferred to an oven and dried to a constant weight at 103°–105°C. The total residue is equal to the difference between the cooled weight of the dish and the original weight of the empty dish. The concentration of total solids is the weight of dry solids

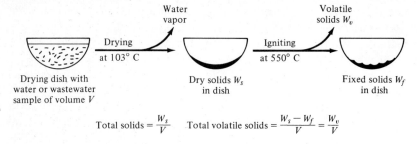

$$\text{Total solids} = \frac{W_s}{V} \qquad \text{Total volatile solids} = \frac{W_s - W_f}{V} = \frac{W_v}{V}$$

**Figure 8.1** Diagram of laboratory procedure to determine total solids and total volatile solids concentrations of a water or wastewater sample.

divided by the volume of the sample, usually expressed in milligrams per liter.

Total volatile solids are determined by igniting the dry solids at 550° ± 50°C in an electric muffle furnace. The residue remaining after burning is referred to as fixed solids, and the loss of weight on ignition is reported as volatile solids. The concentration of total volatile solids is the weight of dry solids minus the weight of fixed solids divided by the volume of the original liquid sample. Volatile solids content also can be expressed as a percentage of the dry solids in the sample.

The term *total suspended solids* refers to the nonfilterable residue that is retained on a glass-fiber disk after filtration of a sample of water or wastewater (Fig. 8.2). A measured portion of a sample is drawn through a glass-fiber filter, retained in a funnel, by applying a vacuum to the suction flask under the filter. The filter with damp suspended solids adhering to the surface

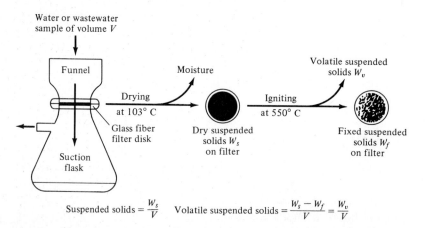

$$\text{Suspended solids} = \frac{W_s}{V} \qquad \text{Volatile suspended solids} = \frac{W_s - W_f}{V} = \frac{W_v}{V}$$

**Figure 8.2** Diagram of laboratory procedure to determine the total suspended-solids and volatile suspended-solids concentrations of a water or wastewater sample.

is transferred from the filtration apparatus to an aluminum or stainless steel planchet as a support. After drying at 103°–105°C in an oven, the filter with the dry suspended solids is weighed. The weight of suspended solids is equal to the difference between this weight and the original weight of the clean filter. The concentration of total suspended solids is the weight of the dry solids divided by the volume of the sample and is usually expressed in milligrams per liter.

Volatile suspended solids are determined by igniting the dry solids at 550° ± 50°C after placing the filter disk in a porcelain dish. The concentration of volatile suspended solids is the weight of dry solids minus the weight of fixed solids divided by the volume of the original liquid sample.

Dissolved solids are the solids that pass through the glass-fiber filter and are calculated from total and suspended solids analyses. Total dissolved solids equals total solids minus total suspended solids. Volatile dissolved solids equals total volatile solids minus volatile suspended solids.


# 8.9  BIOCHEMICAL AND CHEMICAL OXYGEN DEMANDS


## Biochemical Oxygen Demand

*Biochemical oxygen demand* is the quantity of oxygen used by microorganisms in the aerobic stabilization of wastewaters and polluted waters. The standard 5-day BOD value is commonly used to define the strength of municipal wastewaters, to evaluate the efficiency of treatment by measuring oxygen demand remaining in the effluent, and to determine the amount of organic pollution in surface waters.

Laboratory analyses of wastewaters and polluted waters are conducted using 300-ml BOD bottles incubated at a temperature of 20°C. The preparation of BOD tests is diagrammed in Fig. 8.3. To measure the BOD of a wastewater sample, a measured portion is placed in the BOD bottle, seed microorganisms are added if needed, and the bottle is filled with aerated dilution water containing phosphate buffer and inorganic nutrients. The amount of wastewater added to a 300-ml bottle depends on the estimated strength. For example, typical amounts are 5.0 ml for untreated wastewater in the BOD range of 120–420 mg/l and 50 ml for effluents with 12–42 mg/l. Untreated municipal wastewaters and unchlorinated effluents have adequate microbial populations without seeding. Industrial wastewaters and dechlorinated effluents require a prepared seed, usually aged domestic wastewater, to perform the biological reactions. While the wastewater provides the organic matter, the dilution water provides the dissolved oxygen for aerobic decomposition. For a polluted surface water, the sample is adjusted to 20°C and aerated to increase the dissolved oxygen to near saturation. The sample, or a portion of the sample mixed with dilution water, is then placed in the BOD bottle. A

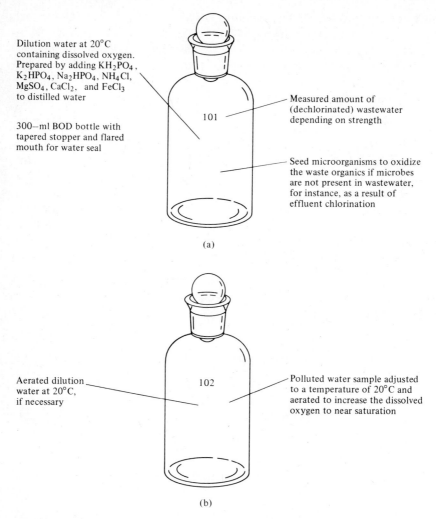

Dilution water at 20°C containing dissolved oxygen. Prepared by adding $KH_2PO_4$, $K_2HPO_4$, $Na_2HPO_4$, $NH_4Cl$, $MgSO_4$, $CaCl_2$, and $FeCl_3$ to distilled water

300—ml BOD bottle with tapered stopper and flared mouth for water seal

Measured amount of (dechlorinated) wastewater depending on strength

Seed microorganisms to oxidize the waste organics if microbes are not present in wastewater, for instance, as a result of effluent chlorination

(a)

Aerated dilution water at 20°C, if necessary

Polluted water sample adjusted to a temperature of 20°C and aerated to increase the dissolved oxygen to near saturation

(b)

**Figure 8.3** Preparation of biochemical oxygen demand (BOD) tests on (a) wastewater sample and (b) polluted surface water.

50% mixture of sample and dilution waters is suitable for a polluted-water BOD in the range of 4–14 mg/l.

The biochemical oxygen demand exerted by a diluted wastewater progresses approximately by first-order kinetics as shown in Fig. 8.4. The initial depletion of dissolved oxygen is the result of carbonaceous oxygen demand resulting from organic matter degradation:

Dissolved oxygen + organic matter

$$\xrightarrow[\text{and protozoans}]{\text{bacteria}} \text{carbon dioxide + biological growths} \qquad (8.3)$$

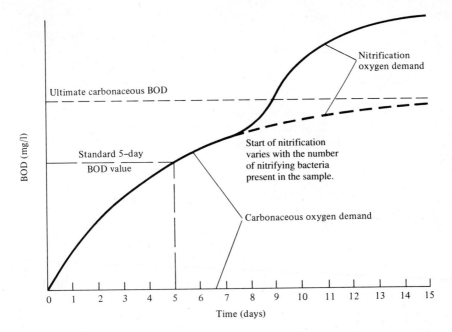

**Figure 8.4** Hypothetical biochemical oxygen demand reaction curve showing the carbonaceous and nitrification reactions.

If present in sufficient numbers, nitrifying bacteria exert a secondary oxygen demand by oxidation of ammonia:

Dissolved oxygen + ammonia nitrogen

$$\xrightarrow[\text{bacteria}]{\text{nitrifying}} \text{nitrate nitrogen} + \text{bacterial growth} \qquad (8.4)$$

Nitrification often lags several days behind the start of carbonaceous oxygen demand.

The carbonaceous oxygen-demand curve can be expressed mathematically as

$$BOD_t = L(1 - 10^{-kt}) \qquad (8.5)$$

where   $BOD_t$ = biochemical oxygen demand at time $t$, mg/l
$\quad\quad\quad L$ = ultimate BOD, mg/l
$\quad\quad\quad k$ = deoxygenation rate constant, day$^{-1}$
$\quad\quad\quad t$ = time, days

The equation for calculating BOD from a seeded laboratory test is

$$BOD = \frac{(D_1 - D_2) - (B_1 - B_2)f}{P} \qquad (8.6)$$

where   BOD = biochemical oxygen demand, mg/l
$D_1$ = dissolved oxygen (DO) of diluted seeded wastewater about 15 min after preparation, mg/l
$D_2$ = DO of wastewater after incubation, mg/l
$B_1$ = DO of diluted seed sample about 15 min after preparation, mg/l
$B_2$ = DO of seed sample after incubation, mg/l
$f$ = ratio of seed volume in seeded wastewater test to seed volume in BOD test on seed
$P$ = decimal fraction of wastewater sample used
$= \dfrac{\text{volume of wastewater}}{\text{volume of dilution water plus wastewater}}$

If the sample is unseeded, the relationship is

$$\text{BOD} = \frac{D_1 - D_2}{P} \tag{8.7}$$

The standard value is the BOD exerted during the first 5 days of incubation. More detailed discussions of BOD testing are presented by Hammer [15].

### ■ EXAMPLE 8.1
BOD tests were conducted on composited samples of a raw wastewater and a treated wastewater after chlorination.

  a. The BOD tests for the raw wastewater were set up by pipetting 5.0 ml into each 300-ml bottle. For one pair of bottles, the test results were: the initial dissolved oxygen (DO) was 8.4 mg/l, and after 5 days of incubation at 20°C the final DO was 3.7 mg/l. Calculate the BOD and estimate the 20-day BOD value assuming a $k$ of 0.10 day$^{-1}$.
  b. The treated wastewater sample was dechlorinated prior to conducting a seeded test. The BOD bottles were set up with 50.0 ml of treated wastewater and 0.5 ml of raw wastewater for seed added to each bottle. For one pair of bottles, the test results were: the initial DO was 7.6 mg/l, and the final DO was 2.9 mg/l. Calculate the BOD.

*Solution*

  a. Using Eq. 8.7,

$$\text{BOD}_5 = \frac{8.4 - 3.7}{5.0/300} = 282 \text{ mg/l}$$

  Using Eq. 8.5,

$$L = \frac{282}{1 - 10^{-0.1 \times 5.0}} = 412 \text{ mg/l}$$

$$\text{BOD}_{20} = 412(1 - 10^{-0.1 \times 20}) = 408 \text{ mg/l}$$

  b. Using Eq. 8.6,

$$\text{BOD} = \frac{(7.6 - 2.9) - (8.4 - 3.7)(0.5/5.0)}{50/300} = 25 \text{ mg/l}$$  ■

## Chemical Oxygen Demand

The *chemical oxygen demand* (COD) of a wastewater or a polluted water is a measure of the oxygen equivalent of the organic matter susceptible to oxidation by a strong chemical oxidant. The organic matter destroyed by the mixture of chromic and sulfuric acids is converted to $CO_2$ and water. The test procedure is to add measured quantities of standard potassium dichromate, sulfuric acid reagent containing silver sulfate, and a measured volume of sample into a flask. After attaching a condenser on top, this mixture is refluxed (vaporized and condensed) for 2 hr. The oxidation of organic matter converts dichromate to trivalent chromium,

$$\text{Organic matter} + Cr_2O_7^{2-} + H^+ \xrightarrow[\text{Ag}^+]{\text{heat}} CO_2 + H_2O + 2Cr^{3+} \qquad (8.8)$$

After cooling, washing down the condenser, and diluting the mixture with distilled water, the excess dichromate remaining in the mixture is measured by titration with standardized ferrous ammonium sulfate. A blank sample of distilled water is carried through the same COD testing procedure as the wastewater sample. The purpose of testing a blank is to compensate for any error that can result because of the presence of extraneous organic matter in the reagents. COD is calculated from the following equation:

$$\text{COD} = \frac{(a - b)[\text{normality of Fe(NH}_4)_2(SO_4)_2]8000}{V} \qquad (8.9)$$

where   COD = chemical oxygen demand, mg/l
       $a$ = amount of ferrous ammonium sulfate titrant added to blank, ml
       $b$ = amount of titrant added to sample, ml
      $V$ = volume of sample, ml
   8000 = multiplier to express COD in milligrams per liter of oxygen

## Applications of BOD and COD Testing

BOD is the most common parameter for defining the strengths of untreated and treated municipal and biodegradable industrial wastewaters. The oxygen requirement and tank sizing for aerobic treatment processes are based on BOD loadings. It is also used in quantifying the quality of an effluent from a treatment plant; the maximum allowable BOD being specified in the waste-water discharge permit. Since oxygen depletion in surface waters results from aquatic microorganisms exerting oxygen demand, the BOD test is a key measurement in the evaluation of water pollution by biodegradable wastes.

Commonly COD is used to define the strength of industrial wastewaters that are either not readily biodegradable or contain compounds that inhibit biological activity. Frequently laboratory wastewater treatability studies are based on COD testing rather than on BOD analyses. The COD test has the advantages of rapid analysis and reproducible results. The BOD test requires incubation for 5 days, and the results of multiple analyses on an industrial wastewater sample often show considerable scatter. Also, COD testing is becoming more popular in all applications of oxygen demand analyses as a result of simplified laboratory techniques.

The relationship between BOD and COD concentrations must be defined for each individual wastewater. Ideally for a wastewater composed of biodegradable organic substances, the COD concentration approximates the ultimate carbonaceous BOD value. Yet this simple relationship is rarely substantiated in testing of municipal wastewaters. Many organic compounds can be oxidized chemically that are only partly biodegradable.

## 8.10  COLIFORM BACTERIA

The coliform group of bacteria is defined as aerobic and facultative anaerobic, nonspore forming, Gram-stain negative rods that ferment lactose with gas production within 48 hr of incubation at 35°C.

### Fermentation Tube Technique

The basic coliform analysis, diagrammed in Fig. 8.5, is the presumptive test based on gas production during the fermentation of lactose (or lauryl tryptose) broth, which contains beef extract, peptone (protein derivatives), and lactose (milk sugar). Ten-milliliter portions of a water sample are transferred using sterile pipettes into prepared fermentation tubes containing inverted glass vials. The inoculated tubes are placed in a warm-air incubator at $35° \pm 0.5°C$. Growth with the production of gas, identified by the presence of a bubble in the inverted vial, is a positive test, indicating that coliform bacteria may be present. A negative reaction, either no growth or growth without gas, excludes the coliform group.

The confirmed test is used to substantiate, or deny, the presence of coliforms in a positive presumptive test. One method is to transfer about one-hundredth of a milliliter on a sterile wire loop from the positive lactose tube to a sterile tube containing brilliant green lactose bile broth. The green dye in this broth inhibits noncoliform growth. If growth with gas occurs within 48 hr at 35°C, the presence of coliforms is confirmed. If no gas is produced the test is negative. Normally, in examining potable water, coliform testing is terminated with this confirmed test.

With a positive confirmed test, a completed coliform test can be conducted by transferring this growth back to lactose broth and onto a nutrient solidified agar medium. If gas is produced in the lactose tube, a portion of the growth from the agar surface is smeared onto a glass slide and prepared

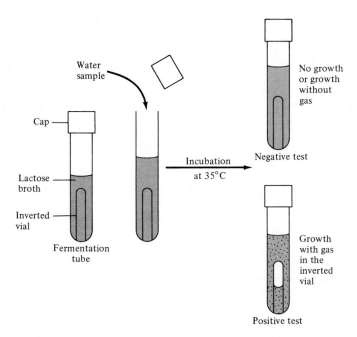

**Figure 8.5** Diagram of basic coliform test using a fermentation tube containing lactose or lauryl tryptose broth.

for observation under a microscope using the Gram-stain technique. The coliform group is present and the completed test is positive if the bacteria observed under the microscope are short rods, without spores present, and Gram-stain negative. If the culture is Gram-stain positive, the completed test is proved negative and coliforms were not present in the water sample.

## Membrane Filter Technique

The membrane filter test for coliform testing is commonly used when a large number of water samples are routinely analyzed. This method, diagrammed in Fig. 8.6, consists of drawing a measured volume of water through a filter membrane with small enough openings to take out bacteria and then placing the filter on a growth medium in a culture dish. This technique assumes that each bacterium retained by the filter grows and forms a small visible colony. The number of coliforms present in a filtered sample is determined by counting the number of typical colonies and expressing this value in terms of number per 100 ml of water.

The apparatus to perform membrane filter coliform testing includes a filtration unit, sterile filter membranes, sterile absorbent pads, culture dishes, nutrient media, and forceps. The forceps for handling the filters are sterilized before each use. Rapid decontamination of filter units between successive filtrations can be accomplished using an ultraviolet sterilizer with exposure

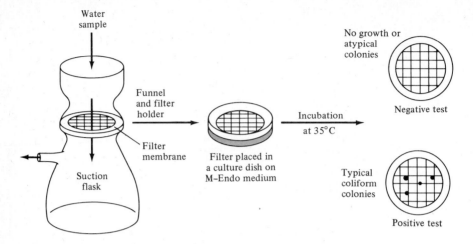

**Figure 8.6** Diagram of membrane filter technique for coliform testing.

to radiation for 2 min. The filter membranes with pore openings of 0.45 $\mu$m are in 2-in.-diameter disks with a grid printed on the surface for ease in counting colonies. In drinking-water testing, a 100-ml sample is filtered through the membrane. A culture dish with a fitted cover is prepared by placing an absorbent pad in the bottom half of the dish and saturating it with 1.0–2.0 ml of M-Endo medium. Using the forceps, the membrane is removed from the filtration unit and placed on the pad in the culture dish. After incubation for 22–24 hr at $35° \pm 0.5°C$, the cover is removed from the dish and the membrane examined for growth. Typical coliform colonies are pink to dark-red color with a green metallic surface sheen. Absence of bacterial growth on the membrane is a negative test. Occasionally, atypical colonies occur as light-colored growth without dark centers or a green sheen.

## Presence–Absence Technique

This test recommended by EPA for drinking-water analysis determines only the presence or absence of coliform bacteria in a water sample without indicating the number of coliforms in a positive test. A 100-ml sample of water is added to a sterile 250-ml bottle containing 50 ml of prepared P–A (presence–absence) broth containing lactose broth, lauryl tryptose broth, and bromcresol purple indicator. The bottle is incubated for 48 hr at 35°C. If a distinct yellow color forms in the broth, the test is positive and the presumption is that the water contained one or more bacteria of the coliform group. Coliform bacteria ferment lactose sugar, forming lactic acid and lowering the pH, which changes the bromcresol purple indicator color to yellow. If the yellow color does not appear, the test is negative and no coliform bacteria were present in the sample.

## Fecal Coliform Testing

The term *total coliforms* in the EPA drinking-water regulations includes any bacteria from feces, soil, or other origin that grow in a lactose (milk sugar) broth, producing acid and gas at 35°C or 95°F (approximate human body temperature) within 48 hr of incubation. Total coliforms is the first test conducted on a water sample by the fermentation tube, membrane filter, or presence–absence technique. The term *fecal coliforms* refers to coliform bacteria, such as *Escherichia coli,* from human or warm-blooded animal feces that grow in a lactose broth, producing acid and gas after 22–26 hr of incubation at the elevated temperature of 44.5°C (112.1°F). Nonfecal coliforms do not grow at this higher temperature.

The common test for fecal coliforms is a second-phase confirmatory test following growth of coliforms in the presumptive total-coliform test. A minute portion of the broth from a positive test, or a positive colony from a membrane filter, is transferred aseptically on a sterile wire loop to a fermentation tube of an EC medium containing tryptose, lactose, bile salts, and chemical buffers. The glass tube has a removable cap and an inverted vial at the bottom of the tube in the broth, as illustrated in Fig. 8.5. A positive test is growth with gas after 22–26 hr at 44.5°C. Gas production is evidence that coliforms have converted the lactose sugar to lactic acid, thus lowering the pH releasing gas. If no gas appears in the inverted vial, the test is negative and no fecal coliform bacteria were present in the positive total-coliform test.

## PROBLEMS

**8.1**  Compare the latency, persistence, and infective dose of *Ascaris* and *Salmonella*.

**8.2**  Discuss the significance of the carrier condition in transmission of enteric diseases. What major waterborne disease in the United States is commonly spread by carriers? How is this disease amplified by beavers?

**8.3**  Why can coliform bacteria be used as indicators of drinking-water quality? Discuss the limitations of coliforms as an indicator. Why is a positive test for fecal coliforms in a public water supply considered more serious than a positive test for total coliforms?

**8.4**  The NOAEL for chronic exposure of rats to drinking water containing cadmium was 0.75 mg/kg. Calculate the ADI in drinking water. Use an uncertainty factor of 1000 and assume 20% of the intake is apportioned to water.

**8.5**  Based on laboratory animal studies, a synthetic organic chemical has an estimated cancer risk of $1 \times 10^{-6}$ after a lifetime consumption of 1 l/d of water containing 1.0 $\mu$g/l of chemical. If the MCL is 0.005 mg/l and consumption is 2 l/d, what is the calculated risk and the number of excess cases of cancer expected among persons exposed?

**8.6**  Discuss the health risk of finding lead in drinking water.

**8.7**  What is the health risk of excess nitrate ion in drinking water?

**8.8**   How does the source of THMs differ from VOCs and other synthetic organic chemicals?

**8.9**   What are the conventional water pollutants often limited by an NPDES permit and their main pollution concern for limitation?

**8.10** The following data are from total solids and total volatile solids tests on a wastewater. Calculate the total and volatile solids concentrations in milligrams per liter.

> Weight of empty dish = 68.942 g
>
> Weight of dish plus dry solids = 69.049 g
>
> Weight of dish plus ignited solids = 69.003 g
>
> Volume of wastewater sample = 100 ml

**8.11** Listed below are total solids and suspended solids data on an industrial wastewater. Calculate the total and volatile solids, suspended solids, and dissolved solids.

*Total solids data*

> Weight of empty dish = 85.337 g
>
> Weight of dish plus dry solids = 85.490 g
>
> Weight of dish plus ignited solids = 85.375 g
>
> Volume of wastewater sample = 85 ml

*Suspended solids data*

> Weight of glass-fiber filter disk = 0.1400 g
>
> Weight of disk plus dry solids = 0.1530 g
>
> Weight of disk plus ignited solids = 0.1426 g
>
> Volume of wastewater filtered = 200 ml

**8.12** A unseeded BOD test is conducted on a polluted surface water by adding 100 ml to a 300-ml BOD bottle and filling with dilution water. The initial dissolved oxygen measured 8.2 mg/l, and the final concentration after 5 days of incubation at 20°C measured 2.9 mg/l. Calculate the BOD.

**8.13** A BOD test was conducted on the unchlorinated effluent of a municipal treatment plant. The wastewater portion added to a 300-ml BOD bottle was 30 ml, and the dissolved oxygen values listed below were measured using a dissolved-oxygen probe. Plot a BOD-versus-time curve and determine the 5-day BOD value.

| Time (days) | DO (mg/l) | Time (day) | DO (mg/l) |
|---|---|---|---|
| 0 | 8.7 | 6.0 | 4.9 |
| 2.0 | 6.7 | 10.0 | 3.9 |
| 4.0 | 5.7 | 14.0 | 0.7 |

**8.14** A BOD test was conducted on a raw domestic wastewater. The wastewater portion added to each 300-ml test bottle was 8.0 ml. The dissolved-oxygen values and incubation periods are listed below. Plot a BOD-versus-time curve and determine the 5-day BOD value.

| Bottle number | Initial DO (mg/l) | Incubation period (days) | Final DO (mg/l) | DO drop (mg/l) | Calculated BOD (mg/l) |
|---|---|---|---|---|---|
| 1 | 8.4 | 0 | 8.4 | | |
| 2 | 8.4 | 0 | 8.4 | | |
| 3 | 8.4 | 1.0 | 6.2 | | |
| 4 | 8.4 | 1.0 | 5.9 | | |
| 5 | 8.4 | 2.0 | 5.2 | | |
| 6 | 8.4 | 2.0 | 5.2 | | |
| 7 | 8.4 | 3.0 | 4.4 | | |
| 8 | 8.4 | 3.0 | 4.6 | | |
| 9 | 8.4 | 5.0 | 3.8 | | |
| 10 | 8.4 | 5.0 | 3.5 | | |

**8.15** A seeded BOD analysis was conducted on a food-processing wastewater. Ten-milliliter portions were used in preparing the 300-ml bottles to determine the dissolved-oxygen demand of the aged, settled wastewater seed. The seeded sample BOD bottles contained 2.7 ml of food-processing wastewater and 1.0 ml of seed wastewater. The results of this series of test bottles are listed below. Calculate the wastewater BOD values and plot a BOD–time curve. What is the 5-day BOD?

| Time (days) | Seed tests $B_1$ (mg/l) | Seed tests $B_2$ (mg/l) | Sample tests $D_1$ (mg/l) | Sample tests $D_2$ (mg/l) |
|---|---|---|---|---|
| 0 | 7.8 | — | 8.1 | — |
| 1.0 | 7.8 | 6.9 | 8.1 | 5.6 |
| 2.0 | 7.8 | 6.6 | 8.1 | 4.3 |
| 3.0 | 7.8 | 6.3 | 8.1 | 3.6 |
| 4.0 | 7.8 | 5.8 | 8.1 | 3.0 |
| 5.0 | 7.8 | 5.7 | 8.1 | 2.5 |
| 6.0 | 7.8 | 5.3 | 8.1 | 2.0 |
| 7.0 | 7.8 | 5.4 | 8.1 | 1.8 |

# REFERENCES

1. R. G. Feachem, D. J. Bradley, H. Garelick, and D. D. Mara, *Sanitation and Disease, Health Aspects of Excreta and Wastewater Management: World Bank Studies in Water Supply and Sanitation 3.* (Chichester: Wiley, 1983).

2. E. C. Lippy and S. C. Waltrip, "Waterborne Disease Outbreaks—1946–1980: A Thirty-Five-Year Perspective," *J. Am. Water Works Assoc.* 76(2) (1984): 60–67.

3. *Drinking Water and Health,* Vol. 1, National Academy of Sciences (Washington, DC: National Academy Press, 1977).

4. *Drinking Water and Health,* Vol. 6, National Academy of Sciences (Washington, DC: National Academy Press, 1986).

5. *Drinking Water and Health,* Vol. 3, National Academy of Sciences (Washington, DC: National Academy Press, 1980).

6. D. F. Striffler, W. O. Young, and B. A. Bart, "The Prevention and Control of Dental Caries: Fluoridation," in *Dentistry, Dental Practice, and the Community* (Philadelphia: Saunders, 1983): 155–199.

7. "Radioactivity in Drinking Water," Environmental Protection Agency, Office of Drinking Water, EPA-570/9-81-002 (January 1981).

8. M. J. Hammer and K. A. Mac Kichan, *Hydrology and Quality of Water Resources* (New York: Wiley, 1981).

9. "Quality Criteria for Water," Environmental Protection Agency (July 1976).

10. D. S. Szumski, D. A. Barton, H. D. Putman, and R. C. Polta, "Evaluation of EPA Un-ionized Ammonia Toxicity Criteria," *J. Water Poll. Control Fed.* 54(3) (1982): 281–291.

11. "Ambient Water Quality Criteria" (for a specific priority pollutant), Environmental Protection Agency, Office of Water Regulations and Standards, Criteria and Standards Division (October 1980).

12. "Methods of Measuring the Acute Toxicity of Effluents to Freshwater and Marine Organisms," 3rd ed., Environmental Protection Agency, Environmental Monitoring and Support Laboratory, EPA-600/4-85-013 (March 1985).

13. "Short-Term Methods for Estimating the Chronic Toxicity of Effluents and Receiving Waters to Freshwater Organisms," 2nd ed., Environmental Protection Agency, Environmental Monitoring Systems Laboratory (March 1989).

14. *Standard Methods for the Examination of Water and Wastewater,* 17th ed. (Washington DC: Am. Public Health Assoc., Am. Water Works Assoc., Water Poll. Control Fed., 1989).

15. M. J. Hammer, *Water and Wastewater Technology,* 2nd ed. (New York: Wiley, 1986) and *Water and Wastewater Technology, SI Version,* 2nd ed. (New York: Wiley, 1986).

# Chapter 9

# Systems for Treating Wastewater and Water

The objective of this chapter is to provide students with an overview of wastewater and water treatment systems relative to water pollution and water quality. Sections on selection of wastewater and water treatment processes are introductory to subsequent chapters that deal separately with physical, chemical, biological, and sludge treatment methods. They also act as summaries for subsequent chapters by drawing the unit processes into integrated treatment systems. *Students should refer to Sections 9.5 and 9.8 while studying Chapters 10–13.*

## Wastewater Treatment Systems

The purpose of municipal wastewater treatment is to prevent pollution of the receiving watercourse. Characteristics of a municipal wastewater depend to a considerable extent on the type of sewer collection system and industrial wastes entering the sewers. The degree of treatment required is determined by the beneficial uses of the receiving stream or lake. Stream pollution and lake eutrophication resulting from municipal wastes are particularly troublesome in water use for water supplies and recreation.

### 9.1 INDUSTRIAL WASTEWATERS

An industry has three possibilities for disposal of process wastewaters: (1) they may be treated separately in an industrial waste treatment plant prior to discharge to a watercourse, (2) raw wastewaters may be discharged to the municipal treatment plant for complete treatment, or (3) industrial wastes can be pretreated at the industrial site prior to discharge in the municipal sewerage system. A careful study must be performed to determine the most feasible method for disposal. Discharge to a municipal plant relieves the industry of wastewater treatment, and the municipality accepts the responsibility for its disposal.

Joint treatment of industrial and domestic wastewaters has several advantages. A municipality can apply for government aid for plant construction, whereas private industry is not eligible for such cost-sharing monies. Only one major treatment plant is required, often resulting in more economical operation. The responsibility for operation is placed with one owner. The majority of industrial wastes are more amenable to biological treatment after dilution with domestic wastewater.

Three categories of wastes should be excluded from the municipal sewers, those which (1) create a fire or explosion hazard (e.g., gasoline or cleaning solvents), (2) impair hydraulic capacity (e.g., paunch manure or sand), and (3) create a hazard to people, the sewer system, or the biological treatment system (e.g., toxic metal ions or hazardous organic wastes).

Large volumes of high-strength wastes must be considered in the design capacity of municipal plants since overloaded treatment plants resulting from industrial expansion become a serious problem. Facilities constructed on funds provided from a 20-yr bond issue frequently reach design loading in 5–10 yr. A municipal sewer code, user fees, and separate contracts between an industry and the city can provide adequate control and sound financial planning, while accommodating industry by joint wastewater treatment. A municipality should enact a sewer ordinance and establish charges for industry to protect the public investment in its treatment plant.

Pretreatment should be considered for industrial discharges that have strengths or characteristics differing significantly from domestic wastewater. Only a fine line of difference may exist between an untreatable trade waste and one that can be jointly treated after removal of toxic wastes, equalization of the wastewater flow, and strength reduction.

## Process Modifications

Pretreatment of wastewaters at the industrial sites are frequently an integral part of plant design. Process changes, equipment modifications, by-product recovery, and wastewater reuse can have an economic advantage. Industries with a large volume of high-strength wastewater should consider recovery processes to reduce the pollution load on the municipal treatment plant. For example, a tannery in the dehairing process can use a proprietary unhairing compound with low sulfides and realize significant BOD reductions in comparison with the usual high-sulfide process. Many industries have installed countercurrent washing or spray-rinsing systems to reduce water consumption. Collecting and processing blood is a good example of by-product recovery by the meat industry of a salable commodity. Reuse of water is practiced extensively by closed systems in paper production. Thus, a paper mill can recycle white water (that passing through a wire screen upon which paper is formed), thus saving costs of both supply and treatment.

## Segregation of Wastewaters

In many older industrial plants, process, cooling, and sanitary wastewaters are mixed in one sewer line. Modern industrial treatment dictates segregation of these waste streams for separate disposal, individual pretreatment, or controlled mixing. Cooling water, although slightly contaminated by leakage, corrosion, and heat, generally contains little or no pollutional organic matter. Therefore, it can be disposed of in a storm sewer or directly in a watercourse provided that thermal pollution is not a problem. Segregated waste streams containing coarse solids (i.e., paunch manure, feathers, corn, onion peelings, and hair) can be screened readily. Combined waste streams having coarse solids mixed with grease and sanitary wastes are more difficult to screen. In treating tannery wastes, the chrome dump waste (the spent chrome tanning liquor) can be proportioned into the tannery wastewater to improve primary sedimentation.

## Equalization

Industries using a diversity of processes may be required to equalize wastes by holding them in a basin for a certain period of time to get a stable effluent easier to treat in a municipal plant. Neutralization of alkaline and acid waste streams, stabilization of BOD, and settling of heavy metals are some objectives of equalization. Textile mill wastes discharged to a municipal plant must be equalized to prevent fluctuations in pH and BOD, which upset the efficiency of a biological treatment system. Unequalized wastewaters high in alkalinity or acidity commonly require neutralization by chemical addition to prevent upsetting the system.

## Waste Strength Reduction

Certain industrial wastes can be pretreated to reduce the organic or inorganic solids prior to disposal in a municipal plant. Tannery or textile wastes may require equalization and sedimentation to decrease settleable solids. Dairy wastes from a bottling plant, a creamery, or a cheese factory located on the sewer line a considerable distance from the municipal plant may require pretreatment by aeration to prevent acidic, malodorous influent wastewater. Reduction in drippage and spillage and by-product recovery in milk processing can reduce the strength of dairy wastes significantly. Whole milk has a BOD of approximately 100,000 mg/l, and whey a BOD of 35,000 mg/l.

Metal-plating wastes containing cyanide, chromium, zinc, copper, and other heavy metals may require treatment in addition to equalization if the toxic ions are not sufficiently diluted with domestic wastewater. Chemical treatment by oxidation or coagulation are commonly used for removal of these inorganic pollutants. Waste streams with high concentrations of refractory pollutants (i.e., salt or ammonia nitrogen) must be regulated at the plant site if the municipal waste-disposal system does not have sufficient assimilative capacity.

## 9.2  INFILTRATION AND INFLOW

Infiltration is groundwater entering sewers and building connections through defective joints and cracks in pipes and manholes. Inflow is water discharged into service connections and sewer pipes from foundation and roof drains, outdoor paved areas, cooling water from air conditioners, and unpolluted discharges from businesses and industries. Excessive infiltration and inflow cause surcharging of sewer lines with possible backup of sanitary wastewaters into basements, hydraulic overloading of treatment facilities, and bypassing of pumping stations or processing plants.

The amount of infiltration depends on condition of the sewer system, groundwater levels, and porosity of the soil profile. Proper design, selection of sewer pipe with watertight joints, and supervision during construction can limit the quantity of seepage. Specifications for new construction permit a maximum infiltration rate of 200–300 gpd/mi/in. of pipe diameter. This quantity of flow is equal to about 3% of the peak hourly, or 7% of the average, domestic wastewater flow rate. Evaluation of an existing sewer system is initiated by measuring both the wet- and dry-weather flows to determine sources and quantities of excessive infiltration. Possible corrective measures include grouting or sealing soils, relining pipe, and sewer replacement.

Inflow is the result of deliberately connecting sources of extraneous water to sanitary sewer systems. Storm water and drainage should be disposed of in storm sewers; however, sanitary lines are often more convenient because of their greater depth of burial and convenient location. Establishing and enforcing a sewer-use regulation that excludes unpolluted waters from separate sanitary collectors is an effective method of preventing excess inflow. A municipal sewer code should require disposal of surface runoff, foundation drainage, air-conditioning effluents, industrial cooling waters, and other clean waters to storm lines or natural drainage courses. Where problems already exist, surveys can be conducted to locate illegal connections and institute corrective measures.

## 9.3  COMBINED SEWERS

Combined wastewater collection systems have only one sewer pipe network to collect domestic wastewater, industrial wastes, and storm runoff water. The dry-weather flow (domestic and industrial wastewater, plus street drainage from washing operations and lawn sprinkling) is intercepted for treatment. If during storms the combined wastewater and runoff water exceed the capacity of the treatment plant, the overflow is bypassed directly to the receiving watercourse without treatment. This excess may result in significant pollution, even during high river stages, and create a health hazard for the downstream water user. Combined sewers are no longer acceptable except in cities having combined systems, where separate systems are not feasible due to a lack of storm sewer outfalls.

In antiquity, sewers were constructed primarily as drains to carry away storm water from urbanized areas. As late as 1850 disposal of sanitary wastes into the sewers of London was prohibited. Development of the water carriage waste-disposal system led to the use of storm drains as combined sewers.

In the United States many cities along rivers built combined sewers to convey both storm runoff and sanitary wastewater directly to the river through a series of lines oriented perpendicular to the river. Required treatment of sanitary wastewater resulted in construction of a collector sewer along the river to intercept the dry-weather flow in combined sewers and convey it to a treatment plant. The flow entering a collector line from a combined sewer is commonly regulated by a gate or weir arrangement. The dry-weather flow from an overlying combined sewer falls into the interceptor below, but the greater wet-weather flows, with their higher velocities, leap across the opening and flow directly to the river.

In many cities, only the older sections are served by combined sewers. In other cities, the sanitary sewers in aging neighborhoods are cracked and the mortar joints deteriorated, so infiltration of groundwater is excessive during wet seasons. Foundation drains and roof downspouts are sometimes connected to the sanitary sewer; hence the design of a municipal plant must consider these hydraulic loads or the high flows eliminated by reconstruction. If drainage from streets is involved, special consideration should be given to the design of bar screens and grit-removal units to prevent carryover of solids damaging to subsequent treatment equipment.

## 9.4 PURPOSES OF WASTEWATER TREATMENT

The location of a typical municipal wastewater treatment plant is illustrated in Fig. 9.1. Wastewaters from households, industries, and combined sewers are collected and transported to the treatment plant with the effluent commonly disposed of by dilution in rivers, lakes, or estuaries. This is normally the only feasible method of disposal and, for some communities on small rivers, the only system that ensures adequate downstream river flow during periods of drought. Other means of disposal include irrigation, evaporation from lagoons, and submarine outfalls extending into the ocean.

Water-quality criteria have been established for receiving waters to assist in defining the degree of treatment required for disposal by dilution. Effluent standards prescribe the required quality of the discharge from each municipality and industry. In general, the minimum extent of processing is secondary treatment. Some cities and industries are required to install tertiary or ad-

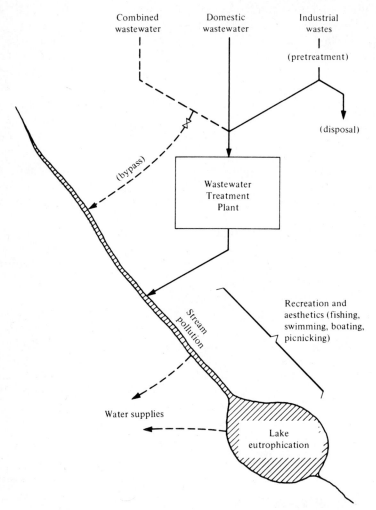

**Figure 9.1** Location of a municipal wastewater treatment plant.

vanced wastewater treatment processes for removal of pollutants that are resistant to conventional treatment, e.g., removal of phosphorus to retard eutrophication of receiving lakes. Stream classification documents, published by each state as required by federal law, categorize surface waters according to their most beneficial present or future use (i.e., for drinking-water supplies, body-contact recreation, etc.). These publications also incorporate stream standards that establish maximum allowable pollutant concentrations for a given stream under defined flow conditions. Effluent standards under the National Pollutant Discharge Elimination System (NPDES) are used for regulatory purposes to achieve compliance with these stream standards.

Rules and regulations of the Environmental Protection Agency define the minimum level of effluent quality that must be attained by secondary treatment of municipal wastewaters. Acceptable secondary effluent is defined in terms of biochemical oxygen demand, suspended solids, fecal coliform bacteria, and pH. The arithmetic mean of BOD and suspended-solids concentrations for effluent samples collected in a period of 30 consecutive days must not exceed 30 mg/l; and, during any 7-consecutive-day period, the average must not exceed 45 mg/l. Furthermore, removal efficiencies shall not be less than 85% (i.e., if an influent concentration is less than 200 mg/l, the effluent cannot exceed 15% of this value). The geometric mean of fecal coliform counts for effluent samples collected in a period of 30 consecutive days must not exceed 200 per 100 ml; and the geometric mean of fecal coliform bacteria for any 7-consecutive-day period must not exceed 400 per 100 ml. The effluent values for pH shall remain within the limits 6.0–9.0. These numerical limitations have been modified for some treatment processes where higher concentrations of BOD, suspended solids, and fecal coliforms do not create a problem in effluent disposal. Trickling-filter plants and lagoons are often permitted to discharge BOD and suspended solids concentrations greater than the maximum allowable average of 30 mg/l. Disinfection for control of bacterial populations may not be required at anytime if no threat to public health exists.

Meeting the 30-mg/l BOD and suspended-solids effluent requirement requires a well-designed plant that is properly operated. Moderately loaded activated-sludge processes, biological towers, and two-stage trickling-filter plants are capable of achieving this degree of treatment. However, some designs that were popular in the past are not efficient enough (e.g., single-stage trickling-filter plants and high-rate activated-sludge systems). Reduction of fecal coliforms to a level of 200 per 100 ml requires disinfection with chlorine. Facultative stabilization ponds handling raw wastewater are not able to meet discharge standards without further treatment. The main problem is the high suspended-solids content of the lagoon water during the summer due to growth of algae. Where the average suspended-solids content cannot exceed 30 mg/l, the best solution in most cases is disposal by land irrigation or evaporation of the liquid from complete retention lagoons to avoid effluent discharge to surface waters.

Refractory contaminants, both inorganic and organic materials, are pollutants resistant to, or totally unaffected by, conventional treatment processes. The mineral quality of wastewater depends largely on the character of the municipal water supply, but during water use numerous substances are added—common salt (sodium chloride) and other dissolved solids. Phosphates, which occur in low concentrations in most natural waters, are increased during domestic use, principally by synthetic detergents. Organic nitrogeneous compounds decompose to ammonia and to a variable extent oxidize to nitrates in treatment. Certain high-molecular-weight materials (e.g., tannins and lignins) and aromatic compounds (e.g., surfactants and dyes) are resistant to biological degradation. Rivers are also polluted by waste effluents from industries and agricultural chemicals in land runoff. The latter includes refractory pesticides such as chlorinated hydrocarbons.

If a refractory pollutant originates from an industrial waste that can be segregated, the pollutant may be removed by pretreatment, or by eliminating the isolated waste stream using land burial, or incineration rather than disposal in a sewer. Highly colored industrial waste streams can be handled by chemical treatment or separate disposal. Common plant nutrients are derived from organic matter and phosphate builders of detergents found in domestic wastewaters. Phosphorus and nitrogen removals in conventional treatment are only 30%–60% effective, depending on the types of processes and concentrations in the raw wastewater. Advanced treatment methods are required for greater nitrogen reduction and phosphorus removal.

Eutrophication of a lake induced by municipal wastewater can be retarded by removing the source of plant nutrients, notably phosphorus. This is accomplished by diversion of the treated effluent around the lake, or by treatment of the wastewater employing advanced treatment processes. No ready solution exists for removing nutrients from runoff, either from natural land or cultivated fields. Before instituting phosphorus and nitrogen removal from a municipal wastewater, therefore, a survey should determine existing levels and identify sources. Although in many cases municipal wastes have been the major origin of phosphorus, various receiving waters have concentrations well above the desirable limit originating from land drainage.

## 9.5  SELECTION OF TREATMENT PROCESSES

Conventional wastewater treatment consists of preliminary processes (pumping, screening, and grit removal), primary settling to remove heavy solids and floatable materials, and secondary biological aeration to metabolize and flocculate colloidal and dissolved organics (Fig. 9.2). Waste sludge drawn from these unit operations is thickened and processed for ultimate disposal.

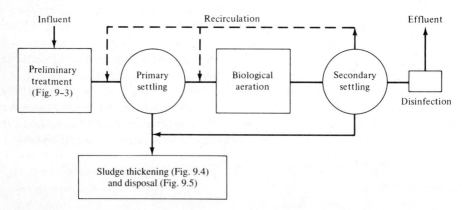

**Figure 9.2** Schematic diagram of conventional wastewater treatment.

## Preliminary Treatment Units

The following preliminary processes are used in municipal wastewater treatment: coarse screening (bar racks), medium screening, comminution, flow measuring, pumping, grit removal, and preaeration. Although not common in pretreatment, flotation, flocculation, and chemical treatment are sometimes dictated by the industrial pollutants in the municipal wastewater. Flotation is used to remove fine suspensions, grease, and fats and is performed either in a separate unit or in a preaeration tank also used for grit removal. If adequate pretreatment is provided by petroleum industries and meat-processing plants, flotation units are not required at a municipal facility. Flocculation with or without chemical additions may be practiced on high-strength municipal wastewaters to provide increased primary removal and prevent excessive loads on the secondary treatment processes. Chlorination of raw wastewater is sometimes used for odor control and to improve settling characteristics of the wastes.

The arrangement of preliminary treatment units varies depending on raw-wastewater characteristics, subsequent treatment processes, and the preliminary steps employed. A few general rules always apply in arrangement of units. Screens are used to protect pumps and prevent solids from fouling grit-removal units and flumes. In small plants a Parshall flume is normally placed ahead of constant-speed lift pumps, but may be located after them in large plants or where variable-speed pumps are used. Grit removal should be placed ahead of the pumps when heavy loads are anticipated, although the grit chamber follows the lift pumps in most separate sanitary wastewater plants. Three possible arrangements for preliminary units are shown in Fig. 9.3.

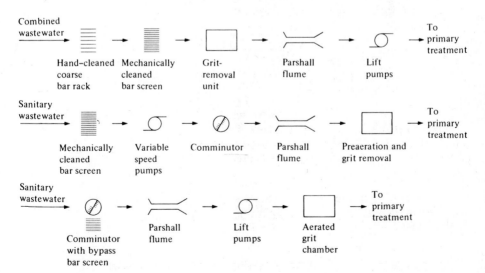

**Figure 9.3** Possible arrangements of preliminary treatment units in municipal wastewater processes.

## Primary Treatment Units

Primary treatment is sedimentation. In common usage, the term usually includes the preliminary treatment processes. Sedimentation of raw wastewater is practiced in all large municipal plants and must precede trickling (biological) filtration. Completely mixed activated-sludge processes can be used to treat unsettled raw wastewater; however, this method is generally restricted to small municipalities because of the costs involved in sludge disposal and operation.

## Secondary Treatment Units

Primary sedimentation removes 30%–50% of the suspended solids in raw municipal wastewater. Remaining organic matter is extracted in biological secondary treatment to the allowable effluent residual using activated-sludge processes, trickling filters, or rotating biological contactors. In the activated-sludge method wastewater is fed continuously into an aerated tank, where microorganisms synthesize the organics. The resulting microbial floc (activated sludge) is settled from the aerated mixed liquor under quiescent conditions in a final clarifier and returned to the aeration tank (Fig. 9.2). The plant effluent is clear supernatant from secondary settling. Advantages of liquid aeration are high-BOD removals, ability to treat high-strength wastewater, and adaptability for future use in plant conversion to advanced treatment. On the other hand, a high degree of operational control is needed, shock loads may upset the stability of the biological process, and sustained hydraulic or organic overloading results in process failure.

Trickling filters and rotating biological contactors have media to support microbial films. These slime growths extract organics from the wastewater as it trickles over the surfaces. Oxygen is supplied from air moving through voids in the media. Excessive biological growth washes out and is collected in the secondary clarifier. In northern climates two-stage shallow trickling filters are needed to achieve efficient treatment. Biological towers may be either single- or two-stage systems. Advantages of filtration are ease of operation and capacity to accept shock loads and overloading without causing complete failure.

## Sludge Disposal

Primary sedimentation and secondary biological flocculation processes concentrate the waste organics into a volume of sludge significantly less than the quantity of wastewater treated. But disposal of the accumulated waste sludge is a major economic factor in wastewater treatment. The construction cost of a sludge processing facility is approximately one-third that of a treatment plant.

Flow schemes for withdrawal, holding, and thickening raw waste sludge from sedimentation tanks are illustrated in Fig. 9.4. The settled solids from clarification of trickling filter effluent are frequently returned to the plant head for removal with the primary sludge (top diagram, Fig. 9.4). Raw settlings may be stored in the primary tank bottom until processed or pumped into a holding tank for storage. The withdrawn sludge may be concentrated in a gravity thickener prior to processing. In the bottom diagram of Fig. 9.4, waste-activated sludge is mixed with primary residue after withdrawal. A holding tank is commonly used in this system arrangement, along with a thickener. The waste activated may be thickened separately or the combined sludges thickened prior to processing.

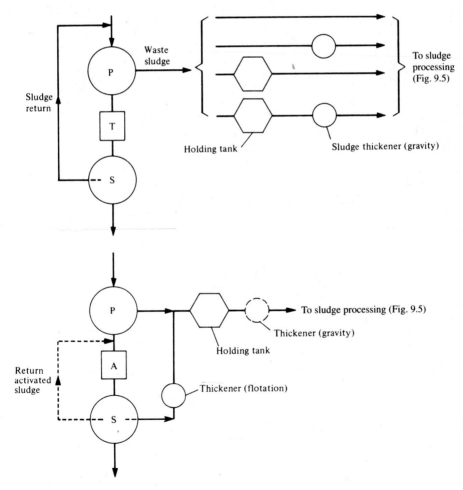

**Figure 9.4** Flow schemes for withdrawal, holding, and thickening of waste sludge. P, primary settling; T, trickling filtration; A, activated-sludge aeration; S, secondary settling.

Alternatives for processing and disposal of raw sludge are shown in Fig. 9.5. Common methods are anaerobic digestion and mechanical dewatering by pressure or vacuum filtration. Centrifugation and wet combustion are less often employed. Conventional methods of disposal include burial in landfill, incineration, production of soil conditioner, and barging to sea. For coastal cities ocean dumping is often least expensive, whereas burial is regularly practiced if landfill area is available. Incineration, although more costly, is frequently the only feasible method of disposal in urbanized areas.

All possible sludge-disposal processes for a municipal treatment plant must be given careful consideration. The method selected should be the most economical process, if it is best, with due regard to environmental conditions. Attention must be given to such factors as trucking sludge through residential areas, future use of landfill areas, groundwater pollution, air pollution, other potential public health hazards, and esthetics.

## Advanced Wastewater Treatment

Phosphorus removal is practiced in processing wastewaters that are discharged to receiving waters subject to eutrophication including lakes, reservoirs, and slow-moving rivers acting like impoundments. Since nitrogen removal is much more difficult and costly, it is rarely performed, although nitrification to reduce the ammonia content is sometimes done to control the oxygen demand and toxicity of the plant effluent. Water reclamation is achieved in varying degrees by many plants depending on the local environmental concerns; however, only a few large-scale plants are reclaiming water to near original quality. Advanced wastewater treatment systems are discussed in Chapter 14.

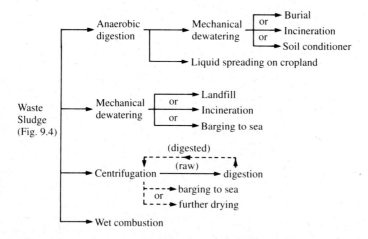

**Figure 9.5** Common methods for processing and disposal of raw waste sludge.

## Size of Municipality

Operational management and control and the necessity for sludge handling dictate the selection of wastewater treatment processes for small communities. Methods that do not require sludge disposal (stabilization ponds) or only occasional sludge withdrawals (extended aeration) are preferable for small villages and subdivisions. Towns large enough to employ a part-time operator frequently use systems that require more operational control and maintenance (i.e., contact stabilization and oxidation ditch plants). Cities with trickling-filter and activated-sludge treatment plants employ several workers.

Listed in Table 9.1 are the common types of wastewater treatment plants built in municipalities of various sizes. Many existing plants do not conform to the listing in Table 9.1. Some are no longer popular and others were selected on the basis of unique local conditions.

## Layout of Treatment Plants

The flow diagram for a two-stage trickling filter plant is shown in Fig. 9.6; an aerial view of the same facility is given in Fig. 9.7. The sequence of treatment units is raw-wastewater pumping station with a mechanically cleaned bar screen, primary settling tank, first-stage trickling filter, intermediate clarifier, second-stage filter, final settling basin, and chlorine contact tank. The stone-media biological filters are covered with fiberglass domes to prevent ice formation and maintain an adequate wastewater temperature

**TABLE 9.1**  COMMON TYPES OF WASTEWATER
TREATMENT PLANTS FOR
MUNICIPALITIES OF VARIOUS SIZES

| |
|---|
| Subdivisions, schools, etc. |
|    Extended aeration (factory built) |
|    Stabilization ponds |
| Towns (population less than 2000) |
|    Extended aeration |
|    Contact stabilization (field erected, factory built) |
|    Oxidation ditch |
|    Stabilization ponds |
| Small cities (2000–8000 population) |
|    Contact stabilization |
|    Oxidation ditch |
|    Completely mixed activated sludge without primary |
|    Primary plus trickling filters |
|    Primary plus rotating biological contactors |
| Cities (population greater than 10,000) |
|    Primary plus trickling filters |
|    Primary plus activated sludge |

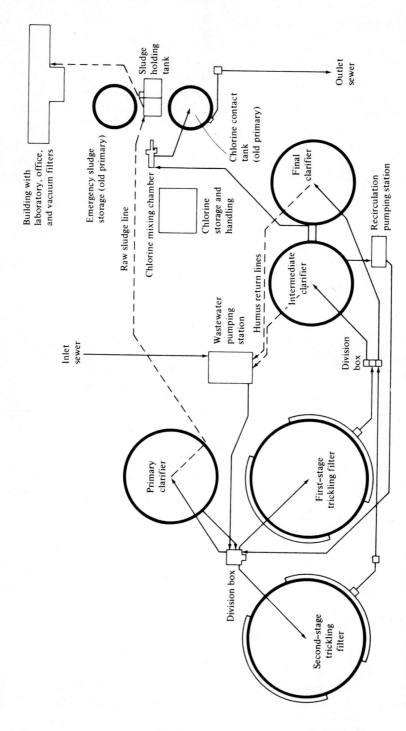

**Figure 9.6** Plant layout for a two-stage trickling-filter wastewater treatment plant, Kearney, NE. Average design flow, 3.0 mgd with 230 mg/l of BOD and 250 mg/l of suspended solids, and maximum wet weather flow of 7.5 mgd. (Courtesy of HDR Engineering Inc.)

**Figure 9.7** Aerial view of two-stage filter plant shown in Fig. 9.6. (Courtesy of HDR Engineering Inc.)

during the winter for biological activity. Wastewater is recirculated through the trickling filters to maintain optimum hydraulic loading and increase the removal of organic matter. Humus washed off the filter media collects in the intermediate and final clarifiers. Underflow from the clarifiers returns this material to the lift station, where it is pumped to the primary settling basin with the raw wastewater. Primary sludge containing raw organics and filter humus is withdrawn from the bottom of the clarifier and discharged to a holding tank. From here it is pumped to vacuum filters for dewatering prior to disposal on agricultural land after composting with feedlot manure.

The sequence of wastewater treatment units in the activated-sludge plant shown in Fig. 9.8 is as follows: a mechanically cleaned bar screen with a grinder that returns shredded solids to the raw wastewater, constant- and variable-speed lift pumps with standby gas engines, a Parshall flume, clarifier-type aerated grit chambers with a separate grit washer, primary clarifiers and a raw-sludge pumping station, and a completely mixed activated-sludge secondary with sludge-recirculating pumps and air blowers. Primary sludge is pumped into sludge holding tanks, from which it is withdrawn for dewatering by vacuum filtration, returning filtrate to the wet well. Chemical feeders are provided to allow conditioning of the sludge with either a polymer or ferric chloride and lime. The filter cake is disposed of by either land burial or composting and spreading on agricultural land. The waste-activated sludge is aerobically digested and applied to cropland near the treatment plant by subsurface injection. In addition to the small sludge holding tank adjacent to the injection pumping station, a sludge lagoon can be used to store digested sludge during winter months when subsurface injection is not possible.

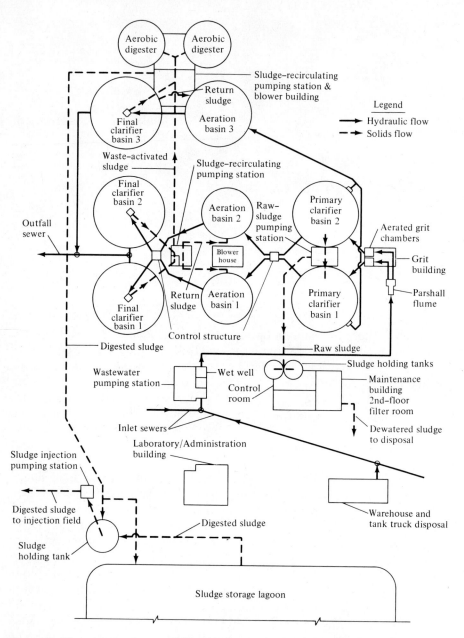

**Figure 9.8** Plant layout for a completely mixed activated-sludge wastewater treatment plant, design flow, 11 mgd, Grand Island, NE. (Courtesy of Black & Veatch, Engineers-Architects.)

# Water Treatment Systems

## 9.6  WATER SOURCES

Typical water sources for municipal supplies, illustrated in Fig. 9.9, are deep wells, shallow wells, rivers, natural lakes, and impounding reservoirs. No two sources of supply are alike, and the same origin may produce water of varying quality at different times. Water treatment processes selected must consider the raw-water quality and differences in quality for each particular water source.

Municipal water-quality factors of safety, temperature, appearance, taste and odor, and chemical balance are most easily and frequently satisfied by a deep-well source. Furthermore, the treatment processes employed are simplest because of the relatively uniform quality of such origin. Excessive concentrations of iron, manganese, and hardness habitually exist in well waters. Some deep-well supplies may contain hydrogen sulfide, others may have excessive concentrations of chlorides, sulfates, and carbonate. Hydrogen sulfide can be removed by aeration or other oxidation processes, but the prevalent sodium and potassium salts of anions bicarbonate, chloride, and

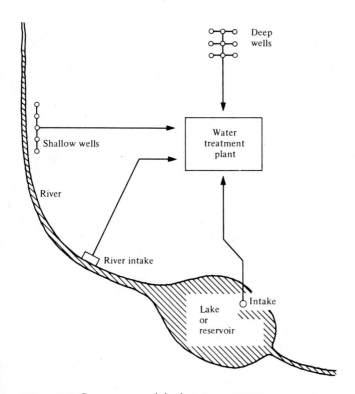

**Figure 9.9** Common municipal water sources.

sulfate are refractory inorganics. Calcium and magnesium can be removed by precipitation softening. Excessive concentration of fluoride, responsible for mottled enamel of the teeth, has in rare cases caused a municipality to abandon a well-supply source. The mineral constituents of a potential groundwater source should be carefully examined in selecting a deep-well water supply.

Shallow wells recharged by a nearby surface watercourse may have quality characteristics similar to deep wells or may relate more closely to the watercourse quality. A sand aquifer adjacent to a river may act as an effective filter for removal of organic matter and as a heat exchanger for leveling out temperature changes of the recharge water seeping into it. A case in point is the municipal water supply for Lincoln, Nebraska, located in the Platte River Valley near Ashland. The wells are located in a sand aquifer adjacent to the river channel. Chemical composition of the well water and river water are nearly identical. River-water temperature varies from 80°F in the summer to 33°F in the winter, whereas the well-water temperature remains within 3° of 56°F year-round. None of the bacterial or organic pollution existing in the river has been identified in the well water. Another shallower well field located adjacent to the Platte River in a coarser sand and gravel aquifer withdraws water of significantly poorer quality after peak pumping periods. The well-water temperature increases to nearly 70°F in late summer and bacterial counts in the raw water rise. Clogging of the well screens, attributed in part to bacterial slime growths, has occurred in some wells.

To predict water quality from shallow wells, careful studies of the aquifer and nature of recharge water are necessary. Quantity of the well water depends upon such things as aquifer permeability, well spacing and depth, seasonal changes in river flow, and pumping rates. The best way to evaluate these variables is by full-scale field pumping tests or by checking similar existing well fields.

Pollution and eutrophication are major concerns in selecting and treating surface-water supplies. Where these are highly contaminated and difficult to treat, municipalities seek either groundwater supplies or alternate less polluted surface sources within a feasible pumping distance. However, in many regions of the United States, adequate groundwater resources are not available and high-quality surface supplies are not within economic reach. Although the majority of municipal water-supply systems are from underground origins, only about one-fourth of the nation's population is served by these sources. In general, larger cities are dependent on surface supplies. Treatment of polluted waters is perhaps the greatest challenge in water-supply engineering, and the importance of pollution control is realized most vividly when one ponders the problems of providing safe, palatable water from contaminated sources.

Water quality in rivers depends on the character of the watershed; pollution caused by municipalities, industries, and agricultural practices; river development such as dams; the season of the year; and climatic conditions. During spring runoff and other periods of high flows, river water may be muddy and high in taste- and odor-producing compounds. During drought flows, pollutants are often present in higher concentrations and odorous conditions can occur. During late summer, algal blooms frequently create problems. River temperature variations depend on latitude and the location of the stream headwaters. In Northern states, river water is warm in the summer and cold in the winter. Water-quality conditions vary from one stream to another and, in addition, each has its own peculiar characteristics. River-water quality is usually deteriorating if the watershed is under development.

Water supplies from rivers normally require the most extensive treatment facilities and the greatest operational flexibility of any source. A river-water-treatment plant must be capable of handling day-to-day variations, and the anticipated quality changes likely to occur within its useful life.

The quality of water in a lake or reservoir depends on the physical, chemical, and biological characteristics (limnology) of the body. Size, depth, climate, watershed, degree of eutrophication, and other factors influence the nature of an impoundment. The relatively quiescent retention of river water produces marked changes in quality brought about by self-purification. Common benefits derived from impounding river water include reduction of turbidity, coliform counts and usually color, and elimination of the day-to-day variations in quality.

Lakes and reservoirs are subject to seasonal changes that are particularly noticeable in eutrophic waters. In summer and winter during stratification, the hypolmnion (unmixed bottom water layer) in an eutrophic lake may contain dissolved iron and manganese and taste and odor compounds. The decrease of oxygen by bacterial activity on the bottom results in dissolution of iron and manganese and production of hydrogen sulfide and other metabolic intermediates. Algal blooms frequently occur in the epilimnion (warmer mixed surface-water layer) of fertile lakes in early spring and late summer. Heavy growths of algae, particularly certain species of blue-greens produce difficult-to-remove tastes and odors. Normally, the best quality of water is found near middepth, below the epilimnion and above the bottom. The intake for a fertile lake or reservoir should preferably be built so that water can be drawn from selected depths. Lakes that stratify experience spring and fall overturns, which occur when the temperature profile is nearly uniform and wind action stirs the lake from top to bottom. If the bottom water has become anaerobic during stratification, water drawn from all depths during the overturn will contain taste- and odor-producing compounds.

The limnology of a lake or reservoir and the present and future level of eutrophication should be thoroughly understood before beginning the design of intake structures or treatment systems.

## 9.7  PURPOSE OF WATER TREATMENT

The purpose of a municipal water-supply system is to provide potable water that is chemically and microbiologically safe for human consumption and has adequate quality for industrial users. For domestic uses water should be free from unpleasant tastes and odors, and improved for human health (i.e., by fluoridation). Since the quality of public supplies is based primarily on drinking-water standards, the special temperature and other needs of some industries are not always met by public supplies. Boiler feedwater, water used in food processing, and process water in the manufacture of textiles and paper have special quality tolerances that may require additional treatment of the municipal water at the industrial site.

## 9.8 SELECTION OF WATER TREATMENT PROCESSES

Watershed management should be considered part of the operation of a water-supply system. Stream flows may be augmented by controlled releases from upstream storage, or water-supply reservoirs can be used for natural purification of river waters. In reservoirs, algicides such as copper sulfate can be introduced to control algal blooms that interfere with coagulation and filtration processes and produce taste and odor problems.

Current pretreatment processes in municipal water treatment are screening, presedimentation or desilting, chemical addition, and aeration. Screening is practiced in pretreating surface waters. Presedimentation is regularly used to remove suspended matter from river water. Chemical treatment, in advance of in-plant coagulation, is most frequently applied to improve presedimentation, to pretreat hard-to-remove substances, such as taste and odor compounds and color, and to reduce high bacterial concentrations. Conventional chemicals used with presedimentation are polyelectrolytes and alum. Aeration is customarily the first step in treatment for the removal of iron and manganese from well waters and a standard way to separate dissolved gases such as hydrogen sulfide and carbon dioxide.

Treatment processes used in water plants depend on the raw-water source and quality of finished water desired. The specific chemicals selected for treatment are based on their effectiveness to perform the desired reaction and cost. For example, activated carbon, chlorine, chlorine dioxide, and potassium permanganate are all used for taste and odor control. Excess chlorination, although least expensive, can create undesired trihalomethanes; activated carbon is the most effective chemical. In surface-water treatment plants equipment for feeding two or three taste- and odor-removal chemicals is usually provided, so the operator can select the most effective and economic chemical applications. There is no fixed rule for color removal applicable to all waters. Alum coagulation with adequate pretreatment, and applying oxidizing chemicals or activated carbon, may provide satisfactory removal. On the other hand, a more expensive coagulant might prove to be more effective and reduce overall chemical costs (i.e., copperas or ferric salts can be substituted for alum in the coagulation process).

Perhaps the most important consideration in designing water treatment processes is to provide flexibility. The operator should have the means to change the point of application of certain chemicals. For example, chlorine

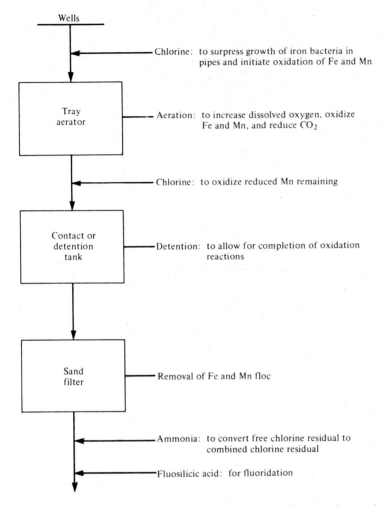

**Figure 9.10** Iron and manganese removal plant using aeration and chlorine for oxidation.

feedlines are normally provided for pre-, intermediate-, and postchlorination. Multiple chemical feeders and storage tanks should be supplied so that various chemicals can be employed in the treatment process. Degradation of the raw-water quality, or changes in costs of chemicals, may dictate a change in the kind of coagulant or auxiliary chemicals used in coagulation. In the case of surface-water treatment plants it is desirable to provide space for the construction of additional pretreatment facilities. The flow in rivers may change due to construction of dams, channel improvements, or upstream water use. The quality of water changes due to human alteration and occupation of the watershed. Concentrations of pollutants from disposal of municipal and industrial wastewaters and agricultural land runoff may increase. Lakes can become more eutrophic.

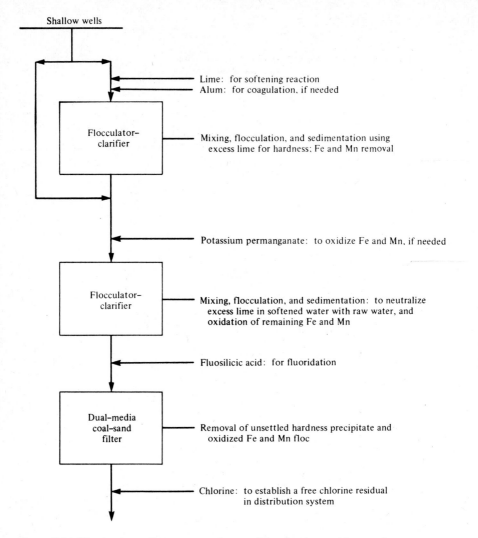

**Figure 9.11** Plant using split treatment for partial softening and iron and manganese removal.

The treatment plant described in Fig. 9.10 was built in the 1930s with provisions for iron and manganese removal and lime softening. Since then plant capacity has been expanded and the treatment process converted to only iron and manganese removal by increasing the aerator capacity, adding chlorine feed after the aerators, and using the settling basin as a detention tank. At present this plant is only capable of performing oxidation by aeration and chlorination for iron and manganese removal and disinfection. Finished water quality is excellent all year round because of the stable character of the well supply.

The treatment plant shown in Fig. 9.11 is a 1968 installation for partial softening and iron and manganese removal. Flocculator–clarifiers can be operated in series or parallel for flexibility in processing. The flow diagram illustrates split-treatment operation. An alternative method is single-step coagulation using selective calcium softening by applying lime and alum. Both dry and liquid chemical feeders are provided with lines to each flocculator–clarifier. No pretreatment facilities are required in the treatment of the existing water quality; however, land area is available for expansion of the plant facilities (Fig. 9.12).

The treatment plant in Fig. 9.13 was constructed about 1950 as a conventional treatment plant for coagulation, sedimentation, and filtration. Since then the lake-water source has become increasingly eutrophic, and additional facilities for taste and odor control have been provided. Activated carbon, chlorine dioxide, and various auxiliary chemicals for improved chemical treatment are now available to the operator during critical periods of poorer raw-water quality. During most of the year the finished water is very palatable, but tastes and odors cannot be completely removed during spring and fall lake overturns.

The river-water processing scheme shown in Fig. 9.14 is a very complex, flexible treatment system for a turbid and polluted stream with highly variable quality. The plant has a complex of uncovered and covered mixing and settling basins with a variety of chemicals that can be fed at several points during water processing. The general treatment scheme consists of plain sedimentation for desilting; mixing followed by sedimentation, using coagulants as necessary; split treatment for partial softening and coagulation in flocculator–clarifiers; blending of the split flows followed by sedimentation; dual-media filtration; and chemical additions for chlorine residual, pH adjustment, and control of scaling. Operation of this plant is varied from day to

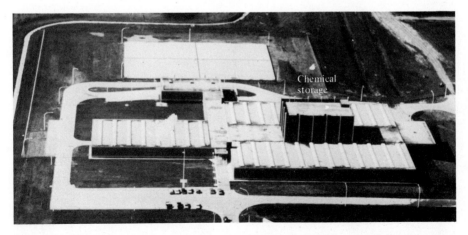

**Figure 9.12** Aerial view of a water treatment plant. The flow diagram for this water utility is given in Fig. 9.11. (Courtesy of Metropolitan Utilities District, Omaha, NE.)

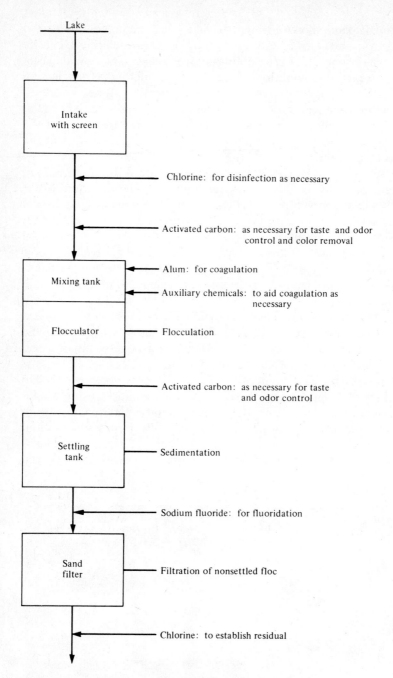

**Figure 9.13** Chemical coagulation treatment plant with special provisions for taste and odor control.

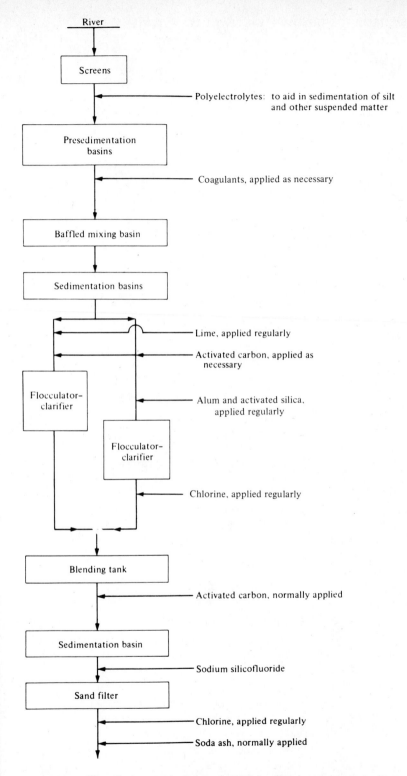

**Figure 9.14** Chemical coagulation and partial-softening treatment plant with provisions for handling high turbidity, tastes and odors, and color.

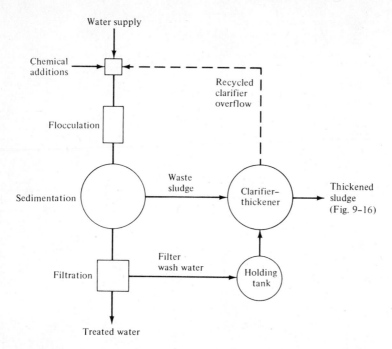

**Figure 9.15** General flow scheme for withdrawal and gravity thickening of water-processing wastes.

day and from season to season depending on the raw-water quality. In general, the most troublesome time of the year is during spring runoff. Little is known about the degree to which refractory inorganic or organic substances can be transported through a complex treatment system. A river-water treatment plant should have process depth to prevent any possibility of short-circuiting in pollution emergencies.

## 9.9  WATER-PROCESSING SLUDGES

Chemical residues from water treatment historically were discharged to surface watercourses without processing. Water-quality regulations now require handling of these wastes to minimize enviromental degradation. The method of eliminating residues is unique to each water utility because of the individuality of each treatment scheme and differing characteristics of the waste slurries generated. The two primary wastes are sludge from the settling basins following chemical coagulation, or precipitation softening, and wash water from backwashing the filters. These residues, containing matter removed from the raw water and chemicals added during processing, are relatively nonputrescible and high in mineral content. Figure 9.15 is a generalized scheme of alternatives for disposing of water-processing wastes. Filter wash

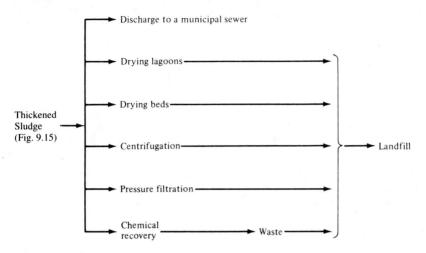

**Figure 9.16** Common methods for dewatering and disposal of water-processing sludges.

water is a very dilute waste discharged for only a few minutes once or twice a day for each filter bed. Therefore, a holding tank is necessary for flow equalization (i.e., to accept the intermittent peak discharges and release a relatively uniform effluent flow). Backwash water is settled in a clarifier-thickener with the overflow recycled to the plant influent; the underflow is withdrawn for further processing. Waste sludge from settling basins may also be concentrated further in a clarifier–thickener.

Disposal alternatives for gravity-thickened precipitates are diagrammed in Fig. 9.16. In some municipalities, slurries can be drained into a sanitary sewer for processing with the municipal wastewater. Air drying is accomplished either in lagoons or on sand-drying beds. Ponding is a popular method for dewatering, thickening, and temporary storage of waste solids. Where suitable land area is available, this technique is both inexpensive and efficient when compared to other methods. Drying beds are more costly and applicable only to small water plants. Centrifugation and pressure filtration are the two most successful mechanical dewatering processes. Recovery of alum and recalcination for lime are feasible in some locations. Often, the economics do not warrant processing of sludge for chemical recovery. Residue from air drying, concentrated cake from dewatering, and waste after chemical recovery may be disposed of by landfill.

# Chapter 10
# Physical Treatment Processes

Physical treatment methods are used in both water treatment and wastewater processing. Except for preliminary steps, most physical processes are associated directly with chemical and biological operations. In water treatment, granular-media filtration (a physical method) must be preceded by chemical coagulation. In wastewater processing, the physical procedures of mixing and sedimentation in activated sludge are related directly to the biology of the system.

## Flow-Measuring Devices

Measurement of flow is essential for operation, process control, and record keeping of water and wastewater treatment plants.

### 10.1 MEASUREMENT OF WATER FLOW

The flow of water through pipes under pressure can be measured by mechanical or differential head meters, such as a venturi meter, flow nozzle, or orifice meter. The drop in piezometric head between the undisturbed flow and the constriction in a differential head meter is a function of the flowrate. A venturi meter (Fig. 10.1), although more expensive than a nozzle or orifice meter, is preferred because of its lower head loss.

Piping changes upstream from a venturi meter produce nonuniform flow, causing inaccuracies in metering. In general, a straight length of pipe equal to 10–20 pipe diameters is recommended to minimize error from flow disturbances created by pipe fittings.

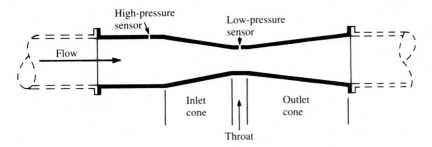

**Figure 10.1** Venturi meter for measuring flow in a pipeline by differential pressure between inlet and throat.

Flow in a venturi meter can be calculated using the equation

$$Q = CA_2 \left[ 2g \left( \frac{P_1 - P_2}{w} \right) \right]^{0.5}$$
(10.1)

where    $Q$ = flow, cfs
         $C$ = discharge coefficient, commonly in the range of 0.90–0.98
         $A_2$ = cross-sectional area of throat, ft²
         $g$ = acceleration of gravity, ft/sec²
$P_1 - P_2$ = differential pressure, psf
         $w$ = specific weight of water, pcf

## 10.2 MEASUREMENT OF WASTEWATER FLOW

The flow of wastewater through an open channel can be measured by a weir or a venturi-type flume, such as the Parshall flume [1] (Fig. 10.2). Parshall flumes have the advantages of a lower head loss than a weir and a smooth hydraulic flow preventing deposition of solids.

Under free-flowing (unsubmerged) conditions, the Parshall flume is a critical-depth meter that establishes a mathematical relationship between the stage $h$ and discharge $Q$. For a flume with a throat width of at least 1 ft but less than 8 ft, flow can be calculated as

$$Q = 4Wh^{1.522W^{0.026}}$$
(10.2)

where    $Q$ = flow, cfs
         $W$ = throat width, ft
         $h$ = upper head, ft

■ **EXAMPLE 10.1**
What is the calculated water flow through a venturi meter with $5\frac{1}{16}$-in. throat diameter when the measured pressure differential is 160 in. $H_2O$? Assume $C = 0.95$ and a water temperature of 50°F.

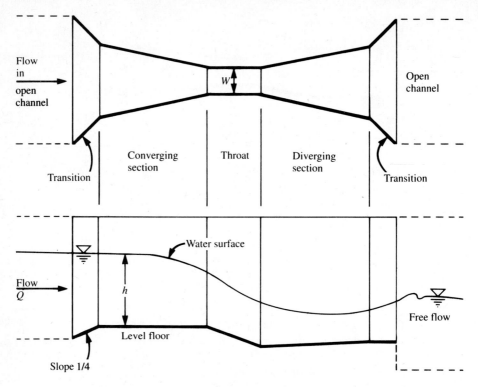

**Figure 10.2** Parshall flume for measuring flow in an open channel by measuring the free-flowing upper head $h$.

*Solution.* Substituting in Eq. 10.1, we obtain

$$Q = 0.95 \times \frac{\pi(0.422)^2}{4} \left( 64.4 \frac{160 \times 62.4}{12 \times 62.4} \right)^{0.5}$$

$$= 3.9 \text{ cfs} = 1750 \text{ gpm} \qquad \blacksquare$$

**■ EXAMPLE 10.2**

What is the calculated wastewater flow through a Parshall flume with a throat width of 2.0 ft at the maximum free-flow head of 2.5 ft?

*Solution.* By Eq. 10.2,

$$Q = 4 \times 2.0 \times 2.5^{1.522 \times 2.0^{0.026}}$$
$$= 33.4 \text{ cfs} = 21.6 \text{ mgd} \qquad \blacksquare$$

# Screening Devices

River water and wastewater in sewers frequently contain suspended and floating debris varying in size from logs to small rags. These solids can clog and damage pumps or impede the hydraulic flow in open channels and pipes. Screening is the first step in treating water containing large solids.

## 10.3 WATER-INTAKE SCREENS

River-water intakes are commonly located in a protected area along the shore to minimize collection of floating debris. Lake water is withdrawn below the surface to preclude interference from floating materials.

Coarse screens of vertical steel bars having openings of 1–3 in. are employed to exclude large materials. The clear openings should have sufficient total area so that the velocity through them is less than 3 fps to prevent pulling through stringy solids. These screens are available with mechanical rakes to take accumulated material from the bars. A coarse screen can be installed ahead of a finer one used to remove leaves, twigs, small fish, and so on. The traveling screen shown in Fig. 10.3 serves this purpose. Trays with sections constructed of wire mesh or slotted metal plates generally have $\frac{3}{8}$-in. openings. As water passes through the upstream side of the screen, the solids are retained and elevated by the trays. As the trays rise into the head enclosure, the solids are removed by means of water sprays. The operation

**Figure   10.3** Traveling   water-intake screen. (Courtesy of Envirex Inc., a Rexnord Co.)

of a traveling screen is intermittent, being controlled by the water head differential across it resulting from clogging.

## 10.4  SCREENS IN WASTEWATER TREATMENT

Coarse screens (bar racks) constructed of steel bars with clear openings not exceeding 2.5 in. are normally used to protect wastewater lift pumps. (Centrifugal pumps used for wastewater works, in sizes larger than 100 gpm, are capable of passing spheres at least 3 in. in diameter.) To facilitate cleaning, the bars are usually set in a channel inclined 22°–45° to the horizontal.

Mechanically cleaned medium screens (Fig. 10.4) with clear bar openings of $\frac{5}{8}$–$1\frac{3}{4}$ in. are commonly used instead of a manually cleaned coarse screen. The maximum velocity through the openings should not exceed 2.5 fps.

Collected screenings are generally hauled away for disposal by land burial or incineration, although they can be shredded in a comminutor and returned to the wastewater flow for removal in a treatment plant.

**Figure 10.4** Mechanically cleaned bar screen for wastewater treatment. (Courtesy of Envirex Inc., a Rexnord Co.)

Fine screens with openings as small as $\frac{1}{32}$ in. have been used in wastewater treatment, but because of their high installation and operation cost, they are rarely employed in handling municipal wastewater. Fine screens are used to take out suspended and settleable solids in pretreating certain industrial waste streams. Typical applications are the removal of coarse solids from cannery and packing-house wastewaters.

## 10.5  COMMINUTORS

Comminutors or shredding devices, like the unit shown in Fig. 10.5, grind solids passing through bar screens to about $\frac{1}{4}$–$\frac{3}{8}$ in. in size. They are installed directly in the wastewater flow channel and provided with a bypass so that the length containing the unit can be isolated and drained for machine maintenance. In small waste treatment plants, comminutors may be installed without being preceded by medium screens. The bypass channel around the comminutor contains a hand-cleaned medium screen (Fig. 10.5).

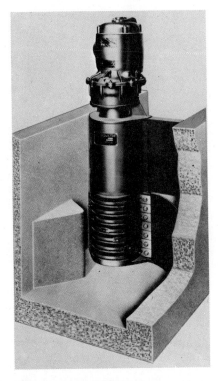

**Figure 10.5** The *Griductor* comminutor has a stationary semicircular screen grid and rotating circular cutter disks and is installed in a wastewater flow channel provided with an emergency bypass containing a fixed-bar screen. (Courtesy of Infilco Degremont, Inc.)

# Hydraulic Characteristics of Reactors

Chemical and biological reactions are performed in tanks often referred to as reactors in process engineering. The common mixed reactor in water treatment is for chemical flocculation, whereas that in wastewater treatment is biological activated sludge. Sedimentation basins in both water and wastewater treatment are used for the physical process of solid–liquid separation. The following sections discuss the hydraulic characteristics of complete mixing, plug flow, and plug flow with longitudinal dispersion.

## 10.6  RESIDENCE TIME DISTRIBUTION

The hydraulic character of a process reactor is defined by the residence time distribution of individual particles of the liquid flowing through the tank. Since routes through the reactor differ in travel times, the effluent age distribution of a nonideal reactor extends from less than to greater than the theoretical detention time, which is defined as

$$t_R = \frac{V}{Q} \tag{10.3}$$

where  $t_R$ = theoretical mean residence time
$V$ = volume of the reactor
$Q$ = rate of flow through the reactor

Evaluation of an actual process tank is performed by introducing a tracer to the influent during steady-state flow and measuring the concentration in the output over an extended period of time. The general shapes of residence time distributions for a dispersed plug-flow reactor are illustrated in Fig. 10.6. If the tracer input is suddenly initiated and then held at a constant application rate, the tracer output curve of concentration in the effluent slowly rises and approaches the influent tracer concentration $C_0$ at a time interval greater than the mean residence time $t_R$. Applying a pulse input of dye solution results in an effluent tracer concentration curve as shown in Fig. 10.6b. The centroid of the curve is located to the left of time $t_R$ as a result of flow dispersion with backmixing and short-circuiting of liquid in the tank because of stagnant pockets.

## 10.7  IDEAL REACTORS

Common ideal reactors are the plug flow reactor, completely mixed reactors, and completely mixed vessels in series. The residence time distributions of these units can be expressed in mathematical equations. Although real process tanks never precisely follow these flow patterns, many designs approxi-

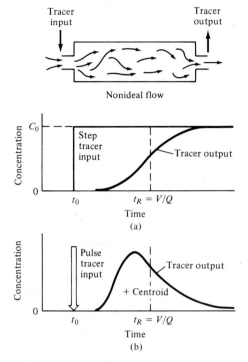

**Figure 10.6** Output tracer distribution curves for flow through a nonideal reactor in response to (a) a continuous tracer input and (b) a pulse tracer input.

mate ideal conditions with negligible error. Laboratory analyses of biological and chemical processes are often conducted using small tanks that are considered to be ideal reactors.

## Plug-Flow Reactor

The ideal plug-flow reactor is a tube with the particles of liquid entering and discharging in the same sequence. The residence time of each particle is equal to the mean residence time $t_R$, as illustrated in Fig. 10.7. If a tracer with concentration $C_0$ is applied to the input, at time $t_R$ the output contains the tracer at the same concentration. A pulse input is observed as a pulse in the output after time $t_R$. Actual treatment units that approach plug flow are long rectangular tanks used as chlorine contact chambers and in the conventional activated-sludge process.

## Completely Mixed Reactor

The contents of a completely mixed reactor are uniformly and continuously redistributed. Particles entering are dispersed immediately throughout the tank and leave in proportion to their concentration in the mixing liquid. Figure 10.8a is the residence time distribution for a step tracer input. The

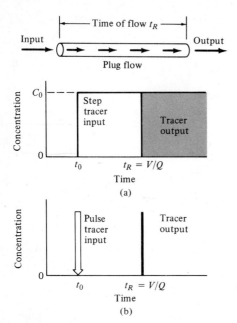

**Figure 10.7** Output residence time distributions for an ideal plug-flow reactor in response to (a) a continuous input and (b) a pulse input.

rate of accumulation of tracer in the reactor, after start of the continuous feed, is equal to the input minus output, which is expressed in the following mass balance equation:

$$V \frac{dC}{dt} = QC_0 - QC \tag{10.4}$$

Rearranging and integrating between the limits of 0 and $C$ and 0 and $t$ gives

$$\int_0^C \frac{dC}{C_0 - C} = \frac{Q}{V} \int_0^t dt \tag{10.5}$$

which, when solved, yields

$$C = C_0(1 - e^{-t/t_R}) \tag{10.6}$$

Figure 10.8b is the residence time distribution for a pulse input of tracer in an amount such that the initial concentration after immediate mixing throughout the tank is $C_0$. The formula for calculating the concentration of tracer in the effluent with passage of time under steady-state flow is

$$C = C_0 e^{-t/t_R} \tag{10.7}$$

Treatment processes that closely simulate this ideal reactor are chemical mixing tanks and completely mixed activated-sludge basins. Studies of wastewater treatability and the evaluation of biological kinetics are commonly performed in laboratory containers that are completely mixed reactors.

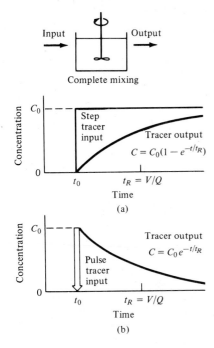

**Figure 10.8** Output residence time distributions for an ideal completely mixed reactor in response to (a) a continuous input and (b) a pulse input.

## Completely Mixed Reactors in Series

The residence time distributions for flow through a series of equal-sized ideal completely mixed reactors are plotted in Fig. 10.9. These output curves for tanks in series are the response to a pulse tracer input to the first tank at time $t_0$. Particles entering each reactor are dispersed immediately and leave in proportion to their concentration in the mixing liquid. For the last tank in a series of $n$, the residence time distribution can be calculated using the generalized equation

$$C_n = C_0 \frac{(t/t_R)^{n-1}}{(n-1)!} e^{-t/t_R} \tag{10.8}$$

where  $C_n$ = effluent tracer concentration from reactor $n$
  $C_0$ = influent tracer concentration in the first reactor
  $n$ = number of reactors in series
  $t$ = time lapsed since tracer input to the first reactor
  $t_R$ = theoretical mean residence time of $n$ reactors

Substituting $n = 1$ in this equation yields Eq. 10.7.

The model of ideal reactors in series is representative in water treatment of long rectangular flocculation tanks separated into a series of compartments by baffles. In wastewater processing, compartmented aeration tanks for high-purity-oxygen activated sludge are commonly three or more reactors in

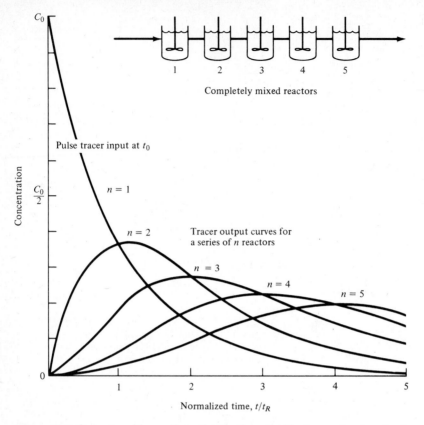

**Figure 10.9** Output residence time distributions for ideal completely mixed reactors in series in response to a pulse tracer input.

series. Several completely mixed reactors in series are frequently used in laboratory studies to simulate dispersed plug flow through a rectangular aeration basin. Observe the similarity in the shapes of the curves in Fig. 10.6b and the distribution diagrams for an $n$ of 4 or 5 in Fig. 10.9.

## 10.8 DISPERSED PLUG FLOW

Longitudinal flow passing through a tank has variations in flow velocities that result in axial intermixing and backmixing. The turbulence of this nonideal flow is in the range between ideal plug flow and ideal mixed flow. Dispersed plug flow, however, does not account for stagnant pockets and short-circuiting of the liquid in the tank.

The dispersion number for a particular reactor is defined as follows:

$$d = \frac{D}{uL} \tag{10.9}$$

where   $d$ = reactor dispersion number, dimensionless
       $D$ = longitudinal dispersion coefficient, ft²/min
       $u$ = mean velocity of flow, fpm
       $L$ = length of the reactor, ft

Figure 10.10 illustrates tracer response curves to a pulse input for dispersed plug flow. The dispersion number for negligible mixing approaches zero, and hence plug flow. With greater intermixing, the dispersion number increases with a completely mixed reactor, represented by a value of infinity.

The spread of a tracer response curve, since it represents a continuous distribution, may be measured by calculating the variance $\sigma^2$ of the distribution about its mean. For large extents of dispersion, which are common in tracer studies of actual processing tanks, the residence time curves are often skewed. This is evidenced by an elongated curve where the concentration of tracer gradually approaches zero. For this case, the relationship between variance and the dispersion number is given by [2]

$$\sigma_\theta^2 = \frac{\sigma^2}{\bar{t}^2} = 2\frac{D}{uL} - 2\left(\frac{D}{uL}\right)^2 (1 - e^{-uL/D}) \tag{10.10}$$

where   $\sigma_\theta^2$ = normalized variance, dimensionless
       $\sigma^2$ = variance, min²
       $\bar{t}$ = mean residence time (time to the centroid of the distribution), min

The variance and dispersion number for an actual reactor are calculated from the effluent concentration-time data from a pulse tracer input. After dye is injected at the inlet of the tank, output samples are collected at time intervals from $\frac{1}{2}$ min to several minutes, with the greatest frequency during discharge of the dye peak. The shape of the residence time distribution is determined by graphing the concentration versus time. If desired, a normalized curve can be plotted by dividing the concentration values by the average of the measurements and corresponding times by the mean time. Variance of a curve is calculated from a finite number of measurements $i$ at equal time intervals using the equation

$$\sigma^2 = \frac{\sum t_i^2 C_i}{\sum C_i} - \bar{t}^2 \tag{10.11}$$

where   $\bar{t} = \dfrac{\sum t_i C_i}{\sum C_i}$

From the second term of this equation, the time to the centroid of the distribution $\bar{t}$ is equal to the sum of time-concentration values divided by the sum of the concentrations. After calculation of the variance $\sigma^2$, the dispersion number $D/uL$ can be computed using Eq. 10.10. Another parameter used to define a residence time distribution is the ratio of time for passage of 90% of the tracer to the time for passage of 10% [3].

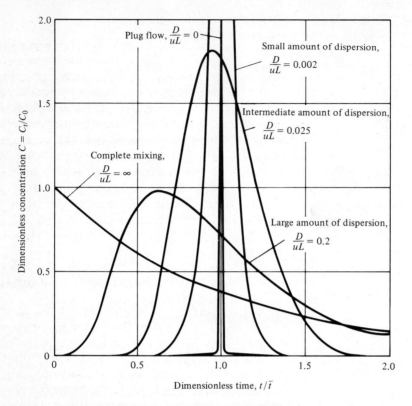

**Figure 10.10** Normalized residence time distributions in response to a pulse tracer input and corresponding dispersion numbers for various degrees of longitudinal dispersion resulting from intermixing and back-mixing of the flow through the tank.

### ■ EXAMPLE 10.3

A completely mixed tank for blending chemicals into water is tested for effectiveness of hydraulic dispersion [4]. A pulse of tracer, equivalent to a concentration of 10 $\mu$g/l $C_0$ in the tank volume, was injected in the influent, followed by analyses of effluent samples for tracer concentrations for a period of three detention times. The experimental data are listed in the first two columns of Table 10.1. The theoretical detention time $t_R$ of the tank is 60 sec. Plot a residence time distribution for the test measurements and compare it to the theoretical curve.

**Solution.** The normalized time $t/t_R$ and tracer concentrations $C/C_0$ are calculated by dividing the experimental $t$ and $C$ values by 60 sec and 10 $\mu$g/l, respectively. The theoretical values are computed using Eq. 10.7. Plots of both the measured and theoretical curves are shown in Fig. 10.11. ■

### ■ EXAMPLE 10.4

Dispersed plug flow through a long narrow chlorination tank was analyzed by injecting a pulse input of a dye tracer in the influent. The tracer response curve is drawn in

**TABLE 10.1** EXPERIMENTAL DATA AND CALCULATED VALUES FOR THE OUTPUT RESIDENCE TIME DISTRIBUTION FOR THE COMPLETELY MIXED REACTOR IN EXAMPLE 10.3

| Experimental values | | | | Theoretical values, from Eq. 10.7 | |
| --- | --- | --- | --- | --- | --- |
| Measured | | Calculated | | | |
| $t$ (sec) | $C$ ($\mu g/l$) | $t/t_R$ | $C/C_0$ | $t/t_R$ | $C/C_0$ |
| 10 | 4.5 | 0.17 | 0.45 | 0.17 | 0.85 |
| 20 | 7.6 | 0.33 | 0.76 | 0.33 | 0.72 |
| 40 | 6.5 | 0.67 | 0.65 | 0.67 | 0.51 |
| 60 | 5.6 | 1.00 | 0.56 | 1.00 | 0.37 |
| 80 | 3.0 | 1.33 | 0.30 | 1.33 | 0.26 |
| 100 | 2.1 | 1.67 | 0.21 | 1.67 | 0.19 |
| 120 | 1.8 | 2.00 | 0.18 | 2.00 | 0.14 |
| 140 | 1.2 | 2.33 | 0.12 | 2.33 | 0.10 |
| 160 | 0.7 | 2.67 | 0.07 | 2.67 | 0.07 |
| 180 | 0.4 | 3.00 | 0.04 | 3.00 | 0.05 |

Fig. 10.12, and the measured $t$ and $C$ values are listed in the first two columns of Table 10.2. The theoretical detention time at the test flowrate was 50 min. Calculate the dispersion number and the time ratios of $t_{50}/t_R$ and $t_{90}/t_{10}$.

**Solution.** Calculated values for $\Sigma\,tC$ and $\Sigma\,t^2C$ are given in Table 10.2. Using Eq. 10.11,

$$\bar{t} = \frac{\Sigma\,t_iC_i}{\Sigma\,C_i} = \frac{1440}{34} = 42 \text{ min}$$

$$\sigma^2 = \frac{69,800}{34} - \left(\frac{1440}{34}\right)^2 = 260$$

The normalized variance from Eq. 10.10 is

$$\sigma_\theta^2 = \frac{260}{(42)^2} = 0.15$$

and the reactor dispersion number $D/uL$, from a trial and solution of Eq. 10.10, equals 0.073.

From the values given in Fig. 10.12 and $t_R = 50$ min,

$$t_{50}/t_R = 39/50 = 0.78$$

and

$$t_{90}/t_{10} = 65/31 = 2.1 \qquad \blacksquare$$

# Mixing and Flocculation

Chemical reactors in water treatment and biological aeration basins are mixed to produce the desired reactions and to keep the solids produced by the process in suspension. Rapid-mixing tanks are used to blend chemicals into

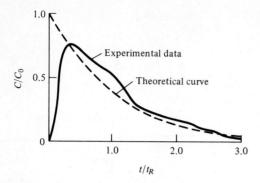

**Figure 10.11** Plot of residence time distribution curves from data in Table 10.1 for Example 10.3.

the water. The gentle mixing in flocculation basins is to promote agglomeration of particles into large floc that can be removed from suspension by subsequent sedimentation.

## 10.9 RAPID MIXING

A rapid mixer provides dispersion of chemicals into water so that blending occurs in 10–30 sec. Most common is mechanical mixing using a vertical-shaft impeller in a tank with stator baffles, as illustrated in Fig. 10.13a. The

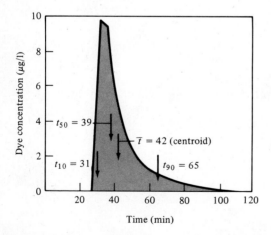

**Figure 10.12** Tracer response curve from injecting a pulse input of dye into the influent of a long narrow chlorination tank. [Adapted from D. M. Marske and J. D. Boyle, "Chlorine Contact Chamber Design—A Field Evaluation," *Water and Sewage Works* 120 (1) (1973): 70.]

**TABLE 10.2**  EXPERIMENTAL DATA AND
CALCULATED VALUES FOR
HYDRAULIC ANALYSIS OF THE LONG
NARROW CHLORINATION TANK IN
EXAMPLE 10.4

| Measured values | | Calculated values | |
|---|---|---|---|
| $t$ (min) | $C$ ($\mu$g/l) | $tC$ | $t^2C$ |
| 25 | 0 | 0 | 0 |
| 30 | 4.0 | 120 | 3,600 |
| 35 | 9.7 | 340 | 11,900 |
| 40 | 5.8 | 232 | 9,300 |
| 45 | 3.6 | 162 | 7,300 |
| 50 | 2.3 | 115 | 5,800 |
| 55 | 1.9 | 105 | 5,700 |
| 60 | 1.4 | 84 | 5,000 |
| 65 | 1.1 | 72 | 4,600 |
| 70 | 0.8 | 56 | 3,900 |
| 75 | 0.6 | 45 | 3,400 |
| 80 | 0.4 | 32 | 2,600 |
| 85 | 0.3 | 26 | 2,200 |
| 90 | 0.2 | 18 | 1,700 |
| 95 | 0.2 | 19 | 1,800 |
| 100 | 0.1 | 10 | 1,000 |
| Total | 34.0 | 1440 | 69,800 |

stators reduce vortexing about the impeller shaft. They are located to disrupt rotational flow by deflecting the water inward toward the impeller shaft. Vortexing hinders mixing and reduces the effective impeller power usage. The turbulent flow pattern in mixing is a function of the tank size and shape; number, shape, and size of impellers; kind and location of stator baffles; and power input. Recommended guidelines for designing a mechanical rapid mixer are the following: a square vessel is superior in performance to a cylindrical vessel, stator baffles are advantageous, a flat-bladed impeller performs better than a fan or propeller impeller, and chemicals introduced at the agitator blade level enhance coagulation [5].

Other methods of rapid mixing include hydraulic action, e.g., injection of chemical into the inlet of a centrifugal pump, jet injection into the water flow in a pipe, and mechanical or static in-line blending. The static mixing elements, shown in Fig. 10.13b, are inserted into a pipe immediately after the point of chemical injection for blending. The kind of mixer configuration selected depends on whether the blending is for dispersion, reaction, or gas–water contact.

## 10.10  FLOCCULATION

Flocculation is agitation of chemically treated water to induce coagulation. In this manner, very small suspended particles collide and agglomerate into larger heavier floc that settle out by gravity. Flocculation is a principal

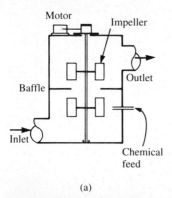

(a)

(b)

**Figure 10.13** (a) Impeller-type mechanical rapid mixer for dispersion of chemicals into water. (b) Static mixing elements that are inserted into a pipe for in-line blending of chemicals. (Courtesy of Koch Engineering Company Inc.)

mechanism in removing turbidity from water. Floc growth depends primarily on two factors: intermolecular chemical forces and the physical action induced by agitation.

The common mechanical mixing devices in water treatment are paddle (reel) flocculators (Figs. 10.14a and 10.23), flat-blade turbines (Fig. 10.14b), and vertical-turbine mixers. The paddle flocculator consists of a shaft with

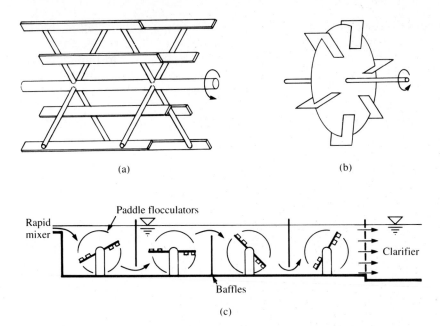

**Figure 10.14** The common mixing devices for flocculation in water treatment are (a) the paddle flocculator and (b) the flat-blade turbine. (c) A typical flocculation tank arrangement is horizontal shaft paddles in a series of compartments separated by baffles to direct water flow through the paddle flocculators.

protruding steel arms that support wooden or steel blades. The paddle shafts can be located transverse to or parallel to the flow, although transverse installation is more common (Fig. 10.14c). The paddle units slowly rotate, usually 1–2 rpm, causing collision among the floc particles that are held in suspension by the gentle agitation. The result is growth of the suspended particles from colloids to settleable floc.

The energy dissipated by the flocculator as shearing forces in the water has been observed to be the controlling factor in floc growth. This energy dissipation is related to the velocity gradient, which represents the changes in velocity from point to point in the water being mixed. The relationship between velocity gradient and energy dissipated is defined as [6]

$$G = \left(\frac{P}{\mu V}\right)^{1/2} \qquad (10.12)$$

where  $G$ = velocity gradient, fps/ft or sec$^{-1}$ (s$^{-1}$)
  $P$ = power input, ft-lb/sec (N·m/s)
  $\mu$ = dynamic viscosity, lb-sec/ft$^2$ (kg·s/m$^2$)
  $V$ = volume, ft$^3$ (m$^3$)

Optimum values exist for flocculation time and velocity gradient. Below a minimum time, no flocculation occurs, and increasing the time beyond maximum floc formation does not significantly improve flocculation. Based on experience, flocculaton tanks with a minimum of three compartments in series reduce the time required for flocculation. In general, the minimum and maximum values of $G$ to promote satisfactory growth of floc are 10 and 75 fps/ft ($s^{-1}$), respectively, with an optimum range in water treatment with paddle flocculators of 30–60 fps/ft [7].

The power dissipated in water by a paddle flocculator can be calculated as

$$P = \frac{C_d A \rho v^3}{2} \tag{10.13}$$

where  $P$ = power dissipated, ft-lb/sec (N·m/s)
  $C_d$ = coefficient of drag
   = 1.8 for flat plates
  $A$ = area of paddles, ft$^2$ (m$^2$)
  $\rho$ = density of water, lb-sec$^2$/ft$^4$ (kg/m$^3$)
  $v$ = velocity of paddles relative to the water, ft/sec (m/s)
   = normally 0.50–0.75 of the velocity of paddle

For a paddle flocculator,

$$v_p = 2\pi r N \tag{10.14}$$

where  $v_p$ = velocity of paddle blade, ft/sec (m/s)
  $r$ = distance from shaft to center of paddle, ft (m)
  $N$ = rotational speed, rev/sec (rev/s)

Then for a flocculation tank with $n$ symmetrical paddle blades at radii $r_1$ and $r_2$ rotating at $N$ rev/sec, the power dissipated is

$$P = \frac{n}{2} C_d A \rho (1 - k)^3 (2\pi N)^3 (r_1^3 + r_2^3) \tag{10.15}$$

where  $k$ = ratio of water velocity to paddle velocity
   = 0.25–0.50 for well-baffled tanks with paddle flocculators, with 0.30 being a common value

The *Recommended Standards for Water Works, Great Lakes Upper Mississippi River Board of State Public Health & Environmental Managers* [8] recommends that rapid mixing for dispersion of chemicals in water be performed with mechanical mixing devices. The mixing detention period should not be more than 30 sec, and the rapid-mix and flocculation basins shall be as close together as possible.

With regard to flocculation, the following recommendations are presented in the *Standards:*

1. Inlet and outlet design shall prevent short-circuiting and destruction of floc.
2. The flow-through velocity shall be not less than 0.5 ft/min nor greater than 1.5 ft/min with a detention time for floc formation of at least 30 min.

3. Agitators shall be driven by variable speed drives with the peripheral speed of paddles ranging from 0.5 ft/sec to 3.0 ft/sec.
4. Flocculation and sedimentation basins shall be as close together as possible. The velocity of flocculated water through pipes or conduits to settling basins shall be not less than 0.5 ft/sec nor greater than 1.5 ft/sec. Allowances must be made to minimize turbulence at bends and changes in direction.

■ **EXAMPLE 10.5**

A 25-mgd water treatment plant has a flocculation tank 64 ft long and 100 ft wide with a water depth of 16 ft. The four horizontal shafts of paddle flocculators are in compartments separated by baffles. All the paddle units have four arms with three blades each with radii of 3.0, 4.5, and 6.0 ft measuring from the shaft to the center of the 0.80-ft-wide boards. The total length of the paddle boards across the tank is 50 ft. Assume a ratio of water velocity to paddle velocity of 0.3, rotational speed of 0.050 rev/sec, $C_d$ equal to 1.8, and water temperature of 50°F. Calculate the velocity gradient.

**Solution.** From a table in the appendix for water at 50°F, $\rho = 1.936$ lb·sec/ft⁴, and $\mu = 2.735 \times 10^{-5}$ lb·sec/ft²:

$$P = \frac{16}{2} 1.8(0.80 \times 50)1.936(1 - 0.3)^3 (2 \times \pi \times 0.050)^3[(3.0)^3 + (4.5)^3 + (6.0)^3]$$

$$= 3960 \text{ ft-lb/sec}$$

$$G = \left(\frac{3960}{2.735 \times 10^{-5} \times 16 \times 64 \times 100}\right)^{1/2} = 38 \text{ fps/ft} \qquad ■$$

# Sedimentation

Sedimentation (clarification) is the removal of solid particles from suspension by gravity. In water treatment, the common application of sedimentation is after chemical treatment to remove flocculated impurities and precipitates. In wastewater processing, sedimentation is used to reduce suspended solids in the influent wastewater and to remove settleable solids after biological treatment.

Design of clarifiers is based on empirical data from the performance of full-scale sedimentation tanks. Mathematical relationships to predict settling of suspensions in actual treatment processes have not been successful in design, and the application of data from laboratory settling column tests has met with only limited success. For these reasons, design criteria are derived from operation of clarifiers settling similar kinds of waters and wastewaters [9].

## 10.11 FUNDAMENTALS OF SEDIMENTATION

*Settling of nonflocculent particles* in dilute suspension, as defined by classical settling theory of discrete particles, has no direct application in water and wastewater sedimentation. It is, however, the basis for the hypothetical relationship between settling velocity and overflow rate. Figure 10.15 illustrates an ideal rectangular clarifier with an inlet zone for transition of influent flow to uniform horizontal flow, a sedimentation zone where the particles settle out of suspension by gravity, an outlet zone for transition of uniform flow in the sedimentation zone to rising flow for discharge, and a sludge zone where the settled particles collect. All influent discrete particles with a settling velocity greater than $V_0$ settle to the sludge zone and are removed. Particles with settling velocities of $V_i$ less than $V_0$ are removed only if they enter the basin within a vertical distance $H_i = V_i t_0$, where $t_0$ is the flowing-through or detention time. The velocity $V_0$ is defined as the overflow rate (gpd/ft$^2$), or as the surface settling rate (ft/sec), and is calculated as

$$V_0 = \frac{Q}{A} \tag{10.16}$$

where   $V_0$ = overflow rate, gpd/ft$^2$ (m$^3$/m$^2$·d)
   $Q$ = average daily flow, gpd (m$^3$/d)
   $A$ = surface area of the clarifier, ft$^2$ (m$^2$)

Overflow rate is an important design criterion in the sizing of sedimentation tanks.

*Settling of flocculent particles* results in coalescence during sedimentation, with the particles growing into larger floc as they descend. Flocculation is caused by differences in settling velocities of particles, resulting in heavy particles overtaking and coalescing with slower ones and velocity gradients within the water producing collisions among particles. The beneficial results are smaller particles growing into faster-settling floc and flocculation sweeping smaller and slower particles from suspension. Flocculent settling is common in clarifying both chemical and biological suspensions. Section 10.12 describes the laboratory column test for flocculent suspensions.

*Zone settling* is when flocculent particles in high concentration settle as a mass with a well-defined, clear interface between the surface of the settling particles and the clarified water above. Because of crowding of the particles,

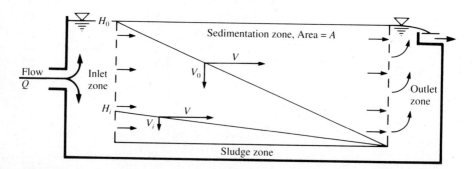

**Figure 10.15** An ideal rectangular clarifier settling discrete particles with an overflow rate of $V_0 = Q/A$.

upward velocity of the displaced water acts to reduce settling velocity. This hindered settling reduces the differences in settling velocities of different size particles and increases the sweeping action of flocculation. Deceleration of settling occurs as hindered settling changes to *compression*. This results when subsiding particles accumulate at the bottom of the tank. Compaction slowly continues from the weight of supported particles above. Hindered settling is usually negligible in water treatment sedimentation and in clarifying of raw wastewaters, because the particles in suspension are dispersed in low concentration. Zone settling does take place in clarifying biological activated sludge in wastewater treatment. Gravity thickening in both water and wastewater sludges exhibits zone settling with hindered settling in the upper layer and compression in the bottom layer. Section 10.13 discusses zone settling.

## 10.12  TYPES OF CLARIFIERS

Figure 10.16 illustrates several types of clarifiers showing the direction of flow through the tanks. In a long rectangular clarifier, the influent is spread across the end of the tank by a baffle structure that dissipates energy to reduce the velocity of the entering water and uniformly distributes the flow. The effluent overflows V-notch weirs mounted along the edges of discharge channels, which convey the water out of the tank. To reduce vertical velocities as the water rises up to the discharge channels, the total effluent weir length is several times greater than the width of the tank. Rectangular clarifiers in water treatment often have long ''finger'' channels extending into the tank (Fig. 10.16a). In contrast, sedimentation tanks in wastewater treatment require scum collectors; therefore, the effluent channels are placed across the width of the tank close to the discharge end (Fig. 10.16b). Sludge is removed by flat scrapers attached to continuous chains that are supported and driven by sprocket wheels mounted on the inside walls of the tank. Rather than submerged chain-and-sprocket mechanisms, in warm climates without snow or ice, the scrapers can be hung in the water from an overhead bridge that spans the width of the tank. The wheels of the bridge roll on steel rails mounted on the side walls of the clarifier.

In a circular clarifier, the influent enters through a vertical riser pipe with outlet ports discharging behind a circular inlet well. This baffle dissipates energy and directs the flow downward and radially out from the center (Fig. 10.16c). The effluent channel may be attached to the outside wall with only a single peripheral weir, or the channel may be located inboard supported from the outside wall with brackets so that water can overflow weirs located on both sides of the channel. Sludge is scraped toward a central hopper by blades attached to arms that rotate around the center of the tank. The drive is a sealed turntable, with gear and pinion running in oil, supported by a bridge or pier. Circular clarifiers are used in both water and wastewater treatment. In wastewater sedimentation, a circular shallow scum baffle is placed in front of the weir channel to prevent the overflow of floating solids,

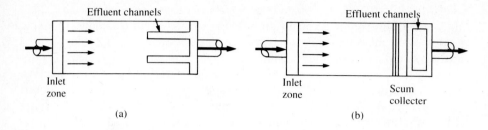

(a)                                                          (b)

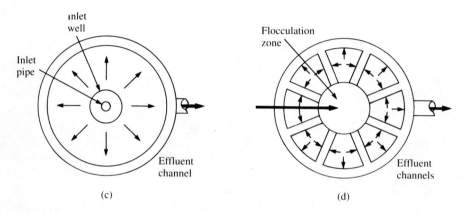

(c)                                                          (d)

**Figure 10.16** Diagrams of various clarifier shapes shown with flow patterns. (a) Rectangular clarifier with horizontal flow to effluent "finger" channels extending into the tank from the outlet end. (b) Rectangular clarifier with a scum collector and effluent channels located at the outlet end. (c) Circular clarifier with central feed well and radial flow to a peripheral effluent channel. (d) Circular flocculator–clarifier with water flowing up to radial effluent channels in the settling zone surrounding the submerged hood of the flocculation zone.

and a rotating skimmer pushes the scum into a hopper that discharges to a scum tank outside of the clarifier.

Figure 10.16d diagrams the plan view of a flocculator–clarifier that is illustrated in Fig. 10.23. The flocculation zone is under a central cone-shaped hood that extends to near the bottom of the tank. As the water flows out from under the hood, it rises vertically to radial effluent channels. This kind of settling tank, common in water softening, is also referred to as an upflow clarifier.

Several parameters are used in sizing of clarifiers. The most common is overflow rate, which is the average flow divided by the total surface area of the tank (Eq. 10.16). Of equal significance is the side-water depth, which is the depth of water in the tank excluding the sludge zone. Detention time, which is the volume divided by the average flow (Eq. 10.3), although less important as a design parameter, is often specified to ensure greater water depth at the overflow rate. Solids loading rates, expressed as the maximum allowable weight of solids per unit area of tank bottom per day, apply only to gravity sludge thickening and final clarifiers of wastewater aeration systems carrying a high concentration of activated sludge.

Conservative overflow rate and sufficient depth are important to compensate for hydraulic instability caused by temperature gradients, inlet energy dissipation, outlet currents, equipment movement, and wind effects. Warm influent to a sedimentation tank containing cooler water can lead to short-circuiting, with the warm water rising to the surface and reaching the effluent channels in a fraction of the theoretical detention time. Similarly, cold water flowing into a tank containing warm water can sink, flow along the bottom, and short-circuit up to the effluent channels. When the inflow and water in the tank are the same temperature, a density current is often caused by the higher turbidity of the influent. The suspension of particles entering, being heavier than the clarified water, descends and moves under the water surface toward the discharge end.

Inlet energy dissipation is important to slow down and distribute the water at the tank entrance. This is particularly critical if the water enters the inlet zone from a pipe conveying the flocculated water at the higher velocity necessary to keep the solids in suspension in the pipeline. Outlet currents are reduced by proper design of the effluent weirs and channels. In preference to flow plates, V-notch weirs provide better lateral distribution of overflow when leveling is not perfect. Channels require sufficient length and adequate spacing to reduce the approach velocity of overflow. Movement of sludge collection mechanisms through the sludge is slow to prevent resuspension of solids. Finally, wind can have a significant effect on large open sedimentation tanks by creating a surface current that moves the upper layer of water downwind.

## 10.13  SETTLING OF FLOCCULENT PARTICLES

Sedimentation of a flocculent suspension results from the combined effects of gravity settling and flocculation. Heavier solids having greater settling velocities overtake and coalesce with smaller lighter solids to form still larger floc with increasing rates of settling. The opportunity for contact among settling solids increases with depth. As a result, removal of suspended solids depends on water depth as well as overflow rate and detention time.

Settling column tests can be used to evaluate the settling characteristics of flocculent suspensions. The standard procedure for a settling column test is to place a uniformly mixed suspension in a column, which has sampling ports at various depths, to cease mixing at the start of the test, and to allow sedimentation under quiescent conditions. Samples from the ports are withdrawn at selected time intervals to determine the concentrations of suspended solids at different depths. The data are used to calculate the percentages of solids removed, and the values are plotted as shown in Fig. 10.17, with each percentage recorded at the proper coordinates of depth and time. Lines representing percentages of removal are drawn through the plotted values. They trace the maximum trajectories of settling paths for specific concentrations in a flocculent suspension. For example, in Fig. 10.17, 45% of the particles in the suspension have average velocities greater than 4 ft/21 min = 0.19 fpm by the time the 4-ft depth has been reached. Computation of removal efficiency is illustrated in Example 10.6 using Eq. 10.17.

Quiescent laboratory settling tests, with the height of the column equal to or greater than the prototype clarifier, have been useful in assessing the settleability of unusual flocculent suspensions. Despite this, the application of the experimental data in design must take into account short-circuiting, inlet and outlet turbulence, and density currents. The settleability of particles may be significantly different as a result of the hydraulic characteristics of an actual basin where the water or wastewater is flowing through the tank, compared to quiescent column settling.

■ **EXAMPLE 10.6**
Using settling trajectories determined by the settling-column analysis shown in Fig. 10.17, find the overall removal for a settling basin 7 ft deep with an overflow rate $v_0$ of $\frac{7}{24}$ fpm.

*Solution.* From Fig. 10.17, 45% of the particles will have settling velocities greater than $\frac{7}{24}$ fpm.

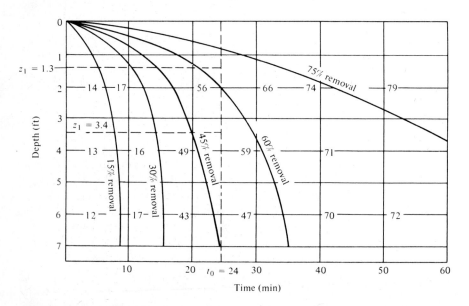

**Figure 10.17** Settling trajectories for a flocculent suspension. (*Note:* Numbers plotted represent fraction removed based on settling column analysis. Percentage removal curves are interpolated using these data.)

Using 15% increments, an additional $60 - 45 = 15\%$ of the particles have an average settling velocity $v \geq 3.4/24$ fpm and a second $75 - 60 = 15\%$ have an average settling velocity $\geq 1.3/24$ fpm. The next increment has a negligible velocity and can therefore be neglected. (*Note*: These values are determined as shown on Fig. 10.17.)

The removal ratio of each percentage of particles is $v/v_0$, as stated previously. The overall removal in percent is therefore

$$R = 45 + \frac{v_1}{v_0}C_1 + \frac{v_2}{v_0}C_2 + \cdots + \frac{v_n}{v_0}C_n \tag{10.17}$$

where in the example $C_1 = C_2 = C_3 = 15\%$. Also, since $v_1 = z_1/t_0$ and $v_0 = z_0/t_0$, $v_1/v_0 = z_1/z_0, \ldots$, where $z_1, z_2, \ldots$, are measured down from the top of the tank and $z_0 = 7$ ft. Thus

$$R = 45 + \frac{3.4}{7} \times (60 - 45) + \frac{1.3}{7} \times (75 - 60) + \cdots$$

$$= 45 + 7.3 + 2.8 + \cdots$$

and the overall removal is approximately 55%. ∎

## 10.14  ZONE SETTLING

Zone settling is when a suspension settles as a blanket without interparticle movement, that is, individual particles of all sizes move downward at the same velocity. The coalescence of the suspended solids occurs during hindered settling of a sufficiently high concentration of flocculent particles.

In observing zone settling, a distinct interface is formed between the mass of settling solids in the sludge zone and the clear supernatant above formed from the water of separation. The settling velocity is the rate of downward movement of the sludge-blanket surface after reaching a constant-velocity characteristic of the suspension. Zone-settling velocities decrease with increasing initial concentrations of the same solids. As shown in Fig. 10.18a, the relationship between $v$ and $C_i$ is such that incremental increases in the suspended-solids content at high concentrations have a greater influence on retarding the settling velocity. The data from this curve can be replotted as solids flux versus initial suspended-solids concentration, as shown in Fig. 10.18b. This plot is accomplished by multiplying the settling velocity times the solids concentration for several points along the $v$ versus $C_i$ curve in Fig. 10.18a to yield the solids flux for a given concentration of suspended solids using

$$G_g = C_i v \tag{10.18}$$

where   $G_g$ = solids flux, lb/ft$^2$/hr (kg/m$^2 \cdot$h)
        $C_i$ = solids concentration, pcf (kg/m$^3$)
        $v$ = settling velocity, ft/hr (m/h)

Zone settling occurs in gravity separation of chemical floc from coagulated water, biological floc from activated sludge, and thickening of waste

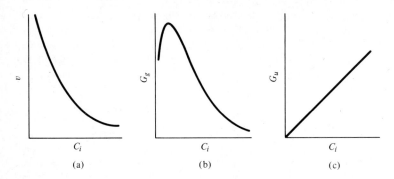

**Figure 10.18** Zone settling and solids flux diagrams. (a) Settling velocity versus solids concentration for coalescing flocculent suspensions. (b) Gravitational flux versus solids concentration calculated from the $v$ versus $C_i$ curve. (c) Underflow solids flux versus solids concentration resulting from the withdrawal of sludge from the bottom of the sedimentation tank.

sludges. Most of the sedimentation tanks used in these processes continuously draw thickened sludge from the bottom of the tank (Fig. 10.18c). Figure 10.19 is a cross-sectional view of a circular final clarifier used in the activated-sludge process. (Figure 10.25 is a picture of this kind of clarifier.) The influent activated sludge enters in the center and the clear effluent is withdrawn from the periphery of the tank. In this continuous-flow system, zone settling is represented by the clear supernatant and hindered settling layer. Compression of solids can occur near the bottom of the tank. Thickened sludge is continuously withdrawn from the tank bottom by uptake pipes. Solids flux resulting from withdrawal of sludge is directly proportional to the concentration of solids in the sludge, as graphed in Fig. 10.18c.

The total solids flux is the sum of the gravitational flux and the underflow flux [10]:

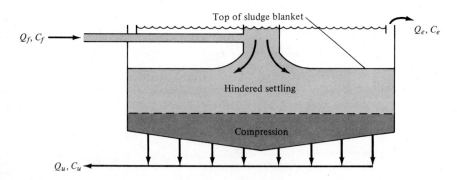

**Figure 10.19** A schematic diagram of zone settling in an activated-sludge final clarifier.

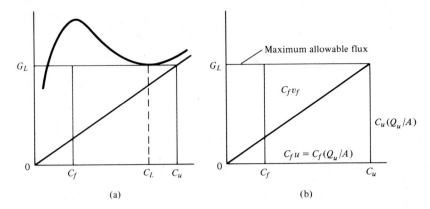

(a)  (b)

**Figure 10.20** Solids flux diagrams for zone settling. (a) Total solids flux as the sum of the laboratory gravitational flux and the underflow flux (Eq. 10.19). (b) Maximum allowable solids flux for the limiting layer of solids concentration $C_L$ as the sum of the gravitational flux of the feed solids and the underflow flux (Eq. 10.20).

$$G = G_g + G_u = C_i v + C_i u \tag{10.19}$$

where   $v =$ settling velocity, $Q_0/A$
$\quad\quad\quad u =$ underflow rate, $Q_u/A$

A total flux curve is shown in Fig. 10.20a. The maximum flux $G_L$ defines the maximum solids loading that can be applied to the clarifier. The selection of $C_L$ as the limiting concentration for the maximum allowable solids flux is because of the difference between laboratory settleability tests and the settling action in a full-scale tank. The experimental solids flux curve is a plot of a series of batch-settling tests performed in a small laboratory container (Section 12.29). For each test, the settling velocity and solids flux of the coalesced solids are plotted against the initial solids concentration of the sample, and consolidation during settlement is not considered. In an actual sedimentation tank operating under continuous flow (Fig. 10.19), the layers of settling solids at greater depths have greater solids concentrations, reduced settling velocities, and lower solids flux. Therefore, the limiting solids flux $G_L$ in a real tank is the value associated with the limiting solids concentration $C_L$. Normal operation encompasses this critical concentration since the feed concentration $C_f$ is less than $C_L$ and the underflow concentration $C_u$ is greater than $C_L$ (Fig. 10.20a).

The maximum allowable flux for a sedimentation tank under steady-state flows is the sum of the gravitational flux of the solids in the feed and the underflow flux:

$$G_L = C_f v_f + C_f u = C_f v_f + C_f(Q_u/A) \tag{10.20}$$

Figure 10.20b is the maximum allowable flux diagram for the total flux curve in Fig. 10.20a. Assuming sedimentation of all the applied solids (100% sus-pended-solids removal), the total solids in the underflow is equal to $G_L$:

$$G_L = C_u u = C_u(Q_u/A) \qquad (10.21)$$

The criteria for design of a sedimentation tank are an overflow rate less than the settling velocity of the feed solids,

$$Q_e/A \leq v_f \qquad (10.22)$$

and a solids loading rate less than the limiting solids flux,

$$C_f(Q_f/A) \leq G_L \qquad (10.23)$$

The maximum allowable solids loading and maximum allowable overflow rate for design of a sedimentation tank can be determined from laboratory settleability data, as illustrated in Example 10.7.

### ■ EXAMPLE 10.7

Settleability of an activated sludge was tested at several suspended-solids concentra-tions between 1000 and 10,000 mg/l using a laboratory settleometer. At each solids concentration, the steady-state settling velocity of the sludge blanket was determined and the solids flux calculated using Eq. 10.18. The results are plotted in Fig. 10.21.

Determine the design criteria of maximum allowable solids loading and maximum allowable overflow rate for an underflow solids concentration of 11.0 g/l and an activated-sludge feed concentration of 3000 mg/l.

*Solution.*   The maximum allowable solids loading for a given underflow solids con-centration can be determined graphically by drawing a line that passes through $C_u$ and tangent to the experimental solids flux curve. The line intersects the ordinate at $G_L$ and is tangent to the flux curve near $C_L$. (The choice of a lower $C_u$, steeper tangent slope, increases the allowable $G_L$ by increasing the rate of underflow $Q_u/A$.) Conversely, a higher $C_u$, flatter tangent slope, decreases $G_L$ and $Q_u/A$.) The tangent drawn in Fig. 10.21a for $C_u = 11.0$ g/l yields $G_L = 7.3$ kg/m²·h.

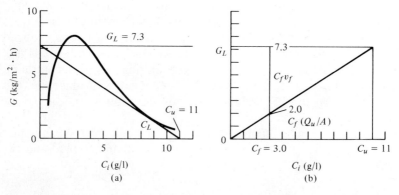

**Figure 10.21** Solids flux data from laboratory settleability analysis of an activated sludge for Example 10.7.

The solids loading diagram in Fig. 10.21b is constructed to graphically determine the maximum allowable overflow rate for an activated-sludge feed concentration of 3000 mg/l. The vertical line drawn at $C_f = 3.0$ g/l relates to Eq. 10.20 such that

$$C_f v_f = 7.3 - 2.0 = 5.3 \text{ kg/m}^2 \cdot \text{h}$$

Therefore, the maximum allowable overflow rate (Eq. 10.22) is

$$\frac{Q_e}{A} = v_f = \frac{5.3 \text{ kg/m}^2 \cdot \text{h}}{3.0 \text{ g/l}} = \frac{5.3 \text{ kg/m}^2 \cdot \text{h}}{3.0 \text{ kg/m}^3} = 1.73 \text{ m}^3/\text{m}^2 \cdot \text{h}$$

From Fig. 10.21b, the rate of underflow is

$$\frac{Q_u}{A} = \frac{2.0 \text{ kg/m}^2 \cdot \text{h}}{C_f} = \frac{2.0 \text{ kg/m}^2 \cdot \text{h}}{3.0 \text{ kg/m}^3} = 0.67 \text{ m}^3/\text{m}^2 \cdot \text{h}$$

The design criteria for $C_u = 11.0$ g/l, $C_f = 3.0$ g/l, and an underflow of 0.67 m³/m²·h are

maximum allowable solids loading = 7.3 kg/m²·h (1.5 lb/ft²/hr)

maximum allowable overflow rate = 1.73 m³/m²·h (1000 gpd/ft²)  ∎

## 10.15 SEDIMENTATION IN WATER TREATMENT

Surface water containing high turbidity (e.g., from a muddy river) may require sedimentation prior to chemical treatment. Presedimentation basins can have hopper bottoms or be equipped with continuous mechanical sludge-removal apparatus. The minimum recommended detention period for presedimentation is 3 hr, although in many cases this is not adequate to settle out fine suspensions that occur at certain times of the year. Chemical feeding equipment for coagulation is frequently provided ahead of presedimentation basins for periods when the raw water is too turbid to clarify adequately by plain sedimentation. Sludge withdrawn from the bottom of presedimentation basins is generally discharged back to the river.

Sedimentation following flocculation depends on the settling characteristics of the floc formed in the coagulation process. A general range for the settling velocities of floc from chemical coagulation is 2–6 ft/hr. Detention periods used in floc sedimentation range from 2 to 4 hr, and overflow rates (surface settling rates) vary from 500 to 1000 gpd/ft² (20–41 m³/m²·d).

The *Recommended Standards for Water Works, Great Lakes Upper Mississippi River Board of State Public Health & Environmental Managers* [8] recommends the following standards for sedimentation design following flocculation: (1) a minimum detention time of 4 hr, which may be reduced by approval when equivalent effective settling is demonstrated; a maximum horizontal velocity through the settling tanks of 0.5 fpm (0.15 m/min) with the tanks designed to minimize short-circuiting; and a maximum rate of flow over the outlet weir of 20,000 gpd/ft of weir length (250 m³/m·d).

Plan-view diagrams of clarifiers used in water treatment are sketched in

Fig. 10.16. Figure 10.22 shows a square clarifier having a sludge scraper equipped with special corner blades. The method of operation for the unit shown in Fig. 10.22 is referred to as cross-flow. Feed is introduced through submerged ports along one side of the tank and flows toward the effluent weir positioned along the opposite side of the clarifier. An alternative method of feed is to pipe influent to the center column of the clarifier, distribute it radially, and collect the effluent over a peripheral weir extending around four sides of the sedimentation tank.

Where quality of the water supply does not vary greatly, such as that from a well supply, flocculator–clarifiers (solids contact units) are an efficient method of chemical treatment. One such unit, shown in Fig. 10.23, combines the processes of mixing, flocculation, and sedimentation in a single-compartmented tank. Raw water and added chemicals are mixed with the slurry

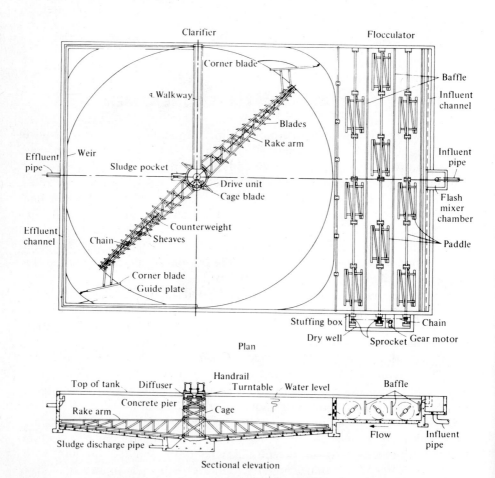

**Figure 10.22** Flocculator and square sedimentation tank for water clarification, illustrating cross-flow operation. (Courtesy of Dorr-Oliver, Inc.)

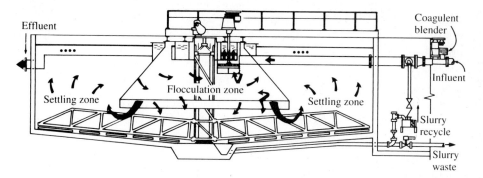

**Figure 10.23** Flocculator–clarifier provides mixing, flocculation, and sedimentation in a compartmented concentric circular tank. (Courtesy of Walker Process Equipment Division of McNish Corp.)

of previously precipitated solids to promote growth of larger crystals and agglomerated clusters that settle more readily.

The *Recommended Standards for Water Works* [8] states that flocculator–clarifiers

> are generally acceptable for combined softening and clarification where water characteristics, especially temperature, do not fluctuate rapidly, flow rates are uniform and operation is continuous. Before such units are considered as clarifiers without softening, specific approval of the reviewing authority shall be obtained. Clarifiers should be designed for the maximum uniform rate and should be adjustable to changes in flow which are less than the design rate and for changes in water characteristics.

Design criteria recommended are as follows: (1) a flocculation and mixing time of not less than 30 min based on total volume of mixing and flocculation zones; (2) minimum detention times of 2 to 4 hr for suspended solids contact clarifiers and softeners treating surface water and 1 to 2 hr for the suspended solids contact softeners treating only groundwater, with the calculated detention time based on the entire volume of the flocculator–clarifier; weir loadings not exceeding 10 gpm/ft (2.1 l/m·s) for units used as clarifiers and 20 gpm/ft (4.1 l/m·s) for units used as softeners; and upflow rates not exceeding 1.0 gpm/ft$^2$ (0.68 l/m$^2$·s) for units used as clarifiers and 1.75 gpm/ft$^2$ (1.19 l/m$^2$·s) for units used as softeners, with the upflow velocity calculated by using the cross-sectional area under the outlet channels at the sludge separation line.

The volume of sludge removed from clarifiers plus the quantity of backwash water from cleaning sand filters generally runs from 4% to 6% of the processed water. The exact amount of wastewater produced depends on chemical treatment processes employed, type of sludge collection apparatus, and kind of filter backwash equipment. Methods of waste disposal are governed by local and state pollution control agency requirements. Possible procedures include direct discharge to a receiving stream or drainage system, lagooning or sludge drying beds, hauling away and spreading on land, discharge to municipal sewer, and dewatering of sludge by pressure filtration or centrifugation.

■ **EXAMPLE 10.8**

Two rectangular clarifiers each 90 ft long, 16 ft wide, and 12 ft deep are used to settle 0.50 mil gal of water in an 8-hr operational period. Calculate the detention time, horizontal velocity, overflow rate, and weir loading, assuming multiple effluent weirs with a total length equal to three tank widths.

*Solution*

$$\text{flow} = 0.50 \text{ mil gal/8 hr} = 8356 \text{ ft}^3/\text{hr} = 139 \text{ cfm}$$

$$\text{detention time} = \frac{2 \times 90 \times 16 \times 12}{8356} = 4.1 \text{ hr}$$

$$\text{horizontal velocity} = \frac{139}{2 \times 16 \times 12} = 0.36 \text{ fpm}$$

$$\text{overflow rate} = \frac{3 \times 500,000}{2 \times 90 \times 16} = 520 \text{ gpd/ft}^2$$

$$\text{weir loading} = \frac{3 \times 500,000}{3 \times 2 \times 16} = 15,600 \text{ gpd/ft}$$ ■

## 10.16 SEDIMENTATION IN WASTEWATER TREATMENT

Clarifiers used to settle raw wastewater are referred to as *primary tanks.* Sedimentation tanks used between trickling filters in a two-stage secondary-treatment system are called *intermediate clarifiers. Final clarifiers* are settling tanks following secondary aerobic treatment units. Placement of sedimentation basins in single-stage trickling-filter plants, two-stage filter plants, and activated-sludge plants are illustrated in Figs. 12.22, 12.23, and 12.32, respectively.

Sedimentation tanks in wastewater treatment have average overflow rates in the range 300–1000 gpd/ft² and detention times between 1 and 3 hr. Side-water depths depend on settling characteristics of the waste and need for sludge storage volume in the tank bottom.

Separate design criteria are recommended for different sedimentation applications [11]. Primary clarifiers normally have an average overflow rate in the range of 600–800 gpd/ft² (24–32 m³/m²·d) with a maximum recommended average rate of 1000 gpd/ft² (41 m³/m²·d) and a maximum rate not exceeding 1500 gpd/ft² (61 m³/m²·d) for peak hourly flow. The side-water depth should be as shallow as practicable, but not less than 7.0 ft (2.1 m). Maximum recommended weir loadings are 10,000 gpd/ft (124 m³/m·d) for plants designed for an average flow of 1 mgd or less and 15,000 gpd/ft (186 m³/m·d) for plants designed for average flows greater than 1 mgd.

Final clarifiers in trickling filter and rotating biological contactor plants normally have an average overflow rate of 800 gpd/ft² (33 m³/m²·d) with a recommended maximum of 1200 gpd/ft² (49 m³/m²·d) for peak hourly flow. The side-water depth is commonly 7.0–8.0 ft (2.1–2.4 m), with 7.0 ft being the recommended minimum. Maximum recommended weir loadings are 10,000 gpd/ft (124 m³/m·d) for plants designed for average flows of 1 mgd or less and 15,000 gpd/ft (186 m³/m·d) for plants designed for average flows greater than 1 mgd. Intermediate clarifiers have the same recommended side-water depths and weir loadings, but the average overflow rate is normally 1000 gpd/ft² (41 m³/m²·d) with a recommended maximum of 1200 gpd/ft² (49 m³/m²·d) for peak hourly flow.

Final settling tanks for activated-sludge processes are sized to account for zone settling of a flocculent suspension. Design detention times, overflow rates, and weir loadings are selected to minimize the problems with solids loadings, density currents, inlet hydraulic turbulence, and occasional poor sludge settleability. Compared to other wastewater sedimentation tanks, activated-sludge final clarifiers are deeper to accommodate the greater depth of hindered solids settling, have a lower overflow rate to reduce carryover of buoyant biological floc, and have inboard weir channels to reduce the approach velocity of the effluent. Typical overflow rates for determining clarifier surface area based on average daily design flow are 600 gpd/ft² (24 m³/m²·d) for plants smaller than 1 mgd and 800 gpd/ft² (33 m³/m²·d) for larger plants. Overflow rates during the peak diurnal flows should not exceed 1200 gpd/ft² (49 m³/m²·d) and 1600 gpd/ft² (65 m³/m²·d) for plants smaller and larger than 1 mgd, respectively. The recommended minimum side-water depth is 10 ft (3.1 m) with greater depths for larger-diameter tanks (e.g., 11 ft at 50-ft diameter and 12 ft at 100-ft diameter). Minimum detention times are 2.0–3.0 hr, depending on the values selected for overflow rate and side-water depth. The maximum recommended weir loadings are 10,000–20,000 gpd/ft (125–250 m³/m·d). The allowable solids loading relates primarily to the depth of the clarifier, type of mechanical sludge-removal equipment, and characteristics of the activated sludge. Appropriate solids loadings for a properly designed clarifier are 50–60 lb/ft²/day (250–290 kg/m²·d) for a good settling sludge, sludge volume index less than 150 ml/g, and 40–50 lb/ft²/day (200–250 kg/m²·d) for a poor settling sludge. Clarifiers with shallow water depths or high overflow rates, in comparison to optimum design, may be limited to solids loadings of 20–30 lb/ft²/day (100–150 kg/m²·d) for adequate solids separation.

The BOD removal from raw domestic wastewater in primary settling is generally 30%–40%, with the commonly assumed design BOD removal being 35% and suspended-solids removal of 50%. Of course, the effectiveness of plain sedimentation depends on the characteristics of the wastewater as well as clarifier design. If a municipal wastewater contains an unusual amount of soluble organic matter from industrial wastes, the BOD removal could be less than 30%; conversely, if industrial wastes contribute settleable solids, the BOD removal could be greater than 40%.

Design of a clarifier, the hydraulic loading, and wastewater characteristics influence the density of the accumulated sludge withdrawn from a clarifier [12]. With proper design and operation, the sludge gravity thickens in the bottom of the tank. Hydraulic disturbances usually result in a more dilute sludge, in addition to reduced BOD and suspended-solids removal from the wastewater. Retaining the sludge solids too long can also upset the clarification process in the primary and final tanks. In primary sedimentation, return of waste-activated sludge to the head of a plant can cause detrimental microbiological activity in the settled raw sludge, leading to foul odors, floating sludge, and thinning of the withdrawn sludge. Shortening the retention time of the sludge in the tank by more frequent withdrawal can help, but the best arrangement is separate disposal of the waste-activated sludge. In final sedimentation of activated sludge, retention of the settled sludge for too long can lead to floating solids and thinning of the sludge. The best solution is to use a scraper mechanism with uptake pipes spaced along the arm to remove the solids from the entire floor of the tank rather than plowing them to a central hopper prior to withdrawal.

Figure 10.24 pictures a typical primary settling tank. Wastewater enters at the center behind a stilling baffle and travels down and outward toward effluent weirs located on the periphery of the tank. The inlet line usually terminates near the surface, but the wastewater must travel down behind the stilling baffle before entering the actual settling zone. A stilling well reduces the velocity and imparts a downward motion to the solids, which drop to the tank floor.

The clarifier in Fig. 10.25 is designed as a final clarifier for an activated-sludge secondary. The liquid flow pattern is the same as that of a primary clarifier, but the sludge collection system is unique. Sludge uptake pipes are attached to and spaced along the scraper mechanism to convey the less dense sludge to a stationary collecting channel constructed around the stilling baffle on the center column. This continuously withdrawn sludge is activated sludge returned to the aeration tank. Heavy sludge plowed to the hopper near the center column is excess activated sludge wasted from the system.

### ■ EXAMPLE 10.9

Determine the dimensions for two rectangular primary clarifiers to settle 3400 $m^3/d$ of domestic wastewater based on the following criteria: an overflow rate of 32 $m^3/m^2 \cdot d$, a side-water depth of 2.4 m, weir loading of 125–250 $m^3/m \cdot d$, and a length/width ratio in the range of 3/1 to 5/1. Calculate the detention time and estimate the BOD removal efficiency after sizing the tanks.

### *Solution*

$$\text{surface area of each tank} = \frac{3400 \text{ m}^3/\text{d}}{2 \times 32 \text{ m}^3/\text{m}^2 \cdot \text{d}} = 53.1 \text{ m}^2$$

**Figure 10.24** Typical primary settling tank for wastewater treatment. (a) Circular settling tank using inboard weir trough. (b) Stilling well and sludge collecting mechanism in circular clarifier. (Courtesy of Walker Process Corp.)

The appropriate width for a standard-sized sludge-removal mechanism is 4.0 m. Therefore, the tank length is 53.1/4.0 = 13.3 m, and the length/width ratio is 13.3/4.0 = 3.3/1, which is within the desired range of 3/1–5/1.

$$\text{detention time} = \frac{2 \times 53.1 \text{ m}^2 \times 2.4 \text{ m} \times 24 \text{ h/d}}{3400 \text{ m}^3/\text{d}} = 1.8 \text{ h}$$

$$\text{maximum weir length} = \frac{3400 \text{ m}^3/\text{d}}{125 \text{ m}^3/\text{m}^2 \cdot \text{d}} = 27.2 \text{ m}$$

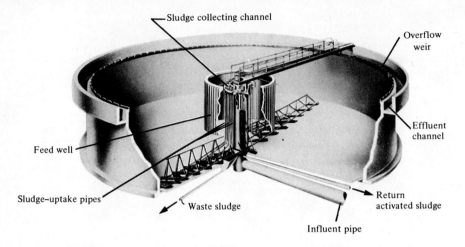

**Figure 10.25** Final clarifier for an activated-sludge secondary with rapid-sludge-removal apparatus. (Courtesy of Dorr-Oliver, Inc.)

$$\text{minimum weir length} = \frac{3400 \text{ m}^3/\text{d}}{250 \text{ m}^3/\text{m}^2\cdot\text{d}} = 13.6 \text{ m}$$

Consider one inboard effluent channel and an end weir across the width of each tank. For this arrangement,

$$\text{weir loading} = \frac{3400 \text{ m}^3/\text{d}}{2 \times 3 \times 4 \text{ m}} = 140 \text{ m}^3/\text{m}\cdot\text{d (OK)}$$

The BOD removal efficiency for an overflow rate of 32 m³/m²·d = 800 gpd/ft²/day is assumed to be 35%. ∎

### ■ EXAMPLE 10.10

Two final clarifiers of the type shown in Fig. 10.25, with 100-ft diameter and 12-ft side-water depth, are provided for an activated-sludge plant designed to treat 12.5 mgd. Calculate the overflow rate and detention time based on design flow. If the aeration tank is operated at a mixed-liquor suspended-solids concentration of 4000 mg/l and a recirculation ratio of 0.5, calculate the solids loading on the clarifier.

*Solution*

$$\text{surface area of clarifiers} = 2 \times \pi(50)^2 = 15,700 \text{ ft}^2$$

$$\text{volume of clarifiers} = 15,700 \times 12 \times 7.48 = 1,410,000 \text{ gal}$$

$$\text{overflow rate} = \frac{12,500,000}{15,700} = 800 \text{ gpd/ft}^2 \text{ (32 m}^3/\text{m}^2\cdot\text{d)}$$

$$\text{detention time} = \frac{1,410,000 \times 24}{12,500,000} = 2.7 \text{ hr}$$

Flow from the aeration tank to the clarifier with a recirculation ratio of 0.5 equals $1.5 \times 12.5 = 18.8$ mgd

$$\text{solids loading} = \frac{18.8 \times 4000 \times 8.34}{15,700} = 40 \text{ lb/ft}^2/\text{day} (195 \text{ kg/m}^2\cdot\text{d}) \qquad \blacksquare$$

## 10.17  GRIT CHAMBERS IN WASTEWATER TREATMENT

Grit includes gravel, sand, and heavy particulate matter such as corn kernels, bone chips, and coffee grounds. For design purposes grit is defined as fine sand, 0.2-mm-diameter particles with a specific gravity of 2.65 and a settling velocity of 0.075 fps.

Grit removal in municipal wastewater treatment protects mechanical equipment and pumps from abnormal abrasive wear, prevents pipe clogging by its deposition, and reduces accumulation in settling tanks and digesters. Grit chambers are commonly placed between lift pumps and primary settling tanks. Wear on the centrifugal pumps is tolerated to obtain the convenience of ground-level construction of grit chambers.

Several types of grit-removal units are used in wastewater treatment. The kind selected depends on the amount of grit in the wastewater, size of the plant, convenience of operation and maintenance, and costs of installation and operation. Standard types are channel-shaped settling tanks, aerated tanks of various shapes with hopper bottoms, clarifier-type tanks with mechanical scraper arms, and cyclone grit separators with screw-type grit washers.

Historically popular but rarely in use today, the channel-type grit chamber is a long, narrow, shallow settling channel with a proportional weir placed in the discharge end. The weir opening is shaped to keep the horizontal velocity relatively constant with varying depths of flow. To prevent scouring of the precipitated grit, the horizontal velocity is controlled at approximately 1 fps.

Square and rectangular hopper-bottom tanks with the inlet and effluent weir on opposite sides of the tank are generally used in small plants. These units are often mixed by diffused aeration to keep the organic solids in suspension while the grit settles out. Design of an aerated hopper-bottom grit tank is based on a detention time of approximately 1 min at peak hourly flow. Settled substances are removed from the hopper bottom by pumping, screw conveyor, bucket elevator, or, if possible, gravity flow.

Clarifier-type tanks are generally square, with influent and effluent weirs on opposite sides. A centrally driven collector arm scrapes the settled grit into an end hopper. The grit is then elevated into a receptacle with a chain and bucket, or screw conveyor. The clarifier may be a shallow tank or a deeper aerated chamber.

Grit taken from these kinds of grit chambers is sometimes relatively high in organic content. A counterflow grit washer that functions like a screw conveyor can be used to wash the grit and return waste organics to the plant influent. Alternatively, a special centrifugal pump can lift the slurry from a grit sump to a centrifugal cyclone. The cyclone separates the grit from the organic material and discharges it to a classifier for washing and draining. Wash water and washed-out organics from the cyclone and classifier return to the wastewater.

Preaeration of raw wastewater prior to primary sedimentation is practiced to restore freshness, scrub out entrained gases, and improve subsequent settling. The process is similar to flocculation if the detention time is not less than 45 min without chemical addition, or about 30 min with chemicals. Normally, the purpose of preaeration is freshening wastewater, not flocculation, and little or no BOD reduction occurs. The detention time in preaeration basins is generally less than 20 min. Air supplied for proper agitation ranges from 0.05 to 0.20 ft³/gal of applied wastewater.

Processes of grit removal and preaeration can be performed in the same basin—a hopper-bottom or clarifier-type tank provided with grit-removal and washing equipment. Aeration mixing improves grit separation while freshening the wastewater.

### ■ EXAMPLE 10.11

Design wastewater flows for a treatment plant are an average daily flow of 280 gpm and a peak hourly rate of 450 gpm. Estimate tank volumes and dimensions for the following grit-removal units: (a) a hopper-bottom tank with grit washer and (b) a clarifier-type unit for grit removal and preaeration.

*Solution*

    **a.** Design criteria for a hopper-bottom grit-removal tank:

        detention time = 1 min at peak flow
        volume of tank required = $450 \times 1.0 = 450$ gal = 60 ft³

    **b.** Design criteria for a preaeration basin with grit removal:

        detention time = 20 min
        volume of basin required = $280 \times 20 = 5600$ gal = 750 ft³

    Use a 12 × 12 ft basin 6 ft deep + freeboard. Provide a revolving clarifier mechanism, a separate grit washer, and diffused aeration along one side of the tank.    ■

## Filtration

Filtration is used to separate nonsettleable solids from water and wastewater by passing it through a porous medium. The most common system is filtration through a layered bed of granular media, usually a coarse anthracite coal underlain by a finer sand.

## 10.18  GRAVITY GRANULAR-MEDIA FILTRATION

Gravity filtration through beds of granular media is the most common method of removing colloidal impurities in water processing and tertiary treatment of wastewater.

The mechanisms involved in removing suspended solids in a granular-media filter are complex, consisting of interception, straining, flocculation, and sedimentation as shown schematically in Fig. 10.26. Initially, surface straining and interstitial removal results in accumulation of deposits in the upper portion of the filter media. Because of the reduction in pore area, the velocity of water through the remaining voids increases, shearing off pieces of captured floc and carrying impurities deeper into the filter bed. The effective zone of removal passes deeper and deeper into the filter. Turbulence and the resulting increased particle contact within the pores promotes flocculation, resulting in trapping of the larger floc particles. Eventually, clean bed depth is no longer available and breakthrough occurs, carrying solids out in the underflow and causing termination of the filter run.

Microscopic particulate matter in raw water that has not been chemically treated will pass through the relatively larger pores of a filter bed. On the

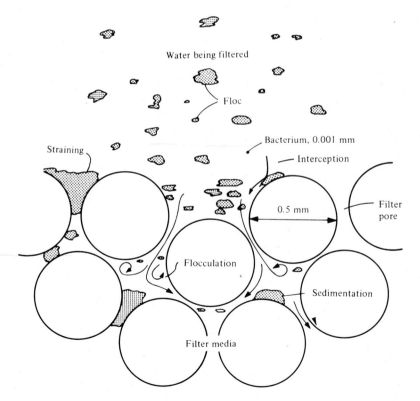

**Figure 10.26** Schematic diagram illustrating straining, flocculation, and sedimentation actions in a granular-media filter.

other hand, suspended solids fed to a filter with excess coagulant carryover from chemical treatment produces clogging of the bed pores at the surface. Optimum filtration occurs when impurities in the water and coagulant concentration cause "in-depth" filtration. The impurities neither pass through the bed nor are all strained out on the surface, but a significant amount of flocculated solids is removed throughout the entire depth of the filter.

Granular-media filtration of surface water for removal of *Giardia* cysts using field-scale pilot filters and testing at full-scale treatment plants have demonstrated the necessity of proper chemical pretreatment for effective removal of microscopic particles [13, 14]. The dual-media pilot filters were operated at a rate of 5 gpm/ft² and filtered low-temperature (near 0°C–8°C) low-turbidity (0.5–1.5 NTU) waters after coagulation with either a cationic coagulant or alum and a polymer aid. The contaminant removals with proper chemical pretreatment were turbidity 84%–96%, coliform bacteria 97%–99.95%, and *Giardia* cysts 100%, meaning no cysts were detected in the filtered water. The removals without chemical pretreatment were turbidity 35%–57%, coliform bacteria 60%, and *Giardia* cysts 80%–91%. Even a lapse in chemical feed for a short time allowed a higher proportion of particles, including cysts, to pass through the filter. To ensure that large enough numbers of coliform bacteria and *Giardia* cysts were present, the raw water was seeded with these microorganisms for several of the test runs. The best indicators of proper chemical pretreatment for removal of *Giardia* cysts and other microscopic particles were the percentage reduction in turbidity and the concentration of turbidity in the filtered water. The survey of full-scale treatment plants processing waters from mountain streams containing *Giardia* cysts verified the importance of proper chemical pretreatment. *Giardia* cysts were found in the finished waters of those plants with either no chemical pretreatment or improper coagulation. In filtering these low-turbidity waters (less than 1 NTU), effective chemical pretreatment is either a finished water turbidity of less than 0.1 NTU or the absence of microscopic particles [14].

## Traditional Filtration

A typical scheme for processing surface supplies to drinking-water quality, shown in Fig. 10.27, consists of flocculation with a chemical coagulant and sedimentation prior to filtration. Under the force of gravity, often by a

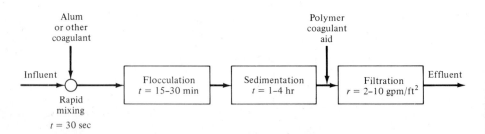

**Figure 10.27** Flow diagram for a traditional surface-water treatment system.

combination of positive head and suction from underneath, water passes downward through the media that collect the floc and particles. When the media become filled or solids break through, a filter bed is cleaned by backwashing where upward flow fluidizes the media and conveys away the impurities that have accumulated in the bed. Destruction of bacteria and viruses depends on satisfactory turbidity control to enhance the efficiency of chlorination.

Filtration rates following flocculation and sedimentation are in the range of 2–10 gpm/ft$^2$ (1.4–6.8 l/m$^2$·s), with 5 gpm/ft$^2$ (3.4 l/m$^2$·s) normally the maximum design rate.

## Direct Filtration

The process of direct filtration, diagrammed in Fig. 10.28, does not include sedimentation prior to filtration. The impurities removed from the water are collected and stored in the filter. Although rapid mixing of chemicals is necessary, the flocculation stage is either eliminated or reduced to a mixing time of less than 30 min. Contact flocculation of the chemically coagulated particles in the water takes place in the granular media [15]. Successful advances in direct filtration are attributed to the development of coarse-to-fine multimedia filters with greater capacity for "in-depth" filtration, improved backwashing systems using mechanical or air agitation to aid cleaning of the media, and the availability of better polymer coagulants.

Surface waters with low turbidity and color are most suitable for processing by direct filtration. Based on experiences cited in the literature, waters with less than 40 units of color, turbidity consistently below 5 units, iron and manganese concentrations of less than 0.3 and 0.05 mg/l, respectively, and algal counts below 2000/ml can be successfully processed [16]. Operational problems in direct filtration are expected when color exceeds 40 units or turbidity is greater than 15 units on a continuous basis. Potential problems can often be alleviated during a short period of time by application of additional polymer. Tertiary filtration of wastewaters containing 20–30 mg/l of suspended solids following biological treatment can be reduced to less than 5 mg/l by direct filtration. For inactivation of viruses and a high

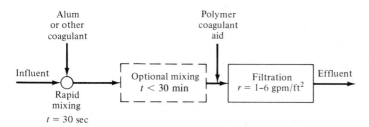

**Figure 10.28** Flow diagram for direct filtration of a surface-water supply or tertiary treatment of wastewater.

degree of bacterial disinfection, filtration of chemically conditioned waste-water precedes disinfection by chlorine.

The feasibility of filtration without prior flocculation and sedimentation relies on a comprehensive review of water quality data. The incidence of high turbidities caused by runoff from storms and blooms of algae must be evaluated. Often, pilot testing is valuable in determining efficiency of direct filtration compared to conventional treatment, design of filter media, and selection of chemical conditioning.

Filtration rates in direct filtration are usually 1–6 gpm/ft$^2$ (0.7–4.1 l/m$^2$·s), somewhat lower than the rates following traditional pretreatment.

## 10.19  DESCRIPTION OF A TYPICAL GRAVITY FILTER SYSTEM

A cutaway view of a gravity filter is shown in Fig. 10.29. During filtration, the water enters above the filter media through an inlet flume. After passing downward through the granular media (24–30 in. in thickness) and the supporting gravel bed, it is collected in the underdrain system and discharged through the underdrain pipe. During backwashing, wash water passing up-

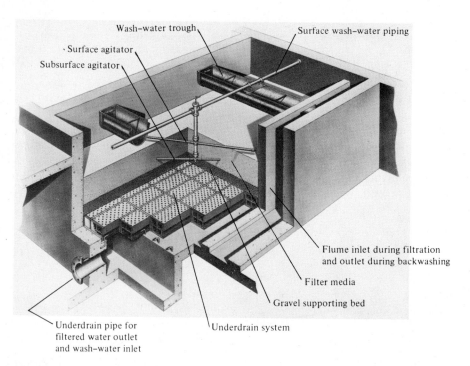

**Figure 10.29** Cutaway view of a gravity filter. (Courtesy of The F. B. Leopold Co. Inc., a subsidiary of Mueller Co.)

ward through the filter carries out the impurities that accumulated in the media. After entering the underdrain pipe, it is distributed by the underdrain system and flows upward, hydraulically expanding the filter media. The turbid wash water is collected in the wash-water troughs that discharge to the outlet flume. While backwashing, the agitator arms rotate and spray water into the expanded bed to loosen any impurities that adhere to the grains of filter media.

Typical construction of a gravity filter system in a water treatment plant is illustrated in Fig. 10.30. The filters are placed on both sides of a pipe gallery that contains inlet and outlet piping, wash-water inlet lines, and wash-water drains. The gallery is decked by an operating floor where control consoles are placed near the filters. A clear well for storage of filtered water is located under a portion of the filter bed area.

## Filter Media

Broadly speaking, filter media should possess the following qualities: (1) coarse enough to retain large quantities of floc, (2) sufficiently fine to prevent passage of suspended solids, (3) deep enough to allow relatively long filter runs, and (4) graded to permit backwash cleaning. These attributes are not all compatible. For example, a very fine sand retains floc, which also tends to shorten the filter run, while for a coarse sand the opposite would be true. Recent trends are toward coarse sands and dual-media beds of anthracite overlying sand so that high rates of filtration can be obtained.

A filter medium is defined by effective size and uniformity coefficient. The effective size is the 10-percentile diameter; that is, 10% by weight of the filter material is less than this diameter. The uniformity coefficient is the ratio of the 60-percentile size to the 10-percentile size. In water treatment [8], the conventional sand medium has an effective size of 0.45–0.55 mm, a uniformity coefficient less than 1.65, and a bed depth of 24–30 in. For dual-media filters, the top anthracite layer has an effective size of 0.8–1.2 mm, a uniformity coefficient of less than 1.85, thickness of a few inches to two thirds of the total filter thickness of 24–30 in., and is underlain by a sand filter layer as described above. The supporting coarse sand layer between the filter sand and the underlying gravel has an effective size of 0.8–2.0 mm and a uniformity coefficient less than 1.7. The conventional supporting gravel layers, beneath the filter to the underdrain, are [8] 2 to 3 in. of $\frac{2}{32}$–$\frac{3}{16}$ in., 2 to 3 in. of $\frac{3}{16}$–$\frac{1}{2}$ in., 3 to 5 in. of $\frac{1}{2}$–$\frac{3}{4}$ in., 3 to 5 in. of $\frac{3}{4}$–$1\frac{1}{2}$ in., and 5 to 8 in. of $1\frac{1}{2}$–$2\frac{1}{2}$ in. The coarsest layer of gravel required is determined by the kind of underdrain and size of openings for passage of filtered and backwash water.

A sand filter bed with a relatively uniform grain size can provide effective filtration throughout its depth. If the grain-size gradation is too great, effective filtering is confined to the upper few inches of sand. This results because the finest sand grains accumulate on the top of the bed during stratification after backwashing. The problem of surface plugging of sand filters led to development of dual-media filters. A dual-media filter consists of a sand

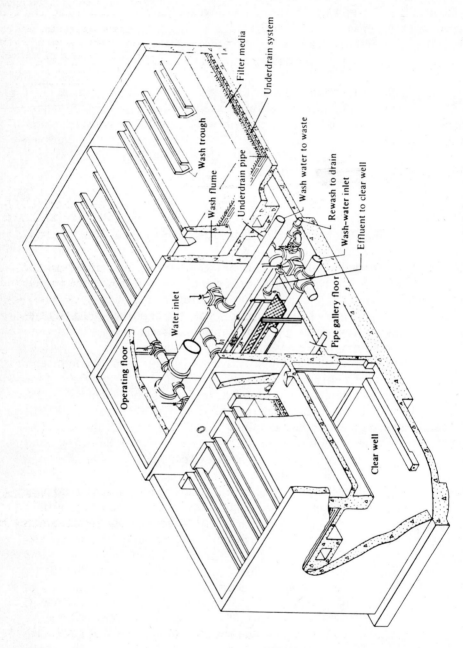

Filter media

Underdrain system

Wash trough

Wash flume

Underdrain pipe

Wash water to waste

Rewash to drain

Wash-water inlet

Effluent to clear well

Pipe gallery floor

Water inlet

Operating floor

Clear well

**Figure 10.30** Typical gravity filter system in water treatment. (Courtesy of the Permutit Co., Inc.)

[specific gravity (sp gr), 2.65] layer topped with a bed of anthracite coal medium (1.4–1.6 sp gr). The coarser anthracite top media layer has pores about 20% larger than the sand medium. These openings are capable of adsorbing and trapping particles so that floc carried over in clarified water does not accumulate prematurely on the filter surface and plug the sand filter.

Unconventional filters described by Cleasby [17] are dual-media with coarse anthracite having an effective size of about 1.5 mm and a low uniformity coefficient to provide a greater volume of voids to collect impurities and extend filter runs of highly turbid surface waters and wastewaters. To avoid problems in backwashing, the recommended effective size of the underlying sand medium is 0.75 to 0.90 mm for anthracite densities of 1.45 to 1.65. The use of a single-medium coarse-sand filter is practiced in Europe. For example, sand with a size range of 0.9–1.5 mm and about 1.0 m depth has been used in surface-water treatment plants. These filters are scoured with concurrent air and water at nonfluidizing velocity followed by a brief wash at high velocity with water alone. Triple-media filters comprising anthracite, sand, and garnet layers have been used for many years in the United States. Both dual- and triple-media filters are substantially better than the conventional sand filter in providing longer filter runs with a corresponding reduction in required backwash water. In comparing these two filters, however, the benefit of adding a third layer has not been well demonstrated [17].

## Underdrain Systems

The purposes for the underdrain of a filter are to support the filter media, collect the filtered water, and distribute the water for backwashing and air for scouring. The number and depths of the gravel layers, both to prevent loss of the media during filtration and to contribute to uniform distribution of backwash water, depend on the size of openings in the underdrain. Larger openings require a deep gravel layer, and hence a deeper filter box, but have reduced head loss during backwashing. Smaller openings, such as the slots in nozzles, require only the finer gravel layers but have higher head loss. The underdrain systems of pipe laterals with orifices and vitrified tile blocks allow only water backwash, other underdrains permit separate air scour and water backwash, and a few are designed for concurrent air-and-water scour and water backwash. Table 10.3 lists some of the common filter underdrain systems. Vitrified tile, plastic block, and nozzles are manufactured, proprietary systems. In contrast, pipe lateral and precast concrete T-Pees can be fabricated on site. The quality of water or wastewater being filtered, type of filter media, backwash and scouring system desired, and cost are factors that influence underdrain design [18].

The earliest underdrain is the pipe lateral with orifices composed of perforated pipes surrounded and overlain by graded gravel layers. The filter bottom in Fig. 10.29 is constructed of vitrified tile block with holes through the top connected to channels that carry away filtered water and distribute the upward flow of wash water. Gravel layers with a gradation between $\frac{1}{8}$ in.

**TABLE 10.3** COMMON FILTER UNDERDRAIN SYSTEMS

| Kind of underdrain | Features |
|---|---|
| Pipe lateral with orifices | Deep gravel layer |
| | Medium head loss |
| | No air scour |
| Pipe lateral with nozzles | Shallow gravel layer |
| | High head loss |
| | Air scour |
| Vitrified tile block | Shallow gravel layer |
| | Medium head loss |
| | No air scour |
| Precast concrete T-Pees | Deep gravel layer |
| | Low head loss |
| | Concurrent air-and-water scour |
| Plastic dual-lateral block | Deep gravel layer |
| | Low head loss |
| | Concurrent air-and-water scour |
| Plastic nozzles | Shallow gravel layer |
| | High head loss |
| | Air scour |

and $\frac{3}{4}$ in. are placed between the block and filter media. Rather than air scouring, surface and subsurface rotating agitators are employed for auxiliary scour in the expanded bed during backwashing. The precast concrete T-Pee underdrain system, as illustrated in Fig. 10.31, is constructed with long reinforced concrete V-blocks that are inverted and supporting graded gravel layers. The advantages of this underdrain are simplicity of design, concurrent air-and-water scouring at nonfluidizing wash-water flow, and very low head loss during backwashing. This underdrain system has been installed in waste-water filtration plants of the Sanitation Districts of Los Angeles County, California (Section 14.5).

The filter underdrains illustrated in Fig. 10.32 consist of plastic nozzles placed through holes in a steel underdrain plate. Each nozzle is constructed of wedge-shaped segments assembled with stainless steel bolts to form media-retaining slots that are tapered larger in the direction of filtering flow. If air scour is to be used before water backwash, the nozzles have 6-in.-long plastic tubes that extend into the collecting well. Air is applied through the air holes that are sized to maintain a 3.5-in. cushion to ensure equal nozzle pressure and uniform air distribution. The 16-in. gravel layer recommended to support a 30-in. dual-media filter bed has graded layers ranging from 1.5–0.75 in. in the bottom layer to $\frac{1}{4}-\frac{3}{32}$ in. in the top layer.

## 10.20 FLOW CONTROL THROUGH GRAVITY FILTERS

Depending on the design, the operation of a gravity filter can be controlled by the following methods: rate of discharge, constant level, influent flow splitting, and declining rate.

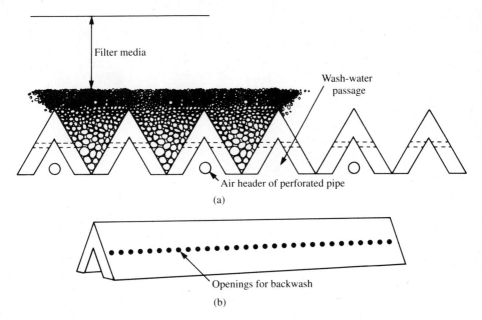

**Figure 10.31** Precast concrete T-Pee underdrain. (a) The T-Pees with openings for air-and-water scouring and water backwash are covered by graded gravel layers supporting the filter media. (b) The concrete T-Pee is a long reinforced-concrete V-block with holes along both sides.

## Rate of Flow Control

Traditional systems for control of gravity filtration regulate the rate of discharge from the filter underdrain. The influent enters each filter below the operating water level and is essentially unrestricted. A step-by-step description of the flow control operation follows the valve numbering illustrated in Fig. 10.33. Initially, valves 1 and 4 are open and 2, 3, and 5 are closed, permitting filtration to proceed. Overflow from the settling basin is applied to the filter, and water passes through the bed into the clear well. When head loss becomes excessive, valves 1 and 4 are closed (3 remains closed), and 2 and 5 are opened to permit backwashing. Clean wash water flows into the filter underdrain system, where it is distributed upward through the filter media. Dirty wash water is collected by troughs and flows into the flume connected to a drain. At the beginning of a filter run, some filtered water can be wasted to flush out the wash water remaining in the bed. This initial wasting is accomplished by opening valve 3 when valve 1 is opened to start filtration (2, 4, and 5 are shut). Then valve 4 is opened while valve 3 is being closed. The sequence is then repeated.

    The piezometric diagrams in Fig. 10.34 illustrate the approximate hydraulic profiles through a gravity filter system operated by rate of flow control. During filtration the depth of water above the surface of the filter media is 3–4.5 ft (0.9–1.4 m). The filter effluent pipe is trapped in the clear well to

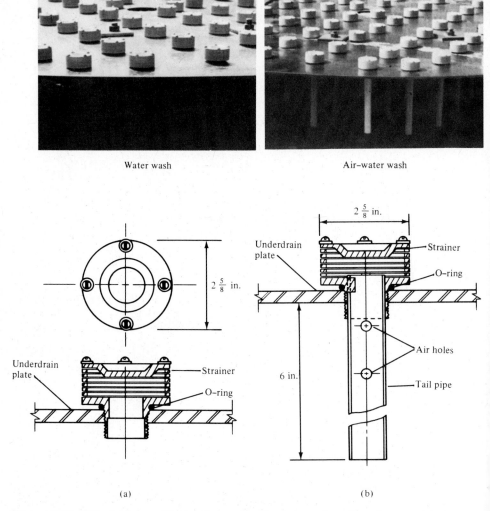

**Figure 10.32** Filter bottom with plastic nozzles fitted through uniformly spaced holes in a steel underdrain plate. (a) Nozzle design for water backwash only. (b) Nozzle design for air–water wash. The tail pipe with air holes restricts the flow of air forming a cushion of air under the plate to ensure uniform air distribution through the nozzles. (Courtesy of General Filter Co., Ames, IA.)

provide a connection to water above the media and to prevent backflow of air to the bottom of the filters. The total head available for filtration is equal to the difference between the elevation of the water surface above the media and the water level in the clear well, commonly 9–12 ft (2.7–3.7 m). A rate of flow controller (a valve controlled by a flowmeter) in the filter effluent pipe regulates the rate of flow through a clean bed. As the pores of the filter media

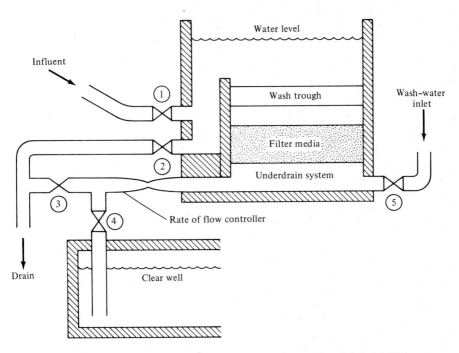

**Figure 10.33** Diagram illustrating the operation of a traditional gravity filter system by control of the rate of flow.

fill, the head loss in the bed increases and the valve automatically opens wider to maintain constant flow. The flow controller valve is wide open when the measured head loss through the filter media is 8–10 ft (2.4–3.0 m) and the bed is cleaned by backwashing. A control console for each filter unit is provided with a head-loss gage, flowmeter, and rate controller.

The advantage of rate of flow control is that the design procedures are well established. The main disadvantages are the high construction and maintenance costs of this relatively complex control system. Also, manual or automatic control of plant influent is necessary to equal the outflow and to maintain constant water levels above the filter beds.

## Constant-Level Control

Constant-level control, which is similar to constant-rate control, employs an effluent control valve that automatically adjusts the filter discharge to maintain a preset water level above the bed in the filter box. If the filtration rate is too high, a dropping water level throttles the valve. Conversely, as the rate decreases, resulting from accumulation of impurities in the bed, the tendency of the water level to rise opens the control valve. For uniform operation, the plant inflow must remain unchanged and the effluent water level in the clear well held constant because of the hydraulic connection to the water above

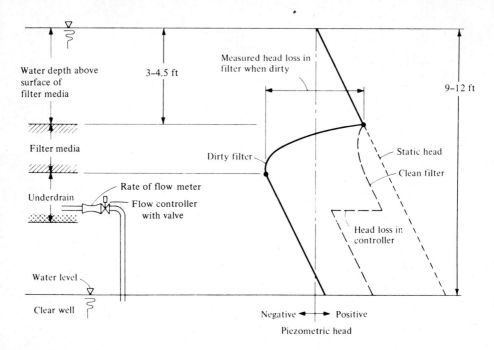

**Figure 10.34** Piezometric head diagrams through a gravity filter system operated by control of the rate of flow.

the bed. In addition to cost and mechanical complexity, a potential disadvantage for this type of control system is the possibility of momentary surges in filtration rates caused by an increased plant inflow or sudden rise in water level above the operating filters when one unit is taken out of service for backwashing. Sudden filtration rate increases through partially dirty filters can result in accumulated impurities being flushed from the bed in the effluent.

## Influent Flow Splitting

The typical arrangement of filtration control through flow splitting for constant-rate filtration is illustrated in Fig. 10.35. While employed selectively in drinking-water filtration, this system commonly is constructed for effluent filtration in advanced wastewater processing.

Plant inflow is split equally to all operating filters through an influent weir box (1) at each unit, passes through the media to the underdrain, and then into the clear well through valve 2. At minimum operating water level, the applied water is conveyed over the bed through the wash troughs and overflows into the water overlying the bed without disturbing the media. Head loss through the clean filter is $h_c$ (Fig. 10.35). After the troughs are submerged, the influent weir discharges directly into the water above the bed. To prevent accidental dewatering of the filter media, the underflow passes over an efflu-

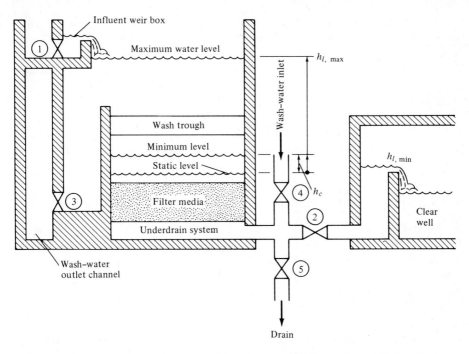

**Figure 10.35** Typical arrangement of flow control by influent flow splitting for constant-rate filtration.

ent weir located above the surface of the bed to enter the clear well. This arrangement eliminates the possibility of creating a negative head in the filter since the hydraulic pressure for filtration is only generated by the depth of water above static level. Constant-rate filtration is achieved without rate of flow controllers if the total plant inflow remains constant. Furthermore, effluent flowmeters are not necessary on the filters because the flow per filter is approximately the total plant inflow divided by the number of units in operation. As impurities fill the media, the water level above the bed rises to maintain the constant filtration rate. Thus, the head loss is apparent by the depth of water above the filter bed, which can be read by either a simple staff gage or an automatic water-level recorder. When the water level reaches the maximum operating level, the filter is backwashed.

The sequence for cleaning the bed is started by closing influent valve 1 and opening valve 3 to the wash-water outlet channel, thus draining out the water overlying the bed and lowering the level to the wash troughs. After the close of valve 2 to the clear well, wash-water valve 4 can be opened, allowing wash water to flush impurities out of the fluidized bed. The clean media are allowed to restratify in quiescent water before closing and opening appropriate valves to restore filtration. The initial filtered water, if contaminated, can be wasted to a drain by opening valve 5. When one unit is taken out of service for backwashing, the water level gradually rises above the remaining beds until sufficient head is achieved to filter the higher flow received. Filtration rates increase slowly and smoothly with minimal effect on the filtered water quality.

The flow-splitting filtration system reduces the need for the piping and hydraulic controls required by traditional rate-of-flow control systems. The only disadvantage is the additional 5–6 ft (1.5–1.8 m) required in the depth of filter boxes to provide available head above the bed for filtration. On the other hand, eliminating the suction head under the bed removes the possibility of a negative pressure in the media with the consequent release of dissolved gases.

## Declining-Rate Filtration

As illustrated in Fig. 10.36, the cross-sectional view of a declining-rate filter appears similar to that of an influent flow-splitting unit. The principal differences are the location and type of influent arrangement and the provision of less available head loss.

Inflow enters a filter through valve 1 below the low water level from a common channel (or large influent pipe) supplying a series of filters. It passes through the filter media, underdrain, and valve 2 into the clear well. Because of the common supply, the depth of water over all filters in the group being served is the same at any time. The filtration rate of any one unit depends on

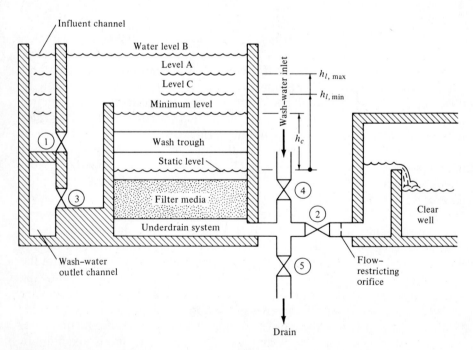

**Figure 10.36** Typical arrangement of a declining-rate filtration system.

the head loss through that bed; consequently, the cleanest filter accepts the greatest flow and the dirtiest the least. As the rates of filtration through some beds decrease, inflow is automatically redistributed to the cleaner beds to compensate for the capacity lost by the dirtier filters. During redistribution of flow, filtration rates shift gradually among the interconnected units as the water level rises slowly, providing the needed additional hydraulic head to maintain a constant overall quantity of water being processed. Eventually, the operating water level reaches the maximum elevation and the dirtiest filter is backwashed to increase its rate of filtration. After this unit returns to service, the flow to all filters redistributes at a lower operating water level.

The general behavior of declining rate filters is depicted in Fig. 10.37 [19]. The upper curve gives the rate of filtration for filter 1 starting at 6 gpm/ft$^2$ and slowly declining to 3 gpm/ft$^2$, which represents the time from starting as a clean bed to backwash. In the lower diagram, the solid line shows the water level gradually rising above all filters as their head losses increase, resulting from accumulation of impurities in the granular media. At point $A$ the dirtiest bed (2) is backwashed and the three remaining filters receive additional flow to compensate for this unit being removed from service for cleaning. Overall filtration capacity is retained by the automatic increase in rates through the operating filters forced by the rise in water level from point $A$ to point $B$. After the clean filter 2 is returned to operation, the common water level drops to level $C$.

The vertical distance from the horizontal axis of Fig. 10.37 to the water-

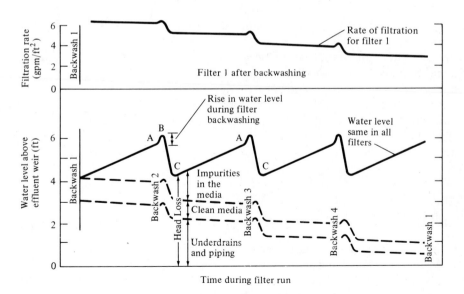

**Figure 10.37** Diagrams for declining-rate filtration showing the filtration rate, water level, and head losses for one filter run in a plant having four filters. [From J. L. Cleasby, "Declining-Rate Filtration," *J. Am. Water Works Assoc.* 73(9) (1981): 485.]

level line is the elevation of water above the effluent weir, which is the total head loss through any filter. The dashed lines apply to filter 1 starting as a clean bed to the time of backwash. Measuring upward from the horizontal axis, the vertical distances between the lines represent head losses in the underdrains and piping, clean media, and head loss caused by the accumulation of impurities in the media. The latter increases as the bed becomes dirtier, and the losses in the clean media and underdrain system decrease relative to the loss caused by solids removal as the rate of filtration through filter 1 decreases. Backwashing of filters 2–4 causes the temporary rises in all of the curves.

Design considerations unique to a declining rate filtration system are predicting water-level variations, preventing excessive rates through clean filters, and the influence of inflow variation on filtration hydraulics. Cleasby and DiBernardo [20] discuss the hydraulic considerations in declining-rate filtration in detail, including sample calculations to predict water-level variations and head losses necessary for design. The minimum water level with a head loss of $h_c$ in Fig. 10.36 results only if all beds are clean. The other water levels labeled $A–C$ in Fig. 10.36, correspond to the labeled points in Fig. 10.37. The maximum level $B$ results when one filter is out of service for backwashing. The range between levels $C$ and $A$ is the normal operating zone of head loss during the time period between backwashing of filters. The recommendation for preventing an excessive filtration rate through a clean filter is to insert a flow-restricting device in the pipeline leading from the filter underdrain to the clear well. This could be an orifice, a fixed-position valve, or a two-position valve.

Advantages attributed to declining-rate filtration compared to traditional system are that effluent quality is improved, head loss is lower for an equal length of filter run, smooth transitions in filtration rate changes are attained, negative pressure cannot occur in the bed, flowmeters and effluent control equipment are not needed, and construction and maintenance costs are lower. The disadvantages are that a greater depth is required in filter boxes, a high-water-level alarm is desirable, and, since the rate through filters cannot be manipulated, the ability of the plant to process increased flows must be considered by the operator [20].

## 10.21 HEAD LOSSES THROUGH FILTER MEDIA

Head loss through a clean granular-media filter is generally less than 3 ft (0.9 m). With accumulation of impurities, head loss gradually increases until the filter is backwashed, usually at 8–10 ft (2.4–3.0 m). In this section the Kozeny equation is presented for calculating head loss through a clean filter. Several other formulas not discussed here have been developed, such as those of Carman and Kozeny [21] and Rose [22]. Formulation of a relationship for head loss through dirty filters is very difficult; although several techniques have been proposed, none is considered to provide substantially accurate predictions.

Poiseuille's equation for laminar flow through a circular capillary tube is

$$\frac{h}{l} = \frac{32\nu v}{gD^2} \qquad (10.24)$$

where   $h/l$ = head loss per unit length, ft $H_2O$/ft (m/m)
   $\nu$ = kinematic viscosity ($\mu/\rho$), ft$^2$/sec (m$^2$/s)
   $v$ = mean velocity of flow, ft/sec (m/s)
   $g$ = acceleration of gravity, ft$^2$/sec (m$^2$/s)
   $D$ = capillary tube diameter, ft (m)

Granular media can be idealized as a bundle of capillary tubes representing the pore passages of the bed. The hydraulic radius for such an ideal bed is equal to the volume of the pores divided by the wetted internal surface area of the tubular passages. For a unit volume of the bed, the pore volume is the porosity and the wetted surface area is the volume of spherical grains times the volume of a sphere ($\pi d^3/6$) divided by the area of a sphere ($\pi d^2$). Therefore,

$$\text{hydraulic radius} = \frac{\varepsilon}{(1 - \varepsilon)(\pi d^3/6)/(\pi d^2)} = \frac{6\varepsilon}{d(1 - \varepsilon)} \qquad (10.25)$$

where   $\varepsilon$ = porosity of the stationary filter bed, dimensionless
   $d$ = diameter of spherical grains, ft (m)

Replacing the 6 with $S$ (the shape factor) and substituting a constant times the square of the hydraulic radius for $D^2$ in Poiseuille's equation yields the Kozeny equation [23] for the rate of head loss for water flow through a clean granular medium:

$$\frac{h}{l} = \frac{J\nu(1 - \varepsilon)^2 V S^2}{g\varepsilon^3 d^2} \qquad (10.26)$$

where   $h/l$ = head loss per unit depth of filter bed, ft $H_2O$/ft (m/m)
   $J$ = constant, approximately 6 for filtration in the laminar flow region, dimensionless
   $\nu$ = kinematic viscosity ($\mu/\rho$), ft$^2$/sec (m$^2$/s)
   $g$ = acceleration of gravity, ft/sec$^2$ (m/s$^2$)
   $\varepsilon$ = porosity of the stationary filter bed, dimensionless
   $V$ = superficial (approach) velocity of water above the bed, ft/sec (m/s)
   $S$ = shape factor ranging from 6.0 for spherical grains to 7.5 for angular grains, dimensionless
   $d$ = mean grain diameter, ft (m)

Equation 10.26 can be easily applied to determine head loss in a homogeneous granular-media bed. Filter beds in water and wastewater, however, are usually graded beds stratified as a result of backwashing with the coarsest grains on the bottom and finest on top. The grain-size distribution in a bed is defined by a sieve analysis. Table 10.4 lists the U.S. sieve series by the sieve designation number (the approximate number of meshes per inch) and the size of openings. For practical purposes, the thicknesses of substantially uniform layers in a stratified bed can be assumed to be proportional to the weights of the portions separated by the sieves. Hence, the total head loss is the sum of the head losses calculated by the Kozeny equation for successive layers based on the weight gradation from a sieve analysis of the filter material. Fair and Hatch [24] developed this concept and proposed the following equation for calculating head loss through each layer of a clean stratified filter bed:

$$\frac{h}{l} = \frac{36k\nu(1-\varepsilon)^2 V}{g\varepsilon^3\psi^2} \sum_{i=1}^{n} \frac{P_i}{d_i^2}$$

$(10.27)$

where    $k$ = constant equal to 5.0, dimensionless

$\psi$ = sphericity of grains, ratio of surface area of equal volume spheres to the actual surface area of grains, dimensionless

$P_i$ = fraction of total weight of filter grains in any layer $i$, dimensionless

$d_i$ = geometric mean diameter of grains in layer $i$, ft (m)

Based on tests, Fair and Hatch [24] used a value for $k$ of 5.0 as opposed to Kozeny's $J$ of 6. The geometric mean grain diameter $d_i$ (the square root of the product of the upper and lower grain sizes) can be computed from either the upper and lower sieve size of any layer $i$ or taken from the gradation curve of a grain size analysis. $P_i$ is the percentage by weight of filter grains trapped between adjacent sieves. The $6/\psi$ is the same as the shape factor $S$. Values of $\psi$ for filter sand are generally in the range of 0.8–0.7, rounded to angular, and for anthracite media from 0.7 to 0.4, angular to jagged.

**TABLE 10.4**   U.S. SIEVE SERIES

| Sieve designation number | Size of opening (mm) | Sieve designation number | Size of opening (mm) |
|---|---|---|---|
| 200 | 0.074 | 20 | 0.84 |
| 140 | 0.105 | (18) | (1.00) |
| 100 | 0.149 | 16 | 1.19 |
| 70 | 0.210 | 12 | 1.68 |
| 50 | 0.297 | 8 | 2.38 |
| 40 | 0.42 | 6 | 3.36 |
| 30 | 0.59 | 4 | 4.76 |

Example 10.12 illustrates the application of Eq. 10.27.

■ **EXAMPLE 10.12**
Calculate the head loss through a clean sand filter with a gradation defined by the sieve analysis given in Table 10.5 at a filtration rate of 2.7 $1/m^2$·s (0.0027 m/s). The bed has a depth of 0.70 m with a porosity of 0.45 and the grains of sand have a sphericity of 0.75. Assume a water temperature of 10°C.

*Solution.* From Table 10.5 the computed value

$$\Sigma \frac{P_i}{d_i^2} = 2.34 \text{ mm}^{-2} = 2.34 \times 10^{-6} \text{ m}^{-2}$$

Appendix Table A.9 gives $\nu$ at 10°C as $1.306 \times 10^{-6}$ $m^2/s$.
Substituting into Eq. 10.27,

$$\frac{h}{l} = \frac{36 \times 5.0 \times 1.306 \times 10^{-6} \text{ m}^2/\text{s} \times (1 - 0.45)^2 \, 0.0027 \text{ m/s} \times 2.34 \times 10^{-6} \text{ m}^{-2}}{9.807 \text{ m/s}^2 \, (0.45)^3 \, (0.75)^2}$$

$$= 0.89 \text{ m/m}$$

head loss through filter = 0.70 m × 0.89 m/m = 0.62 m ■

# 10.22 BACKWASHING AND MEDIA FLUIDIZATION

Filtration can be stopped because of a low rate of filtration, passage of excess turbidity through the bed, or "air binding." As head loss increases across the bed, the lower portion of the filter is under a partial vacuum (Fig. 10.34). This negative head permits the release of dissolved gases, which tend to fill the pores of the filter, causing air binding and reducing the rate of filtration.
Under average operating conditions granular-media filters are back-

**TABLE 10.5** SIEVE ANALYSIS AND COMPUTATION OF HEAD LOSS FOR EXAMPLE 10.12

| Sieve designation number | Size of opening $S$ (mm) | Geometric mean diameter $(S_1 \times S_2)^{0.5}$ (mm) | Fraction of sand retained | $\dfrac{P_i}{d_i^2}$ (mm)$^{-2}$ |
|---|---|---|---|---|
| 12 | 1.68 | 1.41 | 0.02 | 0.01 |
| 16 | 1.19 | 1.00 | 0.25 | 0.25 |
| 20 | 0.84 | 0.70 | 0.47 | 0.96 |
| 30 | 0.59 | 0.50 | 0.24 | 0.96 |
| 40 | 0.42 | 0.35 | 0.02 | 0.16 |
| 50 | 0.297 | | | |
| Total | | | 1.00 | 2.34 |

washed about once in 24 hr at a rate of about 15 gpm/ft² (10 l/m²·s) for a period of 5–10 min. Initial filtered water may be wasted for 3–5 min. A bed is out of operation for 10–15 min to complete the cleaning process. The amount of water used in backwashing varies from 2% to 4% of the filtered water. During backwashing the bed of filter media is expanded hydraulically about 50%, and the released impurities are conveyed in the wash water to the wash troughs.

Problems in backwashing of dual-media filters can result if the cleaning action is limited to wash-water fluidization. Nonuniform expansion and poor scouring can result in mud balls dropping through the coarser coal media and lodging on top of the sand layer. Several devices have been developed to improve backwashing by increasing the scrubbing action in an expanded bed and decreasing the quantity of wash water used. One popular system, shown in Fig. 10.29, consists of revolving agitator arms driven by nozzles that spray wash water under a pressure of 45–75 psi when submerged in the expanded media. Another effective method of cleaning dual-media filters is to use air scouring prior to water backwashing. After lowering the water level below the wash-water troughs, air is introduced to mix and scour the media. Wash water is then used to purge and restratify the media. Where underdrains are specifically designed for air–water backwash, air scour can be performed concurrent with the flow of wash water at a nonfluidization rate of flow. In some underdrains, separate perforated pipes are inserted as headers for injecting the air (Fig. 10.31). The wash cycle begins by draining off the water above the filter. After backwash flow has started, at about one quarter of the fluidizing rate, air is supplied and the simultaneous flow of air and water scours the bed as the wash-water level rises in the filter box. When the water surface approaches the bottom of the wash troughs, air injection is terminated and the backwash rate increases to the desired fluidization velocity to carry the impurities out of the expanded bed.

The granular media are thoroughly mixed in the agitated, turbulent flow of an expanded bed during backwashing. When the upward flow of wash water is stopped, the suspended grains settle down to form a stratified bed with the finest grains of each medium on top. In a mixed-media bed the medium of lowest density settles on top, that is, the anthracite layer above the sand bed.

## Fluidization

Fluidization is defined as upward flow through a granular bed at sufficient velocity to suspend the grains in the water. During the process of fluidization, the upward flow overcomes the gravitational force on the grains, and the energy loss is due to fluid motion. The viscous energy loss is proportional to the velocity of flow, and the kinetic energy loss is proportional to the square of the velocity.

The pressure loss through a fixed bed is a linear function of flowrate at low superficial velocities when flow is laminar. (The superficial velocity is the quantity of flow divided by the cross-sectional area of the filter.) As the flowrate increases further, the resistance of the grains to wash-water flow increases until this resistance equals the gravitational force and the grains are suspended in the water. Any further increase in upward velocity results in additional expansion of the bed while maintaining a constant pressure drop equal to the buoyant weight of the media. The characteristics of an ideal fluidized bed and departures from that behavior because of real conditions are shown in Fig. 10.38 [25].

The frictional drag of grains suspended in upward flowing water is counterbalanced exactly by the pull of gravity. Therefore, the pressure drop after fluidization is equal to the buoyant weight of the grains, which can be calculated as follows:

$$\Delta p = h\rho g = l(\rho_s - \rho)g(1 - \varepsilon) \tag{10.28}$$

where   $\Delta p$ = pressure drop after fluidization, lb (N)
$h$ = head loss (as a water column height), ft (m)
$l$ = height of expanded bed, ft (m)
$\rho$ = mass density of wash water, lb·sec²/ft⁴ (kg/m³)
$\rho_s$ = mass density of solid grains, lb·sec²/ft⁴ (kg/m³)
$\varepsilon$ = porosity of expanded bed, dimensionless
$g$ = acceleration of gravity, ft/sec² (m/s²)

Since the quantity of filter medium remains the same whether the bed is stationary or fluidized, the volume of grains initially can be equated to the volume of grains after expansion.

$$l(1 - \varepsilon) = l_0(1 - \varepsilon_0) \tag{10.29}$$

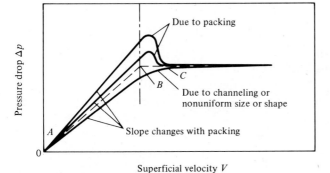

**Figure 10.38** Characteristics of a fluidized bed of granular media. Solid curves: real bed; dashed lines: ideal bed. [From J. L. Cleasby and K. Fan, "Predicting Fluidization and Expansion of Filter Media," *J. Env. Eng. Div., Proc. Am. Soc. Civil Engrs.* 107 (EE3) (1981): 459.]

where    $l_0$ = height of stationary bed
         $\varepsilon_0$ = porosity of stationary bed

## Minimum Fluidizing Velocity

The superficial water velocity required for the onset of fluidization is defined by point $B$ in Fig. 10.38 for an ideal bed composed of unisized spherical particles. For a graded bed, the minimum fluidizing velocity is not the same for all of the grains. Therefore, the change from a stationary to an expanded bed occurs gradually, with complete fluidization at a higher superficial velocity indicated by point $C$. Further increase in flow causes the concentration of particles to decrease and the porosity to approach 1. Ideally, the superficial velocity equals the unhindered terminal settling velocity of a single particle. It can be measured by experimentation in a laboratory using a filter column. At gradually increasing superficial velocities, head losses through the expanding bed are measured and plotted in a form similar to Fig. 10.38. The minimum fluidizing velocity is at the point where the curve becomes horizontal and subsequent head-loss measurements remain constant.

The minimum fluidizing velocity can also be calculated from relationships substantiated by experimental observations. One rational approach equates the head loss determined by the Ergun equation [25, 26] to the head loss using Eq. 10.28 at the point of incipient fluidization. For nonspherical particles, Wen and Yu [27] modified this relationship to account for sphericity and stationary bed porosity. The following equations resulted:

$$\mathrm{Re}_{mf} = [(33.7)^2 + 0.0408\mathrm{Ga}]^{0.5} - 33.7 \tag{10.30}$$

$$\mathrm{Re}_{mf} = \frac{d_{eq}V_{mf}\rho}{\mu} \tag{10.31}$$

$$\mathrm{Ga} = \frac{d_{eq}^3\rho(\rho_s - \rho)g}{\mu^2} \tag{10.32}$$

where    $\mathrm{Re}_{mf}$ = Reynolds number, dimensionless
         $\mathrm{Ga}$ = Galileo number, dimensionless
         $d_{eq}$ = grain diameter of a sphere of equal volume, ft (m)
         $V_{mf}$ = minimum fluidizing velocity (superficial velocity at the point of minimum fluidization), ft/sec (m/s)
         $\mu$ = absolute (dynamic) viscosity of water, lb·sec/ft$^2$ (kg/m·s)
         $\rho$ = mass density of water, lb·sec$^2$/ft$^4$ (kg/m$^3$)
         $\rho_s$ = mass density of grains, lb·sec$^2$/ft$^4$ (kg/m$^3$)

In a bed with a gradation in size of the grains, a coarser particle diameter is used to calculate $V_{mf}$ to ensure fluidizing of the entire bed. Cleasby [25] recommends substituting $d_{90}$ sieve size (90% of the grains by weight are smaller) in Eqs. 10.31 and 10.32; the value for $d_{eq}$ is often not conveniently available. Furthermore, to allow free movement of these coarse grains during backwashing, a 30% margin of safety is recommended so that the minimum backwash rate becomes $1.3V_{mf}$ [25].

This procedure is adequate for selecting the minimum backwash rate for a single-medium filter. For a dual-media bed, the minimum fluidizing velocity for the mixed interfacial region between the two adjacent media may be somewhat lower than the fluidizing velocity of the coarser grains of the upper layer at the interface.

## Expansion of a Fluidized Bed

Calculation of bed expansion relies on selecting the best estimates for sphericity and stationary-bed porosity. For irregular shapes, indirect laboratory methods are required to determine these parameters. A technique based on the hydrodynamic behavior of the grains, proposed by Briggs et al. [28], is the dynamic shape factor defined from one of the following equations:

$$DSF = \left(\frac{V_s}{V_n}\right)^2 \tag{10.33}$$

or

$$DSF = a\left(\frac{V_s}{V_n}\right) + b\left(\frac{V_s}{V_n}\right)^2 \tag{10.34}$$

where   $DSF$ = dynamic shape factor
$\quad\quad\quad V_s$ = unhindered terminal settling velocity of the grains as measured for a representative sample of grains, ft/sec (m/s)
$\quad\quad\quad V_n$ = unhindered terminal settling velocity of equivalent volume spherical grains calculated as follows with a $C_D$ value from Fig. 10.39, ft/sec (m/s)

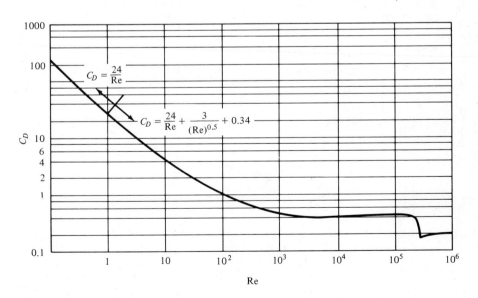

**Figure 10.39**  Drag coefficient $C_D$ for a sphere versus Reynolds number Re.

$$= \left[ \frac{4g(\rho_s - \rho)d}{3C_D\rho} \right]^{0.5} \tag{10.35}$$

$a, b$ = constants depending on Reynolds number of the spherically equivalent grains: $a$ and $b$ are selected such that $a + b = 1$ and $b/a$ is the Reynolds number

An empirical expansion correlation for uniform beds of spherical particles, including a broad range of Reynolds numbers ranging from laminar to turbulent flow, was developed by Richardson and Zaki [29]. From laboratory fluidization experiments, they found that the logarithm of superficial velocity versus the logarithm of porosity graphed as a straight line, as shown in Fig. 10.40. The mathematical expression for this relationship is

$$\frac{V}{V_i} = \varepsilon^n \tag{10.36}$$

where   $V$ = superficial velocity of water above the bed
$V_i$ = intercept velocity at a porosity of 1 (a log porosity of zero)
$\varepsilon$ = porosity of the expanded bed at $V$
$n$ = slope of log $V$ versus log $\varepsilon$ plot (Fig. 10.40)

The slope $n$ is a characteristic value for grains of a particular size, shape, and density.

Cleasby [25] developed empirical relationships for media with irregular shaped grains by incorporating DSF in the determination of $n$ for Eq. 10.36. This was accomplished using experimental data from fluidization of sand and anthracite beds of known physical properties. The following expression was solved for the exponent $\alpha$ to satisfy the expression for each size medium:

$$n_{(actual)} = n_{(spherical)}DSF^\alpha \tag{10.37}$$

A two-variable correlation of $\alpha$ against DSF and $Re_0$ yielded the desired form for $\alpha$:

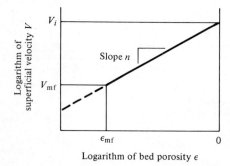

**Figure 10.40** Relationship between superficial velocity and porosity for fluidization of an ideal bed of spherical particles developed by Richardson and Zaki [29].

$$\alpha = -2.2715(DSF)^{0.420}Re_0^{-0.441} \tag{10.38}$$

Thus the resulting equations for the expansion coefficients are

$$n = \left(4.45 + 18\frac{d}{D}\right) Re_0^{-0.1}(DSF)^\alpha \tag{10.39}$$

for $Re_0$ data from 15 to 200, and

$$n = 4.45Re_0^{-0.1}(DSF)^\alpha \tag{10.40}$$

for $Re_0$ data from 200 to 503.

where    $Re_0$ = Reynolds number based on unhindered terminal settling velocity $V_s$ of grains

Values of $n$ calculated using these formulas agreed closely with actual measured values of $n$ at temperatures of both 25°C and 40°C. In these equations, the expansion slope $n$ is related to $Re_0$, since $V_s$ can be determined in a simple settling column and is measured in determining DSF. During fluidization experiments, the intercept velocities $V_i$ when log $\varepsilon = 0$ can be determined by extrapolation. Values of $V_i$ are different from the $V_s$ values in experiments with angular particles. The relation of $V_i/V_s$ to DSF determined from experimental data is

$$\frac{V_i}{V_s} = 0.90(DSF)^{-0.261} \tag{10.41}$$

Predicting expansion of a graded filter bed is possible if a sieve analysis is available. The technique involves dividing the material represented by the sieve analysis curve into several segments and applying the equation to determine the expanded depth of each segment at a particular flowrate. The total expanded depth is then the sum of the depths of equal segments, which are usually taken as five equal portions of 20% by weight. The middle size in each segment is used for the diameter term in the equation.

### ■ EXAMPLE 10.13

Calculate the recommended minimum backwash rate for a sand filter with a $d_{90}$ sieve opening equal to 1.00 mm (0.001 m). This grain diameter designation means that 90% of the filter sand by weight passed through a U.S. sieve number 18, which has a size of opening equal to 1.00 mm. The water temperature is 25°C and the density of the sand is 2.65 g/cm³ (2650 kg/m³).

*Solution.*    Calculation of the Reynolds number using $d_{90}$ in Eq. 10.31:

$$Re_{mf} = \frac{0.001 \text{ m} \times V_{mf} \times 997 \text{ kg/m}^3}{0.890 \times 10^{-3} \text{ kg/m·s}} = 1120V_{mf} \text{ m/s}$$

Values for $\rho$ and $\mu$ are for water at 25°C taken from Table A.9 in the appendix. Computation of the Galileo number using Eq. 10.32:

$$Ga = \frac{(0.001\ \text{m})^3 \times 997\ \text{kg/m}^3 \times [(2650 - 997)\text{kg/m}^3] \times 9.81\ \text{m/s}^2}{(0.890 \times 10^{-3}\ \text{kg/m·s})^2} = 20{,}400$$

Substituting $Re_{mf}$ and Ga into Eq. 10.30 and solving for $V_{mf}$ yields

$$1120 V_{mf} = \{[(33.7)^2 + 0.0408 \times 20{,}400]^{0.5} - 33.7\}\ \text{m/s}$$

$$V_{mf} = 0.0095\ \text{m/s} = 9.5\ \text{mm/s} = 9.5\ \text{l/m}^2\text{·s} = 14\ \text{gpm/ft}^2$$

With addition of a 30% margin of safety, the recommended backwash rate is 1.3 × 9.5 = 12 l/m²·s = 18 gpm/ft². ∎

## 10.23 PRESSURE FILTERS

Pressure filters have the granular media and underdrains contained in a steel tank as illustrated in Fig. 10.41. Water is pumped through the filter under pressure, and the media are washed by reversing flow through the bed, flushing out the impurities. Filtration rates are comparable to gravity filters; however, the maximum head loss can be significantly greater since it is a function of the input pump pressure rather than static water levels. Pressure filters are commonly installed in small municipal softening and iron-removal

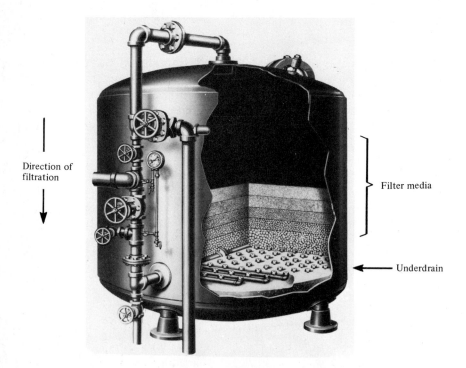

Direction of
filtration

Filter media

Underdrain

**Figure 10.41** Pressure filter (Courtesy of Infilco Degremont, Inc.)

plants, in industrial water treatment processes, and for tertiary filtration of effluent from small wastewater plants.

## 10.24  SLOW-SAND FILTERS

Slow-sand filters are limited to low-turbidity waters not requiring chemical pretreatment. Filtration rates are in the range of 0.05–0.10 gpm/ft² (0.03–0.07 l/m²·s). Beds are cleaned by scraping off a thin top layer of sand.

The filtering action of a slow sand filter is a combination of straining, adsorption, and biological flocculation. Gelatinous slimes of microbial growths form on the surface and in the upper sand layer. Effective bacteria and turbidity reductions and reasonable color removal are realized in slow-sand filters provided that the applied water turbidity does not exceed an upper practical limit of 30–50 mg/l.

Use of slow-sand filters has declined because of their high construction cost, large filter area needed, and unsuitability for treating highly turbid and polluted waters requiring chemical coagulation. No major installations exist in the United States. However, in several countries with moderate climates, where the beds do not require covers, slow-sand filter plants are an effective and efficient method for producing a potable water supply. The only chemical use is with chlorine for postfiltration disinfection and to establish a protective residual in the distribution system.

## PROBLEMS

**10.1**  A venturi meter with a throat diameter of 6.0 in. shows a pressure differential of 128 in. of water. Calculate the water flow, using a discharge coefficient of 0.93.

**10.2**  Calculate the upper head in a Parshall flume with a throat width of 1.5 ft for a flow of 5.56 cfs.

**10.3**  The recommended minimum and maximum measurements for a Parshall flume have been experimentally determined. For a throat width of 1.5 ft, the minimum head is 0.10 ft and the maximum is 2.5 ft. Calculate the flowrates for these heads.

**10.4**  A wastewater bar screen is constructed with 0.25-in.-wide bars spaced 2 in. apart center to center. If the approach velocity in the channel is 2.0 fps, what is the velocity through the screen openings?

**10.5**  Calculate points and sketch ideal output tracer curves (concentration versus normalized time) for a pulse tracer input for (a) a single completely mixed reactor and (b) a series of four equal-sized completely mixed reactors.

**10.6**  Dispersed plug flow through a baffled tank was analyzed by injecting a pulse of dye tracer into the influent and measuring the tracer concentrations in the effluent at time intervals of every 2 min after injection. The results are listed below. Sketch the tracer distribution curve by plotting $C$ (mg/l) versus $t$ (min) and locate the centroid. Calculate the dispersion number $D/uL$. By comparing this calculated value to those given in Fig. 10.10, how is this amount of dispersion described?

| $t$ | $C$ | $t$ | $C$ | $t$ | $C$ | $t$ | $C$ |
|---|---|---|---|---|---|---|---|
| 0 | 0.00 | 6 | 0.10 | 12 | 0.40 | 18 | 0.05 |
| 2 | 0.00 | 8 | 0.35 | 14 | 0.22 | 20 | 0.00 |
| 4 | 0.00 | 10 | 0.65 | 16 | 0.11 | | |

**10.7** Dispersed plug flow through a compartmented aeration tank was analyzed by injecting a pulse of lithium chloride tracer in the influent. From the time and output concentration data listed, plot $C$ (kg/m³) versus $t$ (min), the tracer response curve. Calculate the location of the centroid of the distribution $\bar{t}$, variance of the curve $\sigma^2$, normalized variance $\sigma_\theta^2$, and the reactor dispersion number $D/uL$.

| $t$ | $C$ | $t$ | $C$ | $t$ | $C$ | $t$ | $C$ |
|---|---|---|---|---|---|---|---|
| 0 | 0 | 105 | 89.0 | 210 | 33.5 | 315 | 6.0 |
| 15 | 0 | 120 | 95.0 | 225 | 25.8 | 330 | 4.6 |
| 30 | 0 | 135 | 88.0 | 240 | 20.0 | 345 | 3.5 |
| 45 | 3.5 | 150 | 78.2 | 255 | 15.4 | 360 | 2.6 |
| 60 | 16.5 | 165 | 65.0 | 270 | 12.1 | 375 | 1.7 |
| 75 | 46.5 | 180 | 55.2 | 285 | 9.5 | 390 | 0.7 |
| 90 | 72.0 | 195 | 43.0 | 300 | 7.5 | 405 | 0 |

**10.8** The residence time distribution of a cross-baffled serpentine chlorination tank was determined by injecting a pulse of 200 g of dissolved rhodamine dye into the influent. The tank is 9.4 m long and 6.7 m wide with a longitudinal wall in the center. Each side has three baffle walls extending 2.5 m in from the outside wall and two extending in from the center wall to direct the serpentine flow pattern. The wastewater effluent enters through a channel near the top of the tank and discharges over an outlet weir. At the operating water depth of 2.55 m, the liquid volume is 137 m³ (excluding the volume occupied by the concrete walls). The test was conducted at the peak hourly flow rate of 10,100 m³/d. Effluent sampling times in minutes after injecting the dye and corresponding dye concentrations are listed below. The dye concentrations were determined by measuring absorbance with a spectrophotometer at a wavelength of 550 m$\mu$. From these data, plot the tracer response curve $C$ (mg/l) versus $t$ (min). Calculate and locate the mean residence time $\bar{t}$ and the theoretical mean residence time $t_R$. Calculate the reactor dispersion number $D/uL$. In order to normalize the concentration test data, the value of $C_0$ must be determined. A portion of the dye adsorbed onto the walls of the tank and became trapped in stagnant water in the corners and near the bottom at the inlet and outlet.

Therefore, $C_0$ must be based on the calculated quantity of dye recovered in the effluent rather than on the amount injected. To estimate the milligrams of dye discharged, determine the area under the $C$-versus-$t$ tracer response curve in milligrams·minutes/liter and multiply this value by the flowrate in liters/minute. The $C_0$ value is this amount of dye divided by the volume of water in the tank, which is equal to 137 m³. Calculate the normalized value of the centroid $\bar{t}/t_R$ and the peak normalized concentration $C_{peak}/C_0$.

| $t$ | $C$ | $t$ | $C$ | $t$ | $C$ | $t$ | $C$ |
|---|---|---|---|---|---|---|---|
| 0 | 0.00 | 10 | 0.97 | 19 | 0.92 | 40 | 0.08 |
| 2 | 0.00 | 11 | 1.01 | 20 | 0.83 | 44 | 0.06 |
| 3 | 0.00 | 12 | 1.08 | 22 | 0.72 | 48 | 0.04 |
| 4 | 0.00 | 13 | 1.12 | 24 | 0.60 | 52 | 0.06 |
| 5 | 0.00 | 14 | 1.15 | 26 | 0.49 | 56 | 0.00 |
| 6 | 0.04 | 15 | 1.17 | 28 | 0.42 | 60 | 0.00 |
| 7 | 0.45 | 16 | 1.06 | 30 | 0.33 | 64 | 0.00 |
| 8 | 0.67 | 17 | 1.06 | 32 | 0.22 | 68 | 0.00 |
| 9 | 0.83 | 18 | 0.92 | 36 | 0.11 | | |

**10.9**  A flocculation basin equipped with revolving paddles is 60 ft long (the direction of flow), 45 ft wide, and 14 ft deep and treats 10 mgd. The power input to provide paddle-blade velocities of 1.0 and 1.4 fps, for the inner and outer blades, respectively, is 930 ft·lb/sec. Calculate the detention time, horizontal flow-through velocity, and $G$ (the mean velocity gradient) for a water temperature of 50°F.

**10.10**  A surface-water treatment plant is being designed to process 50 mgd. The preliminary size of the flocculation tank is 96.0 ft long with a series of 6 baffled compartments to create an over–under flow pattern. The tank width is 96.0 ft, and the water depth is 14.5 ft. Each of the six compartments has a horizontal shaft supporting six paddle flocculators with four arms to attach blades; the total number of units in the tank is 36. Up to five blades, each 15 ft by 0.5 ft, can be attached on each arm at radii of 6.5 ft, 5.5 ft, 4.5 ft, 3.5 ft, and 2.5 ft from the centerline of the shaft to the center of the blades. The maximum rotation of the flocculators, driven by variable-speed drives, can provide a velocity of 2.5 fps at a radius of 6.5 ft (center of the outer blade). What is the minimum number of blades needed on the flocculators for a velocity gradient of 60 fps/ft. Assume a ratio of water velocity to paddle velocity of 0.3, $C_d$ equal to 1.8, and a water temperature of 50°F.

**10.11**  The settling velocity of alum floc is approximately 0.0014 fps in water at 10°C. Calculate the equivalent overflow rate in gpd/ft². What is the minimum detention time in hours to settle out alum floc in an ideal basin with a depth of 10 ft?

**10.12** The data from a settling-column analysis of a dilute suspension of flocculating solids are listed below. Plot the removal percentages at the corresponding coordinates of depth and time. Sketch lines represent 15%, 30%, 45%, 60%, and 75% removals through the data. Estimate the overall removal for a settling tank 6.0 ft deep with an overflow rate of 6.0/5.0 fpm.

| $t$ | Percentage removal | | |
|-----|----------|----------|----------|
| (min) | (2.0 ft) | (4.0 ft) | (6.0 ft) |
| 10 | 20 | 15 | 13 |
| 20 | 36 | 30 | 27 |
| 30 | 50 | 39 | 37 |
| 40 | 58 | 50 | 45 |
| 60 | 71 | 60 | 55 |
| 80 | 75 | 67 | 62 |

**10.13** The settlement of flocculent particles after coagulation of a surface water was determined by a settling column test. The procedure was as described in Section 10.13. Plot the data listed below and sketch lines representing 50%, 60%, and 70% removals. For a depth of 4.0 m, estimate the overall removal for a settling time of 120 min. If the calculated percentage of removal (nearly 70%) were considered satisfactory for plant operation, why would the detention (settling) time for design be increased to 180 min or more?

| $t$ | Percentage removal | | | |
|-----|---------|---------|---------|---------|
| (min) | (1.0 m) | (2.0 m) | (3.0 m) | (4.0 m) |
| 20 | 48 | 40 | 33 | 31 |
| 40 | 61 | 49 | 43 | 39 |
| 60 | 65 | 55 | 53 | 52 |
| 80 | 68 | 59 | 57 | 53 |
| 100 | 70 | 65 | 62 | 58 |
| 120 | 74 | 68 | 65 | 61 |
| 150 | 75 | 71 | 67 | 63 |

**10.14** The settleability of an activated sludge was tested at several suspended-solids concentrations between 3200–21,000 mg/l (0.0032–0.021 lb/lb). Data for the settling velocities and corresponding solids concentrations are listed below. Plot settling velocity versus solids concentration. Calculate $G_g$ values and then plot the gravitational flux (lb/ft$^2$/hr) versus the solids concentration (lb/lb). Determine the design criteria of maximum allowable solids loading and maximum allowable overflow rate for an underflow solids concentration of 0.015 lb/lb (1.5%) and an activated-sludge feed concentration of 0.0040 lb/lb (4000 mg/l).

| $v$ (ft/hr) | $C_i$ (lb/lb) | (pcf) | $v$ (ft/hr) | $C_i$ (lb/lb) | (pcf) |
|---|---|---|---|---|---|
| 9.2 | 0.0032 | 0.20 | 1.1 | 0.010 | 0.62 |
| 5.2 | 0.0053 | 0.33 | 0.50 | 0.014 | 0.87 |
| 3.4 | 0.0062 | 0.39 | 0.30 | 0.016 | 1.0 |
| 2.0 | 0.0080 | 0.50 | 0.17 | 0.021 | 1.3 |

**10.15** Two rectangular clarifiers, each 30-ft long, 15-ft wide, and 10-ft deep, settle 0.40 mgd following alum coagulation. The effluent channels have a total weir length of 60 ft. Calculate the detention time, horizontal velocity of flow, and rate of flow over the outlet weir. Do these values meet the *Recommended Standards for Water Works* given in Section 10.15?

**10.16** Lime precipitation of raw wastewater may be practiced for removal of phosphorus (Section 14.9) or as a preliminary process in wastewater reclamation (Section 14.19). Circular primary clarifiers with aerated flocculation wells as illustrated in Fig. 10.42 were installed at a wastewater treatment plant for phosphorus precipitation and enhanced suspended-solids removal. The maximum hourly flow rate for each tank was 1.1 mgd. The lime dosage of 300 mg/l of CaO in a slurry was applied to the influent prior to entering the flocculation well. For this peak hydraulic loading, calculate the detention time based on the entire tank volume, flocculation time based on the volume of the flocculation well, overflow rate, and weir loading. Compare these values to the *Recommended Standards* for flocculator–clarifiers given in Section 10.15.

The performance of the clarifiers was very poor. The overflow was continuously milky, carrying out approximately one third of the applied lime. Subsequent recarbonation was not adequate to neutralize the pH below the range of 8.5–9.5; consequently, calcium carbonate precipitated in the following equal-

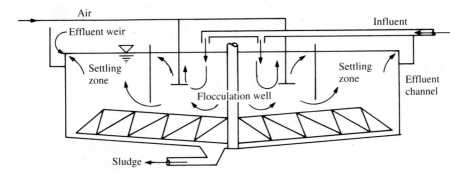

**Figure 10.42** Illustration for Prob. 10.16. A circular primary clarifier with an aerated flocculation well that receives the wastewater after the addition of lime in a rapid-mix tank. The key dimensions are diameter of the tank 40 ft, diameter of the flocculation well 20 ft, side-water depth 10 ft, and submergence of flocculation well 6.0 ft.

ization tank and RBC tanks. The sludge production was significantly less than the anticipated quantity. Why was the operation of the clarifiers unsatisfactory? (Read the discussion of flocculator–clarifiers and examine Fig. 10.23.)

**10.17** A wastewater treatment plant has two primary clarifiers, each 20 m in diameter with a 2-m side-water depth. The effluent weirs are inboard channels set on a diameter of 18 m. For a flow of 12,900 m³/d, calculate the overflow rate, detention time, and weir loading.

**10.18** Calculate the diameter and depth of a circular sedimentation basin for a design flow of 3800 m³/d based on an overflow rate of 0.00024 m/s and a detention time of 3 hr.

**10.19** A rectangular sedimentation basin is to be designed for a flow of 1.0 mgd using a 2 : 1 length/width ratio, an overflow rate of 0.00077 fps, and a detention time of 3.0 hr. What are the dimensions of the basin?

**10.20** A wastewater treatment plant has two rectangular primary settling tanks, each 40 ft long, 12 ft wide, and 7 ft deep. The effluent weir length in each tank is 45 ft. The average daily wastewater flow is 387,000 gal. Calculate the overflow rate and effluent weir loading. What is the estimated BOD removal?

**10.21** List the recommended design criteria for final clarifiers for an activated-sludge process with a design capacity of 20 mgd. Assuming four identical circular clarifiers, calculate the surface area, diameter, side-water depth, detention time, and weir loading based on an inboard weir channel. If the flow into each tank is 6.5 mgd (5.0 mgd plus 30% recirculation flow) and the suspended-solids concentration equals 2500 mg/l, calculate the solids loading. Is this solids loading satisfactory?

**10.22** An aerated clarifier-type unit for grit removal and preaeration is 12 ft square with an 8-ft liquid depth. The wastewater flow is 0.80 mgd with an estimated grit volume of 3 ft³/mil gal. A separate hopper-bottomed grit storage tank has a usable volume of 3 yd³. Compute the detention time in the aerated unit and the estimated length of time required to fill the storage tank with grit.

**10.23** Why must granular-media filtration in water treatment be preceded by chemical coagulation?

**10.24** What are the limitations in applying direct filtration in water processing?

**10.25** What is the major advantage of a dual-media coal-sand filter compared to a conventional sand filter?

**10.26** Figures 10.29, 10.31, and 10.32 illustrate different kinds of underdrains. For backwashing, which one can use air scour, which one air-and-water scour, and which one no air scour?

**10.27** In a gravity filter, how can the head-loss gage record a 9-ft loss when the water depth above the filter media surface is only 3.5 ft?

**10.28** Refer to Fig. 10.34. Assume the water depth above the surface of the filter media is 4 ft, the filter media is 2-ft deep, and the measured head loss as diagrammed in the dirty filter is 8 ft. This traditional flow control system, by regulating the discharge from the underdrain, was installed at a treatment plant where the influent was cold water transported in a pipeline from a mountain reservoir at a higher elevation. Being delivered under pressure, the influent contained dissolved gases released during processing by depressurization and warming of the water. The dissolved gases came out of solution under the negative piezometric head in the filter media and formed tiny bubbles. This "air binding" (Section 10.22) increased the head loss and, at the start of backwashing, disturbed the underdrain media by eruption of the accumulated gas bubbles. A plant expansion is being proposed. What change in design of the filtration system would you recommend to reduce the problem of air binding?

**10.29** What are the principal similarities and differences between declining-rate filtration and influent flow-splitting filtration?

**10.30** Calculate the initial head loss through a filter with 18 in. of uniform sand having a porosity of 0.42 and a grain diameter of $1.6 \times 10^{-3}$ ft. Assume spherical particles and a water temperature of 50°F. The filtration rate is 2.5 gpm/ft$^2$. Use the Kozeny equation.

**10.31** Calculate the initial head loss through a dual-media filter consisting of a 0.30-m layer of uniform anthracite with a grain diameter of 1.0 mm and a 0.30-m layer of uniform sand with a grain diameter of 0.50 mm at a filtration rate of 2.7 l/m$^2$·s. The porosity of both media is 0.42, the shape factor for the anthracite is 7.5, and the shape factor for the sand is 6.0. Assume a water temperature of 10°C.

**10.32** Calculate the head loss through a clean sand filter with a gradation as given by the sieve analysis below. The filtration rate is 2.7 l/m$^2$·s, and the water temperature is 10°C. The filter depth is 0.70 m with a porosity of 0.45, and the sand grains have a sphericity of 0.75.

| Sieve designation number | 12 | 16 | 20 | 30 | 40 | 50 |
|---|---|---|---|---|---|---|
| Fraction of sand retained | 0 | 0.05 | 0.22 | 0.51 | 0.20 | 0.02 |

**10.33** How do the surface and subsurface agitators shown in Fig. 10.29 function?

**10.34** Calculate the recommended minimum backwash rate for a sand filter with a $d_{90}$ sieve opening equal to 0.84 mm. (Ninety percent of the filter sand by weight passed through a sieve number 20.) The water temperature is 10°C and the density of the sand is 2650 kg/m$^3$.

**10.35** A dual-media filter consists of a 0.50-m layer of anthracite and 0.30-m layer of sand. The anthracite medium has a specific gravity of 1.67, porosity of 0.50, and a $d_{90}$ grain size of 1.50 mm. The sand medium has a specific gravity of 2.65, a porosity of 0.40, and a $d_{90}$ grain size of 0.90 mm. (a) Calculate the recommended minimum backwash rate for the filter. (b) Calculate the pressure drop through the filter during fluidization.

# REFERENCES

1. L. D. Benefield and J. F. Judkins, Jr., *Treatment Plant Hydraulics for Environmental Engineers* (Englewood Cliffs, NJ: Prentice-Hall, 1984).
2. O. Levenspiel, *Chemical Reaction Engineering* (New York: Wiley, 1972): 276.
3. A. B. Morrill, "Sedimentation Basin Research and Design," *J. Am. Water Works Assoc.* 24 (1932): 1442–1463.
4. D. M. Marske and J. D. Boyle, "Chlorine Contact Chamber Design—A Field Evaluation," *Water and Sewage Works* 120 (1) (January 1973): 70–77.

5. A. Amirtharajah, "Design of Rapid Mix Units," in *Water Treatment Plant Design,* R. Sanks, editor (Ann Arbor, MI: Ann Arbor Science, 1978): 131–147.

6. T. R. Camp and P. C. Stein, "Velocity Gradients and Internal Work in Fluid Motion," *J. Boston Soc. Civil Engrs.* 30 (1943): 219.

7. G. M. Fair, J. C. Geyer, and D. A. Okun, *Water and Wastewater Engineering,* Vol. 2 (New York: Wiley, 1968).

8. *Recommended Standards for Water Works, Great Lakes Upper Mississippi River Board of State Public Health & Environmental Managers* (Albany, NY: Health Research Inc., 1987).

9. American Water Works Association, *Water Treatment Plant Design,* 2nd ed. (New York: McGraw-Hill, 1990).

10. R. I. Dick and K. W. Young, "Analysis of Thickening Performance of Final Settling Tanks," *Proc. 27th Industrial Waste Conf., Purdue Univ., Eng. Ext. Series* 141 (1972): 33–54.

11. *Recommended Standards for Sewage Works, Great Lakes Upper Mississippi River Board of State Public Health & Environmental Managers* (Albany, NY: Health Research Inc., 1978).

12. "Design Manual, Dewatering Municipal Wastewater Sludges," U.S. Environmental Protection Agency, Office of Research and Development, Center for Environmental Research Information, EPA/625/1-87/014 (September 1987).

13. R. R. Mosher and D. W. Hendricks, "Rapid Rate Filtration of Low Turbidity Water Using Field-Scale Pilot Filters," *J. Am. Water Works Assoc.* 78(12) (1986): 42–51.

14. D. W. Hendricks et al., *Filtration of Giardia Cysts and Other Particles under Treatment Plant Conditions* (Denver: Am. Water Works Assoc. Research Foundation, 1988).

15. R. L. Culp, "Direct Filtration," *J. Am. Water Works Assoc.* 69(7) (1977): 375–378.

16. Committee Report, "The Status of Direct Filtration," *J. Am. Water Works Assoc.* 72(7) (1980): 405–411.

17. J. L. Cleasby, "Unconventional Filtration Rates, Media, and Backwashing Techniques," in *Innovations in the Water and Wastewater Fields,* E. A. Glysson et al., eds. (Boston: Butterworths, 1985): 1–23.

18. R. D. G. Monk, "Design Options for Water Filtration," *J. Am. Water Works Assoc.* 79(9) (1987): 93–106.

19. J. L. Cleasby, "Declining-Rate Filtration," *J. Am. Water Works Assoc.* 73(9) (1981): 484–489.

20. J. L. Cleasby and L. DiBernardo, "Hydraulic Considerations in Declining-Rate Filtration," *J. Env. Eng. Div., Proc. Am. Soc. Civil Engrs.* 106(EE6) (1980): 1043–1055.

21. L. G. Rich, *Unit Operations of Sanitary Engineering* (New York: Wiley, 1961), pp. 137–146.

22. H. E. Rose, "On the Resistance Coefficient-Reynolds Number Relationship for Fluid Flow through a Bed of Granular Material," *Proc. Inst. Mech. Engrs.* 153 (1945): 154–168; 160 (1949): 492–511.

23. T. R. Camp, "Theory of Water Filtration," *J. San. Eng. Div., Proc. Am. Soc. Civil Engrs.* 90 (SA4) (1964): 1–30.

24. G. M. Fair and L. P. Hatch, "Fundamental Factors Governing the Streamline Flow of Water through Sand," *J. Am. Water Works Assoc.* 25 (11) (1933): 1551–1565.

25. J. L. Cleasby and K. Fan, "Predicting Fluidization and Expansion of Filter Media," *J. Env. Eng. Div., Proc. Am. Soc. Civil Engrs.* 107 (EE3) (1981): 455–471.

26. S. Ergun, "Fluid Flow through Packed Columns," *Chem. Eng. Progress* 48(2) (1952): 89–94.

27. C. Y. Wen and Y. H. Yu, "Mechanics of Fluidization," *Chem. Eng. Progress Symposium Series* 62 (1966): 100–111.

28. L. I. Briggs, D. S. McCulloch, and F. Moser, "The Hydraulic Shape of Sand Particles," *J. of Sedimentary Petrol.* 32 (1962): 645–657.

29. J. F. Richardson and W. N. Zaki, "Sedimentation and Fluidization, Part I," *Trans. Institution of Chem. Engrs.* 32 (1954): 35–53.

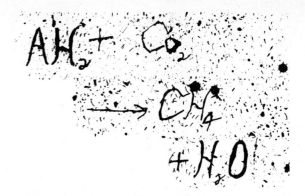

# Chapter 11

# Chemical Treatment Processes

Chemical treatment is the most important step in processing public water supplies. Surface water normally requires chemical coagulation to eliminate turbidity, color, and taste- and odor-producing compounds, while well water supplies are commonly treated to remove dissolved minerals, such as iron and manganese, and hardness. Through a variety of saline-water conversion processes, saltwater can be converted to fresh water. Chlorine is applied for disinfection and to establish a protective residual in the distribution system, and most utilities fluoridate their water as a health benefit.

## Chemical Considerations

The purpose of these initial sections is to refresh students on selected fundamental concepts, including definitions, compounds, units of expression, the bicarbonate–carbonate system, chemical equilibria, and process kinetics. Basic chemistry is also presented as introductory material for each specific unit process discussed in the chapter. These brief comments are not intended to replace formal chemistry courses prerequisite to sanitary engineering.

### 11.1 INORGANIC CHEMICALS AND COMPOUNDS

#### Definitions

Chemical elements and their atomic weights are given in Table A.7 in the Appendix. *Atomic weight* is the weight of an element relative to that of carbon-12, which has an atomic weight of 12. *Valence* is the combining power of an element relative to that of the hydrogen atom, with an assigned value of 1. Thus, an element with a valence of 2 + can replace two hydrogen atoms in a compound or, in the case of a 2 − valence, can react with two hydrogen atoms. Equivalent or combining weight of an element is equal to its atomic weight divided by the valence. For example, the equivalent weight of calcium $(Ca^{2+})$ equals 40.0 g divided by 2, or 20.0 g.

The *molecular weight* of a compound equals the sum of the weights of the combining elements and is conventionally expressed in grams. *Equivalent weight* is the molecular weight divided by the number of positive or negative electrical charges resulting from dissolution of the compound. Consider sulfuric acid, with a molecular weight of 98.1 g. *Ionization* releases two $H^+$ ions and one $SO_4^{2-}$ radical; therefore, the equivalent weight of sulfuric acid is 98.1 divided by 2, or 49.0 g.

## Chemicals Applied in Treatment

The common inorganic compounds used in water and wastewater processing are listed in Table 11.1. Given are the name, formula, common usage, molecular weight, and equivalent weight when appropriate. Common names and purity of commercial-grade chemicals are presented in the sections dealing with specific chemical treatment processes.

## Units of Expression

The concentration of ions or chemicals in solution is normally expressed as weight of the element or compound in milligrams per liter of water, abbreviated as mg/l. Occasionally, the term *parts per million* (ppm) is used rather than mg/l. These are identical in meaning, since 1 mg/1,000,000 ml is essentially the same as 1 part by weight to 1 million parts for low concentrations. Chemical dosages may be expressed in units of pounds per million gallons or rarely as grains per gallon. To convert milligrams per liter to pounds per million gallons, multiply by 8.34, which is the weight of 1 gal of water. In other words,

$$1.0 \text{ mg/l} = \frac{1 \text{ gal by weight}}{1,000,000 \text{ gal}} = 8.34 \text{ lb/mil gal}$$

One pound contains 7000 grains, and 1.0 grain per gallon (gpg) equals 17.1 mg/l.

Elemental concentrations expressed in units of mg/l can usually be interpreted to mean that the solution contains the stated number of milligrams of that particular element. For example, a water containing 1.0 mg/l of fluoride means that there is 1.0 mg of F ion by weight per liter. However, in some cases, the concentration given in milligrams of weight does not relate to the specific element whose concentration is being expressed. For example, hardness, which is a measure of the calcium ion and magnesium ion content

**TABLE 11.1**   COMMON CHEMICALS IN WATER AND WASTEWATER PROCESSING

| Name | Formula | Common application | Molecular weight | Equivalent weight |
|---|---|---|---|---|
| Activated carbon | C | Taste and odor control | 12.0 | n.a.[a] |
| Aluminum sulfate | $Al_2(SO_4)_3$ $\cdot 14.3H_2O$ | Coagulation | 600 | 100 |
| Ammonia | $NH_3$ | Chloramine disinfection | 17.0 | n.a. |
| Ammonium sulfate | $(NH_4)_2SO_4$ | Coagulation | 132 | 66.1 |
| Calcium hydroxide | $Ca(OH)_2$ | Softening | 74.1 | 37.0 |
| Calcium hypochlorite | $Ca(ClO)_2 \cdot 2H_2O$ | Disinfection | 179 | n.a. |
| Calcium oxide | $CaO$ | Softening | 56.1 | 28.0 |
| Carbon dioxide | $CO_2$ | Recarbonation | 44.0 | 22.0 |
| Chlorine | $Cl_2$ | Disinfection | 71.0 | n.a. |
| Chlorine dioxide | $ClO_2$ | Taste and odor control | 67.0 | n.a. |
| Copper sulfate | $CuSO_4$ | Algae control | 160 | 79.8 |
| Ferric chloride | $FeCl_3$ | Coagulation | 162 | 54.1 |
| Ferric sulfate | $Fe_2(SO_4)_3$ | Coagulation | 400 | 66.7 |
| Ferrous sulfate | $FeSO_4 \cdot 7H_2O$ | Coagulation | 278 | 139 |
| Fluosilicic acid | $H_2SiF_6$ | Fluoridation | 144 | n.a. |
| Magnesium hydroxide | $Mg(OH)_2$ | Defluoridation | 58.3 | 29.2 |
| Oxygen | $O_2$ | Aeration | 32.0 | 16.0 |
| Potassium permanganate | $KMnO_4$ | Oxidation | 158 | n.a. |
| Sodium aluminate | $NaAlO_2$ | Coagulation | 82.0 | n.a. |
| Sodium bicarbonate | $NaHCO_3$ | pH adjustment | 84.0 | 84.0 |
| Sodium carbonate | $Na_2CO_3$ | Softening | 106 | 53.0 |
| Sodium chloride | $NaCl$ | Ion exchanger regeneration | 58.4 | 58.4 |
| Sodium fluoride | $NaF$ | Fluoridation | 42.0 | n.a. |
| Sodium fluosilicate | $Na_2SiF_6$ | Fluoridation | 188 | n.a. |
| Sodium hexameta-phosphate | $(NaPO_3)_n$ | Corrosion control | n.a. | n.a. |
| Sodium hydroxide | $NaOH$ | pH adjustment | 40.0 | 40.0 |
| Sodium hypochlorite | $NaClO$ | Disinfection | 74.4 | n.a. |
| Sodium silicate | $Na_4SiO_4$ | Coagulation aid | 184 | n.a. |
| Sodium thiosulfate | $Na_2S_2O_3$ | Dechlorination | 158 | n.a. |
| Sulfur dioxide | $SO_2$ | Dechlorination | 64.1 | n.a. |
| Sulfuric acid | $H_2SO_4$ | pH adjustment | 98.1 | 49.0 |

[a] Not applicable.

of a water, is given in weight units of calcium carbonate. This facilitates treating hardness as a single value rather than two concentrations expressed in different weight units, one for $Ca^{2+}$ and the other for $Mg^{2+}$. The alkalinity of a water may consist of one or more of the following ionic forms: $OH^-$, $CO_3^{2-}$, and $HCO_3^-$. For commonality, the concentrations of these various radicals are given as mg/l as $CaCO_3$. According to *Standard Methods* [1], all nitrogen compounds—ammonia, nitrate, and organic nitrogen—are expressed in units of mg/l as nitrogen and phosphates are given as mg/l as phosphorus.

The term *milliequivalents per liter* (meq/l) expresses the concentration of a dissolved substance in terms of its combining weight. Milliequivalents are calculated from milligrams per liter for elemental ions by Eq. 11.1 and for radicals or compounds by Eq. 11.2:

$$\text{meq/l} = \text{mg/l} \times \frac{\text{valence}}{\text{atomic weight}} = \frac{\text{mg/l}}{\text{equivalent weight}} \qquad (11.1)$$

$$\text{meq/l} = \text{mg/l} \times \frac{\text{electrical charge}}{\text{molecular weight}} = \frac{\text{mg/l}}{\text{equivalent weight}} \qquad (11.2)$$

## Milliequivalents-per-Liter Bar Graph

Results of a water analysis are normally expressed in milligrams per liter and reported in tabular form. For better visualization of the chemical composition, these data can be expressed in milliequivalents per liter to permit graphical presentation, as illustrated in Fig. 11.1. The top row of the bar graph consists of major cations arranged in the order of calcium, magnesium, sodium, and potassium. Anions in the bottom row are aligned in the sequence of carbonate (if present), bicarbonate, sulfate, and chloride. The sum of the positive milliequivalents per liter must equal the sum of the negative values for a water in equilibrium. Hypothetical combinations of positive and negative ions can be written from a bar graph. These combinations are useful in evaluating a water for lime–soda ash softening.

Table 11.2 lists basic data for selected elements, radicals, and compounds. Included are equivalent weights for calculating milliequivalents per liter for use in bar graph presentations and chemical equations.

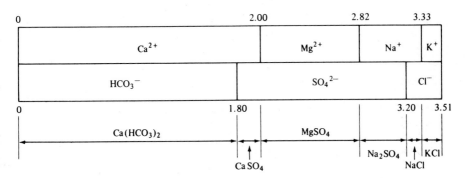

**Figure 11.1** Milliequivalents-per-liter bar graph for water analysis.

**TABLE 11.2**   DATA ON SELECTED ELEMENTS, RADICALS, AND COMPOUNDS

| Name | Symbol or formula | Atomic or molecular weight | Equivalent weight |
|------|-------------------|---------------------------|-------------------|
| Aluminum | $Al^{3+}$ | 27.0 | 9.0 |
| Calcium | $Ca^{2+}$ | 40.1 | 20.0 |
| Carbon | C | 12.0 | |
| Hydrogen | $H^+$ | 1.0 | 1.0 |
| Magnesium | $Mg^{2+}$ | 24.3 | 12.2 |
| Manganese | $Mn^{2+}$ | 54.9 | 27.5 |
| Nitrogen | N | 14.0 | |
| Oxygen | O | 16.0 | |
| Phosphorus | P | 31.0 | |
| Sodium | $Na^+$ | 23.0 | 23.0 |
| Ammonium | $NH_4^+$ | 18.0 | 18.0 |
| Bicarbonate | $HCO_3^-$ | 61.0 | 61.0 |
| Carbonate | $CO_3^{2-}$ | 60.0 | 30.0 |
| Hydroxyl | $OH^-$ | 17.0 | 17.0 |
| Hypochlorite | $OCl^-$ | 51.5 | 51.5 |
| Nitrate | $NO_3^-$ | 62.0 | 62.0 |
| Orthophosphate | $PO_4^{3-}$ | 95.0 | 31.7 |
| Sulfate | $SO_4^{2-}$ | 96.0 | 48.0 |
| Aluminum hydroxide | $Al(OH)_3$ | 78.0 | 26.0 |
| Calcium bicarbonate | $Ca(HCO_3)_2$ | 162 | 81.0 |
| Calcium carbonate | $CaCO_3$ | 100 | 50.0 |
| Calcium sulfate | $CaSO_4$ | 136 | 68.0 |
| Carbon dioxide | $CO_2$ | 44.0 | 22.0 |
| Ferric hydroxide | $Fe(OH)_3$ | 107 | 35.6 |
| Hydrochloric acid | HCl | 36.5 | 36.5 |
| Magnesium carbonate | $MgCO_3$ | 84.3 | 42.1 |
| Magnesium hydroxide | $Mg(OH)_2$ | 58.3 | 29.1 |
| Magnesium sulfate | $MgSO_4$ | 120 | 60.1 |
| Sodium sulfate | $Na_2SO_4$ | 142 | 71.0 |

■ **EXAMPLE 11.1**

The results of a water analysis are: calcium 40.0 mg/l, magnesium 10.0 mg/l, sodium 11.7 mg/l, potassium 7.0 mg/l, bicarbonate 110 mg/l, sulfate 67.2 mg/l, and chloride 11.0 mg/l. Draw a milliequivalents-per-liter bar graph and list the hypothetical combinations. Express the hardness and alkalinity in units of mg/l as $CaCO_3$.

*Solution.*   Using Eqs. 11.1 and 11.2, we have the data in Table 11.3. From the bar graph of these data in Fig. 11.1, the hypothetical chemical combinations are

1.80 meq/l of $Ca(HCO_3)_2$      0.38 meq/l of $Na_2SO_4$

0.20 meq/l of $CaSO_4$            0.13 meq/l of NaCl

0.82 meq/l of $MgSO_4$            0.18 meq/l of KCl

**TABLE 11.3**  WATER ANALYSIS DATA FOR EXAMPLE 11.1

| Component | mg/l | Equivalent weight | meq/l |
|---|---|---|---|
| $Ca^{2+}$ | 40.0 | 20.0 | 2.00 |
| $Mg^{2+}$ | 10.0 | 12.2 | 0.82 |
| $Na^+$ | 11.7 | 23.0 | 0.51 |
| $K^+$ | 7.0 | 39.1 | 0.18 |
| | | Total cations = | 3.51 |
| $HCO^{3-}$ | 110 | 61.0 | 1.80 |
| $SO_4^{2-}$ | 67.2 | 48.0 | 1.40 |
| $Cl^-$ | 11.0 | 35.5 | 0.31 |
| | | Total anions = | 3.51 |

Hardness is the sum of the $Ca^{2+}$ and $Mg^{2+}$ concentrations expressed in mg/l as $CaCO_3$, and alkalinity equals the bicarbonate content. Thus

$$\text{hardness} = 2.82 \text{ meq/l} \times 50 \frac{\text{mg/l of } CaCO_3}{\text{meq/l}} = 141 \text{ mg/l}$$

$$\text{alkalinity} = 1.80 \times 50 = 90 \text{ mg/l} \qquad \blacksquare$$

## 11.2  HYDROGEN ION CONCENTRATION

The hydrogen ion activity (i.e., intensity of the acid or alkaline condition of a solution) is expressed by the term pH, which is defined as follows:

$$pH = \log \frac{1}{[H^+]} \qquad (11.3)$$

Water dissociates to only a slight degree, yielding hydrogen ions equal to $10^{-7}$ mole/l; thus pure water has a pH of 7. It is also neutral since $10^{-7}$ mole/l of hydroxyl ion is produced simultaneously,

$$H_2O \rightleftharpoons H^+ + OH^- \qquad (11.4)$$

When an acid is added to water, the hydrogen ion concentration increases resulting in a lower pH value. Addition of an alkali reduces the number of free hydrogen ions, causing an increase in pH, because $OH^-$ ions unite with $H^+$ ions. The pH scale is acidic from 0 to 7 and basic from 7 to 14.

The chemical equilibrium of water can be shifted by changing the hydrogen ion activity in solution. Thus pH adjustment is used to optimize coagulation, softening, and disinfection reactions, and for corrosion control. In wastewater treatment, pH must be maintained in a range favorable for biological activity.

## 11.3 ALKALINITY AND pH RELATIONSHIPS

Alkalinity is a measure of water's capacity to absorb hydrogen ions without significant pH change (i.e., to neutralize acids). It is determined in the laboratory by titrating a water sample with a standardized sulfuric acid solution. The three chemical forms that contribute to alkalinity are bicarbonates, carbonates, and hydroxides that originate from the salts of weak acids and strong bases. Bicarbonates represent the major form since they originate naturally from reactions of carbon dioxide in water.

Below pH 4.5, dissolved carbon dioxide is in equilibrium with carbonic acid in solution; thus no alkalinity exists. Between pH 4.5 and 8.3, the balance shown in Eq. 11.5 shifts to the right, reducing the $CO_2$ and creating $HCO_3^-$ ions. Above 8.3 the bicarbonates are converted to carbonate ions. Hydroxide appears at a pH greater than 9.5 and reacts with carbon dioxide to yield both bicarbonates and carbonates (Eq. 11.6). The maximum $CO_3^{2-}$ concentration for dilute solutions is in the pH range 10–11. Figure 11.2 shows the relationship between carbon dioxide and the three forms of alkalinity with respect to pH calculated for water having a total alkalinity of 100 mg/l at 25°C. Temperature, total alkalinity, and the presence of other ionic species influence alkalinity–pH relationships; nevertheless, Fig. 11.2 is a realistic representation of most natural waters.

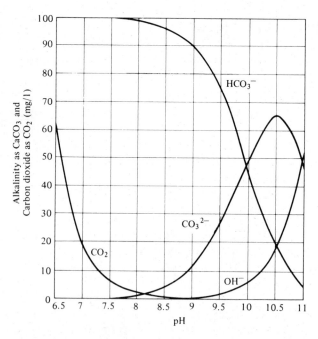

**Figure 11.2** Carbon dioxide and various forms of alkalinity relative to pH in water at 25°C.

$$CO_2 + H_2O \rightleftharpoons H_2CO_3 \rightleftharpoons H^+ + HCO_3^- \tag{11.5}$$

$$CO_2 + OH^- \rightleftharpoons HCO_3^- \rightleftharpoons H^+ + CO_3^{2-} \tag{11.6}$$

Substances that offer resistance to change in pH as acids or bases are added to a solution are referred to as *buffers*. Since the pH falls between 6 and 9 for most natural waters and wastewaters, the primary buffer is the bicarbonate–carbonate system. When acid is added, a portion of the $H^+$ ions is combined with $HCO_3^-$ to form un-ionized $H_2CO_3$; only the $H^+$ remaining free affect pH. If a base is added, the $OH^-$ ions react with free $H^+$, increasing the pH. However, some of the latter are replaced by a shift of $HCO_3^-$ to $CO_3^{2-}$, attenuating the change in hydrogen ion concentration. Both chemical reactions and biological processes depend on this natural buffering action to control pH changes. Sodium carbonate or calcium hydroxide can be added if naturally existing alkalinity is insufficient, such as in the coagulation of water where the chemicals added react with and destroy alkalinity.

Precipitation softening is easily understood by referring to the pH–alkalinity relationship illustrated in Fig. 11.2. Calcium and magnesium ions are soluble when associated with bicarbonate anions. But, if the pH of a hard water is increased, insoluble precipitates of $CaCO_3$ and $Mg(OH)_2$ are formed. This is accomplished in water treatment by adding lime to raise the pH level. At a value of about 10, hydroxyl ions convert bicarbonates to carbonates to allow the formation of calcium carbonate precipitate as follows:

$$\underset{\substack{\text{bicarbonate} \\ \text{hardness}}}{Ca(HCO_3)_2} + \underset{\substack{\text{lime} \\ \text{slurry}}}{Ca(OH)_2} \rightarrow \underset{\substack{\text{solid} \\ \text{precipitate}}}{2CaCO_3} + 2H_2O \tag{11.7}$$

## 11.4  CHEMICAL EQUILIBRIA

Many chemical reactions are reversible to some degree, and the concentrations of reactants and products determine the final state of equilibrium. For the general reaction expressed by Eq. 11.8, increase in either A or B shifts the equilibrium to the right, whereas a larger concentration of either C or D drives it to the left. A reaction in true equilibrium can be expressed by the *mass-action formula*, Eq. 11.9:

$$aA + bB \rightleftharpoons cC + dD \tag{11.8}$$

$$\frac{[C]^c[D]^d}{[A]^a[B]^b} = K \tag{11.9}$$

where   A, B = reactants
         C, D = products
         [ ] = molar concentrations
         $K$ = equilibrium constant

Strong acids and bases in dilute solutions approach 100% ionization, while weak acids and bases are poorly ionized. The degree of ionization of the latter is expressed by the mass-action equation. For example, Eqs. 11.10–11.13 are, for carbonic acid,

$$H_2CO_3 \rightleftharpoons H^+ + HCO_3^- \tag{11.10}$$

$$\frac{[H^+][HCO_3^-]}{[H_2CO_3]} = K_1 = 4.45 \times 10^{-7} \text{ at } 25°C \tag{11.11}$$

$$HCO_3^- \rightleftharpoons H^+ + CO_3^{2-} \tag{11.12}$$

$$\frac{[H^+][CO_3^{2-}]}{[HCO_3^-]} = K_2 = 4.69 \times 10^{-11} \text{ at } 25°C \tag{11.13}$$

The foregoing characterizes homogeneous chemical equilibria where all reactants and products occur in the same physical state. Heterogeneous equilibrium exists between a substance in two or more physical states. For example, at greater than pH 10, solid calcium carbonate in water reaches a stability with the calcium and carbonate ions in solution,

$$CaCO_3 \rightleftharpoons Ca^{2+} + CO_3^{2-} \tag{11.14}$$

Equilibrium between crystals of a compound in the solid state and its ions in solution can be treated mathematically as if the equilibrium is homogeneous in nature. For Eq. 11.14 the expression is as follows:

$$\frac{[Ca^{2+}][CO_3^{2-}]}{[CaCO_3]} = K \tag{11.15}$$

Concentration of a solid substance can be treated as a constant $K_s$ in mass-action equilibrium; therefore, $[CaCO_3]$ can be assumed equal to $KK_s$, and then

$$[Ca^{2+}][CO_3^{2-}] = KK_s = K_{sp} = 5 \times 10^{-9} \text{ at } 25°C \tag{11.16}$$

The constant $K_{sp}$ is called the *solubility-product constant*.

If the product of the ionic molar concentrations is less than the solubility-product constant, the solution is unsaturated. Conversely, a supersaturated solution contains a $[A^+][B^-]$ value greater than $K_{sp}$. In this case, crystals form and precipitation progresses until the ionic concentrations are reduced equal to those of a saturated solution.

Other ions in solution affect the solubility of a substance by either the common-ion effect or the secondary-salt effect. The common-ion effect is a repression of solubility of a substance in the presence of an excess of one of the solubility-product ions. For example, in pure water the solubility of $CaCO_3$ is about 13 mg/l, whereas in a solution containing 100 mg/l of carbonate alkalinity the solubility is only 0.5 mg/l. The secondary-salt effect is the increase in solubility of slightly soluble salts when other salts in solution do not have an ion in common with the slightly soluble substance. For example, $CaCO_3$ is several times more soluble in seawater than in fresh water.

## 11.5 WAYS OF SHIFTING CHEMICAL EQUILIBRIA

Chemical reactions in water and wastewater treatment rely on shifting of homogeneous or heterogeneous equilibria to achieve the desired results. The most common methods for completing reactions are by formation of insoluble substances, weakly ionized compounds, gaseous end products, and oxidation and reduction [2].

The best example of shifting equilibrium to form precipitates is lime–soda ash softening. Calcium is removed from solution by adding lime, as shown in Eq. 11.7. If insufficient alkalinity is available to complete this reaction, sodium carbonate (soda ash) is also applied. However, magnesium hardness must be removed by forming $Mg(OH)_2$ since $MgCO_3$ is relatively soluble. This is affected by addition of excess lime to increase the value of $[Mg^{2+}][OH^-]^2$ above the solubility product of magnesium hydroxide, which equals $9 \times 10^{-12}$.

$$MgCO_3 + Ca(OH)_2 \xrightarrow{\text{excess OH}^-} CaCO_3\downarrow + Mg(OH)_2\downarrow \qquad (11.17)$$

A common example of destroying equilibrium by forming a poorly ionized compound is neutralization of acid or caustic wastes. Here, the combining of hydrogen ions and hydroxyl ions forms poorly ionized water and a soluble salt. For example,

$$2H^+ + SO_4^{2-} + 2Na^+ + 2OH^- \rightarrow 2H_2O + 2Na^+ + SO_4^{2-} \qquad (11.18)$$

Reactions involving production of a gaseous product go to practical completion if the gas escapes from solution. One illustration is breakpoint chlorination, which oxidizes ammonia to nitrogen and nitrous oxide gases.

Oxidation and reduction is a very positive method of sending reactions to completion, since one or more of the ions involved in the equilibrium is destroyed. A practical example in water treatment is removal of soluble iron from solution by oxidation using potassium permanganate. In this reaction (Eq. 11.19), the iron gains one positive charge while the manganese in the permanganate ion is reduced from a valence of $7+$ to a valence of $4+$, forming manganese dioxide.

$$Fe(HCO_3)_2 + KMnO_4 \rightarrow Fe(OH)_3\downarrow + MnO_2\downarrow \qquad (11.19)$$

## 11.6 CHEMICAL PROCESS KINETICS

Chemical reactions are classified on the basis of stoichiometry, which defines the number of moles of a substance entering into a reaction and the number of moles of products, and process kinetics that describe the rate of reaction. The types discussed in this section are irreversible reactions occurring in one phase where the reactants are distributed uniformly throughout the liquid. Heterogeneous processes involve the presence of a solid-phase catalyst, such as adsorption onto a granular medium. In irreversible reactions, the stoichiometric combination of reactants leads to nearly complete conversion to products. Most reactions in water and wastewater are sufficiently irreversible to allow this assumption for purposes of kinetic interpretation of experimental data. The type of reactor containing a reaction influences the degree of completion; therefore, process kinetics must also incorporate the hydraulic characteristics of reactors discussed in Sections 10.6–10.8.

## Reaction Rates

**Zero-Order Reactions**    These reactions proceed at a rate independent of the concentration of any reactant or product. Consider conversion of a single reactant to a single product, represented as follows:

A (reactant) → P (product)

If $C$ represents the concentration of A at any time $t$, the disappearance of A is expressed as

$$\frac{dC}{dt} = -k \tag{11.20}$$

where $\dfrac{dC}{dt}$ = rate of change in concentration of A with time

$k$ = reaction-rate constant

The negative sign in front of $k$ means that the concentration of A decreases with time. Integrating Eq. 11.20 and rearranging yields

$$C = C_0 - kt \tag{11.21}$$

where  $C$ = concentration of A at any time $t$
$C_0$ = initial concentration of A

and $C_0$ becomes the constant of integration if we let $C = C_0$ when $t = 0$.

A plot of a zero-order reaction is shown in Fig. 11.3a by the straight line with a constant slope equal to $-k$.

**First-Order Reactions**    These reactions proceed at a rate directly proportional to the concentration of one reactant. Again consider conversion of a single reactant to a single product:

A (reactant) → P (product)

If $C$ represents the concentration of A at any time $t$, the disappearance of A is expressed

$$\frac{dC}{dt} = -kC \tag{11.22}$$

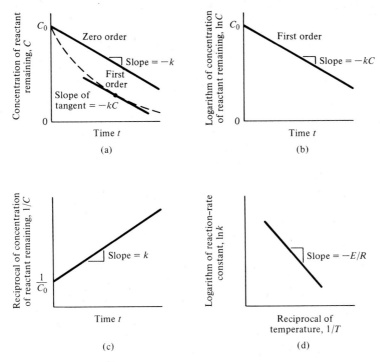

**Figure 11.3** Diagrams illustrating reaction rates for irreversible homogeneous reactions occurring in one phase. (a) Arithmetic graphs of a zero-order reaction as a solid line and first-order reaction as a dashed line. (b) Semilogarithmic plot of a first-order reaction. (c) Plot of a second-order process. (d) Diagram showing the effect of temperature on reaction rate plotting $\ln k$ versus $1/T$ based on the Arrhenius equations (Eqs. 11.26 and 11.27).

Integrating Eq. 11.22 and letting $C = C_0$ at $t = 0$ gives

$$\ln\frac{C_0}{C} = kt \quad \text{or} \quad \log\frac{C_0}{C} = \frac{kt}{2.30} \tag{11.23}$$

Plotting a first-order reaction on arithmetic paper results in a curve, as shown by the dashed line in Fig. 11.3a, with the slope of a tangent at any point equal to $-kC$. With passage of time from the start of the reaction, the rate decreases with the decreasing concentration of A remaining; this is reflected by the reduction in tangent slopes along the curve. First-order reactions can be linearized by graphing on semilogarithmic paper, as illustrated in Fig. 11.3b.

**Second-Order Reactions**  These reactions proceed at a rate proportional to the second power of a single reactant being converted to a single product,

2A (reactant) $\rightarrow$ P (product)

The rate of disappearance of A is described by the rate equation

$$\frac{dC}{dt} = -kC^2 \tag{11.24}$$

The integrated form for a second-order reaction is

$$\frac{1}{C} - \frac{1}{C_0} = kt \tag{11.25}$$

A plot of Eq. 11.25 is shown in Fig. 11.3c. On arithmetic paper, the reciprocal of the concentration of reactant remaining versus time plot is a straight line with a slope equal to $k$.

## Effect of Temperature on Reaction Rate

The rates of simple chemical reactions accelerate with increasing temperature of the liquid in a reactor, provided the higher temperature does not alter a reactant or catalyst. In 1889 Arrhenius examined the available data on the effect of temperature on the rates of chemical reactions and proposed the following equation:

$$\frac{d(\ln k)}{dT} = \frac{E}{RT^2} \tag{11.26}$$

where   $k$ = reaction-rate constant
   $T$ = absolute temperature, K
   $E$ = a constant characteristic of the reaction called activation energy, cal/mole
   $R$ = ideal gas constant (1.987 cal/K/mole)

Integration of this equation between $T_1$ and $T_2$, corresponding to reaction-rate constants $k_1$ and $k_2$, respectively, gives the relationship

$$\ln \frac{k_2}{k_1} = \frac{E(T_2 - T_1)}{RT_1T_2} \tag{11.27}$$

If experimental data are available, then $E/R$ can be determined from the slope of a plot of $\ln k$ versus $1/T$ (Fig. 11.3d). In turn, the activation energy $E$ for the reaction can be calculated by dividing the value of the slope by the ideal gas constant $R$.

The integrated form of Eq. 11.27 is

$$\frac{k_2}{k_1} = (e^{E/RT_1T_2})^{T_2-T_1} \tag{11.28}$$

For water and wastewater reactions near ambient temperature, the quantity $E/RT_1T_2$ located on the right side of Eq. 11.28 can be assumed constant for practical temperature ranges. Replacing the term in brackets with a temperature coefficient $\Theta$ yields

$$\frac{k_2}{k_1} = \Theta^{T_2 - T_1} \tag{11.29}$$

where $T$ = temperature, °C

which is commonly used to adjust the value of a rate constant for a temperature change. Equation 11.29 is applied to both chemical and biological processes, although in some cases linearity may be limited to a narrow temperature range. A common value for the temperature coefficient is 1.072, which doubles the reaction-rate constant with a 10°C temperature rise.

## Mass Balance Analysis

Reaction-rate equations are based on batch reactor analyses in which the reactants are added to a stirred container and the change in concentrations measured with respect to time. In continuous-flow reactors, the extent of a reaction depends on how the hydraulic characteristics of the tank affect reaction time. For steady-state conditions of uniform flow and reactant concentrations, mass balance analyses are used to calculate the changes that occur between the influent and effluent of a reactor. Mathematical expressions can be derived for process reactions to predict the degree of reaction completion in a particular system or to compute the mean residence time (mean hydraulic detention time) needed for a specific degree of reaction completion.

Mass balance analysis of ideal plug flow is the same as reaction analysis of a stirred batch reactor, since the liquid flows through an ideal tubular reactor without longitudinal mixing. The horizontal time scales in Fig. 11.3 represent time of passage, or travel distance along the reactor at a known velocity of flow. The plug-flow equations relating mean hydraulic detention time to reaction-rate constants for zero-, first-, and second-order reactions are given in Table 11.4. These were derived by rearranging and solving Eqs. 11.21, 11.23, and 11.25 for $t$.

TABLE 11.4 MEAN HYDRAULIC DETENTION TIMES (MEAN RESIDENCE TIMES) FOR REACTIONS OF DIFFERENT ORDER IN PLUG-FLOW AND COMPLETELY MIXED REACTORS

| | Equations for mean hydraulic detention times | |
|---|---|---|
| Reaction order | Ideal plug flow | Ideal completely mixed flow |
| 0 | $\frac{1}{k}(C_0 - C_t)$ | $\frac{1}{k}(C_0 - C_t)$ |
| 1 | $\frac{1}{k}\left[\ln\left(\frac{C_0}{C_t}\right)\right]$ | $\frac{1}{k}\left(\frac{C_0}{C_t} - 1\right)$ |
| 2 | $\frac{1}{kC_0}\left(\frac{C_0}{C_t} - 1\right)$ | $\frac{1}{kC_t}\left(\frac{C_0}{C_t} - 1\right)$ |

Analysis of an ideal completely mixed reactor assumes steady-state conditions, so that the reactants flow continuously into the tank and products are continuously discharged at time $t$ later. The concentrations of reactants and products are uniform throughout the mixing liquid, and the effluent has identical concentrations of reactants and products. Considering both the rate of reaction within the reactor and hydraulic characteristics of the system, the mass balance is

$$V\left(\frac{dC}{dt}\right) = QC_0 - QC_t - V(\text{rate of reaction}) \tag{11.30}$$

where the change of mass of reactant A within the reactor equals the mass input minus the mass output and minus the decrease in mass caused by reaction in the reactor.

For a first-order reaction rate (Eq. 11.22), the mass balance becomes

$$V\left(\frac{dC}{dt}\right) = QC_0 - QC_t - V(kC_t) \tag{11.31}$$

Under steady-state conditions, the rate of change in the mass of reactant within the tank is zero. Hence,

$$0 = QC_0 - QC_t - V(kC_t) \tag{11.32}$$

Rearranging Eq. 11.32 and solving for $V/Q$, which equals $t$, gives

$$\frac{V}{Q} = t = \frac{1}{k}\left(\frac{C_0}{C_t} - 1\right) \tag{11.33}$$

where    $V$ = volume of the reactor
$Q$ = rate of flow through the reactor
$t$ = mean hydraulic detention time (mean residence time)
$k$ = reaction-rate constant
$C_0$ = influent concentration of reactant A
$C_t$ = effluent concentration of reactant A

Thus, the reaction-rate constant and mean hydraulic detention time are expressed mathematically with the influent and effluent concentrations of reactant A for a first-order reaction.

The equations for zero- and second-order reactions can be derived in a similar manner. Table 11.4 lists the equations for mean hydraulic detention times of ideal completely mixed-flow reactors based on zero-, first-, and second-order reactions. Mathematical expressions for more complex chemical process kinetics are given by others [3, 4].

## 11.7 COLLOIDAL DISPERSIONS

*Colloidal dispersions* in water consist of discrete particles held in suspension by their extremely small size (1–200 nm), state of hydration (chemical combination with water), and surface electric charge. The size of particles is the most significant property responsible for the stability of a *sol* (a colloidal dispersion in a liquid). With larger particles the ratio of surface area to mass is low, and mass effects, such as sedimentation by gravity forces, predominate. For colloids, the ratio of surface area to mass is high, and surface phenomena, such as electrostatic repulsion and hydration, become important.

There are two types of colloids—hydrophilic and hydrophobic. *Hydrophilic* colloids are readily dispersed in water, and their stability (lack of tendency to agglomerate) depends on a marked affinity for water rather than on the slight charge (usually negative) that they possess. Examples of hydrophilic colloidal materials are soaps, soluble starch, soluble proteins, and synthetic detergents.

*Hydrophobic* colloids possess no affinity for water and owe their stability to the electric charge they possess. Metal oxide colloids, most of which are positively charged, are samples of hydrophobic sols. A charge on the colloid is gained by adsorbing positive ions from the water solution. Electrostatic repulsion between the charged colloidal particles produces a stable sol.

The concept of *zeta potential* is derived from the diffuse double-layer theory applied to hydrophobic colloids (Fig. 11.4). A fixed covering of positive ions is attracted to the negatively charged particle by electrostatic attraction. This stationary zone of positive ions is referred to as the *Stern layer,* which is surrounded by a movable, diffuse layer of counterions. The concentration of these positive ions in the diffuse zone decreases as it extends into the surrounding bulk of electroneutral solution. Zeta potential is the magnitude of the charge at the surface of shear. The boundary surface between the fixed ion layer and the solution serves as a shear plane when the particle undergoes movement relative to the solution. The zeta potential magnitude can be estimated from electrophoretic measurement of particle mobility in an electric field.

A colloidal suspension is defined as stable when the dispersion shows little or no tendency to aggregate (Fig. 11.5a). The repulsive force of the charged double layer disperses particles and prevents aggregation, thus particles with a high zeta potential produce a stable sol. Factors tending to destabilize a sol are van der Waals forces of attraction and Brownian movement. *Van der Waals forces* are the molecular cohesive forces of attraction that increase in intensity as particles approach each other. These forces are negligible when the particles are slightly separated but become dominant when particles contact. *Brownian movement* is the random motion of colloids caused by their bombardment by molecules of the dispersion medium. This movement has a destabilizing effect on a sol because aggregation may result.

*Destabilization* of hydrophobic colloids can be accomplished by adding electrolytes to the solution (Fig. 11.5b). Counterions of the electrolyte suppress the double-layer charge of the colloids sufficiently to permit particles to contact. As the particles meet, van der Waals forces of attraction become dominant and aggregation results. Electrolytes found to be most effective are multivalent ions of opposite charge to that of the colloidal particles. Sols can

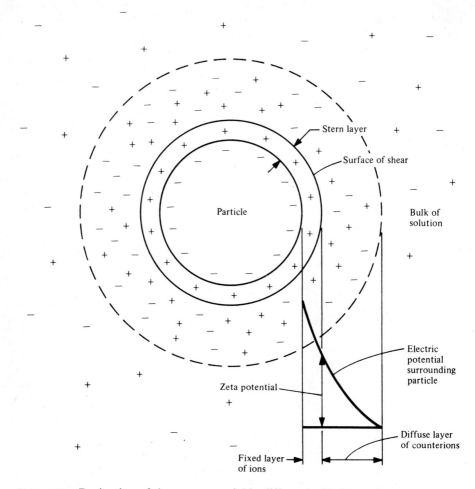

**Figure 11.4** Derivation of the zeta potential in diffuse double-layer theory.

also be destabilized by cationic polymers that act much the same as neutral salts in suppressing the diffuse double layer. The primary mechanism of polymers appears to be bridging. The long polymer molecule attaches to absorbent surfaces of colloidal particles by chemical or physical interactions, resulting in aggregation (Fig. 11.5c). The destabilizing action of hydrolyzed metal ions (i.e., aluminum and iron salts) appears to fall into an intermediate category between simple ions and cationic polymers. Highly charged, soluble hydrolysis products of these metal salts reduce the repulsive forces between colloids by compressing the double-layer charge, bringing on coagulation. Hydrolyzed metal ions are also adsorbed on the colloids, creating bridges between the particles.

In contrast to the electrostatic nature of hydrophobic colloids, the stability of hydrophilic colloids is related to their state of hydration (i.e., their

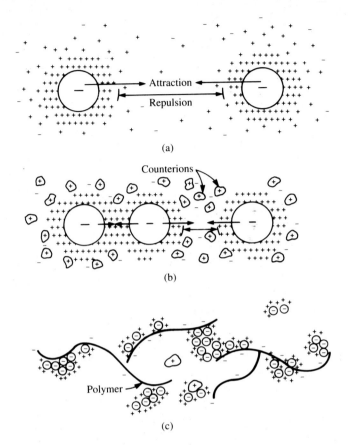

**Figure 11.5** Schematic representations of coagulation and bridging of colloids. (a) A stable suspension of particles where forces of repulsion exceed the forces of attraction. (b) Destabilization and coagulation caused by counterions of a coagulant suppressing the double-layer charges. (c) Agglomeration of destabilized particles by attaching of coagulant ions and bridging of polymers.

marked affinity for water). Chemical coagulation does not materially affect the degree of hydration of colloids. Therefore, hydrophilic colloids are extremely difficult to coagulate, and heavy doses of coagulant salts, often 10–20 times the amount used in conventional water treatment, are needed for destabilization.

## 11.8  COAGULATION PROCESS

Two basic mechanisms have been defined in destabilizing a colloidal suspension to produce floc for subsequent solids-separation processes: *Coagulation* reduces the net electrical repulsive forces at particle surfaces by adding coagulant chemicals, whereas *flocculation* is agglomeration of the destabilized particles by chemical joining and bridging.

The primary purpose of chemical treatment is to agglomerate particulate matter and colloids into floc that can be separated from the water by sedimentation and filtration. In water treatment, coagulation and flocculation are used to destabilize turbidity, color, odor-producing compounds, pathogens, and other contaminants in surface waters. In wastewater reclamation, coagulation precedes tertiary filtration necessary to clarify a biologically treated effluent for effective chemical disinfection.

Traditionally, environmental engineers have not restricted the use of the terms *coagulation* and *flocculation* to describing chemical mechanisms only. Common use of these terms refers to both chemical and physical processes in treatment, possible because the complex reactions that take place in chemical coagulation–flocculation are only partially understood. Even more important is that engineers tend to associate coagulation with the operational units (rapid mixing and flocculation) used in chemical treatment. Hence, the term *coagulation* in common usage refers to the series of chemical and mechanical operations by which coagulants are applied and made effective. These operations are customarily considered to comprise two distinct phases: (1) rapid mixing to disperse coagulant chemicals by violent agitation into the water being treated and (2) flocculation to agglomerate small particles into well-defined floc by gentle agitation for a much longer time [5]. (Refer to Sections 10.9 and 10.10 for a description of rapid-mixing and flocculation processes.)

The common unit operations and chemical additions in the treatment of surface waters for a potable supply are diagrammed in Fig. 11.6. Coagulant chemicals are added by rapid mixing, and auxiliary chemicals are usually blended into the destabilized water prior to or during flocculation. After filtration, the turbidity must be equal to or less than 0.5 NTU to ensure

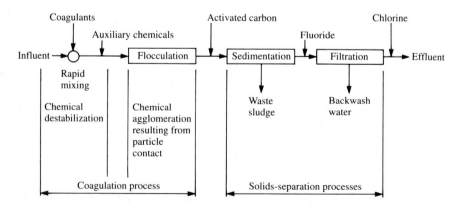

**Figure 11.6** Schematic diagram of the coagulation process as a part of the scheme for treating surface water for a potable supply.

satisfactory disinfection without creating undesirable by-products, such as trihalomethanes. Activated carbon may be applied to adsorb taste- and odor-producing compounds and fluoride to achieve an optimum concentration as a public health measure. If the formation of trihalomethanes is not a problem, prechlorination may be used for disinfection of the raw water and/or for control of tastes and odors. Chemical treatment varies with season of the year, and, for river waters, operational flexibility is required to handle the day-to-day variations.

The removal of contaminants by coagulation depends on their nature and concentration; use of both coagulants and coagulant aids; and other characteristics of the water, including pH, temperature, and ionic strength. Because of the complex nature of coagulation reactions, chemical treatment is based on empirical data derived from laboratory and field studies.

The *jar test* is widely used to simulate a full-scale coagulation–flocculation process to determine optimum chemical dosages. The apparatus consists of six agitator paddles coupled to operate at the same rotational speed, which can be varied from 10 to 100 rpm. The laboratory containers are 1- to 2-l beakers or square jars. The general procedure for conducting a jar test is as follows:

1. Fill six 1- or 2-liter beakers with a measured amount of the water to be treated.
2. Add the coagulant and/or other chemicals to each sample.
3. Flash-mix the samples by agitating at maximum speed (100 rpm) for 1 min.
4. Flocculate the samples at a stirring rate of 20–70 rpm for 10–30 min. Record the time of floc appearance for each beaker.
5. Stop the agitation and record the nature of the floc, clarity of supernatant, and settling characteristics of the floc.

Experiments can be conducted to evaluate the effectiveness of different coagulants and optimum dosage for destabilization, optimum pH, effectiveness of various coagulant aids and best dosage for floc formation, and the most effective sequence of chemical applications [6]. Ordinarily, a full-scale plant provides better results than the laboratory jar test for the same chemical dosages.

# 11.9 COAGULANTS

The most widely used coagulants for water and wastewater treatment are aluminum and iron salts. The common metal salt is aluminum sulfate, which is a good coagulant for water containing organic matter. Iron coagulants are effective over a wider pH range and are generally more effective in removing color from water, but they usually cost more. For some waters, cationic polymers are effective as a primary coagulant, but polymers are more commonly applied as coagulant aids. The final choice of coagulants and chemical aids is based on being able to achieve the contaminant removals and low turbidity desired in the filtered water at least cost.

The following paragraphs discuss various coagulants and present the theoretical chemical reactions. These traditional coagulation reactions yield only approximate results. The molecular and equivalent weights of chemical compounds listed in Tables 11.1 and 11.2 are useful in solving numerical problems.

## Aluminum Sulfate (Filter Alum)

Aluminum sulfate is the standard coagulant used in water treatment. The commercial product strength ranges from 15% to 22% as $Al_2O_3$ with a hydration of about 14 moles of water. A formula used for filter alum is $Al_2(SO_4)_3 \cdot 14.3H_2O$ with a molecular weight of 600. The material is commonly shipped and fed in a dry granular form, although it is available as a powder or liquid alum syrup.

Aluminum sulfate reacts with natural alkalinity in water to form aluminum hydroxide floc.

$$Al_2(SO_4)_3 \cdot 14.3H_2O + 3Ca(HCO_3)_2 \qquad\qquad (11.34)$$
$$\rightarrow 2Al(OH)_3 \downarrow + 3CaSO_4 + 14.3H_2O + 6CO_2$$

Each mg/l of alum decreases water alkalinity by 0.50 mg/l (as $CaCO_3$) and produces 0.44 mg/l of carbon dioxide. Production of carbon dioxide is undesirable since this increases the corrosiveness of water.

If water does not contain sufficient alkalinity to react with the alum, lime or soda ash is fed to provide the necessary alkalinity.

$$Al_2(SO_4)_3 \cdot 14.3H_2O + 3Ca(OH)_2$$
$$\rightarrow 2Al(OH)_3 \downarrow + 3CaSO_4 + 14.3H_2O \qquad\qquad (11.35)$$

$$Al_2(SO_4)_3 \cdot 14.3H_2O + 3Na_2CO_3 + 3H_2O$$
$$\rightarrow 2Al(OH)_3 \downarrow + 3Na_2SO_4 + 3CO_2 + 14.3H_2O \qquad\qquad (11.36)$$

An advantage of using sodium carbonate (soda ash) is that unlike lime it does not increase water hardness, only corrosiveness. Lime, which is more popular, is less expensive than soda ash.

The dosage of alum used in water treatment is in the range 5–50 mg/l. The effective pH range for alum coagulation is 5.5–8.0. Alum is preferred in treating relatively high-quality surface waters because it is the only chemical needed for coagulation.

## Ferrous Sulfate (Copperas)

Commercial ferrous sulfate has a strength of 55% $FeSO_4$ and is supplied as green crystal or granule for dry feeding. Ferrous sulfate reacts with natural alkalinity (Eq. 11.37), but the response is much slower than that between alum and natural alkalinity. Lime is generally added to raise the pH to the point where ferrous ions are precipitated as ferric hydroxide by the caustic alkalinity (Eq. 11.38).

$$2FeSO_4 \cdot 7H_2O + 2Ca(HCO_3)_2 + 0.5O_2$$
$$\rightarrow 2Fe(OH)_3 \downarrow + 2CaSO_4 + 4CO_2 + 13H_2O \tag{11.37}$$

$$2FeSO_4 \cdot 7H_2O + 2Ca(OH)_2 + 0.5O_2 \tag{11.38}$$
$$\rightarrow 2Fe(OH)_3 \downarrow + 2CaSO_4 + 13H_2O$$

Treatment using ferrous sulfate and lime adds some hardness but no corrosiveness to the water. This process is usually cheaper than alum coagulation, but the dosing operation with two chemicals is more difficult. If excess lime is used, the water may require treatment for stabilization.

Chlorinated copperas treatment is a second method of using ferrous sulfate. In this process chlorine is used to oxidize the ferrous sulfate to ferric sulfate.

$$3FeSO_4 \cdot 7H_2O + 1.5\,Cl_2 \rightarrow Fe_2(SO_4)_3 + FeCl_3 + 21H_2O \tag{11.39}$$

Oxidation is generally performed by adding ferrous sulfate to the discharge from a solution feed chlorinator. Theoretically, 1.0 lb of chlorine oxidizes 7.8 lb of copperas. In practice, a chlorine feed slightly in excess of the theoretical amount produces good results.

Ferric sulfate and the ferric chloride react with natural alkalinity or lime, as illustrated by the following reactions with ferric chloride:

$$2FeCl_3 + 3Ca(HCO_3)_2 \rightarrow 2Fe(OH)_3 \downarrow + 3CaCl_2 + 6CO_2 \tag{11.40}$$

$$2FeCl_3 + 3Ca(OH)_2 \rightarrow 2Fe(OH)_3 \downarrow + 3CaCl_2 \tag{11.41}$$

Color in water is generally not affected by copperas and lime treatment, whereas chlorinated copperas is effective in the removal of color.

## Ferric Salts

Ferric sulfate and ferric chloride are available as coagulants under a variety of trade names. The reactions of these salts with natural alkalinity and with lime are noted in Eqs. 11.40 and 11.41. Advantages of the ferric coagulants are that (1) coagulation is possible over a wider pH range, generally pH 4–9 for most waters; (2) the precipitate produced is a heavy quick-settling floc; and (3) they are more effective in the removal of color, taste, and odor compounds.

Ferric sulfate is available in crystalline form and may be fed using dry or liquid feeders. Ferric sulfate, although not as aggressive as ferric chloride or chlorinated copperas, must be handled by corrosive-resistant equipment.

Ferric chloride is supplied in either crystalline or liquid form. Although ferric chloride can be used in water treatment, its most frequent application is in wastewater treatment (e.g., as a waste-sludge-conditioning chemical in combination with lime prior to mechanical dewatering).

## Sodium Aluminate

Sodium aluminate is essentially alumina dissolved in sodium hydroxide. The principal use of sodium aluminate is as an additional coagulant with aluminum sulfate, generally in the treatment of boiler water. The limited employment of sodium aluminate is dictated by its high cost. Sodium aluminate is alkaline in its reactions, instead of acidic like other coagulants. Reactions of sodium aluminate with aluminum sulfate and carbon dioxide are

$$6NaAlO_2 + Al_2(SO_4)_3 \cdot 14.3H_2O$$
$$\rightarrow 8Al(OH)_3 \downarrow + 3Na_2SO_4 + 2.3H_2O \qquad (11.42)$$

$$2NaAlO_2 + CO_2 + 3H_2O \rightarrow 2Al(OH)_3 \downarrow + Na_2CO_3 \qquad (11.43)$$

## Polymers

Synthetic polymers are water-soluble high-molecular-weight organic compounds that have multiple electrical charges along a molecular chain of carbon atoms. If the ionizable groups have a positive charge, the compound is referred to as a cationic polymer. If ionizable groups have a negative charge, it is an anionic polymer. If no charges are exhibited, it is a nonionic polymer.

Cationic polymers can be effective for coagulation, without hydrolyzing metals, by producing destabilization through charge neutralization and interparticle bridging. In contrast to metal coagulants, rapid mixing for low-molecular-weight cationic polymers is performed at reduced turbulence relative to metal coagulants to prevent breakup of the more fragile polymer floc. Feed control of polymer coagulants must be precise since effective dosage has a narrow band; overdosing or underdosing results in restabilization of the colloidal particles. Because polymers do not affect pH, they are advantageous for coagulating low-alkalinity waters. Another potential advantage of polymers relative to metal salts is reduced sludge production. Dosages of cationic polymers for coagulation are commonly 0.5–1.5 mg/l, which is much less than the dosages of metal coagulants.

## 11.10 COAGULANT AIDS

## Polymers

Anionic and nonionic polymers are effective coagulant aids. After destabilizing the colloidal suspension by hydrolyzing metals such as alum, polymers promote larger and tougher floc by a bridging mechanism (Fig. 11.5). Generally, nonionic polymers are more effective in water containing higher concentrations of divalent cations, i.e., $Ca^{2+}$ and $Mg^{2+}$. Common dosages are 0.1–0.5 mg/l, to aid alum coagulation. The combination of polymer and alum can reduce the alum dosage that would be required without the polymer. Since a wide selection of proprietary polymers is available, the brand name of the selected polymer should be determined and verified by laboratory and full-scale testing to ensure it is the best aid at least cost. The effectiveness of a polymer also depends on the point of application. The mixing must be adequate for complete blending but not so turbulent as to hinder particle bridging.

## Activated Silica

Activated silica is sodium silicate that has been activated with sulfuric acid, aluminum sulfate, carbon dioxide, or chlorine. One method of preparing activated silica is to dilute a sodium silicate solution to a level of 1.5% $SiO_2$ and then add enough sulfuric acid to neutralize 85% of the alkalinity. This solution is aged 2 hr before use. Another procedure is to add 1 part of a 1% silicate solution to 4 parts of a 1% alum solution applying the mixture immediately. These and other methods of preparation require proper equipment and close operational control to produce activated silica successfully.

When activated silica is placed in water, it produces a stable sol having a negative charge. As an aid to coagulation, this colloidal dispersion of activated silica has several advantages. The activated silica unites with positively charged metal hydroxides of the coagulant, resulting in a larger, tougher, denser floc and causes more rapid, improved settling, which in turn reduces turbidity carryover to the filter. Activated silica may lower the coagulant dosage required for good coagulation and thus reduce overall chemical costs.

## Clays

Clay improves coagulation by adding weight to the floc, thus increasing the settling rate, and exerts adsorptive action that aids in the floc formation. Bentonite and other clays are available in different degrees of fineness as coagulant aids.

## Acids and Alkalies

Acids and alkalies are added to water to adjust the pH for optimum coagulation. Typical acids used to lower the pH are sulfuric and phosphoric.

Alkalies used to raise the pH are lime, sodium hydroxide, and soda ash. Hydrated lime, with about 70% available CaO, is suitable for dry feeding but costs more than quicklime, which is 90% CaO. The latter must be slaked (combined with water) and fed as a lime slurry. Soda ash is 98% sodium carbonate and can be fed dry but is more expensive than lime. Sodium hydroxide is purchased and fed as a concentrated solution.

### ■ EXAMPLE 11.2

A surface water is coagulated with a dosage of 30 mg/l of ferrous sulfate and an equivalent dosage of lime. (a) How many pounds of ferrous sulfate are needed per mil gal of water treated? (b) How many pounds of hydrated lime are required assuming a purity of 70% CaO? (c) How many pounds of Fe(OH)₃ sludge are produced per mil gal of water treated?

*Solution.* Converting the dosage of ferrous sulfate from mg/l to lb/mil gal, one obtains

$$30 \times 8.34 = 250 \text{ lb/mil gal}$$

By Eq. 11.38 and molecular weights from Tables 11.1 and 11.2,

$$\underset{\text{FeSO}_4 \cdot 7\text{H}_2\text{O}}{2 \times 278 \text{ lb}} + \underset{\text{Ca(OH)}_2}{2 \times 74 \text{ lb}} \rightarrow \underset{\text{Fe(OH)}_3}{2 \times 107 \text{ lb}}$$

Therefore,

$$\frac{556}{250 \text{ lb}} = \frac{148}{Y \text{ lb Ca(OH)}_2} = \frac{214}{Z \text{ lb Fe(OH)}_3}$$

Solving for the lime dosage yields

$$Y \text{ lb Ca(OH)}_2 = \frac{250 \times 148}{556} = 66.5 \text{ lb/mil gal (8.0 mg/l)}$$

$$\text{lb 70\% CaO} = 66.5 \times \frac{56}{74} \times \frac{1}{0.70} = 72 \text{ lb/mil gal (8.6 mg/l)}$$

The $Fe(OH)_3$ sludge production is

$$Z \text{ lb Fe(OH)}_3 = \frac{250 \times 214}{556} = 96 \text{ lb/mil gal (12 mg/l)}$$

*Alternative solution for lime dosage:*
    One equivalent weight of ferrous sulfate (139) reacts with 1 equivalent weight of 70% CaO (28 ÷ 0.70 = 40). Therefore,

$$\text{lime dosage} = 30 \times \frac{40}{139} \times 8.34 = 72 \text{ lb/mil gal} \qquad \blacksquare$$

### ■ EXAMPLE 11.3

Dosage of alum with the alum–lime coagulation of a water is 50 mg/l. It is desired to react only 10 mg/l (as $CaCO_3$) of the natural alkalinity with the alum. Based on the theoretical Eqs. 11.34 and 11.35, what dosage of lime is required, in addition to 10 mg/l of natural alkalinity, to react with the alum dosage?

*Solution.* Using equivalent weights, the alum that reacts with 10 mg/l of natural alkalinity is 10 × (100/50) = 20 mg/l. The amount of alum remaining to react with lime is 50 − 20 = 30 mg/l. The lime dosage required to react with 30 mg/l of alum is 30 × (28/100) = 8.4 mg/l as CaO. ■

# Water Softening

Hardness in water is caused by the ions of calcium and magnesium. Although ions of iron, manganese, strontium, and aluminum also produce hardness, they are not present in significant quantities in natural waters.
    A singular criterion for maximum hardness in public water supplies is

not possible. Water hardness is largely the result of geological formations of the water source. Public acceptance of hardness varies from community to community, consumer sensitivity being related to the degree to which the consumer is accustomed.

Hardness of more than 300–500 mg/l as $CaCO_3$ is considered excessive for a public water supply and results in high soap consumption as well as objectional scale in heating vessels and pipes. Many consumers object to water harder than 150 mg/l, a moderate figure being 60–120 mg/l.

## 11.11 CHEMISTRY OF LIME–SODA ASH PROCESS

The lime–soda water-softening process uses lime, $Ca(OH)_2$, and soda ash, $Na_2CO_3$, to precipitate hardness from solution. Carbon dioxide and carbonate hardness (calcium and magnesium bicarbonate) are complexed by lime. Noncarbonate hardness (calcium and magnesium sulfates or chlorides) requires the addition of soda ash for precipitation.

Chemical reactions in the lime–soda process are

$$CO_2 + Ca(OH)_2 = CaCO_3\downarrow + H_2O \tag{11.44}$$

$$Ca(HCO_3)_2 + Ca(OH)_2 = 2CaCO_3\downarrow + 2H_2O \tag{11.45}$$

$$Mg(HCO_3)_2 + Ca(OH)_2 = CaCO_3\downarrow + MgCO_3 + 2H_2O \tag{11.46}$$

$$MgCO_3 + Ca(OH)_2 = Mg(OH)_2\downarrow + CaCO_3\downarrow \tag{11.47}$$

$$MgSO_4 + Ca(OH)_2 = Mg(OH)_2\downarrow + CaSO_4 \tag{11.48}$$

$$CaSO_4 + Na_2CO_3 = CaCO_3\downarrow + Na_2SO_4 \tag{11.49}$$

These equations give all the reactions taking place in softening water containing both carbonate and noncarbonate hardness, by additions of both lime and soda ash. The carbon dioxide in Eq. 11.44 is not hardness as such, but it consumes lime and must therefore be considered in calculating the amount required. Equations 11.45–11.47 demonstrate removal of carbonate hardness by lime. Note that only 1 mole of lime is needed for each mole of calcium alkalinity, whereas 2 moles are required for each mole of magnesium alkalinity (Eqs. 11.46 and 11.47). Equation 11.48 shows the removal of magnesium noncarbonate hardness by lime. No softening results from this reaction because 1 mole of calcium noncarbonate hardness is formed for each mole of magnesium salt present. Equation 11.49 is for removal of calcium sulfate originally present in the water and also that formed as stated in Eq. 11.48.

Precipitation softening cannot produce a water completely free of hardness because of the solubility of calcium carbonate and magnesium hydroxide. Furthermore, completion of the chemical reactions is limited by physical considerations, such as adequate mixing and limited detention time in settling basins. Therefore, the minimum practical limits of precipitation softening are 30 mg/l of $CaCO_3$ and 10 mg/l of $Mg(OH)_2$ expressed as $CaCO_3$. Hardness levels of 80–100 mg/l are generally considered acceptable for a public water supply, but the magnesium content should not exceed 40 mg/l as $CaCO_3$ in a softened municipal water.

There are several advantages of lime softening in water treatment. The most obvious is that the total dissolved solids may be significantly reduced, hardness is taken out of solution, and the lime added is also removed. When soda ash is applied, sodium ions remain in the finished water; however, noncarbonate hardness requiring the addition of soda ash is generally a small portion of the total hardness. Lime also precipitates the soluble iron and manganese often found in groundwaters. In processing surface waters, excess lime treatment provides disinfection and aids in coagulation for removal of turbidity.

## 11.12 PROCESS VARIATIONS IN LIME–SODA ASH SOFTENING

Three different basic schemes are used to provide a finished water with the desired hardness: excess lime treatment, selective calcium removal, and split treatment.

### Excess Lime Treatment

Carbonate hardness associated with the calcium ion can be effectively removed to the practical limit of $CaCO_3$ solubility by stoichiometric additions of lime (Eq. 11.45). Precipitation of the magnesium ion (Eqs. 11.47 and 11.48) calls for a surplus of approximately 35 mg/l of CaO (1.25 meq/l) above stoichiometric requirements. The practice of excess-lime treatment reduces the total hardness to about 40 mg/l (i.e., 30 mg/l of $CaCO_3$ and 10 mg/l of magnesium hardness).

After excess-lime treatment, the water is scale forming and must be neutralized to remove caustic alkalinity. Recarbonation and soda ash are regularly used to stabilize the water. Carbon dioxide neutralizes excess lime as follows:

$$Ca(OH)_2 + CO_2 = CaCO_3\downarrow + H_2O \tag{11.50}$$

This reaction precipitates calcium hardness and reduces the pH from near 11 to about 10.2. Further recarbonation of the clarified water converts a portion of the remaining carbonate ions to bicarbonate by the reaction

$$CaCO_3 + CO_2 + H_2O = Ca(HCO_3)_2 \tag{11.51}$$

The final pH is in the range 9.5–8.5, depending on the desired carbonate-to-bicarbonate ratio (Fig. 11.2).

The equipment required for generating carbon dioxide gas consists of a furnace to burn the fuel (coke, coal, gas, or oil), a scrubber to remove soot and other impurities from the gas, and a compressor for forcing the gas into the water. Commercial liquid carbon dioxide is also used if the consumption is sufficiently low to be cost effective. Receiving and storing bulk quantities of commercial gas and the ease of feeding reduce equipment and operating costs compared to on-site gas generation for smaller facilities.

A two-stage system is preferred for excess-lime treatment (Fig. 11.7). Lime is applied in first-stage mixing and sedimentation to precipitate both calcium and magnesium. Then carbon dioxide is applied to neutralize the excess lime (Eq. 11.50), and soda ash is added to reduce noncarbonate hardness. Solids formed in these reactions are removed by secondary settling and subsequent filtration. Recarbonation immediately ahead of the filters may be used to prevent scaling of the media (Eq. 11.51).

■ **EXAMPLE 11.4**
Water defined by the following analysis is to be softened by excess lime treatment in a two-stage system (Fig. 11.7):

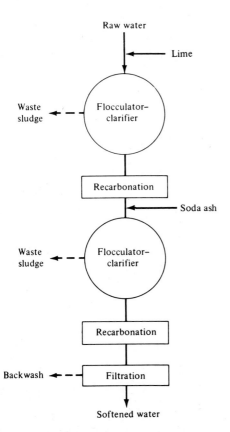

**Figure 11.7** Schematic flow diagram for a two-stage excess-lime softening plant.

$$CO_2 = 8.8 \text{ mg/l as } CO_2 \qquad Alk(HCO_3^-) = 115 \text{ mg/l as } CaCO_3$$
$$Ca^{2+} = 70 \text{ mg/l} \qquad\qquad SO_4^{2-} = 96 \text{ mg/l}$$
$$Mg^{2+} = 9.7 \text{ mg/l} \qquad\qquad Cl^- = 10.6 \text{ mg/l}$$
$$Na^+ = 6.9 \text{ mg/l}$$

The practical limits of removal can be assumed to be 30 mg/l of $CaCO_3$ and 10 mg/l of $Mg(OH)_2$, expressed as $CaCO_3$. Sketch a meq/l bar graph and list the hypothetical combinations of chemical compounds in the raw water. Calculate the quantity of softening chemicals required in pounds per million gallons of water treated and the theoretical quantity of carbon dioxide needed to provide a finished water with one-half of the alkalinity converted to bicarbonate ion. Draw a bar graph for the softened water after recarbonation and filtration.

*Solution.*

| Components | mg/l | Equivalent weight | meq/l |
|---|---|---|---|
| $CO_2$ | 8.8 | 22.0 | 0.40 |
| $Ca^{2+}$ | 70 | 20.0 | 3.50 |
| $Mg^{2+}$ | 9.7 | 12.2 | 0.80 |
| $Na^+$ | 6.9 | 23.0 | 0.30 |
| Alk | 115 | 50.0 | 2.30 |
| $SO_4^{2-}$ | 96 | 48.0 | 2.00 |
| $Cl^-$ | 10.6 | 35.5 | 0.30 |

The meq/l bar graph of the raw water is shown in Fig. 11.8a, and the hypothetical combinations are listed.

| Component | meq/l | Lime | Soda ash |
|---|---|---|---|
| $CO_2$ | 0.4 | 0.4 | 0 |
| $Ca(HCO_3)_2$ | 2.3 | 2.3 | 0 |
| $CaSO_4$ | 1.2 | 0 | 1.2 |
| $MgSO_4$ | 0.8 | 0.8 | 0.8 |
| | | 3.5 | 2.0 |

lime required = stoichiometric quantity + excess lime

$$= 3.5 \times 28 + 35 = 133 \text{ mg/l of CaO} = 1100 \text{ lb/mil gal}$$

soda ash required $= 2.0 \times 53 = 106 \text{ mg/l of Na}_2CO_3 = 900 \text{ lb/mil gal}$

A hypothetical bar graph for the water after addition of softening chemicals and first-stage sedimentation is shown in Fig. 11.8b. The dashed box is the excess-lime addition, 35 mg/l of CaO = 1.25 meq/l. The 0.6 meq/l of $Ca^{2+}$ (30 mg/l as $CaCO_3$) and 0.20 meq/l of $Mg^{2+}$ (10 mg/l as $CaCO_3$) are the practical limits of hardness reduction. The 2.0 meq/l of $Na_2SO_4$ results from the addition of soda ash. Alkalinity consists of

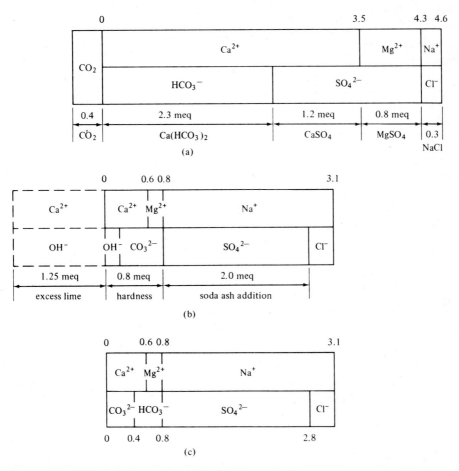

**Figure 11.8** Milliequivalent bar graphs for Example 11.4. (a) Bar graph and hypothetical chemical combinations in the raw water. (b) Bar graph of the water after lime and soda ash additions and settling but before recarbonation. (c) Bar graph of the water after two-stage recarbonation and final filtration.

0.20 meq/l of $OH^-$ associated with $Mg(OH)_2$ and 0.60 meq/l of $CO_3^{2-}$ related to $CaCO_3$.

Recarbonation converts the excess hydroxyl ion to carbonate ion; using the relationship in Eq. 11.50 and 22.0 as the equivalent weight of carbon dioxide is $(1.25 + 0.2)22.0 = 31.9$ mg/l of $CO_2$. After second-stage processing, final recarbonation converts one-half of the remaining alkalinity to bicarbonate ion by Eq. 11.51, giving $0.5 \times 0.8 \times 22.0 = 8.8$ mg/l of $CO_2$. Therefore the total carbon dioxide reacted is $(31.9 + 8.8)8.34 = 340$ lb/mil gal of $CO_2$.

The bar graph of the finished water is shown in Fig. 11.8c.    ■

## Selective Calcium Removal

Waters with a magnesium hardness of less than 40 mg/l as $CaCO_3$ can be softened by removing only a portion of the calcium hardness. The processing scheme can be a single-stage system of mixing, sedimentation, recarbonation, and filtration. Enough lime is added to the raw water to precipitate calcium hardness without providing any excess for magnesium removal. Soda ash may be required, depending on the amount of noncarbonate hardness. Recarbonation is usually practiced to reduce scaling of the filter media and produce a stable effluent.

■ **EXAMPLE 11.5**

Determine the chemical dosages needed for selective calcium softening of the water described in Example 11.4. Draw a bar graph of the processed water.

*Solution.* The hypothetical combinations of concern in calcium precipitation are $CO_2$, $Ca(HCO_3)_2$, and $CaSO_4$ (Fig. 11.8a).

| Component | meq/l | Lime | Soda ash |
|-----------|-------|------|----------|
| $CO_2$ | 0.4 | 0.4 | 0 |
| $Ca(HCO_3)_2$ | 2.3 | 2.3 | 0 |
| $CaSO_4$ | 1.2 | 0 | 1.2 |
| | | 2.7 | 1.2 |

lime required $= 2.7 \times 28 = 76$ mg/l of CaO $= 630$ lb/mil gal

soda ash required $= 1.2 \times 53 = 64$ mg/l of $Na_2CO_3 = 530$ lb/mil gal

Final hardness is all the $Mg^{2+}$ in the raw water plus the practical limit of $CaCO_3$ removal (30 mg/l or 0.60 meq/l), $(0.8 + 0.6)50 = 70$ mg/l as $CaCO_3$. The bar graph of the softened water is shown in Fig. 11.9. Recarbonation would be desirable to stabilize the water by converting a portion of the carbonate alkalinity to bicarbonate. ■

## Split Treatment

Split treatment consists of treating by excess lime a portion of the raw water and then neutralizing the excess lime in the treated flow with the remaining portion of raw water. When split treatment is used, any desired hardness

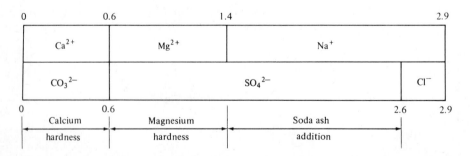

**Figure 11.9** Bar graph of the softened water after selective calcium removal.

level above 40 mg/l is attainable. Since hardness levels of 80–100 mg/l are generally considered acceptable, split treatment can result in considerable chemical savings. Recarbonation after the first stage is not customarily required; however, it may be needed after the second stage before filtration. Split treatment is particularly advantageous on well waters. In softening surface waters, where taste, odor, and color may be problems, two stages of processing for the total flow are usually preferred over split treatment.

The flow pattern of a typical two-stage split-treatment plant, shown in Fig. 11.10, and the following description are from Cleasby and Dillingham [7]. If $X$ is the ratio of the bypassed flow to the total quantity $Q$, then bypassed flow is $XQ$, and that through the first stage is $Q - XQ$ or $(1 - X)Q$. The magnesium content leaving the first stage (designated $Mg_1$) will be less than 10 mg/l (as $CaCO_3$). Magnesium in the bypass will be the same as that in the raw water (designated as $Mg_r$). Finished water total hardness is dictated by what is considered acceptable to the consumer, about 80–100 mg/l. Calcium in the finished water should not exceed 30–40 mg/l (as $CaCO_3$) for good operation. Therefore, permissible magnesium in finished water (designated $Mg_f$) is about 50 mg/l (as $CaCO_3$). Some cities produce a water of 40 mg/l of Mg to reduce problems with hard magnesium silicate in high-temperature

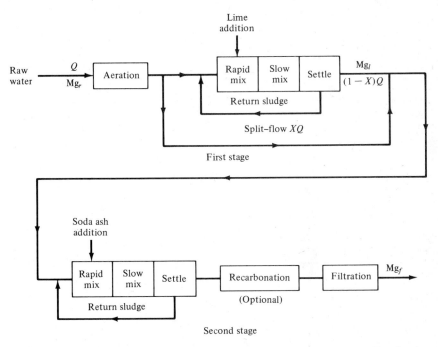

**Figure 11.10** Flow diagram for a typical split-treatment lime–soda ash softening plant. [From J. L. Cleasby and J. H. Dillingham, "Rational Aspects of Split Treatment," *Proc. Am. Soc. Civil Engrs., J. San. Eng. Div.* 92 (SA2) (1966): 1–7.]

(180°F) services. The bypass flow fraction can be calculated for any desired level of magnesium by using the formula

$$X = \frac{Mg_f - Mg_1}{Mg_r - Mg_1} \qquad\qquad (11.52)$$

## Coagulation and Softening

Addition of a coagulant or a coagulant aid may result in more efficient removal of the hardness precipitates formed in lime–soda softening. Alum is the prevalent coagulant. Activated silica has been effective when applied in softening waters high in calcium.

Lime softening is often used to treat surface waters. Improved coagulation and turbidity removal usually result from lime–soda softening, as compared to simple coagulation, because of the greater quantity of precipitate formed in chemical treatment processes.

■ **EXAMPLE 11.6**

Consider the split-treatment softening of water described by the following analysis. Criteria for the finished water are a maximum permissible magnesium hardness of 40 mg/l as $CaCO_3$ and calcium hardness in the range 40–60 mg/l.

$CO_2$ = 0.5 meq/l

$Ca^{2+}$ = 3.5 meq/l $\quad$ $HCO_3^-$ = 3.2 meq/l

$Mg^{2+}$ = 1.8 meq/l $\quad$ $SO_4^{2-}$ = 2.2 meq/l

$Na^+$ = 0.5 meq/l $\quad$ $Cl^-$ = 0.4 meq/l

*Solution.* Solving for the fraction of bypassed flow using Eq. 11.52 gives

$$X = \frac{40 - 10}{(1.8)(50) - 10} = \frac{30}{80} = 0.375$$

The first-stage flow is therefore

$$1 - X = 1 - 0.375 = 0.625$$

The lime and soda ash required for treatment are as follows:

| Component | meq/l | Lime | | Soda ash |
|---|---|---|---|---|
| | | First-stage flow | Bypassed flow | |
| $CO_2$ | 0.5 | 0.625 × 0.5 = 0.313 | 0.375 × 0.5 = 0.188 | 0 |
| $Ca(HCO_3)_2$ | 3.2 | 0.625 × 3.2 = 2.000 | 0.375 × 3.2 = 1.200 | 0 |
| $CaSO_4$ | 0.3 | 0 | 0 | 0.3 |
| $MgSO_4$ | 1.8 | 0.625 × 1.0 = 0.625 | 0 | 1.0 |
| | | 2.938 | 1.388 | 1.300 |

The lime dose for $MgSO_4$ is for only 1.0 meq/l of the 1.8 meq/l since 0.8 meq/l of Mg (40 mg/l of magnesium hardness) is the allowable concentration in the finished water.

The lime addition to the first stage is

$$2.938 + 1.388 = 4.326 \text{ meq/l} = 121 \text{ mg/l of CaO}$$

The soda ash addition to the second stage is

$$1.300 \text{ meq/l} = 69 \text{ mg/l of Na}_2\text{CO}_3$$

The hypothetical bar graphs of the water at various stages of treatment are shown in Fig. 11.11. The first bar graph showing the hypothetical combinations is for the raw water. All of the lime is applied to this water in the first stage. Since only $0.625Q$ of the influent is treated, the lime addition relative to the full bar graph is the actual amount of 4.326 meq/l divided by 0.625 to equal 6.922 meq/l. Reacting with this applied lime are 0.5 meq/l of $CO_2$ (Eq. 11.44), 3.2 of $Ca(HCO_3)_2$ (Eq. 11.45), and 1.8 of $MgSO_4$ (Eq. 11.48). The excess lime remaining after these reactions equals $6.922 - (0.50 + 3.20 + 1.80) = 1.422$ meq/l, which is greater than the required minimum of 1.25 meq/l for $Mg(OH)_2$ precipitation. The 0.3 meq/l of $CaSO_4$ do not react, and the 1.8 meq/l of $MgSO_4$ are converted to $CaSO_4$ (Eq. 11.18). The practical limits of removal are assumed to be 0.60 meq/l of $CaCO_3$ and 0.20 meq/l of $Mg(OH)_2$. Figure 11.11b is a bar graph for the settled effluent from the first stage.

The next step is the reaction between the first-stage effluent and the untreated bypassed flow. Figure 11.11c is a bar graph of the combined flow without considering any reactions. It was drawn by adding Fig. 11.11a times 0.375 to Fig. 11.11b times 0.675. The 1.014 meq/l of $Ca(OH)_2$ reacts with the $CO_2$ and $Ca(HCO_3)_2$ to form $CaCO_3$, which precipitates, leaving 0.60 meq/l of $CaCO_3$ in solution. Figure 11.11d is a hypothetical bar graph after reaction of the excess lime in the first-stage effluent with the $CO_2$ and $Ca(HCO_3)_2$ in the bypassed flow.

In the second stage, 1.30 meq/l of soda ash ($Na_2CO_3$) are added and react with 1.30 meq/l of $CaSO_4$ to precipitate 1.30 meq/l of $CaCO_3$ in accordance with Eq. 11.49. The net effect in the bar graph is the replacement of 1.30 meq/l of Ca with Na. The finished water after sedimentation and filtration is graphed in Fig. 11.11e.

The finished water has the following calculated alkalinity and hardness values:

$$\text{alkalinity} = 0.973 \times 50 = 49 \text{ mg/l}$$

$$\text{calcium hardness} = 0.973 \times 50 = 49 \text{ mg/l}$$

$$\text{magnesium hardness} = 0.800 \times 50 = 40 \text{ mg/l}$$

$$\text{total hardness} = 49 + 40 = 89 \text{ mg/l} \qquad \blacksquare$$

## 11.13  CATION EXCHANGE SOFTENING

The hardness-producing elements of calcium and magnesium are removed and replaced with sodium by a cation resin. Ion exchange reactions for softening may be written

$$Na_2R + \left.\begin{matrix} Ca \\ Mg \end{matrix}\right\} \left\{\begin{matrix} (HCO_3)_2 \\ SO_4 \\ Cl_2 \end{matrix}\right. \rightarrow \left.\begin{matrix} Ca \\ Mg \end{matrix}\right\} R + \left\{\begin{matrix} 2NaHCO_3 \\ Na_2SO_4 \\ 2NaCl \end{matrix}\right. \qquad (11.53)$$

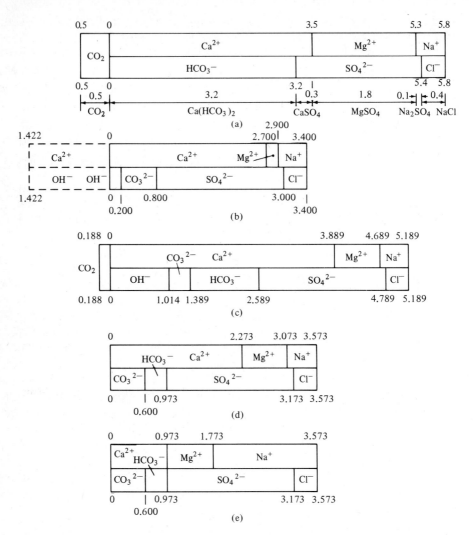

**Figure 11.11** Milliequivalent bar graphs for Example 11.6. (a) Bar graph and hypo-thetical chemical combinations in the raw water. (b) Bar graph of the water after first-stage treatment with lime. (c) Unreacted combination of the first-stage effluent times 0.625 plus the bypassed flow times 0.375. (d) Hypothetical graph after reaction of the excess lime in the combined flows. (e) Bar graph of finished water after second-stage treatment with soda ash, sedimentation, and final filtration.

where R represents the exchange resin. They show that if a water containing calcium and magnesium is passed through an ion exchanger, these metals are taken up by the resin, which simultaneously gives up sodium in exchange.

After the ability of the bed to produce soft water has been exhausted, the unit is removed from service and backwashed with a solution of sodium chloride. This removes the calcium and magnesium in the form of their soluble chlorides and at the same time restores the resin to its original sodium condition. The bed is rinsed free of undesirable salts and returned to service. The governing reaction may be written

$$\left.\begin{array}{c}Ca\\Mg\end{array}\right\}R \;+\; 2NaCl \rightarrow Na_2R \;+\; \left.\begin{array}{c}Ca\\Mg\end{array}\right\}Cl_2 \tag{11.54}$$

A majority of ion exchange softeners are the pressure type, with either manual or automatic controls. They normally operate at rates of 6–8 gpm/ft$^2$ of surface filter area. A water meter is usually employed on the inlet or outlet side. For manual-type operations, this meter can be set to turn on a light or sound an alarm at the end of the softening run.

About 8.5 lb of salt is required to regenerate 1 ft$^3$ of resin and remove approximately 4 lb of hardness in a commercial unit. The reduction is directly related to the amount of cations present in the raw water and the amount of salt used to regenerate the resin bed.

## 11.14 RADIONUCLIDES REMOVAL BY LIME–SODA ASH SOFTENING

The natural radionuclides of interest in water are within the decay series of uranium (U-238), actinium (U-235), and thorium (Th-232). The U-238 decay series (i.e., Rn-222, Ra-226, and U) are of primary health concern because of their relatively high exposure associated with drinking water [8]. Radon-222 gas is a health risk because of the alpha dose to the lungs. Although little is ingested from drinking water, Rn-222 gas can be released to the air during normal domestic water use. Radon-222 sources are almost exclusively confined to groundwaters. Radium, uranium, and artificial radionuclides are a health risk by ingestion. Radium is found primarily in groundwaters, uranium in both groundwaters and surface waters, and artificial radionuclides in surface waters.

The most significant exposures to radioactivity in drinking water occur in small public and private shallow groundwater supplies. The removal from large water supplies can be achieved by conventional treatment methods. Aeration is effective for removal of Rn-222 because of its high volatility and short half-life of 3.8 days. Traditional surface-water treatment by coagulation and filtration is effective for removal of artificial radionuclides for the degree of contamination normally expected.

The best process for removal of radium and uranium from groundwaters is precipitation by lime–soda ash softening at pH > 10. The removal efficiency is generally 80%–90% [9]. The recommended disposal of lime sludge containing the radionuclides is by spreading on land or by landfill.

# Iron and Manganese Removal

Iron and manganese in concentrations greater than 0.3 mg/l of iron and 0.05 mg/l of manganese stain plumbing fixtures and laundered clothes. Although discoloration from precipitates is the most serious problem associated with water supplies having excessive iron and manganese, foul tastes and odors can be produced by growth of iron bacteria in water distribution mains. These filamentous bacteria, using reduced iron as an energy source, precipitate it, causing pipe incrustations. Decay of the accumulated bacterial slimes creates offensive tastes and odors.

Dissolved iron and manganese are often found in groundwater from wells located in shale, sandstone, and alluvial deposits. Impounded surface-water supplies may also have troubles with iron and manganese. An anaerobic hypolimnion (stagnant bottom-water layer) in a reservoir dissolves precipitated iron and manganese from the bottom muds, and during periods of overturn these minerals are dispersed throughout the entire depth.

## 11.15 CHEMISTRY OF IRON AND MANGANESE

Iron (II) ($Fe^{2+}$) and manganese (II) ($Mn^{2+}$) are chemically reduced, soluble forms that exist in a reducing environment (absence of dissolved oxygen and low pH). These conditions exist in groundwater and anaerobic reservoir water. When it is pumped from underground or an anaerobic hypolimnion, carbon dioxide and hydrogen sulfide are released, raising the pH. In addition, the water is exposed to air, creating an oxidizing environment. The reduced iron and manganese start transforming to their stable, oxidized, insoluble forms of iron (III) ($Fe^{3+}$) and manganese (IV) ($Mn^{4+}$).

The rate of oxidation of iron and manganese depends on the type and concentration of the oxidizing agent, pH, alkalinity, organic content, and presence of catalysts [10].

Oxygen, chlorine, and potassium permanganate are the most frequent oxidizing agents. The natural reaction by oxygen is enhanced in water treatment by using spray nozzles or waterfall-type aerators. Chlorine and potassium permanganate ($KMnO_4$) are the chemicals commonly used in iron and manganese-removal plants. Oxidation reactions using potassium permanganate are

$$3Fe^{2+} + MnO_4^- \rightarrow 3Fe^{3+} + MnO_2 \qquad (11.55)$$

$$3Mn^{2+} + 2MnO_4^- \rightarrow 5MnO_2 \qquad (11.56)$$

Rates of oxidation of the ions depend on the pH and bicarbonate ion concentration. The pH for oxidation of iron should be 7.5 or higher: manganese oxidizes readily at pH 9.5 or higher. Organic substances (i.e., humic or tannic acids) can create complexes with iron (II) and manganese (II) ions holding them in the soluble state to higher pH levels. If a large concentration of organic matter is present, iron can be held in solution at pH levels of up to 9.5.

Copper ions and silica have a catalytic effect on the oxidation of iron and manganese [11]. The presence of about 0.1 mg/l of copper increases the rate of iron oxidation 5–6 times. Silica increases oxidation rates of both metals. Manganese oxides are catalytic in the oxidation of manganese. Tray aerators frequently contain coke or stone contact beds through which the water percolates. These media develop and support a catalytic coating of manganese oxides.

## 11.16  PREVENTATIVE TREATMENT

When an industry or city is confronted with iron and manganese problems, solutions are difficult and probably costly. Treatment of the water supply is the only permanent answer. Control and preventative measures can be employed with expectation of reasonable success.

Polyphosphates have been effective in sequestering iron and manganese in some well-water supplies [10]. When applied at the proper dosage, before oxidation of the iron and manganese occurs, polyphosphates tend to hold the metals in solution and suspension, preventing destabilization and, thus, stopping agglomeration of the individual tiny particles of iron and manganese oxides. The concept is that the sequestered metals will pass through the distribution system without creating discolored water. Nevertheless, oxidized particles often settle out and collect in water mains at times when the velocities of flow in the pipes are low and in storage reservoirs during quiescent periods. Then, when the velocities of flow increase and when water in storage is agitated, these particles are resuspended in the water at much higher concentrations than in the raw water. The scouring and mixing flows can be caused by the dramatic increase in water consumption (e.g., because of lawn watering in the spring of the year), operation of well or booster pumps in the distribution system that have not been used for several weeks or months, and sudden withdrawals of water from hydrants.

Growth of iron bacteria also gravitates the adverse effects of iron and manganese [12]. Some of the iron bacteria are autotrophs that oxidize soluble iron (II) to insoluble iron (III) for energy and use carbon dioxide as a carbon source. Filamentous *Crenothix* grow in sheaths impregnated with ferric hydroxide that form reddish colored slimes. *Crenothix* infestations in distribution systems are often widespread, since these bacteria produce and release vast numbers of spores that can be carried throughout the pipe network. *Leptothrix,* a member of the genus *Sphaerotilus,* grow on iron(II) and organic matter. Their growth may be accompanied by a musty odor, which changes to a foul odor upon their death. Also associated with decay of the slime growth, sulfate-reducing bacteria can release hydrogen sulfide, giving the water a rotten-egg odor. Manganese oxidation is also attributed to several species of iron bacteria.

No easy or inexpensive way exists for controlling the chemical and bacterial oxidation of iron and manganese. Periodic flushing of small distribution pipes can be effective in removing accumulations of oxide particles; however, elimination of iron bacteria is generally impossible. Heavy chlorination of isolated sections of water mains followed by flushing may be effective for a limited time. If chlorine is continuously added for disinfection, the rate of oxidation of iron and manganese is increased and the problem of colored water is likely to become more severe. The only permanent solution is treatment to remove iron and manganese from the raw water before distribution in the pipe network.

## 11.17 IRON AND MANGANESE REMOVAL PROCESSES

### Aeration–Sedimentation–Filtration

The simplest form of oxidation treatment uses plain aeration. The units most commonly employed are the tray type, where a vertical riser pipe distributes the water on top of a series of trays, from which it then drips and spatters down through a stack of three or four of them. Soluble iron is readily oxidized by the following reaction:

$$2Fe(HCO_3)_2 + 0.5O_2 + H_2O = 2Fe(OH)_3 + 4CO_2 \qquad (11.57)$$

Manganese cannot be oxidized as easily as iron, and aeration alone is generally not effective. If, however, the pH is increased to 8.5 or higher (by the addition of lime, soda ash, or caustic soda), and if aeration is accompanied by contact with coke beds coated with oxides in the aerator, catalytic oxidation of the manganese occurs.

In plants using the aeration–sedimentation–filtration process, most of the oxidized iron and manganese is removed by a granular-media filter. Flocculant metal oxides are not heavy enough to settle out in the settling basin. The main function of the basin is to allow sufficient reaction time for oxidation to proceed to near completion.

### Aeration–Chemical Oxidation–Sedimentation–Filtration

This sequence of processes is the usual method for removing iron and manganese from well water without softening treatment. Contact tray aeration is designed to displace dissolved gases (i.e., carbon dioxide) and initiate oxidation of the reduced iron and manganese.

Either chlorine or potassium permanganate can chemically oxidize the manganese. When chlorine is utilized, a free available chlorine residual is maintained throughout the treatment process. The rate of manganese(II) oxidation by chlorine depends on the pH, the chlorine dosage, mixing conditions, and other factors. Copper sulfate is a catalyst in the oxidation of manganese.

Theoretically, 1 mg/l of potassium permanganate will oxidize 1.06 mg/l of iron or 0.52 mg/l of manganese (Eqs. 11.55 and 11.56). In actual practice, however, the permanganate necessary for oxidation of the soluble manganese is less than the theoretical requirement. One main advantage of potassium permanganate oxidation is the high rate of the reaction, many times faster than for chlorine. Also, the rate of reaction is relatively independent of the hydrogen ion concentration within a pH range of 5–9.

Filtration following chemical oxidation and sedimentation is very important. Practice has shown that filters pass manganese unless grains of the media are coated with metal oxides. This covering develops naturally during filtration of manganese-bearing water. The coating serves as a catalyst for the oxidation and removal of manganese.

## Water Softening

Lime–soda softening takes out iron and manganese. If split treatment is employed, potassium permanganate can oxidize the iron and manganese in water bypassing the first-stage excess-lime treatment. Lime–soda softening should be given careful consideration as a possible process for treating a hard water requiring iron and manganese elimination.

Lime treatment has been used to remove organically bound iron and manganese from surface water. The process scheme aeration–coagulation–lime treatment–sedimentation–filtration can treat surface waters containing color, turbidity, and organically bound iron and manganese.

## Manganese Zeolite Process

Manganese zeolite is made by coating natural greensand (glauconite) zeolite with oxides [13]. Manganese dioxide removes soluble iron and manganese until it becomes degenerated (Eq. 11.58). The filter is regenerated by using potassium permanganate (Eq. 11.59).

$$Z\text{---}MnO_2 + \begin{cases} Fe^{2+} \\ Mn^{2+} \end{cases} \to Z\text{---}Mn_2O_3 + \begin{cases} Fe^{3+} \\ Mn^{3+} \\ Mn^{4+} \end{cases} \tag{11.58}$$

$$Z\text{---}Mn_2O_3 + KMnO_4 \to Z\text{---}MnO_2 \tag{11.59}$$

Manganese zeolite filters are generally pressure filters.

Disadvantages of the regenerative-batch process are the possibility of soluble-manganese leakage when the bed is nearly degenerated and the waste of excess potassium permanganate needed to regenerate the greensand. These two drawbacks have been substantially overcome by continuously supplying a feed of potassium permanganate solution ahead of a dual-media filter of anthracite and manganese zeolite. The anthracite filter media remove most insolubles, thereby reducing the problem of plugging the greensand. A continuous feed of permanganate reduces the frequency of greensand regeneration. When the permanganate feed is less than the reduced iron and manganese in the water, excess iron and manganese are oxidized by the greensand. If a surplus is applied, it regenerates the greensand.

■ **EXAMPLE 11.7**

A well-water supply contains 3.2 mg/l of iron and 0.8 mg/l of manganese at pH 7.8. Estimate the dosage of potassium permanganate required for iron and manganese oxidation.

*Solution.*   The theoretical potassium permanganate dosages given in Section 11.17 are 1.0 mg/l per 1.06 mg/l of iron and 1.0 mg/l per 0.52 mg/l of manganese. Therefore,

$$\text{KMnO}_4 \text{ required} = \frac{3.2 \times 1.0}{1.06} + \frac{0.8 \times 1.0}{0.52} = 4.6 \text{ mg/l} \qquad ■$$

# Chemical Disinfection and By-product Formation

Chlorine is the common chemical used for disinfection of water and wastewater. One alternative to chlorine is chlorine dioxide, although it is usually applied for taste and odor control. Another alternative is ozone, but because of high cost it is rarely applied solely for disinfection. Other benefits attributed to ozonation are trihalomethane reduction, taste and odor control, and improved coagulation.

## 11.18 CHEMISTRY OF CHLORINATION

Chlorine gas is soluble in water (7160 mg/l at 20°C and 1 atm) and hydrolizes rapidly to form hypochlorous acid:

$$\text{Cl}_2 + \text{H}_2\text{O} \rightleftharpoons \text{HOCl} + \text{H}^+ + \text{Cl}^- \qquad (11.60)$$

Hydrolysis goes virtually to completion at pH values and concentrations normally experienced in water treatment and waste treatment operations.

Figure 11.12 shows the relationship between HOCl and OCl$^-$ at various pH levels. Hypochlorous acid ionizes according to the following equation:

$$\text{HOCl} \rightleftharpoons \text{H}^+ + \text{OCl}^- \qquad (11.61)$$

$$\frac{[\text{H}^+][\text{OCl}^-]}{[\text{HOCl}]} = K_i \qquad (11.62)$$

The dissociation rate from hypochlorous acid to hypochlorite ion is sufficiently rapid so that equilibrium is maintained even though the former is being continuously consumed. If a reducing agent is put into water containing

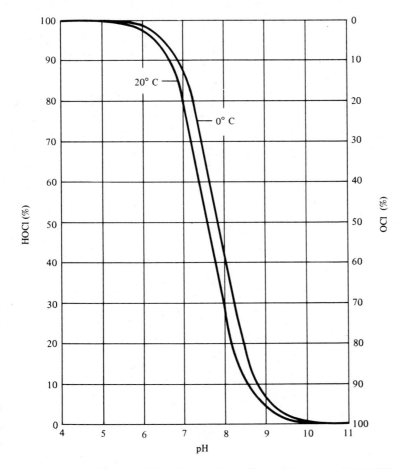

**Figure 11.12** Effect of pH on the portions of hypochlorous acid (HOCl) and hypochlorite ion (OCl⁻) present in water.

free available chlorine, the unconsumed residual redistributes itself between $HOCl$ and $OCl^-$.

Chlorine reacts with ammonia in water to form chloramines as follows:

$$HOCl + NH_3 \rightarrow H_2O + NH_2Cl \quad \text{(monochloramine)} \qquad (11.63)$$

$$HOCl + NH_2Cl \rightarrow H_2O + NHCl_2 \quad \text{(dichloramine)} \qquad (11.64)$$

$$HOCl + NHCl_2 \rightarrow H_2O + NCl_3 \quad \text{(trichloramine)} \qquad (11.65)$$

The chloramines formed depend on the pH of the water, the amount of ammonia available, and the temperature. In the pH range 4.5–8.5, monochloramine and dichloramine are formed. At room temperature, monochloramine exists alone above pH 8.5 and dichloramine occurs alone at pH 4.5. Below pH 4.4 trichloramine is produced.

*Free available residual chlorine* is that residual chlorine existing in water as hypochlorous acid or hypochlorite ion. *Combined available residual chlorine* is that residual existing in chemical combination with ammonia (chloramines) or organic nitrogen compounds. *Chlorine demand* is the difference between the amount added to a water and the quantity of free and combined available chlorine remaining at the end of a specified contact period.

When chlorine is added to water containing reducing agents and ammonia, residuals develop that yield a curve similar to Fig. 11.13. Chlorine reacts first with reducing agents present and develops no measurable residual, as shown by the portion of the curve extending from *A* to *B*. The chlorine dosage at *B* is the amount required to meet the demand exerted by the reducing agents (those common to water and wastewater include nitrites, ferrous ions, and hydrogen sulfide).

The addition of chlorine in excess of that required up to point *B* results in the formation of chloramines. Monochloramines and dichloramines are usually considered together because there is little control over which will be formed. The quantities of each are determined primarily by pH. Chloramines thus established show an available chlorine residual and are effective as disinfectants. When all the ammonia has been reacted with, a free available chlorine residual begins to develop (point *C* on the curve). As the free available chlorine residual increases, the previously produced chloramines are oxidized. This results in the creation of oxidized nitrogen compounds, such as nitrous oxide, nitrogen, and nitrogen trichloride, which in turn reduce the chlorine residual, as seen on the curve between *C* and *D*.

Once most of the chloramines are oxidized, additional chlorine applied to the water creates an equal residual, as indicated by the rising curve at point *D*. Point *D* is generally referred to as the *breakpoint*; beyond it, all

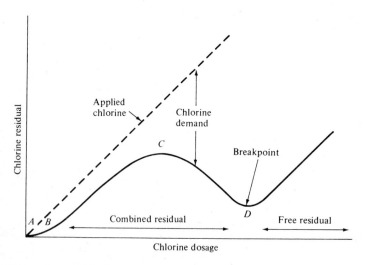

**Figure 11.13** Chlorine-residual curve for breakpoint chlorination.

added residual is free available chlorine. Some resistant chloramines can still be present beyond $D$, but their relative importance is small.

Hypochlorites (salts of hypochlorous acid) may be used for chlorination at small installations such as swimming pools and in emergencies. Since hypochlorites are more expensive, liquid chlorine is applied in most water treatment plants in the United States. Calcium hypochlorite, $Ca(OCl)_2$, is available commercially in granular and powdered forms that contain about 70% available chlorine. Sodium hypochlorite (NaOCl) is handled in liquid form at concentrations between 5% and 15% available chlorine. These salts in water solution yield the hypochlorite ion directly.

Feeding of chlorine involves controlled dissolution of the gas into a carrier water supply for delivery to the point of application and blending with the water or wastewater being chlorinated. Direct feed of chlorine gas into a pipe or channel is not practiced for safety reasons, e.g., the danger is pipes leaking gas outside the controlled environment of the chlorine room. Figure 11.14 is a schematic diagram of a vacuum chlorinator and ejector suitable for a small system. Chlorine is shipped as a liquid in pressurized steel cylinders ranging in size from 100 lb to 1 ton. As gas is released, the liquid vaporizes, yielding about 450 volumes of gas per volume of liquid. To start the flow of gas from the cylinder, water under pressure is pumped at high velocity through the ejector throat to draw a vacuum on the regulator. The safety features of the regulator are that the gas cannot be released except under vacuum, and the regulator case is vented to release the gas outside the room

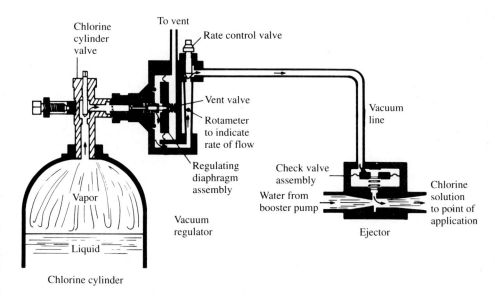

**Figure 11.14** Gas chlorinator with a vacuum regulator mounted on a cylinder and an ejector to dissolve the chlorine in water for conveying a strong solution to the point of application. (Courtesy of Capital Controls, Inc.)

if the diaphragm leaks. The rate of flow is controlled by the rate-control valve on the regulator, and the flow can be observed by the rotameter. The ejector dissolves the gas in water, and this concentrated solution is piped to the water being treated. Moist chlorine gas is extremely corrosive, so piping and dosing equipment are nonmetallic or a special alloy. The yellow-green chlorine gas is poisonous, causing respiratory and eye irritation at very low concentrations and physiologic damage at high doses. Chlorine feeding rooms and storage areas must be kept cool and ventilated.

A manual-control chlorinator, as illustrated in Fig. 11.14, is appropriate for a small treatment plant or well-water supply discharging water at a nearly constant rate of flow. For large facilities, controlling the chlorine rate of feed based on water flowrate or chlorine residual is important for reliable performance and economical operation. The automatic proportional-control system illustrated in Fig. 11.15 adjusts the feed rate to maintain a constant preset dosage for all rates of flow. The chlorine feeder is responsive to signals from both the flowmeter transmitter and the chlorine residual analyzer. In this manner rate of flow measurement is the primary feed regulator, and residual monitoring trims the dosage. At some installations, it may be satisfactory to proportion feed to flow and thus apply a constant preset dosage without residual monitoring. However, this type of regulator is only satisfactory where chlorine demand and flow are reasonably constant and an operator is available to make adjustments as necessary.

## 11.19  CHLORINE DIOXIDE

Chlorine dioxide has had limited application in the United States as a water disinfectant; meanwhile, it has been used for taste and odor control. It is manufactured at a water treatment plant by mixing solutions of sodium chlorite and chlorine in controlled proportions as shown by Eq. 11.66 [14]:

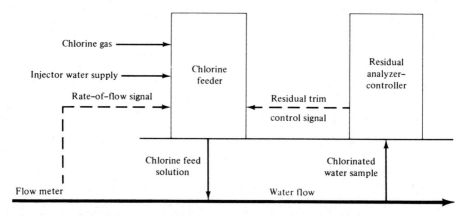

**Figure 11.15** Automatic proportional control system for feeding to a constant preestablished chlorine residual.

$$2NaClO_2 + Cl_2 \rightarrow 2ClO_2 + 2NaCl \tag{11.66}$$

By maintenance of the pH in the reactor at about 3.5, the production of $ClO_2$ is optimized with minimum residuals of unreacted chlorine or chlorite. A mineral acid such as HCl or $H_2SO_4$ can be added to reduce the pH in the reactor if necessary. Under good control, chlorine dioxide yields of 85%–90% are expected with solution concentrations between 500 and 2000 mg/l.

Disinfection using chlorine dioxide has several advantages. It is a strong bactericide and viricide over a wide pH range and forms a residual capable of persisting in the distribution system. Equally important in treatment of surface waters, chlorine dioxide reacts with neither nitrogenous compounds to form chloramines nor humic acids to form trihalomethanes. The greatest potential disadvantage is creation of chlorate and chlorite residuals, which are toxins. For this reason, the recommended limit for these residuals in drinking water is 1.0 mg/l. Second, the high cost of sodium chlorite makes $ClO_2$ disinfection more expensive than chlorination.

## 11.20  OZONE

Ozone is a strong oxidizing gas that reacts with most organic and many inorganic molecules. It is more reactive than chlorine. The reactions are rapid in inactivating microorganisms, oxidizing iron, manganese, sulfide, and nitrite, and slower in oxidizing organic compounds like humic and fulvic substances, pesticides, and volatile organic compounds. Unlike chlorine, it does not react with water to produce disinfecting species but decomposes in water to produce oxygen and hydroxyl free radicals. Since it does not produce a disinfecting residual, chlorine is added to treated potable water before distribution for a protective residual. The half-life of ozone in water is approximately 10–30 min and shorter above pH 8; therefore, it must be generated on site.

An ozonation system consists of (1) air preparation or oxygen feed, (2) electric power supply, (3) ozone generation, (4) ozone contacting, and (5) ozone contactor exhaust gas destruction. Ambient air is dried to prevent fouling of the ozone production tubes and to reduce corrosion. A common system uses desiccant dryers in conjunction with compression and refrigerant dryers. The voltage or frequency of the electrical supply is varied to control the rate of ozone production, which requires a specialized power source normally supplied by the generator manufacturer. Generators for water treatment usually employ a corona discharge cell. The cell consists of two electrodes separated by a discharge gap and a dielectric material across which high-voltage potentials are maintained. As the air or oxygen flows between the electrodes, ozone is produced. If ambient air is used, the concentration of ozone is 1%–3.5% by weight. This is an adequate concentration to dissolve enough ozone to attain the concentration–contact time necessary in water processing. When pure oxygen is used, the concentration of ozone is approximately doubled. The most common ozone contactor has two compartments in series with porous diffusers under a 16-ft water column, as sketched in Fig. 11.16. The water flows downward in each compartment counter to the rising fine bubbles of ozonated air. The covered tank increases the partial pressure of ozone and collects the head gases for disposal. Ozone in the exhaust gas must be destroyed or removed by recycling prior to venting, since the concentration of ozone exceeds air quality standards. When the ozone is generated from air, reducing ozone by the process of thermal–catalytic destruction is less expensive than recirculating the exhaust gas through the air preparation system.

Ozone is rarely applied solely for disinfection because of the high cost relative to chlorine. In most cases, its application is for a combination of operations incorporating taste, odor, or color control, which is a common application; oxidation of humic organic substances that are precursors in the formation of trihalomethanes; destabilization of colloids; and inactivation of microorganisms. The only by-products that have been identified with ozonation of water are detectable levels of aldehydes. The increased interest in application of ozone in potable water treatment relates to this absence of health-related by-products and the potential for destruction of many trace organic compounds.

## 11.21 CHLORINATION BY-PRODUCTS

Chlorination inactivates pathogenic microorganisms and oxidizes many organic molecules to carbon dioxide. In the treatment of surface waters and groundwaters containing humic substances, however, it also produces chlorinated by-products and incompletely oxidized compounds that are potential toxins. Studies involving chlorination of humic and fulvic acid precursors, isolated from waters, have improved the understanding of by-product formation during disinfection of water supplies. Many of the specific chemical

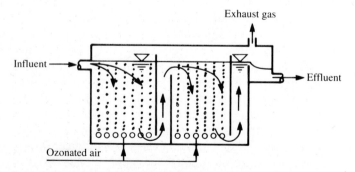

**Figure 11.16** Two-compartment ozone contactor with porous diffusers.

structures have been characterized and vary with chlorine-to-carbon (Cl/C) ratio, pH, time of reaction, and other factors [15]. The principal by-products at high Cl/C molar ratios of 3 : 1 or 4 : 1 are volatile hydrophobic compounds, mainly chloroform. In addition, a large variety of nonvolatile hydrophilic compounds are produced, including chlorinated and unchlorinated aromatic and aliphatic compounds. These appear to increase at lower pH and Cl/C molar ratios less than 1 : 1. At lower Cl/C ratios, which more closely approximate typical drinking-water disinfection conditions, the humic-acid precursors appear to support the formation of unchlorinated by-products, such as monobasic and dibasic aliphatic acids, to a greater extent than do fulvic acid precursors. Increasing the Cl/C ratio appears to drive both kinds of precursors toward chloroform production and a larger fraction of identifiable products. Nevertheless, the by-products represent only a small fraction of the initial organic matter [15].

An extensive study on the occurrence of disinfection by-products in 35 water treatment plants processing surface waters revealed that trihalomethanes (mainly chloroform, bromodichloromethane, and dibromochloromethane) accounted for about 50% of the total by-products on a weight basis [16]. Haloacetic acids were the next most significant fraction accounting for about 25%, and aldehydes accounted for about 7%. Of the remaining by-products, none was present in a significant concentration. The median total trihalomethane concentration was 39 $\mu$g/l, and the median haloacetic acids concentration was 19 $\mu$g/l [16].

The evidence that trihalomethanes are capable of inducing cancer gives rise to a complex set of issues in risk assessment [15]. Convincing evidence that trihalomethanes are intrinsically carcinogenic to humans is lacking. Some epidemiologic studies have associated increased cancer risk with chlorination of drinking waters, but most studies have been seriously hampered by universal exposures, such as diet and smoking, and geographic migration in large populations. On the other hand, trihalomethanes are not the only mutagenic and potentially carcinogenic by-products. Therefore, calculation of the levels of risk to humans, who consume drinking water containing these compounds, must rely on data that demonstrate the carcinogenicity of individual trihalomethanes in experimental animals [15].

The carcinogenicity of chloroform in laboratory animals was studied by several researchers. The results of one study are shown in Table 11.5 [15]. Male rats were administered chloroform for 104 weeks in drinking water containing 0, 200, 400, 900, and 1800 mg/l. Based on water intakes and body weights, the average doses were 0, 19, 38, 81, and 160 mg/kg body weight per day. The rats consuming the high-dose rate had a 14% incidence of adenomas and adenocarcinomas of the renal tubules, which is the 7/50 tumor rate shown in Table 11.5. The control group had a 1% (4/301) incidence of tumors. Based on these laboratory data, the carcinogenic risk for chloroform estimated for humans from the linearized multistage mathematical model is $5.16 \times 10^{-8}$ for a lifetime risk and $8.9 \times 10^{-8}$ cancer risk for the upper 95% confidence level. This risk estimate expresses the probability of cancer in a

**TABLE 11.5** TUMOR INCIDENCE IN MALE RATS FED CHLOROFORM
IN DRINKING WATER

| Animal | Sex | Tumor site | Dose (mg/kg·bw/d) | Tumor rates |
|--------|-----|-----------|-------------------|-------------|
| Osborne–Mendel rats | Male | Kidney | 0 | 4/301 |
| | | | 19 | 4/313 |
| | | | 38 | 4/148 |
| | | | 81 | 3/48 |
| | | | 160 | 7/50 |

*Source: Drinking Water and Health, Disinfectants and Disinfection By-products,* National Research Council (Washington, DC: National Academy Press, 1987), p. 131.

person weighting 70 kg who has consumed 1 l of water per day containing 1 $\mu$g/l of chloroform for a lifetime of 70 years. At 2 l/d with a maximum contaminant level of 100 $\mu$g/l, the risk at the upper 95% confidence level increases to $1.78 \times 10^{-5}$, and, therefore, lifetime exposure would be expected to produce one excess case of cancer for every 56,000 persons exposed for their lifetimes. If the maximum contaminant level were lowered to 25 $\mu$g/l, the risk decreases to $4.45 \times 10^{-5}$ and lifetime exposure would be expected to produce one excess case of cancer for every 225,000 persons exposed. A trihalomethane standard of 25 $\mu$g/l or less would severely limit the use of free chlorine for drinking-water disinfection.

## 11.22 CONTROL OF TRIHALOMETHANES

Chlorination is the most common process for disinfecting and establishing a protective residual in water treatment because of its relatively low cost, ease of application, proven reliability, and residual detectability. At the same time, chlorine reacts with humic substances commonly found in raw surface waters to form trihalomethanes, principally chloroform and bromodichloromethane. Decaying vegetation produces the humic substances—humic and fulvic acids—referred to as precursors.

Alternatives for reducing the production of trihalomethanes are to (1) change the point of chlorine application, (2) improve the removal of precursors prior to chlorination, and (3) use an alternative disinfectant. Moving the point of chlorine application to later stages in water treatment is the easiest method for reducing trihalomethane formation. Prior to the mid-1970s, the common practice for taste and odor control and disinfection was breakpoint chlorination of raw surface waters as the first step in processing. Now, chlorine is not added until after coagulation and sedimentation or, if previous control of microbial populations is not necessary, after filtration. The benefits of delayed chlorination are reduction in the required dosage and prior removal of precursors.

Minimizing precursor concentrations prior to chlorination complements the alternative of delaying and reducing chlorine additions. Optimizing chemical coagulation also has the benefits of reducing turbidity and removing other organic compounds. Powdered activated carbon applied in the early stages of treatment may adsorb humic substances. Where trihalomethane formation cannot be controlled by improved clarification and filtration, predisinfection using chlorine dioxide or ozone may be considered. Both of these chemicals provide taste and odor control and disinfection, and ozone can improve flocculation and filtration.

The alternative disinfectants are ozone, chlorine dioxide, and chloramines. If ozone is used as the primary disinfectant, secondary chlorination is needed to provide a disinfectant residual in the water entering the distribution system. The major disadvantage of ozonation is high cost. Chlorine dioxide does not produce trihalomethanes, but a major portion can revert to the chlorite ion, which is a toxin. Chloramines are weak disinfectants and, except for establishing a residual to control bacteria, are of little benefit in inactivating viruses and protozoal cysts.

# Disinfection of Potable Water

The Environmental Protection Agency has established maximum contaminant level goals of zero for *Giardia lamblia* and viruses of fecal origin for public water supplies. Since routine tests cannot be used to determine the presence of these microorganisms, treatment techniques have been established to ensure their removal and inactivation during water processing. Furthermore, because *G. lamblia* and viruses represent the most persistent pathogens, treatment for their removal ensures the absence of other pathogens. Although some water supplies may not contain significant numbers of these pathogens, demonstrating their absence by water-quality monitoring is not feasible and cannot be used in lieu of applying the specified treatment techniques. The three categories of water supplies are (1) surface water open to the atmosphere and subject to surface runoff, (2) groundwater under the direct influence of surface water (i.e., containing algae, insects, or other macroorganisms, or experiencing significant and relatively rapid shifts in water characteristics), and (3) groundwater.

## 11.23 CONCEPT OF THE *C·t* PRODUCT

Chemical inactivation of a specific specie of microorganism is a function of disinfectant concentration and contact time. Other important factors are the kind of disinfectant, temperature, pH, viability of the microorganisms, and presence of suspended organic matter.

The $C \cdot t$ concept is expressed as

$$k = C^n \times t \quad (\text{or } C^n \cdot t) \tag{11.67}$$

where  $k$ = a constant for a specific microorganism exposed under specific conditions

$n$ = a constant

$C$ = disinfectant concentration, mg/l

$t$ = contact time required to inactivate a specified percentage of the microorganisms, min

To apply this equation, results of several individual experiments with different disinfectant concentrations under identical conditions are plotted on double-logarithmic paper. For a selected degree of inactivation (e.g., 2.0 log or 99%), the $t$-versus-$C$ plot produces a straight line with a slope of $n$. The $C \cdot t$ value remains constant when $n = 1$, regardless of disinfection concentration; also, the concentration and contact time are of equal importance. The $n$ values determined from experimental analyses on a variety of microorganisms were in a broad range (0.5–2.0) with the majority between 0.8 and 1.2, averaging close to 1.0 [17]. Thus, for engineering practice, the $C \cdot t$ values are based on the assumption that $n = 1.0$.

The kind of disinfectant (chlorine, chlorine dioxide, chloramine, and ozone) has individual characteristics that result in different $C \cdot t$ values for the same conditions. All are influenced by water temperature such that a two- to threefold increase in inactivation rates results from a 10°C increase. Of the chlorine disinfectants, free chlorine is most influenced by pH because of the dissociation of HOCl to OCl$^-$ (Fig. 11.12). Hypochlorous acid, the stronger disinfectant specie, is present in a proportion of about 98% below pH 6 and the hypochlorite ion, the weaker disinfectant specie, is present in a proportion of 99% above pH 10. At intermediate pH values, the two chemical species are in equilibrium, and as one is consumed more rapidly in reactions the proportions remain the same, with the concentration of each being decreased accordingly. In general, the order of resistance to chemical disinfection from least to greatest is bacteria, viruses, protozoal cysts, helminth eggs. Each category has a variety of species encompassing a wide diversity of sizes, life cycles, and other biological characteristics, including resistance to chemical disinfectants. Even within the same specie, resistance can vary, e.g., between those cultured in the laboratory and those found naturally in the environment. Protection by clumping, or being protected by organic matter, affects the kinetics of disinfection, thereby extending the required contact time. Consequently, turbidity reduction in water treatment and suspended-solids removal in wastewater treatment are important before applying disinfecting chemicals.

Extensive evaluations of the inactivation of microorganisms by chemicals have been documented in several studies [15, 18]. The $C \cdot t$ values for 99% inactivation in Table 11.6 and summarized in the following statements are from Hoff [18]. The $C \cdot t$ value for 99% inactivation of *Escherichia coli* by free

**TABLE 11.6**    SUMMARY OF *C·t* VALUE RANGES FOR 99% INACTIVATION OF VARIOUS MICROORGANISMS BY DISINFECTANTS AT 5°C

| | Disinfectant | | | |
|---|---|---|---|---|
| Microorganism | Free chlorine pH 6–7 | Preformed chloramine pH 8–9 | Chlorine dioxide pH 6–7 | Ozone pH 6–7 |
| *E. coli* | 0.034–0.05 | 95–180 | 0.4–0.75 | 0.02 |
| *Polio 1* | 1.1–2.5 | 770–3700 | 0.2–6.7 | 0.1–0.2 |
| *Rotavirus* | 0.01–0.05 | 3800–6500 | 0.2–2.1 | 0.006–0.06 |
| *G. lamblia* cysts | 47–>150 | — | — | 0.5–0.6 |
| *G. muris* cysts | 30–630 | — | 7.2–19 | 1.8–2.0 |

*Source:* J. C. Hoff, Project Summary, "Inactivation of Microbial Agents by Chemical Disinfectants," EPA/600/S2-86/067 (September 1986), p. 5.

chlorine at 5°C and pH 6 averaged 0.045 mg/l·min. The *C·t* values for 99% inactivation of poliovirus 1 by free chlorine at 5°C and pH 6–7 averaged 1.1 and 2.0 mg/l·min in different studies; at pH 10 the average was 10.5 mg/l·min. The *C·t* values for 99% inactivation of *G. lamblia* (infective in humans) by free chlorine at 5°C and pH 6 were 65–150 mg/l·min. For *G. muris* (infective in mice), the 99% values in different studies ranged from 68 mg/l·min at 3°C and pH 6.5 to greater than 150 mg/l·min at 5°C and pH 6.

Animal infectivity data for cysts of *G. lamblia* were collected in studies by Hibler [19]. Isolates of *G. lamblia* acquired from several human sources were maintained by passage in Mongolian gerbils. Clean cysts from these animals at a concentration of 1000 cysts/ml were exposed to selected free-chlorine concentrations for specified *C·t* values from 0.5 to 5°C and at pH values of 6, 7, and 8. At specified time intervals for each temperature and pH condition, the chlorine activity was chemically arrested and the cyst suspension was concentrated and washed. Five gerbils were fed 50,000 cysts that had been exposed to the chlorine solution. An equal number of positive control animals were each orally inoculated with 50 unchlorinated cysts maintained at the same temperature and pH. After 6–7 days, the gerbils were examined to determine the number per group infected. A separate infectivity study of gerbils demonstrated that approximately five cysts constituted an infective dose, which is 0.01% of the 50,000 dose of treated cysts fed to the test animals. If all five gerbils became infected, the *C·t* during chlorination of the cyst culture produced less than 99.99% (4.0 log) inactivation, and if no gerbils were infected the inactivation was greater than 99.99%. If some of the group were infected and others were not, the *C·t* value was assumed to have produced 99.99% inactivation for the specified temperature and pH conditions.

A regression analysis of the animal infectivity data yielded the equation

$$C \cdot t = 0.985 C^{0.176} \text{pH}^{2.75} T^{-0.147} \tag{11.68}$$

where    $C \cdot t$ = chlorine concentration times contact time for 99.99% inactivation of *G. lamblia* cysts, mg/l·min

$C$ = free chlorine concentration, mg/l

$T$ = temperature, °C

Equation 11.68 was derived from experimental results for $C$ in the range of 0.44–4.23 mg/l, pH of 6–8, and $T$ of 0.5–5°C. Extrapolating these values to calculate estimated $C \cdot t$ values can be performed by assuming first-order kinetics. For example, $C \cdot t$ values for 99.9%, 99%, and 90% would be three quarters, one half, and one quarter of the calculated value for 99.99%, respectively. For temperature correction, the reaction rate can be assumed to double (a twofold increase) for a 10°C temperature increase. For example, the $C \cdot t$ values at 15°C and 25°C would be one half and one quarter of the calculated value at 5°C, respectively. For intermediate temperatures, use Eq. 11.29 to calculate the change in reaction rate. Extrapolated values are subject to error and should be considered only as estimated values.

### ■ EXAMPLE 11.8

Using Eq. 11.68, calculate the $C \cdot t$ value for *G. lamblia* inactivation for a chlorine concentration of 2.0 mg/l, pH 7.0, and temperature of 5.0°C. Extrapolate this value to 99% and 99.9% inactivation and compare the results with the values given in Tables 11.6 and 11.7, respectively.

**TABLE 11.7**    *C·t* VALUES FOR 99.9% (3.0 LOG) INACTIVATION OF *GIARDIA LAMBLIA* CYSTS BY FREE CHLORINE AT VARIOUS TEMPERATURES AND pH VALUES

| Free residual chlorine (mg/l) | pH | Water temperatures | | | |
|---|---|---|---|---|---|
| | | 0.5°C (mg/l·min) | 5°C (mg/l·min) | 10°C (mg/l·min) | 20°C (mg/l·min) |
| ≤0.4 | 6.5 | 163 | 117 | 88 | 44 |
| | 7.0 | 195 | 139 | 104 | 52 |
| | 7.5 | 237 | 166 | 125 | 62 |
| | 8.0 | 277 | 198 | 149 | 74 |
| 1.0 | 6.5 | 176 | 125 | 94 | 47 |
| | 7.0 | 210 | 149 | 112 | 56 |
| | 7.5 | 253 | 179 | 134 | 67 |
| | 8.0 | 304 | 216 | 162 | 81 |
| 2.0 | 6.5 | 197 | 138 | 104 | 52 |
| | 7.0 | 236 | 165 | 124 | 62 |
| | 7.5 | 286 | 200 | 150 | 75 |
| | 8.0 | 346 | 243 | 182 | 91 |
| 3.0 | 6.5 | 217 | 151 | 113 | 57 |
| | 7.0 | 261 | 182 | 137 | 68 |
| | 7.5 | 316 | 221 | 166 | 83 |
| | 8.0 | 382 | 268 | 201 | 101 |

*Source:* Adapted from *Guidance Manual for Compliance with the Surface Water Treatment Requirements for Public Water Systems* (Environmental Protection Agency, 1990).

*Solution.*

For 99.99%,

$$C \cdot t = 0.985(2.0)^{0.176}(7.0)^{2.75}(5.0)^{-0.147}$$

$$= 185 \text{ mg/l·min}$$

For 99%,

$$C \cdot t = (1/2)185 = 93 \text{ mg/l·min}$$

The value of 93 mg/l·min is within the range of $C \cdot t$ values of 47–>150 mg/l·min given in Table 11.6 for 99% inactivation of *G. lamblia* by free chlorine at 5°C in the pH range of 6–7.

For 99.9%,

$$C \cdot t = (3/4)185 = 139 \text{ mg/l·min}$$

This value is slightly (16%) less than the $C \cdot t$ value of 165 mg/l·min given in Table 11.7 for 2.0 mg/l, pH 7.0, and 5°C.  ■

## 11.24 SURFACE-WATER DISINFECTION

The EPA rule for surface-water treatment requires at least 99.9% (3.0 log) removal and/or inactivation of *G. lamblia* cysts and at least 99.99% (4.0 log) removal and/or inactivation of enteric viruses. These performance standards are generally achieved in well-operated conventional plants processing water by coagulation, sedimentation, filtration, and disinfection. Since *G. lamblia* cysts are very resistant to chlorine, the intent of this rule is to promote coagulation–filtration treatment for physical removal of a substantial portion of the *G. lamblia* cysts.

Treatment of surface water and groundwater under the direct influence of surface water is required to include filtration unless the source water meets stringent water-quality criteria. The fecal coliform concentration in the raw water must be equal to or less than 20 per 100 ml, or total coliform concentration equal to or less than 100 per 100 ml, in at least 90% of the samples tested. The turbidity level cannot exceed 5 NTU except for an unexpected event, but the number of events cannot exceed two in the preceding 12 months. A comprehensive watershed control program is mandated to minimize the potential for contamination by *G. lamblia* cysts and viruses, and a defined water-quality monitoring schedule must be instituted.

The disinfection requirement for unfiltered public water supplies is at least 99.9% (3.0 log) inactivation of *G. lamblia* cysts and 99.99% (4.0 log) inactivation of fecal viruses. The calculated $C \cdot t$ value must be equal to or greater than those listed in Table 11.7. Only the $C \cdot t$ for inactivation of *G. lamblia* cysts is necessary since the $C \cdot t$ for a 4.0-log inactivation of viruses is a much lower value, equivalent to approximately a 0.5-log inactivation of cysts. Disinfection by chemicals other than chlorine must be demonstrated to be equally effective. The free-plus-combined chlorine residual in the water entering the distribution system cannot be less than 0.2 mg/l for more than 4 hr, and the chlorine residual in the distribution system must be detectable in at least 95% of the samples tested each month. Where the residual is undetectable, a measurement of heterotrophic bacteria by a plate count of equal to or less than 500 per ml is deemed to be equivalent to a detectable chlorine residual.

The procedure for calculating the $C \cdot t$ for a treatment system is described in the *Guidance Manual* [20]. The $C \cdot t$ is the summation of the calculated $C \cdot t$ values before the water arrives at the first customer with the $C$ being the free chlorine residual measured at the end of each chlorination segment, in milligrams per liter, and the $t$ being the calculated contact time of the segment, in minutes. For example, if chlorine were added at the pumping station as the water entered a pipeline and again as the water discharged into a contact tank, the $C \cdot t$ would be the chlorine residual measured in the discharge of the pipeline multiplied by the contact time in transit plus the residual at the discharge of the contact tank multiplied by the $t_{10}$ time. The contact time in the pipeline is the theoretical detention time, assuming plug flow. In chlorine contact tanks or storage reservoirs, the contact time can be determined by tracer studies. (Refer to Sections 10.6–10.8.) The tracer tests should be conducted for at least four different flow rates. The residual tracer concentrations $C$ measured at time intervals are normalized by dividing by the applied tracer concentration $C_0$ and plotted versus time, as in Fig. 11.17a. The contact time $t_{10}$ is the time for 10% of the tracer to pass through the contact tank or reservoir. Figure 11.17b is a plot of $t_{10}$ times versus flow rate that can be used to determine $t$ for daily $C \cdot t$ calculations. The contact time $t$ used in computing the $C \cdot t$ is the peak hourly flow for that day.

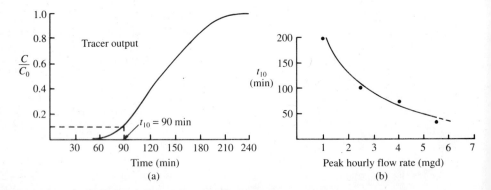

**Figure 11.17** Tracer analyses of a chlorine contact tank to determine $t_{10}$ times at peak hourly flow for calculating $C \cdot t$ values. (a) Normalized tracer output for a test applying a continuous tracer input. (b) A plot of the $t_{10}$ times for four tracer tests at different flowrates to draw a curve to determine $t$ for calculating $C \cdot t$ values.

Surface-water plants that provide filtration are required to achieve at least 99.9% removal and inactivation of *Giardia lamblia* and 99.99% removal and inactivation of viruses by the total treatment system, including sedimentation after coagulation, if provided, granular-media filtration, and chemical disinfection. For filtration to be considered effective in removal of *G. lamblia* cysts, the turbidity in the filtered water must be equal to or less than 0.5 NTU in at least 95% of the measurements taken each month. The recommended removals of pathogens achievable in conventional treatment (coagulation, flocculation, sedimentation, and filtration) are a 2.5-log reduction of *G. lamblia* cysts and a 2.0-log reduction of viruses [20]. Therefore, only 0.5 log of *G. lamblia* and 2.0 log of viruses need to be inactivated by chemical disinfection. For direct filtration (coagulation and filtration excluding sedimentation), the recommended removals are a 2.0-log reduction of cysts and a 1.0-log reduction of viruses, which leaves 1.0 log of cysts and 3.0 log of viruses for inactivation. Table 11.8 lists the $C \cdot t$ values for 90% (1.0 log) inactivation of *G. lamblia* cysts by various chemical disinfectants; the $C \cdot t$ values for 0.5-log inactivation are equal to one half the 1.0-log inactivation values. Recommended $C \cdot t$ values for achieving different levels of virus inactivation are given in Table 11.9. For free chlorine, $C \cdot t$ values recommended for *G. lamblia* inactivation also provide adequate virus inactivation. For other disinfectants, this may not be true.

The residual chlorine (free chlorine plus chloramine) in the water entering the distribution system cannot be less than 0.2 mg/l for more than 4 hr, and a residual in the distribution system must be detectable in at least 95% of the

**TABLE 11.8**   $C \cdot t$  VALUES FOR 90% (1.0 LOG) INACTIVATION OF *GIARDIA LAMBLIA* CYSTS[a]

| | | Water Temperature | | | |
| --- | --- | --- | --- | --- | --- |
| | pH | 0.5°C (mg/l·min) | 5°C (mg/l·min) | 10°C (mg/l·min) | 15°C (mg/l·min) |
| Free Chlorine[b] | 6 | 49 | 35 | 26 | 19 |
| | 7 | 70 | 50 | 37 | 28 |
| | 8 | 101 | 72 | 54 | 36 |
| | 9 | 146 | 146 | 78 | 59 |
| Preformed Chloramine | | 1300 | 730 | 620 | 500 |
| Chlorine Dioxide | | 21 | 8.4 | 7.4 | 6.3 |
| Ozone | | 0.97 | 0.63 | 0.48 | 0.32 |

*Source:* Adapted from *Guidance Manual for Compliance with the Surface Water Treatment Requirements for Public Water Systems* (Environmental Protection Agency, 1990).

[a] $C \cdot t$ values for 0.5-log inactivation are one half those shown in table.

[b] Free chlorine values are based on a residual of 2.0 mg/l.

**TABLE 11.9** $C \cdot t$ VALUES FOR INACTIVATION OF VIRUSES AT VARIOUS TEMPERATURES AND pH 6–9[a]

| | Log Inactivation | Water temperature | | | |
|---|---|---|---|---|---|
| | | 0.5°C (mg/l·min) | 5°C (mg/l·min) | 10°C (mg/l·min) | 15°C (mg/l·min) |
| Free chlorine | 2.0 | 6 | 4 | 3 | 2 |
| | 3.0 | 9 | 6 | 4 | 3 |
| Preformed chloramine | 2.0 | 1200 | 860 | 640 | 430 |
| | 3.0 | 2100 | 1400 | 1100 | 710 |
| Chlorine dioxide | 2.0 | 8.4 | 5.6 | 4.2 | 2.8 |
| | 3.0 | 25.6 | 17.1 | 12.8 | 8.6 |
| Ozone | 2.0 | 0.9 | 0.6 | 0.5 | 0.3 |
| | 3.0 | 1.4 | 0.9 | 0.8 | 0.5 |

*Source:* Adapted from *Guidance Manual for Compliance with the Surface Water Treatment Requirements for Public Water Systems* (Environmental Protection Agency, 1990).

[a] $C \cdot t$ values for free chlorine, ozone, and chlorine dioxide include safety (uncertainty) factors. The chloramine values are based on laboratory data using preformed chloramine to inactivate hepatitis A and do not include a safety factor.

samples tested each month. Where the residual is undetectable, a measurement of heterotrophic plate count equal to or less than 500 per ml is deemed equivalent to a detectable residual.

### ■ EXAMPLE 11.9

A conventional surface-water plant with coagulation, flocculation, sedimentation, and filtration produces a filtered water with a turbidity less than 0.4 NTU, pH 7, and temperature of 5°C on a day when the peak hourly flow is 3.0 mgd. After filtration, the water is chlorinated in a baffled reservoir with hydraulic characteristics as shown in Fig. 11.17. What is the required disinfection of the filtered water if free chlorine is used?

*Solution.*    For conventional treatment, a 2.5-log removal of *G. lamblia* is achieved by filtration, leaving 0.5-log inactivation required by disinfection.

The required $C \cdot t$ value for 1.0-log inactivation from Table 11.8 is 50 mg/l·min at pH 7 and 5°C. (Note that this value is based on a free residual of 2.0 mg/l. If a $C \cdot t$ value for inactivation were given for a lower chlorine residual, it would be lower. In other words, $C \cdot t$ values at a residual of 2.0 mg/l are conservative for postchlorination, since levels above about 0.5 mg/l in drinking water are intolerable for most customers.)

$$C \cdot t \text{ for 0.5-log inactivation} = (1/2)50 = 25 \text{ mg/l·min}$$

From Fig. 11.17b, for a peak-hourly flowrate of 3.0 mgd, the $t_{10}$ is 90 min. Therefore, the required free chlorine residual in the effluent from the baffled reservoir is

$$C = \frac{25 \text{ mg/l·min}}{90 \text{ min}} = 0.28 \text{ mg/l}$$

This chlorine residual also satisfies the requirement of a minimum residual of 0.2 mg/l for water entering the distribution system.    ■

## 11.25 GROUNDWATER DISINFECTION

Environmental Protection Agency regulations require that all groundwaters serving as public water supplies be disinfected unless a variance is granted by the state. (A variance exempts a public water supply from compliance with a treatment technique, or maximum contaminant level, for justifiable reasons, provided the exemption does not result in risk to health.) Since groundwaters under the direct influence of surface waters are included in the surface-water disinfection rule, groundwaters in this regulation are presumed to be free of *G. lamblia* cysts but can contain viruses and bacteria. Microbial cysts are large enough to be removed by filtration through soils, while viruses can be carried for significant distances by groundwater movement. The maximum contaminant level goal for viruses of fecal origin is zero. Since routine testing for viruses is not feasible, adequate protection is based on establishing appropriate treatment techniques and monitoring for coliform bacteria.

Groundwater supplies from deep wells not requiring treatment are commonly pumped directly into the distribution system without chlorination. To prevent microbial contamination of the groundwater prior to withdrawal, the surface and subsurface areas surrounding the wells are protected. The most convincing substantiation of the safety of protected groundwaters is the absence of outbreaks of waterborne disease. For systems supplied by groundwater, the major cause of outbreaks is from contamination entering the distribution pipe network [21].

Some groundwater systems add chlorine to establish a "protective" residual of 0.2–0.4 mg/l. While this level of chlorine residual may inactivate low numbers of pathogens in backflow of contaminated water entering a pipe network, the disinfecting action is very limited for resistant microorganisms. Bacteria are inactivated by the lowest $C \cdot t$ values for disinfection, viruses are more resistant, and protozoal cysts are extremely resistant and require the highest $C \cdot t$ values. A rational assessment of the degree of protection of health provided by maintaining a chlorine residual in a distribution system supplied by groundwater is very difficult. (A chlorine residual in systems-supplied surface waters is important to suppress the growth of numerous species of heterotrophic bacteria that survive treatment and regrow in the pipe network.) If the water in the system is contaminated with only 0.1% of wastewater containing low concentrations of bacterial and viral pathogens, a chlorine residual of 0.2–0.4 mg/l provides some protection. A chlorine residual of 0.2–0.4 mg/l, however, is no guarantee of protection if the percentage of wastewater is greater or concentrations of pathogens is high. The chlorine demand of the contaminating wastewater will rapidly neutralize the chlorine residual, and the pathogens will survive.

A major disadvantage of residual chlorination of uncontaminated groundwater entering a distribution system is interference with coliform testing as an indicator of fecal pollution entering the system. If the supply is unchlorinated, the presence of coliforms in the water is a warning of the possibility of backflow contamination. Conversely, if the supply has a chlorine residual, the coliforms can be inactivated and not indicate the presence of more persistent viruses and protozoal cysts. The water utility personnel may never suspect that backflow at a low level of contamination is occurring in the absence of positive coliform tests.

The best protection against contamination of water in a distribution system is to prevent backflow by installation of backflow preventers in high-risk service connections (e.g., hospitals and mortuaries) and enforcement of a plumbing code for all residential, commercial, and industrial buildings. In case of an unprotected system outlet, back-siphonage is prevented by maintaining adequate pressure in the supply mains to prevent reversal of flow. The risk of backflow is remote in a properly designed and operated water distribution system. A chlorine residual is not a substitute for a well-maintained system with backflow protection.

The general criteria for a state to issue a variance in lieu of requiring disinfection of groundwater supplies are (1) absence of outbreaks of waterborne disease, or approved system modifications if an outbreak has occurred, (2) a satisfactory sanitary-survey report every 5 years that evaluates the vulnerability of the groundwater to fecal contamination, (3) approved well construction, (4) a cross-connection control program to prevent backflow or back-siphonage of contaminated waters into the distribution system, and (5) compliance with coliform testing regulations.

Groundwater supplies vulnerable to fecal contamination or systems that do not qualify for issuance of a variance are required to disinfect for virus inactivation. The appropriate level of disinfection, between 90% and 99.99% virus inactivation, depends on the specific conditions of the site and well construction as determined by the reviewing authority. The $C \cdot t$ values for inactivation of 99% (2.0 log) and 99.9% (3.0 log), which include safety factors (uncertainty factors), are listed in Table 11.9 for various temperatures for a pH range of 6–9. The calculation of $C \cdot t$ values follows the procedures specified in the surface-water regulations. Chlorine residual must be detectable in at least 95% of the samples tested in the distribution system.

■ **EXAMPLE 11.10**

The groundwater from a well discharging 400 gpm at 10°C is to be disinfected with free chlorine for a 99.9% inactivation of viruses. (a) Calculate how many feet of 6-in.-diameter pipeline are needed to provide the required contact time for inactivation with a chlorine residual of 0.4 mg/l. (b) Estimate the volume required for a baffled contact tank, assuming a $t_{10}$ equal to 25% of the theoretical detention time and a chlorine residual of 0.4 mg/l.

*Solution.* From Table 11.9, the $C \cdot t$ for a 3.0-log inactivation by free chlorine at 10°C is 4 mg/l·min.

Contact time for a 0.4-mg/l residual = 4/0.4 = 10 min

  **a.** Velocity of flow in 6-in. pipe

$$= \frac{400 \times 0.134}{\pi(0.25)^2} = 270 \text{ ft/min}$$

  Length of pipeline for contact time

$$= 270 \times 10 = 2700 \text{ ft}$$

  **b.** Theoretical detention time required for contact

  tank = 10/0.25 = 40 min

  Volume of baffled tank = 400 × 40 = 16,000 gal

$$= 2140 \text{ ft}^3$$

These calculations show the problem of providing an adequate $C \cdot t$ for disinfecting the discharge of a well that pumps water directly into a distribution system. ■

# Disinfection of Wastewater

The degree of wastewater disinfection depends on the uses of the receiving watercourse or direct reuse of the wastewater. In general, conventionally treated wastewaters discharged to recreational waters with adequate dilution are given only plain chlorination. Reclaiming water for reuse involving human contact or protection of sensitive aquatic organisms for a food source (e.g., shellfish) requires filtration prior to chemical disinfection.

## 11.26 CONVENTIONAL EFFLUENT DISINFECTION

The kinds and numbers of pathogens in wastewater depend on the health of the contributing population. Although the concentrations of fecal microorganisms are reduced progressively by each stage in wastewater treatment, the effluent still contains a remnant of those contributed to the wastewater in human excreta. The removal of microorganisms by conventional treatment processes are summarized in Table 11.10 [22]. Helminth eggs are most likely to be removed by primary sedimentation because of their large size. In contrast, bacteria and protozoa removals are generally greater in trickling filtration. Apparently they adhere to the fixed-film biological growths coating the filter media. Removals in the activated-sludge processing rank low because of the variability in operational control. If the suspended biological floc under aeration settle out of suspension, removals of all the kinds of microorganisms can be as high as 99%. When the activated-sludge floc are carried out in the effluent, however, removals of microorganisms are greatly reduced. Even though sedimentation and biological treatment are able to remove 99% to 99.9% of microorganisms, the effluent still can contain significant concentrations of fecal viruses, bacteria, protozoal cysts, and helminth eggs.

Wastewater effluents can be chlorinated to inactivate pathogens in an attempt to protect public health where discharges enter surface waters used for body-contact recreation or as municipal water supplies. Disinfection is accomplished by rapid mixing of the chlorine solution with the wastewater followed by contact with the chloramines formed when the chlorine reacts with ammonia present in the wastewater (Eqs. 11.63 and 11.64). Acceptable disinfection is defined by the reduction of fecal coliforms to either less than

**TABLE 11.10**    REMOVAL OF MICROORGANISMS IN CONVENTIONAL
WASTEWATER TREATMENT

| Kind of microorganism | Primary sedimentation (%) | Trickling filtration (%) | Activated sludge (%) | Postchlorination (%) |
|---|---|---|---|---|
| Bacteria | 0–90 | 0–99 | 0–99 | 99–99.99 |
| Viruses | 0–90 | 0–90 | 0–90 | — |
| Protozoa | 0–90 | 0–99 | 0–90 | — |
| Helminths | 0–99 | 0–90 | 0–90 | 0 |

*Source:* Adapted from R. G. Feachem et al., *Sanitation and Disease, Health Aspects of Excreta and Wastewater Management* (Chichester: Wiley, 1983).

1000 per 100 ml or less than 200 per 100 ml, depending on the state standard. Some standards are more stringent and require pretreatment by granular-media filtration. The requirement for effluent disinfection may also be seasonal, depending on recreational water use. Since biologically treated wastewater contains approximately 1,000,000 coliforms per 100 ml, oxidative reduction of fecal coliforms from this large number to 200–1000 per 100 ml destroys most bacteria (Table 11.10). Nevertheless, viruses and protozoal cysts are more resistant to chlorination than bacteria, and helminth eggs are unharmed. Furthermore, the presence of suspended solids inhibits disinfection by harboring viruses within floc material shielding them from the action of the chlorine.

An efficient chlorination system provides initial contact of the chlorine solution with the wastewater during mixing followed by contact time with chloramines in a plug-flow tank for a minimum of 30 min at the peak hourly flow. Rapid blending can be accomplished by applying the chlorine into a pressure pipe conveying turbulent flow or into a channel immediately upstream from a mechanical mixer. Adequate plug flow can be achieved by a long narrow tank. Baffled rectangular tanks allow greater short-circuiting and are not as effective. A well-designed unit provides adequate disinfection of a secondary effluent with a dosage of 8–15 mg/l.

Control of chlorine dosage is extremely important for proper operation. Automatic residual monitoring and feedback control is necessary to prevent either inadequate disinfection or excessive chlorination, resulting in discharge of an effluent toxic to aquatic life. To protect receiving streams, some regulatory agencies have specified maximum chlorine residuals in undiluted effluents of 0.1–0.5 mg/l. Dechlorination may be required to detoxify a discharge after chlorination. This is usually done by adding sulfur dioxide in aqueous solution at the discharge end of the chlorination tank. The oxidation–reduction reaction (Eq. 11.69) between the chloramine residual and the sulfur dioxide is very rapid. Any excess sulfur dioxide applied reduces dissolved oxygen (Eq. 11.70).

$$NH_2Cl + SO_2 + 2H_2O = NH_4^+ + SO_4^{2-} + Cl^- + 2H^+ \qquad (11.69)$$

$$2SO_2 + O_2 + 2H_2O = 2SO_4^{2-} + 4H^+ \qquad (11.70)$$

## 11.27 TERTIARY EFFLUENT DISINFECTION

The inability of gravity sedimentation to remove small suspended solids following biological aeration is a major limitation of conventional treatment. Tertiary granular-media filtration preceded by chemical coagulation can be used for effluent clarification to remove tiny particles, like helminth eggs, and colloidal solids that interfere with the disinfecting action of chlorine.

Tertiary treatment and disinfection processes are similar to those applied in surface–water treatment for potable supplies: (1) rapid mixing of chemicals, flocculation, sedimentation, and granular-media filtration or (2) direct filtration after chemical mixing without sedimentation. The floccula-tion–sedimentation–filtration system is recommended for tertiary treatment of a conventional biologically treated effluent with BOD and suspended solids of 30 mg/l. For direct filtration, the biologically treated wastewater must be of consistently higher quality, which can be achieved only by extended aeration in an activated sludge process. The state of California regulatory guideline for direct filtration is a settled wastewater turbidity of less than 10 NTU. (Refer to Section 14.5).

The effluent chlorination system following filtration must be properly designed and operated to ensure effective disinfection. Rapid mixing is used to blend the chlorine solution with the filtered wastewater, and chlorine contact is by plug flow with a minimum of short-circuiting and backmixing in a long narrow tank. Common design specifications are a theoretical deten-tion time of at least 2.0 hr and an actual modal contact time of at least 1.5 hr. A chlorine residual in the range of 5–10 mg/l is maintained by automatic residual monitoring and feedback control. This provides a chloramine $C \cdot t$ in the range of 360–720 mg/l·min based on a $t_{10}$ equal to 60% of the theoretical detention time. If dechlorination of the effluent is necessary, a solution of sulfur dioxide can be added at the discharge end of the chlorination tank.

The standard for effective disinfection for wastewater reuse in the state of California for unrestricted landscape irrigation, recreational impound-ments, and irrigation of food crops eaten raw is a mean of 2.2 total coliforms per 100 ml and a turbidity less than 2 NTU. The allowable limits for specific uses specified by the state of Arizona are a mean of 25 fecal coliforms per 100 ml, 125 enteric viruses per 40 liters, and a turbidity of 5 NTU for open access (unrestricted) landscape irrigation; 200 fecal coliforms per 100 ml, 1 enteric virus per 40 liters, and a turbidity of 1 NTU for full-body contact recreation; and 2.2 fecal coliforms per 100 ml, 1 enteric virus per 40 liters, and a turbidity of 1 NTU for irrigating food eaten raw. These restrictions on coliforms, viruses, and turbidity establish a high microbiologic quality for reuse of wastewater. (Refer to Section 14.16.)

# Taste and Odor

Surface waters contain tastes and odors associated with decaying organic matter, biological growths, and chemicals originating from industrial waste discharges. Land drainage from snowmelt and spring rains contains pollut-ants that are difficult to remove, especially at the low water temperatures that occur in northern climates. During the summer, lake and reservoir supplies may be plagued by blooms of algae that impart compounds producing fishy, grassy, or other foul odors. Actinomycetes are moldlike bacteria that create an earthy odor. Well waters may contain dissolved gases, such as hydrogen sulfide, and inorganic salts or metal ions that flavor the water. Tastes in groundwater can be identified, but defining specific odor-bearing substances in surface waters is usually an impossible task. Therefore, the best practical treatment for each water supply is determined by experiment and experience.

## 11.28  CONTROL OF TASTE AND ODOR

Water treatment plant designs should allow maximum flexibility of operations for control of tastes and odors. Problems may vary considerably during different seasons of the year, and future adverse changes in water quality of a supply are difficult to predict. Breakpoint chlorination and treatment with activated carbon are common control techniques for surface-water supplies, while aeration is applied most frequently in groundwater processing. Preventative measures, such as selection of a source with the best water quality, should be given primary consideration. Regulatory controls should be used to reduce contaminants from waste discharges that enter surface and underground waters. Algal blooms may be suppressed in reservoirs by regular application of copper sulfate.

### Oxidative Methods

Chlorine dioxide, potassium permanganate, and ozone are strong oxidants capable of destroying many odorous compounds. These chemicals rather than heavy chlorination are favored since they do not form trihalomethanes. (Refer to Section 11.21.) Occasionally, a combination of chemicals can be used to achieve maximum control. The use of ozone is limited in the United States.

### Powdered Activated Carbon

Adsorption on activated carbon is the most effective means of taste and odor removal. It is primarily a surface phenomenon; that is, one substance is attracted to the surface of another. The larger the surface area of an adsorber, the greater its power. Carbon for water treatment is rated in terms of square meters of surface area per gram. One pound of activated carbon has an estimated surface area of 100 acres. Besides controlling tastes and odors, powdered carbon aids in sludge stabilization, improved floc formation, and reduction in non-odor-causing organics.

Carbon is fed to water either as a dry powder or as a wet slurry. The latter has the advantage of being cleaner to handle and assures complete effectiveness by thoroughly wetting the carbon. Although granular carbon adsorption beds are used extensively to purify product water in the food and beverage industries, their application in municipal water processing is limited by economic considerations.

Activated carbon can be introduced at any stage of water processing ahead of filtration. Although adsorption is nearly instantaneous, a contact time of 15 min or more is desirable before sedimentation or filtration. The best point of application is generally determined by trial and error based on previous experience. Carbon is often fed during flocculation or just prior to filtration. Since carbon adsorbs chlorine, these two chemicals should not be applied simultaneously or in close sequence.

## Aeration

Air stripping is effective for removing dissolved gases and highly volatile odorous compounds. Aeration as a first step in processing well water may achieve any of the following: removal of hydrogen sulfide, reduction of dissolved carbon dioxide, and addition of dissolved oxygen for oxidation of iron and manganese. Aeration is rarely effective in processing surface waters, since the odor-producing substances are generally nonvolatile.

# Fluoridation

The incidence of dental caries relative to the concentration of fluoride in drinking water is well established [23]. Low levels of fluoride result in increasing incidence of caries, while excessive fluoride results in mottled tooth enamel. At the optimum concentration for the local climate, consumers realize maximum reduction in tooth decay with no esthetically significant dental fluorosis. Recommended limits given in Table 11.11 are based on air temperature, since this influences the amount of water ingested by people. Medical studies have also indicated that fluoride benefits older persons by reducing the prevalence of osteoporosis and hardening of the arteries. Most public water supplies add fluoride ion to achieve an optimum level as a public health measure.

**TABLE 11.11**   RECOMMENDED FLUORIDE LIMITS FOR PUBLIC DRINKING-WATER SUPPLIES

| Annual average of maximum daily air temperatures based on temperature data obtained for a minimum of 5 yr (°F) | Fluoride ion concentrations (mg/l) | | |
|---|---|---|---|
| | Recommended limits | | |
| | Lower | Optimum | Upper |
| 50.0–53.7 | 0.9 | 1.2 | 1.7 |
| 53.8–58.3 | 0.8 | 1.1 | 1.5 |
| 58.4–63.8 | 0.8 | 1.0 | 1.3 |
| 63.9–70.6 | 0.7 | 0.9 | 1.2 |
| 70.7–79.2 | 0.7 | 0.8 | 1.0 |
| 79.3–90.5 | 0.6 | 0.7 | 0.8 |

*Source:* National Primary Drinking Water Regulations, U.S. Environmental Protection Agency.

# 11.29 FLUORIDATION

The fluoride compounds most commonly applied in water treatment are sodium fluoride, sodium silicofluoride, and fluosilicic acid (Table 11.12). Sodium silicofluoride is commercially available in various gradations for application by dry feeders. Sodium fluoride is also popular, particularly the crystalline type where manual handling is involved. Fluosilicic acid is a strong corrosive acid that must be handled with care. It is preferred in waterworks where it can be applied by liquid feeders without prior dilution.

Design of a fluoridation system varies with size and type of water facility, chemical selection, and availability of operating personnel. Small utilities often choose liquid feeders to apply solutions of NaF or $Na_2SiF_6$ that are prepared in batches. The solution tank may be placed on a platform scale for the convenience of weighing during preparation and feed. Saturator tanks containing a bed of sodium fluoride crystals yield a solution of about 4.0% (18,000 mg/l of F). Larger water plants use either gravimetric dry feeders to apply chemicals or solution feeders to inject full-strength $H_2SiF_6$ directly from the shipping drum. Automatic control systems use flow meters and recorders to adjust feed rate.

Application of fluoride is best in a channel or water main coming from the filters, or directly to the clear well. If applied prior to filtration, losses could occur due to reactions with other chemicals, such as coagulation with heavy alum doses or lime softening. If no treatment plant exists, fluoride can be injected into mains carrying water to the distribution system. This may be a single point or several separate fluoride feeding installations where wells supply water at different points in the piping network.

■ **EXAMPLE 11.11**
The fluoride ion concentration in a water supply is increased from 0.30 mg/l to 1.00 mg/l by applying 98% pure sodium silicofluoride. How many pounds of chemical are required per million gallons of water?

**TABLE 11.12**  COMMON CHEMICALS USED IN THE FLUORIDATION OF DRINKING WATER

|  | Sodium fluoride | Sodium silicofluoride | Fluosilicic acid |
|---|---|---|---|
| Formula | NaF | $Na_2SiF_6$ | $H_2SiF_6$ |
| Fluoride ion (%) | 45 | 61 | 79 |
| Molecular weight | 42 | 188 | 144 |
| Commercial purity (%) | 90–98 | 98–99 | 22–30 |
| Commercial form | Powder or crystal | Powder or fine crystal | Liquid |

*Solution.*    From Table 11.11, $Na_2SiF_6$ is 61% F.

$$\text{dosage} = \frac{(1.00 - 0.30) \times 8.34}{0.98 \times 0.61} = 9.77 \text{ lb/mil gal}$$    ∎

## 11.30 DEFLUORIDATION

Consumption of water containing excess fluorides over a period of several years results in dental fluorosis, identified by a permanent gray-to-black discoloration of tooth enamel. Studies have indicated that children consuming water containing 5 mg/l develop mottling to the extent of pitted enamel, resulting in loss of teeth. Although expensive, defluoridation is cost effective when the expenses of dental care and loss of teeth are considered. A few communities have been able to find alternative water sources to solve their fluoride problem without special treatment.

Current methods for defluoridation use either activated alumina or bone char [24]. Fluoride ion is removed by filtering water through these insoluble, granular media. After saturation with fluoride ion, bone char is regenerated by backwashing with a 1% solution of caustic soda and then rinsing the bed. Reactivation of alumina also involves rinsing the bed with a caustic solution.

# Corrosion and Corrosion Control

Internal corrosion of piping and valves is a serious problem in many water distribution systems. A national survey of the chemical composition of water in 130 systems throughout the United States revealed that 17% had highly corrosive water and approximately 50% more had moderately aggressive water [25]. In addition to economic loss, corrosive waters have the potential of degrading water quality by the dissolution of metals from the distribution system and household plumbing.

## 11.31 ELECTROCHEMICAL MECHANISM OF IRON CORROSION

The primary mechanism in metal dissolution is a chemical process accompanied by the passage of an electric current. In aerated water, the following complimentary electrode process is capable of absorbing electrons and therefore acting as a cathode.

$$O_2 + 2H_2O + 4e^- \rightarrow 4OH^-$$    (11.71)

As shown in Fig. 11.18, the principal reaction at the lower potential, the anode, is dissolution of the iron-releasing iron ions into solution. These are oxidized in the presence of hydroxide ions and dissolved oxygen to form ferric oxides that are only slightly soluble. Thus, rather than the ferrous ions being carried away in the water, leaving a clean surface, as in the case of corrosion in deaerated water, the iron oxides collect around the anode. Nodules created by this precipitation reduce the diffusion of oxygen and strengthen the anodic character of the surface covered by the oxide nodules. The result is pitting and the ultimate perforation of the pipe wall by corrosive action.

Corrosion is limited by formation of protective films that coat interior pipe surfaces. In potable water systems the carbonate–carbon dioxide equilibrium can be shifted and controlled by chemical treatment to deposit and maintain a calcium carbonate film. The mechanically applied cement lining in new ductile iron pipe is also preserved by this chemical stabilization. After establishing a water slightly oversaturated with calcium carbonate, polyphosphates can be applied at a dosage of about 2 mg/l to inhibit crystal nucleation. Thereby, the protective carbonate scale does not dissolve and the excess calcium ions cannot precipitate.

Temperature variations, dissolved salts, and the presence of microorganisms also influence corrosion. In general, inhibition of corrosion is more difficult at high temperature and low resistivity. Chlorides are considered the most corrosive salts since, in theory, chloride ions accumulate at anodes assisting the migration of iron ions toward the interface and preventing local precipitation of hydroxides. Heating a water that is in carbonate–carbon

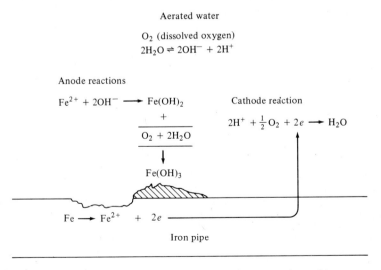

Aerated water

$O_2$ (dissolved oxygen)
$2H_2O \rightleftharpoons 2OH^- + 2H^+$

Anode reactions

$Fe^{2+} + 2OH^- \longrightarrow Fe(OH)_2$          Cathode reaction

$+$                                                $2H^+ + \frac{1}{2}O_2 + 2e \longrightarrow H_2O$

$\overline{O_2 + 2H_2O}$

$\downarrow$

$Fe(OH)_3$

$Fe \longrightarrow Fe^{2+} \quad + \quad 2e$

Iron pipe

**Figure 11.18** Electrochemical mechanism for corrosion of iron exposed to aerated water. Metal dissolution is the result of iron oxidation at the anode coupled with reduction of oxygen at the cathode.

dioxide equilibrium causes scale formation. Conversely, if a stable water is cooled, it becomes unbalanced and corrosive. Although microorganisms can influence corrosion, their reactions are usually related to unesthetic conditions, such as nodules formed by iron bacteria and decomposition of slime growths releasing foul odors.

## 11.32  WATER STABILIZATION

Formation of a protective film containing a mixed precipitate of calcium carbonate and iron oxides on unlined ductile iron and steel pipe depends on the following chemical conditions. First, the pH must be compatible with carbonate–carbon dioxide equilibrium of the water at service temperature. A calculated oversaturation equaling 4–10 mg/l of $CaCO_3$ is recommended to eliminate aggressive carbon dioxide. Second, the calcium and alkalinity values should be about equal, with minimums of 40 mg/l as $CaCO_3$ for waters with low sulfate and chloride contents [26]. Another source recommends minimums of 70 mg/l [27]. Greater concentrations of calcium and alkalinity provide more protection. Third, the water must contain sufficient dissolved oxygen (4–5 mg/l) to oxidize the nonprotective ferrous hydroxide to ferric hydroxide.

Several indices have been developed to quantify the aggressiveness of a water based on the following relationship:

$$CO_2 + CaCO_3 + H_2O \rightleftharpoons Ca(HCO_3)_2 \tag{11.72}$$

Excess free carbon dioxide shifts the reaction to the right with dissolution of calcium carbonate, while less than equilibrium carbon dioxide causes a shift to the left and formation of scale. Mathematical definition of this reversible reaction has been attempted by several researchers [28]. Although no index is universally applicable because of differing corrosion environments, the original and most commonly adopted index was developed by Langelier [29, 30].

The Langelier saturation index (SI) predicts the stability of a water to deposit or dissolve calcium carbonate in the pH range of 6.5–9.5. The index equates the pH, calcium content, alkalinity, and ionic strength as follows:

$$SI = pH - pH_s = pH - [(pK_2' - pK_s') + pCa^{2+} + pAlk] \tag{11.73}$$

where
- $pH$ = measured pH of the water
- $pH_s$ = pH at $CaCO_3$ saturation (equilibrium)
- $pK_2' - pK_s'$ = constants based on ionic strength and temperature
- $pCa^{2+}$ = negative logarithm of the calcium ion concentration, moles/liter
- $pAlk$ = negative logarithm of the total alkalinity, equivalents/liter

A positive value for the index signifies the water is oversaturated and can precipitate calcium carbonate. A negative number indicates a corrosive water. The saturation index measures the tendency of a water to dissolve or precipitate calcium carbonate but relates to neither the rate at which stability is attained nor the capacity.

The value of $pK_2' - pK_s'$ based on temperature and ionic strength is determined from Table 11.13. The ionic strength of a water is calculated as

$$\text{Ionic strength} = 0.5(C_1Z_1^2 + C_2Z_2^2 + \cdots + C_nZ_n^2) \tag{11.74}$$

where   $C$ = concentration of an ionic species, moles/liter
        $Z$ = valence of the individual ion

An unstable water can be treated with chemicals to achieve the proper amounts and balance between calcium and alkalinity at the proper pH. Application of lime increases pH, neutralizes free carbon dioxide, and adds calcium. Soda ash increases pH, neutralizes carbon dioxide, and adds alkalinity. Sodium bicarbonate adds alkalinity with some influence on pH. Carbon dioxide reduces pH and increases alkalinity. Normally, only one of these chemicals is needed to adjust pH and stabilize an untreated or treated water. An exception is remineralization of an extremely soft water or desalted seawater, which requires two chemicals in order to increase both calcium content and alkalinity while controlling pH. A common combination is lime and carbon dioxide.

The Langelier index, like any other method of defining calcium carbonate solubility, is only an approximate measure of the water stability in a system. For example, physical factors like the velocity of flow and residence time of water in the system affect the rate and amount of deposition. Conditions along the pipe wall influence the effectiveness of a coating. Thus, the precise chemical treatment must be based on close field and experimental observations of the system.

■ **EXAMPLE 11.12**

A desalted seawater contains 140 mg/l sodium and 216 mg/l chloride. For remineralization, 56.1 mg/l of CaO (40 mg/l Ca) are added to the water followed by carbon dioxide to convert the hydroxide ion from the lime to bicarbonate alkalinity. A trace amount of the alkalinity remains as carbonate ion. Based on the composition of the treated water given below, calculate the hardness and alkalinity as $CaCO_3$, pH, and the Langelier index at 25°C.

| Component | mg/l | Molecular weight | mol/l | Equivalent weight | meq/l |
|-----------|------|------------------|-------|-------------------|-------|
| $Ca^{2+}$ | 40 | 40.1 | 0.00100 | 20.0 | 2.00 |
| $Na^+$ | 140 | 23.0 | 0.00610 | 23.0 | 6.10 |
| $CO_3^{2-}$ | 2 | 60.0 | 0.00003 | 30.0 | 0.06 |
| $HCO_3^-$ | 118 | 61.0 | 0.00194 | 61.0 | 1.94 |
| $Cl^-$ | 216 | 35.5 | 0.00610 | 35.5 | 6.10 |

| Ionic strength | Total dissolved solids (mg/l) | 25°C $pK'_2$ | 25°C $pK'_s$ | 25°C $pK'_2-pK'_s$ | $pK'_2-pK'_s$ 0°C | 10°C | 20°C | 30°C | 40°C | 50°C | 60°C | 70°C | 80°C | 90°C |
|---|---|---|---|---|---|---|---|---|---|---|---|---|---|---|
| 0.000 |     | 10.26 | 8.32 | 1.94 | 2.45 | 2.23 | 2.02 | 1.86 | 1.68 | 1.52 | 1.36 | 1.23 | 1.08 | 0.95 |
| 0.001 | 40  | 10.26 | 8.19 | 2.07 | 2.58 | 2.36 | 2.15 | 1.99 | 1.81 | 1.65 | 1.49 | 1.36 | 1.21 | 1.08 |
| 0.002 | 80  | 10.25 | 8.14 | 2.11 | 2.62 | 2.40 | 2.19 | 2.03 | 1.85 | 1.69 | 1.53 | 1.40 | 1.25 | 1.12 |
| 0.003 | 120 | 10.25 | 8.10 | 2.15 | 2.66 | 2.44 | 2.23 | 2.07 | 1.89 | 1.73 | 1.57 | 1.44 | 1.29 | 1.16 |
| 0.004 | 160 | 10.24 | 8.07 | 2.17 | 2.68 | 2.46 | 2.25 | 2.09 | 1.91 | 1.75 | 1.59 | 1.46 | 1.31 | 1.18 |
| 0.005 | 200 | 10.24 | 8.04 | 2.20 | 2.71 | 2.49 | 2.28 | 2.12 | 1.94 | 1.78 | 1.62 | 1.49 | 1.34 | 1.21 |
| 0.006 | 240 | 10.24 | 8.01 | 2.23 | 2.74 | 2.52 | 2.31 | 2.15 | 1.97 | 1.81 | 1.65 | 1.52 | 1.37 | 1.24 |
| 0.007 | 280 | 10.23 | 7.98 | 2.25 | 2.76 | 2.54 | 2.33 | 2.17 | 1.99 | 1.83 | 1.67 | 1.54 | 1.39 | 1.26 |
| 0.008 | 320 | 10.23 | 7.96 | 2.27 | 2.78 | 2.56 | 2.35 | 2.19 | 2.01 | 1.85 | 1.69 | 1.56 | 1.41 | 1.28 |
| 0.009 | 360 | 10.22 | 7.94 | 2.28 | 2.79 | 2.57 | 2.36 | 2.20 | 2.02 | 1.86 | 1.70 | 1.57 | 1.42 | 1.29 |
| 0.010 | 400 | 10.22 | 7.92 | 2.30 | 2.81 | 2.59 | 2.38 | 2.22 | 2.04 | 1.88 | 1.72 | 1.59 | 1.44 | 1.31 |
| 0.011 | 440 | 10.22 | 7.90 | 2.32 | 2.83 | 2.61 | 2.40 | 2.24 | 2.06 | 1.90 | 1.74 | 1.61 | 1.46 | 1.33 |
| 0.012 | 480 | 10.21 | 7.88 | 2.33 | 2.84 | 2.62 | 2.41 | 2.25 | 2.07 | 1.91 | 1.75 | 1.62 | 1.47 | 1.34 |
| 0.013 | 520 | 10.21 | 7.86 | 2.35 | 2.86 | 2.64 | 2.43 | 2.27 | 2.09 | 1.93 | 1.77 | 1.64 | 1.49 | 1.36 |
| 0.014 | 560 | 10.20 | 7.85 | 2.36 | 2.87 | 2.65 | 2.44 | 2.28 | 2.10 | 1.94 | 1.78 | 1.65 | 1.50 | 1.37 |
| 0.015 | 600 | 10.20 | 7.83 | 2.37 | 2.88 | 2.66 | 2.45 | 2.29 | 2.11 | 1.95 | 1.79 | 1.66 | 1.51 | 1.38 |
| 0.016 | 640 | 10.20 | 7.81 | 2.39 | 2.90 | 2.68 | 2.47 | 2.31 | 2.13 | 1.97 | 1.81 | 1.68 | 1.53 | 1.40 |
| 0.017 | 680 | 10.19 | 7.80 | 2.40 | 2.91 | 2.69 | 2.48 | 2.32 | 2.14 | 1.98 | 1.82 | 1.69 | 1.54 | 1.41 |
| 0.018 | 720 | 10.19 | 7.78 | 2.41 | 2.92 | 2.70 | 2.49 | 2.33 | 2.15 | 1.99 | 1.83 | 1.70 | 1.55 | 1.42 |
| 0.019 | 760 | 10.18 | 7.77 | 2.41 | 2.92 | 2.70 | 2.49 | 2.33 | 2.15 | 1.99 | 1.83 | 1.70 | 1.55 | 1.42 |
| 0.020 | 800 | 10.18 | 7.76 | 2.42 | 2.93 | 2.71 | 2.50 | 2.34 | 2.16 | 2.00 | 1.84 | 1.71 | 1.56 | 1.43 |

*Source:* T. E. Larson and A. M. Buswell, "Calcium Carbonate Saturation Index and Alkalinity Interpretations," *J. Am. Water Works Assoc.* 34 (1942): 1667. Copyright 1942 by the American Water Works Association, Inc.

*Solution.* The hardness and alkalinity both equal

$$2.0 \text{ meq/l} \times 50 = 100 \text{ mg/l as } CaCO_3$$

For carbonate–bicarbonate equilibrium at 25°C using Eq. 11.13,

$$\frac{[H^+][CO_3^{2-}]}{[HCO_3^-]} = k_2 = 4.69 \times 10^{-11}$$

Rearranging and substituting values for the carbonate and bicarbonate concentrations in moles per liter, one finds

$$[H^+] = \frac{4.69 \times 10^{-11} \times [1.94 \times 10^{-3}]}{[3.0 \times 10^{-5}]} = 3.0 \times 10^{-9}$$

Then, calculating the pH from Eq. 11.3,

$$pH = \log \frac{1}{[H^+]} = \log \frac{1}{[3.0 \times 10^{-9}]} = 8.5$$

The ionic strength using Eq. 11.74 is calculated as follows:

$$
\begin{aligned}
Ca^{2+} &= 0.5 \times 0.00100 \times 4 = 0.00200 \\
Na^+ &= 0.5 \times 0.00610 \times 1 = 0.00305 \\
CO_3^{2-} &= 0.5 \times 0.00003 \times 4 = 0.00006 \\
HCO_3^- &= 0.5 \times 0.00194 \times 1 = 0.00097 \\
Cl^- &= 0.5 \times 0.00610 \times 1 = \underline{0.00305} \\
&\qquad\qquad\qquad\qquad\qquad\; 0.00913
\end{aligned}
$$

From Table 11.13 for 25°C, $pK_2' - pK_s' = 2.3$.

$$pCa^{2+} = -\log 0.00100 = 3.0$$

$$pAlk = -\log \left( \frac{1.94 + 0.06}{1000} \right) = 2.7$$

Substituting into Eq. 11.73 yields the Langelier index:

$$SI = 8.5 - (2.3 + 3.0 + 2.7) = +0.5$$

These calculations show that a desalted water treated with CaO and $CO_2$ to a pH of 8.5 meet the criteria for forming a protective film of $CaCO_3$. The saturation index is positive, signifying oversaturation. The 2 mg/l of carbonate ion is equivalent to 5 mg/l of oversaturation expressed as $CaCO_3$; the recommended range is 4–10 mg/l. Finally, the calcium concentration and alkalinity are balanced, and both equal 100 mg/l as $CaCO_3$, which is greater than the recommended minimum of 70 mg/l. ∎

## 11.33 CORROSION OF LEAD PIPE AND SOLDER

Lead in excess of 5 $\mu$g/l in raw water sources is rare. In tap water, it is a corrosion by-product from customer service lines, household piping, and plumbing fixtures. The sources are solder containing 50% lead and 50% tin joining copper pipes (which is no longer being used), old lead goosenecks that were used to connect the service line to the water main, and brass fixtures containing 3–8% lead. The highest lead concentration is in water that has been in contact with the service line and plumbing for several hours, while the content in flowing water is undetectable. Therefore, first-flush sampling is recommended in monitoring for compliance with the allowable limit of 15 $\mu$g/l at the customer's tap. Aging of copper plumbing reduces the dissolution of lead from solder to a level below 15 $\mu$g/l usually after 5 years. Chemical characteristics that increase dissolution of lead are low alkalinity, acidic pH, and high temperature. Recommended methods of controlling excessive lead concentrations are corrosion-control treatment and replacement of long lead service lines.

Lead corrosion control is complicated because of the many interdependent reactions that occur in the formation of protective films on lead surfaces. The effect of pH on lead solubility is very strong [31]. Based on this, one recommendation to reduce lead corrosion has been to increase the pH to 8.0 or above. Nevertheless, dissolution in actual service lines is often lower than the theoretically predicted concentration because of protective coatings and other factors. The formation of an effective film of lead carbonates (e.g., $PbCO_3$) depends on pH and alkalinity, which has resulted in a second recommendation of chemical addition for supplementing alkalinity. But because the relationship is complex, increasing the carbonate content (except in very low alkalinity waters) can be complicated by the precipitation of calcium carbonate. Although $CaCO_3$ deposition is accepted practice to reduce iron corrosion, the adequacy of protection from lead leaching has been questioned by the argument that sloughing off of the deposit or incomplete coating of the pipe could result in periodic dissolution of lead in high concentrations. Despite this, in well waters high in calcium and alkalinity, service lines with lead goosenecks and copper lines joined with lead solder have shown negligible lead content in first-flush samples.

Orthophosphate addition is a recommended remedial measure to reduce lead levels in tap water. Lead forms several orthophosphate deposits [e.g., $Pb_5(PO_4)_3OH$]. Film deposition depends on orthophosphate concentration, alkalinity, pH, and temperature. Although the formation rate is slow, orthophosphate addition plus pH control can reduce leaching of lead. Polyphosphate addition is controversial; the chemical differences between ortho- and polyphosphates in relationship to formation of protective film is poorly understood [31].

Verifying the effectiveness of a lead corrosion control program is time-consuming and complicated by changing conditions. The most likely measures are adjustment of pH, addition of phosphate, supplementing alkalinity, or some combination of these. The only way to measure the success of corrective treatment is to monitor changes in lead concentrations in selected first-flush tap waters. Because formation of protective films is a slow process, changing field conditions can occur that are independent of the treatment. For example, the most significant source of lead at the tap is lead-based solder in copper service lines [32]. Aging of the solder rapidly decreases lead dissolution. In fact, the most important measure to control lead in drinking water has been to require the use of lead-free solder and plumbing fixtures in public water supplies.

## 11.34  CATHODIC PROTECTION

Cathodic protection prevents corrosion by lowering the electrode potential so that the metal structure or pipe being protected acts as a cathode. Electrons are supplied to the protected metal faster than they are lost. The two methods of applying cathodic protection to iron are (1) using reactive, consumable anodes made of magnesium or aluminum and (2) applying an external electromotive force.

Attaching a mass of magnesium metal to iron creates an electric cell that sacrifices the magnesium anode, preventing dissolution of the iron. Buried iron and steel water mains can be protected from corrosion of the pipe exterior by this method. In water distribution, application is limited to reducing deterioration of older uncoated pipe or new piping laid in very corrosive soils. Modern ductile iron pipes are coated with coal tar or enamel on the exterior for protection under normal circumstances. If the soils are considered corrosive, the pipes are placed in sheaths of polyethylene tubing that are taped together as each pipe is laid to provide a continuous watertight sleeve.

In the second method, an external current is supplied by rectifying alternating current to direct current and impressing a potential difference of several volts between the structure being protected and an incorrodible material. Buried tanks are protected from external corrosion by placing the anodes in the ground surrounding the tank. Cathodic protection is commonly used to prevent corrosion of the interior surfaces of steel tanks by suspending anodes in the contained water. The electric current is impressed between the tank and the anodes.

## 11.35  CORROSION OF SEWER PIPES

Corrosion in sewers is the destruction of pipe materials by chemical action. Sewer corrosion can result from biological production of sulfuric acid and is caused by strong industrial wastes unless they are neutralized prior to disposal. Most municipal sewer ordinances prohibit industrial wastes having a pH less than 5.5 or higher than 9.0 or having other corrosive effects.

Crown corrosion in sanitary sewers is most prevalent in warm climates where they are laid on flat grades or where the sulfur content of the wastewater is high. Biological activity in wastewater in a sewer creates anaerobic conditions, producing hydrogen sulfide. Condensation moisture on the crown and walls of the sewer pipe absorbs hydrogen sulfide and oxygen from the atmosphere in the sewer. The sulfur-oxidizing bacteria *Thiobacillus* form sulfuric acid in the moisture of condensation:

$$H_2S + O_2 \xrightarrow{\text{Thiobacillus}} H_2SO_4 \qquad (11.75)$$

In concrete sewers, sulfuric acid reacts with lime to form calcium sulfate, which lacks structural strength. If the concrete is sufficiently weakened, the pipe can collapse under heavy overburden loads.

The best protection for sanitary sewers is a corrosion-resistant pipe material such as vitrified clay or plastic. In large sewers, where size and economics dictate concrete pipe, sacrificial concrete can be placed in the crown of the pipe. Crown corrosion can be retarded by ventilation or by chlorinating the wastewater to control hydrogen sulfide generation. Recent advances for the protection of concrete pipe include the development of synthetic coatings and linings.

# Reduction of Dissolved Salts

The common processes for desalination of seawater are distillation and reverse osmosis; for desalting brackish groundwater, the processes are reverse osmosis and electrodialysis. The selection of process is based on size of plant, sources of energy, and capital and operating costs.

## 11.36 DISTILLATION OF SEAWATER

Typical seawater has a salinity (total dissolved solids) of 35,000 mg/l, of which 30,000 mg/l is NaCl. The generally accepted quality standards for drinking water are 500 mg/l of total dissolved solids and 200 mg/l of chloride. Distillation is cost competitive for desalination of feedwater with a high salt content since the process operates virtually independent of influent solids concentration. Moreover, a product purity of less than 100 mg/l is easily attained [33].

Distillation involves heating feedwater to the boiling point and then into steam to form water vapor, which is then condensed to yield a salt-free water. The principal commercial processes are multistage flash distillation and thin-film multiple-effect evaporation. The prefix "multi" in the names means that a series of evaporation–condensation units is employed to obtain multiple reuse of the energy content of the heated steam. There may be as many as 15–25 stages.

*Multistage flash distillation* is illustrated schematically in Fig. 11.19. Seawater entering the plant is initially heated in a heat recovery unit in which the hot desalted product water and waste brine discharge are cooled. The warmed seawater is then blended with recycled brine and passed through a series of condenser tubes in the evaporator chambers for further heating. In

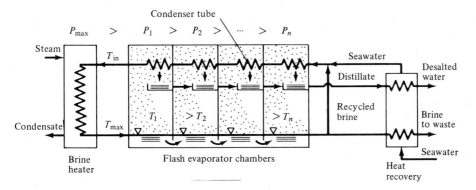

**Figure 11.19** Multistage flash distillation.

the process of condensing the steam to distillate (the desalted water), the temperature of the brine is increased in stages to temperature $T_{in}$ as it enters the brine heater. In this unit, the temperature of the brine feed is raised by external thermal energy to just below the saturation temperature $T_{max}$ under a pressure $P_{max}$. The hot brine is then discharged through the series of stages each at a reduced pressure compared with the previous stage. The pressure $P_{max} > P_1$, $P_1 > P_2$, and so forth. Because of the reduction in pressure, a portion of the heated feed flashes to vapor in each stage to obtain equilibrium with the vapor condition prevailing in each individual stage. This results in a temperature drop in each stage (e.g., $T_{max}$ drops to $T_1$, and $T_1$ drops to $T_2$) arriving at the minimum temperature $T_n$ in the last stage. The total tempera-ture drop is usually from a $T_{max}$ of 250°F to a $T_n$ of 100°F. The fraction of recirculating brine that can be flashed with each cycle is restricted to 0.10–0.15. Therefore, to produce the desired rate of distillate requires a minimum brine recirculaton rate in the range of 10–6.6 times the production rate of desalted water. The brine wasted from the recycling feed line may contain approximately 70,000 mg/l for seawater input of 35,000 mg/l.

The controlling parameters for output of desalted water and energy con-sumption are the temperature drop allowed in each stage, the difference between the brine inlet temperature to the first stage and the discharge temperature at the last (overall flash range), and the stage heat transfer coefficients. The principal unavoidable heat losses result from imperfect heat transfer by the condenser tubes and heat exchangers. Other losses include poor venting resulting in vapor blanketing of the condenser tube surfaces and tube fouling as a result of scale formation. Energy consumption in distillation is always well in excess of the ideal theoretical minimum. For a particular installation, efficiency is related to design factors such as the number of stages.

*Thin-film multiple-effect evaporation* is the second process widely used for distillation of seawater. The steam generated in one effect condenses on the outside of long vertical tubes in the next effect, evaporating more water from a film of brine that runs down the inside of the tube. Tracing this on Fig. 11.20, prime steam enters the shell of the first effect where it condenses on the outside of the tubes. The latent heat of condensation furnishes the energy required to evaporate a portion of the brine feed within the tubes. The partially concentrated brine proceeds to the second effect, which operates at a slightly lower pressure. The vapor leaving the first effect condenses on the tubing of the second effect, causing further evaporation of water from the brine. This process continues from effect to effect until the lowest pressure vapor is condensed in a final condenser by giving up its latent heat to circulating cooling water. The combined condensate from all effects constitutes the product water from the plant. Using this same principle, several variations may be designed into the multiple-effect process. One modification is the use of horizontal rather than vertical tubes with the steam inside the tubes and the evaporating brine flowing in a film on the outside tube surfaces.

## 11.37  REVERSE OSMOSIS

Reverse osmosis is the most common process for reducing the salinity of brackish groundwater. It is the forced passage of water through a membrane against the natural osmotic pressure to accomplish separation of water from a solution of dissolved salts. The process of osmosis is illustrated in Fig. 11.21, where a thin membrane separates waters with different salt concentrations. In direct osmosis, water naturally flows from the side of lower salt concentration through the membrane to the solution of higher concentration, attempting to equalize the salt content; the membrane allows water flow while blocking the passage of salt ions. If pressure is applied to the side of

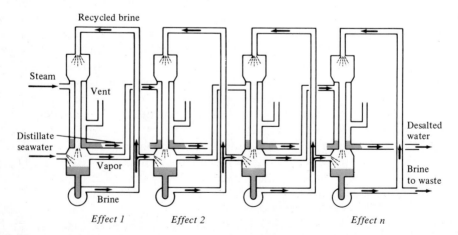

**Figure 11.20** Thin-film multiple-effect evaporation with long vertical condenser tubes.

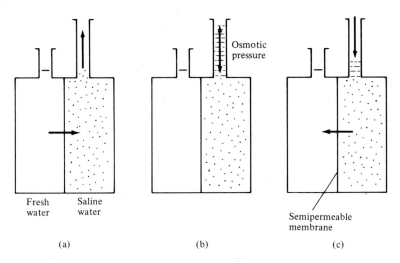

Fresh water    Saline water

Osmotic pressure

Semipermeable membrane

(a)    (b)    (c)

**Figure 11.21** Illustrations describing the process of reverse osmosis to remove dissolved salts from water. (a) Direct osmosis. (b) Osmotic equilibrium. (c) Reverse osmosis.

higher salt content, this flow of water can be prevented at a pressure termed the *osmotic pressure* of the salt solution. In reverse osmosis, the water is forced by high pressure from a salt solution through the membrane into fresh water, separating desalted water from the saline solution. The rate of flow through a reverse-osmosis membrane is directly proportional to the effective pressure (i.e., to the difference between the applied and osmotic pressures). In practice, operating pressures vary between 350 and 1500 psi, with a typical range of 600–800 psi. The quantity of product water is 70%–90% for a feed of brackish groundwater and about 30% for a feed of seawater.

The two common membrane materials are cellulose acetate and aromatic polyamide. Cellulose acetate membranes have a high flowrate per unit area and are commonly used to form tubes of spiral-wound flat sheets. In contrast, polyamide membranes have a lower specific flowrate and are manufactured in the form of hollow fibers to achieve the maximum surface area per unit volume, which is about 15 times that of spiral-wound membranes. By assembling the membranes in modular units, a large membrane surface area can be compacted into a cylindrical pressure vessel fitted with an inlet for the saline feedwater and outlets for the product water (permeate) and reject brine.

A *spiral-wound module* (Fig. 11.22) is made up of large membrane sheets covering both sides of a flat sheet of porous backing material that collects the permeate (product water). The membranes are sealed on the two long edges and one end to form an envelope enclosing the permeate collector. The other end of the membrane envelope is sealed to a perforated tube, which receives and carries away the permeate from the collectors. Several of these membrane envelopes, with mesh spacers in between for brine flow, are rolled

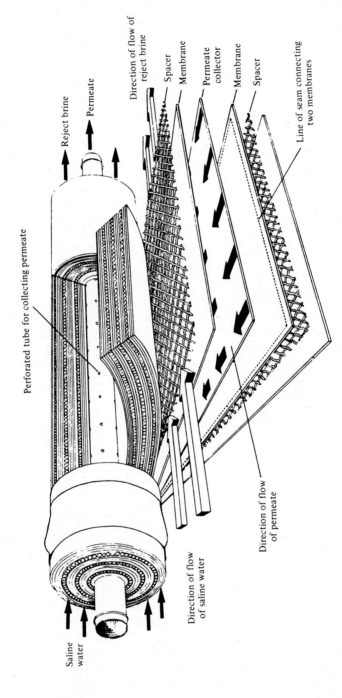

Reject brine

Permeate

Direction of flow of reject brine

Spacer

Membrane

Permeate collector

Membrane

Spacer

Line of seam connecting two membranes

Perforated tube for collecting permeate

Direction of flow of permeate

Direction of flow of saline water

Saline water

**Figure 11.22** Spiral-wound module for reverse osmosis. (Courtesy of Degremont, 183 Avenue de Juin 1940—92508 Rueil Malmaison CEDEX—France.)

up to form a spiral-type module. Operation of a spiral-wound module is diagrammed in Fig. 11.22. Saline water enters the end of the module through the voids between the membrane envelopes provided by the spacers. Under high pressure, water is forced from the brine in the spacer voids through the membranes and conveyed by the enclosed porous permeate collectors to the perforated tube in the center of the module. Brine reject discharges from the spacer voids at the outlet end of the module.

A *hollow-fiber module* is a pressure vessel containing a very large number of microfiber membranes densely packed in a U-bundle with their openings secured in an end block of the module. The hollow fibers, finer than a human hair, have an outside diameter of 85–100 $\mu$m and an inside diameter of 42 $\mu$m. Because of their small diameter and thick wall, these tubes can withstand the high reverse-osmosis pressure required to force water from the surrounding brine into the hollow cores of the fibers. The process flow is shown in Fig. 11.23. Saline water enters the module through a central perforated feed tube and flows radially through the fiber bundle toward the outer shell of the cylinder. Under high pressure, water enters the hollow fibers and exits from their open ends at the discharge end of the module. Reject brine arriving at the outer shell of the cylinder is collected by a flow screen and conveyed from the module.

The potential for membrane fouling and scaling must be considered in both the design and operation of a reverse-osmosis system. The feedwater must be clear and low in chemical ions that can cause scale (e.g., calcium). In groundwater, silt and iron oxides are the two common contaminants resulting from poor well construction. If the gravel pack and screen are not properly designed and placed, fine sand and silt from the aquifer can be carried out in the well water. Since many groundwaters are aggressive, the well casing, pump column, and transmission piping must be constructed of noncorrosive materials to prevent formation of rust particles.

The three most common chemical scalants are calcium sulfate, calcium carbonate, and silicon dioxide. As shown schematically in Fig. 11.24, the reject brine is in contact with the membrane. Therefore, the scaling potential is determined by the concentration of ions in the brine, which in turn is controlled by the recovery of product water. For example, if the concentration of calcium ion in feedwater is 200 mg/l and the recovery of product water is 75% with 7 mg/l, the reject brine contains 779 mg/l of calcium ion (4 times the concentration in the feedwater). The calcium concentration can be lowered by reducing the permeate recovery and wasting more of the feedwater as brine. If a high recovery is necessary, such as 90%, the feedwater can be pretreated by softening to reduce the calcium content. Regardless of the methods used to reduce the calcium concentration, sodium hexametaphosphate or a polymeric antiscalant is commonly applied to inhibit the formation of calcium sulfate [34].

Calcium carbonate precipitation is controlled by acidifying the feedwater to convert bicarbonate ion to carbon dioxide, thus making the water corrosive. Since the reverse-osmosis process is a closed system, the carbon dioxide

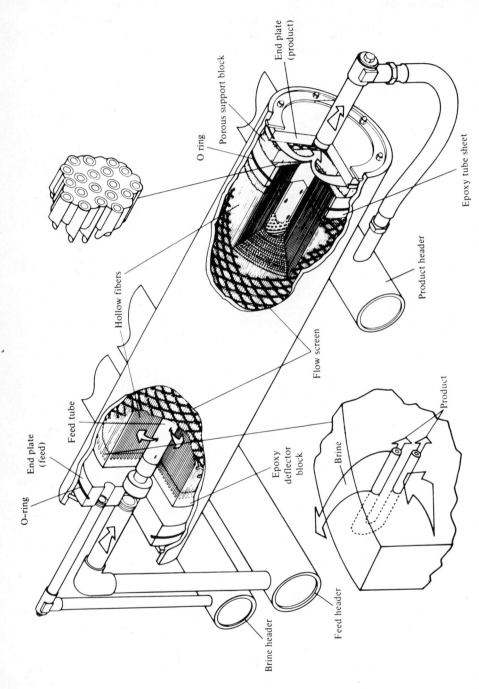

**Figure 11.23** Hollow-fiber module for reverse osmosis. [Courtesy of Permasep Products, E. I. du Pont de Nemours & Co. (Inc.).]

End plate (product)

Porous support block

O ring

Epoxy tube sheet

Product header

Hollow fibers

Flow screen

Feed tube

End plate (feed)

O-ring

Epoxy deflector block

Brine

Product

Brine header

Feed header

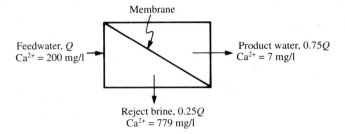

Membrane

Feedwater, $Q$
$Ca^{2+} = 200$ mg/l

Product water, $0.75Q$
$Ca^{2+} = 7$ mg/l

Reject brine, $0.25Q$
$Ca^{2+} = 779$ mg/l

**Figure 11.24** Schematic diagram of a reverse-osmosis module illustrating the concentration of calcium ion in the reject brine relative to the feedwater and product water for 75% recovery.

cannot escape and appears in the permeate and reject brine. Silica must be removed by pretreatment of the feedwater by chemical coagulation and filtration. High feedwater temperature deteriorates membrane material; hence, hot groundwaters require cooling prior to processing.

The product water from reverse osmosis is highly corrosive with low pH and high concentration of carbon dioxide. The three methods of stabilization are degasification (decarbonation), addition of lime or soda ash for neutralization and increase in alkalinity, and blending with raw water. Degasification can be performed in a packed column with the water sprayed in at the top and percolating through the media against a countercurrent of air. Carbon dioxide is reduced to less than 10 mg/l and the pH raised to near neutrality. Addition of lime slurry [$Ca(OH)_2$] neutralizes the carbon dioxide, increases the calcium concentration, and raises the pH. Soda ash ($Na_2CO_3$) addition increases alkalinity and raises the pH. It also adds sodium ion, which is undesirable for both corrosion control and for consumption as drinking water by persons with hypertension (high blood pressure). Nevertheless, soda ash is often applied in small systems since it is much easier to feed; a solution of soda ash is a clear liquid, whereas lime is a milky slurry. Blending permeate (after degasification if performed) with raw water is often an economical method of partial or, in some cases, complete stabilization. The blending ratio depends on the chemical characteristics of the two waters. Since the fluoride ion content is reduced in desalting, supplemental fluoridation of the stabilized water to the optimum level is recommended.

Reverse osmosis in advanced wastewater treatment at Water Factory 21 in Orange County, California, is discussed in Section 14.20.

This limited discussion and the calculations in Example 11.13 are only an overview of reverse osmosis. The presentation by Ko and Guy [35] on brackish and seawater desalting expands the topics of membranes, modules, pretreatment, and design and operation. In the same book, Ridgway [36] reports the results of research studies on biological fouling of membrane surfaces.

■ **EXAMPLE 11.13**

A reverse-osmosis plant treats warm brackish groundwater with total dissolved solids of 2600–2700 mg/l and pH 6.8–7.2. Since the well water is free of silt, iron, manganese, and low in silica, no granular-media filtration is required. First step in pretreatment is to cool the water in a heat exchanger to 35°C, when necessary, for longer membrane life. Next, the water is acidified to pH 5.8 with 150 mg/l of sulfuric acid to prevent $CaCO_3$ scale formation, and 10 mg/l of hexametaphosphate are added to prevent $CaSO_4$ formation. Cartridge filters remove particles down to 5 microns in size; the average life of the replaceable filter elements is 12 weeks. To force the pretreated water through the membranes, a high-pressure pump at each reverse-osmosis unit increases the pressure to 370 psi. Each unit has 13 modules, with 9 in the first stage and 4 in the second stage. The feedwater applied to the first stage produces 50% permeate and 50% brine. The first-stage brine is applied to the second stage to again produce 50% permeate and 50% brine, which is rejected. Therefore, the total recovery of product water is 75% of the feedwater, and the reject brine is 25%. The product water is stripped to remove carbon dioxide in packed countercurrent columns, raising the pH from 5.8 to about 7.0. It is then stabilized by adding approximately 10 mg/l of soda ash to raise the pH to 8.2–8.5. The fluoride ion concentration is increased from 0.2 mg/l to the optimum of 0.8 mg/l by adding fluosilicic acid. The finished water has a total dissolved-solids concentration of 250–350 mg/l, alkalinity of 80–100 mg/l, and calcium ion concentration of approximately 7 mg/l.

Trace the chemical changes that occur in the water during treatment. The milli-equivalents-per-liter bar graph of the untreated groundwater is shown in Fig. 11.25a.

*Solution.* Adding 150 mg/l of sulfuric acid for acidification to a pH of 5.8 converts bicarbonate ion to carbon dioxide and increases the sulfate ion content as shown in Fig. 11.25b.

If the water recovery is 75%, the concentrations of ions in the reject brine are about 4 times the concentrations in the acidified feedwater. Therefore, the contents are calcium ion 860 mg/l (0.0215 moles/l), sulfate ion 2270 mg/l (0.0236 moles/l), alkalinity 228 mg/l (4.56 meq/l), total dissolved solids 10,200 mg/l, and ionic strength 0.226. The scaling potential of $CaCO_3$ can be estimated by calculating the Langelier saturation index by using Eq. 11.73.

$$SI = 5.8 - [2.1 + p(1/0.0215) + p(1000/4.56)]$$

$$= -0.3 \quad \text{(non-scale-forming water)}$$

(Table 11.13 lists $pK_2' - pK_s'$ values for ionic strengths less than 0.020 and total dissolved solids less than 800 mg/l. The value of 2.10 used in the above calculation was taken from a Langelier diagram in Reference 37 for a total dissolved-solids content of 10,200 mg/l.)

The potential for $CaSO_4$ scale is estimated by calculating the product of the ionic molar concentrations of the calcium and sulfate ions and comparing the result to the solubility-product constant. For the acidified brine,

$$[Ca^{2+}][SO_4^{2-}] = 0.0215 \times 0.0236 = 0.51 \times 10^{-3}$$

The estimated $K_{sp}$ for a brine after addition of sodium hexametaphosphate is $1.0 \times 10^{-3}$; hence, this brine is unsaturated and non-scale-forming.

Figure 11.25c is the approximate bar graph of the finished water with a total

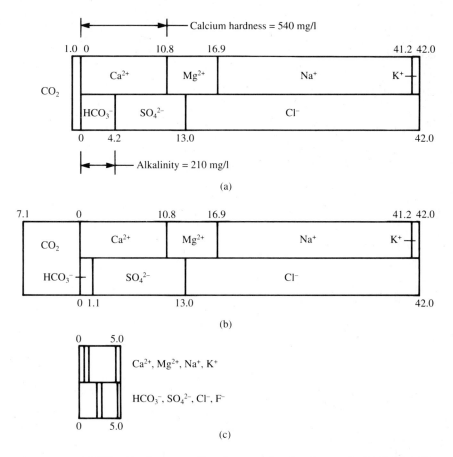

**Figure 11.25** Milliequivalents-per-liter bar graphs for Example 11.13. (a) Untreated groundwater with a total dissolved solids concentration of 2600 mg/l. (b) Feedwater after acidification to pH 5.8. (c) Product water after post treatment with a total-dissolved-solids concentration of 350 mg/l.

dissolved-solids concentration of 350 mg/l after carbon dioxide stripping, soda ash addition, and fluoridation.    ∎

## 11.38 ELECTRODIALYSIS

The principle of electrodialysis is separation of ions from water by pulling them through selective ion-permeable membranes by electrical attraction [38]. Figure 11.26 illustrates conventional dialysis as a stack of membranes interposed in the path of a direct current generated by the end electrodes. The selectivity of the membranes alternate: $A$ membranes are permeable only to anions such as $Cl^-$, and $B$ membranes are permeable only to cations such as $Na^+$. To minimize cell resistance to current flow, the space between

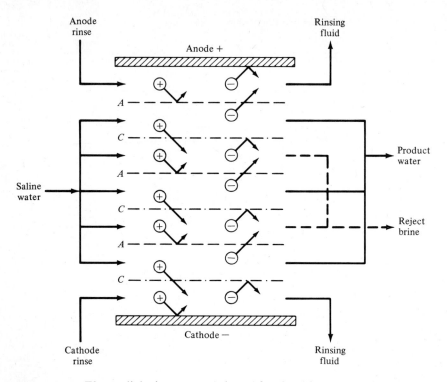

**Figure 11.26** Electrodialysis process scheme for desalting water. Membranes labeled *A* are permeable to anions only, and those labeled *C* are permeable to cations only.

membranes may be only 1 mm. Saline water passes between the membranes perpendicular to the path of the current at sufficiently high velocities to induce mixing. The direct current passing through the saline water draws anions toward the anode from one compartment to the adjacent one through the *A* membranes. In this compartment, the anions are trapped because of the *C* membrane between them and the anode. Cations are drawn in a similar manner in the opposite direction. From this movement of ions, the water in alternate compartments is reduced in ionic concentration while the water in the other compartments increases in salinity. The two flows become, respectively, the desalted product water and the reject brine.

# Nitrate Removal

The nitrate ion ($NO_3^-$) is not removed from water by chemical coagulation or from wastewater by conventional biological treatment. Nitrate in surface waters rarely exceeds the maximum contaminant level of 10 mg/l because of uptake as a plant nutrient in photosynthesis and biological conversion to nitrogen gas in an anoxic environment. However, nitrate is a stable ion in groundwater contaminated by leaching of unused nitrogen fertilizer and nitrate from decomposing organic matter. The most feasible method for removing nitrate from groundwater is anion exchange, as discussed in the following sections. The removal from wastewater by denitrification is discussed in Sections 14.13–14.14.

# 11.39 ANION EXCHANGE FOR NITRATE REMOVAL

The common anion exchange medium for removal of the nitrate ion is a strongly basic resin that substitutes the chloride ion for adsorbed anions. The resin beads are spherical in the range of 0.4–1.2 mm diameter with a specific gravity of 1.3–1.4. With this fine a resin, the bed is susceptible to plugging and loss of efficiency caused by mineral deposits, particulates, and bacterial growths. Thus, the groundwater must be free of silt, iron, manganese, and be nonscaling. If a bed is fouled, cleaning by backwashing is not possible since the sand, metal oxides, and most other contaminants have greater resistance to fluidization than the resin beads.

Aging of a strongly basic resin results in decreased capacity and increased leakage as the sites on the resin change from strongly basic to a mixture of weakly and strongly basic sites. The life of a resin should be at least 6 years with a maximum loss of 25% exchange capacity in 2 years. The factors influencing resin life are type of resin, regeneration temperature, regeneration level (salt applied per unit bed volume), operating water temperature, water flowrate, bed depth, and chemical characteristics of the water.

Sketched breakthrough curves for an anion exchange resin bed used for nitrate removal are shown in Fig. 11.27. The concentrations of ions in the feedwater are shown on the right edge of the diagram, where all of the ions are breaking through the bed. The sulfate ion is removed in preference to all other anions. The bicarbonate ion is removed initially and then slowly increases in leakage through the resin until the concentration is unchanged from the feedwater. The release of chloride ion by the resin to the filtered water is highest in concentration at the beginning of the run when sulfate, bicarbonate, and nitrate are being replaced. Then the concentration gradually decreases to a relatively constant release to sustain sulfate and nitrate removal. The nitrate ion continuously leaks through the resin bed at a low concentration throughout the filtration until a sudden increase at breakthrough. The initial concentration during breakthrough is much greater than the concentration in the feedwater. As seen from the decreasing release of $Cl^-$ and the delayed rise of the $SO_4^{2-}$ curve, the $NO_3^-$ previously absorbed by the resin is being displaced by the $SO_4^{2-}$ ion. The maximum practical operating range is from less than full regeneration (to reduce salt consumption) on the left to near the point of nitrate breakthrough.

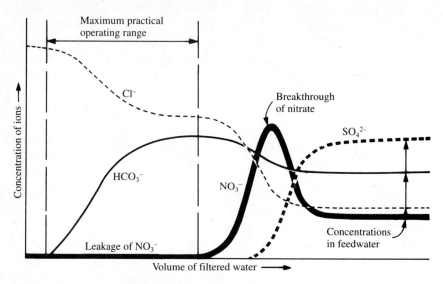

**Figure 11.27** Typical breakthrough curves for an anion exchange resin bed. (Based on data from Boyle Engineering Corp.)

## 11.40 ANION EXCHANGE TREATMENT SYSTEM

The common arrangement for a nitrate-removal plant consists of three ion exchange vessels, a brine tank, and piping for filtration and regeneration [39]. Figure 11.28 diagrams the flow with the two vessels in service and the third unit on the right being regenerated. Feedwater enters the operating vessels at the top, passes down through the resin beds, and leaves through the underdrain manifolds at the bottom. Since the nitrate–nitrogen concentration is reduced to 1–2 mg/l, well below the maximum contaminant level of 10 mg/l, the treated water is blended with untreated water prior to discharge to the distribution system. The ion exchange vessels are operated in rotational sequence with two units always in service. This provides a treated water with more uniform chemical characteristics since the lag unit is treated water in the first half of the operating range and the lead unit in the second half of the operating range, as identified in Fig. 11.27. Three vessels are also necessary in case one is taken out of service for maintenance.

The regeneration cycle consists of three stages: filling of the resin bed with brine, slow rinsing, and backwashing. After the vessel is taken out of service, diluted brine is pumped into the ion exchange vessel through an internal distributor located just above the surface of the resin bed. The dose is about 1.5 bed volumes. The downward flow of the brine, with limited mixing with the water in the vessel above the bed, exits through the underdrain manifold and discharges to waste.

The slow-rinse water, which is treated water, enters the vessel at the top

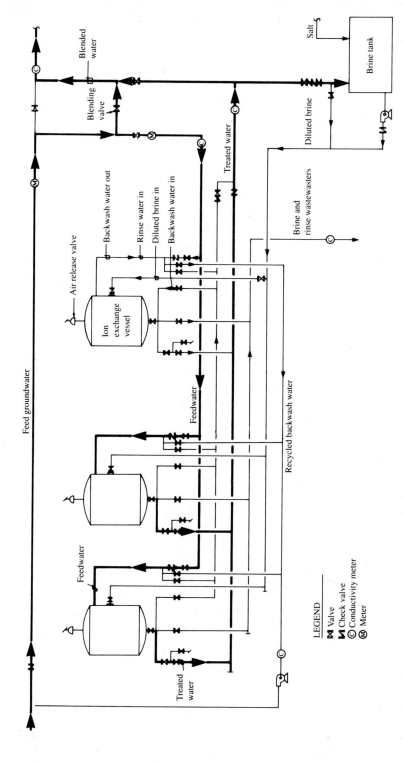

**Figure 11.28** Flow diagram of an anion exchange nitrate-removal plant with two vessels filtering water and the third vessel on the right undergoing regeneration. (Courtesy of Boyle Engineering Corp.)

through the same distributor used for the untreated water supply during filtration. It exits through the underdrain to waste.

Backwashing is started by upflow of treated water through one half of the underdrain manifold on one side, followed by a pause to allow the resin to settle. The same is done through the other half of the underdrain manifold and a second time through the first half. The wash water exits out the manifold at the top of the vessel. This backwashing technique declassifies the resin bed, preventing the resin beads from settling into graded layers with the smallest beads on top and largest beads on the bottom. If this stratification were allowed to occur repeatedly after backwashing, the large beads at the bottom would remain in the nitrate form after bed regeneration and cause increased nitrate in the product water when the bed is returned to service.

McFarland, California, has two nitrate-removal plants, each with a nominal design flow of 1.0 mgd, located at two different well sites [40]. The groundwater is clear (free of sand and silt), slightly saline (total dissolved salts of 500–600 mg/l), moderately hard (about 80 mg/l of calcium hardness), moderately high in sulfate (about 80 mg/l), free of iron and manganese, and contaminated with approximately 15 mg/l of nitrate nitrogen.

The plants use three ion exchange vessels, as diagrammed in Fig. 11.28, each with a diameter of 6 ft containing a resin bed 3 ft deep. Approximately one half of the water bypasses the vessels through the blending valve for mixing with the treated water. With 14–16 mg/l in the untreated water and 1.5–2.5 mg/l in the treated water, the blended water contains 7–9 mg/l of nitrate–nitrogen. Pressure from the well pump forces the water through the ion exchange vessels and blending valve and into a hydropneumatic tank. (The distribution system does not have elevated storage tanks.) Pressure switches in the hydropneumatic tank control the operation of the well pump and, therefore, the operation of the plant. Nitrate and sulfate concentrations in the treated water are automatically monitored, and the plant shuts down if processing fails.

Each bed is regenerated after filtering approximately 150,000 gal (250 bed volumes) of water. Wastewater flows total 3%–4% of the water processed through the resin beds, which is about 50% rinse water, 35% backwash water, and 15% dilute brine. The amount of sodium chloride for regeneration is 3–4 lb per 1000 gal of water processed through the resin. The regeneration wastewater is discharged to the sanitary sewer and blended with the municipal wastewater. After treatment in aerated lagoons, the municipal wastewater is applied on land for irrigation of cotton.

# Volatile Organic Chemical Removal

The two processes for removal of volatile organic chemicals (VOCs) are stripping by aeration and granular activated carbon (GAC) adsorption. Because of their volatility, these chemicals are rarely found in surface waters. However, VOCs are stable in groundwaters contaminated by leaching of chemicals from industrial discharges, improper chemical use, and spillage. The maximum contaminant levels for VOCs are vinyl chloride, 2 $\mu$g/l; 1,2-dichloroethane, benzene, carbon tetrachloride, and trichloroethylene, 5 $\mu$g/l; 1,1-dichloroethylene, 7 $\mu$g/l; p-chlorobenzene, 75 $\mu$g/l; and 1,1,1-trichloroethane, 200 $\mu$g/l. Because of these extremely low allowable concentrations, air stripping in a countercurrent packed tower is the only aeration method satisfactory for drinking-water treatment. In cold climates the process may not be feasible because of poor removal at low temperatures and the possibility of ice formation on the tower packing. The more costly process of GAC adsorption may replace air stripping or be applied as a second-stage following partial removal by aeration.

## 11.41  DESIGN OF AIR-STRIPPING TOWERS

The water is sprayed on the top of the packing and passes down through the column while air is blown countercurrent up through the tower. The packing can be random or stacked lightweight plastic media. As the water spreads over the surfaces of the packing, a large area of water is exposed for mass transfer to the flow of air. The VOCs can move freely toward equilibrium between liquid and gas phases. For air-stripping very dilute solutions, this equilibrium can be expressed by Henry's law as follows:

$$C_{GM}^E = HC_{LM}^E \qquad (11.76)$$

where $C_{GM}^E$ = gas-phase molar concentration in equilibrium with the liquid-phase concentration, kmol/m$^3$

$\quad\quad\quad C_{LM}^E$ = liquid-phase molar concentration in equilibrium with the gas-phase concentration, kmol/m$^3$

$\quad\quad\quad\quad H$ = Henry's law constant, mass concentration/mass concentration (dimensionless)

For efficient air stripping, the equilibrium between the liquid and gas phases is continuously destabilized by replenishing the air exhausted from the top with contaminant-free air entering at the bottom. By the time the water discharges from bottom of the column, the contaminant in the liquid phase is reduced to a very low concentration. The higher the value of Henry's law constant the more readily a VOC is air stripped from water.

The rate of mass transfer of a VOC from water to air is proportional to the difference between the equilibrium concentration in solution and the existing concentration in solution.

$$J = K_L a(C_{LM}^E - C_{LM}) \qquad (11.77)$$

where $\quad\quad J$ = rate of mass transfer per unit volume of packing, kmol/m$^3$·s

$\quad\quad\quad K_L a$ = overall mass transfer coefficient, s$^{-1}$

$C_{LM}^E$ = molar concentration in liquid phase in equilibrium with the gas-phase concentration, kmol/m$^3$

$C_{LM}$ = average molar concentration in liquid phase, kmol/m$^3$

The value of $K_L a$ depends on the geometry of the tower and packing, operation of the air-stripping system (e.g., the air-to-water ratio), and temperature.

The design of an air-stripping packed column for steady-state mass transfer is based on the following relationship:

$$Z = (HTU)(NTU) \qquad (11.78)$$

where    $Z$ = depth of packing, m
         HTU = height of a transfer unit, m
         NTU = number of transfer units (dimensionless)

The height of a transfer unit (HTU) characterizes the mass transfer efficiency from the liquid to the gas phase.

$$HTU = L/\rho_L K_L a \qquad (11.79)$$

where    $L$ = water mass loading rate, kg/m$^2$·s
         $\rho_L$ = water density, kg/m$^3$
         $L/\rho_L$ = volumetric loading rate, m$^3$/m$^2$·s
         $K_L a$ = overall mass transfer coefficient, s$^{-1}$

The number of mass transfer units (NTU) corresponds to the difficulty in removing the VOC from the liquid phase.

$$NTU = \left(\frac{S}{S-1}\right) \ln \frac{(C_{in}/C_{out})(S-1) + 1}{S} \qquad (11.80)$$

where    $S$ = stripping factor, mol/mol (dimensionless)
         $C_{in}$ = VOC concentration in influent, kg/m$^3$
         $C_{out}$ = VOC concentration in effluent, kg/m$^3$

The stripping factor is defined as

$$S = H(Q_A/Q_W) \qquad (11.81)$$

where    $H$ = Henry's law constant, mass concentration/mass concentration (dimensionless)
         $Q_A$ = volumetric airflow rate, m$^3$/s
         $Q_W$ = volumetric water loading rate, m$^3$/s

Several mathematical models to calculate mass transfer coefficients have been proposed based on the two-film theory, which assumes that overall resistance to mass transfer is the sum of liquid- and gas-phase resistances. Lamarche and Droste [41] evaluated theses models in packed-column air stripping of 6 VOCs. They also presented a laboratory technique for determining the value of Henry's law constant for VOCs at different temperatures.

The procedure for designing an air-stripping tower starts with the selection of a packing. A stripping factor is selected between 2 and 5, if high removal efficiency is required, and, based on this value, the air-to-water ratio is calculated. An allowable air pressure drop is selected. Data on the air pressure drop through a particular packing are generally available from the manufacturer. Operation at a high-pressure drop allows a smaller volume of packing, reducing the construction cost but increasing the operational costs. Various combinations of pressure drops and air-to-water ratios can be calculated to find the most cost-effective choice. Henry's law constant for the anticipated operating temperature is taken from the literature or determined by laboratory analysis. Selection of a mass transfer coefficient should preferably be from pilot-plant studies or based on experience in full-scale performance. Using these data, along with influent and effluent VOC concentrations, the required depth of packing can be calculated from Eqs. 11.78–11.81. The surface area of the packing is calculated from the quantity of water to be treated and the water loading.

■ **EXAMPLE 11.14**

Determine the depth of packing and surface area for a countercurrent air-stripping tower to reduce the trichloroethylene from 200 $\mu$g/l to 2 $\mu$g/l (99% removal). The design water flowrate is 76 l/s, and the lowest operating temperature anticipated is 10°C based on groundwater temperature and cooling in the tower. Henry's law constant for trichloroethylene is 0.30 at 10°C. The manufacturer of the proprietary random packing recommends a mass transfer coefficient of 0.017 s$^{-1}$ based on pilot studies and an allowable pressure drop of 50 N/m$^2$·m.

*Solution.* After discussions with the client and manufacturer of the packing, the designer selected a loading of 10 l/m$^2$·s (kg/m$^2$·s) and a stripping factor of 3.6. Using Eq. 11.81,

air-to-water ratio = 3.6/0.30 = 12 m$^3$/m$^3$

volumetric airflow rate = 0.076 × 12 = 0.91 m$^3$/s

Using Eq. 11.79,

HTU = 10/(1000)(0.017) = 0.59 m

Using Eq. 11.80,

$$ NTU = \left(\frac{3.6}{3.6 - 1}\right) \ln \frac{(200/2)(3.6 - 1) + 1}{3.6} = 5.9 $$

Using Eq. 11.78,

Z (depth of packing) = 0.59 × 5.9 = 3.5 m

surface area of packing = 76/10 = 7.6 m$^2$     ■

# Synthetic Organic Chemical Removal

Synthetic organic chemicals (SOCs) include pesticides (herbicides and insecticides), volatile organic chemicals, and trihalomethanes. Trace concentrations of pesticides are found in runoff from agricultural lands, occasionally in groundwaters under agricultural lands, and groundwaters contaminated by seepage from improper disposal of industrial wastes and spillage of chemicals. Conventional water treatment provides limited removal of organic chemicals. If adsorbed on particles or associated with large hydrophobic molecules, they can be taken out by coagulation–sedimentation–filtration. However, dissolved organic chemicals rarely adsorb to metal hydroxides and polymers, resulting in negligible removal. Adjustment of pH, changing coagulants or coagulant aids, and application of powdered activated carbon are options to be considered for improved treatment. In surface-water treatment, these process variables are generally successful for greater removal of natural organic precursors to reduce subsequent formation of trihalomethanes.

## 11.42  ACTIVATED CARBON ADSORPTION

Activated carbon can be made from a variety of carbonaceous raw materials. Processing is dehydration and carbonization by slow heating in the absence of air followed by chemical activation to produce a highly porous structure. Powdered activated carbon (PAC) for water treatment, which has good characteristics for adsorption of taste and odor compounds, is commonly made from lignin or lignite. Granular activated carbon (GAC) made from coal has the best physical properties of density, particle size, abrasion resistance, and ash content. These characteristics are essential, since GAC is subject to filter backwashing, conveyance as a slurry, and heat reactivation.

The activation process in manufacture creates a highly porous surface on the carbon particles with macropores and micropores down to molecular dimensions. Organic contaminants are adsorbed by attraction to and accumulation in pores of appropriate size, thus the pore structure is extremely important in determining adsorptive properties for particular compounds. In general, GAC most readily adsorbs branch-chained high-molecular–weight organic chemicals with low solubility. These include pesticides, volatile organic chemicals, and trihalomethanes. Macropores are large enough for colonies of bacteria to grow and proliferate if biodegradable organic compounds are in the water. The benefits of microbial growth or potential risks to water quality are not well understood. GAC is reactivated thermally at a furnace temperature and retention time based on the volatility of the adsorbed chemicals. Two percent to 5% of the carbon is lost during each reactivation and must be replaced with fresh carbon.

Powdered activated carbon is a fine powder applied in a water slurry, which can be added at any location in the treatment process ahead of filtration. At the point of application, the mixing must be adequate to ensure dispersion and the contact time long enough for adsorption. The dosage for normal taste and odor control is usually up to 5 mg/l with a contact time of 10–15 min. Although PAC is an effective adsorber of organic compounds that cause taste and odor, this success is not repeated by the adsorption of

SOCs. Poor adsorption is attributed to the pore structure of the PAC, short contact time between the carbon particles and the dissolved organic chemicals, and interference by adsorption of other organic compounds.

Efficiency in removal of SOCs requires a granular activated carbon filter to ensure close contact between the water and carbon for a sufficient time for adsorption to occur. Design of a GAC system requires pilot-plant tests for selection of the best carbon, determination of the required contact time, the effects of influent water quality variations, and to establish the carbon loss during reactivation. The analytic methodology includes data on (1) adsorbability of the various organic compounds present, (2) performance of several different GACs in their removal, (3) information on the kind of process for reactivation, (4) frequency of reactivation, and (5) balancing costs of increased contact time against savings from less frequent reactivation [42].

## 11.43 GRANULAR ACTIVATED CARBON SYSTEMS

Granular activated carbon can be used in place of anthracite and sand media in conventional filters to increase removal of compounds that cause taste and odor in surface waters. The filter-adsorber beds may be dual-media GAC sand or single-medium GAC. Graese, Snoeyink, and Lee [43] reviewed the performance of treatment plants with GAC filter adsorbers. The conclusions were that, if properly designed, they were as effective as conventional filtration media for removing turbidity and can consistently reduce tastes and odors from water supplies for extended periods of time before reactivation. GAC filter adsorbers do not perform well in removing compounds, such as trihalomethanes and volatile organic compounds, that are less strongly adsorbed.

A fixed-bed contactor has a GAC bed that remains stationary (fixed) during operation. Although the bed can be designed for downflow or upflow, downflow operation with provision for backwashing is more common. The GAC is not reactivated until chemical breakthrough, then the entire bed is removed and replaced. The design of a fixed-bed contactor is similar to a gravity granular-media filter (Sections 10.18–10.22) or a pressure filter (10.23). In a surface-water treatment plant, the conventional filter is retained for removal of turbidity, and the fixed-bed contactor added as a second stage. A postfilter contactor designed specifically to adsorb SOCs (without the necessity of turbidity removal) provides better use of the adsorptive capacity of the GAC and allows longer contact times. Adsorptive capacity can be significantly reduced by organic contaminants in an unfiltered water. Also, GAC suitable for filtration may not be optimum for adsorbing the contaminating organic chemicals. Contact time is expressed as empty bed contact time calculated by dividing the volume of the bed by the flowrate. In conventional filtration, the empty-bed contact time is usually 3–9 min, while in a GAC contactor it is 15–30 min or greater. A contactor following filtration also reduces the quantity of backwash water required and can be designed for easier removal of spent carbon and replacement. A gravity contactor is appropriate for removal of chemicals from a groundwater supply in a large plant. Pretreatment may be necessary to remove contaminants that can interfere with filtration through the GAC bed, for example, iron oxide deposits and growth of iron bacteria. For individual wells, the contactor may be a pressure vessel with discharge pressure from the well pump forcing the water through the bed.

A countercurrent pulsed-bed (moving-bed) contactor operates with the water flowing upward through the bed while spent carbon is removed from the bottom as an equal amount of fresh carbon is added to the top. In effect, the pulsed-bed system performs like a series of fixed-bed contactors with nearly spent carbon contained in the first and fresh carbon in the last. This is the most efficient use of the adsorptive capacity of the carbon. Since the column of GAC is fluidized by the upward flow of water to reduce the clogging, this arrangement is most appropriate for water or wastewater containing suspended solids that do not have to be removed in treatment. The granular-carbon columns and carbon regeneration system installed in Water Factory 21 for reclamation of wastewater are discussed in Sections 14.6 and 14.7.

## PROBLEMS

**11.1** (a) Using atomic weights from the table of elements given in Table A7, calculate the molecular and equivalent weights of alum (aluminum sulfate), ferric sulfate, and soda ash (sodium carbonate). The formulas of these compounds are given in Table 11.1. (b) Using atomic weights, compute the equivalent weights of the ammonium ion, bicarbonate ion, calcium carbonate, and carbon dioxide. Values are given in Table 11.2.

**11.2** (a) Water contains 50 mg/l of calcium ion and 15 mg/l of magnesium ion. Express the hardness as mg/l of $CaCO_3$. (b) Alkalinity in water consists of 150 mg/l of bicarbonate ion and 15 mg/l of carbonate ion. Express the alkalinity in units of mg/l of $CaCO_3$.

**11.3** Draw a milliequivalent per liter bar graph and list the hypothetical combinations for the following analysis of a soft water:

$Ca^{2+}$ = 36 mg/l     $HCO_3^-$ = 208 mg/l
$Mg^{2+}$ = 14 mg/l      $SO_4^{2-}$ =  14 mg/l
$Na^+$ = 43 mg/l       $Cl^-$ =  44 mg/l
$K^+$ =  7 mg/l

**11.4** Draw a milliequivalents-per-liter bar graph and list the hypothetical combinations for the following analysis of a groundwater:

$$Ca^{2+} = 94 \text{ mg/l} \qquad HCO_3^- = 317 \text{ mg/l}$$
$$Mg^{2+} = 24 \text{ mg/l} \qquad SO_4^{2-} = 67 \text{ mg/l}$$
$$Na^+ = 14 \text{ mg/l} \qquad Cl^- = 24 \text{ mg/l}$$

**11.5** Draw a milliequivalents-per-liter bar graph for the following water analysis:

| | |
|---|---|
| calcium hardness = 185 mg/l | alkalinity = 200 mg/l |
| magnesium hardness = 50 mg/l | sulfate ion = 58 mg/l |
| sodium ion = 23 mg/l | chloride ion = 36 mg/l |
| potassium ion = 20 mg/l | pH = 7.7 |

**11.6** What is the dominant form of alkalinity in a natural water at pH 7? What are the forms present at pH 10.5?

**11.7** Based on laboratory studies, the rate constant for a chemical coagulation reaction was found to be first-order kinetics with a $k$ equal to 75 per day. Calculate the detention times required in an ideal completely mixed reactor and an ideal plug-flow reactor for 80% reduction, $C_0 = 200$ mg/l, and $C_t = 40$ mg/l.

**11.8** The kinetics of a chemical reaction were analyzed by laboratory experiment. Lime was added to a water sample to precipitate reactant A as product P. While continuously stirring the water, portions were withdrawn at 10-min intervals to measure the amount of A remaining. The data collected were as follows: $t = 0$, $C_0$ of A = 100 mg/l; $t = 10$ min, $C$ of A remaining = 55 mg/l; $t = 20$ min, $C = 22$ mg/l; $t = 30$, $C = 8$; and $t = 40$, $C = 5$. Plot graphs as shown in Fig. 11.3a and 11.3b. What are the kinetics of the reaction? What is the value of the reaction rate constant?

**11.9** The number of coliform bacteria is reduced from an initial concentration of 2,000,000 per 100 ml to 400 per 100 ml in a long, narrow chlorination tank under a steady wastewater flow with a hydraulic detention time of 30 min. Assuming first-order kinetics and ideal plug flow, calculate the reaction-rate constant.

**11.10** Alternative reactor systems are being considered to reduce the reactant in a steady liquid flow from an initial concentration of 100 mg/l to a final concentration of 20 mg/l. Assuming a first-order reaction-rate constant of 0.80 $day^{-1}$, calculate the hydraulic detention time required for each of the following reactor systems: (a) one plug-flow reactor, (b) one completely mixed reactor, and (c) two equal-volume completely mixed reactors in series.

**11.11** Define the meanings of the terms *coagulation* and *flocculation* in reference to destabilization of colloidal suspensions. When these terms are used by an environmental engineer in reference to water treatment processes, what are their meanings?

**11.12** The results from a jar test for coagulation of a turbid alkaline raw water are given in the table. Each jar contained 1000 ml of water. The aluminum sulfate solution used for chemical addition had such strength that each milliliter of the solution added to a jar of water produced a concentration of 8.0 mg/l of aluminum sulfate. Based on the jar test results, what is the most economical dosage of aluminum sulfate in mg/l?

| Jar | Aluminum sulfate solution (ml) | Floc formation |
|-----|-------------------------------|----------------|
| 1 | 1 | None |
| 2 | 2 | Smoky |
| 3 | 3 | Fair |
| 4 | 4 | Good |
| 5 | 5 | Good |
| 6 | 6 | Very heavy |

If another jar had been filled with freshly distilled water and dosed with 5 ml of aluminum sulfate solution, what would have been the degree of floc formation?

**11.13** In the coagulation reaction, commercial aluminum sulfate (alum) reacts with natural alkalinity or can be reacted with lime or soda ash if the water is deficient in alkalinity. Based on Eqs. 11.34–11.36, calculate the milligrams-per-liter amounts of alkalinity, lime as CaO, and soda ash as $Na_2CO_3$ that react with 1.0 mg/l of alum.

**11.14** A ferrous sulfate dosage of 40 mg/l and an equivalent dosage of lime are used to coagulate a water. (a) How many pounds of ferrous sulfate per million gallons are used? (b) How many pounds of hydrated lime per million gallons are used, assuming a purity of 70% CaO? (c) How many pounds of ferric hydroxide are theoretically produced per million gallons of water treated?

**11.15** In chlorinated copperas coagulation, how many milligrams of ferrous sulfate are oxidized by 1.0 mg of chlorine?

**11.16** Treatment of a water supply requires 60 mg/l of ferric chloride as a coagulant. The natural alkalinity of the water is 40 mg/l. Based on theoretical chemical reactions, what dosage of lime as CaO is required to react with the ferric chloride after the natural alkalinity is exhausted?

**11.17** The data listed below are from a pilot-plant study to determine turbidity and *Giardia* cyst removal from a cold low-turbidity water (less than 1°C and 0.5 NTU in winter) by direct filtration using a cationic polymer as the coagulant [14, Chapter 10]. The filter was a dual-media coal-sand bed, 2 ft by 2 ft square, operated for most test runs at a loading of approximately 12 $m^3/m^2$·h (4.9 gpm/ $ft^2$) for durations in the range of 3–22 h. During selected filter runs, *Giardia* cysts and coliform bacteria were injected into the raw water for 40–60 min and tested for presence in the filtered water. Calculate the percentages for *Giardia* cyst and coliform removals. Plot turbidity versus polymer dosage for both the raw water and filtered water on the same diagram. What is the least dosage of polymer for maximum turbidity removal? What appears to be an acceptable turbidity in the effluent to ensure 98%–99% (virtually 100%) *Giardia* cyst removal?

| Filter loading $(m^3/m^2 \cdot h)$ | Water temp. (°C) | Polymer dosage (mg/l) | Turbidity | | Giardia lamblia | | Coliforms | |
|---|---|---|---|---|---|---|---|---|
| | | | Inf. (NTU) | Eff. (NTU) | Inf. (Cysts/l) | Eff. (Cysts/l) | Inf. (Org/100 ml) | Eff. (Org/100 ml) |
| 12.3 | 2.7 | 0 | 0.46 | 0.30 | | | | |
| 8.0 | 0.3 | 0 | 0.45 | 0.22 | 340 | 69 | | |
| 12.1 | 1.9 | 5 | 0.48 | 0.30 | | | | |
| 12.2 | 0.3 | 10 | 0.60 | 0.05 | | | | |
| 12.3 | 0.3 | 10 | 0.47 | 0.03 | | | | |
| 11.4 | 0.3 | 12 | 0.45 | 0.03 | | | 360 | 9 |
| 12.3 | 0.3 | 12 | 0.48 | 0.04 | | | | |
| 8.7 | 1.9 | 13 | 0.61 | 0.06 | 270 | 0.4 | 3900 | 580 |
| 12.3 | 0.3 | 13 | 0.47 | 0.03 | | | | |
| 12.2 | 0.3 | 14 | 0.49 | 0.03 | | | | |
| 11.9 | 8.3 | 18 | 0.65 | 0.07 | 3.2 | 0 | 8700 | 190 |
| 12.5 | 7.6 | 24 | 0.43 | 0.02 | 410 | 0 | 5800 | 3 |
| 12.3 | 0.2 | 24 | 0.48 | 0.02 | | | | |

**11.18** Presedimentation reduces the turbidity of a raw river water from 1500 mg/l suspended solids to 200 mg/l. How many pounds of dry solids are removed per million gallons? If the settled sludge has a concentration of 8% solids and a specific gravity of 1.03, calculate the sludge volume produced per million gallons of river water processed.

**11.19** Sketch a preliminary process flow diagram for a water treatment plant to clarify and disinfect a turbid surface water at a design flow of 50 mgd. Use two identical, parallel, and separate processing lines with rapid mixing, flocculation, sedimentation, filtration, and clear-well storage. The flocculation and sedimentation processes for each line are in the same large rectangular concrete tank with paddle flocculators in baffled compartments ahead of the sedimentation section with effluent "finger" channels extending into the tank from the outlet end (Figs. 10.14 and 10.16). For each line, use four gravity dual-media coal-sand filters with deep filter boxes to prevent "air binding" and flow control by influent flow splitting for constant-rate filtration (Fig. 10.35). The preferred filter bottom is the precast concrete T-Pee underdrain illustrated in Fig. 10.31.

On the flow diagram, show the chemicals to be added with alternate points of application. The raw water has a turbidity ranging from 10 to 40 NTU, and in the spring the water contains natural organic matter that creates taste and odor and forms trihalomethanes with prechlorination. The fluoride concentration is less than optimum. The treatment plant must meet the EPA rule for surface-water disinfection as discussed in Section 11.24. Sketch a longitudinal cross-sectional view of the flocculation–sedimentation tank. List the design criteria for sizing the flocculation section and specifying paddle flocculators, sizing the sedimentation section and effluent channels, and sizing the filters. Sketch a plan view of the clear well to ensure compliance with the EPA disinfection rule.

**11.20** The water defined by the analysis given below is to be softened by excess lime treatment. (a) Sketch an meq/l bar graph. (b) Calculate the softening chemicals required. (c) Draw a bar graph for the softened water after recarbonation and filtration assuming 80% of the alkalinity is in the bicarbonate form.

$$CO_2 = 8.8 \text{ mg/l} \qquad Alk(HCO_3^-) = 135 \text{ mg/l}$$
$$Ca^{2+} = 40.0 \text{ mg/l} \qquad SO_4^{2-} = 29.0 \text{ mg/l}$$
$$Mg^{2+} = 14.7 \text{ mg/l} \qquad Cl^- = 17.8 \text{ mg/l}$$
$$Na^+ = 13.7 \text{ mg/l}$$

**11.21** Settled water after excess lime treatment, before recarbonation and filtration, contains 35 mg/l of CaO excess lime in the form of hydroxyl ion, 30 mg/l of $CaCO_3$ as carbonate ion, and 10 mg/l as $CaCO_3$ of $Mg(OH)_2$ in the form of hydroxyl ion. First-stage recarbonation precipitates the excess lime as $CaCO_3$ for removal by sedimentation, and second-stage recarbonation converts a portion of the remaining alkalinity to bicarbonate ion. Calculate the carbon dioxide needed to neutralize the excess lime and convert one-half of the alkalinity in the finished water to the bicarbonate form. Assume an excess of 20% of the calculated $CO_2$ is required to account for unabsorbed gas escaping from the recarbonation chamber.

**11.22** For the water analysis given in Prob. 11.5, calculate the additions of CaO, $Na_2CO_3$, and $CO_2$ needed for excess lime softening. Sketch a bar graph for the finished water after two-stage precipitation softening by excess lime treatment with intermediate and final recarbonation (Fig. 11.9). Assume that three quarters of the alkalinity in the finished water is in the bicarbonate form.

**11.23** For the water analysis given in Prob. 11.5, calculate the lime dosage required for selective calcium removal. The process flow scheme is a single-stage system of mixing, sedimentation, and filtration. Draw a bar graph for the finished water. Is this softening process recommended for this water?

**11.24** Assume that the water described in Prob. 11.5 is processed by precipitation softening in a split-treatment system, as illustrated in Fig. 11.10. Assume a permissible magnesium hardness in the finished water ($Mg_f$) of 30 mg/l ($X = 0.5$). Calculate the chemical doses. Draw initial, intermediate, and final bar graphs.

**11.25** Compute the lime dosage needed for selective calcium-removal softening of the water described by the following analysis. What is the finished water hardness?

$$Ca^{2+} = 63 \text{ mg/l} \qquad CO_3^{2-} = 16 \text{ mg/l}$$
$$Mg^{2+} = 15 \text{ mg/l} \qquad HCO_3^- = 189 \text{ mg/l}$$
$$Na^+ = 20 \text{ mg/l} \qquad SO_4^{2-} = 80 \text{ mg/l}$$
$$K^+ = 10 \text{ mg/l} \qquad Cl^- = 10 \text{ mg/l}$$

**11.26** Consider the split-treatment softening of water described by the analysis below. Use the criteria for the finished water as given in Example 11.6. Draw a bar graph for the finished water after second-stage treatment with soda ash, sedimentation, and final filtration.

$$CO_2 = 15 \text{ mg/l as } CO_2 \qquad HCO_3^- = 200 \text{ mg/l as } CaCO_3$$
$$Ca^{2+} = 60 \text{ mg/l} \qquad SO_4^{2-} = 96 \text{ mg/l}$$
$$Mg^{2+} = 24 \text{ mg/l} \qquad Cl^- = 35 \text{ mg/l}$$
$$Na^+ = 46 \text{ mg/l}$$

**11.27** Sketch a preliminary process flow diagram for a split-treatment lime–soda ash water treatment plant to soften a design flow of 60 mgd. Use six equal-sized flocculator–clarifiers (Fig. 10.23). The first-stage flow for excess-lime treatment is expected to be no greater than 50% of the raw water. Use eight gravity dual-media coal-sand filters with traditional flow control using rate-of-flow controllers (Fig. 10.33). The preferred backwashing system is air scouring prior to water backwash. The clear-well capacity is 6.0 mil gal. The raw water has a hardness of approximately 230 mg/l, iron in the range of 0.2–0.3 mg/l, and less than optimum concentration of fluoride. On the flow diagram, show the chemicals being added and their points of application. Sketch a cross-sectional view of the filter box showing the wash-water troughs, filter media, and underdrain system. List the design criteria for sizing the flocculator–clarifiers and filters.

**11.28** The ionic characteristics of a fossil groundwater in an arid region are listed below. Draw a milliequivalents-per-liter bar graph and calculate total hardness and alkalinity. One recommendation for improving the quality of the water for domestic use is lime–soda ash softening to reduce hardness and total dissolved solids (TDS). Calculate the lime and soda ash additions for excess lime treatment and draw the final bar graph after recarbonation. Calculate the theoretical total dissolved solids content by summing the weights of the ions (or hypothetical combinations) in the softened water. Was the recommendation of lime–soda ash softening appropriate?

$$Ca^{2+} = 108 \text{ mg/l} \qquad HCO_3^- = 146 \text{ mg/l} \qquad TDS = 900 \text{ mg/l}$$
$$Mg^{2+} = \phantom{0}44 \text{ mg/l} \qquad SO_4^{2-} = 110 \text{ mg/l}$$
$$Na^+ = 138 \text{ mg/l} \qquad Cl^- = 366 \text{ mg/l}$$

**11.29** Sketch a meq/l bar graph of the water described in Prob. 11.20 after it is softened to zero hardness by cation exchange softening.

**11.30** Consider the ion exchange softening of water described in Example 11.4. If 0.3 lb of NaCl is required to regenerate the resin bed per 1000 grains of hardness removed, calculate the salt required per million gallons of water softened. Sketch a meq/l graph for the ion-exchange-softened water. How does finished water from ion exchange softening differ from finished water produced in lime–soda softening?

**11.31** A small community has used an unchlorinated well-water supply containing approximately 0.3 mg/l of iron and manganese for several years without any apparent iron and manganese problems. A health official suggested that the town install chlorination equipment to disinfect the water and provide a chlorine residual in the distribution system. After initiating chlorination, consumers complained about water staining washed clothes and bathroom fixtures. Explain what is occurring due to chlorination.

**11.32** The iron-and-manganese removal process for the well supply of a small community is mechanical aeration, the addition of potassium permanganate followed by detention in a contact tank, pressure filtration, and postchlorination. The construction specifications called for manganese-treated greensand; however, the actual filter medium provided was plain sand. Customers often complained that the treated water caused staining of bathroom fixtures and laundry. The common response of the plant operator was to increase the chemical dosage, which did not seem to improve the situation. The operator even tried prechlorination of the water in combination with potassium permanganate addition, but that appeared to increase the staining characteristics of the treated water. Discuss the most probable cause of the poor-quality finished water and your recommendations for improvement.

**11.33** Untreated well water contains 1.2 mg/l of iron and 0.8 mg/l of manganese at a pH of 7.5. Calculate the theoretical dosage of potassium permanganate required for iron and manganese oxidation.

**11.34** Iron and manganese are removed by aeration, chlorine oxidation, sedimentation, and sand filtration from a well-water supply. The plant processes 3000 $m^3/d$ with application of 2.5 mg/l chlorine. Calculate the usage in kilograms per month for each of the following chemicals: chlorine gas, 70% granular calcium hypochlorite, and 12% sodium hypochlorite solution. At the same pH of the chlorinated water, do all of these chemicals form the same kind of chlorine residual?

**11.35** The results of a chlorine demand test on a raw water at 20°C are given in the following table.

| Sample | Chlorine dosage (mg/l) | Residual chlorine after 10 min of contact (mg/l) |
|--------|------------------------|--------------------------------------------------|
| 1 | 0.20 | 0.19 |
| 2 | 0.40 | 0.37 |
| 3 | 0.60 | 0.51 |
| 4 | 0.80 | 0.50 |
| 5 | 1.00 | 0.20 |
| 6 | 1.20 | 0.40 |
| 7 | 1.40 | 0.60 |
| 8 | 1.60 | 0.80 |

(a) Sketch the chlorine demand curve.
(b) What is the breakpoint chlorine dosage?
(c) What is the chlorine demand at a chlorine dosage of 1.20 mg/l?

**11.36** The practice of combined residual chlorination involves feeding both chlorine and anhydrous ammonia. Calculate the stoichiometric ratio of chlorine feed to ammonia feed for combined chlorination.

**11.37** List the possible applications of ozone in water treatment. If ozone is applied for disinfection, may the use of chlorine be eliminated?

**11.38** What is the suspected health risk of trihalomethanes (THMs) in drinking water, and how was this risk demonstrated? What is the origin of THMs in treated water? If the finished water from a river-water treatment plant contains an excessive concentration of THMs during spring runoff, what remedial actions can be taken to reduce THM formation?

**11.39** Define the meaning of the $C \cdot t$ product. What factors affect the $C \cdot t$ value used in design and operation of a drinking-water disinfection system? With reference to Table 11.6, what kind of microorganism is most readily inactivated by free chlorine? What kind is the most difficult to inactivate? List the tabulated disinfectants in the order of most effective to least effective in disinfecting action.

**11.40** Using Eq. 11.68, calculate the $C \cdot t$ value for *G. lamblia* inactivation at a free chlorine concentration of 0.80 mg/l, pH equal to 8.0, and temperature of 10°C. Correct this value for 5°C, using Eq. 11.29, and extrapolate to 99% inactivation. How does this result compare to the value in Table 11.6? Extrapolate the calculated value at 10°C to 99.9% inactivation and compare the result with the value in Table 11.7.

**11.41** A surface-water treatment plant at a winter resort city has been designed to process cold low-turbidity water by direct filtration based on the pilot-plant study described in Prob. 11.17. The filtered water has a turbidity of less than 0.1 NTU, temperature of 0.5°C, and pH of 7.4 during the period of highest water consumption in the winter. Tracer analyses of the clear-well reservoir are illustrated in Fig. 11.17. At the critical hourly flow of 3.0 mgd, $t_{10} = 90$ min. To comply with the EPA surface-water disinfection regulation, what free chlorine residual must be maintained in filtered water at the outlet of the clear well?

**11.42** A surface-water treatment plant with coagulation, sedimentation, filtration, and effluent chlorination is being evaluated for compliance with the EPA surface-water disinfection regulation. The critical time of the year is during peak demand in the summer when the finished water temperature is 15°C or greater, pH is 7.5 or less, and the turbidity is 0.3–0.4 NTU. Prechlorination cannot be practiced because of trihalomethane formation. Data from the tracer analysis at peak hourly flow through the clear well, which has been modified by installation of baffles to reduce short-circuiting, are given in Prob. 10.6. After the clear well and prior to the first service connection, the finished water is transported through a pipeline for 4000 ft at a velocity of 5 fps. What chlorine residual must be present in the water at the outlet of the pipeline?

**11.43** A groundwater supply has been designated as vulnerable to fecal contamination, and the state has specified a disinfection level of 99.9% (3.0 log) virus inactivation. The peak-hourly pumping rate is 2000 gpm from the well field through an 4200-ft pipe with a 16-in. diameter to a reservoir in the town. The water temperature is 5°C. What chlorine residual is required in the water at the outlet of the pipeline?

**11.44** What determines the occurrence and concentrations of different kinds of pathogens present in a municipal wastewater? In general terms, to what degree are different kinds of pathogens removed in conventional wastewater treatment without effluent chlorination? with effluent chlorination?

**11.45** A conventional activated-sludge treatment plant with effluent chlorination consistently produces a treated wastewater with a BOD less than 20 mg/l, suspended solids less than 30 mg/l, and fecal coliform count less than 200 organisms per 100 ml. The effluent has been described by the plant superintendent as clear "sparkling" water. You have been asked to evaluate the feasibility of using the effluent to irrigate the city's public park with playgrounds and to apply to the state department of environmental control for permission to reuse the wastewater for this application. To introduce the idea and to allay the fears of an environmental citizens group, you have been asked to present a preliminary assessment of the situation to the city council. Outline the subjects that you would discuss in your presentation. (Refer to Sections 8.1, 11.26, 11.27, 14.5, 14.16, and 14.18.)

**11.46** What dosage of commercial fluosilicic acid is needed to increase the fluoride ion concentration from 0.3 to 1.0 mg/l? Use fluosilicic acid data from Table 11.12 and express the answers as mg/l and lb/mil gal.

**11.47** A 4.0% sodium fluoride solution is applied to increase the fluoride concentration from 0.4 to 1.0 mg/l in a municipal water supply. (a) What is the application rate of NaF solution in gal/mil gal? (b) How many pounds of commercial-grade sodium fluoride are needed per million gallons?

**11.48** Calculate the saturation index of the softened water from Example 11.6 (Fig. 11.11e) at 20°C and pH of 10.2.

**11.49** Calculate the saturation index from the water analysis in Prob. 11.25 for pH 8.0 at 50°F.

**11.50** What is the health risk of dietary intake of lead? (Refer to Section 8.4.) Why are first-flush samples collected from consumer's faucets used to assess lead contamination of drinking water? What are recommended methods of controlling excessive lead concentrations?

**11.51** List three possible methods for controlling crown corrosion in a large concrete sanitary sewer.

**11.52** A distilled seawater contains 25 mg/l of sodium chloride. (Negligible amounts of magnesium, calcium, potassium, and sulfate are also present.) To stabilize this corrosive water, sufficient lime is applied to add 15 mg/l of calcium ion, and the hydroxide ion is converted to bicarbonate ion by applying carbon dioxide. (The equilibrium pH for stabilization is generally 8.5–8.7.) The desalinated stabilized water is transported to a blending plant where brackish groundwater is added prior to delivering to the water distribution system. The criterion for blending is a product water with not more than 250 mg/l chloride, 250 mg/l sulfate, or 100 mg/l of sodium. Based on the analysis of the groundwater listed below, what percentage of the blended water can be groundwater? Sketch milliequivalent-per-liter bar graphs for the distilled seawater, stabilized distilled water, and blended water.

$Ca^{2+}$ = 220 mg/l         $Na^+$ = 580 mg/l      $SO_4^{2-}$ = 420 mg/l

$Mg^{2+}$ = 74 mg/l      $HCO_3^-$ = 210 mg/l         $Cl^-$ = 1030 mg/l

**11.53** Review Example 11.13. Evaluate the feasibility of operating this reverse-osmosis plant at a water recovery of 90%. (The $pK_2' - pK_S'$ value for a total dissolved-solids concentration above 10,000 mg/l can be assumed to remain essentially constant at 2.1.) If 90% recovery were to be performed, what process changes would be required?

**11.54** In Prob. 11.28, lime–soda ash softening of the groundwater did not produce a treated water of drinking quality. Sketch a preliminary process flow diagram to treat this groundwater for potable use. On the flow diagram, show the chemicals being added and their points of application. Consider also the disposal of any process wastewaters.

**11.55** An anion exchange treatment system as illustrated in Fig. 11.28 is used to remove nitrate from well water. The raw water supply is 0.70 mgd, of which

75% is treated and 25% bypassed to produce a blended product water. The chemical characteristics of the untreated, treated, and blended waters are given in the following table, with the ionic concentrations expressed in milligrams per liter, except for pH.

| Component | Untreated | Treated | Blended |
|---|---|---|---|
| $Ca^{2+}$ | 96 | 94 | 95 |
| $Mg^{2+}$ | 2 | 2 | 2 |
| $Na^+$ | 68 | 68 | 68 |
| $K^+$ | 4 | 4 | 4 |
| $HCO_3^-$ | 146 | 34 | 62 |
| $SO_4^{2-}$ | 120 | Trace | 30 |
| $Cl^-$ | 88 | 260 | 217 |
| $NO_3^-$ as N | 16 | 2 | 5.5 |
| pH | 7.2 | 7.4 | 7.4 |

The average wastewaters generated each day in producing the blended product water of 700,000 gpd are as follows: 3400 gal of diluted brine, 12,300 gal of rinse water, and 8000 gal of backwash water. The diluted brine solution applied for regeneration contains 860 gal of saturated brine with approximately 360 g/l of NaCl. The volume of resin in each ion exchange vessel is 85 ft³, and the volume of water treated before regeneration is 165,000 gal (260 bed volumes).
(a) What are the changes in the ionic characteristics of the well water resulting from passage through the anion exchange resin. Relate these changes to the breakthrough curves shown in Fig. 11.27.
(b) How many milliequivalents-per-liter of chloride were added to the blended product water, and how many milliequivalents-per-liter each of bicarbonate, sulfate, and nitrate were removed?
(c) How many gallons of saturated brine were used in regeneration per 1000 gal of blended product water produced? Convert this value to pounds of NaCl per 1000 gal of blended product water and milliequivalents of chloride per liter of blended water.
(d) Calculate the brine use factor equal to the equivalents of chloride in the brine regenerant to equivalents of nitrate removed.
(e) Calculate the milliequivalents-per-liter of unreacted chloride appearing in the process wastewater, and the percentage of the equivalents of chloride in the blended water relative to the equivalents of chloride brine regenerant.
(f) How many pounds of NaCl are used in regeneration per cubic foot of resin?

**11.56** Determine depths of packing and surface areas between 4m and 8m for a countercurrent air-stripping tower to treat well water to reduce tetrachloroethylene from 100 $\mu$g/l to 2 $\mu$g/l, trichloroethylene from 25 to 2 $\mu$g/l, and cis-1,2 dichloroethylene from 70 to 2 $\mu$g/l. The design flow rate is 44 l/s, and the lowest temperature anticipated is 7°C, based on groundwater temperature and cooling in the tower. The following data are based on pilot-scale packed-column tests: (1) Henry's law constants for tetrachloroethylene, trichloroethylene, and cis-1,2-dichloroethylene are 0.30, 0.21, and 0.094 at 7°C, respectively. (2) The mass transfer coefficient for the most efficient and cost-effective commercial packing is 0.015 s⁻¹. (3) The optimum air-to-water ratio is 20:1.

**11.57** Outlined below is the sequence of unit operations and chemical additions used in the treatment of a well-water supply. Briefly state the function or purpose of each unit process and the reason for each chemical addition.

**1.** Mixing and flocculation with the addition of lime.
**2.** Sedimentation.
**3.** Recarbonation.
**4.** Granular-media filtration.
**5.** Postchlorination.

**11.58** Outlined below is the sequence of unit operations and chemical additions used in the treatment of a well-water supply. Briefly state the function or purpose of each unit process and the reason for each chemical addition.

**1.** Prechlorination at the wells.
**2.** Aeration over a tray aerator.
**3.** Rechlorination.
**4.** Detention in a settling basin.
**5.** Granular-media filtration.
**6.** Addition of anhydrous ammonia.

**11.59** Outlined below is the sequence of unit operations and chemical additions used in the treatment of a well-water supply. Briefly state the function or purpose of each unit process and the reason for each chemical addition.

**1.** Prechlorination at the wells.
**2.** Mixing–flocculation–sedimentation in flocculator–clarifiers using split treatment with lime and alum added to one leg and potassium permanganate to the other leg.
**3.** Granular-media filtration.
**4.** Postchlorination.

**11.60** Consider the following sequence of unit operations and chemical additions used in the treatment of a river-water supply. Briefly state the function or purpose of each unit process and the reason for each chemical addition.

**1.** Presedimentation with polymer addition.
**2.** Activated carbon available when needed.
**3.** Mixing and flocculation with the addition of alum and polymer.
**4.** Sedimentation.
**5.** Addition of activated carbon.
**6.** Granular-media filtration.
**7.** Postchlorination.

**11.61** Outlined below is the sequence of unit operations and chemical additions used in the treatment of a reservoir-water supply. Briefly state the function or purpose of each unit process and the reason for each chemical addition.

**1.** Intermittent applications of copper sulfate to reservoir during summer and fall.
**2.** Chlorine dioxide available when needed.
**3.** Mixing and flocculation with the addition of alum and polymer.
**4.** Sedimentation.
**5.** Addition of activated carbon.

**6.** Granular-media filtration.

**7.** Postchlorination.

**11.62** Outlined below is the sequence of unit operations and chemical additions used in the treatment of a brackish groundwater. Briefly state the function or purpose of each unit process and the reason for each chemical addition.

**1.** Acidification with sulfuric acid.

**2.** Addition of sodium hexametaphosphate.

**3.** Cartridge filtration.

**4.** High-pressure pumps.

**5.** Reverse-osmosis modules.

**6.** Degasification using stripping towers.

**7.** Addition of sodium hydroxide.

**8.** Addition of chlorine.

# REFERENCES

1. American Public Health Association, American Water Works Association, and Water Pollution Control Federation, *Standard Methods for the Examination of Water and Wastewater,* 17th ed. (Washington, DC: American Public Health Association, 1989).
2. C. N. Sawyer and P. L. McCarty, *Chemistry for Environmental Engineering,* 3rd ed. (New York: McGraw-Hill, 1978).
3. V. L. Snoeyink and D. Jenkins, *Water Chemistry* (New York: Wiley, 1980).
4. W. J. Weber, Jr., *Physicochemical Processes for Water Quality Control* (New York: Wiley-Interscience, 1972).
5. Committee Report, "Coagulation as an Integrated Water Treatment Process," *J. Am. Water Works Assoc.* 81(11) (1989): 72–78.
6. H. E. Hudson, Jr., "Jar Testing and Utilization of Jar Test Data," *Water Clarification Processes Practical Design and Evaluation* (New York: Van Nostrand Reinhold Co., 1981), Chap. 3.
7. J. L. Cleasby and J. H. Dillingham, "Rational Aspects of Split Treatment," *Proc. Am. Soc. Civil Engrs., J. San. Eng. Div.* 92 (SA2) (1966): 1–7.
8. J. D. Lowry and S. B. Lowry, "Radionuclides in Drinking Water," *J. Am. Water Works Assoc.* 80(7) (1988): 50–64.
9. E. M. Aieta, J. E. Singley, A. R. Trussell, K. W. Thorbjarnarson, and M. J. McGuire, "Radionuclides in Drinking Water: An Overview," *J. Am. Water Works Assoc.* 79(4) (1987): 144–152.
10. Committee Report, "Research Needs for the Treatment of Iron and Manganese," *J. Am. Water Works Assoc.* 79(9) (1987): 119–122.
11. H. H. Chambers and R. S. Ingols, "Copper Sulfate Aids in Manganese Removal," *Water and Sewage Works* 103 (1956): 248.
12. W. S. Holden, *Water Treatment and Examination* (Baltimore: Williams and Wilkins, 1970).
13. W. A. Welch, "Potassium Permanganate in Water Treatment," *J. Am. Water Works Assoc.* 55(6) (1963): 735.
14. Committee Report, "Disinfection," *J. Am. Water Works Assoc.* 74(7) (1982): 376–379.

15. National Research Council, *Drinking Water and Health, Disinfectants and Disinfection By-products,* Vol. 7 (Washington, DC: National Academy Press, 1987).

16. S. W. Krasner, M. J. McGuire, J. G. Jacangelo, N. L. Patania, K. M. Reagan, and E. M. Aieta, "The Occurrence of Disinfection By-products in U.S. Drinking Water," *J. Am. Water Works Assoc.* 81(8) (1989): 41–53.

17. C. N. Hass and S. B. Karra, "Kinetics of Microbial Inactivation by Chlorine, Review of Results in Demand Free Systems," *Water Res.* 18 (1984): 1443–1449.

18. J. C. Hoff, Project Summary, "Inactivation of Microbial Agents by Chemical Disinfectants," Environmental Protection Agency, Water Engineering Research Laboratory, EPA/600/S2-86/067 (September 1986).

19. C. P. Hibler, C. M. Hancock, L. M. Perger, J. G. Wegrzn, and K. D. Swabbly, *Inactivation of Giardia Cysts with Chlorine at 0.5°C to 5.0°C* (Denver: Research Foundation, Am. Water Works Assoc., 1987).

20. Environmental Protection Agency, *Guidance Manual for Compliance with the Surface Water Treatment Requirements for Public Water Systems* (Washington DC: EPA, 1990).

21. E. C. Lippy and S. C. Waltrip, "Waterborne Disease Outbreaks 1964–1980: A Thirty-Five-Year Perspective," *J. Am. Water Works Assoc.* 76(2) (1984): 60–67.

22. R. G. Feachem, D. J. Bradley, H. Garelick, and D. D. Mara, *Sanitation and Disease, Health Aspects of Excreta and Wastewater Management; World Bank Studies in Water Supply and Sanitation 3* (Chichester: Wiley, 1983).

23. D. F. Striffler, W. O. Young, and B. A. Burt, "The Prevention and Control of Dental Caries: Fluoridation," in *Dentistry, Dental Practice, and the Community* (Philadelphia: Saunders, 1983), pp. 155–199.

24. E. Bellack, *Fluoridation Engineering Manual,* U.S. Environmental Protection Agency, Office of Water Programs, Water Supply Programs Division (1972).

25. J. R. Millette, A. F. Hammonds, M. F. Pansing, E. C. Hanson, and P. J. Clark, "Aggressive Water: Assessing the Extent of the Problem," *J. Am. Water Works Assoc.* 72(5) (1980): 262–266.

26. D. T. Merrill and R. L. Sanks, "Corrosion Control by Deposition of $CaCO_3$ Films," *J. Am. Water Works Assoc.* 69(11) (1977): 592–599; 69 (12) (1977): 634–640; 70(1) (1978): 12–18.

27. Degremont, *Water Treatment Handbook,* 5th ed. (New York: A Halsted Press Book, Wiley, 1979), pp. 385, 386.

28. J. E. Singley, "The Search for a Corrosion Index," *J. Am. Water Works Assoc.* 73(11) (1981): 579–582.

29. W. F. Langelier, "The Analytical Control of Anticorrosion Water Treatment," *J. Am. Water Works Assoc.* 28 (1936): 1500.

30. T. E. Larson and A. M. Buswell, "Calcium Carbonate Saturation Index and Alkalinity Interpretations," *J. Am. Water Works Assoc.* 34 (1942): 1667–1684.

31. M. R. Schock, "Understanding Corrosion Control Strategies for Lead," *J. Am. Water Works Assoc.* 81(7) (1989): 88–100.

32. R. G. Lee, W. C. Becker, and D. W. Collins, "Lead at the Tap: Sources and Control," *J. Am. Water Works Assoc.* 81(7) (1989): 52–62.

33. A. Porteous, *Saline Water Distillation Processes* (London: Longman, 1975).

34. R. L. Reitz, "Pretreatment for Potable Water Production by Membrane Processes," in *AWWA Seminar Proceedings, Membrane Processes: Principles and Practices* (Denver: Am. Water Works Assoc., 1988), pp. 1–11.

35. A. Ko and D. B. Guy, "Brackish and Seawater Desalting," in *Reverse Osmosis Technology, Application for High-Purity-Water Production* (New York: Marcel Dekker, 1988), pp. 185–277.

36. H. F. Ridgway, Jr., "Microbial Adhesion and Biofouling of Reverse Osmosis Membranes," in *Reverse Osmosis Technology, Application for High-Purity-Water Production* (New York: Marcel Dekker, 1988), pp. 429–481.

37. Degremont, *Water Treatment Handbook,* 5th ed. (New York: Halsted Press, 1979), p. 878.

38. L. H. Shafer and M. S. Mintz, "Electrodialysis," in *Principles of Desalination,* 2nd ed., Part A (New York: Academic Press, 1980), Chap. 6.

39. R. P. Lauch and G. A. Guter, "Ion Exchange for the Removal of Nitrate from Well Water," *J. Am. Water Works Assoc.* 78(5) (1986): 83–88.

40. G. A. Guter, Project Summary, "Nitrate Removal from Contaminated Water Supplies: Vol. I. Design and Initial Performance of a Nitrate Removal Plant," Environmental Protection Agency, Water Engineering Research Laboratory, EPA/600/S2-86/111 (April 1987).

41. P. L. Lamarche and R. L. Droste, "Air-Stripping Mass Transfer Correlation for Volatile Organics," *J. Am. Water Works Assoc.* 81(1) (1989): 78–89.

42. Culp/Wesner/Culp, *Handbook of Public Water Systems* (New York: Van Nostrand Reinhold, 1986).

43. S. L. Graese, V. L. Snoeyink, and R. G. Lee, "Granular Activated Carbon Filter-Adsorber Systems," *J. Am. Water Works Assoc.* 79(12) (1987): 64–74.

# Chapter 12

# Biological Treatment Processes

Biological treatment is the most important step in processing municipal wastewaters. Physical treatment of raw wastewater by sedimentation removes only about 35% of the BOD, owing to the high percentage of nonsettleable solids (colloidal and dissolved) in domestic wastewaters. Chemical treatment is not favored because of high chemical costs and inefficiency of dissolved BOD removal by chemical coagulation and precipitation. Advanced physical treatment methods, such as carbon adsorption and reverse osmosis, can remove dissolved BOD and other contaminants, but are very costly to construct, operate, and maintain. A modern treatment plant uses a variety of physical, chemical, and biological processes to provide the best, most economical treatment.

## Biological Considerations

Biological treatment systems are "living" systems that rely on mixed biological cultures to break down waste organics and remove organic matter from solution. Domestic wastewater supplies the biological food, growth nutrients, and inoculum. A treatment unit provides a controlled environment for the desired biological process. Historically, civil engineers designed treatment systems on the basis of empirical rules. This practice has led to failures in sanitary design—not unsuccessful in the sense of collapse of a structure but deficient in that the biological process did not function properly. Understanding the biological processes involved in wastewater treatment is essential to a design engineer.

## 12.1 BACTERIA AND FUNGI

*Bacteria* (singular, bacterium) are the simplest forms of plant life that use soluble food and are capable of self-reproduction. Bacteria are fundamental microorganisms in the stabilization of organic wastes and therefore of basic importance in biological treatment. Individual bacterial cells range in size from approximately 0.5 to 5 $\mu$m in rod, sphere, and spiral shapes and occur in a variety of forms: individual, pairs, packets, and chains.

Bacteria reproduce by binary fission (the mature cell divides into two new cells). In most species, the process of reproduction—growth, maturation, and fission—occurs in 20–30 min under ideal environmental conditions. Certain bacterial species form spores as a means of survival under adverse environmental conditions. Their tough coating is resistant to heat, lack of moisture, and loss of food supply. Fortunately, only one spore-forming bacterium, *Bacillus anthracis,** is pathogenic to man. As the result of stringent public health measures, incidents of anthrax in man are rare.

Based on nutritive requirements, bacteria are classified as heterotrophic or autotrophic bacteria, although several species may function both heterotrophically and autotrophically.

*Heterotrophic bacteria* use organic compounds as an energy and carbon source for synthesis. Another term used instead of heterotroph is *saprophyte,* which refers to an organism that lives on dead or decaying organic matter. The heterotrophic bacteria are grouped into three classifications, depending on their action toward free oxygen. *Aerobes* require free dissolved oxygen to live and multiply. *Anaerobes* oxidize organic matter in the complete absence of dissolved oxygen. Pasteur referred to anaerobiosis as "life without air." *Facultative bacteria* are a class of bacteria that use free dissolved oxygen when available but can also respire and multiply in its absence. *Escherichia coli,* a common coliform, is a facultative bacterium.

*Autotrophic bacteria* use carbon dioxide as a carbon source and oxidize inorganic compounds for energy. Autotrophs of greatest significance in sanitary engineering are the nitrifying, sulfur, and iron bacteria. Nitrifying bacteria perform the following reactions:

$$NH_3 \text{ (ammonia)} + \text{oxygen} \xrightarrow{\textit{Nitrosomonas}} NO_2^- \text{ (nitrite)} + \text{energy} \qquad (12.1)$$

$$NO_2^- \text{ (nitrite)} + \text{oxygen} \xrightarrow{\textit{Nitrobacter}} NO_3^- \text{ (nitrate)} + \text{energy} \qquad (12.2)$$

Autotrophic sulfur bacteria, *Thiobacillus,* convert hydrogen sulfide to sulfuric acid (Eq. 12.3). This bacterial production of sulfuric acid occurs in the moisture of condensation on side walls and crowns of sewers conveying septic wastewater. Since thiobacilli can tolerate pH levels less than 1.0,

---

* Bacteria are named using a binominal system; that is, each species is given a name consisting of two words. The first word is the genus and the second the name of the species.

sanitary sewers constructed on flat grades in warm climates should be built using corrosion-resistant materials.

$$H_2S \text{ (hydrogen sulfide)} + \text{oxygen} \rightarrow H_2SO_4 + \text{energy} \qquad (12.3)$$

True iron bacteria are autotrophs that oxidize inorganic ferrous iron as a source of energy. These filamentous bacteria occur in iron-bearing waters and deposit the oxidized iron, $Fe(OH)_3$, in their sheath (Eq. 12.4). All species of the iron bacteria *Leptothrix* and *Crenothrix* may not be strictly autotrophic; however, they are truly iron-accumulating bacteria and thrive in water pipes conveying water containing dissolved iron and form yellow or reddish colored slimes. When mature bacteria die, they may decompose, imparting foul tastes and odors to water.

$$Fe^{2+} \text{ (ferrous)} + \text{oxygen} \rightarrow Fe^{3+} \text{ (ferric)} + \text{energy} \qquad (12.4)$$

*Fungi* (singular, fungus) is a common term used to refer to microscopic nonphotosynthetic plants, including yeasts and molds. The most important group of yeasts for industrial fermentations are the genus *Saccharomyces*. *Saccharomyces cerevisiae* is the common yeast used by bakers, distillers, and brewers. *Saccharomyces cerevisiae* is single celled, commonly 5–10 $\mu$m in size, and reproduces by budding, in which large, mature cells divide, each producing one or more daughter cells. Under anaerobic conditions, this yeast produces alcohol as an end product. *Saccharomyces cerevisiae* is facultative and performs the following reactions:

*Anaerobic*:          Sugar $\rightarrow$ alcohol $+ CO_2 +$ energy          (12.5)

*Aerobic*:    Sugar $+$ oxygen $\rightarrow CO_2 +$ energy          (12.6)

Energy yield in the aerobic reaction is much greater than in the anaerobic fermentation.

Molds are saprophytic or parasitic filamentous fungi that resemble higher plants in structure, composed of branched, filamentous, threadlike growths called hyphae. Molds are nonphotosynthetic, multicellular, heterotrophic, aerobic, reproduced by spore formation, and grow best in low-pH solutions (pH 2–5) high in sugar content. Molds are undesirable growths in activated sludge and can be created by low-pH conditions. The operation of an activated-sludge wastewater treatment system relies on gravity separation of microorganisms from the wastewater effluent. A large growth of molds creates a filamentous activated sludge that does not settle easily.

## 12.2 ALGAE

*Algae* (singular, alga) are microscopic photosynthetic plants. The process of photosynthesis is illustrated by the equation

$$CO_2 + 2H_2O \underset{\text{dark reaction}}{\overset{\text{sunlight}}{\rightleftharpoons}} \text{new cell tissue} + O_2 + H_2O \qquad (12.7)$$

The overall effect of this reaction is to produce new plant life, thereby increasing the number of algae. By-product oxygen results from the biochemical conversion of water.

Algae are autotrophic, using carbon dioxide (or bicarbonates in solution) as a carbon source. The nutrients of phosphorus (as phosphate) and nitrogen (as ammonia, nitrite, or nitrate) are necessary for growth. Certain species of blue-green algae are able to fix atmospheric nitrogen. In addition, certain trace nutrients are required, such as magnesium, sulfur, boron, cobalt, molybdenum, calcium, potassium, iron, manganese, zinc, and copper. In natural waters, the nutrients most frequently limiting algal growth are inorganic phosphorus and nitrogen.

Energy for photosynthesis is derived from sunlight. Photosynthetic pigments biochemically convert the energy in sun's rays to useful energy for plant synthesis. The most common pigment is chlorophyll, which is green in color. Other pigments or combinations of pigments result in algae of a variety of colors, such as blue-green, yellowish green, brown, and red. In the prolonged absence of sunlight, the algae perform a dark reaction—for practical purposes the reverse of synthesis. In the dark reaction, the algae degrade stored food or their own protoplasm for energy to perform essential biochemical reactions for survival. The rate of this endogeneous reaction is significantly slower than the photosynthetic reaction.

Algae grow in abundance in stabilization ponds rich in inorganic nutrients and carbon dioxide released from bacterial decomposition of waste organics. Green algae *Chlorella* are commonly found in oxidation ponds. Certain genera of algae are identified with clean water, such as *Navicula*. Descriptions and pictorial representations of algae occurring in water and wastewater are in *Standard Methods for the Examination of Water and Wastewater* [1].

## 12.3  PROTOZOANS AND HIGHER ANIMALS

Protozoans are single-celled animals that reproduce by binary fission. The protozoans of significance in biological treatment systems are strict aerobics found in activated sludge, trickling filters, and oxidation ponds. These microscopic animals have complex digestive systems and use solid organic matter as an energy and carbon source. Protozoans are a vital link in the aquatic chain since they ingest bacteria and algae.

Protozoans with cilia may be categorized as free swimming and stalked. Free-swimming forms move rapidly in the water, ingesting organic matter at a very high rate. The stalked forms attach by a stalk to particles of matter and use cilia to propel their head about and bring in food. Another group of protozoans move by flagella. Long hairlike strands (flagella) move with a whiplike action, providing motility. *Amoeba* move and ingest food through the action of a mobile protoplasm.

Rotifers are the simplest multicelled animals. They are strict aerobes and metabolize solid food. A typical rotifer uses the cilia around its head for catching food. The name *rotifer* is derived from the apparent rotating motion of the cilia on its head. Rotifers are indicators of low pollutional waters and are regularly found in streams and lakes.

Crustaceans are multicellular animals with branched swimming feet or a shell-like covering, with a variety of appendages (antennae). The two most common crustaceans of interest are *Daphnia* and *Cyclops*. Crustaceans are strict aerobes and ingest microscopic plants. The zooplankton population in a lake includes a wide selection of crustaceans that serve as food for fishes.

## 12.4 METABOLISM, ENERGY, AND SYNTHESIS

*Metabolism* (catabolism) is the biochemical process (a series of biochemical oxidation–reduction reactions) performed by living organisms to yield energy for synthesis, motility, and respiration to remain viable. In standard usage, metabolism implies both catabolism and anabolism, that is, both degradation and assimilative reactions.

The metabolism of autotrophic bacteria is illustrated in Eqs. 12.1 through 12.4. In these reactions, the reduced inorganic compounds are oxidized, yielding energy for synthesis of carbon from carbon dioxide, producing organic cell tissue. (In the case of algae, Eq. 12.7, the carbon source is carbon dioxide but the energy is from sunlight.)

In heterotrophic metabolism, organic matter is the substrate (food) used as an energy source. However, the majority of organic matter in wastewater is in the form of large molecules that cannot penetrate the bacterial cell membrane. The bacteria, in order to metabolize high-molecular-weight substances, must be capable of hydrolyzing the large molecules into diffusible fractions for assimilation into their cells. Therefore, the first biochemical reactions are hydrolysis* of complex carbohydrates into soluble sugar units, protein into amino acids, and insoluble fats into fatty acids. Under aerobic conditions, the reduced soluble organic compounds are oxidized to end products of carbon dioxide and water (Eq. 12.8). Under anaerobic conditions, soluble organics are decomposed to intermediate end products, such as organic acids and alcohols, along with the production of carbon dioxide and water (Eq. 12.9). Many intermediates, such as butyric acid, mercaptons (organic compounds with —SH radicals), and hydrogen sulfide have foul odors.

Under anaerobic conditions, if excess organic acids are produced, the pH of the solution will drop sufficiently to "pickle" the fermentation process. This is the principle used for preservation of silage. Bacteria produce an overabundance of organic acids in the anaerobic decomposition of the green fodder stored in the silo, inhibit further bacterial decomposition, and preserve

---

* Hydrolysis is the addition of water to split a bond between chemical units.

the food value of the fodder. However, if proper environmental conditions exist to prevent excess acidity from the production of organic acid intermediates, populations of acid-splitting methane-forming bacteria will develop and use the organic acids as substrate (Eq. 12.10). The combined biological processes of anaerobic decomposition of raw organic matter to soluble organic intermediates and the gasification of the intermediates to carbon dioxide and methane is referred to as digestion.

$$\textit{Aerobic}: \quad \text{Organics} + \text{oxygen} \rightarrow CO_2 + H_2O + \text{energy} \qquad (12.8)$$

$$\textit{Anaerobic}: \quad \text{Organics} \qquad \rightarrow \text{intermediates} + CO_2 + H_2O + \text{energy} \qquad (12.9)$$

$$\text{Organic acid intermediates} \quad \rightarrow CH_4 + CO_2 + \text{energy} \qquad (12.10)$$

The growth and survival of nonphotosynthetic microorganisms are dependent on their ability to obtain energy from the metabolism of substrate. Biochemical metabolic processes of heterotrophs are energy-yielding oxidation–reduction reactions in which reduced organic compounds serve as hydrogen donors and oxidized organic or inorganic compounds act as hydrogen acceptors. *Oxidation* is the addition of oxygen, removal of hydrogen, or removal of electrons. *Reduction* is the removal of oxygen, addition of hydrogen, or addition of electrons.

The simplified diagram of substrate dehydrogenation shown in Fig. 12.1 is intended to illustrate the general relationship between energy yields of aerobic and anaerobic metabolism. Enzymatic processes of hydrogen transfer and methods of biologically conserving energy released are beyond the scope of this discussion. For students to understand fully the mechanisms illustrated in Fig. 12.7, a knowledge of the biochemistry of microorganisms is necessary [2].

Energy stored in organic matter ($AH_2$) is released in the process of biological oxidation by dehydrogenation of substrate followed by transfer of hydrogen, or electrons, to an ultimate acceptor. The higher the ultimate hydrogen acceptor is on the energy (electromotive) scale, the greater will be the energy yield from oxidation of 1 mole of a given substrate. Aerobic metabolism using oxygen as the ultimate hydrogen acceptor yields the greatest amount of energy. Aerobic respiration can be traced on Fig. 12.1 from reduced organic matter ($AH_2$) at the bottom, through the hydrogen and electron carriers, to oxygen. Facultative respiration, using oxygen bound in nitrates and sulfates, yields less energy than aerobic metabolism. The least energy yield results from strict anaerobic respiration, where the oxidation of $AH_2$ is coupled with reduction of B (an oxidized organic compound) to $BH_2$ (a reduced organic compound). The preferential use of hydrogen acceptors based on energy yield in a mixed bacterial culture is illustrated by the following equations:

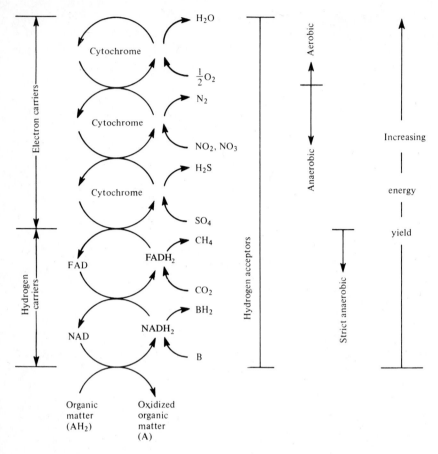

**Figure 12.1** General scheme of substrate dehydrogenation for energy yield. (FAD: flavin adenine dinucleotide; NAD: nicotinamide adenine dinucleotide.)

| | | |
|---|---|---|
| *Aerobic* | $AH_2 + O_2 \rightarrow CO_2 + H_2O$ + energy | (12.11) |
| | $AH_2 + NO_3^- \rightarrow N_2 + H_2O$ + | (12.12) |
| *(Facultative)* | $AH_2 + SO_4^{2-} \rightarrow H_2S + H_2O$ + | (12.13) |
| | $AH_2 + CO_2 \rightarrow CH_4 + H_2O$ + | (12.14) |
| *Anaerobic* | $AH_2 + B \rightarrow BH_2 + A$ + energy | (12.15) |

energy yield decreasing

Hydrogen acceptors are used in the sequence of dissolved oxygen first, followed by nitrates, sulfates, and oxidized organic compounds, in this general order. Thus hydrogen sulfide odor formation follows nitrate reduction and precedes methane formation.

The biochemical reactions in Fig. 12.1 are performed by oxidation–reduction enzymes. Enzymes are organic catalysts that perform biochemical reactions at temperatures and chemical conditions compatible with biological life.

The coenzyme component of the enzyme determines what chemical reaction will occur. Coenzymes nicotinamide adenine dinucleotide (NAD) and flavin adenine dinucleotide (FAD) are responsible for hydrogen transfer. Cytochromes are respiratory pigments that can undergo oxidation and reduction and serve as hydrogen carriers.

Synthesis (anabolism) is the biochemical process of substrate utilization to form new protoplasm for growth and reproduction. Microorganisms process organic matter to create new cells. The cellular protoplasm formed is a combination of hundreds of complex organic compounds, including proteins, carbohydrates, nucleic acids, and lipids. Major elements in biological cells are carbon, hydrogen, oxygen, nitrogen, and phosphorus. On a dry-weight basis, protoplasm is 10%–12% nitrogen and approximately 2.5% phosphorus; the remainder is carbon, hydrogen, oxygen, and trace elements.

Relationships between metabolism, energy, and synthesis are important in understanding biological treatment systems. The primary product of metabolism is energy, and the chief use of this energy is for synthesis. Energy release and synthesis are coupled biochemical processes that cannot be separated. The maximum rate of synthesis occurs simultaneously with the maximum rate of energy yield (maximum rate of metabolism). Therefore, in heterotrophic metabolism of wastewater organics, maximum rate of removal of organic matter for a given population of microorganisms occurs during maximum biological growth. Conversely, the lowest rate of removal of organic matter occurs when growth ceases.

The major limitation of anaerobic growth is energy, owing to the fact that in anaerobic decomposition a low energy yield per unit of substrate results from an incomplete reaction (Fig. 12.2). In other words, the limiting factor in anaerobic metabolism is a lack of hydrogen acceptors. When the supply of biologically available energy is exhausted, the processes of metabolism and synthesis cease.

Aerobic metabolism is the antithesis of anaerobiosis, biologically avail-

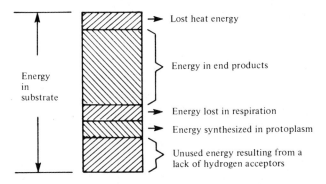

Energy in substrate

Lost heat energy

Energy in end products

Energy lost in respiration

Energy synthesized in protoplasm

Unused energy resulting from a lack of hydrogen acceptors

**Figure 12.2**  Energy conversions in anaerobic metabolism.

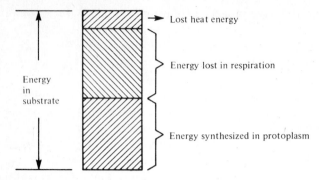

**Figure 12.3** Energy conversions in aerobic metabolism.

able carbon being the limiting factor (Fig. 12.3). Abundance of oxygen creates no shortage of hydrogen acceptors. But the supply of substrate carbon is rapidly exhausted through respiration of carbon dioxide and synthesis into new cells.

The energy conversion diagrams shown schematically in Figs. 12.2 and 12.3 illustrate the major features of anaerobic and aerobic metabolism. An anaerobic process has the following characteristics: incomplete metabolism, small quantity of biological growth, and production of high-energy products such as acetic acid and methane. An aerobic process results in complete metabolism and synthesis of the substrate, ending in a large quantity of biological growth.

## 12.5 ENZYME KINETICS

The reaction between an enzyme E and substrate S is postulated to be

$$E + S \underset{k_2}{\overset{k_1}{\rightleftharpoons}} ES \overset{k_3}{\rightarrow} E + \text{products} \tag{12.16}$$

In the first step, E combines with S to form ES by a reversible reaction where $k_1$ is the rate constant for formation of ES, and $k_2$ is the rate constant for dissociation of ES to E and S. After combining to form ES, S is converted to products in the second step and E is released for combination with more S. The rate of conversion of ES to products is represented by $k_3$. An enzyme-catalyzed reaction can be experimentally performed by placing a small quantity of an enzyme in a substrate solution and measuring the decomposition of substrate with time. For example, sucrose is split into its two separate sugar rings of glucose and fructose by the enzyme *invertase* that performs this hydrolysis. The reaction in the form of Eq. 12.16 is written

Sucrose + *invertase* $\rightleftharpoons$ sucrose-*invertase*

$\rightarrow$ glucose + fructose + *invertase*

The first satisfactory mathematical analysis of the effect of substrate concentration on the rate of enzyme-catalyzed reactions was made in 1913 by Michaelis and Menten [3]. The derivation of their equation is based on Eq. 12.17, that is, the rate of decomposition of substrate is proportional to the concentration of the intermediate enzyme–substrate complex. For the reversible reaction $E + S \rightleftharpoons ES$, the dissociation constant of ES, defined as $K_m$, can be written as

$$K_m = \frac{(E - ES)S}{ES} \tag{12.17}$$

Rearranging the equation,

$$(ES) = \frac{(E)(S)}{K_m + S} \tag{12.18}$$

with $k_3$ the rate constant for decomposition of ES, the measured rate of decomposition of substrate $r$ equals $k_3 (ES)$, and

$$r = \frac{k_3(E)(S)}{K_m + S} \tag{12.19}$$

The maximum rate of decomposition $r_m$ occurs when ES is at its maximum concentration, that is, when all of the enzyme is combined with substrate and $ES = E$. Therefore,

$$r_m = k_3(ES) = k_3(E) \tag{12.20}$$

Substituting $r_m$ for $k_3(E)$ in Eq. 12.19 produces the Michaelis–Menten equation:

$$r = r_m \left( \frac{S}{K_m + S} \right) \tag{12.21}$$

or

$$K_m = S \left( \frac{r_m}{r} - 1 \right) \tag{12.22}$$

Since $K_m$ and $r_m$ are constants, Eq. 12.21 is a rectangular hyperbola, as shown in Fig. 12.4. Data from enzyme-catalyzed experiments by Michaelis and

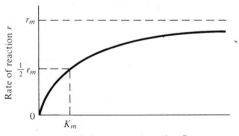

**Figure 12.4** Theoretical curve of reaction rate versus substrate concentration for an enzyme-catalyzed reaction (Eq. 12.21).

Menten graphed to form this diphasic curve. From Eq. 12.22, when $r_m/r = 2$, the measured rate $r$ is one-half the value of the limiting rate $r_m$ and $K_m = S$. Therefore, the substrate concentration for attaining the half-maximum reaction rate is a characteristic constant $K_m$ of an enzyme-catalyzed reaction, which is termed the *saturation constant*.

The substrate saturation phenomenon is a unique characteristic of enzymatic reactions and is the mathematical basis of kinetics in microbiology. For further explanation of this process, consider a batch experiment starting with a large amount of substrate relative to the concentration of enzyme. From Eq. 12.21, for a large S, $K_m$ can be neglected and $r = r_m$. This can also be seen in Fig. 12.4. The magnitude of $r_m$ depends on the maximum rate at which the enzyme present can decompose the substrate. The rate is zero order; that is, the reaction proceeds at a rate independent of the concentration of substrate. As the quantity of substrate decreases, the concentration enzyme remaining constant, the rate of reaction becomes increasingly dependent on the remaining substrate concentration. The rate at $\frac{1}{2}r_m$ corresponds to a substrate concentration equal to the saturation constant $K_m$. Below this substrate concentration, some of the enzyme molecules are not combined with substrate, resulting in a first-order reaction, which proceeds at a rate directly proportional to the substrate concentration. When S is much smaller than $K_m$, S can be neglected in Eq. 12.21 and $r = r_m S/K_m$.

## 12.6 GROWTH KINETICS OF PURE BACTERIAL CULTURES

A pure culture can be grown in a laboratory reactor by inoculating a sterile liquid medium with a small number of viable bacteria of a single species. The characteristic growth pattern for bacteria in such a batch culture is sketched in Fig. 12.5. After a short lag period for adaptation to the new environment, the bacteria reproduce by binary fission, exponentially increasing the number of viable cells and biomass in the culture medium. The existence of excess substrate promotes this maximum rate of growth. The rate of metabolism in the *exponential growth phase* is limited only by the microorganisms' ability to process the substrate. With $X$ representing the concentration of biomass and $\mu$ a proportionality constant, the biomass growth rate can be expressed as

$$\left(\frac{dX}{dt}\right)_g = \mu X \tag{12.23}$$

Dividing both sides of Eq. 12.23 by $X$ yields

$$\mu = \frac{(dX/dt)_g}{X} \tag{12.24}$$

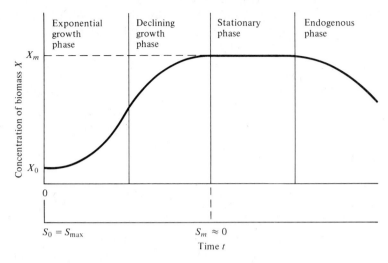

**Figure 12.5** Characteristic growth phases of a pure culture of bacteria.

where $\mu$ = specific growth rate (rate of growth per unit of biomass), time$^{-1}$
$(dX/dt)_g$ = biomass growth rate, mass/unit volume·time
$X$ = concentration of biomass, mass/unit volume

The *declining growth phase* is caused by an increasing shortage of substrate. The rate of reproduction decreases until the number of viable bacteria is stationary, which occurs when the rate of reproduction is equal to the rate of death. The total biomass exceeds the mass of viable cells since many of the microorganisms stopped reproducing owing to substrate-limiting conditions. By laboratory experimentation Monod [4] studied the growth of bacteria in batch cultures. He found that growth was a function of both the concentration of microorganisms and the concentration of the growth-limiting substrate. The mathematical relationship proposed by Monod between the residual concentration of the growth-limiting substrate and the specific growth rate of biomass is the hyperbolic equation

$$\mu = \mu_m \left( \frac{S}{K_S + S} \right) \tag{12.25}$$

where $\mu$ = specific growth rate, time$^{-1}$
$\mu_m$ = maximum specific growth rate (at a concentration of the growth-limiting substrate at or above saturation), time$^{-1}$
$S$ = concentration of growth-limiting substrate in solution, mass/unit volume
$K_S$ = saturation constant (equal to the limiting substrate concentration at one half the maximum growth rate), mass/unit volume

As shown in Fig. 12.6, Eq. 12.25 is similar in form to the Michaelis–Menten equation for enzyme-catalyzed reactions (Eq. 12.21 and Fig. 12.4). The constant $K_S$, similar in function to $K_m$ in Eq. 12.22, is equal to the concentration of substrate $S$ when the specific growth rate $\mu$ equals one-half of the maximum growth rate $\mu_m$.

The specific growth rate relationship in Eq. 12.25 can be substituted for the proportionality constant in Eq. 12.23. The result is the following expression for biomass growth rate in a substrate-limiting solution:

$$\left(\frac{dX}{dt}\right)_g = \frac{\mu_m X S}{K_S + S} \tag{12.26}$$

where
$(dX/dt)_g$ = biomass growth rate, mass/unit volume·time
$\mu_m$ = maximum specific growth rate, time$^{-1}$
$X$ = concentration of biomass, mass/unit volume
$S$ = concentration of growth-limiting substrate, mass/unit volume
$K_S$ = saturation constant, mass/unit volume

The *growth yield Y* is defined as the incremental increase in biomass resulting from metabolism of an incremental amount of substrate [5]. The growth yield in a batch culture (Fig. 12.5) is the biomass increase during the exponential and declining growth phases $(X_m - X_0)$ relative to the substrate used $(S_0 - S_m)$. Since growth is limited by depletion of the substrate, $S_m$ is assumed to be zero. Therefore,

$$X_m - X_0 = Y S_0 \tag{12.27}$$

If a series of batch cultures are grown starting with different initial substrate concentrations, a plot of $X_m$ values versus their respective $S_0$ values is a

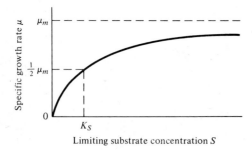

**Figure 12.6** Specific growth rate versus substrate concentration for the exponential and declining growth phases of a bacterial culture.

straight line with slope $Y$ (Fig. 12.7). For bacterial cultures, when conditions are maintained constant, the growth yield is a constant reproducible value [4].

Growth yield can also be expressed in derivative form as

$$\left(\frac{dX}{dt}\right)_g = Y\left(\frac{dS}{dt}\right)_u \tag{12.28}$$

Substituting this into Eq. 12.26 results in an equation that defines the rate of substrate utilization in a solution in which the biomass growth rate is limited by the low concentration of substrate:

$$\left(\frac{dS}{dt}\right)_u = \frac{\mu_m X S}{Y(K_S + S)} \tag{12.29}$$

where   $(dS/dt)_u$ = substrate utilization rate, time$^{-1}$
  $\mu_m$ = maximum specific growth rate, time$^{-1}$
  $X$ = concentration of biomass, mass/unit volume
  $S$ = concentration of growth-limiting substrate, mass/unit volume
  $Y$ = growth yield, mass/mass
  $K_S$ = saturation constant, mass/unit volume

In the *endogenous growth phase* (Fig. 12.5), viable bacteria are competing for the small amount of substrate still in solution. The rate of metabolism is decreasing at an increasing rate, resulting in a rapid decrease in the number of viable cells. Starvation occurs such that the rate of death exceeds the rate of reproduction. The total biomass decreases as cells utilize their own protoplasm as an energy source. Cells become old, die, and lyse, releasing nutrients back into solution. The action of cell lysis decreases both the number and mass of microorganisms.

The rate of biomass decrease during endogenous respiration is proportional to the biomass present. Thus,

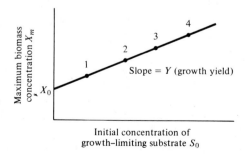

**Figure 12.7** Growth yield for a series of four batch cultures (Fig. 12.11) is determined by plotting $X_m$ versus $S_0$.

$$\left(\frac{dX}{dt}\right)_d = -k_d X \tag{12.30}$$

where   $(dX/dt)_d$ = biomass decay rate, mass/unit volume·time
$\quad\quad\quad k_d$ = microbial decay coefficient, $\text{time}^{-1}$
$\quad\quad\quad X$ = concentration of biomass, mass/unit volume

To determine *net biomass growth rate* during the endogenous growth phase, Eq. 12.30 is combined with Eq. 12.26 and Eq. 12.30 with Eq. 12.28, resulting in the following equations:

$$\left(\frac{dX}{dt}\right)_g^{\text{net}} = \frac{\mu_m X S}{K_S + S} - k_d X \tag{12.31}$$

$$\left(\frac{dX}{dt}\right)_g^{\text{net}} = Y\left(\frac{dS}{dt}\right)_u - k_d X \tag{12.32}$$

The Monod relationship (Eq. 12.25) modified to yield *net specific growth rate* is

$$\mu_{\text{net}} = \mu_m \frac{S}{K_S + S} - k_d \tag{12.33}$$

and the observed growth yield $Y_{\text{obs}}$ accounting for the effect of endogenous respiration on the net biomass growth rate from the relationship in Eq. 12.28 is

$$Y_{\text{obs}} = \frac{(dX/dt)_g^{\text{net}}}{(dS/dt)_u} \tag{12.34}$$

## 12.7 BIOLOGICAL GROWTH IN WASTEWATER TREATMENT

Both fixed-film growth and suspended-solids growth systems are used in biological treatment of wastewaters. In a fixed-growth process, organic matter is removed from wastewater as it flows over a biological film (slime layer) attached to a filter medium. Trickling filters use a variety of media, including stones, crushed rock, small plastic cylinders, plastic vertical-sheet packing, and racks of redwood slats. A rotating biological contactor consists of a series of large-diameter plastic disks that slowly rotate in tanks conveying wastewater flow. In a suspended-growth process, active biological solids are mixed with the wastewater and held in suspension by aeration as the organic matter is taken out of solution by the microbial floc. The common name for this suspended-growth process is activated sludge.

Wastewater treatment relies on a mixed biological culture consisting of a variety of bacteria and protozoans. Microorganisms in the raw wastewater provide continuous inocula for the treatment process. The heterogeneous substrate content of a wastewater is expressed either as BOD or COD. The purpose of the treatment unit is to hold the biological culture in a controlled environment to promote growth of the microorganisms for extraction of colloidal and dissolved organics from solution.

The batch-culture growth pattern shown in Fig. 12.5 is not directly applicable to biological treatment processes that are continuous-flow systems. For example, an activated-sludge system is fed continuously, and excess microorganisms are withdrawn, either continuously or intermittently, to maintain the desired mass of microorganisms for metabolizing incoming organic wastes. A schematic diagram (Fig. 12.8) illustrates the flow pattern for organic matter and microorganisms in an activated-sludge system. Influent wastewater is aerated with a mixed culture of microorganisms for a sufficient period of time to permit synthesis of the waste organics into biological cells. The microorganisms are then settled out of solution, removed from the bottom of the settling tank, and returned to the aeration tank to metabolize additional waste organics. Unused organic matter and nonsettleable microorganisms pass out in the system effluent. Metabolism of the organic matter results in an increased mass of microorganisms in the system. Excess microorganisms are removed (wasted) from the system to maintain proper balance between food supply and mass of microorganisms in the aeration tank. This balance is referred to as the *food-to-microorganism ratio* (F/M).

The F/M ratio maintained in the aeration tank defines the operation of an activated-sludge system. At a high F/M ratio, microorganisms are in the exponential growth phase, characterized by excess food and maximum rate of metabolism (Fig. 12.9). Although the exponential growth phase is desirable for maximum rate of organic matter removal, distinct disadvantages make it undesirable for operation of an activated-sludge system. The microorganisms are in dispersed growth such that they do not settle out of solution by gravity. Consequently, the settling tank is not effective in separating microorganisms from the effluent for return to the aeration tank. Second, there is excess unused organic matter in solution which cannot be removed by sedimentation

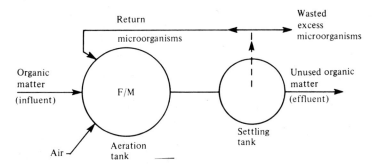

**Figure 12.8** Schematic diagram of a continuous-flow activated-sludge process. (F/M: food/microorganism ratio.)

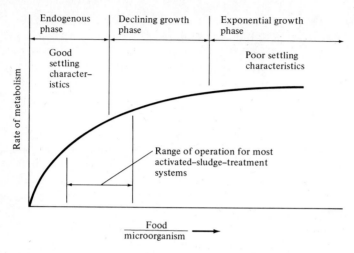

**Figure 12.9** Rate of metabolism versus increasing food/microorganism ratio.

and passes out in the effluent. Operation at a high F/M ratio results in poor BOD removal efficiency.

At a low F/M ratio, overall metabolic activity in the aeration tank may be considered as endogenous. Although initially there is rapid growth when the influent food and return microorganisms are mixed, competition for the small amount of food made available to the large mass of microorganisms results in near-starvation conditions for the majority of microorganisms within a short period of time. Under these conditions, continued aeration results in autooxidation of the microbial mass through cell lysis and resynthesis and also through the predator–prey activity where bacteria are consumed by the protozoans. Although the rate of metabolism is relatively low in the endogenous phase, metabolism of the organics is nearly complete and the microorganisms flocculate rapidly and settle out of solution by gravity. The good settling characteristics exhibited by activated sludge in the endogenous phase make operation in this growth period desirable where a high BOD removal efficiency is desired. Figure 12.9 summarizes the previous discussion and shows the range of operation for most activated-sludge treatment systems to be between the declining growth phase and the endogenous phase.

## 12.8 FACTORS AFFECTING GROWTH

Several factors affect the growth of microorganisms. The most imortant are temperature, pH, availability of nutrients, oxygen supply, presence of toxins, types of substrate, and, in the case of photosynthetic plants, sunlight. Growth with respect both to aerobic and anaerobic conditions and to the need for essential nutrients was discussed in Section 12.4.

Bacteria are classified as psychrophilic, mesophilic, or thermophilic, depending on their optimum temperature range for growth. Of least significance to sanitary engineers are the *psychrophilic* (cold-loving) bacteria, which grow best at temperatures slightly above freezing (4°–10°C).

*Thermophilic* (heat-loving) bacteria like an optimum temperature range of 50°–55°C. They are significant in sludge composting and attempts have been made to use a thermophilic temperature range for the anaerobic digestion of waste sludge. Thermophilic digestion has not been successful in practice because thermophilic bacteria are sensitive to small temperature changes, and it is difficult to maintain the required high operating temperature in a digestion tank.

The *mesophilic* (moderation-loving) bacteria grow best in the temperature range 20°–40°C. The vast majority of biological treatment systems operate in the mesophilic temperature range. Anaerobic digestion tanks are normally heated to near the optimum level of 35°C (95°F). Aeration tanks and trickling filters operate at the temperature of the wastewater as modified by that of the air. Generally, this is within the 15°–25°C range. A high wastewater temperature increases biological activity in the treatment process but rarely causes any severe operating problems. At high temperatures, odor problems may be more pronounced at a wastewater plant.

The rate of biological activity in the mesophilic range between 5° and 35°C doubles for every 10°–15°C temperature rise (Fig. 12.10). The common mathematical expression for relating the change in the reaction-rate constant with temperature, developed in Section 11.6, is

$$K = K_{20}\Theta^{T-20} \qquad\qquad (12.35)$$

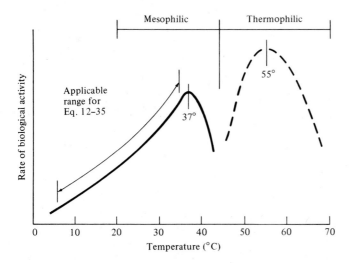

**Figure 12.10** General effect of temperature on biological activity.

where $K$ = reaction-rate constant at temperature $T$
$K_{20}$ = reaction-rate constant at 20°C
$\Theta$ = temperature coefficient
$T$ = temperature of biological reaction, °C

The value of $\Theta$ is 1.072 if the rate of biological activity doubles with a 10°C temperature rise; the value of $\Theta$ is 1.047 if the rate doubles in 15°C. Above 40°C, mesophilic bacterial metabolism drops off sharply and thermophilic growth starts. Thermophilic bacteria have a range of approximately 45°C–75°C, with an optimum near 55°C.

Cold wastewater can reduce BOD removal efficiency of biological processes. The efficiency of trickling filters is definitely decreased during cold weather and increased during warm periods. Trickling filters operating at 5°C–10°C have poor BOD removals. However, low-loaded extended aeration systems operating at the same temperatures showed good efficiencies. Extended aeration at a reduced BOD loading and resultant long aeration time compensates for the low metabolism rate of microorganisms. The biological treatment system most affected by cold winter temperature is the stabilization pond. Heat in the wastewater is not adequate to prevent formation of an ice cover in northern climates during winter.

The hydrogen ion concentration of the culture medium has a direct influence on microbial growth. Most biological treatment systems operate best in a neutral environment. The general range for operation of activated-sludge systems is between pH 6.5 and 8.5. At pH 9.0 and above, microbial activity is inhibited. Below 6.5, fungi are favored over bacteria in the competition for food. The methane-forming bacteria in anaerobic digestion have a much smaller pH tolerance range. General limits for anaerobic digestion are pH 6.7–7.4 with optimum operation at pH 7.0–7.1.

Biological treatment systems are adversely affected by toxic substances. Industrial wastes from metal-finishing industries usually contain toxic metal ions, such as nickel and chromium. Phenol is an extremely toxic compound found in chemical industry wastes. These and other inhibiting compounds are commonly removed by pretreatment at the industrial site prior to disposal of the industrial wastes to a municipal sewer.

Environmental conditions that adversely affect the desired microbial growth in an activated-sludge aeration tank can cause production of sludge with poor settling characteristics. This condition, resulting in excessive carry-over of activated-sludge floc in the clarifier effluent (referred to as sludge bulking), is associated with filamentous growths and pinpoint floc with poor settleability.

## 12.9 POPULATION DYNAMICS

Previous sections described the important characteristics of each group of microorganisms (bacteria, fungi, algae, and protozoans) independently. In biological waste treatment systems, however, the naturally occurring cultures are mixtures of bacteria growing in mutual association and with other microscopic plants and animals. A general knowledge of the relationships, both cooperative and competitive, between various microbial populations in mixed cultures is essential to understanding biological treatment processes.

When organic matter is made available to a mixed population of microorganisms, competition arises for this food between the various species. Primary feeders that are most competitive become the dominant microorganisms. Under normal operating conditions, bacteria are the dominant primary feeders in activated sludge (Fig. 12.11). Saprobic protozoans, those that feed on dead organic matter (e.g., *Euglena*), are not effective competitors against bacteria.

Species of dominant primary bacteria depend chiefly on the nature of the organic waste and environmental conditions in aeration tanks. Conditions adverse to bacteria, such as acid pH, low dissolved oxygen, and nutrient shortage, can produce a predominance of filamentous fungi, resulting in

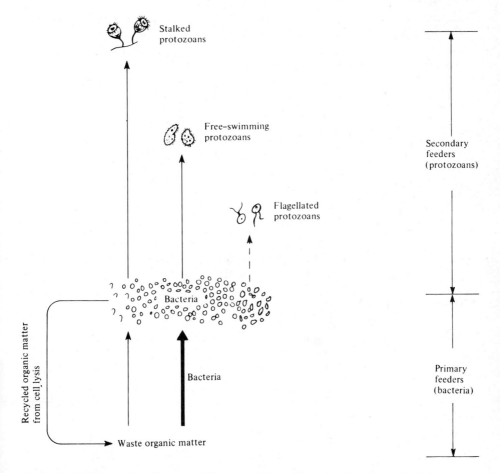

**Figure 12.11** Schematic diagram of the population dynamics in activated sludge.

sludge bulking. These abnormal circumstances are rare in municipal activated-sludge systems treating wastewater composed chiefly of domestic wastewater. However, when bulking in an activated-sludge system does occur, sanitary engineers must be prepared to find the cause and recommend corrective action (Section 12.29).

Primary bacteria in an activated-sludge system are maintained in the declining or endogenous growth phases. Under these conditions, the primary bacteria die and lyse, releasing their cell contents to solution. In this process, raw organic matter is synthesized and resynthesized by various groups of bacteria.

Holozoic protozoans, which feed on living organic matter, are common in activated sludge. They grow in association with the bacteria in a prey–predator relationship; that is, the bacteria (plants) synthesize the organic matter, and the protozoans (animals) consume the bacteria (Fig. 12.11). For a single reproduction a protozoan consumes thousands of bacteria, with two major beneficial effects of the prey–predator action. Removal of the bacteria stimulates further bacterial growth, resulting in accelerated extraction of organic matter from solution. Second, the flocculation characteristics of activated sludge are improved by reducing the number of free bacteria in solution, and a biological floc with improved settling characteristics results.

Competition for food also occurs between the secondary feeders. In a solution with high bacterial populations, free-swimming protozoans are dominant, but when food becomes scarce, stalked protozoans increase in numbers. Stalked protozoans do not require as much energy as free-swimming protozoans; therefore, they compete more effectively in a system with low bacterial concentrations. The photomicrographs shown in Fig. 12.12 illustrate the appearance of a healthy activated sludge.

The process of anaerobic digestion is carried out by a wide variety of bacteria, which can be categorized into two main groups, acid-forming bacteria and methane-forming bacteria. (Protozoans do not function in the digester's strict anaerobic environment.) The acid formers are facultative or anaerobic bacteria that metabolize organic matter, forming organic acids as an end product, along with carbon dioxide and methane (associated with oxidation of fats to organic acids). Acid-splitting methane formers use organic acids as substrate and produce gaseous end products of carbon dioxide and methane. These methane bacteria are strict anaerobes inactivated by the presence of dissolved oxygen and inhibited by the presence of oxidized compounds. The growth medium must contain a reducing agent such as hydrogen sulfide. Acid-splitting methane bacteria are sensitive to pH changes and other environmental conditions.

A simplified diagram (Fig. 12.13) portrays the relationship between the two bacterial stages in anaerobic digestion of organic matter. Both major groups of bacteria must cooperate to perform the overall gasification of organic matter. The first stage creates organic acids for the second stage, where these organic acids are converted to gas, preventing excess acid accumulation. In addition to producing food for the methane bacteria, acid

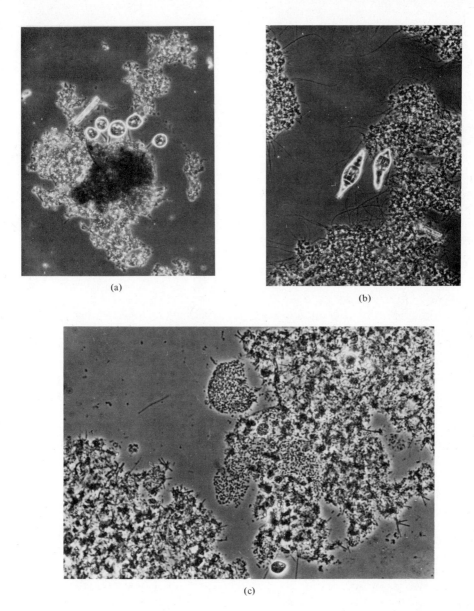

**Figure 12.12** Photomicrographs of activated sludge. (a) Activated-sludge floc with stalked protozoans (100×). (b) Rotifers in activated sludge (100×). (c) Activated-sludge floc showing clusters of bacterial cells. Lower center: flagellated protozoan (400×).

formers also reduce the environment to one of strict anaerobiosis by using the oxidized compounds and excreting reducing agents.

Problems in operating anaerobic treatment systems result when an imbalance occurs in the population dynamics. For example, if a sudden excess of

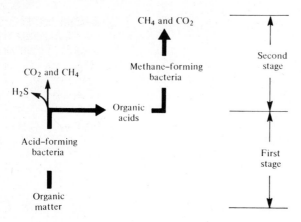

**Figure 12.13** Simplified diagram of population dynamics in anaerobic digestion.

organic matter is fed to a digester, acid formers very rapidly process this food, developing excess organic acids. The methane formers, whose population had been limited by a previous lower organic-acid supply, are unable to metabolize the organic acids fast enough to prevent a drop in pH. When the pH drops, the methane bacteria are affected first, further reducing their capacity to break down the acids. Under severe or prolonged overloading, the contents of the digester "pickles" in excess acids, and all bacterial activity is inhibited. In addition to organic overloading, the digestion process can be upset by a sudden increase in temperature, a significant shift in the type of substrate, or additions of toxic or inhibiting substances from industrial wastes.

A unique relationship exists between bacteria and algae in wastewater stabilization ponds (Fig. 12.14). The bacteria metabolize organic matter for reproduction, releasing soluble nitrogen and phosphorus nutrients and carbon dioxide. Algae use these inorganic compounds, along with the energy from sunlight, for synthesis, releasing oxygen. The resulting dissolved oxygen in the pond water is taken up by the bacteria, thus closing the cycle. This type of association between organisms is referred to as *symbiosis,* a relationship where two or more species live together for mutual benefit such that the association stimulates more vigorous growth of each species than if the growths were separate. The growth of algae replaces in part the organic matter decomposed by the bacteria. For this reason, ponds do not always appear to provide satisfactory removal of organic matter. A variety of predators (protozoans, rotifers, and higher animals) that feed on the plant growth (algae and bacteria) are also present in pond water.

At the liquid depth commonly used in stabilization-pond design, bottom waters may become anaerobic while the surface remains aerobic. In terms of general oxygen conditions, these lagoons are commonly referred to as *facultative lagoons* (see Fig. 12.14). During periods when the dissolved oxy-

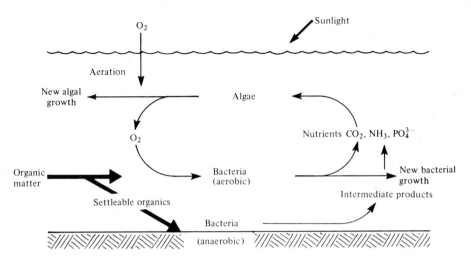

**Figure 12.14** Schematic diagram of microbiological activity in a wastewater stabilization pond showing the symbiotic relationship between bacteria and algae and decomposition of organic matter by both aerobic and anaerobic bacteria.

gen is less than saturation level, the surface water is aerated through wind action. During the winter, both bacterial metabolism and algal synthesis are slowed by cold temperatures. The lagoon generally remains aerobic, even under a transparent ice cover. If the sunlight is blocked by a snow cover, the algae cannot produce oxygen, and the lagoon becomes anaerobic. The result is odorous conditions during the spring thaw until the algae become reestablished. This may take from a few days to weeks, depending on climatic conditions and the amount of organic matter accumulated in the lagoon during the winter.

# Characteristics of Wastewater

*Wastewater* is defined as liquid wastes collected in a sewer system and conveyed to a treatment plant for processing. In most communities, storm runoff water is collected in a separate storm sewer system and conveyed to the nearest watercourse for disposal without treatment. Several large cities have a combined wastewater collection system where both storm water and sanitary wastes are collected in the same pipe system. The dry-weather flow in the combined sewers is collected for treatment, but during storms the wastewater flow in excess of plant capacity is bypassed directly to the receiving watercourse.

Sanitary or domestic wastewater refers to liquid material collected from residences, business buildings, and institutions. The term *industrial wastes* refers to that from manufacturing plants. *Municipal wastewater* is a general term applied to liquid treated in a municipal treatment plant. Municipal wastes from towns frequently contain industrial effluents from dairies, laundries, bakeries, and factories, and those in a large city may have wastes from major industries, such as chemical manufacturing, breweries, meat processing, metal processing, or paper mills.

## 12.10  FLOW AND STRENGTH VARIATIONS

The quantity of municipal wastewater flow varies from 50 to over 250 gal per capita per day (gpcd), depending on sewer uses of the community. A common value for sanitary flow is 120 gpcd (450 l/d). Per capita contribution of organic matter in domestic wastewater is approximately 0.24 lb (109 g) of suspended solids per day and 0.20 lb (91 g) of BOD per day in communities where a substantial portion of the household kitchen wastes are discharged to the sewer system through garbage grinders. These values are the population equivalents used to convert total pounds of solids or BOD of industrial wastes to equivalent population.

The actual quantities of flow and organic matter can differ significantly from these common values. The frequency distributions for per capita contributions of flow, suspended solids, and BOD based on analyses of data from 100 cities in Illinois, Indiana, Ohio, and Minnesota are graphed in Fig. 12.15 [6]. The data show wide variations. Larger cities have greater waste contributions per capita. For cities with populations of greater than 100,000, mean flow was 194 gpcd, mean suspended solids were 0.40 lb/capita/day, and BOD was 0.30 lb/capita/day. For cities with populations of less than 10,000, the values were 140 gpcd, 0.15 lb of suspended solids, and 0.14 lb of BOD. Geographic location and climate influence waste contributions, particularly flow as related to the amount of infiltration and inflow. From a study of 700 Texas communities having populations of less than 10,000, the average contributions per capita per day were 89 gal, 0.21 lb of suspended solids, and 0.16 lb of BOD [7].

The flows and loads entering a wastewater treatment plant during the year vary with climatic conditions, industrial production, domestic water use, and other factors. In a typical industrial community in a moderate climate, the summer average daily flow frequently exceeds winter flows by 20%–30%. Munksgaard and Young [8] analyzed operational data from 11 municipal plants in the range of 0.25–40 mgd to determine peak flows and loads. All the cities had separate sanitary sewers receiving less than 20% of

---

**Figure 12.15** Histograms of average wastewater flow and quantities of suspended solids and BOD contributed per capita based on data from 100 cities in Illinois, Indiana, Ohio, and Minnesota. [From D. H. Stoltenberg, "Midwestern Wastewater Characteristics," *Public Works* 111(1) (1980): 52–53.]

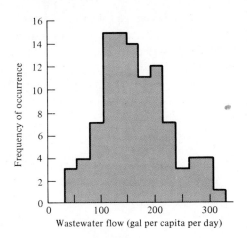

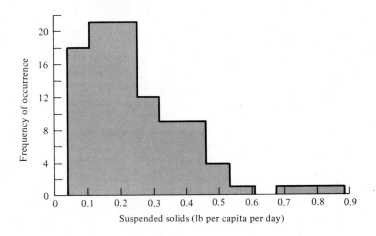

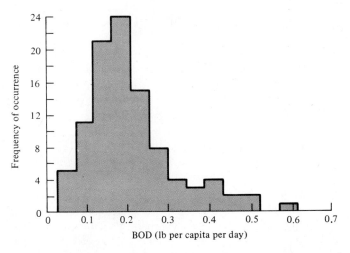

the wastewater from industrial sources. The wastewater flows included normal infiltration expected in humid climatic areas with an annual precipitation in the range of 20–40 in. The peaking ratio equations derived for the average annual peak month and average annual peak day are

$$Q_m = Q\left(\frac{1.26}{Q^{0.0101}}\right) \qquad (12.36)$$

$$B_m = B\left(\frac{1.91}{B^{0.0430}}\right) \qquad (12.37)$$

$$S_m = S\left(\frac{2.18}{S^{0.0517}}\right) \qquad (12.38)$$

$$Q_d = Q\left(\frac{1.96}{Q^{0.0360}}\right) \qquad (12.39)$$

$$B_d = B\left(\frac{4.08}{B^{0.0732}}\right) \qquad (12.40)$$

$$S_d = S\left(\frac{5.98}{S^{0.0716}}\right) \qquad (12.41)$$

where

$Q$ = average annual wastewater flow, mgd
$B$ = average annual BOD load, lb/day
$S$ = average annual suspended solids load, lb/day
$Q_m, B_m, S_m$ = average flow, BOD, and suspended solids values during the peak month
$Q_d, B_d, S_d$ = average flow, BOD, and suspended solids values for the peak day

Diurnal flow variations depend primarily on the size of a municipality and industrial flows. Hourly flowrates range from a minimum of 20% to a maximum of 250% or more of the average daily rate for small communities and from 50% to 200% for larger cities. The wastewater flow diagrams in Fig. 12.16 exemplify hourly flow variations for two municipalities of different sizes.

### ■ EXAMPLE 12.1

The sanitary sewer system in a municipality located in a humid climate receives over three-quarters of the wastewater discharges from domestic and commercial sources. The average annual wastewater flow is 10.0 mgd, containing 16,700 lb of BOD and 20,000 lb of suspended solids. Calculate the average daily wastewater flow, BOD load, and suspended solids load during the peak month of the year.

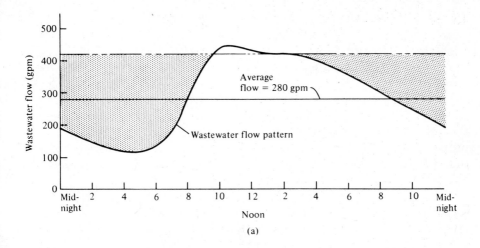

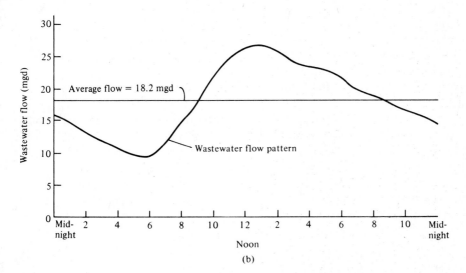

**Figure 12.16** Diagrams of municipal wastewater flow showing hourly variations. (a) Flow diagram from a town with a population of 4500. (Shaded area is the typical recirculation flow for a high-rate trickling-filter plant at $R = 0.5$.) (b) Flow diagram from a city with a population of 150,000.

*Solution.*   Applying Eqs. 12.36 through 12.38,

$$Q_m = 10\left(\frac{1.26}{10^{0.0101}}\right) = 12.3 \text{ mgd}$$

$$B_m = 16{,}700\left(\frac{1.91}{16{,}700^{0.0430}}\right) = 21{,}000 \text{ lb/day}$$

$$S_m = 20{,}000\left(\frac{2.18}{20{,}000^{0.0517}}\right) = 26{,}100 \text{ lb/day} \qquad\blacksquare$$

## 12.11 COMPOSITION OF WASTEWATER

The data in Table 12.1 represent the approximate composition of domestic wastewater before and after treatment. BOD and suspended solids (nonfiltrable residue) are the two most important parameters used to define the characteristics of a domestic wastewater. A suspended-solids concentration of 240 mg/l is equivalent to 0.24 lb of suspended solids in 120 gal, and 200 mg/l of BOD is equivalent to 0.20 lb of BOD in 120 gal. Reduction of suspended solids and BOD in primary sedimentation is approximately 50% and 35%, respectively. Approximately 70% of the suspended solids are volatile, defined as those lost upon ignition at 550°C.

Total solids (residue on evaporation) include organic matter and dissolved salts; the concentration of the latter is dependent to a considerable extent on the hardness of the municipal water. The concentration of nitrogen in domestic waste is directly related to the concentration of organic matter (BOD). Approximately 40% of the total nitrogen is in solution as ammonia. If raw wastewater has been retained for a long time in collector sewers, a greater percentage of ammonia nitrogen results from deamination of the proteins and urea in wastewater. Ten milligrams per liter of phosphorus is approximately equivalent to a 3-lb phosphorus contribution per capita per year. Approximately 2 lb of this is from phosphate builders used in synthetic detergents.

The surplus of nitrogen and phosphorus in biologically treated wastewater reveals that domestic wastewater contains nutrients in excess of biological needs. The approximate BOD/nitrogen/phosphorus (BOD/N/P) weight ratio required for biological treatment is 100/5/1. The exact BOD/N/P ratio needed for treatment depends on the process and the biological availability of the nitrogen and phosphorus compounds in the wastewater. A minimum of 100/6/1.5 is commonly related to treatment of unsettled sanitary wastewater, while 100/3/0.7 is generally adequate for wastewater where the nitrogen and

**TABLE 12.1** APPROXIMATE COMPOSITION OF AN AVERAGE DOMESTIC WASTEWATER (mg/l)

|  | Before sedimentation | After sedimentation | Biologically treated |
|---|---|---|---|
| Total solids | 800 | 680 | 530 |
| Total volatile solids | 440 | 340 | 220 |
| Suspended solids | 240 | 120 | 30 |
| Volatile suspended solids | 180 | 100 | 20 |
| BOD | 200 | 130 | 30 |
| Ammonia nitrogen as N | 15 | 15 | 24 |
| Total nitrogen as N | 35 | 30 | 26 |
| Soluble phosphorus as P | 7 | 7 | 7 |
| Total phosphorus as P | 10 | 9 | 8 |

phosphorus are in soluble forms. The average domestic wastewater listed in Table 12.1 has a ratio of 100/17/5 before sedimentation and 100/19/6 after sedimentation, both of which are in excess of the minimum 100/6/1.5. For biological treatment of industrial wastewaters deficient in nutrients, soluble phosphorus can be supplied by adding $H_3PO_4$ and soluble nitrogen by adding $NH_4NO_3$.

Biodegradable organic matter in wastewater is generally classified in three categories: carbohydrates, proteins, and fats. *Carbohydrates* are hydrates of carbon with the empirical formula $C_nH_{2n}O_n$ or $C_n(H_2O)_n$. The simplest carbohydrate unit is known as a *monosaccharide,* although few monosaccharides occur naturally. Glucose is a common monosaccharide in the structure of polysaccharides. *Disaccharides* are composed of two monosaccharide units. Sucrose (table sugar) is glucose plus fructose. Common milk sugar is lactose consisting of glucose plus galactose. *Polysaccharides* are long chains of monosaccharides, such as cellulose, starch, and glycogen. Cellulose is the common polysaccharide in wood, cotton, paper, and plant tissues. Starches are primary nutrient polysaccharides for plant growth and are abundant in potatoes, rice, wheat, corn, and other plant forms.

*Proteins* in simple form are long-chain molecules composed of amino acids connected by peptide bonds and are important in both the structural (e.g., muscle tissue) and dynamic aspects (e.g., enzymes) of living matter. Twenty-one common amino acids when linked together in long peptide chains form a majority of simple proteins found in nature. A mixture of proteins as a bacterial substrate is an excellent growth medium, since proteins contain all the essential nutrients. On the other hand, pure carbohydrates are unsuitable as a growth medium since they do not contain the nitrogen and phosphorus essential for synthesis.

*Lipids,* together with carbohydrates and proteins, form the bulk of organic matter of living cells. The term refers to a heterogeneous collection of biochemical substances having the mutual property of being soluble to varying degrees in organic solvents (e.g., ether, ethanol, hexane, and acetone) while being only sparingly soluble in water. Lipids may be grouped according to their shared chemical and physical properties as fats, oils, and waxes. A simple fat when broken down by hydrolytic action yields fatty acids. In sanitary engineering, the word *fats* in current usage apparently conveys the meaning of lipids. The term *grease* applies to a wide variety of organic substances in the lipid category that are extracted from aqueous solution or suspension by trichlorotrifluoroethane.

Actually not all biodegradable organic matter can be classed into these three simple groupings. Many natural compounds have structures that are combinations of carbohydrates, proteins, and fats, such as lipoproteins and nucleoproteins.

Approximately 20%–40% of the organic matter in wastewater appears to be nonbiodegradable. Several organic compounds, although biodegradable in the sense that specific bacteria can break them down, must be considered by sanitary engineers as partially biodegradable because of time limitations in waste treatment processes. For example, lignin, a polymeric noncarbohy-drate material associated with cellulose in wood fiber, is for all practical purposes nonbiodegradable. Cellulose itself is not readily available to the general population of domestic wastewater bacteria. Saturated hydrocarbons are a problem in treatment because of their physical properties and resistance to bacterial action. Alkyl benzene sulfanate (ABS synthetic detergent) is only sparingly biodegradable in wastewater treatment.

■ **EXAMPLE 12.2**
Using values of 0.24 lb of suspended solids and 0.20 lb of BOD per 120 gal of domestic wastewater, calculate the suspended solids and BOD concentration in milligrams per liter.

*Solution*

$$BOD = \frac{0.20 \text{ lb}}{120 \text{ gal}} \times \frac{1,000,000 \text{ gal}}{8.34 \text{ lb}} = 200 \text{ mg/l}$$

$$\text{suspended solids} = \frac{0.24 \text{ lb}}{120 \text{ gal}} \times \frac{1,000,000 \text{ gal}}{8.34 \text{ lb}} = 240 \text{ mg/l} \quad ■$$

■ **EXAMPLE 12.3**
What is the BOD equivalent population for an industry that discharges 0.10 mgd of wastewater with an average BOD of 450 mg/l? What is the hydraulic equivalent population of this wastewater?

*Solution*

$$\left(\begin{array}{c} \text{BOD equivalent} \\ \text{population} \end{array}\right) = \frac{0.10 \text{ mil gal} \times 450 \text{ mg/l} \times 8.34 \frac{\text{lb/mil gal}}{\text{mg/l}}}{0.20 \text{ lb/person}} = 1900$$

$$\left(\begin{array}{c} \text{hydraulic equivalent} \\ \text{population} \end{array}\right) = \frac{100,000 \text{ gal/day}}{120 \text{ gal/person}} = 830 \quad ■$$

# Trickling (Biological) Filters

Trickling filters are fixed-growth biological beds where wastewater is spread on the surface of media supporting microbial growths. Although the term *filter* is the accepted designation for this unit, no physical filtration occurs; contaminants are removed by biological action. The media placed in a tank under the wastewater distributor can be crushed rock, plastic-sheet packing formed into modules, or random plastic packing of various shapes. Trickling filters are preceded by primary clarifiers to remove settleable solids and are followed by final clarifiers to collect microbiological growths that slough from the media. The primary reasons for the popularity of trickling filters are their simplicity, low operating cost, and production of a waste sludge that is easy to process.

## 12.12 BIOLOGICAL PROCESS IN TRICKLING FILTRATION

The biological slime layers on filter media consist of bacteria, protozoans, and fungi. Frequently, noticeable populations of larger organisms such as sludge worms and rotifers are present. The top surface of a bed exposed to sunlight is coated with algae, and the lower portion of a deep filter can support nitrifying bacteria. Microorganisms near the surface of the bed where the food concentration is high are in a rapid growth phase, while the microorganisms near the bottom are in a state of starvation.

Figure 12.17 is a schematic diagram illustrating the general biological process at the surface of the medium. Although classified as aerobic treatment, the microbial film is aerobic to a depth of only 0.1–0.2 mm. The zone

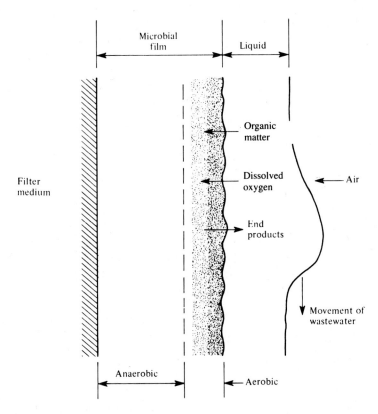

**Figure 12.17** Schematic diagram showing the form of the biological process in a trickling filter.

adjacent to the medium is anaerobic. As the wastewater flows over the microbial film, the soluble organic matter is metabolized and the colloidal organic matter is adsorbed on the film. Dissolved oxygen taken up in the liquid layer is replenished by reoxygenation from the surrounding air. Continuous passage of air through the bed is essential to prevent undesirable anaerobic conditions. The actual biological process is more complex than this simplified description and is poorly understood. Physical characteristics, such as media configuration, bed depth, and hydraulic loading, strongly influence the process. Several forms of manufactured media provide a high specific surface area with a corresponding large percentage of void volume. This permits substantial biological growth without inhibiting passage of air through the bed. A uniform medium also allows even loading distribution, and lightweight packing permits the construction of deep beds.

Problems with the biological process that can occur include poor effluent quality and emission of offensive odors, both associated with organic loading, industrial wastes, and cold-weather operation. The anaerobic zone of the film adjacent to the medium surface can produce odorous metabolic end products. If they are not oxidized while moving through the aerobic zone, they can be released and carried out in the airflow through the bed. Some industrial wastes have characteristic odors that are not easily oxidized by the microbial film. Cooling of the applied wastewater reduces the biological activity, allowing more organic matter to pass through the bed. In northern climates, the walls of deep filters can be insulated and covers can be placed over the tops to reduce cooling. Filter flies, *Psychoda,* are a nuisance problem around filters during warm weather. They breed on the inside of the retaining walls and on the surface of the media around the margin of the bed.

## 12.13 TRICKLING-FILTER OPERATION AND FILTER MEDIA REQUIREMENTS

A cutaway view of a shallow trickling filter is shown in Fig. 12.18. The major components are a rotary distributor, underdrain system, and filter medium. The distributor spreads the wastewater at a uniform hydraulic load per unit area on the surface of the bed. The arms are driven by the reaction of wastewater flowing out of the distributor nozzles, which usually requires a pressure head of 24 in. measured at the center column that supports the arms. The underdrain system, often vitrified clay blocks with entrance holes to drainage channels, carries away the effluent and permits circulation of air through the bed. The need for free passage of air controls the size of openings in the underdrain. The filter media provide a surface for biological growth and voids for passage of air and water.

Traditionally, the common filter media have been crushed rock, slag, or field stone, since these are durable, insoluble, resistant to spalling, and often locally available. Stones, however, have the disadvantage of occupying the majority of the volume in a filter bed, thus reducing the void spaces for

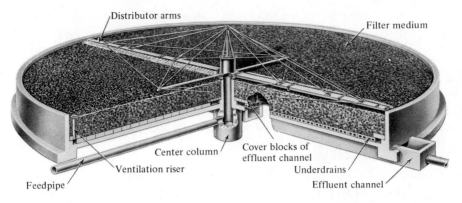

Distributor arms

Filter medium

Center column

Cover blocks of
effluent channel

Ventilation riser

Underdrains

Feedpipe

Effluent channel

**Figure 12.18** Cutaway view of a trickling filter. (Courtesy of Dorr-Oliver, Inc.)

passage of air and limiting the surface area per unit volume for biological growth. Several forms of chemical-resistant plastic media are available that have much greater surface area per unit volume and a large percentage of free space. For shallow filters, the common types are random packing and high-density cross-flow media manufactured in handleable modules that can be cut and fitted into a circular tank. The common random packing is small (2–4 in.) cylinders with perforated walls and internal ribs made of plastic, as shown in Fig. 12.19. The specific surface area is 31–40 ft$^2$/ft$^3$ (100–130 m$^2$/ m$^3$) with a void space of 91%–94%. The packing is placed by dumping it into a filter tank on top of the underdrains. Because of the random placement, the bed is efficient in distributing the applied wastewater to the media surfaces.

Biological towers are trickling filters with deep beds (10–20 ft) either in circular tanks with rotary distributors or rectangular tanks with fixed-nozzle

**Figure 12.19** Random packing for both shallow and deep trickling filters. Each element is a plastic cylinder (3.5 × 3.5 in.) with perforated walls and internal ribs. (Courtesy of Norton Co., Industrial Ceramics Div., Worcester, MA.)

distributors. The media are modules of corrugated PVC plastic sheets 2 ft wide, 4 ft long, and 2 ft deep stacked on top of each other in a crisscross pattern. The light weight and strength of the modules allows self-supporting stacks in towers 20 ft or more in depth. Vertical-flow packing is constructed of corrugated sheets bonded between flat sheets, as shown in Fig. 12.20a. By preventing clear vertical openings, the wastewater passing down through the packing is distributed over the surfaces of the media. The specific surface area varies from 26 to 43 ft$^2$/ft$^3$ (90–140 m$^2$/m$^3$), depending on the manufacturer, and the void space is about 95%. Cross-flow packing is made of ridged corrugated sheets with the ridges on adjacent sheets at 45° or 60° angles to each other and bonded together where the ridges contact (Fig. 12.20b). As wastewater flows down through the media, each contact point permits flow splitting and combining (Fig. 12.20c). Cross-flow wets the surfaces completely and slows the flowrate, resulting in an increased hydraulic residence time in the bed. The specific surface of low-density cross-flow media is 27 ft$^2$/ft$^3$ (90 m$^2$/m$^3$), and high density is 42 ft$^2$/ft$^3$ (140 m$^2$/m$^3$).

## 12.14  TRICKLING-FILTER SECONDARY SYSTEMS

A trickling-filter secondary treatment system includes a final settling tank to remove biological growths that are washed off the filter media. These sloughed solids are commonly disposed of through a drain line from the bottom of the final clarifier to the head end of the plant. This return sludge flow is mixed with the raw wastewater and settled in the primary clarifier.

The raw wastewater in trickling filtration is diluted with recirculated flow so that it is reduced in strength and passes through the filter more than once. A typical filter design is illustrated in Fig. 12.21. A return line from the final clarifier serves a dual function as a sludge return and a recirculation line. The combined flow ($Q + Q_R$) through the filter is always sufficient to maintain a minimum hydraulic loading on the media and sufficient flow to turn the distributor arms.

The two most common recirculation patterns used in trickling filter systems are shown in Fig. 12.22. The recirculation ratio is the ratio of recirculated flow to the quantity of raw wastewater. A common range for recirculation ratio values is 0.5–3.0. Recirculation may be done (1) only during periods of low wastewater flow, (2) at a rate proportional to raw-wastewater flow, (3) at a constant rate at all times, or (4) at two or more constant rates predetermined automatically or by manual control.

The gravity-flow-recirculation sludge-return system illustrated in Figs. 12.21 and 12.22 is common for treatment of average-strength wastewater. The rate of recirculation flow is generally automatically regulated by the rate of raw-wastewater flow into the clear well. In this manner, the rate of return is increased during periods of low raw-wastewater flow and reduced, or even stopped, during high-flow periods. This kind of recirculation operation is shown graphically in Fig. 12.16a. The shaded area represents recirculated

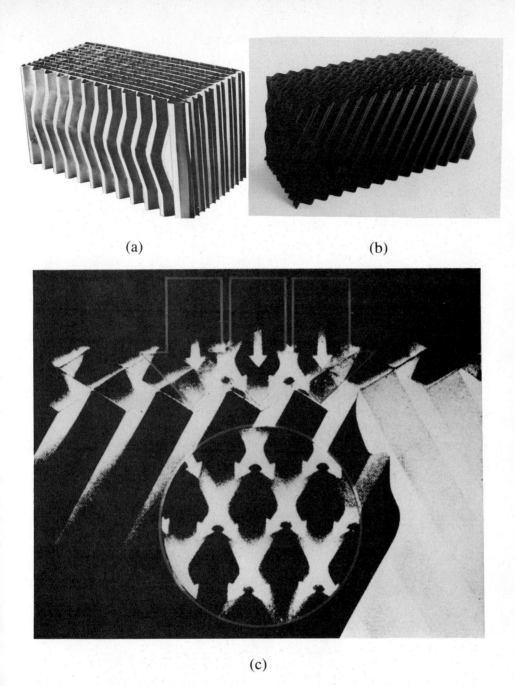

(a)

(b)

(c)

**Figure 12.20** Plastic media for trickling filters are manufactured in modules, commonly 2 ft wide, 2 ft high, and 4 ft long. These self-supporting modules are placed with each layer at right angles to the layer beneath to form an interlocking stack up to 20 ft in depth. (a) A vertical-flow module is constructed with corrugated sheets bonded between flat sheets. (Although installed in many existing filters, *Munters* has discontinued manufacture of vertical-sheet media because of improved performance of cross-flow media.) (b) A cross-flow module is constructed of sheets corrugated at a 60° angle assembled with adjacent sheets in a crossed pattern and bonded at the contact points of the ridges. (c) The cross-flow configuration provides redistribution of the wastewater as it flows through the media, resulting in better spreading over the wetted surfaces and increased contact time. (Courtesy of Munters Corp., Fort Myers, FL.)

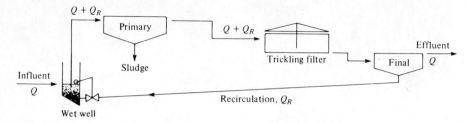

**Figure 12.21** Profile of a typical single-stage trickling-filter plant with recirculation of underflow from the final clarifier to the head of the plant.

flow. The average raw-wastewater flow is 280 gpm. Flow through the plant is 420 gpm equal to $Q + Q_R$, except from 10 A.M. to 3 P.M., when the raw-wastewater flow exceeds 420 gpm. The recirculation ratio for this illustrated flow recirculation pattern is 0.5.

Direct recirculation, depicted in Fig. 12.22b, is frequently used in the treatment of stronger wastewaters, where recirculation ratios of 2–3 are desirable. If high rates are used in pattern (a), the primary clarifier must be sized for the increased flowrate created by the greater volume. In other words, if recirculation flow is routed through a clarifier during the peak hourly flows of the raw wastewater, the clarifier must be increased in size to prevent disturbance of the settled solids and resultant loss of removal efficiency in the sedimentation tank. Consider the flow diagram in Fig. 12.16a, and assume that the shaded area is the flow in the sludge-return line shown in Fig. 12.22b. Then apply a direct recirculation of 420 gpm around the filter using constant-

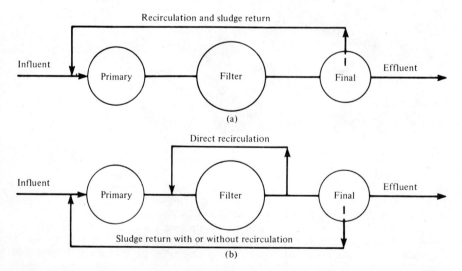

**Figure 12.22** Typical recirculation patterns for single-stage trickling filter plants. (a) Recirculation with sludge return. (b) Direct recirculation around the filter.

speed pumps. The resultant ratio for the trickling filter is 2.0:0.5 from indirect recirculation (from the final to the head of the primary) and 1.5 from direct recirculation.

Two-stage trickling-filter secondary systems have two filters in series, usually with an intermediate settling tank. Two typical flow diagrams for two-stage filter installations are sketched in Fig. 12.23. Two-stage filters are used where a high-quality effluent is required, for treatment of strong wastewater, and to compensate for lower bacterial activity in treating cold wastewater.

Deep biological filters for treating municipal wastewater are generally single-stage units following primary sedimentation, although two stages are installed when processing strong wastewater resulting from industrial discharges. Direct recirculation is employed to maintain the desired flow through the tower. Sometimes a portion of the recirculation flow is drawn from the bottom of the final clarifier to develop a microbial floc in the wastewater circulating through the tower. Waste sludge accumulated in the final clarifier can be returned to the plant influent for settling in the primary clarifier.

## 12.15 EFFICIENCY EQUATIONS FOR STONE-MEDIA TRICKLING FILTERS

The BOD load on a trickling filter is calculated using the raw BOD in the primary effluent applied to the filter, without regard to the BOD in the recirculated flow. BOD loadings are expressed in terms of pounds of BOD applied per unit of volume per day. Common values are 25–45 lb/1000 ft$^3$/day (400–720 g/m$^3$·d) for single-stage filters and 45–65 lb/1000 ft$^3$/day

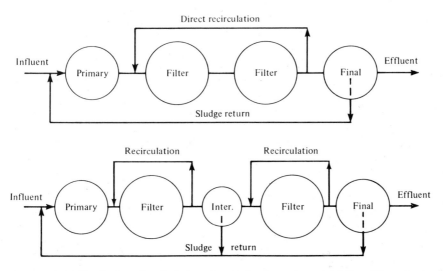

**Figure 12.23** Typical recirculation patterns for two-stage trickling filter plants without and with intermediate sedimentation.

(720–1040 g/m³·d) for two-stage filters based on the total media volume of both filters.

The hydraulic load is computed from the raw-wastewater flow plus recirculated flow. Hydraulic loadings are expressed in terms of average flow in gpm applied per square foot of surface area per day. Recirculation flow is required to maintain an open, well-aerated bed by preventing excessive accumulation of biological growth in the voids and impeding the passage of water and air. The minimum recommended hydraulic loading is 0.16 gpm/ft² (9.4 m³/m²·d). The maximum recommended is 0.48 gpm/ft² (28 m³/m²·d). Above this, the flushing action is excessive and contact time of the wastewater with the filter media becomes too short. The recirculation ratio for this range of hydraulic loadings is usually between 0.5 and 3.0.

Trickling filters have a bed depth of 5–7 ft (1.5–2.1 m) for most efficient BOD removal per unit volume of stone or slag media. In the early development of filters, they were constructed as deep as 8 ft and as shallow as 3 ft. Experience showed that two 3-ft-deep filters in series were no more efficient than one 6-ft-deep filter, and that the filter media below 6 ft in a bed did not result in significant increased BOD removal.

General practice in trickling-filter design has been to use empirical relationships to find the required filter volume for a desired degree of wastewater treatment. Several of these associations have been developed from operational data collected at existing treatment plants. One of the first evolved was the National Research Council (NRC) formula, based on data collected from filter plants at military installations in the United States in the early 1940s [9].

The NRC formula for a single-stage trickling filter is

$$E = \frac{100}{1 + 0.0561 \, (w/VF)^{0.5}} \qquad (12.42)$$

where    $E$ = BOD removal at 20°C, %
         $w$ = BOD load applied, lb/day
         $V$ = volume of filter media, ft³ × 10⁻³
         $F$ = recirculation factor
       $w/V$ = BOD loading, lb/1000 ft³/day

The recirculation factor is calculated from the formula

$$F = \frac{1 + R}{(1 + 0.1R)^2} \qquad (12.43)$$

where $R$ is the recirculation ratio (ratio of recirculation flow to raw-wastewater flow).

The NRC formula for the second stage of a two-stage filter is

$$E_2 = \frac{100}{1 + [0.0561/(1 - E_1)](w_2/VF)^{0.5}} \qquad (12.44)$$

where    $E_2$ = BOD removal of the second stage at 20°C, percent
$E_1$ = fraction of BOD removed in the first stage
$w_2$ = BOD load applied to the second stage, lb/day
$w_2/V$ = BOD loading, lb/1000 ft³/day

The effect of wastewater temperature on stone-filled trickling-filter efficiency may be expressed as follows:

$$E = E_{20}1.035^{T-20} \tag{12.45}$$

where    $E$ = BOD removal efficiency at temperature $T$ in °C
$E_{20}$ = BOD removal efficiency at 20°C

The BOD removal efficiencies computed by the NRC formulas include final settling of the filter effluent. In the empirical development of these formulas, the field procedure used in collecting data sampled the filter influent and final clarifier effluent. Therefore, in evaluating the efficiency of a trickling-filter secondary, the overflow rate and detention time of the final clarifier should be examined for adequacy of design.

For a two-stage filter secondary without an intermediate settling tank (Fig. 12.23), the NRC formulas cannot be used to determine the efficiency of the first stage. In this case it is common to assume that the first-stage efficiency is 50% and find the efficiency of the second stage from Eq. 12.44.

■ **EXAMPLE 12.4**
Calculate the BOD and hydraulic loadings and BOD removal efficiency of a single-stage high-rate trickling filter based on the following data:

wastewater flow pattern = as shown in Fig. 12.16a

recirculation rate = as shown in Fig. 12.16a

settled wastewater BOD (primary effluent) = 130 mg/l

diameter of filter = 12.5 m

depth of media = 2.1 m

*Solution*

raw-wastewater flow = 280 gpm = 1530 m³/d

recirculation flow = 0.50 × 1530 = 765 m³/d

$$\text{BOD load} = 1530 \text{ m}^3/\text{d} \times 130 \text{ mg/l} \times \frac{\text{kg/m}^3}{1000 \text{ mg/l}} = 200 \text{ kg/d}$$

surface area of filter = $\pi(12.5)^2/4$ = 122 m²

volume of media = 122 × 2.1 = 256 m³

$$\text{BOD loading} = \frac{200{,}000 \text{ g}}{256 \text{ m}^3} = 781 \text{ g/m}^3\cdot\text{d} = 48.8 \text{ lb/1000 ft}^3/\text{day}$$

$$\text{hydraulic loading} = \frac{1530 \text{ m}^3 + 765 \text{ m}^3}{122 \text{ m}^2} = 18.8 \text{ m}^3/\text{m}^2 = 0.32 \text{ gpm/ft}^2$$

By Eqs. 12.43 and 12.42,

$$F = \frac{1 + 0.5}{(1 + 0.1 \times 0.5)^2} = 1.36$$

$$E = \frac{100}{1 + 0.0561(48.8/1.36)^{0.5}} = 75\% \qquad \blacksquare$$

### ■ EXAMPLE 12.5

The design flow for a new two-stage trickling-filter plant is 1.2 mgd with an average BOD concentration of 450 mg/l. Determine the dimensions of the sedimentation tanks and trickling filters (surface areas and depths) for the flow scheme shown in Fig. 12.24. Calculate the volume of filter media based on a loading of 50 lb of BOD/1000 ft³/day, and divide the resulting volume equally between the primary and secondary filters. Estimate the BOD concentration in the plant effluent.

*Solution*

**Primary Tank**  Criteria: (1) 500-gpd/ft² overflow rate based on raw $Q$ or 750 gpd/ft² based on $Q$ plus recirculation flow; (2) minimum depth of 7 ft. However, if accumulated sludge is to be retained in the bottom of the tank, increase the depth to accommodate the necessary sludge storage volume.

$$\text{area required} = \frac{1,200,000}{500} = 2400 \text{ ft}^2$$

or

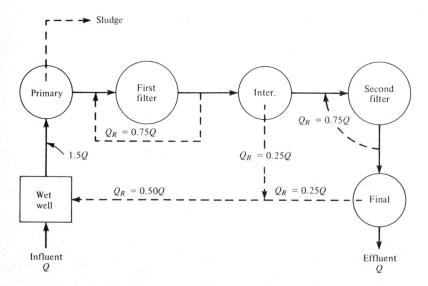

**Figure 12.24**  Flow scheme of the two-stage trickling-filter plant for Example 12.5.

$$\text{area required} = \frac{1.5 \times 1,200,000}{750} = 2400 \text{ ft}^2 \quad \text{(Use)}$$

Estimate the daily sludge accumulation at 4% solids, assuming a sludge solids accumulation equal to 90% of the BOD load:

$$\text{volume} = \frac{0.9 \times 450 \times 1.2 \times 8.34}{0.04 \times 62.4} = 1620 \text{ ft}^3$$

$$\text{depth of sludge} = \frac{1620}{2400} = 0.7 \text{ ft}$$

Provide a side-wall depth of 8 ft plus freeboard.

$$\text{primary BOD removal} = 35\%$$

**Trickling Filters**   Criteria: (1) 50 lb/1000 ft$^3$/day BOD loading; (2) 0.16–0.48 gpm/ft$^2$ hydraulic loading.

$$\text{volume required} = \frac{0.65 \times 450 \times 1.2 \times 8.34}{0.050} = 58,500 \text{ ft}^3$$

$$\text{volume of each filter} = 29,300 \text{ ft}^3$$

Try 6-ft depth:

$$\text{area} = \frac{29,300}{6.0} = 4880 \text{ ft}^2$$

Check the hydraulic loading:

$$\frac{(1.5 + 0.75)1,200,000}{4880 \times 1440} = 0.38 \text{ gpm/ft}^2 \quad \text{(OK)}$$

Use 6-ft-deep filters with a 4880-ft$^2$ area.

**Intermediate Settling Tank**   Criteria: (1) 1000-gpd/ft$^2$ overflow rate; (2) minimum depth of 7 ft.

$$\text{area required} = \frac{1.25 \times 1,200,000}{1000} = 1500 \text{ ft}^2$$

Use a side-wall depth of 7 ft plus freeboard.

**Final Settling Tank**   Criteria: (1) 800-gpd/ft$^2$ overflow rate; (2) minimum depth of 7 ft.

$$\text{area required} = \frac{1,200,000}{800} = 1500 \text{ ft}^2$$

Use a side-wall depth of 7 ft plus freeboard.

*Calculation of BOD removal efficiency*

primary tank $= 35\%$

First-stage filter:

$$\text{BOD loading} = \frac{0.65 \times 450 \times 1.2 \times 8.34}{29.3} = 100 \text{ lb}/1000 \text{ ft}^3/\text{day}$$

$$R = \frac{0.50Q + 0.75Q}{Q} = 1.25$$

$$E = 70\%$$

Second-stage filter:

$$\text{BOD loading} = 0.30 \times 100 = 30 \text{ lb}/1000 \text{ ft}^3/\text{day}$$

$$R = \frac{0.25Q + 0.75Q}{Q} = 1.0$$

$$F = \frac{1 + 1.0}{(1 + 0.1 \times 1.0)^2} = 1.65$$

By Eq. 12.44,

$$E_2 = \frac{100}{1 + [0.0561/(1 - 0.70)](30/1.65)^{0.5}} = 56\%$$

The plant efficiency is

$$E = 100 - 100[(1 - 0.35)(1 - 0.70)(1 - 0.56)] = 92\%$$

The estimated effluent BOD is $0.08 \times 450 = 38$ mg/l.  ■

## 12.16  EFFICIENCY EQUATIONS FOR PLASTIC-MEDIA TRICKLING FILTERS

BOD loadings on biological towers with plastic media are usually 50 lb/1000 ft$^3$/day (800 g/m$^3$·d) or greater with surface hydraulic loadings of 1.0 gpm/ft$^2$ (60 m$^3$/m$^2$·d, 0.68 l/m$^2$·s) or greater. The design loading selected for treatment of a particular wastewater depends on BOD concentration, biodegradability, temperature, type and depth of media, and the ratio and pattern of wastewater circulation.

Plastic media, such as vertical-flow and cross-flow modules (Fig. 12.20) and random packing (Fig. 12.19), are manufactured with uniform geometric features of specific surface and shape. The objectives in design of these media are to provide a surface area that supports a continuous biological growth and distribution of the applied wastewater so that it flows uniformly over the biological growth. Based on this "definable" physical–biological system, several efficiency equations have been developed, modified, and refined in an attempt to find a suitable mathematical model. Nevertheless, no universal model exists to precisely describe substrate removal in a trickling filter. The following equations define a common model used in environmental engineering practice.

The residence time of applied wastewater in a filter bed is an important factor in BOD removal. The mean contact time between the wastewater and the biological film on the surface of the media can be related to filter depth, hydraulic loading, and the geometry of the media as

$$t = \frac{CD}{Q^n} \qquad (12.46)$$

where     $t$ = mean residence time, min
$D$ = depth of media, ft (m)
$Q$ = hydraulic loading, gpm/ft$^2$ (m$^3$/m$^2$·h)
$C, n$ = constants related to specific surface and configuration of media

Exponent $n$ and constant $C$ are determined from pilot plant studies using an experimental filter column packed with synthetic media to the depth of a full-scale unit. A minimum of three ports at different depths are installed to collect samples of the wastewater percolating down through the packing. Residence time is measured by adding a pulse tracer input, such as fluorescent dye, and observing the concentration of tracer in the effluent with respect to time. The mean residence time is when 50% of the tracer has passed out of the filter column. Since the thickness of the biological film affects the residence time, the experimental filter is acclimated to an applied wastewater feed for a sufficient period of time to establish dynamic equilibrium between the organic loading and microbial mass. Values of $n$ and $C$ are determined from a graph of the $t$ and $Q$ data plotted on logarithmic paper. From Eq. 12.46, $t$ is proportional to $Q^{-n}$; therefore,

$$n = -\frac{\Delta \log t}{\Delta \log Q} \qquad (12.47)$$

and $C = t/D$ when $Q = 1.0$.

The removal of soluble BOD in a trickling filter, where the media supports a continuous biological growth and the wastewater is uniformly distributed, is as follows, based on first-order kinetics:

$$\frac{S_e}{S_0} = e^{-KD/Q^n} \qquad (12.48)$$

where     $S_e$ = soluble (filtered) BOD in effluent, mg/l
$S_0$ = soluble (filtered) BOD in influent, mg/l
$e$ = 2.718 (the Napierian base)
$K$ = reaction-rate constant, gal/min/ft$^3$
$D$ = depth of media, ft
$Q$ = hydraulic loading, gpm/ft$^2$
$n$ = empirical flow constant

This equation can be rewritten to include the specific surface area of the media by substituting $k_{20}A_s$ for $K$:

$$\frac{S_e}{S_0} = e^{-k_{20}A_sD/Q^n} \tag{12.49}$$

where   $S_e$ = soluble (filtered) BOD in effluent, mg/l
  $S_0$ = soluble (filtered) BOD in influent, mg/l
  $k_{20}$ = reaction-rate coefficient at 20°C, $(gpm/ft^2)^{0.5}$ $[(l/m^2{\cdot}s)^{0.5}]$
  $A_s$ = specific surface area of media, $ft^2/ft^3$ $(m^2/m^3)$
  $D$ = depth of media, ft (m)
  $Q$ = hydraulic loading, $gpm/ft^2$ $(l/m^2{\cdot}s)$
  $n$ = empirical flow constant, normally selected as 0.5 for vertical-flow and cross-flow media

The reaction-rate coefficient is corrected for temperature by the relationship

$$k = k_{20}\Theta^{T-20} \tag{12.50}$$

where   $k$ = reaction-rate coefficient at temperature $T$, °C
  $k_{20}$ = reaction-rate coefficient at 20°C
  $\Theta$ = temperature coefficient, normally selected as 1.035
  $T$ = wastewater temperature, °C

Equation 12.49 can be rewritten and include Eq. 12.50 as follows:

$$\ln\frac{S_0}{S_e} = \frac{k_{20}A_sD\Theta^{T-20}}{Q_n} \tag{12.51}$$

This is a linear equation convenient for analysis of experimental data. The value of $k_{20}$ is determined graphically by plotting $\ln S_0/S_e$ versus $A_sD\Theta^{T-20}/Q^n$ on arithmetic paper. The value of $k$ is the slope of a straight line originating from the origin and drawn as a best fit through the plotted data.

The BOD concentrations of the applied wastewater before and after dilution with clear recirculation flow is

$$S_0 = \frac{S_p + RS_e}{1 + R} \tag{12.52}$$

where   $S_0$ = soluble BOD in influent after dilution with recirculated flow, mg/l
  $S_p$ = soluble BOD in primary effluent before dilution with recirculated flow, mg/l
  $S_e$ = soluble BOD in effluent, mg/l
  $R$ = recirculation ratio, recirculated flow/primary effluent flow, $Q_R/Q$

Combining Equations 12.49, 12.50, and 12.52:

$$\frac{S_e}{S_p} = \frac{e^{-k_{20}\Theta^{T-20}A_s D/(Q_p(1+R))^n}}{(1+R) - Re^{-k_{20}\Theta^{T-20}A_s D/(Q_p(1+R))^n}} \tag{12.53}$$

where  $Q_p$ = hydraulic loading of primary effluent without recirculation flow, gpm/ft$^2$

This mathematical model is a simplification of complex biological–physical interactions that occur in trickling filters. The values for the reaction-rate coefficient $k_{20}$ vary with media of different configurations with the same specific surface areas. The coefficient also varies with wastewater treatability and depth of media. The approximate range for $k_{20}$ values for vertical-flow media is 0.0008–0.0016 (gpm/ft$^2$)$^{0.5}$ [0.0010–0.0020 (l/m$^2$·s)$^{0.5}$], and for cross-flow media it is 0.0014–0.0023 (gpm/ft$^2$)$^{0.5}$ [0.0017–0.0028 (l/m$^2$·s)$^{0.5}$]. The $k_{20}$ values for cross-flow media are greater than those for vertical-flow media, which is attributed to longer contact time and better wastewater flow distribution. Often manufacturers provide field data to verify reaction-rate coefficients for their media; however, where feasible, pilot studies using selected media are recommended to determine the $k_{20}$ for design.

Example 12.6 illustrates the application of these equations to determine the values of the reaction-rate constant $K_{20}$ and the empirical flow constant $n$ for random packing in a pilot-plant study. Example 12.7 presents the evaluation of data from a field study to determine the $k_{20}$ for high-density cross-flow media in a shallow trickling filter.

■ **EXAMPLE 12.6**

Balakrishnan, Eckenfelder, and Brown [10] demonstrated the feasibility of determining the constants for Eqs. 12.46 and 12.48, using data from a pilot-plant study.

The trickling filter was a 20-in.-diameter, 9-ft-long fiberglass cylinder with an air sparger to provide a uniform distribution of air from the bottom and distribution plates on top for uniform hydraulic loading. The filter was filled to a depth of 8 ft with 1.5-in. in diameter, cylindrical, random packing with a specific surface of 40 ft$^2$/ft$^3$ and 96% void space. Eight sampling ports were located at 1-ft intervals from the top of the filter bed.

The feed was settled domestic wastewater, containing filtered (soluble) BOD concentrations in the range of 65–90 mg/l, applied at hydraulic loadings of 0.20, 0.30, and 0.43 gpm/ft$^2$. After acclimation at each loading rate, samples were collected at various depths in the filter for laboratory analysis. The wastewater samples were settled for 30 min and filtered through Whatman No. 42 filter paper prior to BOD testing.

*Solution.*  Mean residence times of the wastewater in the filter were determined by measuring the time for concentrated doses of salt solution to flush through the column. The results, with and without the presence of biological slime growth, are graphed in

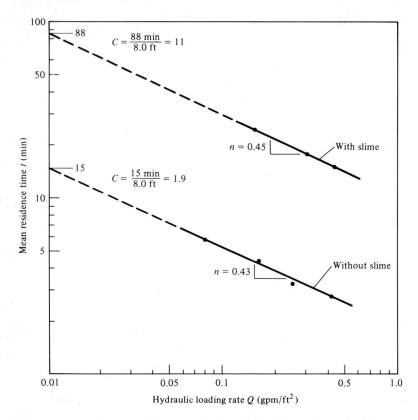

**Figure 12.25** Relationship between hydraulic loading and mean residence time for Example 12.6. [Adapted from S. Balakrishnan, W. W. Eckenfelder, and C. Brown, "Organics Removal by a Selected Trickling Filter Media," *Water and Wastes Eng.* 6(1) (1969): A.22–A.25.]

Fig. 12.25. The constant $n$ for Eq. 12.46 is the negative slope of the line, and $C$ is the intercept of the line at the vertical axis divided by the depth of media. With a biological growth coating the packing, the mean residence time was

$$t = \frac{11D}{Q^{0.045}}$$

The mean residence time of clean packing, without slime, was

$$t = \frac{1.9D}{Q^{0.43}}$$

A comparison of these equations shows the significant effect that slime growth has on the mean residence time.

Soluble BOD removal data are plotted in Fig. 12.26 as the logarithm of BOD remaining, $S_e/S_0$, versus the depth $D$ in the filter packing. From Eq. 12.48,

$$\log\left(\frac{S_e}{S_0}\right) = -\left(\frac{K}{2.3}\right)\left(\frac{D}{Q^n}\right) \tag{12.54}$$

Based on this equation, the slope of each line drawn through $\log(S_e/S_0)$ versus $D$ data for a specific hydraulic loading is defined as

$$\text{slope} = \frac{\log(S_e/S_0)}{D} = -\left(\frac{K}{2.3}\right)Q^{-n} \tag{12.55}$$

The slopes in Fig. 12.26a for hydraulic loadings of 0.20, 0.30, and 0.43 gpm/ft$^2$ are 0.059, 0.051, and 0.045, respectively.

Taking the logarithm of both sides of Eq. 12.55 yields

$$\log(\text{slope}) = n \log Q - \log(K/2.3) \tag{12.56}$$

which is the equation for a straight line of slope $n$. Therefore, the slope of a line drawn through $\log(\text{slope})$ versus $\log Q$ data is the value of the constant $n$. From the logarithmic plot in Fig. 12.26b, $n = 0.39$ (based on the slopes and hydraulic loadings given in Fig. 12.26a).

Finally, the constant $K$ is determined by plotting $\log(S_e/S_0)$ versus $D/Q^n$ and determining the slope of the line drawn through the data, which from Eq. 12.54 equals $-(K/2.3)$. For the $S_e/S_0$ and $D/Q^{0.39}$ data plotted in Fig. 12.26c, the slope of the line is 0.0324, and $K = 2.3 \times 0.0324 = 0.0745$ min$^{-1}$.

The average operating temperature of the pilot filter was 14°C. Correcting $K$ to 20°C using Eq. 12.50, the reaction-rate constant at 20°C is

$$K_{20} = \frac{K}{(1.035)^{T-20}} = \frac{0.0745}{(1.035)^{14-20}} = 0.092 \text{ min}^{-1}$$

Substituting values of $K = 0.092$ and $n = 0.39$ in Eq. 12.48, the soluble BOD removal from a settled domestic wastewater for the specific filter packing tested is

$$\frac{S_e}{S_0} = e^{-0.092D/Q^{0.39}} \tag{12.57}$$

■

## ■ EXAMPLE 12.7

Drury, Carmona, and Delgadillo [11] describe the rehabilitation of an old shallow stone-media trickling filter by installing plastic media to solve the problem of ponding due to plugging of voids in the bed.

High-density cross-flow media with a specific surface area of 138 m$^2$/m$^3$ (42 ft$^2$/ft$^3$) and a depth of only 1.02 m (3.33 ft) was installed in the trickling-filter tank, which was 21.3 m (70 ft) in diameter. Before undertaking costly rehabilitation of the final settling tank, a study was performed to evaluate performance of the media and to establish a suitable loading for operation. The trickling filter was operated at a constant flow rate throughout the day (7 A.M.–midnight) with the primary effluent diluted by trickling-filter effluent returned to the wet well; the recirculation ratio ranged from 0.8 to 1.1. During the 8-hr period from 8 A.M.–4 P.M., the influent and effluent were composited and refrigerated by automatic samplers at 15-min intervals. The effluent sample was allowed to settle for 1 hr in a laboratory container to simulate clarification in a final settling tank, and the supernatant was decanted for analysis. The samples

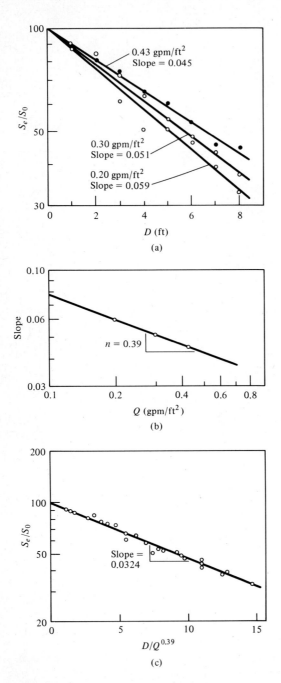

**Figure 12.26** Plotted data for Example 12.6. (a) Relationship between filter depth and fraction of BOD remaining at various hydraulic loads. (b) Diagram for determination of *n*. (c) Diagram for determination of *K*. [Adapted from S. Balakrishnan, W. W. Eckenfelder, and C. Brown, "Organics Removal by a Selected Trickling Filter Media," *Water and Waste Eng.* 6(1) (1969): A.22–A.25.]

**TABLE 12.2** SAMPLING DATA AND CALCULATED VALUES FOR DETERMINATION OF REACTION-RATE COEFFICIENT $k_{20}$ IN EXAMPLE 12.7

| $Q$ (l/m²·s) | $T$ (°C) | Influent BOD (mg/l) | $S_0$ (mg/l) | Effluent BOD (mg/l) | $S_e$ (mg/l) | $\ln \dfrac{S_0}{S_e}$ | $\dfrac{A_s D \Theta^{T-20}}{Q^n}$ [(l/m²·s)$^{-0.5}$] |
|---|---|---|---|---|---|---|---|
| 0.49 | 20.0 | 75 | 44 | 32 | 18 | 0.89 | 201 |
| 0.54 | 20.5 | 50 | 29 | 23 | 13 | 0.80 | 195 |
| 0.49 | 20.0 | 73 | 38 | 27 | 14 | 1.00 | 201 |
| 0.49 | 18.5 | 53 | 22 | 22 | 9.5 | 0.84 | 191 |
| 0.59 | 18.5 | 64 | 34 | 32 | 17 | 0.69 | 174 |
| 0.49 | 19.0 | 45 | 30 | 24 | 12 | 0.92˙ | 195 |
| 0.31 | 23.0 | 80 | 33 | 16 | 9.2 | 1.28 | 280 |
| 0.35 | 21.5 | 66 | 37 | 18 | 12 | 1.13 | 251 |

were tested for total BOD and dissolved BOD after filtration through a glass-fiber filter. The test results for the winter evaluation period are given in Table 12.2.

**Solution.**  For each set of $Q$, $T$, $S_0$, and $S_e$, calculate values for $\ln S_0/S_e$ and $A_s D \Theta^{T-20}/Q^n$. The following are sample calculations for the first set of data:

$$\ln \frac{S_0}{S_e} = \ln \frac{44}{18} = 0.89$$

$A_s = 138$ m²/m³, $D = 1.02$ m, $\Theta = 1.035$, and $n = 0.5$:

$$\frac{A_s D \Theta^{T-20}}{Q^n} = \frac{138 \times 1.02(1.035)^{20.0-20}}{(0.49)^{0.5}} = 201$$

The calculated values for determining $k_{20}$ are listed in Table 12.2 and plotted in Fig. 12.27. The slope of the best-fit line through the plotted points is the reaction-rate coefficient $k_{20}$ equal to 0.0047 (l/m²·s)$^{0.5}$.

This field evaluation demonstrated that high-density cross-flow media can be effectively used in shallow filters at high BOD loadings. The $k_{20}$ was significantly greater than expected, based on other studies conducted of towers with much deeper media. Although the reason for this is not known, the authors speculated that the high BOD removal efficiency of a shallow bed may be related to improved oxygen transfer.

An allowable loading on the filter used in the study with high-density cross-flow media can be calculated based on the following parameters:

Depth of media = 1.02 m

Specific surface area = 138 m²/m³

$k_{20} = 0.0047$ (l/m²·s)$^{0.5}$

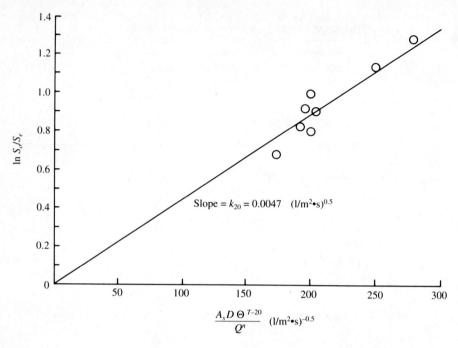

**Figure 12.27** Plot of $\ln S_0/S_e$ versus $A_s D\Theta^{T-20}/Q^n$ to determine $k_{20}$ for Example 12.7.

Wastewater temperature = 10°C

Influent BOD with recirculation flow = 70 mg/l

Soluble influent BOD = 35 mg/l

Allowable effluent BOD = 25 mg/l

Soluble effluent BOD = 12.5 mg/l

Rearranging Eq. 12.51,

$$Q = \left(\frac{k_{20}A_s D\Theta^{T-20}}{\ln(S_0/S_e)}\right)^2 \tag{12.58}$$

$$= \left(\frac{0.0047 \times 138 \times 1.02(1.035)^{10-20}}{\ln(35/12.5)}\right)^2 = 0.208 \ \text{l/m}^2\cdot\text{s}$$

Allowable BOD loading is

$$\left(Q\frac{1}{\text{m}^2\cdot\text{s}}\right)\left(86{,}400\ \frac{\text{s}}{\text{d}}\right)\left(\text{BOD}\ \frac{\text{mg}}{\text{l}}\right)\left(\frac{\text{g}}{1000\ \text{mg}}\right)\left(\frac{1}{A_s}\frac{\text{m}^2}{\text{m}^3}\right) \tag{12.59}$$

$$= \frac{0.208 \times 86{,}400 \times 70}{1000 \times 1.02} = 1200 \ \text{g/m}^3\cdot\text{d} \quad (77 \ \text{lb/1000 ft}^3/\text{day}) \quad \blacksquare$$

## 12.17  COMBINED TRICKLING-FILTER AND ACTIVATED-SLUDGE PROCESSES

Trickling filters alone may not be able to meet stringent effluent standards occasionally specified for treatment. Combining trickling filtration with activated-sludge processes can increase removal efficiency. The combined system consists of a deep trickling filter (biological tower) followed by an aeration tank with sludge recirculation from a final clarifier. The filter media is either cross-flow or vertical-flow plastic modules or redwood packing that consists of stacked racks of wooden lath attached to supporting rails. The aeration tank and final clarifier is an activated-sludge system often referred to as a solids contact process. The final clarifier is either a conventional activated-sludge sedimentation tank or one with a flocculation zone inside a large inlet baffle.

The trickling-filter activated-sludge process can be designed to operate with various flow patterns as illustrated in the composite flow diagram in Figure 12.28. By providing direct recirculation for adequate hydraulic loading, the trickling filter can be operated independent of the second-stage aeration process. In this arrangement, the filter is called a *roughing* filter, and the system is suitable to treat variable-strength wastewaters. The filter absorbs shock loads to stabilize loading on the subsequent activated-sludge process.

The other common flow pattern includes recirculation of a portion of the settled biological solids from the final clarifier to the trickling-filter influent to form an activated-sludge that recycles through both the filter and aeration tank. [This process is referred to by several names: activated biological filtration (ABF), activated biofiltration–activated sludge, and trickling filtration–solids contact process.) This composite system has the characteristics of both fixed-film and suspended-growth processes. The mixed liquor in the aeration tank enhances removal of nonsettleable and dissolved BOD in the filter effluent. Because of good process stability, a high-quality effluent can be consistently produced. In a way, the biofilter in this process is acting as a static aerator; therefore, by increasing the hydraulic loading by recirculation, the transfer of oxygen is increased. However, if excess sludge solids

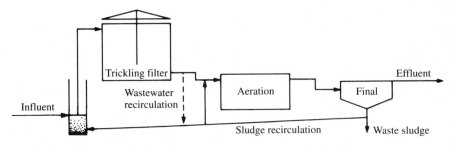

**Figure 12.28** Profile of a combined trickling-filter activated-sludge process.

are recirculated through the biofilter, the oxygen demand exceeds the transfer capability of the media and limits BOD removal.

Equations to model the activated biofiltration–activated sludge process have not been successfully formulated because of the difficulty in analyzing the system. In design, removal by the biofilter must be determined first so that the aeration tank and associated aeration equipment can be properly sized. Arora and Umphres [12] evaluated the operation of 17 activated biofiltration–activated sludge plants with redwood-media filters. The following are the ranges of various design and operational parameters of these treatment systems: percentage of the total volume (filter media plus aeration tank) that is aeration tank volume, 40%–75%; system BOD loading based on total volume, 20–50 lb/1000 ft$^3$/day (320–800 g/m$^3$·d); biofilter loading, 50–110 lb/1000 ft$^3$/day (800–1800 g/m$^3$·d); system food/microorganism ratio, 0.2–1.0 with the median at 0.5; and effluent BOD and effluent suspended solids, 5–25 mg/l. The operating personnel at most of the plants believed the system was more stable than conventional activated sludge and was better able to absorb shock loads.

# Rotating Biological Contactors

A rotating biological contactor (RBC) consists of a shaft of circular plastic disks revolving partly submerged in a contour-bottomed tank. The disks are spaced such that during submergence the wastewater can enter between the surfaces. When rotated out of the tank, air enters the spaces while the liquid trickles out over films of biological growth attached to the media. Alternating exposure to organics in the wastewater and oxygen in the air is similar to dosing a trickling filter with a rotating distributor. Excess microbial solids are hydraulically scoured from the media and carried out in the process effluent for gravity separation in a final clarifier. A rotating biological contactor system has the advantages of low power consumption and good process stability.

## 12.18 DESCRIPTION OF ROTATING BIOLOGICAL CONTACTOR MEDIA AND PROCESS

Most of the first commercially produced RBCs were constructed of alternate flat and corrugated plastic sheets bonded together with radial and circumferential openings for passage of wastewater and air. Bundles of media 10–12 ft in diameter were slid over and attached to a central shaft. Subject to high stress concentrations in the media near the shaft, this type of construction was conducive to fatigue failures. Bundles of media that broke away from the shaft and stopped rotating had to be replaced. Often associated with excessive BOD loading, media failures and broken shafts were attributed in part to plugging of the media voids with biological growth. Replacement of damaged media was very difficult and costly, since the shaft had to be lifted from the tank for removing the large-diameter disks.

The RBC pictured in Fig. 12.29 is designed to reduce the probability of mechanical failures and provide easier access for maintenance. The media bundles are in the shape of truncated pyramids that fit into structural supports mounted on a central shaft. This arrangement limits the stresses on the

(a)

(b)

**Figure 12.29** Rotating biological contactor media. (a) Placement of a segment of polyethylene media in a radial support structure made of stainless steel. (b) An assembled RBC shaft 27 ft long and 12 ft in diameter with a media surface area of 100,000 ft$^2$. (Courtesy of CLOW, Waste Treatment Div.)

polyethylene media and allows withdrawal and replacement of a unit without removing the shaft from the tank. The RBC units can also be shipped as components for assembly and installation on site. In order to reduce the possibility of plugging, the sheets of media are separated by conical spacers, rather than by corrugated sheets, to maximize the open area between layers. The nominal spacing between sheets for BOD removal is $\frac{3}{4}$ in.; the spacing for low BOD loading and nitrification is $\frac{1}{2}$ in. The disks are 12 ft in diameter and operate at 40% submergence. At the operating speed of 1.5 rpm, the peripheral velocity is 57 fpm for a 12-ft-diameter disk.

The arrangement for secondary treatment of municipal wastewater by the RBC process, as shown in Fig. 12.30, is similar to the flow diagram of a trickling-filter plant. The biological process is preceded by primary clarifiers and followed by secondary clarifiers. These settling tanks are sized using the same design criteria as for clarifiers in a trickling-filter plant. Settled solids that accumulate in the final clarifier are returned to the head of the plant for settlement with the raw-wastewater solids; thus, waste sludge is withdrawn only from the primary clarifiers. Since recirculation does not improve performance, the return flow is not designed to recycle wastewater through the RBC units.

Wastewater after sedimentation is applied to the first stage of a series of RBC chambers separated by baffles. In processing a wastewater that is essentially domestic in characteristics, a series of four stages is usually

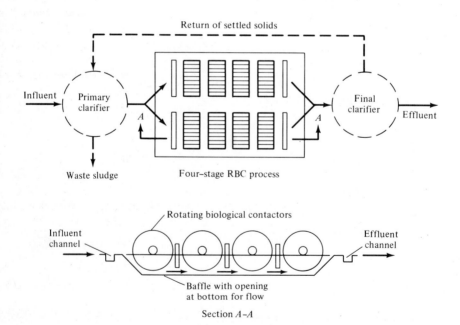

**Figure 12.30** Flow diagram for a domestic wastewater plant using rotating biological contactors for secondary treatment.

employed to ensure adequate BOD reduction; additional stages can be added to initiate nitrification. Each stage acts as a completely mixed chamber, and the slow movement of wastewater through the system simulates plug flow. Biological solids sheared from the disk surfaces are hydraulically transported under the baffles separating the chambers and conveyed from the system suspended in the effluent. In large plants the shafts of biological disks are placed perpendicular to the direction of flow (Fig. 12.30). Small installations have the disks attached to a common shaft with the wastewater flow parallel to the shaft. In this arrangement, bundles of media can be spaced to allow baffling for staging of the disks attached to the common shaft.

Rotating biological contactors are sensitive to cold temperature and must be protected from normal outdoor weather conditions, that is, precipitation, wind, and intensive sunshine. Standard design procedure is either to enclose individual stages under insulated plastic covers or else to house a series of units in a suitable building with adequate ventilation.

## 12.19  DESIGN OF THE RBC PROCESS

The removal of substrate by the RBC process may theoretically be expected to follow first-order kinetics, but this has not been consistently observed during process operation. The heavy biological growths on the first-stage disks often assimilate a major portion of the substrate at a removal rate much higher than exhibited by the remaining stages. Furthermore, the limiting condition can be oxygen transfer rather than substrate concentration. The last stage, supporting a much thinner biological film, receives a small amount of substrate that may appear only as an intermittent loading during the hours of peak wastewater flow. Because of the variations between theory and observation, the design of a RBC process for treating domestic wastewater relies on empirical data from operating plants. If the wastewater is from an industry or a municipality with a substantial portion of industrial discharge, the criteria for design are determined by pilot-plant studies.

The criterion for determining the required area of disk surface is the allowable BOD loading expressed in pounds per 1000 ft$^2$ per day (grams per square meter per day). The original criterion for sizing a RBC process for domestic wastewater was hydraulic loading in terms of gallons per square foot per day; however, this proved to be an unreliable standard for treating municipal wastewaters containing measurable amounts of industrial wastes. Manufacturers' literature usually suggest design criteria based on operational performance of their full-scale installations. Typical recommendations follow for secondary treatment of domestic wastewater to yield an effluent of less than 30 mg/l of BOD and 30 mg/l of suspended solids:

1. Average loading based on total RBC surface area should be 1.5 lb/1000 ft$^2$/day of soluble BOD or 3.0 lb/1000 ft$^2$/day of total BOD (7.5 g/m$^2$·d of soluble BOD or 15 g/m$^2$·d of total BOD).

2. Maximum loading on the first stage should be 6 lb/1000 ft²/day of soluble BOD or 12 lb/1000 ft²/day of total BOD (30 g/m²·d of soluble BOD or 60 g/m²·d of total BOD).

3. A temperature correction for additional RBC surface area of 15% should be made for each 5°F below a design wastewater temperature of 55°F (15% for each 2.8°C below 13°C).

These recommendations are consistent with observed satisfactory performance at larger installations. For small plants, the safe upper limit is an average soluble BOD loading of approximately 1.0 lb/1000 ft²/day (5 g/m²·d). Mathematical models to reliably predict the performance of the RBC process are still in the developmental stage [13].

Many bench-scale studies using RBC units with a disk diameter of 2 ft or less have been conducted to establish design criteria for treatment of strong wastewaters. Often these smaller-diameter units have exhibited a higher treatment efficiency than the full-scale biological disks designed and installed based on the study data. The primary cause appears to be the better oxygen transfer capability of the bench-scale unit, which led to overloading of the full-scale system. For processing combined domestic and industrial wastewaters, a full-scale pilot-plant unit is recommended to increase the reliability of applying treatability data to process design.

The arrangement and staging of biological disks for processing combined wastewaters requires special consideration. A series of equal-sized stages have proven to be an inefficient use of media in the treatment of high-strength wastewaters. Overloading the first-stage RBC leads to excessive biological growths that fill the space between the disks. Results have been the limitation of oxygen transfer and mechanical failures of media and shafts. In contrast, the final stage is ineffective because of underloading. The substrate remaining in the applied wastewater is not adequate to maintain a viable biological film on the media. The recommended solution is to design RBC chambers to provide the greatest disk surface area in the first stage and lesser areas in later stages [14].

### ■ EXAMPLE 12.8

Calculate the RBC area required for secondary treatment of a raw domestic wastewater having 230 mg/l of BOD. The design flow is 2.0 mgd at a temperature of 50°F. The effluent quality specified is 30 mg/l of BOD.

### *Solution*

primary effluent BOD concentration = 0.65 × 230 = 150 mg/l

An appropriate average BOD loading to achieve an effluent BOD of 30 mg/l is 3.0 lb/1000 ft²/day at 55°F. Therefore,

$$\text{RBC area at } 55°F = \frac{2.0 \times 150 \times 8.34}{3.0} = \frac{2500}{3.0} = 834,000 \text{ ft}^2$$

Correcting for a temperature of 50°F by a 15% area increase per 5°F,

RBC area at 50°F = $1.15 \times 834,000 = 959,000 \text{ ft}^2$

Use RBC shafts manufactured with a nominal surface area of 60,000 ft² with 12-ft diameter disks for installation in tanks with a length of 17 ft 4 in. Install four rows of four stages for a total of $16 \times 60,000 = 960,000 \text{ ft}^2$.

BOD loading on the first stage based on a temperature of 55°F should not exceed 12 lb/1000 ft²/day.

$$\text{first-stage loading at } 55°F = \frac{2500 \times 4}{(960/1.15)} = 12.0 \text{ lb/1000 ft}^2/\text{day} \quad \text{(OK)} \quad \blacksquare$$

# Activated Sludge

Activated-sludge processes are used for both secondary treatment and complete aerobic treatment without primary sedimentation. Wastewater is fed continuously into an aerated tank, where the microorganisms metabolize and biologically flocculate the organics (Fig. 12.8). Microorganisms (activated sludge) are settled from the aerated mixed liquor under quiescent conditions in the final clarifier and returned to the aeration tank. Clear supernatant from the final settling tank is the plant effluent.

The primary feeders in activated sludge are bacteria; secondary feeders are holozoic protozoans (Fig. 12.11). Microbial growth in the mixed liquor is maintained in the declining or endogenous growth phase to ensure good settling characteristics (Fig. 12.9). Synthesis of the waste organics results in a buildup of the microbial mass in the system. Excess activated sludge is wasted from the system to maintain the proper food/microorganism ratio (F/M) and sludge age to ensure optimum operation.

Activated sludge is truly an aerobic treatment process since the biological floc are suspended in a liquid medium containing dissolved oxygen. Aerobic conditions must be maintained in the aeration tank; however, in the final clarifier the dissolved-oxygen concentration can become extremely low. Dissolved oxygen extracted from the mixed liquor is replenished by air supplied to the aeration tank.

## 12.20  BOD LOADINGS AND AERATION PERIODS

General loading and operational parameters for the activated-sludge processes used in treatment of municipal wastewater are listed in Table 12.3.

The BOD load on an aeration tank is calculated using the BOD in the influent wastewater without regard to that in the return sludge flow. BOD loadings are expressed in terms of pounds of BOD applied per day per 1000 ft³ of liquid volume in the aeration tank and in terms of pounds of BOD applied/day/lb of mixed-liquor suspended solids (MLSS) in the aeration tank. The latter, the F/M ratio, is expressed by some authors in terms of lb of BOD applied/day/lb of volatile mixed-liquor suspended solids (MLVSS).

**TABLE 12.3** GENERAL LOADING AND OPERATIONAL PARAMETERS FOR ACTIVATED-SLUDGE PROCESSES

| Process | BOD loading | | Sludge age (days) | Aeration period (hr) | Average return sludge rates (%) | BOD efficiency (%) |
|---|---|---|---|---|---|---|
| | lb BOD/ day/ 1000 ft³ [a] | lb BOD/ day/ lb of MLSS | | | | |
| High rate (complete mixing) | 100 up | 0.5–1.0 | 3–10 | 2.5–3.5 | 100 | 70–85 |
| Step aeration | 30–50 | 0.2–0.5 | 5–15 | 5.0–7.0 | 50 | 80–90 |
| Conventional (tapered aeration) | 30–40 | 0.2–0.5 | 5–15 | 6.0–7.5 | 30 | 80–90 |
| Contact stabilization | 30–50 | 0.2–0.5 | 5–15 | 6.0–9.0 | 100 | 75–90 |
| Extended aeration | 10–30 | 0.05–0.2 | 20+ | 20–30 | 100 | 85–95 |
| High-purity oxygen | 120+ | 0.6–1.5 | 5–10 | 1.0–3.0 | 30 | 80–90 |

[a] 1.0 lb/1000 ft³/day = 16 g/m³·d.

The aeration period is the detention time of the raw-wastewater flow in the aeration tank expressed in hours. It is calculated by dividing the tank volume by the daily average flow without regard to return sludge. The activated sludge returned is expressed as a percentage of the raw-wastewater influent. For example, if the return sludge rate is 20% and the raw-wastewater flow into the plant is 10 mgd, the return sludge is 2.0 mgd.

BOD loadings per unit volume of aeration tank vary from greater than 100 to less than 10 lb of BOD/1000 ft³/day, while the aeration periods correspondingly vary from 2.5 to 24 hr. The relationship between volumetric BOD loading and aeration period is directly related to BOD concentration in the wastewater. For example, converting the average BOD concentration of 200 mg/l into units of pounds per 1000 ft³ yields a concentration of

$$200 \text{ mg/l} \times \frac{62.4 \text{ lb/1000 ft}^3}{1000 \text{ mg/l}} = 12.5 \text{ lb/1000 ft}^3$$

Therefore, 200 mg/l wastewater applied to an extended aeration system with a 24-hr (1-day) aeration period results in a BOD loading of 12.5 lb/1000 ft³/day. If a high-rate aeration period of 3.0 hr is considered, the BOD loading becomes

$$12.5 \text{ lb/1000 ft}^3/\text{day} \times \frac{24 \text{ hr}}{3.0 \text{ hr}} = 100 \text{ lb/1000 ft}^3/\text{day}$$

Sludge age (mean cell residence time) relates the quantity of microbial solids in an activated-sludge process to the quantity of solids lost in the effluent and excess solids withdrawn in the waste sludge. Equation 12.60 establishes the sludge age in days on the basis of MLSS in the aeration tank relative to SS discharged in the effluent and SS in the waste sludge withdrawn daily:

$$\text{Sludge age} = \frac{\text{MLSS} \times V}{\text{SS}_e \times Q_e + \text{SS}_w \times Q_w} \tag{12.60}$$

where    sludge age = mean cell residence time, days
$\qquad$ MLSS = mixed-liquor suspended solids, mg/l
$\qquad$ $V$ = volume of the aeration tank, mil gal (m³/d)
$\qquad$ $\text{SS}_e$ = suspended solids in effluent, mg/l
$\qquad$ $\text{SS}_w$ = suspended solids in waste sludge, mg/l
$\qquad$ $Q_e$ = quantity of effluent wastewater, mgd (m³/d)
$\qquad$ $Q_w$ = quantity of waste sludge, mgd (m³/d)

Sludge age is also calculated using the MLVSS (volatile portion of the MLSS) and the VSS (volatile suspended solids) in the effluent and waste sludge. The argument is that the volatile portion of the suspended solids is more representative of the microbial masses, and thus the sludge age expresses the residence time of the microbial cells in the system more realistically.

The suspended-solids concentration maintained in the MLSS of conventional and step aeration processes ranges from 1500 to 3000 mg/l. The concentration held in the operation of a particular system depends on the desired F/M and sludge age for the applied BOD load. High-rate completely mixed processes generally operate with higher MLSS concentrations of 3000–5000 mg/l. Because of the variety of extended aeration processes, MLSS values encompass the entire range of 1000 to greater than 6000 mg/l.

Solids retention in an activated-sludge system is measured in days, whereas the liquid aeration period is in hours. For example, a conventional activated-sludge process with a MLSS of 2500 mg/l in the aeration tank, treating an average domestic wastewater, and operating at a 6-hr aeration period has a sludge age of approximately 7 days. The suspended solids are cycled in the system from final clarifier back to aeration tank, while the liquid flows through the aeration tank and clarifier.

The wide range of aeration periods and BOD loadings used in activated-sludge processes tends to contrast one process with another. Also, the variety of physical features, such as the aeration tank size and shape, used in the various processes tends to accent the differences. Actually all activated-sludge processes are biologically similar, as seen in the generalized activated-sludge diagram of Fig. 12.31. BOD is removed in the process by assimilative respiration of microorganisms, and the new cell growth is reduced by endogenous respiration. Excess microbial growth is withdrawn from the system by wasting activated sludge. Oxygen is added to the process to maintain aerobic biological activity.

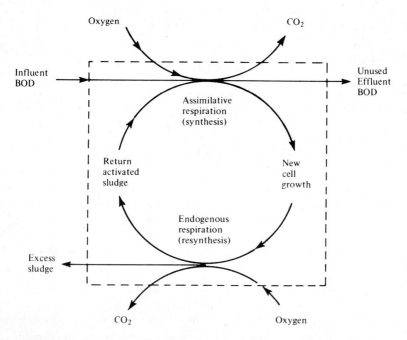

**Figure 12.31** Generalized biological process reactions in the activated-sludge process.

■ **EXAMPLE 12.9**

Data from a field study on a step-aeration activated-sludge secondary are as follows:

$$\text{aeration tank volume} = 120{,}000 \text{ ft}^3 = 0.898 \text{ mil gal}$$

$$\text{settled wastewater flow} = 3.67 \text{ mgd}$$

$$\text{return sludge flow} = 1.27 \text{ mgd}$$

$$\text{waste sludge flow} = 18{,}900 \text{ gpd} = 0.0189 \text{ mgd}$$

$$\text{MLSS in aeration tank} = 2350 \text{ mg/l}$$

$$\text{SS in waste sludge} = 11{,}000 \text{ mg/l}$$

$$\text{influent wastewater BOD} = 128 \text{ mg/l}$$

$$\text{effluent wastewater BOD} = 22 \text{ mg/l}$$

$$\text{effluent SS} = 26 \text{ mg/l}$$

Using these data, calculate the loading and operational parameters listed in Table 12.3 and the excess sludge production in pounds of excess suspended solids per pound of BOD applied.

*Solution*

$$\text{BOD load} = 3.67 \text{ mgd} \times 128 \text{ mg/l} \times 8.34 = 3920 \text{ lb/day}$$

$$\text{MLSS in aeration tank} = 0.898 \text{ mil gal} \times 2350 \text{ mg/l} \times 8.34 = 17{,}600 \text{ lb}$$

$$\text{BOD loading} = 3920/120 = 32.7 \text{ lb/day/1000 ft}^3$$

$$\text{BOD loading} = 3920/17{,}600 = 0.22 \text{ lb/day/lb of MLSS}$$

Using Eq. 12.60,

$$\text{sludge age} = \frac{2350 \times 0.898}{26 \times 3.67 + 11{,}000 \times 0.0189} = 7.0 \text{ days}$$

$$\text{aeration period} = \frac{0.898 \times 24}{3.67} = 5.9 \text{ hr}$$

$$\text{return sludge rate} = \frac{1.27 \times 100}{3.67} = 35\%$$

$$\text{BOD efficiency} = \frac{(128 - 22)100}{128} = 83\%$$

$$\text{sludge production} = \frac{0.0189 \text{ mgd} \times 11{,}000 \text{ mg/l} \times 8.34}{3920}$$

$$= 0.44 \text{ lb SS wasted/lb BOD applied} \qquad ■$$

# 12.21 ACTIVATED-SLUDGE TREATMENT SYSTEMS

Flow diagrams for the most common activated-sludge processes are shown in Fig. 12.32.

The *conventional activated-sludge process* (Fig. 12.32a), an outgrowth of the earliest activated-sludge systems constructed, is used for secondary treatment of domestic wastewater. The aeration basin is a long rectangular tank with air diffusers on one side of the tank bottom to provide aeration and mixing. Settled raw wastewater and return activated sludge enter the head of the tank and flow down its length in a spiral flow pattern. An air supply is tapered along the length of the tank to provide a greater amount of diffused air near the head where the rate of biological metabolism and resultant oxygen demand are the greatest. A conventional activated-sludge aeration tank is shown in Fig. 12.33.

The standard-activated-sludge process uses bubble air diffusers set at a depth of 8 ft or more to provide adequate oxygen transfer and deep mixing. Three of the many different kinds of diffusers manufactured are illustrated in Fig. 12.34. These diffusers can be attached to an air header pipe of a swing diffuser assembly (Fig. 12.33). The swing diffuser arm is jointed so the diffusers can be swung out of the tank for cleaning and maintenance.

Fine-bubble diffusers are ceramic plates, tubes, or domes commonly constructed of an aluminum oxide material. Plate diffusers are installed in recesses or in holders on the bottom of an aeration tank. Tubes are attached to air manifolds mounted on the bottom or to removeable drop pipes. One type of ceramic diffuser installation is illustrated in Fig. 12.35. Individual diffusers are mounted on baseplates attached to piping on the tank bottom. If the diffusers are distributed over the entire floor area, the rising streams of fine bubbles mix the liquid, keeping the microbial floc in suspension. The principal advantage attributed to fine-bubble diffused aeration, as compared to coarse-bubble systems, is the higher oxygen transfer efficiency at a lower power cost. Nevertheless, greater maintenance costs result from the need to periodically clean the diffusers as a result of external biological growths plugging the fine pores. Also, the compressed air supply must be very clean to prevent interior plugging of the diffusers.

The *step-aeration activated-sludge process* (Fig. 12.32b) is a modification of the conventional process. Instead of introducing all raw wastewater at the tank head, raw flow is introduced at several points along the tank length. Stepping the influent load along the tank produces a more uniform oxygen demand throughout. While tapered aeration attempts to supply air to match oxygen demand along the length of the tank, step loading provides a more uniform oxygen demand for an evenly distributed air supply.

The *contact stabilization activated-sludge process* (Fig. 12.32c) provides for reaeration of the return activated sludge from the final clarifier, allowing this method to use a smaller aeration tank. The sequence of aeration-sedimen-

Figure 12.32 Flow diagrams for common activated-sludge processes. (a) Conventional activated-sludge process. (b) Step-aeration activated-sludge process. (c) Contact stabilization without primary sedimentation. (d) High-rate (completely mixed) activated-sludge process. (e) Extended aeration without primary sedimentation.

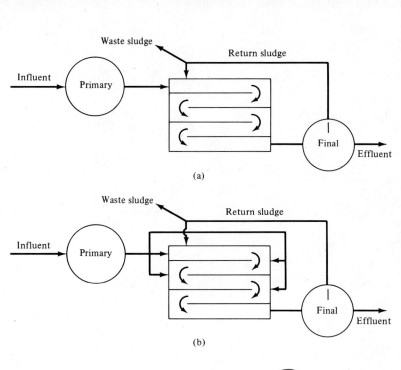

(a)

(b)

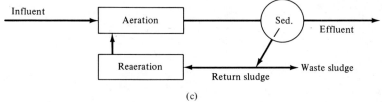

(c)

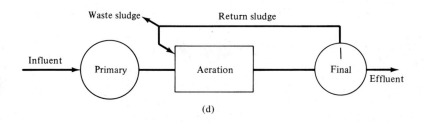

(d)

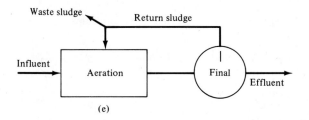

(e)

(a)

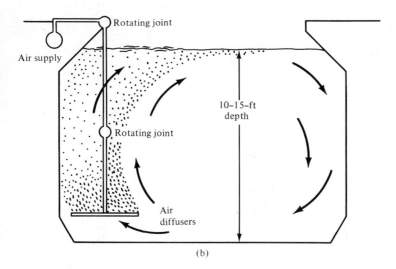

(b)

**Figure 12.33** Conventional activated-sludge process. (a) Long rectangular aeration tank with submerged coarse-bubble diffusers along one side (Santee, CA). (b) Cross-section of a typical aeration tank illustrating the spiral flow pattern created by aeration along one side.

tation–reaeration has been used as a secondary treatment process. However, current use is primarily in complete aerobic treatment without primary sedimentation (Fig. 12.41). Raw wastewater is aerated in a contact tank and then settled. The supernatant from the clarifier is the plant effluent, and the subnatant is reaerated prior to mixing with raw influent in the contact aeration tank.

The *high-rate (completely mixed) activated-sludge process* (Fig. 12.32d)

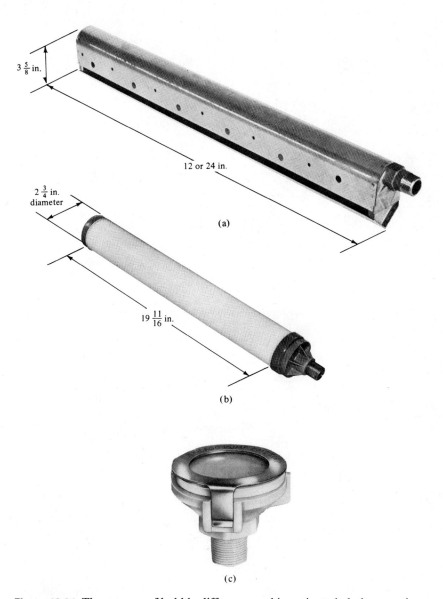

$3\frac{5}{8}$ in.

12 or 24 in.

(a)

$2\frac{3}{4}$ in.
diameter

$19\frac{11}{16}$ in.

(b)

(c)

**Figure 12.34** Three types of bubble diffusers used in activated-sludge aeration tanks. (a) The *Sanitaire* stainless steel, nonclog, wide-band diffuser consists of a balancing nozzle, an air reservoir, air exit ports, and a deflector. Located below the open bottom of the air reservoir, the deflector directs the liquid being aerated along the inverted air reservoir's outer walls and air exiting through the ports in the sides of the diffuser body is sheared into small bubbles. (Courtesy of SANITAIRE, Water Pollution Control Corp.) (b) The *Pearlcomb* fine-bubble, wide-band diffuser is a hollow cylinder manufactured of a porous, polymeric material. (Courtesy of FMC Corp., Material Handling Systems Div.) (c) The *Discfuser* coarse-bubble diffuser has a movable plastic disk that is free to rise and fall within the space provided between the housing and an upper retaining ring. As air discharges through the diffuser, the disk rises, allowing air to escape around the periphery of the disk. (Courtesy of FMC Corp., Material Handling Systems Div.)

(a)

(b)

**Figure 12.35** Fine-bubble aeration with ceramic diffusers. (a) Long rectangular aeration tank with uniform mixing and oxygenation of the activated sludge by diffusers attached to the tank bottom. (b) A grid of individual diffusers attached to air pipes mounted on the floor of the tank. (Courtesy of SANITAIRE, Water Pollution Control Corp.)

operates with the highest BOD load per unit volume of aeration tank of any activated-sludge system. The BOD loading is approximately 3 times that of the conventional process, and the aeration period is proportionately shorter. High-rate mixed liquor is in the declining growth phase rather than the endogenous stage, making the activated sludge more difficult to settle. The D-O Aerator illustrated in Fig. 12.36 is a high-rate secondary aeration basin

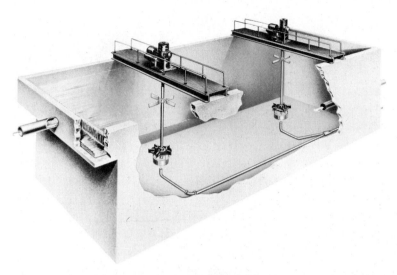

**Figure 12.36** High-rate activated-sludge aeration tank with a large-bubble air sparge ring mounted under mixing impellers. (Courtesy of Dorr-Oliver, Inc.)

that uses a combination of compressed-air aeration and mechanical mixing. Air is introduced through holes in a sparge ring located at the base of the mixer shaft. Turbines churn the mixed liquor, shearing the bubbles and mixing air with the tank contents.

Mechanical aeration systems installed in activated-sludge tanks are surface aerators where air is entrained from the atmosphere (Fig. 12.37) and submerged turbine aerators that have air introduced under the lower turbine (Fig. 12.36). In other applications, floating surface aerators are commonly used in lagoons (Fig. 12.58), platform-mounted surface aerators in extended aeration processes (Fig. 12.39), and horizontal rotors in an oxidation ditch (Fig. 12.40). In large, completely mixed aeration basins, with BOD loadings in the range of conventional and step-aeration processes, a mechanical impeller is placed above an uptake tube for deep mixing along with surface aeration. As illustrated in Fig. 12.37, the liquid in the tank is drawn up through the tube and discharged radially from the impeller on top. A splash plate is mounted above the aerator to deflect the aerial spray downward.

The *extended aeration activated-sludge process* (Fig. 12.32e) is commonly used to treat small wastewater flows from schools, housing developments, trailer parks, institutions, and small communities. The aeration period is 24 hr or greater, the longest of any activated-sludge process. Because of low BOD loading, the extended aeration system operates in the endogenous growth phase.

The extended aeration process can accept periodic (intermittent) loadings without becoming upset. The stability of the process results from the large aeration volume and complete mixing of the tank contents. Final settling

(a)

(b)

**Figure 12.37** Mechanical surface aerator with an uptake tube for deep mixing. (a) Empty aeration tank. (The scaffolding in the center was placed temporarily in the tank for maintenance of the bridge.) (b) Operating aeration tank (Lincoln, NE).

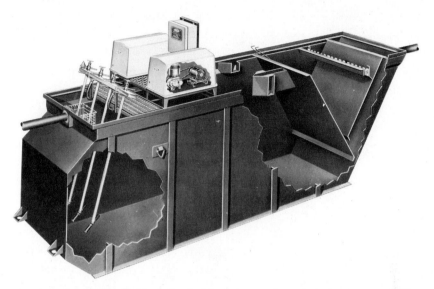

**Figure 12.38** Factory-built extended aeration activated-sludge system with diffused aeration and a slot returning settled sludge from the final settling basin. (Courtesy of Smith & Loveless, Inc.)

tanks are conservatively designed using low overflow rates and long detention times. Overflow rates generally range from 200 to 600 gpd/ft$^2$ for aeration tank volumes ranging from 5000 to 150,000 gal. Excess sludge is not generally wasted continuously from an extended aeration system. Instead, the mixed liquor is allowed to increase in suspended-solids concentration and a large volume of the aeration tank contents are periodically pumped to disposal. The MLSS concentration in the aeration chamber of operating plants varies from 1000 to 10,000 mg/l. Under normal loading conditions, the MLSS concentration increases at a rate of 40–60 mg/l/day when treating domestic wastewater.

The high demand for small extended aeration treatment plants in recent years has stimulated the development of a wide variety of manufactured units. The Oxigest unit in Fig. 12.38 is fabricated and assembled in the factory and shipped to the field site for installation. The Accelo-Biox plant shown in Fig. 12.39 can be factory-built or mechanical aeration equipment installed in cast-in-place reinforced-concrete tanks.

Another extended-aeration system with mechanical aeration is the oxidation ditch with a horizontal rotor, as illustrated in Fig. 12.40. Unsettled wastewater is fed to mixed liquor in the ditch, where it is mechanically aerated using a caged rotor. The ditch effluent is clarified and the settled sludge returned to maintain a desired MLSS. An oxidation ditch is basically an extended aeration process; however, a shorter aeration period can be used when adequate rotor aeration and efficient suspended-solids separation in

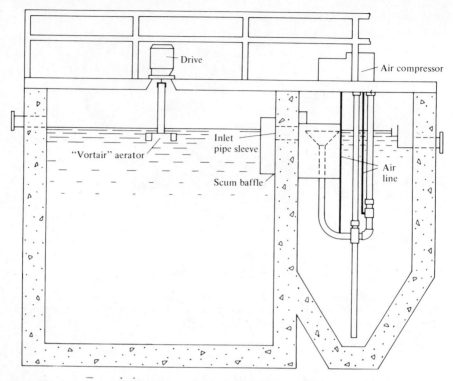

**Figure 12.39** Extended-aeration activated-sludge system with a mechanical aerator and an airlift pump for returning settled sludge from the final clarifier. (Courtesy of Infilco Degremont, Inc.)

the final clarifier are provided. The suggested BOD loading is less than 20 lb/1000 ft³/day with an aeration period greater than 12 hr.

For larger installations, circular steel wastewater treatment plants are commonly used. The design has a center chamber for sedimentation and a segmented cylindrical-shaped outer chamber for aeration. The field-erected Stepaire unit drawn in Fig. 12.41 can be operated using the extended-aeration process, step-aeration process, or contact stabilization.

### ■ EXAMPLE 12.10

A contact-stabilization plant, similar to the one diagrammed in Fig. 12.41, has compartments with the following liquid volumes:

$$\text{aeration chamber} = 85 \text{ m}^3$$

$$\text{reaeration chamber} = 173 \text{ m}^3$$

$$\text{aerobic digester} = 153 \text{ m}^3$$

$$\text{sedimentation tank} = 122 \text{ m}^3 \text{ (30.7-m}^2 \text{ surface area)}$$

If the plant is designed for an equivalent population of 2000 persons, calculate the BOD loading, aeration periods, and detention times.

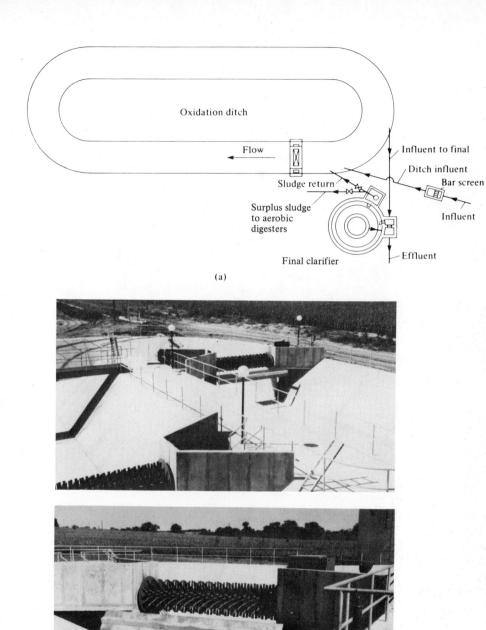

Oxidation ditch

Flow

Sludge return

Surplus sludge
to aerobic
digesters

Final clarifier

Influent to final

Ditch influent

Bar screen

Influent

Effluent

(a)

(b)

**Figure 12.40** Oxidation-ditch activated-sludge process. (a) Flow diagram of an oxidation-ditch treatment plant. (b) Pictures of an oxidation ditch and horizontal-rotor aerators. (Courtesy of Lakeside Equipment Corp.)

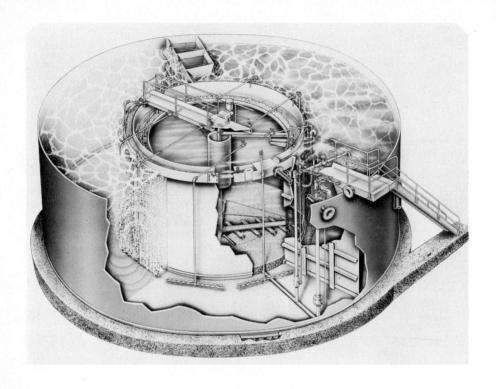

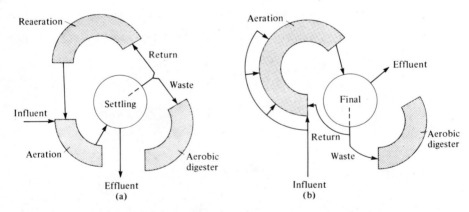

**Figure 12.41** Field-erected circular steel wastewater treatment plant for extended aeration, step aeration, or contact stabilization processes. Photograph shows a cutaway view of the aeration tank and clarifier. (a) Contact stabilization. (b) Extended or step aeration. (Courtesy of FMC Corp., Material Handling Systems Div.)

*Solution*

$$\text{hydraulic load} = 2000 \times 450 \, \text{l/person}$$

$$= 900{,}000 \, \text{l/d} = 900 \, \text{m}^3/\text{d}$$

$$\text{BOD load} = 2000 \times 91 \, \text{g/person}$$

$$= 182{,}000 \, \text{g/d}$$

$$\text{BOD loading on aeration tanks} = \frac{182{,}000}{85 + 173} = 705 \, \text{g/m}^3\cdot\text{d}$$

$$\text{aeration period (based on raw-wastewater flow)} = \frac{85 \times 24}{900} = 2.3 \, \text{h}$$

$$\text{reaeration period (based on raw-wastewater flow)} = \frac{173 \times 24}{900} = 4.6 \, \text{h}$$

The detention time for sedimentation (assuming 100% recirculation flow) is

$$\frac{122 \times 24}{2 \times 900} = 1.6 \, \text{h}$$

The overflow rate on final clarifier (based on effluent flow) is

$$\frac{900}{30.7} = 29.3 \, \text{m}^3/\text{m}^2\cdot\text{d} \qquad \blacksquare$$

### ■ EXAMPLE 12.11

Calculate the dimensions for equal-size aeration basins for secondary treatment of a wastewater flow of 18.2 mgd with 19,700 lb of BOD: (a) for a conventional activated-sludge process with a maximum loading of 35 lb of BOD/1000 ft³/day and minimum aeration period of 6 hr; (b) for a high-rate activated-sludge process with a maximum loading of 100 lb of BOD/1000 ft³/day and minimum aeration period of 2.5 hr.

*Solution*

**a.** Conventional process:

$$V \, (\text{based on BOD loading}) = \frac{19{,}700 \times 1000}{35} = 563{,}000 \, \text{ft}^3$$

$$V \, (\text{based on aeration period}) = \frac{18{,}200{,}000 \times 6}{24 \times 7.48} = 608{,}000 \, \text{ft}^3$$

Use 608,000 ft³ with an aeration period of 6.0 hr at 32 lb of BOD/1000 ft³/day loading. Use ten aeration tanks, with 13 ft liquid depth and 24 ft width.

$$\text{length of each tank} = \frac{608{,}000}{10 \times 13 \times 24} = 196 \, \text{ft}$$

**b.** High-rate process:

$$V \, (\text{based on BOD loading}) = \frac{19{,}700 \times 1000}{100} = 197{,}000 \, \text{ft}^3$$

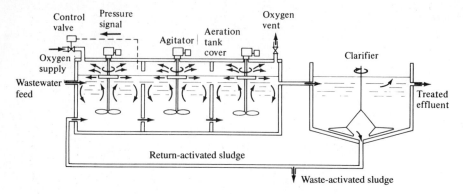

**Figure 12.42** Schematic diagram of a high-purity-oxygen activated-sludge process with surface aerators in three stages. (Courtesy of UNOX System, Lotepro Corp.)

$$V \text{ (based on aeration period)} = \frac{18,200,000 \times 2.5}{24 \times 7.48} = 253,000 \text{ ft}^3$$

Use 253,000 ft³ with an aeration period of 2.5 hr at 78 lb of BOD/1000 ft³/day loading. Use three circular aeration tanks with 16 ft liquid depth.

$$\text{diameter of each tank} = \left(\frac{253,000 \times 4}{3 \times 16 \times \pi}\right)^{0.5} = 82 \text{ ft}$$

■

## 12.22 HIGH-PURITY-OXYGEN ACTIVATED SLUDGE

The primary distinguishing feature of this process is the application of high-purity oxygen instead of air for aerobic treatment by activated sludge. The major deterrent to oxygen use until recently was the production cost. Currently, high-purity oxygen gas can be efficiently generated by cryogenic air separation or pressure-swing adsorption processes. The former is used in large installations, whereas the latter is a reliable method for oxygen production in smaller quantities.

The major components of such a system are an oxygen gas generator, a covered compartmented aeration tank, a final clarifier, pumps for recirculating activated sludge, and sludge disposal facilities. Preliminary treatment must include screening and degritting of the wastewater, with primary sedimentation optional, primarily depending on the size of the treatment plant. The aeration tank is divided into three or four stages by means of baffles to stimulate plug flow and is covered with a gas-tight enclosure (Fig. 12.42). Raw wastewater, return activted sludge, and oxygen gas under a slight pressure are introduced to the first stage and flow concurrently through succeeding sections. Oxygen can be mixed with the tank contents by injection through a hollow shaft to a rotating sparger device, or a surface aerator installed on

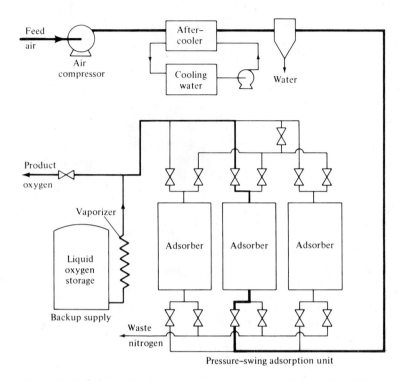

**Figure 12.43** Schematic diagram of a pressure-swing adsorption (PSA) oxygen-generating system for on-site production of high-purity oxygen gas. (Courtesy of UNOX System, Lotepro Corp.)

top of the mixer turbine shaft to contact oxygen gas with the mixed liquor. Successive aeration chambers are connected to each other so that the liquid flows through submerged ports, and head gases pass freely from one stage to the next with only a slight pressure drop. Exhausted waste gas is a mixture of carbon dioxide, nitrogen, and about 10%–20% of the applied oxygen. Effluent mixed liquor is settled in either a scraper type or rapid-sludge-return clarifier, and the activated sludge is returned to the aeration tank.

Oxygen is supplied from on-site gas generators and liquid oxygen is shipped to the treatment plant. Transported liquid oxygen is too costly for regular application; therefore, it is stored for use only as a backup supply or to meet unexpected peak demands. For large installations, standard cryogenic air separation, involving liquefaction of air followed by fractional distillation, is employed to separate the major components of nitrogen and oxygen. The pressure-swing adsorption system (PSA) is more suitable for the majority of treatment plant sizes because of its efficiency and reliability. The main items of equipment in a PSA unit are a feed air compressor, adsorber vessels with manifold pipes and sequencing valves, and a cycle control system (Fig. 12.43). Air is compressed in vessels filled with granular adsorbent that collects

**Figure 12.44** High-purity-oxygen aeration system with mixer drive motors mounted on the tank covers. The three vessels on the right are pressure-swing-adsorption units, and the single tank in the background on the left stores liquid oxygen for a backup supply. (Courtesy of UNOX System, Lotepro Corp.)

carbon dioxide, water, and nitrogen gas under high pressure leaving a gas relatively high in oxygen. The adsorber is regenerated by depressurizing and purging. Three vessels are installed for continuous and constant flow of oxygen gas; two units alternate between operation and regeneration while the third is a standby unit. Figure 12.44 shows covered aeration tanks with mixer motors mounted on the deck, three pressure-swing adsorption vessels, and a liquid oxygen storage tank. Cryogenic oxygen has a purity of 99% or more. PSA oxygen purity can be controlled by system operation and is generally in the 70%–90% range.

Compared to air activated-sludge processes, high-purity-oxygen activated sludge has several advantages that are attributed to its higher oxygenation capacity. If all of the nitrogen in air is displaced by oxygen, the partial pressure of oxygen is 100% resulting in a fivefold increase in the saturation value of dissolved oxygen in water. High efficiency is possible at increased BOD loads and reduced aeration periods by maintaining the food/microorganism ratio with MLSS concentrations of 4000–8000 mg/l. Waste-sludge production is reportedly less than that generated in air activated-sludge systems. This is ascribed to the greater oxidation of organic matter at high-DO levels maintained in the mixed liquor, and improved settleability, resulting in a return sludge containing 2%–3% solids. Even though the process simulates plug flow, shock organic loads do not produce instability, since extra oxygen is supplied to the first stage automatically on demand. Emission of foul odors is virtually eliminated because of the highly aerobic environment and reduced volume of exhaust gases. Covered tanks also help to reduce the cooling of the wastewater during cold-weather operation.

## 12.23 DESIGN CRITERIA FOR HIGH-PURITY-OXYGEN AERATION

Results from pilot-plant studies of oxygen-activated sludge at five different locations are listed in Table 12.4. These installations were operated on a diurnal flow cycle similar to that of the actual wastewater flow pattern for the municipal or combined municipal–industrial waste at the site. At BOD loadings of 160 lb/1000 ft$^3$/day (2.56 kg/m$^3$·d) and mean aeration periods of 1.8 hr, BOD and suspended solids effluent concentrations were consistently less than the maximum allowable criterion of 30 mg/l. Resulting BOD removal efficiency ranged from 87% to 97%, depending on the influent concentration. Loadings in terms of food/microorganism ratio were 0.5–0.6 lb of BOD/day/lb of MLVSS. An F/M ratio of 0.5 is equivalent to treating a 150-mg/l of BOD wastewater with an aeration period of 1.8 hr, holding 4000 MLVSS in the basin. Dissolved oxygen of 3–9 mg/l was held in the mixed liquor, resulting in oxygen utilization of approximately 90%. Oxygen consumption per unit of BOD removed varies over a considerable range, depending on the food/microorganism ratio as shown in Fig. 12.45. At an F/M of 0.5, the oxygen uptake is from 1.0 to 1.5 lb of oxygen per pound of BOD removed.

The settleability of the suspended solids in oxygen-activated sludge, with typical SVI values near 50, is better than in an aerated mixed liquor. Consequently, the thickness of return sludge is generally about 2%, and a normal return sludge rate is 30%. Design parameters for final settling tanks do not differ significantly from those recommended for conventional activated-sludge clarifiers. The maximum allowable recommended overflow rate is 1200 gpd/ft$^2$ during peak hourly flow, generally resulting in an overflow rate of 600–800 gpd/ft$^2$ based on average daily flow. Solids loading on a clarifier ranges from 30 to 60 lb/day/ft$^2$ as a result of the high suspended-solids concentration in the mixed liquor flowing into the basin. Therefore, a rapid-sludge-return clarifier with a hydraulic pickup arm is used to aid in thickening the settled activated sludge. The extent of excess solids production is a function of the sludge age, BOD loading, level of dissolved oxygen held in the mixed liquor, and characteristics of the wastewater. For domestic wastes being treated under normal operating conditions, the excess volatile solids production is 30%–50% of the applied BOD.

■ **EXAMPLE 12.12**
A municipality has an average daily wastewater flow of 280 gpm with a peak hourly rate of 450 gpm. The average BOD concentration is 200 mg/l except during several weeks when a seasonal industry increases the mean BOD to 250 mg/l. A high-purity-oxygen system as in Fig. 12.42 without primary clarification of the raw wastewater is

**TABLE 12.4**   PILOT-PLANT DATA FROM HIGH-PURITY-OXYGEN ACTIVATED-SLUDGE EVALUATIONS AT FIVE LOCATIONS

| Parameter | Municipal | | | Combined municipal–industrial | |
|---|---|---|---|---|---|
| | Primary effluent | Primary effluent | Unsettled wastewater | Primary effluent | Unsettled wastewater |
| BOD loading | | | | | |
| (lb/day/1000 ft$^3$) | 239 | 155 | 162 | 171 | 193 |
| Aeration period (hr) | 0.8 | 1.8 | 1.7 | 2.0 | 1.8 |
| Influent concentrations | | | | | |
| BOD (mg/l) | 128 | 184 | 184 | 227 | 229 |
| COD (mg/l) | 222 | 375 | 380 | 467 | 377 |
| SS (mg/l) | 93 | 168 | 183 | 76 | 236 |
| Effluent concentrations | | | | | |
| BOD (mg/l) | 17 | 18 | 6 | 8 | 12 |
| COD (mg/l) | 76 | 87 | 67 | 81 | 66 |
| SS (mg/l) | 19 | 16 | 17 | 15 | 28 |
| Removal efficiencies | | | | | |
| BOD (%) | 87 | 90 | 97 | 96 | 95 |
| COD (%) | 66 | 77 | 82 | 82 | 82 |
| SS (%) | 90 | 90 | 91 | 80 | 88 |
| Food/microorganism ratio | | | | | |
| (lb of BOD/day/lb of | | | | | |
| MLVSS) | 1.6 | 0.6 | 0.6 | 0.5 | 0.6 |
| Mixed liquor | | | | | |
| DO (mg/l) | 8 | 7 | 6 | 8 | 6 |
| MLSS (mg/l) | 3630 | 6080 | 6200 | 7010 | 7350 |
| MLVSS (mg/l) | 2500 | 3860 | 4650 | 5400 | 5450 |
| SVI (ml/g) | 33 | 40 | 55 | 42 | 48 |
| Return sludge rate | | | | | |
| (%) | 11 | 30 | 27 | 30 | 26 |
| Return sludge density | | | | | |
| (%) | 3.6 | 2.6 | 2.7 | 3.1 | 3.2 |
| Clarifier overflow rate | | | | | |
| (gpd/ft$^2$) | 1100 | 650 | 750 | 600 | 650 |
| Oxygen utilization | | | | | |
| (%) | 93 | 90 | 91 | 90 | 93 |

*Source: Oxygen Activated Sludge Wastewater Treatment Systems,* Environmental Protection Agency [15].

being considered. Calculate (a) the volume of aeration tank capacity required, and (b) surface area and depth for a final settling tank. Recommended design criteria are a maximum BOD loading of 160 lb/1000 ft$^3$/day, a largest food/microorganism ratio of 0.5 lb of BOD/day/lb of MLVSS, an operating MLSS concentration of 5500 mg/l (MLVSS of 4200 mg/l), and a highest overflow rate of 1200 gpd/ft$^2$ during peak flow.

**Solution**

$$\textbf{a.} \quad \left( \begin{array}{c} \text{aeration tank volume} \\ \text{required at 250 mg/l of BOD} \end{array} \right) = \frac{280 \times 1440 \times 250 \times 8.34}{1{,}000{,}000 \times 0.160} = 5250 \text{ ft}^3$$

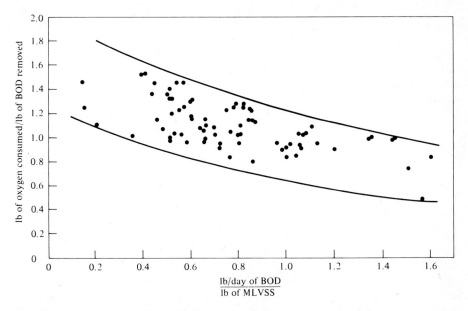

**Figure 12.45** Oxygen consumption in high-purity-oxygen activated sludge, per pound of BOD removed, as a function of the food/microorganism ratio expressed as lb BOD/day/lb volatile MLSS. [From *Oxygen Activated Sludge Wastewater Treatment Systems,* Environmental Protection Agency, Technology Transfer (August 1973).]

$$\left(\begin{array}{c}\text{aeration period}\\\text{at average flow}\end{array}\right) = \frac{5250 \times 7.48 \times 24}{280 \times 1440} = 2.34 \text{ hr} \quad \text{(OK)}$$

$$\left(\begin{array}{c}\text{BOD load at}\\\text{200 mg/l of BOD}\end{array}\right) = \frac{280 \times 1440 \times 200 \times 8.34}{1,000,000 \times 5.25}$$

$$= 128 \text{ lb/1000 ft}^3/\text{day} \quad \text{(OK)}$$

Check the F/M ratio for both waste loadings (128 and 160 lb/1000 ft³/day) at a MLVSS concentration of 4200 mg/l.

$$\text{At} \frac{128 \text{ lb of BOD/day}}{1000 \text{ ft}^3} : \frac{F}{M} = \frac{128/1000}{\dfrac{4200 \times 62.4}{1,000,000}} = 0.49 \frac{\text{lb of BOD/day}}{\text{lb of MLVSS}} \quad \text{(OK)}$$

$$\text{At} \frac{160 \text{ lb of BOD/day}}{1000 \text{ ft}^3} : \frac{F}{M} = \frac{160/1000}{\dfrac{4200 \times 62.4}{1,000,000}} = 0.61 \quad \text{(slightly greater than 0.5)}$$

Therefore, use a three-stage aeration basin with a total volume of 5250 ft³.

**b.** $\left(\begin{array}{c}\text{final clarifier surface area}\\\text{required based on peak flow}\end{array}\right) = \frac{450 \times 1440}{1200} = 540 \text{ ft}^2$

$$\text{overflow rate at average flow} = \frac{280 \times 1440}{540} = 747 \text{ gpd/ft}^2$$

Assume the additional final clarifier design parameters of 8.0 ft minimum depth and 2.5 hr as the minimum detention time. Then the clarifier depth required for an overflow rate of 747 gpd/ft² is

$$\text{depth} = \frac{747 \text{ gal}}{\text{day} \times \text{ft}^2} \times \frac{\text{ft}^3}{7.48 \text{ gal}} \times 2.5 \text{ hr} \times \frac{\text{day}}{24 \text{ hr}} = 10.4 \text{ ft} \qquad \blacksquare$$

## 12.24 KINETICS MODEL OF THE ACTIVATED-SLUDGE PROCESS

The principles of growth kinetics from pure culture microbiology can be applied in suspended-growth wastewater treatment even though the processes differ significantly in relation to the method of growth (batch as compared to continuous flow), microbial population (pure culture as opposed to mixed culture), and substrate content (uniform as opposed to a variety of organics). The shape of the wastewater processing curve in Fig. 12.9 is similar to Monod's growth rate curve (Fig. 12.6) and the Michaelis–Menten rate curve for an enzyme-catalyzed reaction (Fig. 12.4). When plotted against substrate concentration, the reaction rates of enzyme, pure culture, and mixed culture reactions all follow the shape of a rectangular hyperbola where the substrate at one-half the maximum reaction rate is a characteristic constant of the reaction.

The following mathematical equations apply to completely mixed activated-sludge systems operating in a substrate limiting condition, i.e., at a low F/M ratio. The fundamental relationships of growth kinetics used in the derivation of these equations are presented in Section 12.6.

The flow scheme for a completely mixed activated-sludge process is shown in Fig. 12.46,

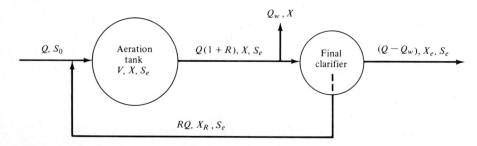

**Figure 12.46** Flow scheme for the completely mixed activated-sludge system used in the derivation of the kinetics equations.

where
$Q$ = rate of influent flow
$Q_w$ = rate of excess sludge wasting from aeration tank
$Q - Q_w$ = rate of effluent flow
$R$ = recirculation ratio $(Q_R/Q)$
$RQ$ = rate of sludge recirculation
$Q(1 + R)$ = rate of flow from aeration tank
$V$ = volume of aeration tank
$X$ = concentration of biomass in aeration tank (MLVSS)
$X_R$ = concentration of biomass in recirculating sludge (VSS)
$X_e$ = concentration of biomass in effluent (VSS)
$S_0$ = concentration of substrate in influent flow (soluble BOD or COD)
$S_e$ = concentration of substrate in effluent flow, recirculating sludge, and aeration tank (soluble BOD or COD)

The following conditions are assumed in the formulation of the mass balance equations:

1. Flows, biomass concentrations, and substrate concentrations are in a steady state.
2. All substrates are soluble (filtered BOD or COD).
3. The substrate concentration in the aeration tank equals the substrate concentration in the effluent after treatment.
4. Biological activity occurs only in the aeration tank.
5. No microorganisms are present in the influent wastewater.
6. The mean cell residence time is calculated based on the biomass in the aeration tank.
7. Excess activated sludge is wasted from the aeration tank rather than from the sludge recirculation line.
8. The aeration tank is complete mixing.

The aeration period (liquid detention time) is defined as

$$\theta = \frac{V}{Q} \tag{12.61}$$

where $\theta$ = aeration period, time

The mean cell residence time (sludge age) is the biomass (MLVSS) in the aeration tank divided by the sum of the biomasses in the waste sludge and effluent.

$$\theta_c = \frac{VX}{Q_w X + (Q - Q_w)X_e} \tag{12.62}$$

where $\theta_c$ = mean cell residence time, time

The definition of specific growth rate $\mu$ from Eq. 12.24 is the rate of growth per unit of biomass (time$^{-1}$). The inverse of $\mu$ is the biomass divided by the rate of growth, $X/(dX/dt)_g$. Therefore, under steady-state conditions the mean cell residence time is

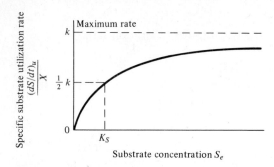

**Figure 12.47** Specific substrate utilization rate versus substrate concentration surrounding the microorganisms based on Eq. 12.66.

$$\theta_c = \frac{X}{(dX/dt)_g} = \frac{1}{\mu} \tag{12.63}$$

A mass balance for biomass around the entire activated-sludge system shown in Fig. 12.46 is as follows:

$$\begin{pmatrix} \text{net rate of change} \\ \text{of biomass} \\ \text{in system} \end{pmatrix} = \begin{pmatrix} \text{net rate of growth} \\ \text{of biomass} \\ \text{in system} \end{pmatrix} - \begin{pmatrix} \text{rate of loss} \\ \text{of biomass} \\ \text{from system} \end{pmatrix}$$

Using the notation in Fig. 12.46,

$$\left(\frac{dX}{dt}\right) V = \left(\frac{dX}{dt}\right)_g V - [Q_w X + (Q - Q_w)X_e] \tag{12.64}$$

At steady state, the rate of biomass growth equals the rate of biomass loss; hence, $(dX/dt)V = 0$. Setting Eq. 12.64 equal to zero and substituting the endogenous rate of growth from Eq. 12.32 for $(dX/dt)_g$ yields

$$\frac{1}{\theta_c} = Y\frac{(dS/dt)_u}{X} - k_d \tag{12.65}$$

Lawrence and McCarty [16] related the rate of substrate utilization both to the concentration of microorganisms in the aeration tank and to the concentration of substrate surrounding the organsims. The equation is

$$\left(\frac{dS}{dt}\right)_u = k\frac{XS_e}{K_S + S_e} \tag{12.66}$$

where   $k$ = maximum rate of substrate utilization per unit mass of biomass, time$^{-1}$
   $X$ = concentration of biomass, mass/unit volume
   $S$ = concentration of substrate surrounding the microorganisms, mass/unit volume
   $K_S$ = saturation constant, equal to the substrate concentration when $(dS/dt)_u = k/2$, mass/unit volume.

Equation 12.66, graphed in Fig. 12.47, indicates that the functional relationship between substrate utilization rate and substrate concentration is continuous over the entire range of substrate concentrations. It is similar in form to Eq. 12.29 with $k$ replacing $\mu_m/Y$.

Substituting Eq. 12.66 into Eq. 12.65 incorportes the term $S_e$,

$$\frac{1}{\theta_c} = Y\frac{kS_e}{K_S + S_e} - k_d \tag{12.67}$$

Rearranging, the formula for $S_e$ (effluent substrate concentration) is

$$S_e = \frac{K_S(1 + k_d\theta_c)}{\theta_c(Yk - k_d) - 1} \tag{12.68}$$

The specific substrate utilization rate is defined as the substrate utilization rate divided by the concentration of biomass in the aeration tank. Hence,

$$U = \frac{(dS/dt)_u}{X} \tag{12.69}$$

where $\quad U$ = specific substrate utilization rate, time$^{-1}$

It can be calculated from experimental data using the following formula:

$$U = \frac{Q(S_0 - S_e)}{VX} = \frac{S_0 - S_e}{\theta X} \tag{12.70}$$

Substituting the mathematical expression for specific substrate utilization rate (Eq. 12.69) into Eq. 12.65 gives the relationship

$$\frac{1}{\theta_c} = YU - k_d \tag{12.71}$$

where $\quad \theta_c$ = mean cell residence time, time
$\qquad Y$ = growth yield, biomass increase-substrate metabolized, unitless
$\qquad U$ = specific substrate utilization rate, time$^{-1}$
$\qquad k_d$ = microbial decay coefficient, time$^{-1}$

A mass balance for substrate entering and leaving the aeration tank as diagrammed in Fig. 12.46 is

$$\begin{pmatrix} \text{net rate of change} \\ \text{of substrate} \\ \text{in aeration tank} \end{pmatrix} = \begin{pmatrix} \text{rate of substrate} \\ \text{entering} \\ \text{aeration tank} \end{pmatrix} - \begin{pmatrix} \text{rate of substrate} \\ \text{utilization} \\ \text{in aeration tank} \end{pmatrix}$$

$$- \begin{pmatrix} \text{rate of substrate} \\ \text{leaving} \\ \text{aeration tank} \end{pmatrix}$$

Using the notation in Fig. 12.46,

$$\left(\frac{dS}{dt}\right)V = QS_0 + RQS_e - \left(\frac{dS}{dt}\right)_u V - Q(1 + R)S_e \tag{12.72}$$

At steady state, the rate of substrate entering the aeration tank equals the rate of substrate removal; hence, $(dS/dt)V = 0$. Setting Eq. 12.72 equal to zero and solving for the substrate utilization rate results in

$$\left(\frac{dS}{dt}\right)_u = \frac{Q(S_0 - S_e)}{V} = \frac{S_0 - S_e}{\theta} \tag{12.73}$$

Dividing both sides by $X$, Eq. 12.73 becomes

$$\frac{(dS/dt)_u}{X} = \frac{Q(S_0 - S_e)}{XV} \tag{12.74}$$

Substituting this expression for specific substrate utilization into Eq. 12.65 and rearranging, the equation for $V$ (volume of the aeration tank) is

$$V = \frac{\theta_c YQ(S_0 - S_e)}{X(1 + k_d\theta_c)} \tag{12.75}$$

Equating the terms for $(dS/dt)_u$ from Eqs. 12.73 and 12.66, and dividing by $X$ yields

$$\frac{S_0 - S_e}{\theta X} = k\frac{S}{K_S + S} \tag{12.76}$$

By inverting and linearizing Eq. 12.76,

$$\frac{X\theta}{S_0 - S_e} = \left(\frac{K_S}{k}\right)\left(\frac{1}{S_e}\right) + \frac{1}{k} \tag{12.77}$$

Substituting Eq. 12.70 for the left side of Eq. 12.77 gives

$$\frac{1}{U} = \left(\frac{K_S}{k}\right)\left(\frac{1}{S_e}\right) + \frac{1}{k} \tag{12.78}$$

where 
$U$ = specific substrate utilization rate, time$^{-1}$
$K_S$ = saturation constant, mass/unit volume
$k$ = maximum rate of substrate utilization per unit mass of biomass, time$^{-1}$
$S_e$ = concentration of substrate in effluent, mass/unit volume

This presentation was limited to the kinetics model for completely mixed activated sludge. For a more detailed mathematical analysis of this system and discussions of models for other biological processes, the reader is referred to books by Benefield and Randall [17] and Grady and Lim [18].

## 12.25 LABORATORY DETERMINATION OF KINETIC CONSTANTS

The numerical values for the following kinetic constants must be determined by laboratory experiments before the model described in Section 12.24 can be used for design of completely mixed activated sludge:

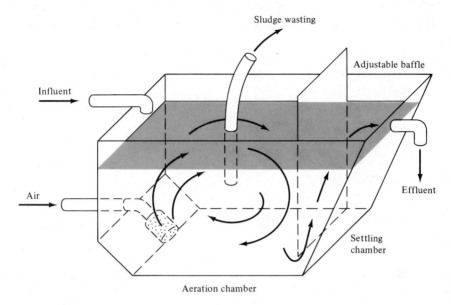

**Figure 12.48** Bench-scale, continuous-flow, activated-sludge unit for laboratory testing of wastewater to determine kinetic constants.

$Y$ = growth yield, mg VSS/mg BOD (or mg COD)
$k_d$ = microbial decay coefficient, d$^{-1}$
$K_S$ = saturation constant, mg/l of BOD (or COD)
$k$ = maximum rate of substrate utilization per unit mass of biomass, d$^{-1}$

For design, additional required data include settling characteristics of the activated sludge and oxygen uptake rates during aeration of the wastewater.

A bench-scale activated-sludge unit as illustrated in Fig. 12.48 is commonly used for laboratory evaluations. Operating conditions are the same as those assumed in the derivation of the mass balance equations, which are listed at the beginning of Section 12.24. Essential for satisfactory results are a continuous influent flowrate with a constant substrate concentration and complete mixing in the aeration tank. To collect sufficient data, the unit is operated at several mean cell residence times in the range of 3–20 days. Wasting the same quantity of suspended solids from the aeration tank maintains a constant $\theta_c$, concentration of MLVSS, and F/M ratio. Ideally, the waste activated sludge is pumped out continuously; however, withdrawing batches of mixed liquor from the aeration tank at uniform time intervals may be practiced. Settleability of the activated sludge is tested by withdrawing a sample of the aerating mixed liquor and placing it in a settleometer; it may then be replaced or discarded as waste sludge. The length of time required for a test period, after steady-state conditions are established, is at least twice the mean cell residence time with four residence times preferred.

Temperature, pH, and dissolved oxygen concentration are held constant throughout the series of test runs.

Laboratory tests and operating conditions are generally recorded daily. The data required for calculating the operating parameters are as follows:

$$Q = \text{rate of influent flow, l/d}$$
$$Q_w = \text{rate of sludge wasting, l/d}$$
$$Q - Q_w = \text{rate of effluent flow, l/d}$$
$$V = \text{volume of aeration tank, l}$$
$$X = \text{concentration of MLVSS in the aeration tank, mg/l}$$
$$X_e = \text{concentration of VSS in effluent, mg/l}$$
$$S_0 = \text{soluble BOD (or COD) in influent, mg/l}$$
$$S_e = \text{soluble BOD (or COD) in effluent, mg/l}$$

From these data, the aeration period $\theta$ is calculated using Eq. 12.61, the specific substrate utilization rate $U$ is calculated from Eq. 12.70, and the mean cell residence time $\theta_c$ from Eq. 12.62.

Values of $1/\theta_c$ and $U$ for each test period are plotted as shown in Fig. 12.49. Based on Eq. 12.71, the slope of the line is equal to the growth yield $Y$ and the intercept with vertical axis is the microbial decay coefficient $k_d$.

A plot of $1/U$ and $1/S_e$ for each test period is shown in Fig. 12.50. Based on Eq. 12.78, the slope of the line is equal to the saturation constant over the maximum rate of substrate utilization per unit mass of biomass $K_S/k$, and the intercept with the vertical axis is equal to $1/k$.

The general range of values for kinetic constants for completely mixed activated sludge treating municipal wastewaters is listed in Table 12.5.

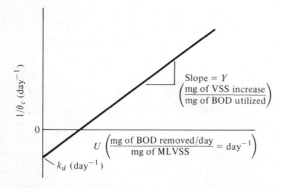

**Figure 12.49** Plot of the inverse of the mean cell residence time versus the specific substrate utilization rate to determine $Y$ and $k_d$ based on Eq. 12.71.

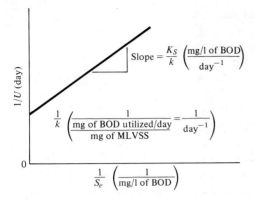

**Figure 12.50** Plot of the inverse of the specific substrate utilization rate versus the inverse of concentration of substrate in the effluent to determine $k$ and $K_S$ based on Eq. 12.78.

**TABLE 12.5**   GENERAL RANGES OF MAGNITUDE FOR KINETIC CONSTANTS FOR COMPLETELY MIXED ACTIVATED-SLUDGE PROCESSES TREATING MUNICIPAL WASTEWATER AT APPROXIMATELY 20°C

| Constant | Units | Range |
|---|---|---|
| $Y$ | mg VSS/mg BOD | 0.4–0.8 |
| $Y$ | mg VSS/mg COD | 0.3–0.4 |
| $k_d$ | $d^{-1}$ | 0.04–0.08 |
| $K_S$ | mg/l of BOD | 25–100 |
| $K_S$ | mg/l of COD | 25–100 |
| $k$ | $d^{-1}$ | 4–8 |

■ **EXAMPLE 12.13**

A municipal wastewater was tested to determine the kinetic constants using a laboratory apparatus similar to the bench-scale unit shown in Fig. 12.48. The volume of the aeration chamber was 10 l. Wastewater feed was established at a constant rate of 34.3 l/d to provide a 7.0-hr aeration period for all of the test runs. A measured volume of sludge was wasted once a day from the tank. Determine the values for $Y$, $k_d$, $k$, and $K_S$ from the following laboratory data.

| $Q$ (l/d) | $S_0$ (mg/l) | $Q_w$ (l/d) | $X$ (mg/l) | $S_e$ (mg/l) | $X_e$ (mg/l) |
|------|------|------|------|------|------|
| 34.3 | 126 | 0.25 | 1730 | 5.2 | 9.4 |
| 34.3 | 126 | 0.35 | 1500 | 7.3 | 8.0 |
| 34.3 | 126 | 0.80 | 968 | 10.5 | 8.4 |
| 34.3 | 126 | 0.90 | 848 | 11.5 | 7.9 |

**Solution.** The following calculations are for the first test period. Using Eq. 12.62,

$$\theta_c = \frac{101 \times 1730 \text{ mg/l}}{0.25 \text{ l/d} \times 1730 \text{ mg/l} + (34.3 \text{ l/d} - 0.25 \text{ l/d}) 9.4 \text{ mg/l}} = 23 \text{ d}$$

The biomass growth rate per liter of aeration tank volume is

$$\left(\frac{dX}{dt}\right)_g = \frac{Q_w X + (Q - Q_w)X_e}{V}$$

$$\left(\frac{dX}{dt}\right)_g = \frac{0.25 \text{ l/d} \times 1730 \text{ mg/l} + (34.3 \text{ l/d} - 0.25 \text{ l/d}) 9.4 \text{ mg/l}}{101} = 75 \text{ mg/l·d of VSS}$$

Checking the $\theta_c$ by Eq. 12.63 gives

$$\theta_c = \frac{1730 \text{ mg/l}}{75 \text{ mg/l}} = 23 \text{ d}$$

The soluble BOD utilization rate using Eq. 12.73 is

$$\left(\frac{dS}{dt}\right)_u = \frac{(34.3 \text{ l/d})(126 \text{ mg/l} - 5.2 \text{ mg/l})}{101} = 414 \text{ mg/l·d}$$

The specific substrate utilization rate as defined by Eq. 12.69 is

$$U = \frac{414 \text{ mg/l·d}}{1730 \text{ mg/l}} = 0.24 \text{ d}^{-1}$$

The following values are calculated for plotting the data:

$$\frac{1}{\theta_c} = \frac{1}{23 \text{ d}} = 0.043 \text{ d}^{-1}$$

$$\frac{1}{U} = \frac{1}{0.24 \text{ d}^{-1}} = 4.2 \text{ d}$$

$$\frac{1}{S_e} = \frac{1}{5.2 \text{ mg/l}} = 0.19 \text{ (mg/l)}^{-1}$$

The calculated data for all of the test runs are listed in Table 12.6.

The plot in Fig. 12.51 is to determine the values of $Y$ and $k_d$. From the slope of the line, the growth yield $Y = 0.35$ mg VSS/mg BOD. The intercept on the vertical axis is a microbial decay coefficient of $k_d = 0.04$ d$^{-1}$.

The plot in Fig. 12.52 is to determine the values of $k$ and $K_S$. The intercept on

**TABLE 12.6    DATA CALCULATED FOR EXAMPLE 12.13 AND PLOTTED IN FIG. 12.51 AND 12.52 TO DETERMINE KINETIC CONSTANTS**

| $\theta_c$ (d) | $\left(\dfrac{dX}{dt}\right)_g$ (mg/l·d) | $\left(\dfrac{dS}{dt}\right)_u$ (mg/l·d) | $U$ (d$^{-1}$) | $1/\theta_c$ (d$^{-1}$) | $1/U$ (d) | $1/S_e$ [(mg/l)$^{-1}$] |
|---|---|---|---|---|---|---|
| 23 | 75 | 414 | 0.24 | 0.0444 | 4.2 | 0.19 |
| 19 | 80 | 407 | 0.27 | 0.053 | 3.7 | 0.14 |
| 9.2 | 106 | 396 | 0.41 | 0.11 | 2.4 | 0.095 |
| 8.3 | 103 | 392 | 0.46 | 0.12 | 2.2 | 0.087 |

the vertical axis is at $1/k = 0.25$ d; therefore, the maximum rate of substrate utilization per unit mass of biomass $k$ is 4.0 d$^{-1}$. The slope of the line $K_S/k$ is 22.5; hence,

$$K_S = 22.5 \text{ d·mg/l} \times 4.0 \text{ d}^{-1} = 90 \text{ mg/l of BOD} \qquad \blacksquare$$

## 12.26 APPLICATION OF THE KINETICS MODEL IN PROCESS DESIGN

The equations developed in Section 12.24 can be applied in the design of a completely mixed activated-sludge process based on kinetic constants determined by laboratory testing. Since the model is based on steady-state conditions that do not exist in full-scale systems, the selection of design parameters such as the mean cell residence time must account for the diurnal and random variations in wastewater loadings. Allowances must also be made for characteristics of the actual wastewater not taken into consideration in the theoretical equations. Rather than consisting solely of a soluble sub-

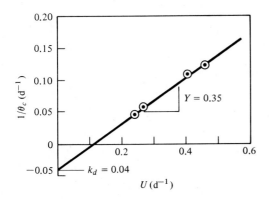

**Figure 12.51** Plot of $1/\theta_c$ versus $U$ to determine $Y$ and $k_d$ for Example 12.13.

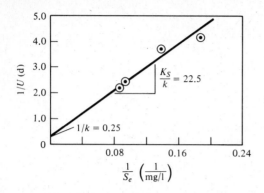

**Figure 12.52** Plot of $1/U$ versus $1/S_e$ to determine $k$ and $K_S$ for Example 12.13.

strate, municipal wastewater contains suspended solids and an abundance of biological organisms. For example, the BOD of a treatment plant effluent includes oxygen demand from both the soluble organic matter and the volatile suspended solids. During the laboratory evaluation of wastewater treatability, both total BOD and filtered BOD analyses can be performed on the effluent from the bench-scale unit to correlate total and soluble substrate. Also, an evaluation of sludge settleability is necessary to establish design criteria for the final clarifier to ensure good gravity separation of the biological suspended solids.

The first step in process design is to select the desired concentration of effluent soluble BOD based on the allowable total effluent BOD and the anticipated performance in final clarification. Soluble BOD removal efficiency of the system is calculated by the following formula:

$$E = \frac{(S_0 - S_e)100}{S_0} \tag{12.79}$$

where  $E$ = efficiency of soluble BOD removal, %
$S_0$ = influent soluble BOD concentration, mg/l
$S_e$ = effluent soluble BOD concentration, mg/l

The recommended loading criterion for completely mixed activated sludge is the mean cell residence time defined by Eq. 12.62:

$$\theta_c = \frac{VX}{Q_w X + (Q - Q_w)X_e}$$

where  $\theta_c$ = mean cell residence time, d
$V$ = volume of aeration tank, m³
$Q$ = influent wastewater flow, m³/d
$Q_w$ = rate of excess sludge wasting, m³/d
$X$ = concentration of MLVSS in aeration tank, mg/l
$X_e$ = concentration of VSS in effluent, mg/l

The high-rate activated-sludge process designed for a short $\theta_c$ of 3–5 d has a low substrate removal efficiency and is likely to exceed the effluent standards of 30 mg/l of total BOD and 30 mg/l suspended solids. Under a conventional loading rate, $\theta_c$ is in the range of 5–15 d and provides a process efficiency resulting in an effluent of satisfactory quality. The extended aeration process that operates at solids retention times greater than 15 d is usually employed only at small treatment plants. The advantages of this process are an ability to absorb wide load variations and the low production of waste sludge solids. Thus, selection of the mean cell residence time in design takes into consideration the factors of process efficiency, treatment reliability, and excess sludge production.

Another common loading criterion is the food/microorganism ratio (F/M), which is defined for the kinetics model as

$$F/M = \frac{QS_0}{VX} = \frac{S_0}{\theta X} \tag{12.80}$$

where  F/M = food/microorganism ratio, g/d of soluble BOD applied per g of MLVSS in the aeration tank

Rearranging Eq. 12.70 allows one to express the specific substrate utilization rate as

$$U = \frac{QS_0}{VX} \times \frac{S_0 - S_e}{S_0} \tag{12.81}$$

Then substituting the appropriate terms from Eq. 12.79 and 12.80 gives a relationship between the food/microorganism ratio and the specific substrate utilization rate:

$$U = \frac{(F/M)E}{100} \tag{12.82}$$

Replacing $U$ in Eq. 12.71 with Eq. 12.82 relates $\theta_c$ and F/M such that

$$\frac{1}{\theta_c} = \frac{Y(F/M)E}{100} - k_d \tag{12.83}$$

where  $\theta_c$ = mean cell residence time, d
  $Y$ = growth yield, unitless
  F/M = food/microorganism ratio, g/d of soluble BOD per g of MLVSS
  $E$ = soluble BOD removal, %
  $k_d$ = microbial decay coefficient, $d^{-1}$

After establishing the desired effluent quality $S_e$ and mean cell residence time $\theta_c$, or the F/M, the required aeration tank volume can be calculated from Eq. 12.75:

$$V = \frac{\theta_c YQ(S_0 - S_e)}{X(1 + k_d\theta_c)}$$

The choice of design flow $Q$ and soluble influent BOD $S_0$ depends on the anticipated flow and strength variations, as discussed in Sections 12.10 and 12.11. Values of $Y$ and $k_d$ are determined from laboratory testing. Thus, the only remaining design parameter is $X$, the mixed-liquor volatile suspended solids to be maintained in the aeration tank.

Selection of an MLVSS concentration is based on a number of considerations, the most important of which are (1) the ability of the final clarifier to provide gravity separation of the activated-sludge suspended solids and (2) the oxygen transfer capacity of the aeration system. At a low design MLVSS the aeration period is long, resulting in an extended process time. A concentration of MLVSS that is too high produces a high-rate process characterized by poor soluble BOD removal efficiency and high suspended-solids concentration in the effluent resulting from the limited aeration period and poor settleability of the microbial floc. In general, conventional secondary activated-sludge systems processing municipal wastewaters operate in the MLSS range of 1500–3000 mg/l with 70%–80% being volatile solids. In a typical design, therefore, the optimum MLVSS is within the range of 1200–2400 mg/l.

Waste sludge production in terms of volatile solids can be calculated based on the kinetics model. The observed growth yield $Y_{obs}$, as defined in Eq. 12.34, can be obtained by first substituting Eq. 12.32 for $(dX/dt)_g^{net}$ in the numerator and then replacing $(dS/dt)_u$ with $UX$ from Eq. 12.69:

$$Y_{obs} = \frac{YU - k_d}{U} \tag{12.84}$$

Substituting the relationship from Eq. 12.71 for $U$ in Eq. 12.84 gives

$$Y_{obs} = \frac{Y}{1 + \theta_c k_d} \tag{12.85}$$

where    $Y_{obs}$ = observed growth yield, g of MLVSS/g of soluble BOD

Using this equation, Lawrence and McCarty [16] expressed the production of excess biomass in the waste-activated sludge as

$$P_x = \frac{YQ(S_0 - S_e)}{1 + \theta_c k_d} \tag{12.86}$$

where    $P_x$ = volatile solids in waste sludge, g/d

### ■ EXAMPLE 12.14

A completely mixed activated-sludge process is being designed for a wastewater flow of 10,000 m³/d (2.64 mgd) using the kinetics equations. The influent BOD of 120 mg/l is essentially all soluble and the design effluent soluble BOD is 7 mg/l, which is based on a total effluent BOD of 20 mg/l. For sizing the aeration tank, the mean cell residence time is selected to be 10 d and the MLVSS 2000 mg/l. The kinetic constants from a bench-scale treatability study are as follows: $Y = 0.60$ mg VSS/mg BOD, $k_d = 0.06$ d$^{-1}$, $K_S = 60$ mg/l of BOD, and $k = 5.0$ d$^{-1}$.

*Solution.*    From Eq. 12.79 the required soluble BOD efficiency is

$$E = \frac{(120 - 7)100}{120} = 94\%$$

Rearranging Eq. 12.83 one obtains the food/microorganism ratio for $\theta_c = 10$ d and $E = 94\%$:

$$F/M = \frac{(1/\theta_c + k_d)100}{YE} = \frac{(1/10 + 0.06)100}{0.60 \times 94} = 0.28 \frac{\text{g of soluble BOD}}{\text{g of MLVSS}}$$

The volume of the aeration tank, based on Eq. 12.75, is

$$V = \frac{10 \times 0.60 \times 10,000(120 - 7)}{2000 (1 + 0.06 \times 10)} = 2100 \text{ m}^3 \text{ (74,800 ft}^3\text{)}$$

$$\theta = \frac{2100 \times 24}{10,000} = 5.0 \text{ h}$$

From Eq. 12.86 the excess volatile solids in the waste sludge are

$$P_x = \frac{0.60 \times 10,000(120 - 7)}{1 + 10 \times 0.06} = 420,000 \text{ g/d} = 420 \text{ kg/d}$$

## 12.27  OXYGEN TRANSFER AND OXYGENATION REQUIREMENTS

In activated sludge processes, oxygen is supplied to the microorganisms by dispersing air into the mixed liquor by either diffused-air or mechanical surface aeration. Diffused-air systems use a variety of diffusers such as those illustrated in Figs. 12.34 and 12.35. The two kinds of mechanical aerators, differentiated by the plane of rotation, are horizontal rotors (Fig. 12.40) and impellers mounted on vertical shafts (Figs. 12.36 and 12.37).

The commonly accepted oxygen transfer scheme is diagrammed in Fig. 12.53. Oxygen is dissolved in solution and then extracted from solution by

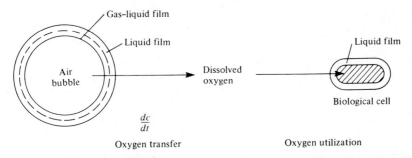

**Figure 12.53** Schematic diagram of oxygen transfer in activated sludge.

the biological cells. Direct oxygen transfer from bubble to cell is possible if the microorganisms are adsorbed on the bubble surface. Bennett and Kempe [19] demonstrated direct oxygen transfer in a laboratory fermenter using a culture of *Pseudomonas ovalis* converting glucose to gluconic acid. The extent of direct oxygen transfer in activated-sludge systems is not known; however, it is generally felt to be secondary to oxygen transfer through the intermediate dissolved-oxygen phase.

The rate of oxygen transfer from air bubbles to dissolved oxygen in an aeration tank is expressed as follows:

$$\frac{dc}{dt} = \alpha K_L a(\beta C_s - C_t) \tag{12.87}$$

where $dc/dt$ = rate of oxygen transfer, mg/l/hr
$\alpha$ = oxygen transfer coefficient of the wastewater
$\beta$ = oxygen saturation coefficient of the wastewater
$K_L a$ = oxygen transfer coefficient, hr$^{-1}$
$C_s$ = oxygen concentration at saturation, mg/l
$C_t$ = oxygen concentration in the liquid, mg/l
$\beta C_s - C_t$ = dissolved-oxygen deficit, mg/l

Equation 12.87 without the $\alpha$ and $\beta$ coefficients applies to clean water. The factors $\alpha$ and $\beta$ depend on the characteristics of the wastewater being aerated, primarily the concentration of dissolved solids; $K_L a$ depends on the temperature and the aeration system features, such as the type of diffuser, depth of aerator, type of mixer, and tank geometry. In general, the rate of oxygen transfer increases with decreasing bubble size, longer contact time, and added turbulence. Methods for determining the coefficients $K_L a$, $\alpha$, and $\beta$ are discussed in Section 12.28.

The rate of dissolved-oxygen utilization by microorganisms in an activated-sludge system can be determined by placing a sample of mixed liquor in a closed container and measuring the dissolved-oxygen depletion with respect to time. The slope of the resultant curve $r$ is the oxygen utilization rate. Figure 12.54 is a dissolved-oxygen depletion curve for a mixed liquor from a high-rate activated-sludge aeration tank. The $r$ value depends on the microorganisms' ability to metabolize the waste organics based on such factors as the food/microorganism ratio, mixing conditions, and temperature. A general range for $r$ in the mixed liquor of conventional and high-rate activated-sludge systems is 30–100 mg/l/hr.

Under steady-state conditions of oxygen transfer in an activated-sludge system, the rate of oxygen transfer to dissolved oxygen ($dc/dt$) is equal to the rate of oxygen utilization ($r$). Substituting $r$ in Eq. 12.87 for $dc/dt$ and rearranging yields the following relationship:

$$\alpha K_L a = \frac{r}{\beta C_s - C_t} \tag{12.88}$$

where $r$ = oxygen utilization rate by microorganisms in activated sludge, mg/l/hr

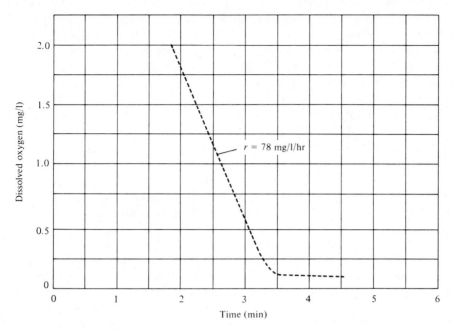

**Figure 12.54** Oxygen utilization curve for a sample of mixed liquor from a high-rate activated-sludge aeration basin.

The rate of aerobic microbial metabolism is independent of the dissolved-oxygen concentration above a critical (minimum) value. Below the critical value, the rate is reduced by the limitation of oxygen required for respiration. Critical dissolved-oxygen concentrations reported in the literature for acti-vated-sludge systems range from 0.2 to 2.0 mg/l, depending on the type of activated-sludge process and characteristics of the wastewater. The most frequently referenced critical dissolved-oxygen value for conventional and high-rate aeration basins is 0.5 mg/l.

The aerator power required to satisfy microbial oxygen demand and to provide adequate mixing in an aeration tank depends on the type of activated-sludge process, BOD loading, and oxygen transfer efficiency of the aeration equipment. In the design of any activated-sludge system, the power require-ments should be based on proved performance of the aeration equipment. The capacity of the aeration equipment must furnish sufficient air to meet peak BOD loads without the dissolved-oxygen concentration dropping below the critical level for aerobic metabolism.

Aeration systems are compared on the basis of mass of gaseous oxygen transferred to dissolved oxygen per unit of energy expended, pounds of oxygen per horsepower-hour (kilogram per kilowatt-hour). Oxygen transfer efficiency is expressed as the percentage of the mass of oxygen dissolved in the water relative to the applied mass of gaseous oxygen. For the purpose of comparison, the values specified for efficiency are based on operation in clean water with zero dissolved-oxygen concentration and standard conditions of 20°C and 1 atm pressure. Table 12.7 lists oxygen transfer efficiencies and oxygen transfer rates for different kinds of aeration systems. Of course, the selection of an aeration system in process design must also consider other factors, including the flexibility and reliability of operation, effective mixing, and the maintenance of equipment. An economic analysis encompasses capital, operation, and maintenance costs.

The amount of dissolved oxygen needed for treatment of a wastewater depends on the carbonaceous and nitrogenous oxygen demands that are satisfied. For biological oxidation of carbonaceous matter, the oxygen requirement varies from approximately 0.8 to 1.6 times the BOD of the applied wastewater for a corresponding food/microorganism loading range of extended aeration to high-rate aeration. The oxygen requirement for nitrification is 4.6 times the ammonia–nitrogen oxidized to nitrate.

The *Recommended Standards for Sewage Works, Great Lakes–Upper Mississippi River Board of State Public Health & Environmental Managers* [20] considers the following as minimum normal air requirements for diffused-air systems: conventional, step aeration, and contact stabilization 1500 ft³ of air applied per lb of BOD aeration tank load; modified or high-rate 400–1500 ft³/lb of BOD load; and extended aeration 2000 ft³/lb of BOD load. These demands assume that the aeration equipment is capable of transferring at

**TABLE 12.7** OXYGEN TRANSFER DATA FOR AIR AERATION
SYSTEMS IN CLEAN WATER AT
15 FT SUBMERGENCE

| System | Oxygen transfer efficiency (%) | Oxygen transfer rate (lb/hp·hr)[a] |
|---|---|---|
| Fine bubble diffusers, total floor coverage | 20–32[b] | 4.0–6.5 |
| Fine bubble diffusers, side-wall installation | 11–15[b] | 2.2–3.0 |
| Jet aerators (fine bubble) | 22–27[b] | 4.0–5.0 |
| Static aerators (medium-size bubble) | 12–14[b] | 2.3–2.8 |
| Mechanical surface aerators | — | 2.5–3.5[c] |
| Coarse-bubble diffusers, wide-band pattern | 6–8[c] | 1.2–1.6 |
| Coarse-bubble diffusers, narrow-band pattern | 4–6[c] | 0.8–1.2 |

*Source: Proceedings, Workshop Toward an Oxygen Transfer Standard,* Environmental Protection Agency, EPA 600/9-78-021 (April 1979): 13.

[a] 1.0 lb/hp·hr = 0.61 kg/kW·h.

[b] From manufacturers' bulletins and technical reports.

[c] Common ranges for variations of these systems.

least 1.0 lb of oxygen to the mixed liquor per pound of BOD aeration tank loading. In any case, aeration equipment shall be capable of maintaining a minimum of 2.0 mg/l of dissolved oxygen in the mixed liquor at all times and ensuring thorough mixing of the mixed liquor.

■ **EXAMPLE 12.15**

The following data were collected during field evaluation of a completely mixed activated-sludge secondary treating municipal wastewater. The aeration basin, with a diameter of 80 ft and a liquid depth of 17 ft, was mixed with four turbine mixers mounted above air sparge rings. Twenty-four-hour composite BOD analyses were run on the aeration basin influent, final clarifier effluent, and waste-activated sludge. The oxygen-utilization rate in the aeration basin was measured each hour throughout the 24-hr sampling period and individual values averaged for oxygen utilization rate of the mixed liquor.

$$\text{influent wastewater flow} = 6.52 \text{ mgd}$$
$$\text{waste-activated sludge} = 15,000 \text{ gpd}$$
$$\text{influent BOD} = 125 \text{ mg/l}$$
$$\text{effluent BOD} = 18 \text{ mg/l}$$
$$\text{waste sludge BOD} = 5300 \text{ mg/l}$$
$$\text{air supplied (20°C and 760 mm)} = 1650 \text{ cfm}$$
$$\text{minimum DO in mixed liquor} = 0.8 \text{ mg/l}$$
$$\text{average DO in mixed liquor} = 1.1 \text{ mg/l}$$
$$\text{temperature of mixed liquor} = 24°C$$
$$\text{oxygen utilization rate of mixed liquor} = 74 \text{ mg/l/hr}$$
$$\text{beta factor of mixed liquor} = 0.9$$

Use these data to calculate the following:

**a.** Pounds of BOD load.
**b.** Cubic feet of air applied per lb of BOD load.
**c.** Pounds of oxygen utilized per lb of BOD.
**d.** Oxygen transfer efficiency.
**e.** $\alpha K_L a$.

*Solution*

**a.**   $\text{lb of BOD load} = 6.52 \text{ mgd} \times 125 \text{ mg/l} \times 8.34 = 6800 \text{ lb/day}$

$\text{volume aeration tank} = \pi(40)^2 17 = 85,500 \text{ ft}^3$

$\text{BOD loading} = 79.5 \text{ lb of BOD/1000 ft}^3\text{/day}$

**b.**   $\text{air applied} = 1650 \dfrac{\text{ft}^3}{\text{min}} \times 1440 \dfrac{\text{min}}{\text{day}} = 2,380,000 \text{ ft}^3$

$\dfrac{\text{ft}^3 \text{ of air applied}}{\text{lb of BOD load}} = \dfrac{2,380,000}{6800} = 350 \dfrac{\text{ft}^3}{\text{lb of BOD}}$

c.    lb of oxygen utilized = $r$ × volume of aeration tank × time

$$= 74\frac{\text{mg/l}}{\text{hr}} \times 28.3\frac{1}{\text{ft}^3} \times 85{,}500 \text{ ft}^3$$

$$\times \frac{\text{lb}}{453{,}600 \text{ mg}} \times 24\frac{\text{hr}}{\text{day}}$$

$$= 9420 \text{ lb/day}$$

lb of BOD satisfied = lb of BOD removed − lb of BOD wasted

$$= 6.52(125 - 18)8.34 - 0.015 \times 5300 \times 8.34$$

$$= 5190 \text{ lb/day}$$

$$\frac{\text{lb of oxygen utilized}}{\text{lb of BOD satisfied}} = \frac{9420}{5190} = 1.82$$

$$\frac{\text{lb of oxygen utilized}}{\text{lb of BOD applied}} = \frac{9420}{6800} = 1.39$$

d.    lb of oxygen applied $= 2{,}380{,}000\frac{\text{ft}^3}{\text{day}} \times 0.0174\frac{\text{lb of oxygen}}{\text{ft}^3}$

$$= 41{,}400 \text{ lb/day}$$

oxygen transfer efficiency $= \frac{9420}{41{,}400} \times 100 = 22.8\%$

e. From Eq. 12.88,

$$\alpha K_L a \text{ at } 24°\text{C} = \frac{74}{0.9 \times 8.5 - 1.1} = 11 \text{ hr}^{-1}$$    ∎

## 12.28  DETERMINATION OF OXYGEN TRANSFER COEFFICIENTS

In order to apply Eq. 12.87 in calculating the mass transfer of oxygen in the design of an activated-sludge process, the coefficients $K_L a$, $\alpha$, and $\beta$ must be experimentally determined.

The rate of oxygen transfer in clean water is defined as

$$\frac{dc}{dt} = K_L a(C_s - C_t) \tag{12.89}$$

where    $dc/dt$ = rate of oxygen transfer, mg/l/hr
  $\quad\quad K_L a$ = oxygen transfer coefficient, hr$^{-1}$
  $\quad\quad C_s$ = oxygen concentration at saturation, mg/l
  $\quad\quad C_t$ = oxygen concentration in liquid, mg/l

The rate of oxygen dissolution is proportional to the dissolved-oxygen deficit

$(C_s - C_t)$ and the area of the air–water interface per unit volume of water. $K_L a$ is the overall coefficient that incorporates the interfacial area $a$ of diffusion and the liquid film coefficient $K_L$. The value of $K_L a$ depends on the hydrodynamics and turbulence at the interface between the air bubbles and the liquid; hence, it depends on the aeration system, geometry of the aeration tank, liquid characteristics, and temperature.

The efficiency of an aeration system to transfer oxygen is measured by conducting a non-steady-state test on a full-scale aeration basin or test tank using clean water [21, 22]. The clean aeration tank is filled with tap water at a temperature as close to 20°C as possible. Next, a cobalt chloride catalyst is dissolved in a small amount of warm water and added to the aeration tank; the concentration must be high enough to assure catalyzation of all of the sodium sulfite added. After operating the aerator for 20–30 min to achieve a steady-state mixing condition, the sodium sulfite is added to deoxygenate the water in the aeration tank as follows:

$$2Na_2SO_3 + O_2 \xrightarrow{\text{cobalt}} 2Na_2SO_4 \tag{12.90}$$

The sulfite addition is in excess of the theoretical requirement (7.88 mg/l of pure sodium sulfite per 1.0 mg/l of DO concentration) to allow a time lag for mixing before the dissolved oxygen starts to rise above zero. Simultaneously, sampling is initiated at several points in the aeration tank when the DO concentration begins to rise from zero and is continued at 1–3-min intervals, or at approximately every 1.0 mg/l increase in dissolved oxygen. At least six samples are collected at each point between the levels of 10% and 80% DO saturation. Water for sampling is continuously withdrawn by submersible pumps with sufficient capacity to limit the detention time between pump and sample outlet to 5–10 sec. Although oxygen concentrations are monitored and recorded by DO probes and meters, the standard test for dissolved oxygen is by the Winkler titration method. Three replicate tests are normally conducted to determine the aeration efficiency for each operating condition.

Dissolved-oxygen data from each sampling point are plotted to determine the $K_L a$ value based on the following relationship, derived from Eq. 12.89:

$$K_L a = \ln\left(\frac{C_s - C_2}{C_s - C_1}\right) \Big/ (t_2 - t_1) \tag{12.91}$$

Nonparallel slopes of the plots from different sampling points indicate poor mixing, and $K_L a$ values that differ significantly from the others are discarded. The saturation concentration $C_s$ is the theoretical value at the temperature of the water during the test (Table A.10). In the case of diffused-air systems, a correction factor for submergence of the bubbles is included in the pressure correction; this is normally taken as the pressure at one-half the depth of submergence of the diffusers. The $K_L a$ at test temperature in degrees Celsius is corrected to 20°C by the relationship

$$(K_L a)_{20} = (K_L a)_T \Theta^{20-T} \tag{12.92}$$

where $\Theta$ is commonly assumed to be 1.024. (The observed range is 1.01–1.05.)

The mass of oxygen dissolved in the water contained in a test tank per unit time at standard conditions (20°C, 1 atm pressure, and zero DO) is calculated as

$$N = 10^{-6}(K_L a)_{20}(C_s)_{20} W \tag{12.93}$$

where     $N$ = rate of oxygen dissolution, lb/hr
$(K_L a)_{20}$ = oxygen transfer coefficient at 20°C, $hr^{-1}$
$(C_s)_{20}$ = oxygen concentration at saturation at 20°C, mg/l
$W$ = weight of water in the test tank, lb
$10^{-6} \sim 1$ mg/l $\sim 1$ mg/1,000,000 mg

Oxygen transfer efficiency $E$ in a diffused-aeration system is computed by

$$E = \frac{N \times 100}{A} \tag{12.94}$$

where   $E$ = oxygen transfer efficiency, %
$A$ = oxygen applied (standard conditions), lb/hr

Determination of the applied oxygen requires accurate measurement of the air flowrate and adjustment of the observed rate to standard conditions of 20°C and 1 atm of pressure.

The oxygen transfer rate can be calculated for both diffused-air and mechanical aeration systems by the relationship

$$R_0 = \frac{N}{P} \tag{12.95}$$

where   $R_0$ = rate of oxygen transfer at standard conditions (20°C, 1 atm pressure, and zero DO), lb/hp·hr
$P$ = power input, hp

The rate of oxygen transfer to a wastewater requires determining the alpha and beta coefficients in Eq. 12.87. The alpha coefficient is defined as the ratio of the oxygen transfer coefficient in wastewater to that in clean water,

$$\alpha = \frac{K_L a \text{ in wastewater}}{K_L a \text{ in clean water}} \tag{12.96}$$

The value of $\alpha$ is influenced by many conditions related to both the characteristics of the wastewater (temperature, soluble BOD, and concentration of suspended solids) and the aeration equipment (type of diffuser or mechanical aerator, mixing intensity, and aeration tank configuration). The magnitude can even change between the influent and effluent ends of a plug-flow aeration tank resulting from stabilization of the wastewater. Even though the most

**TABLE 12.8**  SAMPLING DATA AND CALCULATED
VALUES FOR DETERMINATION OF THE
OXYGEN TRANSFER COEFFICIENT IN
EXAMPLE 12.16

| $t$ (min) | $C_s$ (mg/l) | $C_t$ (mg/l) | $C_s - C_t$ (mg/l) | $\ln(C_s - C_t)$ |
|-----------|--------------|--------------|--------------------|------------------|
| 1.2  | 10.4 | 2.3  | 8.1 | 2.1  |
| 2.8  | 10.4 | 4.6  | 5.8 | 1.8  |
| 5.0  | 10.4 | 6.7  | 3.7 | 1.3  |
| 8.0  | 10.4 | 8.2  | 2.2 | 0.8  |
| 12.0 | 10.4 | 9.4  | 1.0 | 0    |
| 16.0 | 10.4 | 10.0 | 0.4 | -0.9 |
| 22.0 | 10.4 | 10.2 | 0.2 | -1.6 |

reliable method of measuring $\alpha$ is under field design conditions, it is often determined using a bench-scale aeration tank. Different laboratory units are designed to simulate diffused, mechanical-surface, and submerged-turbine aeration systems. The procedure involves conducting tests for $K_La$ in the model aeration tank for both tap water and wastewater [21, 22]. Deoxygenation is performed by stripping the liquid in the tank with nitrogen gas. Conducting tests to determine alpha requires considerable expertise in oxygen transfer processes and laboratory techiques.

Alpha coefficients for municipal wastewater are generally in the range of 0.7–0.9; nevertheless, fine-bubble diffusers can have a value as low as 0.4 and mechanical aerators as high as 1.1.

The beta coefficient in Eq. 12.87 is defined as the ratio of the DO saturation concentration in the wastewater to that in clean water,

$$\beta = \frac{\text{DO saturation concentration in wastewater}}{\text{DO saturation concentration in clean water}} \qquad (12.97)$$

The value of $\beta$ is influenced by wastewater constituents, including dissolved salts, organics, and gases. To determine the saturation concentration in a wastewater, a settled sample is aerated by vigorous hand mixing for several minutes in a half-full jar. The temperature and dissolved oxygen are then both measured, usually with a calibrated DO probe, for several minutes to ensure stable readings. Saturation for clean water is the theoretical value for the same temperature and is corrected for the barometric pressure (Table A.10). These two values are used in Eq. 12.97 to calculate $\beta$.

The magnitude of the beta coefficient for municipal wastewater typically equals 0.9 and is seldom less than 0.8.

### ■ EXAMPLE 12.16

A coarse-bubble, diffused aeration system was tested in a large tank to determine the oxygen transfer coefficient by the non-steady-state procedure in clean water. The diffusers were submerged 8.0 ft, and the atmospheric pressure on the day of the test was 720 mm Hg. Dissolved oxygen data from a sampling point at a depth of 4.0 ft are given in columns 1 and 3 of Table 12.8. The water temperature was 17°C.

The test tank was circular with a diameter of 20.0 ft. The air supply during the test was 310 cfm (20°C and 760 mm) and the power input was 15 hp.

Determine the value of $K_L a$ and, based on this value and the above operating data, calculate the oxygen transfer efficiency and oxygen transfer rate.

**Solution.**   The pressure at one-half the depth of submergence of the diffusers is

$$\frac{4.0\,\text{ft} \times 305\,\text{mm/ft}}{13.6\,\text{mm H}_2\text{O}/1.0\,\text{mm Hg}} = 90\,\text{mm Hg}$$

Therefore, the barometric pressure at mid-depth is equal to $720 + 90 = 810$ mm Hg. Using Table A.10,

$$C_s' = C_s \frac{P - p}{760 - p} = 9.7 \times \frac{810 - 15}{760 - 15} = 10.4\,\text{mg/l}$$

This value for the oxygen concentration at saturation (17°C and 4.0 ft submergence) is entered in column 2 of Table 12.8 and the $C_s - C_t$ and $\ln(C_s - C_t)$ calculated.

The $\ln(C_s - C_t)$ versus time data are plotted in Fig. 12.55. The slope of a straight line of best fit through the points between 20% and 80% DO saturation is $K_L a$, which is calculated using Eq. 12.91 as

$$K_L a = \frac{(2.0 - 0)60}{11.8 - 1.5} = 11.7\,\text{hr}^{-1}$$

Correcting $K_L a$ to 20°C by Eq. 12.92,

$$(K_L a)_{20} = 11.7(1.024)^{20-17} = 12.6\,\text{hr}^{-1}$$

The following parameters are calculated in order to use Eq. 12.93:

$$(C_s)_{20} = 9.2 \times \frac{810 - 18}{760 - 18} = 9.8\,\text{mg/l}$$

$$W = \pi(10)^2 \times 8.0 \times 62.4 = 157,000\,\text{lb}$$

Substituting into Eq. 12.93,

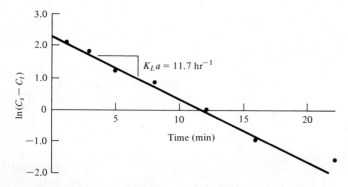

**Figure 12.55** Plot of data from Table 12.8 to determine $K_L a$ for Example 12.16.

$N = 10^{-6} \times 12.6 \times 9.8 \times 157{,}000 = 19.4 \, \text{lb/hr}$

Calculating oxygen transfer efficiency by Eq. 12.94,

$$E = \frac{19.4 \, \text{lb/hr} \times 100}{310 \, \text{ft}^3/\text{min} \times 60 \, \text{min/hr} \times 0.0174 \, \text{lb of oxygen/ft}^3} = \frac{19.4 \times 100}{323} = 6.0\%$$

The rate of oxygen transfer from Eq. 12.95 is

$$R_0 = \frac{19.4}{15} = 1.3 \, \text{lb/hp·hr} \qquad \blacksquare$$

## ■ EXAMPLE 12.17

A step-aeration activated-sludge process is being designed using the diffused-aeration system described in Example 12.16. The design criteria for the period of critical oxygen demand in the aeration basin are as follows:

$$\text{BOD loading} = 3.3 \, \text{lb BOD}/1000 \, \text{ft}^3/\text{hr}$$

$$\text{oxygen transfer requirement} = 1.0 \, \text{lb of oxygen}/1.0 \, \text{lb of BOD applied}$$

$$\text{temperature of mixed liquor} = 14°\text{C}$$

$$\text{minimum allowable DO} = 2.0 \, \text{mg/l}$$

$$\text{beta coefficient} = 0.70$$

$$\text{alpha coefficient} = 0.90$$

$$\text{oxygen transfer coefficient } (K_L a)_{20} = 12.0 \, \text{mg/l/hr}$$

$$\text{oxygen transfer efficiency } E = 6.0\%$$

$$\text{pressure at mid-depth of diffusers} = 810 \, \text{mm Hg}$$

Compare the rate of oxygen transfer $dc/dt$ to the rate of oxygen utilization $r$, and calculate the volume of standard air required per pound of BOD applied. (One cubic foot of air at standard temperature and pressure, 20°C and 760 mm, contains 0.0174 lb oxygen.)

**Solution.** Using Table A.10 for 14°C,

$$C_s = 10.4 \times \frac{810 - 12}{760 - 12} = 11.1 \, \text{mg/l}$$

correcting $(K_L a)_{20}$ for 14°C,

$$K_L a \text{ at } 14°\text{C} = (K_L a)_{20} 1.024^{T-20} = 12.0 \, (1.024)^{-6} = 10.4 \, \text{mg/l/hr}$$

Substituting into Eq. 12.87, the rate of oxygen transfer is

$$\frac{dc}{dt} = 0.70 \times 10.4(0.90 \times 11.1 - 2.0) = 58 \, \text{mg/l/hr}$$

Assuming 1.0 lb of oxygen utilized per 1.0 lb of BOD applied, the rate of dissolved-oxygen utilization is

$$r = \frac{3.3 \, \text{lb}}{1000 \, \text{ft}^3 \cdot \text{hr}} \times \frac{453{,}600 \, \text{mg}}{\text{lb}} \times \frac{\text{ft}^3}{28.31} = 53 \, \text{mg/l/hr}$$

The oxygen transfer rate of the aeration system is adequate since $dc/dt$ exceeds $r$. (Refer to Fig. 12.53.)

The efficiency of oxygen transfer is proportional to the rate of oxygen transfer; therefore,

$$E_{actual} = E\frac{\alpha(K_La)(\beta C_s - C_t)}{(K_La)_{20}(C_s)_{20}} \tag{12.98}$$

$$= 6.0 \times \frac{0.70 \times 10.4(0.90 \times 11.1 - 2.0)}{12.6 \times 9.8} = 2.8\%$$

The volume of standard air required at an actual oxygen transfer efficiency of 2.8% per pound of BOD load, assuming 1.0 lb of oxygen utilized per 1.0 lb of BOD applied, is

$$\frac{1.0 \text{ lb of oxygen}}{0.0174 \text{ lb of oxygen/ft}^3 \times 0.028} = 2000 \text{ ft}^3/\text{lb of BOD applied} \qquad \blacksquare$$

## 12.29 OPERATION AND CONTROL OF ACTIVATED-SLUDGE PROCESSES

Operation of an activated-sludge treatment plant is regulated by (1) the quantity of air supplied to the aeration basin, (2) the rate of activated-sludge recirculation, and (3) the amount of excess sludge withdrawn from the system. Sludge wasting is used to establish the desired concentration of MLSS, food/microorganism ratio, and sludge age.

Field observations for monitoring an aeration system are the rates of wastewater influent, excess sludge wasting, and sludge recirculation; the dissolved-oxygen concentration in the mixed liquor; and the depth of the sludge blanket in the final clarifier. Laboratory tests are used to determine influent and effluent BOD, the concentration of suspended solids in the return sludge, and the concentration of MLSS in the aeration tank. From these data, BOD loadings, the aeration period, the return sludge rate, and the BOD removal efficiency can be calculated. The final clarifier operation is observed by testing for the concentration of suspended solids in the effluent and calculating the overflow rate and solids loading.

An important measurement for operational control is the settleability of the mixed liquor as defined by the sludge volume index (SVI). The SVI is the volume in milliliters occupied by 1 g of suspended solids after 30 min of settling. It is computed as follows:

$$\text{SVI} = \frac{\text{sludge volume after settling (ml/l)} \times 1000}{\text{MLSS (mg/l)}} \tag{12.99}$$

In the standard settleability test, the sludge volume is determined by placing a sample of mixed liquor in the 1-liter graduated cylinder and reading the volume occupied by the settled solids after 30 min. The small diameter of a standard 1-liter laboratory container can slow the settling rate of the suspended solids by bridging across the walls and partial support from the settled solids underneath. To reduce this interference, either a large-diameter cylinder (a settleometer) or a stirred cylinder can be substituted for the standard graduated cylinder. The standard stirred settling apparatus (developed by the Water Research Centre, Great Britain) is a 10-cm tube, graduated with a vertical scale from 0 to 50 cm, fitted with a slow-speed stirrer that rotates at 0.25–1.0 rpm. This index is referred to as the stirred specific volume (SSV), which is expressed in milliliters per gram. In general, SVI or SSV in the range 50–150 indicates a good settling sludge. The MLSS is the concentration of suspended solids in the sample of mixed liquor before settling. The range is normally 1500–3000 mg/l in conventional and step aeration and 3000–5000 mg/l in high-rate completely mixed processes.

The settleability of the mixed liquor governs the rate of activated-sludge recirculation. If the rate of sludge return is less than the rate of accumulation of settled solids, the sludge blanket in the final clarifier gradually rises until suspended solids are carried out in the overflow. If the rate of sludge return exceeds the rate of accumulation of settled solids, clear supernatant is withdrawn with the sludge, diluting the solids content in the recirculated sludge. The settleability can be hypothetically related to the quantity and solids concentration in the return sludge, as depicted in Fig. 12.56, by assuming that the final clarifier responds identically to the laboratory container. This premise appears reasonable with respect to extended aeration and conventional processes, but in high-rate systems considerable deviation can exist between the sludge settleability in an actual clarifier and that measured in laboratory testing. Assuming ideal operation, Eq. 12.100 equates the ratio of return-sludge flow divided by the flow entering a final clarifier to the settled sludge volume over the volume of the laboratory container. This formula can be used to calculate the recirculation ratio $Q_R/Q$ (the rate of sludge return for a given influent flow) and sludge volume after settling:

$$\frac{Q_R}{Q + Q_R} = \frac{V_S}{V} \tag{12.100}$$

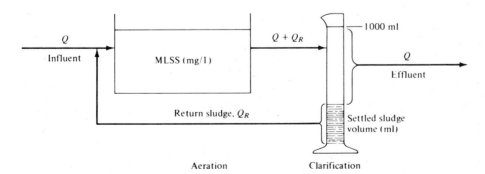

**Figure 12.56** Hypothetical relationship between the settled sludge volume in the SVI or SSV test and the quantity of return sludge in an activated-sludge process.

where    $Q$ = wastewater influent, mgd or gpm
$Q_R$ = return-sludge flow, mgd or gpm
$V_S$ = volume of settled sludge, ml/l
$V$ = volume of laboratory container, l

Assuming the ideal rate of return-sludge flow (Fig. 12.56), the concentration of suspended solids in the return sludge ($SS_R$) can be calculated from the SVI value:

$$SS_R = \frac{1000 \ (mg/g) \times 1000 \ (ml/l)}{SVI \ (ml/g)} = \frac{10^6}{SVI} \ (mg/l) \qquad (12.101)$$

For any recirculation rate, the concentration of suspended solids in the return sludge is calculated as follows, assuming no loss of suspended solids in the effluent:

$$SS_R = \frac{MLSS(Q + Q_R)}{Q_R} \qquad (12.102)$$

The degree of treatment achieved in an activated-sludge process depends directly on the settleability of the suspended solids in the final clarifier. If the biological floc agglomerate and settle rapidly by gravity, the overflow is a clear supernatant. Conversely, poorly flocculated particles (pin floc) and buoyant filamentous growths that do not separate by gravity contribute to BOD and suspended solids in the system effluent. Excessive carryover of floc resulting in inefficient operation is referred to as sludge bulking. This can be caused by any one or a combination of the biological, chemical, and physical factors listed in Table 12.9. If an activated-sludge process is not functioning properly, the loadings on the aeration tank and final clarifier are calculated and compared to established design criteria. Next, operational procedures are reviewed to ensure proper aeration, sludge recirculation, and sludge wasting. Special laboratory tests can be performed to determine detrimental chemical characteristics of the wastewater, such as a lack of nutrients or the presence of toxins. Microscopic examination of the activated sludge can reveal excessive filamentous growth [23].

■ **EXAMPLE 12.18**
The MLSS concentration in an aeration basin is 2200 mg/l and the sludge volume after 30 min of settling in a 1-1 graduated cylinder is 180 ml. Calculate the SVI, required return-sludge ratio, and suspended-solids concentration in the recirculated sludge.

*Solution.*    Using Eqs. 12.99–12.101,

$$SVI = \frac{180 \times 1000}{2200} = 82 \ ml/g$$

**TABLE 12.9**    FACTORS THAT CAN ADVERSELY AFFECT SETTLEABILITY
OF ACTIVATED SLUDGE

Biological factors

Species of dominant microorganisms (filamentous)
Ineffective biological flocculation
Denitrification in final clarifier (floating solids)
Excessive volumetric and food/microorganism loadings
Mixed-liquor suspended-solids concentration
Unsteady-state conditions (nonuniform feed rate and discontinuous wasting of excess activated sludge)

Chemical factors

| Lack of nutrients | Insufficient aeration |
| Presence of toxins | Low temperature |
| Kinds of organic matter | |

Physical factors

Excessive agitation during aeration resulting in shearing of floc
Ineffective final clarification: inadequate rate of return sludge, excessive overflow rate or solids loading, or hydraulic turbulence

$$Q_R/(Q + Q_R) = 180/1000; \text{ thus } Q_R/Q = 0.22.$$

$$SS_R = \frac{1,000,000}{82} = 12,000 \text{ mg/l } (1.2\%) \qquad \blacksquare$$

# Stabilization Ponds

Domestic wastewater can be effectively stabilized by the natural biological processes that occur in shallow ponds. Those treating raw wastewater are referred to as facultative ponds, lagoons, or oxidation ponds. Where small ponds are installed after secondary treatment, they are referred to as tertiary, maturation, or polishing ponds. Their purpose is to further reduce suspended solids, BOD, fecal microorganisms, and ammonia in the plant effluent.

Facultative ponds have a light BOD loading of 0.1–0.3 lb/1000 ft³/day, a normal operating water depth of 5 ft, and a long retention time of 50–150 days. Small ponds may be designed for complete retention with water loss only by evaporation. Tertiary polishing ponds generally have a retention time of only 10–15 days and are shallower with water depths of 2–3 ft for better mixing and sunlight penetration.

A wide variety of microscopic plants and animals find the environment a suitable habitat. Waste organics are metabolized by bacteria and saprobic

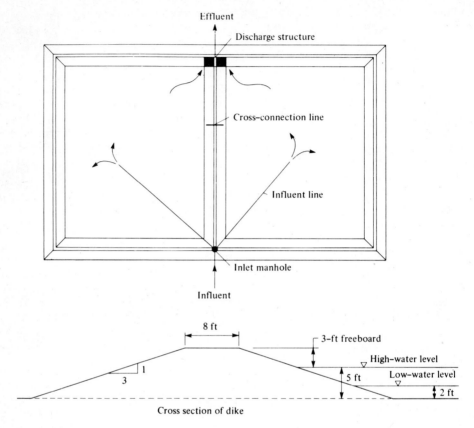

**Figure 12.57** Two-cell stabilization pond.

protozoans as primary feeders. Secondary feeders include protozoans and higher animal forms, such as rotifers and crustaceans. When the pond bottom is anaerobic, biological activity results in digestion of the settled solids. Nutrients released by bacteria are used by algae in photosynthesis. The overall process in a facultative stabilization pond (Fig. 12.14) is the sum of individual reactions of the bacteria, protozoans, and algae.

## 12.30  DESCRIPTION OF A FACULTATIVE POND

A stabilization pond is a flat-bottomed pond enclosed by an earth dike (Fig. 12.57). It can be round, square, or rectangular with a length not greater than three times the width. The operating liquid depth has a range of 2–5 ft with 3 ft of dike freeboard. A minimum of about 2 ft is required to prevent the growth of rooted aquatic plants. Operating depths greater than 5 ft can create odorous conditions because of anaerobiosis on the bottom.

Influent lines discharge near the center of the pond and the effluent usually overflows in a corner on the windward side to minimize short-circuiting. The overflow is generally a manhole or box structure with multiple-valved draw-off lines to offer flexible operation. Where the lagoon area required exceeds 6 acres, it is good practice to have multiple cells which can be operated individually, in series, or in parallel. If the soil is pervious, bottom and dikes should be sealed to prevent groundwater pollution. A commonly used sealing agent is bentonite (clay). Dikes and areas surrounding the ponds are seeded with grass, graded to prevent runoff from entering the ponds, and fenced to preclude livestock and discourage trespassing. In the case of a multiple-pond installation, the sequence of pond operation and liquid operating depth are regulated to provide control of the treatment system. Operating ponds in series generally increase BOD reduction by preventing short-circuiting. Conversely, parallel operation may be desirable to distribute the raw BOD load and avoid potential odor problems.

Where discharges of pond effluent in the winter result in pollution of the receiving stream, the operating level can be lowered before ice formation and gradually increased to 5 ft by retention of winter flows. The elevation can then be slowly lowered in the spring when the dilution flow in the receiving stream is high. Shallow operation can be maintained during the spring with gradually increasing depths to prevent emergent vegetation. In the fall, the level can again be lowered to hold winter flows.

Most stabilization ponds emit odors occasionally. This is the primary reason for locating them as far as practicable from present or future developed areas and on the leeward side so that prevailing winds are in the direction of uninhabited areas. Lagoons treating only domestic wastewater normally operate odor free. Only for a short period of time in the early spring, when the ice melts and the algae are not flourishing, are offensive odors discharged. Lagoons treating certain industrial wastes, in combination with domestic wastewater, are often noted for their persistent obnoxious odors. The cause of these odors is most likely continuous or periodic overloading from industrial waste discharges or the odorous nature of the industrial waste itself. Unpleasant odors can be emitted from lagoons of small municipalities where poultry-processing wastes, slaughterhouse wastes, or creamery wastes are discharged to the municipal sewer without pretreatment.

## 12.31  BOD LOADINGS OF FACULTATIVE PONDS

Many efficiency equations have been proposed for modeling BOD removal in facultative ponds [24]. BOD reduction in a primary cell appears to follow a simple plug-flow hydraulic model with a first-order reaction rate; however, verification by comparison with actual performance data has been only marginally successful. More complex equations that include factors such as water temperature, light intensity, and dispersion are no more successful in predicting the efficiency of stabilization ponds. As a result, design of facultative ponds is commonly dictated by empirical rules based on the observation of pond performance in a region. Design guidelines may specify the maximum allowable BOD loading on primary cells, number of cells required for different-size pond systems, overall wastewater retention time, maximum and minimum water depths, and winter storage for a specified number of months. Example 12.20 illustrates the application of a typical set of design guidelines.

BOD loadings are usually expressed in ponds as BOD applied per acre of surface area per day (kg/ha·d or g/m²·d). In northern states, loadings on primary cells are 25–35 lb BOD/acre/day (2.8–3.9 g/m²·d) to minimize odor nuisance in the spring of the year. In southern states, design loadings are about 40–50 lb BOD/acre/day. Total retention time for primary plus secondary cells is 3–6 months to allow seasonal storage and controlled discharge. Series operation reduces short-circuiting and, as a result, improved BOD and fecal coliform removals.

The degree of stabilization produced in a pond is influenced by climatic conditions. During warm, sunny weather, decomposition and photosynthetic processes flourish, resulting in rapid and complete stabilization of the waste organics. The pond water becomes supersaturated with dissolved oxygen during the afternoon. Suspended solids and BOD in the pond effluent are primarily from the algae. BOD reductions in the summer usually exceed 95%. During cold weather under ice cover, biological activity is extremely slow and the lagoon-treatment process is, for practical purposes, reduced to sedimentation. Anaerobic conditions can occur from a lack of reaeration by wind action and photosynthesis. Suspended solids and BOD in the pond effluent include organics from raw wastes and intermediate organics issuing from incomplete anaerobic metabolism. Under winter ice cover, BOD reductions are generally about 50%.

Warm pond water rich in plant nutrients when exposed to sunlight supports an abundance of algae, giving the water a green color. Since wastewater contains significant concentrations of carbon dioxide, inorganic nitrogen, and soluble phosphorus, the growth of algae is usually limited by shading caused by the high turbidity from the algae in suspension. Higher forms of aquatic animal life, i.e., zooplankton, that graze on the algae are also present. As a result of this natural biological activity, pond water during summer months contains 50–80 mg/l of suspended solids. Although this exceeds the effluent standard of 30 mg/l of suspended solids, most regulatory agencies have established higher limits for stabilization pond effluent, typically 60 mg/l, to allow summer and fall discharge. Where the effluent standard prohibits discharge to a watercourse, irrigation of nearby agricultural land appears to be the best solution. In a semiarid climate where evaporation exceeds rainfall by a wide margin, ponds serving a small community can be constructed large enough to provide complete retention. However, the land area for zero-discharge ponds is often prohibitive.

# 12.32  ADVANTAGES AND DISADVANTAGES OF STABILIZATION PONDS

A general list of items to be considered before selecting the stabilization pond process for treatment of a municipal wastewater is offered. In general, stabilization ponds are suitable for small towns that do not anticipate extensive industrial expansion and where land with suitable topography and soil conditions is available for siting.

### Advantages
1. The initial cost is probably lower than that of a mechanical plant.
2. Operating costs are lower.
3. Regulation of effluent discharge is possible, thus providing control of pollution during critical times of the year.
4. Treatment system is not significantly influenced by a leaky sewer system that collects storm water.

### Disadvantages
1. Extensive land area is required for siting.
2. The assimilative capacity for certain industrial wastes is poor.
3. Potential odor problems are possible.
4. The expansion of town and new developments may encroach on the lagoon site.
5. The effluent quality generally cannot meet the standard for suspended solids of 30 mg/l.

■ **EXAMPLE 12.19**

Design population for a town is 1200 persons, and the anticipated industrial load is 20,000 gpd at 1000 mg/l of BOD from a milk-processing plant. Calculate the surface area required for a stabilization pond system, and estimate the number of days of winter storage available. Assume the following:

a. Wastewater flow of 100 gpcd with 0.17 lb of BOD per capita.
b. Design BOD loading of 25 lb of BOD/acre/day.
c. Water loss from evaporation and seepage of 60 in./yr.
d. Annual rainfall of 20 in./yr.

*Solution*

BOD load (domestic + industrial)
$$= 1200 \times 0.17 + 0.020 \times 1000 \times 8.34 = 371 \text{ lb of BOD/day}$$

stabilization pond area required $= \dfrac{371}{25} = 14.8 \text{ acres (use two ponds)}$

volume available for winter storage between 2- and 5-ft depths
$$= (5 - 2)14.8 \times 43,560 = 1,930,000 \text{ ft}^3$$

water loss per day (evaporation + seepage − rainfall)
$$= \frac{(60 - 20)14.8 \times 43,560}{12 \times 365} = 5890 \text{ ft}^3/\text{day}$$

$$\text{wastewater influent per day} = \frac{1200 \times 100 + 20{,}000}{7.48} = 18{,}700 \, \text{ft}^3/\text{day}$$

$$\text{winter storage available} = \frac{1{,}930{,}000}{18{,}700 - 5900} = 150 \, \text{days} \qquad \blacksquare$$

### ■ EXAMPLE 12.20

The design criteria for stabilization ponds specified by the state regulatory agency are that (1) the BOD loading in the primary cells shall not exceed 25 lb BOD/acre/day, (2) the minimum total water volume, based on influent flow and all cells at a water depth of 5 ft, shall not be less than 120 days, (3) the volume for winter storage between the water depths of 1.5 and 5.0 ft shall be sufficient so that no discharge is necessary for a 4-month period, and (4) the pond system shall have at least two primary cells and one or more secondary cells that cannot receive raw wastewater. Size the ponds for a community with an average daily wastewater discharge of 160,000 gpd with a BOD of 220 mg/l. Based on available data, the net water loss (evaporation plus seepage minus precipitation) during the storage months is 1.0 in./month.

*Solution*

$$\text{area of primary ponds} = \frac{0.160 \times 220 \times 8.34}{25}$$

$$= 11.7 \, \text{acres}$$

construct two primary cells each $= 5.9$ acres

minimum total water volume required $= 120 \times 0.160 = 19.2$ mil gal

volume of primary cells $= 11.7 \times 5 \times 0.326$

$$= 19.1 \, \text{mil gal} \quad (\text{OK})$$

wastewater inflow during 4 months $= 0.160 \times 3.07 \times 4 \times 30 = 58.9$ acre-ft

$$\text{pond area required for storage} = \frac{\text{inflow} - \text{water loss}}{\text{difference in water levels}}$$

$$\text{pond area} = \frac{58.9 - \text{pond area} \times \frac{4}{12}}{5.0 - 1.5}$$

pond area $= 15.4$ acres

area of secondary cell $= 15.4 - 2 \times 5.9 = 3.6$ acres $\qquad \blacksquare$

## 12.33  COMPLETELY MIXED AERATED LAGOONS

Aerated ponds for pretreatment of industrial wastes or first-stage treatment of municipal wastewaters are commonly completely mixed lagoons 8–12 ft deep with floating or platform-mounted mechanical aeration units. A floating aerator consists of a motor-driven impeller mounted on a doughnut-shaped float with a submerged intake cone (Fig. 12.58). Inspection and maintenance is performed using a boat, or by disconnecting the restraining cables and pulling the unit to the edge of the lagoon. Platform-mounted aerators are placed on piles or piers extending into the pond bottom. The impeller is held beneath the liquid surface by a short shaft connected to the motor mounted on the platform. A bridge may be constructed from the lagoon dike to the aerator for ease of inspection and maintenance.

Complete mixing and adequate aeration are essential environmental conditions for a lagoon biota. Selection and design of mixing equipment depend on manufacturer's laboratory test data and field experience. Aerators are spaced to provide uniform blending for the dispersion of dissolved oxygen and suspension of microbial solids. Their oxygen transfer capabilty must be able to satisfy the BOD demand of the waste while retaining a residual dissolved-oxygen concentration. Figure 12.59 illustrates the general relationships between power required for mixing and that required for aeration. Only one detention time exists for a given wastewater strength where both stirring and aeration functions are at optimum. Thus, deviations in loadings should be considered in the design selection and operational control of mechanical aeration units.

Organic stabilization depends on suspended microbial floc developed within the basin, since no provision is made for settling and returning activated sludge. BOD removal is a function of detention time, temperature, and nature of the waste, primarily biodegradability and nutrient content. The common relationships are

$$\frac{L_e}{L_0} = \frac{1}{1 + kt} \tag{12.103}$$

$$k_T = k_{20°C}\Theta^{T-20} \tag{12.104}$$

where   $L_e$ = effluent BOD, mg/l
   $L_0$ = influent BOD, mg/l
   $k$ = BOD-removal-rate constant to base $e$, day$^{-1}$
   $t$ = detention time, days
   $T$ = temperature, °C
   $\Theta$ = temperature coefficient

The value of $k$ relates to degradability of the waste organics, temperature, and completeness of aeration mixing. At 20°C, $k$ values have been found to range from 0.3 to over 1.0; the precise value for a particular waste must be determined experimentally. The coefficient $\Theta$ is a function of biodegradability and generally falls between 1.035 and 1.075, with 1.035 the most common value.

Biological oxygen utilization is equal to assimilative plus endogenous respiration as in the activated-sludge process. However, with the low concentration of microbial suspended solids in the aerating wastewater, oxygen uptake can be simply related to BOD removal by the relationship

lb of oxygen/day = $a$ × lb of BOD removed/day $\qquad$ (12.105)

(a)

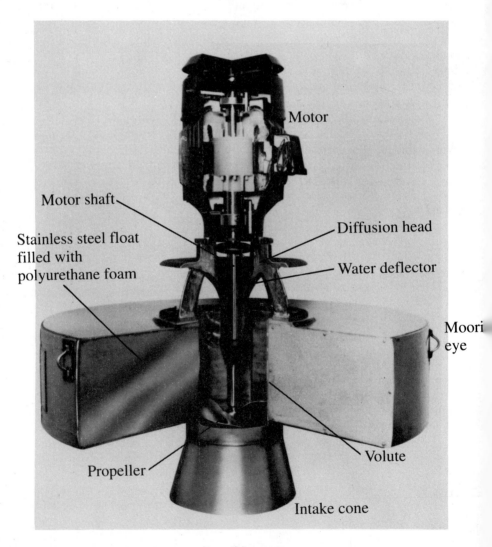

Motor

Motor shaft

Diffusion head

Stainless steel float
filled with
polyurethane foam

Water deflector

Moori
eye

Propeller

Volute

Intake cone

(b)

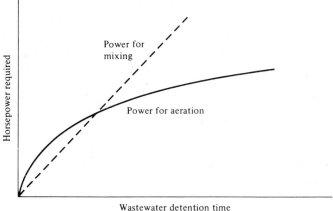

**Figure 12.59** General relationships between the power required for mixing and that needed for aeration relative to wastewater detention time for mechanical aerators in completely mixed lagoons.

The magnitude of $a$ is determined by laboratory or field testing the particular wastewater to be treated. Values are from 0.5 to 2.0, 1.0 being typical.

Oxygen transferred by surface-aeration units can be computed by

$$R = R_0 \frac{\beta C_s - C_t}{9.2} \times 1.02^{T-20}(\alpha) \tag{12.106}$$

where    $R$ = actual rate of oxygen transfer, lb of oxygen/hp·hr
$R_0$ = rate of oxygen transfer of manufacturer's unit under standard conditions (water at 20°C, 1 atm pressure, and zero dissolved oxygen), lb of oxygen/hp·hr
$\beta$ = oxygen saturation coefficient of the wastewater
$C_s$ = oxygen concentration at saturation, mg/l
$C_t$ = oxygen concentration existing in liquid, mg/l
$T$ = temperature of lagoon liquid, °C
$\alpha$ = oxygen transfer coefficient
9.2 = saturation oxygen concentration of pure water at 20°C, mg/l

The manufacturer's oxygen transfer rate $R_0$ is guaranteed performance based on aerator test data stated in terms of standard conditions. The $\alpha$ and $\beta$ factors are discussed in Section 12.28.

---

**Figure 12.58** Floating aerator. (a) Picture of an aerator in operation. (b) Cutaway section showing aerator design with propeller directly connected to motor that is fastened to the float. Water drawn up through the intake cone is deflected by the diffusion head for aeration by dispersion. (Courtesy of Aqua-Aerobic Systems, Inc., Rockford, IL)

A facultative, aerated lagoon results if insufficient mixing permits deposition of suspended solids. BOD removal cannot be predicted with certainty in a nonhomogeneous system. Anaerobic decomposition of the settled sludge can cause emission of foul odors, particularly in treating certain industrial wastes. Facultative conditions often result from overloaded completely mixed lagoons or may derive from poorly designed systems with inadequate mixing. To ensure odor-free operation, pond contents must be thoroughly stirred, with dissolved oxygen available throughout the liquid. Aeration equipment installed should be of proven performance purchased from a reputable manufacturer.

Significant increases in effluent BOD can occur from reducing detention time since the biological process is time dependent. Cooling-water discharges and shock loads of relatively uncontaminated water, for example storm runoff, should be diverted around the lagoon to the secondary ponds. Sudden, large inputs of biodegradable or toxic wastes resulting from industrial spills can also upset the process. Pretreatment and control systems at industrial sites should be furnished to prevent taxing the lagoon's equalizing capacity. Biodegradability studies are essential for municipal wastewater containing measurable amounts of industrial discharges to determine such design parameters as BOD removal-rate constant, influence of temperature, nutrient requirements, oxygen utilization, and sludge production.

### ■ EXAMPLE 12.21

Size an aerated lagoon to treat a domestic plus industrial waste flow of 0.30 mgd with an average BOD of 600 mg/l (1500 lb of BOD/day). The temperature extremes anticipated for the lagoon contents range from 10°C in winter to 35°C in summer. Minimum BOD reduction through the lagoon should be 75%. The surface aerators to be installed carry a manufacturer's guarantee to transfer 2.5 lb of oxygen/hp·hr under standard conditions. During laboratory treatability studies, the wastewater exhibited the following characteristics: $k_{20°C} = 0.68$ per day, $\Theta = 1.047$, $\alpha = 0.9$, and $\beta = 0.8$.

*Solution.*   The required detention time at a critical temperature of 10°C is found using Eqs. 12.103 and 12.104:

$$k_{10°C} = 0.68 \times 1.047^{10-20} = 0.43 \text{ per day}$$

$$\frac{L_e}{L_0} = 1 - 0.75 = \frac{1}{1 + 0.43t}$$

$$t = 7.0 \text{ days}$$

lagoon volume $= 0.30 \text{ mgd} \times 7.0 \text{ days} = 2.1 \text{ mil gal} = 280,000 \text{ ft}^3$

Use a 10-ft depth with earth sides slopes appropriate for soil conditions. Oxygen utilization using $a = 0.8$: At 10°C

$$\text{BOD removal} = 0.75 \times 1500 = 1120 \text{ lb of BOD/day}$$

$$\text{oxygen required} = \frac{1120 \times 0.8}{24} = 37 \text{ lb of oxygen/hr}$$

At 35°C:

$$k = 0.68 \times 1.047^{35-20} = 1.35 \text{ per day}$$

$$\text{BOD removal} = 1500 - \frac{1500}{1 + 1.35 \times 7.0} = 1360 \text{ lb of BOD/day}$$

$$\text{oxygen required} = \frac{1360 \times 0.8}{24} = 45 \text{ lb of oxygen/hr}$$

Aerator power requirements using Eq. 12.106 at a minimum of 2.0 mg/l of dissolved oxygen ($C_s$ = 11.3 mg/l at 10°C and 7.1 mg/l at 35°C):

$$R_{10°C} = \frac{2.5(0.8 \times 11.3 - 2.0)}{9.2} 1.02^{10-20} \times 0.9 = 1.4 \text{ lb of oxygen/hp·hr}$$

$$\text{power required} = \frac{37 \text{ lb of oxygen/hr}}{1.4 \text{ lb of oxygen/hp·hr}} = 26 \text{ hp}$$

$$R_{35°C} = \frac{2.5(0.8 \times 7.1 - 2.0)}{9.2} 1.02^{35-20} \times 0.9 = 1.2 \text{ lb of oxygen/hp·hr}$$

$$\text{power required} = \frac{45}{1.2} = 38 \text{ hp}$$

Power requirements at 35°C control design: Use four 10-hp surface aerators. ∎

# Odor Control

Increased urbanization has resulted in wastewater treatment plants being situated in close proximity to housing areas and commercial developments. This has caused complaints about odors and, in serious situations, led to lawsuits against municipalities operating the disposal systems. Although the problem of foul odors emitting from treatment plants is not new, only in recent years have political and legal pressures forced processing facilities to consider abatement.

## 12.34 SOURCES OF ODORS IN WASTEWATER TREATMENT

Principal odors are hydrogen sulfide and organic compounds generated by anaerobic decomposition. The latter include mercaptans, indole, skatole, amines, fatty acids, and many other volatile organics. Often, industrial wastes in a municipal sewer create odors inherent in the raw materials being processed or the manufactured products (poultry processing, slaughtering and rendering, tanning, and manufacture of volatile chemicals). With the exception of hydrogen sulfide, a specific odor-producing substance is very difficult to identify. Weather conditions, such as temperature and wind velocity, influence the intensity and prevalence of emissions.

Frequently, the initial evolution of malodors is from septic wastewater in the sewer collection system. Flat sewer grades, warm temperatures, and high-strength wastes lead to anaerobiosis. The first sources at the treatment plant are the wet well and grit chamber. Turbulent flow and preaeration of raw waste strip dissolved gases and volatile organics, discharging them into the atmosphere. Odors also may arise from the liquid held in primary clarifiers, particularly if excess activated sludge is returned to the head of the plant, resulting in an active microbial seed being mixed with the settleable raw organic matter. Sludge taken from these tanks has an obnoxious smell. Pumping it into uncovered holding tanks releases the scent previously confined under the water cover. Polymers do not neutralize the olfactory compounds prior to vacuum filtration; therefore, the air drawn through the sludge cake picks up volatile compounds and carries them to the atmosphere. The use of ferric chloride and lime for conditioning chemicals significantly reduces odors, but for most municipal wastes, polymers provide more economical operation. The process of anaerobic digestion takes place in enclosed tanks while digested sludge is dewatered either mechanically or on drying beds. The smell of well-digested sludge is earthy, but if the digestion process is not complete, intermediate aromatic compounds may be released during drying.

Secondary biological processes also yield odors, particularly stone-media trickling filters. Although referred to as aerobic devices, filters are actually facultative, since the microbial films are aerobic on the surface and anaerobic adjacent to the medium. Because of this potential for anaerobic decomposition, filters under heavy organic loading can reek. Odors are not as likely to be created in biological towers because of thinner biological films and improved aeration. Activated-sludge processes yield a relatively inoffensive musty odor carried by the air passing through the mixed liquor. Foul smells are rare since microbial flocs in the aeration basin are completely surrounded by liquid containing dissolved oxygen.

## 12.35  METHODS OF ODOR CONTROL

Modern treatment plant design should incorporate the concept of odor prevention [25]. This involves a comprehensive understanding of the potential problems possible in the application of certain processes for handling a wastewater. For example, the exhaust from vacuum or pressure filtration of a raw sludge coagulated with a polymer is likely to be offensive, particularly if the municipal wastewater contains industrial wastes. Individual unit processes can be designed to minimize the possibility of anaerobiosis, such as the vent and underdrain design of a trickling filter to allow free circulation of air. In siting plants, it is wise to provide a reasonable buffer zone to prevent encroachment of activities that will be offended by the essence of a treatment plant.

The first step in analyzing an existing problem is to determine the cause of odorousness and attempt to isolate the sources. Special attention must

**Figure 12.60** Fiberglass cover enclosing a trickling filter to contain odors and reduce cooling of the wastewater in winter.

be paid to industrial wastes entering the sewer system. Overloading often increases malodors; however, expansion of facilities is no guarantee that the situation will change dramatically. Foul emissions can be given off by properly loaded units if the design is poor, if they are not maintained properly, or when the waste includes organics with an inherent smell. Adding secondary treatment to an existing primary plant can unwittingly create obnoxious conditions that did not previously exist. In one case history, waste sludge from a new activated-sludge secondary was disposed of by return to the head of the plant. This resulted in anaerobic biological decomposition of the settled sludge, causing upset of the primary tanks and operational problems in vacuum filtration, with both leading to foul odors.

Chemicals can sometimes be used to oxidize odorous compounds, particularly hydrogen sulfide. Chlorination of the wastewater in main sewers, or prior to primary settling, may prove beneficial. Using lime and ferric chloride as chemical sludge conditioners reduces bacterial activity and oxidizes many products of anaerobic decomposition. In some cases the only feasible solution has been to provide sealed enclosures to prevent odors from reaching the atmosphere (Fig. 12.60). Air confined under clarifier and grit basin covers must be continuously purified to remove corrosive gases, such as hydrogen sulfide. Air must be circulated through a trickling-filter bed for ventilation. This can be accomplished by forcing fresh air under the dome and collecting it from the underdrain system. Recirculated air can be cleansed using wet scrubbers with a chemical solution or activated-carbon absorption beds. Solutions of either permanganate or hypochlorite have been effective oxidizing agents for some odors, but carbon filters generally perform better. Sometimes carbon beds are used in series with wet scrubbers. In activated-sludge plants, air withdrawn from under the covers of primary clarifiers and grit chambers can be effectively scrubbed by using it as a portion of the air supply to the aeration tanks.

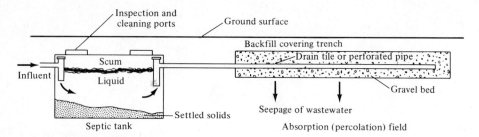

**Figure 12.61** A typical septic tank–absorption field system for on-site disposal of household wastewater.

# Individual On-Site Wastewater Disposal

## 12.36 SEPTIC TANK–ABSORPTION FIELD SYSTEM

Approximately 30% of the population in the United States live in unsewered areas and rely on on-site systems for wastewater treatment and disposal. Almost one-third of the housing units use septic tanks or cesspools; the majority of the remainder, usually in remote locations, use pit privies.

The installation of a septic tank and absorption field, sketched in Fig. 12.61, has the advantages of low cost and underground disposal of effluent. The septic tank is an underground concrete box sized for a liquid detention time of approximately 2 days. With garbage grinders and automatic washers, the recommended minimum capacity is 750 gal ($2.8 \text{ m}^3$) for a two-bedroom house, 900 gal ($3.4 \text{ m}^3$) for three bedrooms, 1000 gal ($3.8 \text{ m}^3$) for four bedrooms, and 250 gal ($1.0 \text{ m}^3$) for each additional bedroom. Inspection and cleaning ports are accessible for maintenance by removing the earth cover of about 1 ft. Inlet and outlet pipe tees, or baffles, prevent clogging of the openings with scum that accumulates on the liquid surface. The functions of a septic tank are settling of solids, flotation of grease, and anaerobic decomposition and storage of sludge. Retention of large solids and grease is essential to prevent plugging of the absorption field. Under normal loading, the accumulated sludge (septage) must be pumped from the tank every 3–5 yr.

A typical absorption field consists of looped or lateral trenches 18–24 in. wide and at least 18 in. deep. Drain tile or perforated pipe in an envelope of gravel distributes the wastewater uniformly over the trench bottom. Organics decompose in the aerobic-facultative environment of the bed and the water seeps downward into the soil. Air enters the gravel bed through the backfill covering the trenches and by ventilation through the drain tile from the house plumbing stack. The percolation area required depends on soil permeability. For a four-bedroom dwelling, the area needed is in the range 300–1300 $\text{ft}^2$ (30–120 $\text{m}^2$). Trench area for a particular location can be determined by subsurface soil exploration and percolation tests; however, most state environmental control agencies and county health departments have guidelines for installations based on local conditions.

The most frequent complaints in the operation of septic tank–absorption field systems involve plumbing stoppages and odorous seepage appearing at or near the ground surface. Plugging of the influent is often the result of excessive accumulation of solids due to either overloading or neglecting to pump out the sludge every few years. When cleaning a tank, a small amount of the black digesting sludge should be left in the tank to ensure adequate bacterial seeding to continue solids digestion. Obstruction of percolation can result for any of the following reasons: construction in soils of low permeability that are inadequate for the seepage, high water-table conditions that saturate the soil profile during the wet-weather season, inadequate design resulting in hydraulic and organic overloading, and improper cleaning of the septic tank.

Soils with a high infiltration rate, although advantageous for effluent disposal, often do not have the capacity to absorb contaminants. The result can be serious groundwater pollution, either limited to the area immediately surrounding an absorption field or widespread depending on the density of housing units. Nearby wells can be polluted with disease-producing microorganisms, particularly hepatitis viruses. The most common pollutants, however, are mobile substances like detergents and ions, such as chloride and nitrate.

# Marine Wastewater Disposal

## 12.37  OCEAN OUTFALLS

For coastal communities, disposal of conventionally treated wastewater through a marine outfall should be considered as an alternate to tertiary processing. An ocean outfall is a pipeline that extends thousands of feet from the shore to relatively deep water. At the end of the pipe, a diffuser discharges the wastewater through a series of ports spaced to provide initial dilution. After reaching the sea surface, or an intermediate equilibrium level, this mixed plume tends to move with the prevailing currents to provide secondary dispersion. Data collected for design of an outfall include physical, chemical, biological, and geological conditions. Of particular concern is a comprehensive oceanographic study to predict dilution and diffusion of the wastewater.

The general water-quality objective in marine disposal is to maintain the indigenous marine life and a healthy and diverse marine community. Relevant considerations are contamination of shellfish with pathogens, toxicity of aquatic life, accumulation of sediments that impair benthic life, and esthetics of the ocean surface. Governmental regulations require treatment of wastewater prior to marine disposal. In general, the specified effluent water quality dictates biological wastewater treatment and limitations on toxins. For further study, the reader is referred to the book by Grace [26], which discusses the planning, design, and construction of marine outfall systems.

## PROBLEMS

**12.1**  (a) How do autotrophic bacteria gain energy?  (b) Why do some bacteria convert ammonia to nitrate (Eqs. 12.1 and 12.2), whereas others reduce nitrate to nitrogen gas (Eq. 12.12)?  (c) Hydrogen sulfide emitted by septic waste-water and sludge is too weak an acid to cause significant corrosion, yet when hydrogen sulfide is present, concrete deteriorates and iron corrodes. Explain this phenomenon.  (d) What promotes the growth of iron bacteria in water pipes distributing well water? How do these bacteria contribute foul tastes and odors to the water?

**12.2**  List the major nutritional and environmental conditions necessary to culture algae in a laboratory container?

**12.3**  A fresh wastewater containing nitrate ions, sulfate ions, and dissolved oxygen is placed in a sealed jar without air. In what sequence are these oxidized compounds reduced by the bacteria? Why is dissolved oxygen the preferred hydrogen acceptor? When do obnoxious odors appear?

**12.4**  Discuss the relationships among metabolism, energy, and synthesis and the effect of these on growth under aerobic and anaerobic conditions. Include comments on growth rates, extent of metabolism, and limiting factors under the two environments.

**12.5**  Why is bacterial synthesis for the same quantity of substrate greater in an aerobic environment than under anaerobiosis?

**12.6**  What are the relationships between biomass and substrate in the exponential growth and the declining growth phases of a pure bacterial culture? List the characteristics of the endogenous growth phase.

**12.7**  What is the mathematical relationship for growth kinetics of a pure bacterial culture proposed by Monod and graphed in Fig. 12.6? How does this relate to enzymatic reactions as defined by Michaelis and Menten?

**12.8**  A series of fermentation tubes containing varying concentrations of glucose in a nutrient broth were inoculated with a pure bacterial culture. The concentrations of cells in the broth media were determined after 16 hr of incubation at 37°C. Rates of growth and initial glucose concentrations are listed below. Plot growth rate expressed as cell divisions per hour versus initial glucose concentration. Estimate the maximum growth rate and the saturation constant (glucose concentration at one half the maximum growth rate). Write an equation in the form of Eq. 12.25, and draw the curve for this equation on the graph with the plotted data. Does the growth of this pure culture appear to be a hyperbolic function as defined by the Monod relationship?

| Glucose (moles $\times$ 10$^{-4}$) | Cell divisions (per hr) | Glucose (moles $\times$ 10$^{-4}$) | Cell divisions (per hr) |
|---|---|---|---|
| 0.1 | 0.23 | 0.8 | 0.94 |
| 0.1 | 0.28 | 1.6 | 1.06 |
| 0.2 | 0.32 | 3.2 | 1.15 |
| 0.4 | 0.71 | | |

**12.9**  How is growth yield determined?

**12.10**  (a) How does temperature affect biological processes?   (b) The rate of BOD reduction in aeration of a synthetic wastewater decreased 25% when the temperature of the laboratory fermentation tank was lowered from 20°C to 15°C. Using Eq. 12.35, calculate the temperature coefficient. How many degrees of temperature drop is required to reduce the rate of BOD reduction by one half?

**12.11**  State the two main reasons why an activated-sludge system is operated at a relatively low food/microorganism ratio.

**12.12**  Why are bacteria rather than protozoans the primary feeders in activated sludge?

**12.13**  Why is impending failure of an anaerobic digestion process forecast by an increase in the percentage of carbon dioxide in the gas produced?

**12.14**  (a) Describe the role of algae in biological stabilization of wastewater in a stabilization pond.   (b) What is wrong with the following statement? "The key to waste stabilization in a pond is the algal growth; overloading kills the algae, whereas they thrive on a reasonable supply of organic matter."

**12.15**  The average annual wastewater flow of a municipality is 24 mgd with an average BOD concentration of 180 mg/l and suspended-solids concentration of 200 mg/l. Based on the equations in Section 12.10, estimate the average annual peak month and average annual peak day of wastewater flows, BOD loads, and suspended-solids loads.

**12.16**  The average wastewater flow and strength anticipated during the maximum month of the year are often used for design. If the annual average wastewater flow and characteristics are estimated to be 32,000 m$^3$/d with 250 mg/l of BOD and 280 mg/l of suspended solids, what are the values likely to be for the peak month using the equations in Section 12.10?

**12.17**  Wastewater from soluble coffee manufacturing is treated jointly with domestic wastewater in an activated-sludge process without primary settling. The mixture of wastewater entering the plant is 40% coffee wastewater by volume and 60% domestic wastewater. The characteristics of the coffee wastewater are 840 mg/l of BOD, 6 mg/l of total nitrogen, and 2 mg/l of total phosphorus. The domestic wastewater characteristics are as listed in Table 12.1.   (a) If the required BOD/N/P weight ratio is 100/6.0/1.5 for activated-sludge processing, are the nitrogen and phosphorus concentrations in the combined wastewater adequate?   (b) If the nutrient content is not adequate, how many milligrams per liter of pure $NH_4NO_3$ and $H_3PO_4$ need to be added to the wastewater?

**12.18** The wastewater from a synthetic textile manufacturing plant is discharged to a municipal treatment plant for processing with domestic wastewater. The characteristics of the synthetic textile wastewater are 1500 mg/l BOD, 2000 mg/l SS, 30 mg/l nitrogen, and no phosphorus. The characteristics of the domestic wastewater are listed in Table 12.1. If the required BOD/N/P weight ratio is 100/5/1, what is the minimum quantity of domestic wastewater required per 1000 gal of textile wastewater to provide adequate nutrients for biological treatment?

**12.19** The wastewater from the manufacture of synthetic chemicals produces a wastewater flow of 779 $m^3$/d containing 4300 mg/l of BOD, 1200 mg/l of suspended solids, 70 mg/l of nitrogen, and negligible phosphorus. You are asked to recommend dosages of pure $NH_4NO_3$ and $H_3PO_4$ to be applied for activated-sludge treatment. After the biological process is stabilized, how would you determine if your recommended dosages were correct?

**12.20** The wastewater from a candy manufacturer is 97,700 gpd with a BOD concentration of 1560 mg/l. Calculate the BOD equivalent population and the hydraulic equivalent population.

**12.21** A meat-processing plant discharges 10,000 $m^3$/d of wastewater containing 1300 mg/l of BOD, 960 mg/l of suspended solids, 2500 mg/l of COD, and 460 mg/l of grease. Calculate the BOD equivalent population and the hydraulic equivalent population.

**12.22** The municipal wastewater flow from a city with a population of 150,000 is 76,000 $m^3$/d with a BOD equal to 320 mg/l. Compute the BOD equivalent population for the wastewater flow.

**12.23** The domestic and industrial waste from a community consists of 100 gpcd from 7500 persons; 65,000 gpd from a milk-processing plant with a BOD of 1400 mg/l; and 90,000 gpd containing 450 lb of BOD from potato-chip manufacturing. Calculate the combined wastewater flow, BOD concentration in the composite waste, and BOD equivalent population.

**12.24** The combined wastewater flow from a community includes domestic waste from a sewered population of 2000 and industrial wastes from a dairy and a poultry-dressing plant. The poultry plant discharges 125 $m^3$/d and 136 kg of BOD/d. The dairy produces a flow of 190 $m^3$/d with a BOD concentration of 900 mg/l. Estimate the total combined wastewater flow from the community and the BOD concentration in the composite discharge.

**12.25** A town with a sewered population of 4000 has a daily wastewater flow (including industrial wastewaters) of 1600 $m^3$ and an average BOD of 280 mg/l. The industrial discharges to the municipal sewers are 60 $m^3$ at 1800 mg/l BOD from a meat-processing plant and 100 $m^3$ at 400 mg/l BOD from a soup-canning plant. Determine the contribution of domestic flow in liters per person and the BOD in grams per person based on the town's wastewater excluding the industrial wastewaters.

**12.26** The municipal wastewater flow from a town is 1890 $m^3$/d with an average BOD of 280 mg/l. Assuming 35% BOD removal in the primary, calculate the size required for one single-stage, high-rate trickling filter. Use a depth of 2.1 m and a BOD loading of 480 g/$m^3$·d. Compute the BOD concentration in the effluent at 22°C using the NRC formula and temperature correction relationship in Eq. 12.25.

**12.27** A single-stage trickling-filter plant (Fig. 12.21) is proposed for treating a dilute wastewater with a BOD concentration of 170 mg/l. The plant is located in a warm climate and the minimum wastewater temperature anticipated is 16°C. Using a recirculation ratio of 0.5, what is the maximum allowable BOD loading for a stone-media filter to achieve an average effluent BOD of 30 mg/l.

**12.28** The sizing of primary clarifiers is generally based on the average weekday flow during the time of the year of greatest flow. If the plant is a trickling-filter system, gravity return of underflow from the final clarifier to the wet well is performed for recirculation during periods of low influent flow (Fig. 12.21). This recirculation flow, usually $0.5Q$ or less as diagrammed in Fig. 12.16a, is necessary to maintain rotation of the distributor arm at night and to provide adequate hydraulic loading of the filter media. Should the size of primary clarifiers be increased to account for this low-flow recirculation? Explain.

**12.29** Estimate the effluent BOD for the two-stage trickling-filter plant designed in Example 12.5 for a wastewater flow of 1.0 mgd with a BOD concentration of 350 mg/l.

**12.30** Determine the NRC BOD removal efficiency for a two-stage trickling-filter plant based on the following: primary clarification with 35% BOD reduction, first-stage filters loaded at 90 lb/1000 ft³/day, intermediate settling, second-stage filters sized identical to the first-stage units, an operating recirculation ratio of 1.0 for all filters, final clarification of the effluent, and a wastewater temperature of 18°C.

**12.31** An existing single-stage trickling-filter plant cannot meet the effluent limitation of 30 mg/l BOD during cold weather. In operation under the design loading with the wastewater temperature at 15°C, the plant influent BOD is 240 mg/l, the primary effluent BOD equals 155 mg/l, and the plant effluent averages 55 mg/l. The proposed modification is the addition of a second-stage trickling filter and final clarifier to reduce the effluent BOD from 55 to 30 mg/l with a waste-water temperature of 15°C. Under normal operation, the recirculation flow in the second-stage filtration is to be one-half of the average wastewater flow entering the plant. Calculate the design BOD loading that should be used to determine the volume of stone media needed in the proposed second-stage filter.

**12.32** The following are design data for the town of Nancy, with a sewered population of 7600. Design flows are as follows: the average daily flow is 0.84 mgd, the peak hourly flow is 1.25 mgd, and the minimum hourly flow is 0.12 mgd. Design average BOD equals 1740 lb/day and average suspended solids equal 1530 lb/day. Calculate the following: equivalent population based on 0.20 lb of BOD per capita; design flows in units of gpm, ft³/sec, m³/min, and m³/d; mean BOD and SS concentrations in mg/l.

**12.33** Consider the feasibility of treating the wastewater from the town of Nancy (Prob. 12.32) in a single-stage stone-media trickling-filter plant as diagrammed in Fig. 12.21. The treated wastewater is required to meet the effluent standards of 30 mg/l of BOD and 30 mg/l of suspended solids at a wastewater temperature of 15°C. Assume a primary BOD removal efficiency of 35%. Size the trickling filters based on a BOD loading of 35 lb/1000 ft³/day, recirculation ratio of 0.5, and a stone-media depth of 7.0 ft. Calculate the volume of filter media required, filter surface area, hydraulic loading with recirculation in gallons per minute per square foot, and effluent BOD concentratin using the NRC formula.

**12.34** Consider the feasibility of treating the wastewater from the town of Nancy (Prob. 12.32) in a two-stage stone-media trickling-filter plant with intermediate clarification. The treated wastewater is required to meet the effluent standards of 30 mg/l of BOD and 30 mg/l of suspended solids at a wastewater temperature of 15°C. Assume a primary BOD removal efficiency of 35%. Calculate the volume of filter media required at a BOD loading of 35 lb/1000 ft³/day and divide the resulting volume equally between the first-stage and second-stage filters. Use a filter media depth of 7.0 ft. Assuming a recirculation ratio of 0.5 for both filters, determine the effluent BOD concentration using the NRC formulas.

**12.35** Consider the feasibility of treating the wastewater from the town of Nancy (Prob. 12.32) in a single-stage trickling-filter plant as diagrammed in Fig. 12.21, using high-density cross-flow media with a specific surface area of 42 ft²/ft³. Use the same filter surface area as calculated for stone-media filtration in Prob. 12.33 ($A = 4600$ ft²) and a recirculation ratio of 0.5. The design wastewater temperature is 15°C. As determined from laboratory testing, the soluble (filtered) BOD of the primary effluent is 100 mg/l or less compared to the unsettled unfiltered BOD of 160 mg/l. The soluble effluent BOD concentration of trickling-filter plants treating similar wastewaters is approximately 50% of the unfiltered BOD concentration. From pilot-plant studies and full-scale experience, the manufacturer of the cross-flow media recommends a reaction-rate coefficient at 20°C of 0.0030 (gpm/ft²)$^{0.5}$ for a media depth of 6.6 ft (2.0 m). From these data, calculate the soluble effluent BOD and double this value for the estimated effluent BOD.

**12.36** A single-stage, trickling-filter plant consists of a primary clarifier, a trickling filter 70 ft in diameter with a 7-ft depth of random packing, and a final clarifier. The raw-wastewater flow is 0.80 mgd with 200 mg/l of BOD and a temperature of 15°C. The constants for the random plastic media are an $n$ of 0.44 and $K_{20}$ of 0.090 min$^{-1}$. Assuming 35% BOD removal in the primary, calculate the effluent BOD concentration (a) without recycle and (b) with an indirect recirculation to the wet well of 0.40 mgd and direct recirculation of 0.80 mgd.

**12.37** A single-stage high-rate trickling-filter plant treating wastewater with 220 mg/l of BOD cannot meet the effluent standard of less than 30 mg/l of BOD. Primary settling removes 35% of the raw-wastewater BOD and the 7-ft-deep filters have an efficiency of 75% under a loading of 40 lb/1000 ft³/day of BOD and a hydraulic loading of 0.18 gpm/ft² without recirculation. This provides average effluent BODs of 30–35 mg/l during the summer and 40–45 mg/l during the winter. One proposal is to replace the stone media with the random packing described in Example 12.6, which has constants $n = 0.39$ and $K_{20} = 0.120$ min$^{-1}$; cover the filters with fiberglass domes to prevent the wastewater from cooling below 16°C in winter; and modify the plant to provide both direct and indirect recirculation. Calculate the effluent BOD at 16°C for the filters packed with the random plastic media at $R = 1.0$.

**12.38** Using Eq. 12.57 developed in Example 12.10, plot the filtered BOD remaining versus depth in a filter with 20 ft of random packing. Assume an applied filtered BOD of 100 mg/l and a hydraulic loading of 0.35 gpm/ft².

**12.39** A pilot-scale study was conducted to determine the reaction-rate coefficient for cross-flow media treating a settled municipal wastewater. The tower was

1.2 m by 1.2 m square with a 6.0-m depth of media, which had a specific surface area equal to 98 $m^2/m^3$. From the following data, plot a graph as shown in Fig. 12.27 and determine $k_{20}$. Assume $n = 0.5$ and $\Theta = 1.035$.

| | | Influent | | Effluent | |
|---|---|---|---|---|---|
| $Q$ ($l/m^2 \cdot s$) | $T$ (°C) | BOD (mg/l) | $S_0$ (mg/l) | BOD (mg/l) | $S_e$ (mg/l) |
| 1.03 | 18 | 92 | 54 | 25 | 14 |
| 0.64 | 23 | 90 | 52 | 18 | 10 |
| 0.47 | 21 | 104 | 54 | 17 | 9 |
| 1.06 | 19 | 72 | 48 | 23 | 12 |
| 0.97 | 24 | 77 | 43 | 20 | 10 |
| 0.70 | 20 | 85 | 45 | 17 | 11 |

**12.40** Calculate the required surface area and the minimum required recirculation ratio based on the following: cross-flow media with $k_{20} = 0.0018$ (gpm/ft²)⁰·⁵, $A_s = 42$ ft²/ft³, $n = 0.5$, and $D = 20$ ft; settled wastewater flow $Q_p = 0.50$ mgd; minimum wetting rate $= 0.7$ gpm/ft²; influent BOD $= 162$ mg/l, $S_p = 80$ mg/l, and temperature $= 15°C$; and effluent BOD $= 20$ mg/l and $S_e = 10$ mg/l. At the calculated surface area, what is the BOD loading in pounds per 1000 ft³ per day?

**12.41** Calculate the RBC surface area required for secondary treatment of wastewater for the town of Nancy (Prob. 12.32). Assume 35% BOD removal in the primary, a required effluent quality of 30 mg/l of BOD, and wastewater temperature of 50°F.

**12.42** A community with a wastewater flow during the peak month of the year of 2.0 mgd at 15°C is considering constructing a new treatment plant. The composition of the raw wastewater is as shown in Table 12.1, and the effluent limits are 30 mg/l of BOD, 30 mg/l of suspended solids, and fecal coliform count of less than 200 per 100 ml. Sketch flow diagrams for the treatment systems listed below showing the arrangement of unit processes (i.e., tanks, pumping stations, and division boxes) and major pipelines (i.e., for wastewater, recycle, and sludge). Assume sludge stabilization is by anaerobic digestion and dewatering on drying beds. Duplication of treatment units is necessary, so if any unit is out of service for maintenance the wastewater can still be processed through the plant, although the effluent quality may be reduced. No pipelines are allowed to bypass raw or settled wastewater to the plant outlet. Standby generators are provided for emergency operation during electrical power outage. (For examples of plant layouts, refer to Figs. 9.6 and 9.8.) List numerical design guidelines for sizing each treatment unit, except pumps and sludge processing.
   a. Preliminary treatment with constant-speed pumps.
   b. Preliminary treatment with variable-speed pumps.
   c. Two-stage trickling-filter plant with stone-media filters.
   d. Trickling-filter plant with single-stage biological towers.
   e. Rotating biological contactor plant.

**12.43** The following are average operating data from a conventional activated-sludge secondary: wastewater flow = 7.7 mgd, volume of aeration tanks = 300,000 ft$^3$ = 2.24 mil gal, influent total solids = 600 mg/l, influent suspended solids = 120 mg/l, influent BOD = 173 mg/l, effluent total solids = 500 mg/l, effluent suspended solids = 22 mg/l, effluent BOD = 20 mg/l, mixed-liquor suspended solids = 2500 mg/l, recirculated sludge flow = 2.7 mgd, waste sludge quantity = 54,000 gpd, and suspended solids in waste sludge = 9800 mg/l. Based on these data, calculate the following: aeration period; BOD loading in lb/1000 ft$^3$/day; F/M ratio in lb BOD/day/lb MLSS; total solids, suspended solids, and BOD removal efficiencies; sludge age; and, return sludge rate.

**12.44** A conventional activated-sludge system treats 11,000 m$^3$/d of wastewater with a BOD of 180 mg/l in an aeration tank with a volume of 3400 m$^3$. The operating conditions are an effluent suspended solids of 20 mg/l, a MLSS concentration maintained in the aeration tank of 2500 mg/l, and an activated-sludge wasting rate of 160 m$^3$/d containing 8000 mg/l. From these data, calculate the aeration period, volumetric BOD loading, F/M loading, and sludge age.

**12.45** Determine the activated-sludge aeration volume required to treat 2.64 mgd with a BOD of 120 mg/l based on the criteria of a maximum BOD loading of 40 lb/1000 ft$^3$/day and a minimum aeration period of 5.0 hr. Assuming an operating F/M of 0.20 lb BOD/day per lb of MLSS, calculate the MLSS to be maintained in the aeration tank. Estimate the operating sludge age, assuming an effluent suspended solids of 30 mg/l and the daily amount of waste-sludge solids from Fig. 13.1. Determine the diameter and side-water depth of two identical final clarifiers of the type shown in Fig. 10.25 for this activated-sludge system.

**12.46** Determine the volume of three identical activated-sludge tanks following primary clarification to aerate 36,000 m$^3$/d with a BOD concentration of 180 mg/l at a BOD loading of 560 g/m$^3$·d. What is the aeration time? For an F/M of 0.35, what MLSS concentration should be maintained in the aeration tanks? Estimate the operating sludge age, assuming an effluent suspended solids of 30 mg/l and the daily amount of waste-activated sludge solids from Fig. 13.1. If the waste sludge has a concentration of 10,000 mg/l, what is the calculated volume of sludge to be wasted each day? Determine the dimensions of three identical rectangular clarifiers (length, width, and side-water depth) with a 2 : 1 length-to-width ratio.

**12.47** A treatment plant has two oxidation-ditch activated-sludge systems as illustrated in Fig. 12.40. Each ditch has a liquid volume of 35,000 ft$^3$ and is equipped with a horizontal-rotor aerator with capacity to transfer 2300 lb of oxygen per day at normal submergence. Each system has a clarifier of the type illustrated in Fig. 10.25 with a diameter of 30 ft, a 9.0 ft side-water depth, and a single weir set on a diameter of 30 ft. The effluent chlorination tank has a volume of 2200 ft$^3$. The design flow is 0.54 mgd: 0.35 domestic and commercial, 0.02 industrial, and 0.17 infiltration and inflow. The design BOD and suspended solids are both 740 lb/day. During heavy rains, the peak hydraulic loading anticipated is 800 gpm for 2–3 hr. Calculate the   (a) BOD concentration at design flow and   (b) at design flow, the aeration period, volumetric BOD loading, and F/M at 2500 mg/l. How do these compare with values listed in Table 12.3?   (c) How many pounds of oxygen can be transferred to the mixed liquor per pound of BOD aeration tank loading?   (d) Calculate the overflow rate and weir loading at the peak hydraulic loading. How do these values compare with recommended design values?   (e) What is the detention in the chlorination tank at peak hydraulic flow? at twice the design flow? How do these values compare with the recommended detention time for chlorination of wastewater effluent?

**12.48** The high-purity-oxygen process illustrated in Fig. 12.42 is being considered for treatment of an unsettled domestic wastewater flow of 1.40 mgd with an average BOD of 200 mg/l. Compute the aeration volume required based on a maximum allowable BOD loading of 130 lb/1000 $ft^3$/day and minimum aeration period of 1.8 hr. For this system, what is the F/M ratio in terms of lb of BOD/day/lb of MLVSS, assuming a MLSS of 5500 mg/l, which is 75% volatile? Estimate the oxygen consumed based on the data given in Fig. 12.45.

**12.49** Size an extended aeration system for the town of Nancy (Prob. 12.32). Provide two identical activated-sludge aeration tanks with diffused aeration and final clarifiers. The treated wastewater is required to meet the effluent standards of 30 mg/l of BOD and 30 mg/l of suspended solids at a wastewater temperature of 15°C. The sludge is to be stabilized by aerobic digestion and dewatered on drying beds.

**12.50** Refer to the instructions given in Prob. 12.42.    (a) Sketch a flow diagram showing the arrangement of unit processes and major pipelines for a treatment system of primary sedimentation followed by a step-aeration secondary and sludge stabilization by anaerobic digestion.    (b) Sketch a flow diagram for a treatment system of extended aeration with aerobic sludge digestion. For both systems, list numerical design guidelines for sizing the unit processes, except preliminary treatment and sludge processing.

**12.51** List the major assumptions made in the derivation of the mathematical model for biological kinetics of the activated-sludge process. What are the limitations in applying the kinetics equations given in Section 12.24?

**12.52** A municipal wastewater containing both domestic and food-processing wastewaters was tested to determine the kinetic constants using a laboratory apparatus similar to the bench-scale unit shown in Fig. 12.48. The volume of the aeration chamber was 10 l, and the wastewater feed was established at a constant rate of 30.0 l/d to provide an 8.0-hr aeration period for all of the test runs. A measured volume of sludge was wasted once a day from the tank. Determine $Y$, $k_d$, $k$, and $K_S$ from the following laboratory data using the procedure in Example 12.16:

| $Q$ (l/d) | $S_0$ (mg/l) | $Q_w$ (l/d) | $X$ (mg/l) | $S_e$ (mg/l) | $X_e$ (mg/l) |
|---|---|---|---|---|---|
| 30.0 | 150 | 0.31 | 2460 | 2.5 | 7.7 |
| 30.0 | 150 | 0.53 | 1690 | 3.3 | 6.5 |
| 30.0 | 150 | 0.98 | 1320 | 4.4 | 5.6 |
| 30.0 | 150 | 1.27 | 1080 | 5.9 | 5.0 |

**12.53** Example 12.14 illustrates the application of the kinetics model in sizing the aeration volume required for a completely mixed activated-sludge process. What operating parameters must be assumed in the design procedure?

**12.54** A completely mixed activated-sludge process is being designed for a wastewater flow of 3.0 mgd using the kinetics equations. The influent BOD of 180 mg/l is essentially all soluble and the design effluent soluble BOD is 10 mg/l. For sizing the aeration volume, the mean cell residence time is selected to be 8.0 days and the MLVSS 2500 mg/l. The kinetic constants from a bench-scale treatability study are as follows: $Y = 0.60$ lb VSS/lb BOD, $k_d = 0.06$ day$^{-1}$, $K_s = 60$ mg/l, and $k = 5.0$ day$^{-1}$.

**12.55** The aeration tank for a completely mixed aeration process is being sized for a design wastewater flow of 7500 m$^3$/d. The influent BOD is 130 mg/l with a soluble BOD of 90 mg/l. The design effluent BOD is 20 mg/l with a soluble BOD of 7.0 mg/l. Recommended design parameters are a sludge age of 10 d and volatile MLSS of 1400 mg/l. Selection of these values takes into account the anticipated variations in wastewater flows and strengths. The kinetic constants from a bench-scale treatability study are $Y = 0.60$ mg VSS/mg soluble BOD and $k_d = 0.06$ per day.

**12.56** A step-aeration activated-sludge system at a loading of 40 lb of BOD/1000 ft$^3$/day requires an air supply of 1200 ft$^3$/lb of BOD applied to maintain an adequate dissolved-oxygen level. The measured average oxygen utilization of the mixed liquor is 36 mg/l/hr. Calculate the oxygen transfer efficiency.

**12.57** An air supply of 1000 ft$^3$ of air is required per pound of BOD applied to a diffused aeration basin to maintain a minimum DO of 2.0 mg/l. Assuming that the installed aeration equiment is capable of transferring 1.0 lb of oxygen to dissolved oxygen per pound of BOD applied, calculate the oxygen transfer efficiency of the system. (One cubic foot of air at standard temperature and pressure contains 0.0174 lb of oxygen.)

**12.58** A 10-hp surface aerator was tested in a tank filled with 9200 ft$^3$ of tap water at 22°C by the non-steady-state procedure. The dissolved-oxygen saturation was assumed to be the standard value of 8.8 mg/l from Table A.10. Based on the following time and dissolved-oxygen data, determine the value of $K_L a$ corrected to 20°C. Also calculate the oxygen transfer rate in pounds per horsepower-hour.

| $t$ (min) | $C_t$ (mg/l) | $t$ (min) | $C_t$ (mg/l) |
|------|------|------|------|
| 0 | 0 | 8.0 | 5.5 |
| 2.0 | 3.0 | 11.0 | 6.3 |
| 4.0 | 4.3 | 14.0 | 7.1 |
| 6.0 | 5.0 | 17.0 | 7.6 |

**12.59** A 40-hp surface aerator was tested in a 120-ft-diameter tank with a water depth of 8.0 ft. Cobalt chloride catalyst and sodium sulfite were added to the tap water in the tank to remove the dissolved oxygen. The water temperature was 21.9°C, and the dissolved-oxygen saturation was 8.7 mg/l. During the test, the average electric power usage was 33.1 kW, which at 90% efficiency is equivalent to 40.0 hp. From the following time and dissolved-oxygen data, determine the value of $K_L a$ corrected to 20°C, and calculate the oxygen transfer rate in pounds per horsepower-hour.

| $t$ (min) | $C_t$ (mg/l) | $t$ (min) | $C_t$ (mg/l) |
|---|---|---|---|
| 0 | 0 | 23 | 6.5 |
| 2 | 1.0 | 28 | 7.1 |
| 4 | 1.9 | 33 | 7.5 |
| 6 | 2.7 | 38 | 7.7 |
| 8 | 3.1 | 43 | 7.9 |
| 13 | 4.9 | 48 | 8.0 |
| 18 | 5.9 | 58 | 8.3 |

**12.60** A sludge volume index test was conducted on a sample of mixed liquor from an aeration tank. The suspended-solids concentration of the sample was determined as 3000 mg/l and the volume of the settled solids after 30 min of undisturbed settling in a 1-1 graduated cylinder was 540 ml. Calculate the SVI. Does this value indicate a good settling sludge?

**12.61** The following field measurements were made in a final clarifier following a conventional activated-sludge aeration tank. The average water depth from the tank bottom to the overflow weir was 3.0 m. During the critical period of maximum hydraulic loading, the thickness of the settled sludge layer on the bottom of the tank was 0.80 m. What recirculation ratio would you recommend for operation of the system?

**12.62** The mixed liquor near the discharge end of a spiral-flow aeration tank was tested for suspended-solids concentration and settleability. The MLSS concentration was 2200 mg/l and the volume occupied by the MLSS after 30-min settling in a 1-1 graduated cylinder was 200 ml. Compute the SVI, theoretical recirculation ratio for system operation, and the estimated solids concentration in the return activated sludge.

**12.63** A step-aeration process is operating at an MLSS of 3000 mg/l and an SVI of 80 ml/g. If the recirculated activated sludge is 0.45 mgd with an influent wastewater flow of 1.10 mgd, what is the suspended-solids concentration in the return sludge?

**12.64** Calculate the surface area required for a stabilization pond to serve a domestic population of 1000. Assume 80 gpcd at 210 mg/l of BOD. Use a design loading of 20 lb of BOD/acre/day. If the average liquid depth is 4 ft, calculate the retention time of the wastewater based on influent flow. The effluent is spread on grassland by spray irrigation at a rate of 2.0 in./week (54,300 gal/acre/wk). Compute the land area required for land disposal. In these computations assume no evaporation or seepage losses from the ponds.

**12.65** Facultative stabilization ponds with a total surface area of 6.0 ha (1 ha = 10,000 m²) serve a community with a waste discharge of 530 m³/d at a BOD of 280 mg/l. Calculate the BOD loading and days of winter storage available between the 0.6- and 1.5-m depths assuming a daily water loss of 0.30 cm by evaporation and seepage.

**12.66** Stabilization pond computations are required for the town of Nancy (Prob. 12.32).   (a) Calculate the lagoon area required for a design loading of 40 lb of

BOD/acre/day.    (b) Determine the percentage of the average design flow that appears as effluent from the lagoons assuming a water loss from the ponds of 60 in./yr (seepage plus evaporation minus precipitation).    (c) Using the water-balance data from part (b), calculate the BOD removal efficiency if the average BOD concentrations in the influent and effluent are 250 mg/l and 25 mg/l, respectively.    (d) How many acres of cropland are needed to dispose of the effluent by spray irrigation if the application rate is 2.0 in./week year-round?

**12.67** Design a layout of stabilization ponds for the town of Nancy (Prob. 12.32) based on the following criteria: (1) BOD loading in the primary cells cannot exceed 25 lb/acre/day, (2) minimum total water volume with the ponds full to a 5.0-ft depth in primary cells and 8.0-ft depth in secondary cells cannot be less than 120 days times the design flow, (3) volume for winter storage above water depths of 1.5 ft should be sufficient so that no discharge is necessary for a 4-month period when the net water loss (evaporation plus seepage minus precipitation) is 1.0 in./month, and (4) system should have at least two primary cells and at least two secondary cells that cannot receive raw wastewater.

**12.68** The wastewater flow from a small town is 240 m$^3$/d with a BOD of 180 mg/l and SS of 210 mg/l. Size and sketch a layout for a stabilization pond arrangement consisting of two identical primary cells, which can be operated in parallel, and one secondary cell. The primary cells can have a maximum water depth of 1.5 m, and the secondary cell can have a maximum depth of 2.5 m. The minimum operating water depth in all cells is 0.5 m. Determine the dimensions of the ponds based on the following criteria: (1) The BOD loading on the primary cells cannot exceed 4.0 g/m$^2$·d. (2) The water storage capacity considering all three cells must be at least 120 d between the minimum and maximum water levels with allowance for a water loss of 2.0 mm/d by evaporation and seepage.

**12.69** A completely mixed aerated lagoon is being considered for pretreatment of a strong industrial wastewater with $k$ = 0.70 at 20°C and $\Theta$ = 1.035, using a detention time of 4 days. What is the BOD reduction at 20°C based on Eqs. 12.103 and 12.104? If the wastewater temperature is 10°C, compute the detention time required to achieve the same degree of treatment.

**12.70** A manufacturer's specified oxygen transfer capacity of a surface-aeration unit is 3.0 lb of oxygen/hp·hr. Using Eq. 12.106, calculate the oxygen transfer capability of this unit for an $\alpha$ = 0.9, $\beta$ = 0.8, a temperature of 20°C, and a dissolved-oxygen level of 2.0 mg/l in the lagoon water.

**12.71** A completely mixed aerated lagoon with a volume of 1.0 mil gal is to treat a daily wastewater flow of 250,000 gal with a BOD of 400 mg/l. The liquid temperature ranges from 4°C in the winter to 30°C in the summer. There are four 5.0-hp surface aerators rated at 2.0 lb of oxygen/hp·hr. The wastewater characteristics are $\Theta$ = 1.035, $\alpha$ = 0.9, $\beta$ = 0.9, $k$ = 0.80 per day at 20°C, and $a$ = 1.0 lb of oxygen utilized/lb of BOD removed. The designer states that this system will remove at least 400 lb of BOD/day and maintain a dissolved-oxygen concentration greater than 2.0 mg/l. Verify these claims by appropriate calculations.

**12.72** Design a layout for an aerated lagoon followed by two secondary facultative ponds to treat a combined food-processing and domestic wastewater. The design flow is 200,000 gpd containing 1100 lb of BOD. The water temperatures in the aerated lagoon are expected to range from 2°C in winter to 25°C in summer. Determine the size and shape of the aerated lagoon and the number and horsepower of the surface aerators based on the following parameters: $\alpha = 0.9$, $\beta = 0.8$, $k = 0.90$ per day at 20°C, $a = 1.0$ lb of oxygen required per lb of BOD removed, a minimum of 80% BOD removal, and an aerator oxygen transfer rate $R_0 = 2.33$ lb of oxygen/hp·hr. The BOD loading on the two facultative ponds receiving the aerated wastewater is limited to 20 lb BOD/acre/day, and the storage volume of the ponds between the minimum water depth of 1.5 ft and maximum of 5.0 ft must be equal to or greater than 4 months of wastewater discharge at design flow.

# REFERENCES

1. *Standard Methods for the Examination of Water and Wastewater, 17th Edition* (Am. Water Works Assoc., Am. Public Health Assoc., and Water Poll. Control Fed., 1989).
2. G. J. Torlora, B. R. Funke, and C. L. Case, *Microbiology* (Menlo Park, CA: Benjamin/Cummings, 1982).
3. L. Michaelis and M. L. Menten, "Die Kinetik der Invertinwirkung," *Biochemische Zeitschrift*, Neunundvierzigster Band (Vol. 49) (Berlin: Springer-Verlag, 1913), pp. 333–369.
4. J. Monod, "The Growth of Bacterial Cultures," *Ann. Rev. Microbiol.* 3 (1949): 371–393.
5. S. J. Pirt, *Principles of Microbe and Cell Cultivation* (New York: Halsted Press Book, Wiley, 1975).
6. D. H. Stoltenberg, "Midwestern Wastewater Characteristics," *Public Works* 3(1) (1980): 52–53.
7. N. W. Classen, "Per Capita Wastewater Contributions," *Public Works* 98(5) (1967): 81–83.
8. D. G. Munksgaard and J. C. Young, "Flow and Load Variations at Wastewater Treatment Plants," *J. Water Poll. Control Fed.* 52(8) (1980): 2131–2144.
9. "Sewage Treatment at Military Installations," Report of the Subcommittee on Sewage Treatment in Military Installations, National Research Council, *Sewage Works J.* 18(5) (1946): 787–1028.
10. S. Balakrishnan, W. W. Eckenfelder, and C. Brown, "Organics Removal by a Selected Trickling Filter Media," *Water and Wastes Eng.* 6(1) (1969): A.22–A.25.
11. D. D. Drury, J. Carmona III, and A. Delgadillo, "Evaluation of High Density Cross Flow Media for Rehabilitating an Existing Trickling Filter," *J. Water Poll. Control Fed.* 58(5) (1986): 364–367.
12. M. L. Arora and M. B. Umphres, "Evaluation of Activated Biofiltration and Activated Biofiltration/Activated Technologies," *J. Water Poll. Control Fed.* 59(4) (1987): 183–190.
13. Proceedings, First National Symposium on Rotating Biological Contactor Technology (Pittsburgh: University of Pittsburgh, 1980).
14. M. J. Hammer, "Rotating Biological Contactors Treating Combined Domestic and Cheese Processing Wastewaters," *Proc. 23rd Industrial Waste Conf.*, Purdue Univ. (Ann Arbor, MI: Ann Arbor Science, 1982).
15. *Oxygen Activated-Sludge Wastewater Treatment Systems* U.S. Environmental Protection Agency, Technology Transfer (August 1973).

16. A. W. Lawrence and P. L. McCarty, "Unified Basis for Biological Treatment Design and Operation," *Proc. Am. Soc. Civil Engrs., J. San. Eng. Div.* 96 (SA3 (1970): 757–778.
17. L. D. Benefield and C. W. Randall, *Biological Process Design for Wastewater Treatment* (Englewood Cliffs, NJ: Prentice-Hall, 1980).
18. C. P. Grady, Jr., and H. C. Lim, *Biological Wastewater Treatment* (New York: Dekker, 1980).
19. G. F. Bennett and L. L. Kempe, "Oxygen Transfer in Biological Systems," *Proc. 20th Industrial Waste Conf., Purdue Univ. Ext. Service* 49(4) (1965): 435–447.
20. *Recommended Standards for Sewage Works, Great Lakes–Upper Mississippi River Board of State Public Health & Environmental Managers* (Albany, NY: Health Research Inc., Health Education Services Div., 1978).
21. American Society of Civil Engineers, *ASCE Standard: Measurement of Oxygen Transfer in Clean Water*, ISBN 0-87262-430-7 (New York, NY: Am. Soc. Civil Eng., 1984).
22. "Development of Standard Procedures for Evaluating Oxygen Transfer Devices," U.S. Environmental Protection Agency, Municipal Environmental Research Laboratory, EPA 600/2-83-002 (Cincinnati, OH: 1983).
23. "The Causes and Control of Activated Sludge Bulking and Foaming," U.S. Environmental Protection Agency, Center for Environmental Research Information, EPA 625/8-87-012 (Cincinnati, OH: July 1987).
24. E. J. Middlebrooks, "Design Equations for BOD Removal in Facultative Ponds," *Water Science Tech.* 19(12)(1987): 187–193.
25. "Odor and Corrosion Control in Sanitary Sewerage Systems and Treatment Plants," U.S. Environmental Protection Agency, Technology Transfer, Center for Environmental Research Information, EPA/625/1-85/018 (Cincinnati, OH: 1985).
26. R. A. Grace, *Marine Outfall Systems, Planning, Design, and Construction* (Englewood Cliffs, NJ: Prentice-Hall, 1978).

# Chapter 13

# Processing of Sludges

Combinations of physical, chemical, and biological processes are employed in handling sludges. While the purpose in treating water and wastewater is to remove impurities from dilute solution and consolidate them into a smaller volume of liquid, the objective of processing sludge is to extract water from the solids and dispose of the dewatered residue. Furthermore, the relationships among the processes must be understood since disposal schemes involve sequencing of operations. For example, gravity thickening, anaerobic digestion, chemical conditioning, and mechanical dewatering form a physical–biological–chemical–physical sequence. Discussions on the characteristics of wastewater sludges and water treatment plant residues are isolated in different sections. Generalized process flow diagrams are also separated, although descriptions of individual unit operations are directed toward both water and wastewater sludges. Comments relating to the applicability of each process are included.

## Sources, Characteristics, and Quantities of Waste Sludges

Mathematical relationships for estimating the specific gravity and computing the sludge volume appear first since they are fundamental calculations applicable to all sludges. Then the characteristics and methods for estimating sludge quantities are presented separately for wastewater and water treatment plant residues.

### 13.1 WEIGHT AND VOLUME RELATIONSHIPS

The majority of sludge solids from biological wastewater processing are organic with a 60%–80% volatile fraction. The concentration of suspended solids in a liquid sludge is determined by straining a measured sample through a glass-fiber filter. Nonfilterable residue, expressed in milligrams per liter, is the solids content. Since the filterable portion of a sludge is very small, sludge solids are often determined by total residue on evaporation (i.e., the total deposit remaining in a dish after evaporation of water from the sample and subsequent drying in an oven at 103°C). Volatile solids are found by igniting the dried residue at 550°C in a muffle furnace. Loss of weight upon ignition is reported as milligrams per liter of volatile solids, and the inerts remaining after burning as fixed solids. Waste from chemical coagulation of a surface water contains both organic matter removed from the raw water and mineral content derived from the chemical coagulants. Most solids are nonfilterable and have a volatile fraction of 20%–40%. Precipitate from treated well water is essentially mineral.

The specific gravity of solid matter in a sludge can be computed from the relationship

$$\frac{W_s}{S_s \gamma} = \frac{W_f}{S_f \gamma} + \frac{W_v}{S_v \gamma} \tag{13.1}$$

where   $W_s$ = weight of dry solids, lb
$S_s$ = specific gravity of solids
$\gamma$ = unit weight of water, lb/ft$^3$ (lb/gal)
$W_f$ = weight of fixed solids (nonvolatile), lb
$S_f$ = specific gravity of fixed solids
$W_v$ = weight of volatile solids, lb
$S_v$ = specific gravity of volatile solids

The specific gravity of organic matter is 1.2–1.4, while the solids in chemically coagulated water vary from 1.5 to 2.5. The value for a solids slurry is calculated from

$$S = \frac{W_w + W_s}{(W_w/1.00) + (W_s/S_s)} \tag{13.2}$$

where   $S$ = specific gravity of wet sludge
$W_w$ = weight of water, lb
$W_s$ = weight of dry solids, lb
$S_s$ = specific gravity of dry solids

Consider a waste biological sludge of 10% solids with a volatile fraction of 70%. Their specific gravity can be estimated using Eq. 13.1 by assuming values of 2.5 for the fixed matter and 1.0 for the volatile residue.

$$\frac{1.00}{S_s} = \frac{0.30}{2.5} + \frac{0.70}{1.0} = 0.82$$

$$S_s = \frac{1}{0.82} = 1.22$$

Then the specific gravity of the wet sludge, by Eq. 13.2, is 1.02:

$$S = \frac{90 + 10}{(90/1.00) + (10/1.22)} = 1.02$$

These calculations demonstrate that for organic sludges of less than 10% solids the specific gravity may be assumed to be 1.00 without introducing significant error. Example 13.1 illustrates that even for mineral residue a high concentration of precipitate is required to increase the specific gravity of a slurry above 1.0.

The volume of waste sludge for a given amount of dry matter and concentration of solids is given by

$$V = \frac{W_s}{(s/100)\gamma S} = \frac{W_s}{[(100 - p)/100]\gamma S} \tag{13.3}$$

where
$V$ = volume of sludge, ft$^3$ (gal) [m$^3$]
$W_s$ = weight of dry solids, lb [kg]
$s$ = solids content, %
$\gamma$ = unit weight of water, 62.4 lb/ft$^3$ (8.34 lb/gal) [1000 kg/m$^3$]
$S$ = specific gravity of wet sludge
$p$ = water content, %

In this formula the volume of a sludge is indirectly proportional to the solids content. Thus, if a waste is thickened from 2% to 4% solids, the volume is reduced by one-half, and if consolidation is continued to a concentration of 8%, the quantity of wet sludge is only one fourth of the original amount. During this concentration process, water content is reduced from 98% to 92%. In applying Eq. 13.3, specific gravity of the sludge $S$ is normally taken as 1.0 and therefore not included in computations, as demonstrated in Example 13.2.

### ■ EXAMPLE 13.1

Coagulation of a surface water using alum produces 10,000 lb (4540 kg) of dry solids/day, of which 20% are volatile. Both the settled sludge following coagulation and filter backwash water are concentrated in clarifier–thickeners to a solids density of 2.5%. Centrifugation can be used to increase the concentration to 20%, a consistency similar to soft wet clay, or the clarifier–thickener underflow can be dewatered to a 40% cake by pressure filtration. (a) Estimate the specific gravities of the thickened sludge, concentrate from centrifugation, and filter cake. (b) Calculate the daily sludge volumes from each process.

### *Solution*

**a.** Applying Eq. 13.1, one has

$$\frac{1.00}{S_s} = \frac{0.80}{2.50} + \frac{0.20}{1.00} = 0.52 \quad S_s = \frac{1}{0.52} = 1.9$$

From Eq. 13.2,

$$S \text{ (thickened sludge)} = \frac{97.5 + 2.5}{(97.5/1.0) + (2.5/1.9)} = 1.0$$

$$S\text{ (centrifuge discharge)} = \frac{80 + 20}{(80/1.0) + (20/1.9)} = 1.1$$

$$S\text{ (filter cake)} = \frac{60 + 40}{(60/1.0) + (40/1.9)} = 1.2$$

**b.** Substituting these values into Eq. 13.3 with $W_s = 10,000$ lb/day gives

$$V\text{ (thickened sludge)} = \frac{10,000}{(2.5/100)8.34 \times 1.0} = 48,000 \text{ gpd}$$

$$= \frac{4540 \text{ kg}}{(2.5/100)1000 \text{ kg/m}^3 \times 1.0} = 182 \text{ m}^3/\text{d}$$

$$V\text{ (centrifuge discharge)} = \frac{10,000}{(20/100)8.34 \times 1.1} = 5400 \text{ gpd } (20.6 \text{ m}^3/\text{d})$$

$$V\text{ (filter cake)} = \frac{10,000}{(40/100)8.34 \times 1.2} = 2500 \text{ gpd } (9.5 \text{ m}^3/\text{d}) \quad \blacksquare$$

■ **EXAMPLE 13.2**

Estimate the quantity of sludge produced by a trickling-filter plant treating 1.0 mgd of domestic wastewater. Assume a suspended-solids concentration of 220 mg/l in the raw wastewater, a solids content in the sludge equivalent to 90% removal, and a sludge of 5.0% concentration withdrawn from the settling tanks.

*Solution*

$$\text{solids in the sludge} = 1.0 \times 220 \times 8.34 \times 0.90 = 1650 \text{ lb/day}$$

$$\text{volume of sludge (using Eq. 13.3)} = \frac{1650}{0.05 \times 62.4} = 530 \text{ ft}^3/\text{day} \qquad \blacksquare$$

## 13.2  CHARACTERISTICS AND QUANTITIES OF WASTEWATER SLUDGES

The purpose of primary sedimentation and secondary aeration is to remove waste organics from solution and concentrate them in a much smaller volume to facilitate dewatering and disposal. The concentration of organic matter in wastewater is approximately 200 mg/l (0.02%), while that in a typical raw-waste sludge is about 40,000 mg/l (4%). Based on these approximate values, treatment of 1.0 mil gal of wastewater produces about 4000 gal of sludge. This raw, odorous and putrescible residue must be further processed and reduced in volume for land disposal or incineration. Common methods include mechanical thickening, biological digestion, and dewatering after chemical conditioning.

The quantity and nature of sludge generated relates to character of the raw wastewater and processing units employed. Daily sludge production may fluctuate over a wide range, depending on size of municipality, contribution of industrial wastes, and other factors. Both maximum and average daily sludge volumes are considered in designing facilities. A limited quantity of solids can be stored temporarily in clarifiers and aeration tanks to provide short-term equalization of peak loads. Mechanical thickening and dewatering units may be sized to handle sludge quantities as high as double the daily average. Other processes, such as conventional anaerobic digestion, have substantial equalizing capacity and are designed on the basis of maximum average monthly loading. Often, selection of conservative design parameters and liberal estimates of sludge yield take into account anticipated quantity variations. In this manner, designers disguise the fact that maximum sludge yield is being considered in sizing unit processes. For example, in designing a trickling-filter plant, the dry solids may be calculated assuming 0.24 lb/capita/day, excluding household garbage grinders, when the actual amount realized in the treatment process is closer to 0.15 lb/capita/day. Furthermore, the required digester volume may be computed using a conservative figure of 5 ft$^3$/population equivalent.

*Primary sludge* is a gray-colored, greasy, odorous slurry of settleable solids, accounting for 50%–60% of the suspended solids applied, and tank skimmings. Scum is usually less than 1% of the settled sludge volume. Primary precipitates can be dewatered readily after chemical conditioning because of their fibrous and coarse nature. Typical solids concentrations in raw primary sludge from settling municipal wastewater are 6%–8%. The portion of volatile solids varies from 60% to 80%.

*Trickling-filter humus* from secondary clarification is dark brown in color, flocculent, and relatively inoffensive when fresh. The suspended particles are biological growths washed from the filter media. Although they exhibit good settleability, the precipitate does not compact to a high density. For this reason and because sloughing is irregular, underflow from the final clarifier containing filter humus is returned to the wet well for mixing with the inflowing raw wastewater. Thus humus is settled with raw organics in the primary clarifier. The combined sludge has a solids content of 4%–6%, which is slightly thinner than primary residue with raw organics only.

*Waste-activated sludge* is a dark-brown, flocculent suspension of active microbial masses inoffensive when fresh, but it turns septic rapidly because of biological activity. Mixed-liquor solids settle slowly, forming a rather bulky sludge of high water content. The thickness of return activated sludge is 0.5%–2.0% suspended solids with a volatile fraction of 0.7–0.8. Excess activated sludge in most processes is wasted from the return sludge line. A high water content, resistance to gravity thickening, and the presence of active microbial floc make this residue difficult to handle. Routing of waste activated to the wet well for settling with raw wastewater is not recommended. Carbon dioxide, hydrogen sulfide, and odorous organic compounds are liberated from the settlings in the primary basin as a result of anaerobic decomposition, and the solids concentration is rarely greater than 4%. Waste-activated sludge can be thickened effectively by flotation or centrifugation; however, chemical additions may be needed to ensure high solids capture in the concentrating process.

*Anaerobically digested sludge* is a thick slurry of dark-colored particles and entrained gases, principally carbon dioxide and methane. When well digested, it dewaters rapidly on sand-drying beds, releasing an inoffensive odor resembling that of garden loam. Substantial additions of chemicals are needed to coagulate a digested sludge prior to mechanical dewatering, owing to the finely divided nature of the solids. The dry residue is 30%–60% volatile, and the solids content of digested liquid sludge ranges from 6% to 12%, depending on the mode of digester operation.

*Aerobically digested sludge* is a dark-brown, flocculent, relatively inert waste produced by long-term aeration of sludge. The suspension is bulky and difficult to thicken, thus creating problems of ultimate disposal. Since decanting clear supernatant can be difficult, the primary functions of an aerobic digester are stabilization of organics and temporary storage of waste sludge. The solids concentration in thickened, aerobically digested sludge is generally in the range 1.0%–2.0% as determined by digester design and operation. The thickness of aerobically digested sludge can be less than that of the influent, since approximately 50% of the volatile solids are converted to gaseous end products. Stabilized sludge, expensive to dewater, is often disposed of by spreading on land for its fertilizer value. For these reasons, aerobic digestion is generally limited to treatment of waste activated from aeration plants without primary clarifiers.

*Mechanically dewatered sludges* vary in characteristics based on the type of sludge, chemical conditioning, and unit process employed. The density of dewatered cakes ranges from 15% to 40%. The thinner cake is similar to a wet mud, while the latter is a chunky solid. The method of ultimate disposal and economics dictate the degree of moisture reduction necessary.

Waste solids production in primary and secondary processing can be estimated using the following formulas:

$$W_s = W_{s_p} + W_{s_s} \tag{13.4}$$

where     $W_s$ = total dry solids, lb/day
           $W_{s_p}$ = raw primary solids, lb/day
           $W_{s_s}$ = secondary biological solids, lb/day

$$W_{s_p} = f \times SS \times Q \times 8.34 \tag{13.5}$$

where     $W_{s_p}$ = primary solids, lb of dry weight/day
           $f$ = fraction of suspended solids removed in primary settling
        SS = suspended solids in unsettled wastewater, mg/l
          $Q$ = daily wastewater flow, mgd
    8.34 = conversion factor, lb/mil gal per mg/l

$$W_{s_s} = k \times BOD \times Q \times 8.34 \tag{13.6}$$

where    $W_{s_s}$ = biological sludge solids, lb of dry weight/day
         $k$ = fraction of applied BOD that appears as excess biological growth in waste-activated sludge or filter humus, assuming about 30 mg/l of BOD and suspended solids remaining in the secondary effluent
         BOD = concentration in applied wastewater, mg/l
         $Q$ = daily wastewater flow, mgd

The first expression simply states that the total weight of dry solids produced equals the sum of the primary plus secondary residues. Settleable matter removed in primary clarification can be considered to be a function of the suspended-solids concentration (Eq. 13.5). For typical municipal wastes, the value for $f$ is between 0.4 and 0.6. The settleable fraction of suspended solids in a fresh domestic wastewater is about 0.5, but septic conditions and industrial waste contributions are likely to decrease the portion of settlings in a wastewater. For example, many food-processing discharges are high in colloidal matter and exhibit BOD/suspended solids ratios of 2:1 or greater. Thus, a combined wastewater may exhibit an $f$ value that is considerably less than the average 0.5 for domestic waste.

Organic matter entering secondary biological treatment is colloidal in nature and best represented by its BOD value. Most is synthesized into flocculent biological growths that entrain nonbiodegradable material. Therefore, excess activated-sludge solids from aeration and humus from biological filtration can be estimated by Eq. 13.6, which relates residue production to BOD load. The coefficient $k$ is a function of process food/microorganism ratio and biodegradable (volatile) fraction of the matter in suspension. For trickling-filter humus, $k$ is assumed to be in the range 0.3–0.5, with the lower value for light BOD loadings and the larger number applicable to high-rate filters and rotating biological contactors. The $k$ for secondary activated-sludge processes can be estimated using Fig. 13.1 by entering the diagram along the ordinate with a known food/microorganism ratio.

Excess solids production for activated-sludge processes treating unsettled wastewater can be estimated using Eq. 13.6 (based on influent BOD), without considering suspended solids input, by increasing the calculated quantity by 100%. Thus, for aeration systems without primary clarifiers, the $k$ factor is the value determined from Fig. 13.1 multiplied by 2.0.

The design of a sludge-handling system is based on the volume of wet sludge as well as dry solids content. Once the dry weight of residue has been estimated, the volume of sludge is calculated by applying Eq. 13.3.

The foregoing formulations are reasonable for sludge quantities from processing domestic wastewater at average daily design flow. Real sludge yields may differ considerably from anticipated values when treating a municipal discharge containing substantial contributions of industrial wastes and when loading or operational conditions create unanticipated peak sludge volumes.

Peak loads must be assessed for each treatment plant design based on

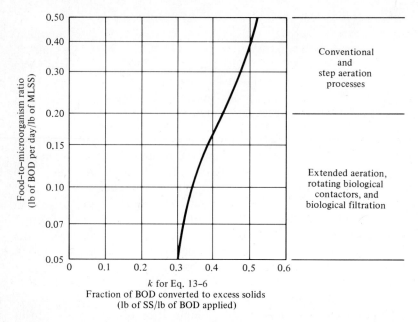

**Figure 13.1** Hypothetical relationship between the food-to-microorganism ratio and the coefficient $k$ in Eq. 13.6.

local conditions such as seasonal industrial discharges, anticipated trends in per capita solids contribution, and the type of unit operations employed in wastewater processing. A rule-of-thumb approach assumes that the maximum weekly dry solids yield will be approximately 25% greater than the yearly mean. Variations in daily sludge volumes may be considerably greater, owing to changes in moisture content. For example, if the concentration of settlings drawn from primary clarifiers shifts from 6% to 4%, the total quantity of wet sludge increases 50% for the same amount of dry solids. These deviations are normally taken into account by selecting conservative design criteria in sizing biological digesters. Where mechanical dewatering is practiced, increased quantities of sludge can be handled by applying higher chemical dosages and by extending the time of operation each day. Perhaps the most difficult parameter to predict is the compactness of waste-activated sludge. Bulking can easily reduce the solids content by one-half, say from 15,000 mg/l to 7500 mg/l, thus doubling the volume. Processing of this larger quantity must be evaluated by the designer. One guideline is to size mechanical thickeners at 200% of the estimated volume during normal operation in order to ensure consolidation of the diluted slurry. Also, provisions should be made for the addition of coagulants to aid in concentrating waste-activated sludge.

### ■ EXAMPLE 13.3

One million gallons of municipal wastewater with a BOD of 260 mg/l and suspended solids of 220 mg/l are processed in a primary plus secondary activated-sludge plant. Estimate the quantities and solids contents of the primary, waste-activated, and mixed sludges. Assume the following: 60% SS removal and 40% BOD reduction in clarification, a water content of 94.0% in the raw sludge, an operating F/M ratio of 1:3 in the aeration basin, and a solids concentration of 15,000 mg/l in the waste-activated sludge. If the sludges are blended and gravity thickened to a solids content of 6.0%, calculate the consolidated sludge volume, assuming 95% solids capture.

*Solution.* The primary sludge solids and volume based on Eqs. 13.5 and 13.3 are, respectively,

$$W_{s_p} = 0.60 \times 220 \times 1.0 \times 8.34 = 1100 \text{ lb}$$

$$V = 1100/[(100 - 94)/100]8.34 = 2200 \text{ gal}$$

To determine $k$ for Eq. 13.6, enter Fig. 13.1 with F/M $= 0.33$, and read 0.48 lb of SS produced/lb of BOD applied.

The waste-activated sludge, based on Eq. 13.6, is

$$W_{s_s} = 0.48 \times 260 \times 1.0 \times 0.60 \times 8.34 = 620 \text{ lb}$$

$$V = \frac{620}{0.015 \times 8.34} = 5000 \text{ gal}$$

The blended sludge volume, solids content, and solids concentration are, respectively,

$$V = 2200 + 5000 = 7200 \text{ gal}$$

$$W_s = 1100 + 620 = 1720 \text{ lb}$$

$$s = \frac{1720 \times 100}{7200 \times 8.34} = 2.9\%$$

The thickened sludge volume is

$$V = \frac{0.95 \times 1720}{0.06 \times 8.34} = 3300 \text{ gal} \qquad \blacksquare$$

## ■ EXAMPLE 13.4

The rational formulas for waste solids production (Eqs. 13.4–13.6) can be used to predict future sludge yields, provided the characteristics of the wastewater are not expected to significantly change.

Listed in columns 1–5 of Table 13.1 are average monthly operating data for a conventional activated-sludge plant treating a municipal wastewater containing discharges from several industries. The average suspended solids removal in primary sedimentation is 53% ($f = 0.53$) and the BOD removal is 22%, leaving an average of 490 mg/l in the settled wastewater. The current solids concentration in the combined waste sludge varies from 4.0% to 4.5%. The sludge is dewatered by belt filter presses 5 days per week using a sludge holding tank for storing weekend production.

Develop equations to calculate the average annual weekday sludge solids production and the average weekday solids production during the maximum month. Calculate these values and sludge volumes at 4.0% solids concentration for a future raw-wastewater flow of 3.0 mgd with 600 mg/l suspended solids and 580 mg/l BOD.

**TABLE 13.1** OPERATING DATA AND CALCULATED VALUES OF SOLIDS PRODUCTION FOR EXAMPLE 13.4

| Month | Flow (mgd) | Measured values | | | Calculated values | |
|---|---|---|---|---|---|---|
| | | Suspended solids in unsettled wastewater (mg/l) | BOD in settled wastewater (mg/l) | Sludge solids produced from wastewater (mg/l) | Sludge solids production from Eq. 13.7 (mg/l) | Percentage difference of columns 5 and 6 (%) |
| Feb. | 2.3 | 480 | 480 | 340 | 410 | −17 |
| Mar. | 2.5 | 620 | 460 | 530 | 480 | 10 |
| Apr. | 2.5 | 640 | 500 | 660 | 500 | 32 |
| May | 2.3 | 780 | 600 | 460 | 610 | −24 |
| June | 2.5 | 740 | 400 | 580 | 520 | 11 |
| July | 2.4 | 870 | 550 | 450 | 640 | −30 |
| Aug. | 2.4 | 600 | 430 | 450 | 460 | −2 |
| Sept. | 2.5 | 460 | 480 | 520 | 400 | 30 |
| Mean | 2.4 | 650 | 490 | 500 | 500 | |

*Solution.*    The average solids production expressed in milligrams per liter of wastewater processed is

$$W_s = W_{s_p} + W_{s_s} = f \times SS + k \times BOD$$

We substitute the mean measured values from Table 13.1 and solve for $k$:

$$500 \text{ mg/l} = 0.53 \times 650 + k \times 490$$

$$k = 0.32$$

The equation for average annual weekday sludge solids production in milligrams per liter of wastewater processed is then

$$W_s = 0.53 \times SS + 0.32 \times BOD \tag{13.7}$$

Using this equation, the sludge solids production are calculated for each month and listed in column 6 of Table 13.1. The percentage difference between the measured and calculated solids production values is computed and listed in column 7 by subtracting the numbers in column 6 from those in column 5 and dividing the resultants by the numbers in column 6 and multiplying by 100 to yield percentages.

Based on Eq. 13.7, the equation for average annual weekday sludge solids production in pounds based on a 5-day work week is

$$W_{s_d} = (7/5)(Q)(8.34)(0.53 \times SS + 0.32 \times BOD)$$

The results in column 7 of Table 13.1 show that the average weekday solids production during the maximum month is 30% greater than the average; therefore,

$$W_{s_m} = 1.3 W_{s_d}$$

For a future raw-wastewater flow of 3.0 mgd with 600 mg/l suspended solids and 580 mg/l of BOD (settled BOD of $0.78 \times 580 = 450$ mg/l), the average annual values are

$$W_{s_d} = (7/5)(3.0)(8.34)(0.53 \times 600 + 0.32 \times 450) = 16,000 \text{ lb}$$

$$V_{s_d} = \frac{16,000}{0.04 \times 8.34} = 48,000 \text{ gal}$$

The average weekday values during the maximum month are $W_{s_m} = 20,000$ lb and $V_{s_m} = 61,000$ gal. ∎

## 13.3 CHARACTERISTICS AND QUANTITIES OF WATER-PROCESSING SLUDGES

The characteristics, methods of processing, and disposal of water treatment plant wastes depend on both the kind of water treatment and issuance of a wastewater discharge permit. Regulations for disposal require compliance with surface-water-quality standards and the application of best available technology in treatment of the wastes [1]. Water treatment residues are derived from sedimentation and filtration of chemically conditioned water. Surface supplies yield wastes containing colloidal matter removed from the raw water and chemical flocs, while groundwater-processing precipitates are mineral with little or no organic material. Sludges vary widely in composition depending on character of the water source and chemicals added during treatment. A typical method of handling a turbid river supply includes presedimentation for reduction of settleable solids, lime softening, alum coagulation, and filtration for removal of colloids, plus the addition of activated carbon for taste and odor control. The presedimentation deposit is silt plus detritus; settling-basin sludge is a mixture of inerts, organics, and chemical precipitates, including metal hydroxides; and filter backwash water contains floc from agglomerated colloids and unspent coagulant hydroxides. Lake and reservoir waters are often dosed with alum and flocculation aids, plus activated carbon. Settlings during the summer may include significant quantities of algae. Precipitates from lime–soda ash softening are predominantly calcium carbonate and magnesium hydroxide with traces of other minerals, such as oxides of iron and manganese.

Sludge storage capacity and the time intervals between withdrawals are governed by the installation design, the type of water processing, and operations management. Clarifiers equipped with mechanical scrapers discharge sludge either continuously at a low rate or intermittently, often daily. Settled sludge is allowed to accumulate and consolidate in plain rectangular or hopper-bottomed basins. These tanks are cleaned at time intervals varying from a few weeks to several months by draining and removing the compacted sludge. Backwashing of filters produces a high flow of dilute wastewater for a few minutes usually once a day for each filter. Obviously, any system for handling water treatment wastes must consider temporary storage and thickening of wash water.

*Alum-coagulation sludge* is dramatically influenced by the gelatinous nature of the aluminum hydroxides formed in the reaction with raw-water alkalinity. Particles entrained in the floc and other coagulation precipitates do not suppress the jellylike consistency that makes an alum slurry difficult to dewater. Coagulation settlings and backwash water can normally be gravity thickened to about 2%–3%, although polymers may be needed to achieve this consolidation. Studies have shown that centrifugal dewatering can concentrate this waste to a truckable 10%–15% with a consistency similar to a soft wet clay. Pressure filtration will produce a 30%–40% cake that breaks easily. Complete dehydration by drying or freezing results in a granular material that does not revert to its original gelatinous form if again mixed with water. Enmeshed water of hydration, not water of suspension, causes the original jelly consistency. Iron coagulants yield slightly denser sludges that are somewhat easier to handle.

Surface-water wastes are highly variable, owing to changes in raw-water quality. High turbidities during spring runoff and periods of high rainfall result in a decreased percentage of aluminum hydroxide solids. The result is a precipitate that settles better and is easier to dewater. Water temperature changes affect algal growth in surface supplies, the rate of chemical reactions in treatment, and filterability of the sludge. In designing a waste-handling system, changes in raw-water quality and accompanying variations in sludge characteristics must be investigated by long-term studies of daily and seasonal records.

Sludge production from surface-water treatment can be estimated from chemical additions and raw-water characteristics. Based on empirical data [1], 1 mg of commercial alum applied as a coagulant produces 0.44 mg of aluminum precipitate, which is nearly double the 0.26 theoretical amount (Eq. 11.34). The observed relationship between turbidity of the raw water, expressed in nephelometric turbidity units (NTU), and weight of impurities removed varies from 0.5 to 2.0 mg/NTU with an average value of 0.74. The following equations for estimating dry sludge solids production from alum coagulation are based on these data:

Total sludge solids (lb/mil gal)
$$= 8.34(0.44 \times \text{alum dosage} + 0.74 \times \text{turbidity}) \qquad (13.8)$$

Sedimentation basin solids (lb/mil gal)
$$= \text{total sludge solids} - \text{solids to filters} \qquad (13.9)$$

The sludge withdrawn from sedimentation basins is 1%–2%, and the solids content of filter wash water is less than 0.1%. Gravity thickening of settled sludge and backwash water from alum coagulation can be thickened in a clarifier–thickener to 3%–6%.

*Coagulation–softening sludges* result from processing hard, turbid surface waters, such as those found in Midwestern rivers. A typical treatment plant flow arrangement is presedimentation followed by two-stage or split treatment; lime softening and coagulation with alum or iron salts; and filtration. Solids concentration in settled sludges varies with turbidity in the raw water, ratio of calcium to magnesium in the softening precipitate, type and dosage of metal coagulant, and filter aids used. In general, filter wash water gravity thickens to about 4%, alum–lime sludges have densities of up to 10%, and lime–iron precipitates range between 10% and 25%. The quantity of sludge produced is difficult to predict because the chemical treatment varies with hardness and turbidity of the river water. In-plant modifications to help even out fluctuations in sludge yield may include applying polymers in coagulation, closer pH control to reduce the amount of magnesium hydroxide produced in softening, and providing flexibility to thicken sludges separately or combined.

*Lime–soda ash softening sludges* produced in treating groundwaters contain calcium carbonate and to a lesser extent magnesium hydroxide. Aluminum hydroxides and other coagulant aids may be present if added in water processing. The quantity of $Mg(OH)_2$ depends on magnesium hardness in the raw water and the softening process employed. In general, dry solids are 85%–95% $CaCO_3$. The residue is stable, dense, inert, and relatively pure, since groundwater does not contain colloidal inorganic or organic matter.

Settled solids concentrations range from 2% to greater than 15%, based on the ratio of calcium to magnesium precipitates, unit operations of the softening process, and method of sludge withdrawal [1].

Calcium carbonate compacts readily while magnesium hydroxide, like aluminum hydroxide, is gelatinous and does not consolidate as well nor dewater as easily. Slurry wasted from flocculator–clarifiers (upflow units) has a solids content in the range 2%–5%. Stored sludges in manually cleaned sedimentation tanks of straight-line systems compact to 10% solids concentration or greater. Stoichiometrically, from Eq. 13.10, 3.6 lb of calcium carbonate is precipitated for each pound of lime applied. However, in actual practice the dry solids yield from softening is closer to 2.6 lb/lb of lime applied owing to incomplete chemical reaction, impurities in commercial-grade lime, and precipitation of variable amounts of magnesium.

$$CaO + Ca(HCO_3)_2 = 2CaCO_3 + H_2O \tag{13.10}$$

*Iron and manganese oxides* removed from groundwater by aeration and chemical oxidation are flocculent particles with poor settleability. The amount of sludge produced in the removal of these metals without simultaneous precipitation softening is relatively small. The majority of hydrated ferric and manganic oxides pass through sedimentation tanks, are trapped in the filters, and appear in the dilute backwash water.

*Filter wash water* is a relatively large volume of wastewater with a low solids concentration of 100–1000 mg/l. The exact amount of water used in backwashing is a function of the type of filter system, cleansing technique, and quality and source of the raw water being treated. Generally, 2%–3% of the water processed in a plant is used for filter washing. The fraction of total waste solids removed by filtration depends on efficiency of the coagulation and sedimentation stages, type of treatment system, and characteristics of the raw water. The amount may be a substantial portion, say 30%, of the dry solids resulting from treatment.

### ■ EXAMPLE 13.5

A reservoir water supply with a turbidity of 10 units in the summer is treated by applying an alum dosage of 30 mg/l. For each million gallons of water processed, estimate the total solids production, volume of settled sludge, and quantity of filter wash water. Assume a settled sludge concentration of 1.5% solids, a filter solids loading of 26 lb/mil gal, and a backwash water with 400 mg/l of suspended solids; also assume that Eqs. 13.8 and 13.9 are applicable. Compute the composite sludge volume after the two wastes are gravity thickened to 3.0% solids.

*Solution.*    By Eqs. 13.8 and 13.9,

$$\text{total sludge solids} = 8.34(0.44 \times 30 + 0.74 \times 10) = 172 \text{ lb/mil gal}$$

$$\text{solids to filters} = 26 \text{ lb/mil gal}$$

$$\text{sedimentation basin sludge} = 172 - 26 = 146 \text{ lb/mil gal}$$

Applying Eq. 13.3 yields

$$V \text{ of settled sludge} = \frac{146}{0.015 \times 8.34} = 1170 \text{ gal}$$

$$V \text{ of wash water} = \frac{26}{(400/1,000,000)8.34} = 7800 \text{ gal}$$

$$V \text{ of thickened sludge} = \frac{172}{0.030 \times 8.34} = 690 \text{ gal} \qquad \blacksquare$$

## ■ EXAMPLE 13.6

The lime–soda ash softening process described in Example 11.5 requires a lime dosage of 2.7 meq/l CaO and a soda ash addition of 1.2 meq/l of $Na_2CO_3$. Based on the appropriate chemical reactions, calculate the calcium carbonate residue produced in the softening of $10^6$ $m^3$ of water, assuming the practical limit of $CaCO_3$ precipitation is 0.60 meq/l (30 mg/l).

*Solution.* Based on Eq. 11.50, the 0.4 meq/l of $CO_2$ is precipitated by addition of 0.4 meq/l of lime to form 0.4 meq/l of $CaCO_3$:

$$Ca(OH)_2 + CO_2 = CaCO_3 \downarrow + H_2O$$

From Eq. 11.45, the 2.3 meq/l of $Ca(HCO_3)_2$ reacts with 2.3 meq/l of lime to form 4.6 meq/l of $CaCO_3$:

$$Ca(HCO_3)_2 + Ca(OH)_2 = 2CaCO_3 \downarrow + 2H_2O$$

Finally, 1.2 meq/l of soda ash precipitates 1.2 meq/l of $CaCO_3$ by the reaction of Eq. 11.49:

$$CaSO_4 + Na_2CO_3 = CaCO_3 \downarrow + Na_2SO_4$$

The residue is theoretically equal to the stoichiometric quantities of $CaCO_3$ formed minus the practical limit of treatment (solubility):

$$0.4 + 4.6 + 1.2 - 0.6 = 5.6 \text{ meq/l}$$

$$5.6 \times 50 = 280 \text{ mg/l of } CaCO_3$$

and

$$280 \text{ g/m}^3 \times 10^6 \text{ m}^3 \times 10^{-3} \text{ kg/g} = 280,000 \text{ kg/} 10^6 \text{ m}^3 \qquad \blacksquare$$

# Arrangement of Unit Processes in Sludge Disposal

Many processes for sludge handling are applied to both wastewater sludges and water treatment plant residues. Individual unit operations are discussed in the latter sections of this chapter with comments relative to their method of operation and application. Under this heading, selection and arrangement of units are outlined to illustrate how they integrate with each other. Referring to these discussions while studying the individual operations will help relate them to complete sludge-disposal schemes.

## 13.4  SELECTION OF PROCESSES FOR WASTEWATER SLUDGES

Techniques selected for processing waste sludges are a function of the type, size, and location of the wastewater plant, the unit operations employed in treatment, and the method of ultimate solids disposal [2, 3]. The system adopted must be able to accept the primary and secondary sludges produced and economically convert them to a residue that is environmentally acceptable for disposal.

Methods for storage, treatment, and disposition are listed in Table 13.2.

**TABLE 13.2**  PROCESSES FOR STORAGE, TREATMENT, AND DISPOSING OF WASTEWATER SLUDGES

Storage prior to processing
  In the primary clarifiers
  Separate holding tanks
Thickening prior to dewatering or digestion
  Gravity settling
  Dissolved air flotation
  Centrifugation
Conditioning prior to dewatering
  Stabilization by anaerobic digestion
  Stabilization by long-term aeration
  Chemical coagulation
  Heat treatment or wet oxidation
Mechanical dewatering
  Pressure filtration
  Centrifugation
  Vacuum filtration
  Heat drying
Composting
Air drying of digested sludge
  Sand drying beds
  Shallow lagoons
Disposal of liquid digested sludge
  Spreading on agricultural land
  Piping into the ocean
Disposal of dewatered solids
  Burial in sanitary landfill
  Incineration
  Production of soil conditioner

Settled raw solids may be stored in the bottom of primary clarifiers during the day, or perhaps over a weekend, and then pumped directly to a processing unit. In small plants, sludge is transferred to anaerobic digesters either once or twice a day. Vacuum filters and belt filter presses require a steady flow of sludge during the operating period, which may extend from 4 to 24 hr/ day. Separate holding tanks can be used to receive and blend primary and secondary sludges. The former may be pumped intermittently using an automatic time-clock control, while the latter flow may be either continuous or periodic. Aerated holding tanks for accumulating and biologically stabilizing waste from complete aeration plants are called aerobic digesters.

The economics of chemical conditioning and biological treatment are directly related to the sludge density. As a general rule, the solids content of settled waste must be at least 4% for feasible handling. Primary precipitates and mixtures of primary and secondary settlings are amenable to compaction by sedimentation; therefore, gravity thickeners are used to increase the concentration of sludges withdrawn from either clarifiers or holding tanks. Because of the flocculent nature of waste-activated sludge, separate thickening is performed by either dissolved air flotation or centrifugation. The float or concentrate is then blended with primary waste and directed to holding tanks or the next dewatering or treatment step.

The intention of anaerobic digestion is to convert bulky, odorous, and putrescible raw sludge to a well-digested material that can be rapidly dewatered without emission of noxious odors. In addition to stabilizing and gasifying the organic matter, the volume of residue is significantly reduced by withdrawal of supernatant from digesters to thicken the sludge. Aerobic digestion is almost exclusively used to treat excess sludge from plants without primary clarifiers. Although it stabilizes the organic matter, solids thickening and dewatering are troublesome, owing to the bulky nature of overaerated sludge. Chemical coagulation with polymers, or metal coagulants and lime, is required for mechanical dewatering of both raw and digested wastes.

Heat treatment of raw sludge for a short period of time under pressure results in coagulation of the solids. This conditioning sterilizes, deodorizes, and prepares the waste for mechanical dewatering without addition of chemicals. Processes using heat treatment apply steam to heat the reactor vessel to about 300°F under a pressure of 150 psi or greater. Sludge from the reactor is discharged to a decant tank, from which the underflow is withdrawn for dewatering. The supernatant contains high concentrations of water-soluble organic compounds and must therefore be returned to the treatment plant for processing. This is a serious drawback in handling some sludges, since it can result in cycling of solids through the heating process back to the treatment plant, which extracts them and returns them again to the heat treatment. Wet oxidation can be achieved under high pressure at elevated temperatures. Liquid sludge with compressed air is fed into a pressure vessel, where the organic matter is stabilized. Inert solids are separated from the effluent by dewatering in lagoons or by mechanical means. Pyrolysis (heating without oxygen) and freezing have also been investigated as sludge-conditioning methods. Significant development is still required before either process can be applied effectively.

Belt filter presses, vacuum filters, and centrifuges are employed in dewatering chemically conditioned sludges. Belt presses are effective in dewatering a wide range of waste concentrations from raw primary to thin aerobically digested sludges. In operating performance, the main advantages are low energy consumption, high cake density, and clarity of filtrate. In the past, vacuum filtration was widely adopted for dewatering both raw and anaerobically digested sludges. However, in comparison to belt filter presses, vacuum filters are costly to operate because of higher energy consumption, time-consuming procedures for start-up and shutdown, and the additional cost of disposal of a wetter cake. For very small treatment plants, any kind of mechanical dewatering may be uneconomical because of the minimum available sizes of dewatering units. The smallest belt filter press can dewater the sludge production from a population of 5000, the smallest vacuum filter from a population of 10,000. The minimum size of centrifuges restricts their application to even larger treatment plants generally serving a population greater than 100,000. Although continuous improvement in machine design is broadening the field of centrifugation, the most prevalent current practice is the dewatering of anaerobically digested sludge in large cities. Mechanical heat drying is associated with the production of soil conditioner, and incineration is used in locations where less costly disposal of dewatered solids is not possible.

The oldest technique for drying digested sludge is on open sand beds contained by short concrete walls. Well-digested slurry is drained or pumped onto the surface of the bed to a depth of 8–12 in. Moisture leaves by evaporation and seepage; the latter is collected in underdrain piping for return to the plant influent. In current design, the beds have vertical draw-off pipes to decant supernatant after the solids settle and paved surfaces that gently slope to a narrow strip of sand bed with an underdrain located along the center line of the bed. This construction allowed cleaning with a small front-end loader, since manual cleaning with shovels is no longer feasible. Shallow lagoons can also be used for air drying of digested sludge. This permits the use of front-end loaders or buckets on draglines, but operations can be disrupted by inclement weather. Towns in agricultural regions may dispose of liquid digested sludge by spreading on farmland. Often, land disposal is originated by the plant operator to avoid the problem of handling the dried humus cake. Led by this precedent, engineers are now designing land application schemes rather than open drying beds for new installations.

Dewatered solids may be disposed of in sanitary landfills. Dried anaerobic cake and incinerator ash are relatively inoffensive, but cake composed of raw solids is putrescible and potentially pathogenic. The latter must be covered every day to prevent nuisances and health hazards. Incineration is much more costly than land disposal, yet in highly urbanized areas it is often the only feasible alternative. Rather than burning, digested sludge may be dried and used as a soil conditioner.

The following process flow diagrams graphically illustrate the selection and arrangement of common sludge-processing layouts. Figure 13.2 is typical for communities with a population of less than 10,000. Raw solids and filter humus are settled and stored in the primary clarifiers. Once or twice a day, sludge and scum are pumped to anaerobic digesters, and supernatant is withdrawn and returned to the treatment plant influent. Stabilized and thickened sludge accumulates in the digesters for withdrawal when weather conditions permit disposal. The residue may be eliminated by spreading the liquid on agricultural land, or by drying on sand beds or in lagoons followed by hauling to land burial.

Many activated-sludge plants that use a flow scheme returning secondary biological floc to the plant influent experience difficulties resulting from anaerobiosis and thinning of settled sludge in the primary clarifiers. As a result, the two waste streams are often separated and thickening units employed to increase solids concentrations. In Fig. 13.3a the waste-activated sludge is thickened independently by dissolved air flotation, a process that gives reliable and effective results. Water of separation is returned for reprocessing, while the float is pumped to mixed holding tanks along with the primary sludge. The tanks are sized so that sludge can accumulate when sludge is not being dewatered, and mixing impellers are installed to ensure a homogeneous feed for chemical conditioning. Dewatered solids are eliminated by sanitary landfill or incineration. An alternative arrangement, Fig. 13.3b, blends waste-activated and raw sludges in a gravity thickener. Consolidation of this waste mixture may yield only marginal results because of carryover of flocculent solids, thus providing poor solids capture. Superna-

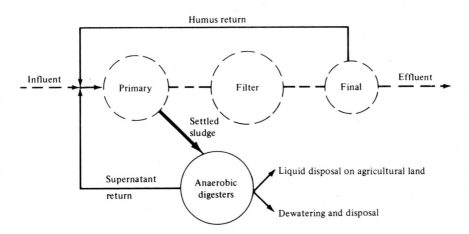

**Figure 13.2** Typical sludge-handling scheme for a trickling-filter plant serving a community of fewer than 10,000 people.

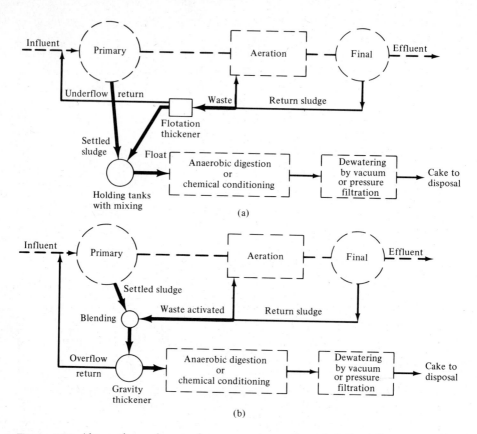

**Figure 13.3** Alternative schemes for processing activated-sludge plant wastes by thickening in advance of conditioning and dewatering. (a) Separate flotation thickening of waste-activated sludge before mixing with primary settlings. (b) Gravity thickening of combined raw primary and waste-activated sludges.

tant from the thickener is returned to the plant inlet, and underflow is pumped to anaerobic digesters. The digested sludge is dewatered and the cake spread on agricultural land.

Aerobic processing of wastewater without primary sedimentation often utilizes aerobic digestion for stabilization (Fig. 13.4). Decanting chambers for return of supernatant, installed in the digesters of package plants, rarely provide efficient solids separation. However, tanks with separate controls for aeration and decanting can be effective in concentrating aerobically digested solids. The most common method for eliminating the liquid stabilized sludge is by spreading on farmland.

Designing a sludge-processing system requires a thorough understanding of the characteristics of the waste being produced and the most feasible method for solids disposal. Selection of the latter is dictated by local conditions and practices. Intermediate steps of thickening, treatment, and dewater-

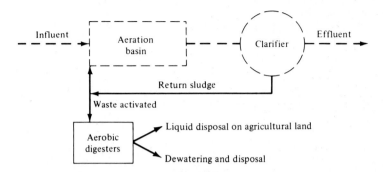

**Figure 13.4** Common disposal methods for waste-activated sludge from small treatment plants without settling prior to aeration.

ing must be integrated so that each relates to the prior operation and prepares the residue for subsequent handling. A scheme should have flexibility to allow alternative modes of operation, since actual conditions may differ from those assumed at the time of design. Too often built-in rigidity, by limiting piping and pumping facilities, does not permit plant personnel to vary operations to meet changing conditions. The practice of sanitary design requires both knowledge and foresight to consider all available options.

The following illustrate typical sludge processing and disposal problems that have confronted plant operators:

1. Returning excess activated sludge to the head of the treatment plant with no alternative for separate thickening or disposal. The result can be upset primary clarifiers, reduced efficiency, thin sludge, and higher chemical costs for conditioning.
2. Employing gravity thickeners to concentrate combined primary and waste-activated sludges (Fig. 13.2) or waste-activated sludge alone. Besides limited thickening of the sludges, the thickener overflow returns to the head of the plant a substantial quantity of suspended solids that are then recycled through the system.
3. Dewatering of sludge with a high proportion of waste-activated solids, or a sludge containing a significant amount of fine solids from an industrial wastewater, by a vacuum filter with a coil-spring medium or by a solid-bowl scroll centrifuge. Because of poor solids capture, the filtrate carries a large portion of solids and BOD back to the head of the plant, resulting in higher chemical consumption in sludge conditioning and long operating time for sludge dewatering.
4. Returning poor-quality digester supernatant to the head of a treatment plant increases the BOD load and recycling of suspended solids.
5. Excessive concentrations of heavy metals in biologically stabilized sludges that limit their application on agricultural land.
6. Being forced to incinerate sludge when an alternate method would be much less expensive if facilities were available.
7. Providing only sand-drying beds for dewatering anaerobic or aerobically digested sludges at small plants with no provision for spreading on adjacent

grassland or farmland. Plant operators usually modify the piping and buy a tank wagon or truck for liquid disposal because of the labor involved in scooping up dried cake by hand.

# 13.5  SELECTION OF PROCESSES FOR WATER TREATMENT SLUDGES

Historically, settled coagulation wastes and backwash waters had been disposed of without treatment. The practice was justified on the basis that the returned material had originally been present in the river or lake. However, this argument is not considered valid, since the discharge also contains the chemicals used in processing the water, and the source of supply may be groundwater. More stringent effluent regulations now apply to waste discharges from water purification and softening facilities. New water plant designs provide for zero discharge by recycling of wastewaters, reclaiming chemicals, and ultimate disposal of the useless or unmarketable residue remaining. Existing plants lack reclamation facilities and generate appreciable amounts of waste that must be hauled to a landfill. Special consideration may be given to old plants with unique problems; otherwise, little or no return discharge is permitted unless it meets effluent standards.

Water treatment processes can be modified to change the characteristics and reduce the quantities of wastes. Polymers as coagulant aids lower the required dosages of alum and auxiliary chemicals. The result is less sludge that is easier to dewater because of the reduced content of hydroxide precipitate. Polymers also enhance presedimentation of turbid river waters, thus controlling carryover of solids to subsequent chemical coagulation. The addition of specially manufactured clays can be used to aid flocculation of relatively clear surface supplies by producing a denser floc that settles more rapidly. Alum substitution is being considered as a cost-effective technique for modifying sludge-handling processes at surface-water plants. Groundwater softening plants can also change their mode of operation to lessen the volume of waste sludge. Emphasis is placed on preventing magnesium hydroxide precipitation, since it inhibits dewaterability and processing for recovery of lime. Also, hardness reduction can be limited to produce less solid material while still supplying a moderately soft water acceptable to the general public.

The common processes for storage, treatment, and disposal of water treatment sludges are listed in Table 13.3. Each waterworks is unique in that local conditions and existing facilities tend to dictate techniques applied in waste sludge disposal. Settled solids from coagulation may be stored in plain sedimentation basins for periods of up to several months, then removed by taking the tanks out of service and emptying the contents.

Modern clarifiers equipped with mechanical scrapers have limited storage capacity, however, and must discharge sludge at regular intervals, usually

**TABLE 13.3**   PROCESSES FOR STORAGE, TREATMENT, AND DISPOSING OF WATER TREATMENT SLUDGES

Storage prior to processing
  Sedimentation basins
  Separate holding tanks
  Flocculator–clarifier basins
Thickening prior to dewatering
  Gravity settling
Chemical conditioning prior to dewatering
  Polymer application
  Lime addition to alum sludges
Mechanical dewatering
  Centrifugation
  Pressure filtration
Air drying
  Shallow lagoons
  Sand drying beds
Disposal of dewatered solids
  Sanitary landfill
  Spreading lime sludges on agricultural land
Chemical recovery
  Recalcination of lime precipitates
  Alum recovery

daily. Separate holding tanks can be installed to accumulate this slurry prior to dewatering. Filter backwash can be stored in clarifier–flocculators that serve as both temporary holding tanks and wash-water settling basins. Equalization and settling are generally the only prerequisites if sludges are discharged to a sewer for processing at the municipal wastewater treatment plant.

A typical system for thickening coagulation waste and filter wash water is shown schematically in Fig. 13.5. The two primary sources of waste are sludge from the clarifier, following chemical coagulation, and wash water from backwashing filters. The latter is discharged to a clarifier holding tank for gravity separation of the suspended solids and flow equalization. Settled solids consolidate to a sludge volume less than one tenth of the wash-water volume. Supernatant is withdrawn slowly and recycled to the plant inlet. After a sufficient portion has been drained, the holding tank is able to receive the next backwash surge. Settled sludges from both the wash-water tanks and in-line clarifiers are given second-stage consolidation in a clarifier–thickener. Polymer is normally added to enhance solids capture, overflow is recycled, and thickened sludge withdrawn for further dewatering and disposal.

Mechanical dewatering of chemical sludges can be accomplished by centrifugation, pressure filtration, and in some instances by vacuum filtration (Fig. 13.6). A major advantage of centrifugation is operational flexibility.

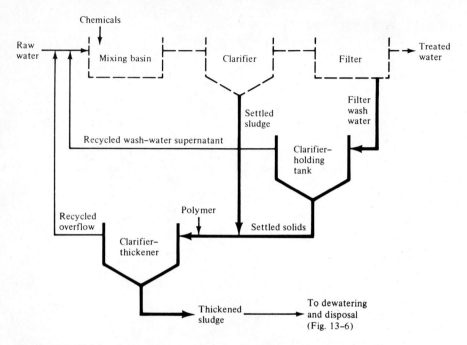

**Figure 13.5** Thickening coagulation waste and filter wash water in preliminary handling of water treatment plant sludges.

Machine variables, such as speed of rotation, allow a range of moisture content in the discharged solids varying from a dry cake to a thickened slurry. Feed rates, sludge solids content, and chemical conditioning are also variables that influence performance. Polymers or other coagulants are normally applied with the slurry feed to enhance solids capture. Lime sludges compact readily, producing a cake with 60%–70% solids from a feed of 10%–20%. In recalcination plants, centrifuges can be operated to selectively thicken calcium carbonate precipitate with the bulk of undesirable magnesium hydroxide appearing in the liquid discharge. Alum sludges do not dewater as readily and discharge with a toothpaste consistency suitable for either further processing or transporting by truck to a disposal site. Gravity-thickened sludge containing about one-half aluminum hydroxide slurry can be concentrated to 10%–15% solids, while one-quarter hydrate slurry can be dewatered to 20% or greater. Removal of solids by centrifugation varies over a broad range, depending on operating situations and chemical conditioning; in general, solids recovery and density of cake are related to polymer dosage.

Pressure filtration is particularly advantageous for dewatering alum sludges if a high solids concentration in the filter cake is desired. Aluminum hydroxide wastes are often conditioned with lime to improve their filterability. The filter medium is precoated with either diatomaceous earth or fly ash before applying sludge solids. Precoat protects against blinding of the filter cloth by fines and ensures easy cake discharge without sticking. With proper

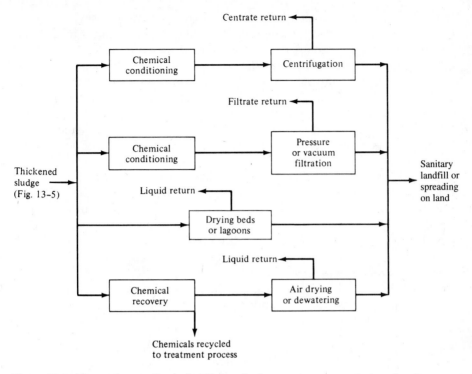

**Figure 13.6** Alternative methods for disposal of water treatment sludges by dewatering, drying, or chemical recovery of thickened chemical coagulation wastes.

chemical conditioning, alum sludges can be pressed to a solids content of about 40%, which can be handled as a chunky solid rather than the paste consistency associated with a 15% density. Belt filter presses can dewater alum sludges to approximately 15% and lime sludges to 40% or greater with polymer conditioning.

Lagooning is an accepted method for dewatering, thickening, and temporary storage of waste sludge where suitable land area is available. The diked pond area needed relates to character of the sludge, climate, design features such as underdrains and decanters, and method of operation. Clarified overflow may be returned to the treatment plant, particularly if filter wash water is directed to the lagoons without prior thickening. Sludges from lime softening consolidate to about 50% solids, which can be removed by a scraper or dragline and hauled to land burial. Alum sludge dewaters more slowly to a density of only 10%–15%. Although the surface may dry to a hard crust, the underlying sludge turns to a viscous liquid upon agitation. This slurry must be removed, usually by dragline, and spread on the banks to air dry prior to hauling. Freezing enhances the dewatering of alum sludge by breaking down its gelatinous character. Neither lime nor alum sludges make a good, stable landfill. Air drying at small water plants can be done on sand beds with tile underdrains. Repeated sludge applications over a period of several months can be made in depths up to several feet. Dewatering action is by drainage and air drying, although operation may include decanting supernatant. Dried cake is removed by either hand shoveling or mechanical means.

Recovery of chemicals is considered where dewatering and disposal by landfill are not feasible. Recovery of lime or alum must be justified both economically and technologically by careful studies based on local conditions. Major steps in recalcination are gravity thickening, mechanical dewatering with classification of solids to remove magnesium hydroxide and other impurities, and heating the dried solids to produce lime and carbon dioxide. The process appears to be most adaptable at large plants where the precipitate is largely calcium carbonate with little or no magnesium or extraneous material. Softening–coagulation sludges from treatment of turbid surface waters are difficult, if not impossible, to recalcine due to insolubles, such as clay, enmeshed in the sludge. Recovery of alum is not widely practiced because of low process efficiency and the poor quality of regenerated chemical. The major problem is separation of undesirable impurities from the aluminum hydroxide. From all chemical recovery processes, unwanted liquid and solid residues remain that require handling and disposal.

# Gravity Thickening

Gravity thickening is the simplest and least expensive process for consolidating waste sludges. Thickeners in wastewater treatment are employed most successfully in consolidating primary sludge separately or in combination with trickling-filter humus. When raw primary and waste-activated sludges are blended and concentrated, results are often marginal because of poor solids capture. Water treatment wastes from both sedimentation and filter backwashing can be compacted effectively by gravity separation.

## 13.6 GRAVITY SLUDGE THICKENERS IN WASTEWATER TREATMENT

The tank of a gravity thickener resembles a circular clarifier except that the depth/diameter ratio is greater and the hoppered bottom has a steeper slope. The three settling zones are the clear supernatant on top, feed zone characterized by hindered settling, and compression near the bottom where consolidation occurs. Figure 13.7 is a schematic diagram with the main components labeled. Influent sludge is applied continuously. However, if this is not practical, sludge application can be intermittent at frequent intervals. The circular inlet baffle is partly submerged but not so deep as to disturb the sludge blanket. The discharge weir is peripheral for maximum length. Although not always installed, a skimmer can be used to push scum over the weir with the overflow or down a pipe for separate collection. Thickened sludge is scraped

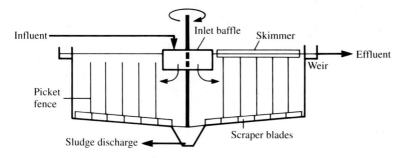

**Figure 13.7** Schematic diagram of a gravity sludge thickener.

to a central outlet for continuous or intermittent discharge. Withdrawal is often governed by the operation of the feed pump to the next processing step, e.g., anaerobic digestion. Attached to the scraper arms are picket fences consisting of vertical rods or pickets that are slowly pulled along. By gentle agitation of the settled solids, consolidation is increased by dislodging gas bubbles and preventing the bridging of solids. A properly designed picket fence is considered essential in thickening of organic waste [4].

The principal design criterion is solids loading expressed in units of pounds of solids applied per square foot of bottom area per day (lb/ft$^2$/day or kg/m$^2$·d). Typical loading values and thickened sludge concentrations based on operational experience are listed in Table 13.4. These data assume good operation and chemical additions, such as chlorine, if necessary to inhibit biological activity. Solids recovery in a properly functioning unit is 90%–95%, with perhaps the exception of a unit handling primary plus waste activated where it is difficult to achieve this degree of solids capture. Most continuous-flow thickeners are designed with a side water depth of approximately 10 ft to provide an adequate clear-water zone, sludge blanket depth, and space for temporary storage of consolidated waste. Sludge blanket depths (feed plus compaction zones) should be 3 ft or greater to ensure maximum compaction, using a suggested solids retention time of 24 hr. This is estimated

**TABLE 13.4**  GRAVITY THICKENER DESIGN LOADINGS AND UNDERFLOW CONCENTRATIONS FOR WASTEWATER SLUDGES

| Type of sludge | Average solids loading (lb/ft$^2$/day)[a] | Underflow concentration (% solids) |
|---|---|---|
| Primary | 20 | 8–10 |
| Primary plus filter humus | 10 | 6–8 |
| Primary plus activated sludge | 8 | 4–6 |

[a] 1.0 lb/ft$^2$/day = 4.88 kg/m$^2$ · d.

by dividing the volume of the sludge blanket by the daily sludge withdrawal; values vary from 0.5 to 2 days, depending on operation. Overflow rates should be 400–900 gpd/ft$^2$ (16–37 m$^3$/m$^2$·d) and are defined by the quantity of sludge plus supplementary dilution water applied.

Gravity thickeners are normally sized to handle the maximum seasonal or monthly sludge yield anticipated. Peak daily sludge production often requires storage in the thickener or other sludge-processing units. Low liquid overflow rates result in malodors from septicity of the thickener contents. A common remedy is to feed dilution water to the thickener along with the sludge to increase hydraulic loading. An alternative is to apply chlorine to reduce bacterial activity. The design of pumps and piping should be sufficiently flexible to allow regulation of the quantity of dilution water and have the capacity to transport viscous, thickened sludges.

■ **EXAMPLE 13.7**

The daily quantity of primary sludge from a trickling-filter plant contains 1130 lb of solids at a concentration of 4.5%. Size a gravity thickener based on a solids loading of 10 lb/ft$^2$/day. Calculate the daily volumes of applied and thickened sludges, assuming an underflow of 8.0% and 95% solids capture. What is the flow of dilution water required to attain an overflow rate of 400 gpd/ft$^2$? If the blanket of consolidated sludge in the tank has a depth of 3.0 ft, estimate the solids retention time.

*Solution*

$$\text{tank area required} = \frac{1130}{10} = 113 \text{ ft}^2$$

$$\text{diameter} = \left(\frac{113 \times 4}{\pi}\right)^{1/2} = 12.0 \text{ ft}$$

Use a depth of 10.0 ft.

$$\text{volume of applied sludge} = \frac{1130}{0.045 \times 62.4} = 402 \text{ ft}^3/\text{day}$$

$$= 3010 \text{ gpd}$$

$$\text{overflow rate of applied sludge} = \frac{3010}{113} = 27 \text{ gpd/ft}^2$$

$$\text{supplemental dilution flow to attain 400 gpd/ft}^2 = (400 - 27)113 = 42{,}000 \text{ gpd}$$

$$\text{volume of thickened sludge} = \frac{1130 \times 0.95}{(8.0/100)62.4} = 215 \text{ ft}^3/\text{day}$$

$$\text{solids retention time} = \frac{3 \times 113 \times 24}{215} = 38 \text{ hr} \quad ■$$

## 13.7  GRAVITY SLUDGE THICKENERS IN WATER TREATMENT

The performance of thickeners handling water treatment plant wastes varies with the character of the water being treated and the chemicals applied. Alum sludges from surface-water coagulation settle to a density in the range of

2%–6% solids. Coagulation–softening mixtures from the treatment of turbid river waters gravity thicken approximately as follows: alum–lime sludge, 4%–10%; iron–lime settlings 10%–20%; alum–lime filter wash water, about 4%; and iron–lime backwash up to 8%. The density achieved in gravity thickening relates to the calcium–magnesium ratio in the solids, quantity of alum, nature of impurities removed from the raw water, and other factors. Calcium carbonate residue from groundwater softening consolidates to 15%–25% solids. In most cases, special studies have to be conducted at a particular waterworks to determine settleability of solids in waste sludges and wash water. Flocculation aids are used to improve clarification in most cases.

Relatively dense chemical slurries are thickened in tanks similar to the one shown in Fig. 13.7. Thin sludges and backwash waters may be concentrated in clarifier–thickeners that have an inlet well equipped with mixing paddles, where the feed can be flocculated with polymers or other coagulants. Holding tanks are used to dampen hydraulic surges of filter wash water. These units can be plain tanks with mixers or clarifiers equipped for removing settled solids and decanting clear supernatant.

Evaluating the performance of a thickener often involves mass balance calculations. Overflow plus underflow solids equals influent solids. Also, the sum of overflow and underflow volumes is equal to the quantity of applied sludge and supplementary dilution water. These values can be calculated using Eq. 13.3, as illustrated in Example 13.8.

### ■ EXAMPLE 13.8

An alum–lime slurry with 4.0% solids content is gravity thickened to 20% with a removal efficiency of 95%. Calculate the quantity of underflow per 1.0 m$^3$ of slurry applied and the concentration of solids in the overflow. Assume a specific gravity of 2.5 for the dry solids.

*Solution*

$$\text{solids applied} = 1.0 \text{ m}^3 \times 1000 \text{ kg/m}^3 \times 0.04 = 40 \text{ kg}$$

$$\text{underflow solids} = 0.95 \times 40 = 38 \text{ kg}$$

The specific gravity of underflow using Eq. 13.2 is

$$S = \frac{80 + 20}{(80/1.0) + (20/2.5)} = 1.14$$

$$\text{volume of underflow (Eq. 13.3)} = \frac{38}{0.20 \times 1000 \times 1.14} = 0.17 \text{ m}^3$$

$$\text{volume of overflow} = 1.0 - 0.17 = 0.83 \text{ m}^3$$

From Eq. 13.3, the concentration of solids in the overflow is

$$s = \frac{0.05 \times 40 \times 100}{0.83 \times 1000 \times 1.0} = 0.24\% = 2400 \text{ mg/l} \qquad ■$$

# Flotation Thickening

Air flotation is most applicable in concentrating waste-activated sludges and pretreatment of industrial wastes to separate grease or fine particulate matter. Fine bubbles to buoy up particles may be generated by air dispersed through a porous medium, by air drawn from the liquid under vacuum, or by air forced into solution under elevated pressure followed by pressure release. The latter, called *dissolved-air flotation,* is the process employed most frequently in thickening sludges because of its reliable performance.

## 13.8 DESCRIPTION OF DISSOLVED-AIR FLOTATION

The major components of a typical flotation system are sludge pumps, chemical feed equipment to apply polymers, an air compressor, a control panel, and a flotation unit. Figure 13.8 is a schematic diagram of a dissolved-air system. Influent enters near the tank bottom and exits from the base at the opposite end. Float is continuously swept from the liquid surface and discharged over the end wall of the tank. Effluent is recycled at a rate of 30%–150% of the influent flow through an air dissolution tank to the feed inlet. In this manner, compressed air at 60–80 psi is dissolved in the return flow. After pressure release, minute bubbles with a diameter about 80 $\mu$m pop out of solution. They attach to solid particles and become enmeshed in sludge flocs, floating them to the surface. The sludge blanket, varying from 8 to 24 in. thick, is skimmed from the surface. Flotation aids are introduced in a mixing chamber at the tank inlet.

The operating variables for flotation thickening are air pressure, recycle ratio, detention time, air/solids ratio, solids and hydraulic loading rates, and application of chemical aids. The operating air pressure in the dissolution tank influences the size of bubbles released. If too large, they do not attach readily to sludge particles, while too fine a dispersion breaks up fragile floc.

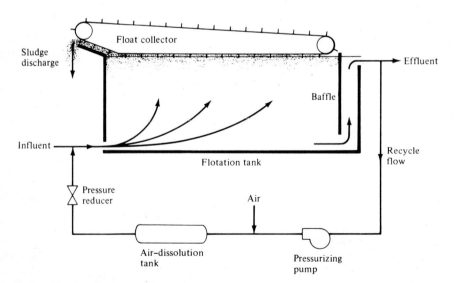

**Figure 13.8** Schematic diagram of a dissolved-air flotation system.

Generally, a bubble size less than 100 $\mu$m is best; however, the only practical way to establish the proper rise rate is by conducting experiments at various air pressures.

Recycle ratio is interrelated with feed solids concentration, detention time, and air/solids ratio. Detention time in the flotation zone is not critical, providing that particles rise rapidly enough and the horizontal velocity does not scour the bottom of the sludge blanket. An air/solids ratio of 0.01–0.03 lb of air/lb of solids is sufficient to achieve acceptable thickening of waste-activated sludge. Optimum recycle ratio must be determined by on-site studies.

Operating data from plant-scale units indicate solids loadings of 2–4 lb/ft²/hr, with hydraulic flows of about 1 gpm/ft², can produce floats of 4%–8% solids. Without polymer addition, solids capture is 70%–90%. However, removal efficiency increases to a mean of 97%, with a polymer dosage of approximately 10 lb/ton of dry suspended solids. This is the reason most wastewater installations use flotation aids.

## 13.9 DESIGN OF DISSOLVED-AIR FLOTATION UNITS

Wherever possible laboratory and pilot-scale tests are recommended to help determine specific design criteria for a given waste. Notwithstanding, the suggested design criteria for flotation thickening of typical waste-activated sludges are listed in Table 13.5. A conservative solids design loading is 2 lb/ft²/hr (10 kg/m²·h) with the use of flotation aids. From actual operating data, at least 3 lb/ft²/hr can be expected, and most thickeners have a built-in

**TABLE 13.5**  DESIGN PARAMETERS FOR DISSOLVED-AIR FLOTATION OF WASTE-ACTIVATED SLUDGE WITH ADDITION OF POLYMER FLOTATION AIDS

| Parameter | Typical design value | Anticipated results |
|---|---|---|
| Solids loading (lb/ft²/hr)[a] | 2 | 3–5 |
| Float concentration (%) | 4 | 5–6 |
| Removal efficiency | 90–95 | 97 |
| Polymer addition (lb/ton of dry solids) | 10 | 5–10 |
| Air/solids ratio (lb of air/lb of solids) | 0.02 | |
| Effluent recycle ratio (% of influent) | 40–70 | |
| Hydraulic loading (gpm/ft²) | 0.8 maximum | |

[a] 1.0 lb/ft²/hr = 4.88 kg/m² · h.

capacity for 4–5 lb/ft$^2$/hr loadings. While a 4% minimum float concentration is specified for design purposes, 5%–6% solids can normally be expected. Flotation without polymers generally results in a concentration that is about 1 percentage point less than with chemical aids. Removal efficiency varies from 90% to 98% with polymer addition. The maximum hydraulic loading for design is set at 0.8 gpm/ft$^2$ (0.54 l/m$^2$·s); this is equivalent to applying a waste with a solids concentration of 5000 mg/l at a loading of 2 lb/ft$^2$/hr. Lesser solids levels or higher hydraulic loadings result in lower removal efficiencies and/or float densities.

The typical design values recommended in Table 13.5 apply to anticipated average sludge production. This procedure provides a significant safety factor and permits flexibility in operations. Peak solids loads at municipal treatment plants can usually be accommodated, since these conservative design criteria allow a maximum loading of nearly 100% greater than the average without a serious drop in performance. Perhaps the most critical condition is during a period of sludge bulking when the waste mixed liquor is more difficult to thicken and maximum hydraulic loading is applied to the flotation unit.

Sizing of flotation units for an existing plant can be calculated from available data on sludge quantities, characteristics, and solids concentrations. For new plant design, raw wastewater is often assumed to contain 0.24 lb of dry solids/capita/day. A portion of these solids is removed in primary settling, and a conservative estimate for secondary activated-sludge production is 0.12 lb/capita/day. The actual amount is likely to be closer to one-half of this value, because of biological decomposition. Solids yield in an activated-sludge process without primary settling may be safely assumed to be 0.20 lb/capita/day for domestic wastewater. If the waste sludge from such a system is aerobically digested, the concentration of solids is reduced by about 35%.

Operating hours of a flotation unit depend on size of plant and the working schedule. Although a unit does not require continuous operator attention, periodic checks of a system are scheduled. Generally, a 48 hr/week is adequate for plants with capacities of less than 2 mgd. For systems of 2–5 mgd, two shifts 5 days per week establishes an operating period of 80 hr/week. Treatment plants handling more than 20 mgd have operators on duty continuously, and thickening units are run on a schedule appropriate for sludge dewatering and disposal.

■ **EXAMPLE 13.9**

A dissolved-air flotation thickener is being sized to process waste-activated sludge based on the design criteria given in Table 13.5. The average waste flow is 33,600 gpd at 15,000 mg/l (1.5%) suspended solids, and the maximum daily quantity contains 50% more solids at a reduced concentration of 10,000 mg/l. What is the peak daily hydraulic loading that can be processed? Base all computations on a 14-hr/day operating schedule.

*Solution.*    The flotation tank surface area required for the average daily flow at a design loading of 2.0 lb/ft$^2$/hr for a 14-hr/day schedule is

$$\text{area} = \frac{33,600 \times 0.015 \times 8.34}{2.0 \times 14} = \frac{4200}{28} = 150 \, \text{ft}^2$$

Check the solids loading and overflow rate at maximum daily sludge production:

$$\text{maximum solids loading} = \frac{1.5 \times 4200}{150 \times 14} = 3.0 \, \text{lb/ft}^2/\text{hr} \quad (\text{OK})$$

$$\text{maximum sludge volume} = \frac{1.5 \times 4200}{0.01 \times 8.34} = 75,500 \, \text{gpd}$$

$$\text{maximum hydraulic loading} = \frac{75,500}{150 \times 14 \times 60} = 0.60 \, \text{gpm/ft}^2 \quad (\text{OK})$$

$$\text{peak hydraulic loading based on } 0.80 \, \text{gpm/ft}^2 = 0.80 \times 150 \times 14 \times 60$$

$$= 100,000 \, \text{gpd} \qquad \blacksquare$$

# Biological Sludge Digestion

Biological digestion of sludge from wastewater treatment is widely practiced to stabilize the organic matter prior to ultimate disposal. Anaerobic digestion is used in plants employing primary clarification followed by either trickling-filter or activated-sludge secondary treatment. Aerobic digestion stabilizes waste-activated sludge from aeration plants without primary settling tanks. The fundamental differences between aerobic and anaerobic digestion are illustrated in Figs. 12.1 and 12.2. The end product of aerobic digestion is cellular protoplasm, and growth is limited by depletion of the available carbon source. The end products of anaerobic metabolism are methane, unused organics, and a relatively small amount of cellular protoplasm. Growth is limited by a lack of hydrogen acceptors. Anaerobic digestion is basically a destructive process, although complete degradation of the organic matter under anaerobic conditions is not possible.

## 13.10  ANAEROBIC SLUDGE DIGESTION

Anaerobic digestion consists of two distinct stages that occur simultaneously in digesting sludge (Fig. 12.13). The first consists of hydrolysis of the high-molecular-weight organic compounds and conversion to organic acids by acid-forming bacteria (Eq. 12.3). The second stage is gasification of the organic acids to methane and carbon dioxide by the acid-splitting methane-forming bacteria (Eq. 12.4).

Methane bacteria are strict anaerobes and very sensitive to conditions of their environment. The optimum temperature and pH range for maximum growth rate are limited. Methane bacteria can be adversely affected by excess concentrations of oxidized compounds, volatile acids, soluble salts, and metal cations and also show a rather extreme substrate specificity. Each species is restricted to the use of only a few compounds, mainly alcohols and organic acids, whereas the normal energy sources, such as carbohydrates and amino acids, are not attacked. An enrichment culture developed on a feed of acetic or butyric acid cannot decompose propionic acid. The sensitivity exhibited by methane bacteria in the second stage of anaerobic digestion, coupled with the rugged nature of the acid-forming bacteria in the first stage, creates a biological system where the population dynamics are easily upset. Any shift in environment adverse to the population of methane bacteria causes a buildup of organic acids, which in turn further reduces the metabolism of acid-splitting methane formers.

Pending failure of the anaerobic digestion process is evidenced by a decrease in gas production, a lowering in the percentage of methane gas produced, an increase in the volatile acids concentration, and eventually a drop in pH when the accumulated volatile acids exceed the buffering capacity created by the ammonium bicarbonate in solution. Therefore, the operation of a digester can be monitored by any of the following methods: plotting the daily gas production per unit raw sludge fed, the percentage of carbon dioxide in the digestion gases, or the concentration of volatile acids in the digesting sludge. A reduction in gas production, an increase in carbon dioxide percentage, and a rise in volatile acids concentration all indicate reduced activity of the acid-splitting methane-forming bacteria. Digester failure may be caused by any of the following: a significant increase in organic loading, a sharp decrease in digesting sludge volume (i.e., when digested sludge is withdrawn), a sudden increase in operating temperature, or the accumulation of a toxic or inhibiting substance.

General conditions for mesophilic sludge digestion are given in Table 13.6.

## 13.11 SINGLE-STAGE FLOATING-COVER DIGESTERS

The cross section of a floating-cover digestion tank is shown in Fig. 13.9. Raw sludge is pumped into the digester through pipes terminating either near the center of the tank or in the gas dome. Pumping sludge into the dome helps to break up the scum layer that forms on the surface.

Digested sludge is withdrawn from the tank bottom. The contents are heated in the zone of digesting sludge by pumping them through an external heater and returning the heated slurry through the inlet lines. The tank contents stratify with a scum layer on top and digested thickened sludge on the bottom. The middle zones consist of a layer of supernatant (water of separation) underlain by the zone of actively digesting sludge. Supernatant is drawn from the digester through any one of a series of pipes extending out of the tank wall. Digestion gas from the gas dome is burned as fuel in the external heater or wasted to a gas burner.

The weight of the cover is supported by sludge, and the liquid forced up

**TABLE 13.6**   GENERAL CONDITIONS FOR
SLUDGE DIGESTION

| | |
|---|---|
| Temperature | |
| Optimum | 98°F (35°C) |
| General range of operation | 85°–95°F |
| pH | |
| Optimum | 7.0–7.1 |
| General limits | 6.7–7.4 |
| Gas production | |
| Per pound of volatile solids added | 8–12 ft³ |
| Per pound of volatile solids destroyed | 16–18 ft³ |
| Gas composition | |
| Methane | 65%–69% |
| Carbon dioxide | 31%–35% |
| Hydrogen sulfide | Trace |
| Volatile acids concentration as acetic acid | |
| Normal operation | 200–800 mg/l |
| Maximum | Approx. 2000 mg/l |
| Alkalinity concentrations as CaCO₃ | |
| Normal operation | 2000–3500 mg/l |

between the tank wall and the side of the cover provides a gas seal. Gas rises out of the digesting sludge, moves along the ceiling of the cover, and collects in the gas dome. The cover can float on the surface of the sludge between the landing brackets and the height of the overflow pipe. Rollers around the circumference of the cover keep it from binding against the tank wall.

Three functions of a single-stage floating-cover digester are (1) anaerobic

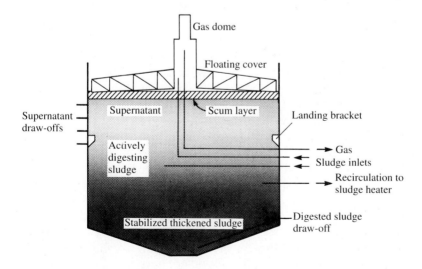

**Figure 13.9**  Cross-sectional diagram of a floating-cover anaerobic digester.

digestion of the volatile solids, (2) gravity thickening, and (3) storage of the digested sludge. A floating-cover feature of the tank provides for a storage volume equal to approximately one third that of the tank. The unmixed operation of the tank permits gravity thickening of sludge solids and withdrawal of the separated supernatant. Anaerobic digestion of the sludge solids is promoted by maintaining near optimum temperature and stirring the digesting sludge through the recirculation of heated sludge. However, the rate of biological activity is inhibited by the lack of mixing; on the other hand, good mixing would prevent supernatant formation. Therefore, in single-tank operation, the biological process is compromised to allow both digestion and thickening to occur in the same tank.

In the operation of an unmixed digester, raw sludge is pumped to the digester from the bottom of the settling tanks once or twice a day. Supernatant is withdrawn daily and returned to the influent of the treatment plant. It is normally returned by gravity flow to the wet well during periods of low raw-wastewater flow, or, in the case of an activated-sludge plant, it may be pumped to the head end of the aeration basin. Because of the floating cover, supernatant does not have to be drawn off simultaneously with the pumping of raw sludge into the digester.

Digested sludge is stored in the tank and withdrawn periodically for disposal. Spreading of liquid sludge on grassland or cropland is common practice in agricultural regions. In some plants, it is dried on sand beds or in lagoons and hauled to land burial. In either case, weather often dictates the schedule for digested sludge disposal. In northern climates, the cover is lowered as close as possible to the corbels (landing brackets) in the fall of the year to provide maximum volume for winter sludge storage.

## 13.12 HIGH-RATE (COMPLETELY MIXED) DIGESTERS

The biological process of anaerobic digestion is significantly improved by complete mixing of the digesting sludge either mechanically or by use of compressed digestion gases. Mechanical mixing is normally accomplished by an impeller suspended from the cover of the digester (Fig. 13.10a). Three common methods of gas mixing are the injection of compressed gas through a series of small-diameter pipes hanging from the cover into the digesting sludge (Fig. 13.10b); the use of a draft tube in the center of the tank, with compressed gas injected into the tube to lift recirculating sludge from the bottom and spill it out on top (Fig. 13.10c); and supplying compressed gas to a number of diffusers mounted in the center at the bottom of the tank (Fig. 13.10d).

A completely mixed digester may have either a fixed- or a floating-cover tank. Digesting sludge is displaced when raw sludge is pumped into a fixed-cover digester. By use of a floating cover, tank volume is available for the storage of digesting sludge, and withdrawals do not have to coincide with the introduction of raw sludge.

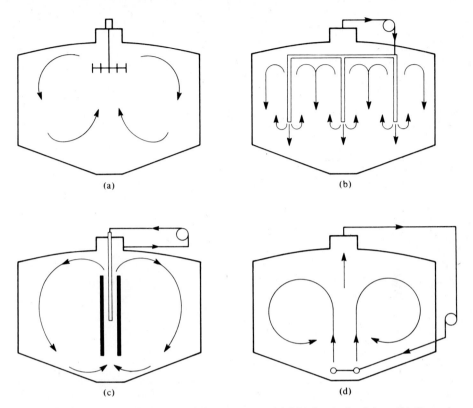

(a)

(b)

(c)

(d)

**Figure 13.10** High-rate digester-mixing systems. (a) Mechanical mixing. (b) Gas mixing using a series of gas discharge pipes. (c) Gas mixing using a central draft tube. (d) Gas mixing using diffusers mounted on the tank bottom.

The homogeneous nature of the digesting sludge in a high-rate digester does not permit formation of supernatant. Therefore, thickening cannot be performed in a completely mixed digester. High-rate digestion systems normally consist of two tanks operated in series (Fig. 13.11). The first stage is a completely mixed, heated, floating, or fixed-cover digester fed as continuously as possible, whose function is anaerobic digestion of the volatile solids. The second stage may be heated or unheated, and it accomplishes gravity thickening and storage of the digested sludge. Two-stage systems may consist of two similar floating-cover tanks with provisions for mixing in one tank.

## 13.13 VOLATILE SOLIDS LOADINGS AND DIGESTER CAPACITY

Typical ranges of loadings and detention times employed in the design and operation of heated anaerobic digestion tanks treating domestic waste sludge are listed in Table 13.7. Values given for volatile solids loading and digester

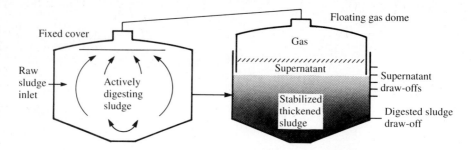

**Figure 13.11** Sketch of a two-stage anaerobic digestion system. The first stage is a completely mixed high-rate digester with a fixed cover. The second stage is a thickening and storage tank covered with a dome for collecting and storing gas.

capacity for conventional, single-stage digesters are based on the total sludge volume available in the tank (i.e., the volume with the floating cover fully raised). Figures given for high-rate digestion apply only to the volume needed for the first-stage tank. There are no established design standards for the tank capacity required in second-stage thickening and supernatant separation.

The loading applied to a digester is expressed in terms of pounds of volatile solids applied per day per cubic foot of digester capacity. Detention time is the volume of the tank divided by the daily raw-sludge pumpage. Digester capacity in Table 13.7 is given in terms of cubic feet of tank volume provided per design population equivalent of the treatment plant.

The *Recommended Standards of the Great Lakes–Upper Mississippi River Board of State Public Health & Environmental Managers* [5] recommend a maximum of 0.08 lb/ft$^3$/day of VS (1.3 kg/m$^3$·d) for high-rate digestion and a maximum of 0.04 lb (0.6 kg) of VS loading for single-stage operation. These loadings assume that the raw sludge is derived from domestic waste-

**TABLE 13.7**   LOADINGS AND DETENTION TIMES FOR HEATED ANAEROBIC DIGESTERS

| | Conventional single-stage (unmixed) | First-stage high-rate (completely mixed) |
|---|---|---|
| Loading (lb/ft$^3$/day of VS)[a] | 0.02–0.05 | 0.1–0.2 |
| Detention time (days) | 30–90 | 10–15 |
| Capacity of digester (ft$^3$/population equivalent)[b] | | |
| Primary only | 2–1 | 0.4–0.6 |
| Primary and secondary | 4–6 | 0.7–1.5 |
| Volatile solids reduction (%) | 50–70 | 50 |

[a] 1.0 lb/ft$^3$/day = 16.0 kg/m$^3$·d.
[b] 1.0 ft$^3$ = 0.0283 m$^3$.

water, the digestion temperature is in the range of 85°–95°F (29°–35°C), volatile solids reduction is 40%–50%, and the digested sludge is removed frequently from the digester.

The capacity required for a single-stage floating-cover digester can be determined by the formula

$$V = \frac{V_1 + V_2}{2} \times T_1 + V_2 \times T_2 \qquad (13.11)$$

where    $V$ = total digester capacity, $ft^3$
$V_1$ = volume of average daily raw-sludge feed, $ft^3/day$
$V_2$ = volume of daily digested sludge accumulation in tank, $ft^3/day$
$T_1$ = period required for digestion, days (approximately 25 days at a temperature of 85°–95°F)
$T_2$ = period of digested sludge storage, days (normally 30–120 days)

Figure 13.12 is a pictorial representation of Eq. 13.11.

Predicting daily volumes of raw sludge produced and the digested sludge accumulated as required in Eq. 13.11 is often difficult. Therefore, the capacity of conventional digesters frequently is based on empirical values relating digester capacity to the equivalent population design of the plant (Table 13.7). Values of 5 and 6 $ft^3/capita$ (0.14 and 0.17 $m^3/capita$) are frequently used for high-rate trickling-filter plants and activated-sludge plants, respectively.

The minimum detention time for satisfactory high-rate digestion at 95°F is in the range 7–10 days. In general, this limiting period depends on the minimum time required to digest the grease component of raw sludge. Also, too short a detention time results in depletion of the methane-bacteria populations since they are washed out of the digester. The maximum volatile solids loadings, at a 10-day detention time, vary from 0.2 to 0.3 $lb/ft^3/day$ (3.2–4.8 $kg/m^3\cdot d$) for adequate volatile solids destruction and gas production. For larger treatment plants with uniform loading conditions, design values of a 10-day minimum detention time and maximum 0.2 $lb/ft^3/day$ of VS loading appear to be satisfactory. Digesters for small treatment plants with wider

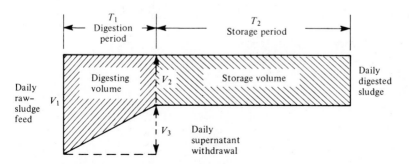

**Figure 13.12** Pictorial presentation of Eq. 13.11.

variations in daily sludge production should be planned using more conservative loading rates.

The capacities required for a high-rate digestion system can be determined by the following equations:

$$V_I = V_1 \times T \tag{13.12}$$

where  $V_I$ = digester capacity required for first-stage high-rate, ft$^3$
$V_1$ = volume of average daily raw sludge feed, ft$^3$/day
$T$ = period required for digestion, days

and

$$V_{II} = \frac{V_1 + V_2}{2} \times T_1 + V_2 \times T_2 \tag{13.13}$$

where  $V_{II}$ = digester capacity required for second-stage digested sludge thickening and storage, ft$^3$
$V_1$ = volume of digested sludge feed = volume of average daily raw sludge, ft$^3$/day
$V_2$ = volume of daily digested sludge accumulation in tank, ft$^3$/day
$T_1$ = period required for thickening, days
$T_2$ = period of digested sludge storage, days

Figure 13.13 explains Eqs. 13.12 and 13.13.

### ■ EXAMPLE 13.10
A high-rate trickling-filter plant treats a domestic wastewater flow of 0.48 mgd. Characteristics of the wastewater are identical to those in Table 12.1. Determine the digester capacities required for a single-stage floating-cover digestion system. Digested sludge is to be dried on sand beds, and the longest anticipated storage period required is 90 days.

*Solution*

Equivalent population = 4000 at 0.24 lb per capita

Assume the following:

water content in raw sludge = 96%

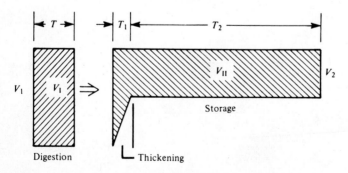

**Figure 13.13** Diagrams for Eqs. 13.12 and 13.13.

volatile solids in raw sludge solids = 70%

water content of digested sludge = 94%

volatile solids reduction = 50%

The volume of raw sludge, from Eq. 13.3, is

$$V = \frac{4000 \times 0.24}{[(100 - 96)/100]62.4} = 385 \text{ ft}^3/\text{day}$$

The volume of digested sludge is

$$V = \frac{0.30(4000 \times 0.24) + 0.70 \times 0.50(4000 \times 0.24)}{[(100 - 94)/100]62.4} = 167 \text{ ft}^3/\text{day}$$

Substituting into Eq. 13.11 yields

$$V = [(385 + 167)/2]25 + 167 \times 90 = 21,900 \text{ ft}^3$$

Check the volatile solids loading:

$$\frac{0.70 \times 4000 \times 0.24}{21,900} = 0.031 \text{ lb/ft}^3/\text{day of VS}$$

Verify the digester capacity per capita:

$$\frac{21,900}{4000} = 5.5 \text{ ft}^3/\text{population equivalent}$$

(This value is in the range of the empirical design figure of 4–6 ft$^3$/day/population equivalent given in Table 13.7.)    ■


## 13.14  AEROBIC SLUDGE DIGESTION

The function of aerobic digestion is to stabilize waste sludge solids by long-term aeration, thereby reducing the BOD and destroying volatile solids. The most common application of aerobic digestion is in handling waste-activated sludge. Customary methods for disposal of the digested sludge are spreading on farmland, lagooning, and drying on sand beds.

Aerobic digestion is accomplished in one or more tanks mixed by diffused aeration. Since dilute solids suspensions have a low rate of oxygen demand, the need for effective mixing rather than microbial metabolism usually governs the air supply required. The volume of air supplied for aerobic digestion is normally in the range of 15–30 cfm/1000 ft$^3$ of digester.

Design criteria vary with the type of activated-sludge system, BOD loading, and the means provided for ultimate disposal of the digested sludge. Small activated-sludge plants without primary sedimentation are generally provided with 2–3 ft$^3$ (57–85 l) of aerobic digester volume per design population equivalent of the plant. For stabilizing waste-activated sludge with a suspended solids concentration of 1.0% or less, the volatile solids loading should be limited to 0.04 lb/ft³/day (0.64 kg/m³·d), and the aeration period 200–300 degree-days computed by multiplying the digesting temperature in degrees Celsius times the sludge age. This equates to a minimum aeration period of 10 days at 20°C or 20 days at 10°C. Volatile solids and BOD reductions at these loadings are in the range of 30%–50%, and the digested sludge can be disposed of without causing odors or other nuisance conditions.

Long-term aeration of waste-activated sludge creates a bulking material that resists gravity thickening. The solids concentration of aerobically digested sludge is usually in the range of 1.0%–2.0%. The maximum concentration in a well-operated system is not likely to exceed 2.5%. This poor settleability frequently creates problems in disposing of the large volume of sludge produced. Thickening by flotation, pressure filtration, or other mechanical methods is too expensive for incorporation in small treatment plants. Therefore, plant design should take care of storage and elimination of the relatively large volume of aerobically digested sludge.

An aerobic digester is operated as a semibatch process with continuous feed and intermittent supernatant and digested sludge withdrawals. The contents of the digester are continuously aerated during filling and for a specified period after the tank is full. Aeration is then discontinued, allowing the stabilized solids to settle. Supernatant is decanted and returned to the head of the plant, and a portion of the gravity-thickened sludge removed for disposal. In practice, aeration and settlement may be a daily cycle with feed applied early in the day and clarified water decanted later in the day. Digested solids are withdrawn when the sludge in the tank does not gravity thicken to provide a supernatant with adequate clarity.

■ **EXAMPLE 13.11**
Refer to the data noted in Example 12.13. Calculate the cubic meters of aerobic digester volume provided per design population equivalent, and estimate the volatile solids loading on the aerobic digester.

*Solution.* The data from Example 12.13 are

   volume of aerobic digester = 153 m³

   design population of plant = 2000

Therefore,

$$\text{volume provided} = \frac{153,000}{2000} = 76 \frac{\text{liters of digester volume}}{\text{design population equivalent}}$$

   BOD load on plant = 182 kg/d

Assuming a BOD loading of 0.20 g of BOD/g of MLSS applied to the aeration tank, the estimated excess sludge produced per day from Fig. 13.1 is 0.42 g of SS/g of BOD load. For the digester volume of 153 m³ and assuming 70% of the SS as volatile, the estimated volatile solids loading applied to the aerobic digester is $2.0(182 \times 0.70 \times 0.42)/153 = 0.70$ kg/m$^3$·d. ∎

## 13.15  OPEN-AIR DRYING BEDS

Historically, small communities have dewatered digested sludge on open beds because of their simplicity, rather than operating more complex mechanical systems. The disadvantages include poor drying during damp weather, potential odor problems, large land area required, and labor for removing the dried cake. In construction of outmoded drying beds, the entire surface area was a level layer of coarse sand supported on a bed of graded gravel. Tile or perforated pipe underdrains were spaced about 20 ft apart in the bottom gravel layer to collect and return drainage to the treatment plant influent. Cleaning dried sludge cake from the beds was a laborious job. The cake had to be lifted from the surface of the sand by hand shoveling and taken off in wheelbarrows for loading in a truck. Attempts to use mechanical equipment resulted in excessive loss of sand and disturbance of the gravel underdrain. In modern design, drying beds are constructed to permit the use of tractors with front-end loaders to scrape up the sludge cake and dump it into a truck. The main features as illustrated in Fig. 13.14 are (a) watertight walls extending 18–24 in. above the surface of the bed, (b) an end opening in the wall, sealed by inserting planks, for entrance of a front-end loader, (c) centrally located drainage trenches filled with a coarse sand bed supported on a gravel filter with a perforated pipe underdrain, (d) paved areas on both sides of the trenches with a 2%–3% slope for gravity drainage, and (e) a sludge inlet at one end and supernatant draw-off at the opposite end. For this illustrated bed, the width is 40 ft and the length 100 ft. The operating procedure is to apply digested sludge to a depth of 12 in. or more, draw off supernatant after the solids settle, and allow the sludge to dry. A well-digested sludge forms a cake 3–5 in. thick that is thoroughly dry, black in color, and has cracks resulting from horizontal shrinkage. Paved drying beds have been constructed with more limited drainage than shown in Fig. 13.14, however, climatic conditions must be very favorable since the major water loss is by evaporation.

Rational design for sludge beds is difficult, owing to the multitude of variables that affect drying rate. These include climate and atmospheric conditions, such as temperature, rainfall, humidity, and wind velocity; sludge characteristics, including degree of stabilization, grease content, and solids concentration; depth and frequency of sludge application; and condition of the sand stratum and drainage piping. The bed area furnished for desiccating anaerobically digested sludge is from 1 to 2 ft$^2$/BOD design population equivalent of the treatment plant. Solids loadings average about 20 lb/ft$^2$/yr (100 kg/m$^2$·y) in northern states, while unit loading may be as high as 40 lb/ft$^2$/yr in southern climates. Drying time ranges from several days to weeks, depending on drainability of the sludge and suitable weather conditions for evapora-

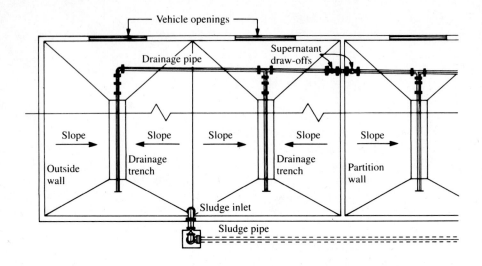

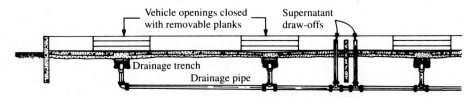

**Figure 13.14** Open-air beds for drying digested sludge with paved surfaces to allow mechanical removal of dried cake. Water is separated by decanting supernatant, draining to trenches, and evaporation. (Courtesy of HDR Engineering, Inc.)

tion. Dewatering may be improved and exposure time shortened by chemical conditioning, such as addition of a polymer.

Air drying of digested sludge may be practiced in shallow lagoons where permitted by soil and weather conditions. Water removal is by evaporation, and the groundwater table must remain below the bottom of the lagoon to prevent contamination by seepage. Sludge is normally applied to a depth of about 2 ft and residue removed by a front-end loader after an extended period of consolidation. Because of long holding times, odor problems are more likely to occur. Design data and operational techniques are defined by local experience.

## 13.16 COMPOSTING

The objectives of sludge composting are to biologically stabilize putrescible organics, destroy pathogenic organisms, and reduce the volume of waste [6]. The optimum moisture content for a compost mixture is 50%–60%; less than

40% may limit the rate of decomposition, while over 60% is too wet to stack in piles. Volatile solids reduction during composting is similar to biological digestion averaging about 50%. The compost product is a moist, friable humus with a water content less than 40%. For most efficient stabilization and pasteurization, the temperature in the compost piles should rise to 130°–150°F (55°–65°C) but not above 176°F (80°C). Moisture content, aeration rates, size and shape of pile, and climatic conditions affect composting temperature. The finished compost, although too low in nutrients to be classified a fertilizer, is an excellent soil conditioner. When mixed with soil, one advantage of the added humus content is increased capacity for retention of water.

The main products of biological metabolism in aerobic composting are carbon dioxide, water, and heat. Anaerobic composting produces intermediate organics, such as organic acids, and gases including carbon dioxide and methane. Since anaerobic decomposition has a higher odor potential and releases less heat, most systems are designed for aerobic composting. Nevertheless, all forms of composting have the potential of nuisance problems such as odors and dust.

Dewatered sludge cake, usually with a moisture content in the range of 70%–85%, is too wet to maintain adequate porosity for aeration. If it is not mixed with another substance, a pile of sludge cake tends to slump and compact to a dense mass with a wet, anaerobic interior and a dried exterior crust. Figure 13.15 is a generalized diagram for composting organic sludges. Dewatered cake is mixed with either an organic amendment (e.g., dried manure, straw, or sawdust) or a recoverable bulking agent (e.g., wood chips) to reduce the unit weight and increase air voids. Finished compost may also be recycled and added to the wet cake. Although composting can be performed in an enclosed reactor, the common processes use outdoor piles either exposed or sheltered under a roofed structure. Compost may be placed in either windrows agitated by periodic turning for remixing and aeration or static piles with forced aeration. The choice between these two processes is

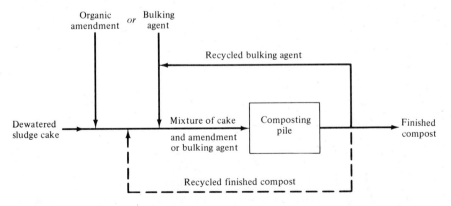

**Figure 13.15** Generalized diagram for composting dewatered wastewater sludges.

based on several factors, including climate, environmental considerations, the availability of a bulking agent, and economics.

In the windrow system, mixed compost material is arranged in long parallel rows. These windrows are turned at regular intervals by mobile equipment to restructure the compost. The piles may be triangular or trapezoidal in shape and may vary in height and width, as determined by the equipment used for turning and the characteristics of the composting material. The height of windrows is usually 4–8 ft and the width 8–12 ft.

Windrow composting is used in agricultural regions where manure from confined feeding of cattle is available for an amendment. The manure is aged and dried by stacking in the feedlot. The wastewater sludge is unstable (raw) filter cake collected immediately after mechanical dewatering with polymer conditioning. Combined in approximately equal portions, the wet cake and dried manure are mixed using a modified manure spreader with the back beaters reversed so that the compost is deposited on the ground in a row rather than being thrown upward by the back beaters for widespread distribution. A large machine straddling the rows, equipped with an auger-type agitator between the outboard wheels, forms the shaped windrows; periodic turning is performed by the same machine. With weekly turning, stabilization requires 4–6 weeks in good weather. In northern climates, the windrows may freeze on the outside and be covered with snow for several weeks, preventing turning and slowing the rate of decomposition. The finished compost is stored and applied at appropriate times on grassland and cropland.

In the aerated static-pile process, oxygen is supplied by mechanically drawing air through the pile. Porosity is maintained by wood chips or similar recyclable bulking agent, which also reduces the initial water content by absorption. The ratio of sludge to wood chips on a volumetric basis is in the range of 1:2 to 1:3. After preparing a base of wood chips over perforated aeration piping, the mixture is placed on a pile 8–10 ft high. This is layered with finished compost to form a cover. Air is then drawn through the pile for a period of about 3 weeks by a blower operating intermittently to prevent excessive cooling. Exhaust air is vented and deodorized through a pile of finished compost. After stabilization, the mixture is cured and dried for several weeks either in the original pile or after moving to a stockpile. The wood chips are separated from the compost by vibrating screens for reuse.

# Vacuum Filtration

Rotary vacuum filters installed in the 1970s or earlier are still in use for dewatering both raw and digested wastewater sludges. In retrofitting sludge-handling facilities and in new installations, vacuum filters are being replaced by belt filter presses that are more economical to operate, have higher solids capture, and produce drier cake [2].

## 13.17  DESCRIPTION OF ROTARY VACUUM FILTRATION

The principal components of a vacuum filter system are illustrated in Fig. 13.16. Positive-displacement pumps draw sludge from clarifiers or holding tanks and discharge it into a conditioning tank. Here the waste is mixed with chemical coagulants metered by solution feeders and is then applied through a feed chute to a vat under the filter. The cylindrical drum, covered with a porous medium, is partially submerged in the liquid sludge. As it slowly rotates, vacuum applied immediately under the filter medium draws solids to form a cake on the surface. Suction continues to dewater the solids adhering to the belt as it rotates out of the liquid; then vacuum is stopped while the belt rides over a small-diameter roller for removal of the cake, and the medium is washed by water sprays before reentering the vat. Collecting channels behind the belt in the drum surface are connected by pipes to a combination vacuum receiver and filtrate pump. The principal purpose of the receiver is air–liquid separation. Air taken from the top is discharged through a wet-type vacuum pump, while water from the bottom is removed by a filtrate pump.

Three categories of rotary vacuum filters are defined by the type of medium used and mechanism for cake discharge. A belt-type unit (Fig. 13.16) can be fitted with a variety of synthetic and natural-fiber filter cloths of

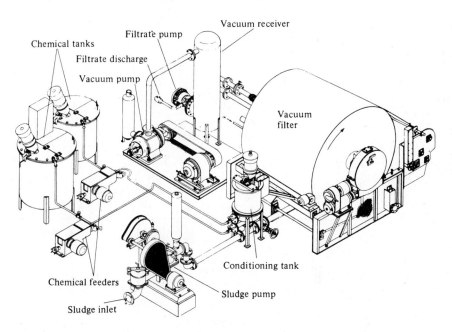

**Figure 13.16** Rotary vacuum-filter system. (Courtesy of Eimco Process Equipment Co.)

differing porosities, such as wool, nylon, Orlon, or Dacron. Each material has advantages and disadvantages related to durability, cake production, solids pickup, and response to washing. A tightly woven medium removes a high percentage of fines but can also lead to premature blinding of the pores and prevent further dewatering. Conversely, an open weave may produce a filtrate high in suspended solids. Careful consideration must be given to proper selection of a fabric for each application. For cake discharge, the medium leaves the drum surface at the end of the drying zone and passes over a small-diameter roller. Normally, a curved bar is placed under the belt between the drum and discharge roller to support the cloth weighted down with solids. This support bar also aids discharge by breaking the cake free from the belt. The cloth then passes under a wash roller, rises vertically, traveling over the top of a takeup roller, and returns to the filter drum. The wash roller is immersed in a trough of water, and spray jets rinse the belt before it is drawn over the takeup roller.

A coil filter has a medium consisting of two layers of stainless-steel helically coiled springs about 0.4 in. in diameter. They are placed around the filter drum in corduroy fashion with the upper layer resting on the bottom springs, which are held in place by grooved division strips attached to the drum surface. As shown in Fig. 13.17, the top layer of coil springs separates from the drum and travels over a small-diameter roller for removal of dewatered cake. Release is rarely a problem, since fork tines positioned between

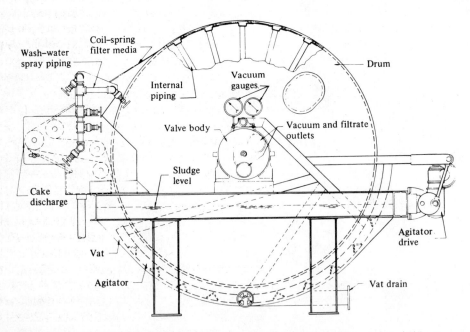

**Figure 13.17** Cross-sectional view of a coil-type vacuum filter. (Courtesy of Komline-Sanderson Engineering Corp.)

the springs ensure complete removal of the solids mat. The lower coils pass over a separate roller, and both layers are washed by spray nozzles and reapplied to the drum by grooved aligning rollers. The coil-spring medium is very effective in drying fibrous wastewater sludge, but slurries with fine particles that resist flocculation dewater poorly.

A drum filter differs from the previous two types described in that the cloth covering does not leave the drum for solids discharge or washing. Cake is scraped from the fabric on the cylinder surface after being loosened by compressed air blown through the medium from the inside. In handling waste sludges, dewatering may have to be stopped periodically to wash the drum cloth to prevent blinding. For this reason, belt and coil filters are preferred.

The objectives of vacuum filtration are to obtain an acceptable filter yield, relatively clear filtrate, and high solids concentration in the cake and to minimize operational costs. Filter yield, expressed in pounds of dry sludge solids discharged per square foot of filter area per hour, varies from 2 to 15 lb/ft²/hr. The yield generally does not include the weight of conditioning chemicals added to the sludge. Output increases with rising dosage of coagulants; for a given chemical conditioning, yield increases with solids concentration in the feed sludge. Solids capture varies from 85% to 95%, depending on the type of filter covering, the character and density of the applied sludge, and chemical conditioning. The cake solids content is affected by the same factors, as well as by the machine variables of vacuum pressure, drum submergence, and speed of rotation. Optimum suction relates to cake compressibility—a relatively incompressible solids layer dewaters better under high vacuum, while compressing an organic cake may decrease its porosity and filterability. Drum submergence and speed can be varied to adjust the form and drying cycles for optimum performance. All these parameters must be viewed as economical factors to determine cost-effective operation. Higher chemical consumption can increase yield and reduce operating time, thus raising chemical costs while decreasing labor and power. Disposal costs may be directly related to cake density. For example, the expense of hauling a relatively wet sludge to a distant landfill may be more costly than operating the filter to produce a drier cake.

## 13.18  VACUUM FILTRATION OF WASTEWATER SLUDGES

Efficient dewatering requires a minimum sludge concentration of about 4% solids for acceptable yield and reasonable chemical conditioning. Many trickling-filter plants pump settled sludge directly from primary clarifiers to vacuum filtration. Gravity thickening can be used to increase the performance of vacuum filters. Unthickened waste-activated sludge is not dewatered separately because of its low solids concentration and high percentage of fine particles. It may be vacuum-filtered after mixing with raw primary, but the blended sludges should be thickened for economical operation. One successful scheme for pretreatment of wastes from an aeration plant is separate flotation thickening of the waste-activated sludge followed by blending with primary wastes in a mixed holding tank.

Vacuum filters are manufactured with surface areas ranging from 50 to more than 300 ft$^2$. The smallest unit can dewater sludge from 1 mgd of domestic wastewater in an operating period of 6–8 hr/day. Sizing of filters is based on an anticipated yield, which is related to sludge characteristics and number of hours of operation. Thirty hours per week are commonly assumed for small plants, while at larger ones operations may extend for two shifts plus cleanup, for a total of 20 hr/day.

Chemical conditioning of wet sludge is necessary to achieve satisfactory yield and a clear supernatant. Fine particles in untreated sludge tend to blind the medium by plugging the pores. Coagulation agglomerates the very fine particles, thus reducing filter resistance and clarifying the filtrate. Minimum chemical conditioning requires greater than 90% solids capture in dewatering. Recycling of matter in the filtrate can lead to excessive solids circulating within the treatment plant, so their continuous and adequate removal is essential to efficient operation.

Common chemicals in conditioning are ferric chloride and lime, or polymers. All are mixed with water and applied by solution feeders, with typical results for various types of sludge listed in Table 13.8. Chemical dosages are expressed as percentages of the dry solids filtered; for example, 2% conditioner means 2 lb of coagulant per 100 lb of dry solids in the filter cake. The tabulated values illustrate the relative chemical conditioning for various residues and are not intended to be used for design or operation. The effectiveness of proprietary polymers differs considerably, and actual polymer additions range from approximately 0.2 to 2 times the dosages listed in Table 13.8.

The choice of chemicals is based on economics, the desired operating

**TABLE 13.8**    TYPICAL RESULTS FROM VACUUM FILTRATION OF CHEMICALLY CONDITIONED WASTEWATER SLUDGES

| Type of sludge | Filter yield (lb/ft$^2$/hr)[a] | Cake solids (%) | Chemical dosage | | |
|---|---|---|---|---|---|
| | | | Ferric chloride (%) | Lime (%) | Polymers[b] (%) |
| Raw primary | 6–10 | 25–40 | 1–4 | 6–8 | 1–4 |
| Primary plus filter humus | 4–8 | 20–35 | 2–4 | 6–8 | 3–6 |
| Primary plus waste activated | 3–5 | 15–25 | 3–6 | 5–10 | 5–10 |
| Anaerobically digested | 5–10 | 20–35 | 3–6 | 5–10 | 4–10 |

[a] 1.0 lb/ft$^2$/hr = 4.88 kg/m$^2$·h

[b] Polymer dosages are given as relative numbers; actual dosages range from 0.2 to 2 times these values.

conditions, and the method of ultimate cake disposal. The advantages of ferric chloride and lime are disinfection and stabilization, thus reducing health hazards and odors. Conversely, they are more difficult to handle and feed and usually more costly. Polymers are easy to apply and frequently more economical but do not provide disinfection. A minimum cake density of 20%–25% solids is generally satisfactory for hauling to landfill. Nevertheless, if long-distance trucking is involved, a lower moisture content reduces the total weight of cake transported.

■ **EXAMPLE 13.12**

A wastewater treatment plant produces 50,000 gpd of sludge containing 5.5% solids. What size vacuum filters are required, assuming a yield of 5.0 lb/ft²/hr and 12 hr of operation per day? Chemical conditioning is 8.0% lime and 2.0% ferric chloride. The cake is 23% solids, including organics plus chemical additions. Calculate the chemical dosages per ton of dry solids and weight of filter cake produced per day.

*Solution*

$$\text{dry sludge solids} = 50,000 \times 8.34 \times 0.055 = 22,900 \text{ lb/day}$$

$$\text{filter area required} = \frac{22,900}{5.0 \times 12} = 382 \text{ ft}^2$$

Use two 200-ft² filters, each 8 ft in diameter by 8 ft wide:

$$\text{lime dosage} = 0.08 \times 2000 = 160 \text{ lb/ton of dry solids}$$

$$\text{ferric chloride dosage} = 0.02 \times 2000 = 40 \text{ lb/ton of dry solids}$$

$$\text{total cake solids} = \text{sludge solids} + \text{chemical additions}$$

$$= 22,900 + (0.08 + 0.02)22,900 = 25,200 \text{ lb/day}$$

$$\text{weight of wet filter cake} = \frac{25,200}{0.23 \times 2000} = 54.8 \text{ tons/day} \qquad ■$$

# Pressure Filtration

Sludges can be dewatered by pressure filtration using either a belt filter press or a plate-and-frame filter press. The belt filter press consists of two continuous porous belts that pass over a series of rollers to squeeze water out of the sludge layer compressed between the belts. A filter press consists of a series of recessed plates with cloth filters and intervening frames held together to form enclosed filter chambers. Sludge pumped under high pressure into these chambers forces water out through the cloth filters, filling the chamber with dewatered cake. At the end of the feed and pressure cycles, the plates are separated to remove the sludge cake. This type of pressure filter is noted for producing a dry cake.

## 13.19 DESCRIPTION OF BELT FILTER PRESS DEWATERING

A belt filter press compresses the sludge between two endless porous belts tensioned over a series of rollers to squeeze out the water. The basic operational steps of the process are illustrated in Fig. 13.18. Before wet sludge is distributed on the top of the upper belt, it is conditioned with polymer to aggregate the solids. Initial dewatering takes place in the gravity drainage zone, where the belt is supported horizontally on an open framework or grid that allows separated water to drain freely through the belt into a collection pan. Most machines use adjustable plastic vanes supported just above the belt surface in the upper drainage zone to open channels in the sludge, aiding the release of free water. Rather than an open framework, some manufacturers use closely spaced, small-diameter rollers to support the belt in the drainage zone. Approximately one half of the water is removed in the gravity zone; thus, the solids content is doubled and the sludge volume halved. After dropping onto the lower belt, the sludge is gradually compressed between the two belts as they come together in the low-pressure, cake-forming zone. This wedge zone terminates with the two belts wrapping over the first of a series of rollers. Some machines have uniform diameter rollers, while on others the subsequent rollers decrease in diameter to gradually increase pressure on the cake. As the belts pass over these rollers, the confined sludge layer is subjected to both compression and shearing action caused by the outer belt being a greater distance from the center of the roller than the inner belt. Depending on the manufacturer, the rollers may be perforated stainless steel cylinders or plain carbon steel with a coating for protection against corrosion. The belt tension, alignment, and drive rollers

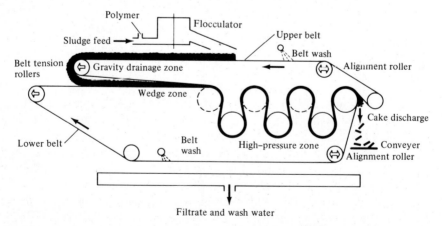

**Figure 13.18** Schematic diagram of a belt filter press. (Courtesy of Ashbrook-Simon-Hartley.)

have a rubber coating to increase frictional resistance and prevent slippage. The cake is scraped from the belts by doctor blades held against the belts.

Belts are made from several fabrics of synthetic fibers. Monofilament polyester woven fabrics with visible clear openings are used in dewatering wastewater sludges. As a result, solids pressed tightly on the surface can penetrate the pores of the fabric and belts require washing with a high-pressure water spray. To confine and collect the wash water, the nozzles are housed in enclosures through which the belts pass. These wash boxes are located on both the upper and lower belts after cake discharge. Belt tensioning is performed by pressing one roller against each belt using hydraulic or pneumatic cylinders connected to the roller shaft. A tension that is too low results in belt slippage, while excess pressure can extrude sludge from between the belts. Each belt is kept centered on its rollers, using a steering assembly consisting of a device that senses the track of the belt and signals adjustment by an alignment roller. By actuating a hydraulic cylinder attached to one end of the shaft, this roller swivels, causing the track of the belt to move along the axis of the roller.

Several manufacturers provide features that differ from the standard belt filter press. Among the various flocculators, the simplest consists of injecting polymer into the sludge feed pipe prior to a static mixing box that distributes the flocculated sludge onto the belt. More common is a horizontal mixing drum fitted with internal paddles that flocculate the sludge prior to discharge. A few proprietary units employ a rotating drum made of a filter screen, which allows thickening in addition to flocculation. While the aggregated solids are retained on the screen and fed to the press, the separated water is discharged from the enclosure surrounding the drum screen. Several manufacturers have the option of adding a separate belt for the gravity drainage zone to allow a longer time for water separation through a more porous belt than those used for the pressure zones (Fig. 13.19). One press has a separate external belt that can exert very high pressure against the two belts with the sludge layer between as they pass over the last roller. Most belt filter press machines require considerable head room since the flocculator is commonly mounted over the gravity drainage zone above the pressure rollers. In contrast, other units attain a lower profile by applying the feed sludge onto the lower belt.

## 13.20 APPLICATION OF BELT FILTER DEWATERING

The most significant variables that affect dewatering performance of a belt filter press are the sludge characteristics, polymer conditioning, sludge feed rate, belt tension, and belt speed. The characteristics of greatest importance in wastewater sludges are the solids concentration, the nature of the solids, and prior biological or chemical conditioning. A press is limited to a hydraulic capacity essentially independent of solids concentration less than about 4%. Most manufacturers suggest a maximum hydraulic loading of 50 gpm (11.4 $m^3$/h) per meter of belt width. At solids contents greater than about 6%, the

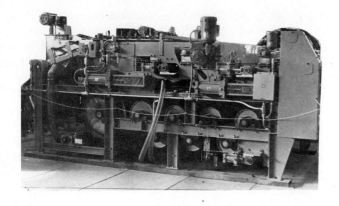

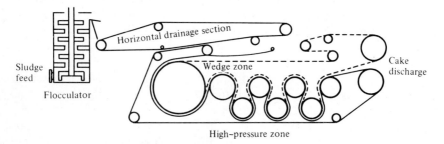

**Figure 13.19** Belt filter press on which feed sludge is mechanically mixed with polymer in a conditioning tank that discharges onto a separate upper belt for gravity drainage. The two belts compressing the sludge pass through a serpentine section of high-pressure rollers that progressively decrease in diameter. (Courtesy of Komline-Sanderson Engineering Corp.)

capacity of a press is restricted by solids loading. The nature of the solids influence both polymer flocculation and mechanical dewatering. Fibrous solids, commonly associated with primary clarifier settlings, are much easier to dewater than the fine, bulky biological solids wasted from secondary activated-sludge processing.

A polymer is applied to flocculate a sludge, forming aggregates of the particles to allow easier release of the water. The dosage of polymer applied per mass of solids dewatered depends essentially on the sludge characteristics and machine loading, which is in turn directly related to belt speed (detention time in the gravity dewatering zone). Besides inadequate flocculation, the maximum sludge feed rate can be governed by poor solids recovery, too wet a cake, or excessive polymer dosage. If sludge loading is too high, the detention time in the gravity drainage section cannot provide sufficient water release. The result is the extrusion of fine solids through the fabric and from between the belts at the edges. Excessive belt tension can also cause extrusion of solids. In actual plant operation, the acceptable loading is based on the economy of operation, the two major costs being the polymer consumption and hours of operation. A logical sequence for adjusting processing variables at a given sludge loading rate is to select a polymer dosage, adjust the belt speed, set the belt tension, and then readjust these three settings to achieve the desired cake dryness and solids recovery with the minimum polymer dosage.

The main performance parameters of a belt filter press are the hydraulic and solids loading rates, polymer dosage, solids recovery, cake dryness, wash-water consumption, and wastewater discharge. Hydraulic loading is expressed in gallons per minute of sludge feed per meter of belt width (cubic meters per meter per hour). Solids loading is expressed as the pounds of total dry solids feed per meter per hour (kilograms per meter per hour). The polymer dosage is calculated as the pounds applied per ton of total dry solids in the sludge feed (kilograms per tonne). Although the fraction of solids recovery is the quantity of dry solids in the cake divided by the dry solids in the feed sludge, it is often calculated based on the suspended solids in the wastewater (filtrate plus wash water) as follows:

$$\text{solids recovery} = \frac{\left(\begin{array}{c}\text{total solids}\\\text{in feed sludge}\end{array}\right) - \left(\begin{array}{c}\text{suspended solids}\\\text{in wastewater}\end{array}\right)}{\text{total solids in feed sludge}} \tag{13.14}$$

Cake dryness is expressed as the percentage of dry solids by weight in the cake. For easy comparison with hydraulic sludge loading, wash-water consumption and wastewater discharge are usually expressed in units of gallons per minute per meter of belt width (cubic meters per meter per hour). Example 13.13 illustrates the calculation of these parameters.

### ■ EXAMPLE 13.13

A belt filter press with an effective belt width of 2.0 m is used to dewater an anaerobically digested sludge. The machine settings during operation are a sludge feed rate of 18.2 m³/h (80 gpm), polymer dosage of 1.8 m³/h (8.0 gpm) containing 0.20% powdered polymer by weight, belt speed of 6.1 m/min, belt tension of 4.7 kN/m of roller, and wash-water application of 15.4 m³/h (68 gpm) at 550 kN/m². Based on laboratory analyses, total solids in the feed sludge equal 3.5%, total solids in the cake are 32%, wastewater from belt washing contains 2600 mg/l suspended solids, and filtrate production measures 17.7 m³/h (78 gpm) with a suspended solids concentration of 500 mg/l. From these data calculate the hydraulic loading rate, solids loading rate, polymer dosage, and solids recovery. Comment on the production water usage and wastewater generated relative to the hydraulic sludge feed.

*Solution*

$$\text{hydraulic loading rate} = \frac{18.2}{2} = 9.1 \text{ m}^3/\text{m·h (40 gpm/m)}$$

$$\text{solids loading rate} = \frac{18.2 \text{ m}^3/\text{h} \times 1000 \text{ kg/m}^3 \times 0.035}{2 \text{ m}}$$

$$= 320 \text{ kg/m·h (700 lb/m/hr)}$$

$$\text{polymer dosage} = \frac{0.5 \times 1.8 \text{ m}^3/\text{h} \times 1000 \text{ kg/m}^3 \times 0.002}{(320 \text{ kg/m·h})/(1000 \text{ kg/t})}$$

$$= 5.7 \text{ kg/t} (11.4 \text{ lb/ton})$$

wastewater suspended solids

$= $ wash-water solids $+$ filtrate solids

$$= \frac{15.4 \text{ m}^3/\text{h} \times 2600 \text{ g/m}^3 + 17.7 \text{ m}^3/\text{h} \times 500 \text{ g/m}^3}{2 \text{ m} \times 1000 \text{ g/kg}}$$

$$= 24 \text{ kg/m·h} (54 \text{ lb/m/hr})$$

(Note that approximately 80% of the waste solids are in the wash water.)

$$\text{solids recovery} = \left(\frac{320 - 24}{320}\right) 100 = 93\%$$

Wash-water consumption equals 7.7 m³/m·h and the polymer feed is 0.9 m³/m·h for a total of 8.6 m³/m·h, and hence the process water added very nearly equals the 9.1-m³/m·h sludge feed. Wastewater production is 16.5 m³/m·h, composed of 7.7 m³/m·h wash water and 8.8 m³/m·h filtrate from the sludge and polymer solution water. This equals 1.8 times the sludge feed rate of 9.1 m³/m·h.  ∎

## 13.21 SIZING OF BELT FILTER PRESSES

Belt widths of presses range from 0.5 to 3 m, with the most common sizes between 1.0 and 2.5 m. Some manufacturers supply only 1.0- and 2.0-m machines, while others build 1.5- and 2.5-m units. The selection during design of a sludge dewatering facility depends on such factors as the size of the plant, the desired flexibility of operations, anticipated conditions of dewatering, and economics. Typical results from filter pressing of wastewater sludges are listed in Table 13.9. The solids loading rates relate to both the feed solids concentration and hydraulic loading. For example, in the first line, a 4% feed at 50 gpm/m yields a solids loading of 1000 lb/m/hr. Also, the cake solids percentage decreases and the polymer dosage increases with greater dilution of the sludges and for those containing waste-activated sludge. Since performance of filter pressing depends on the character of the sludge, sizing of presses for new treatment plants without an existing sludge to test must be based on operating experience at other installations. While such data are more reliable than the values given in Table 13.9, new installations should be conservatively sized to account for the probable inaccuracy in projecting performance.

Design of a belt filter press installation at an existing facility can be reliably done by conducting field testing using a narrow-belt machine enclosed in a mobile trailer. Most manufacturers use a full-scale 0.5- or 1.0-m press that is representative of their larger machines. During the preliminary

**TABLE 13.9    TYPICAL RESULTS FROM BELT FILTER PRESS DEWATERING OF POLYMER FLOCCULATED WASTEWATER SLUDGES**

| Type of sludge | Feed solids (%) | Hydraulic loading (gpm/m)[a] | Solids loading (lb/m/hr)[b] | Cake solids (%) | Polymer dosage (lb/ton)[c] |
|---|---|---|---|---|---|
| Anaerobically digested primary only | 4–8 | 40–50 | 1000–1600 | 25–35 | 3–6 |
| Anaerobically digested primary plus waste activated | 2–5 | 40–50 | 500–1000 | 15–26 | 6–12 |
| Aerobically digested without primary | 1–3 | 30–45 | 200–500 | 11–22 | 8–14 |
| Raw primary and waste activated | 3–6 | 40–50 | 800–1200 | 16–25 | 4–10 |
| Thickened waste activated | 3–5 | 40–50 | 800–1000 | 14–20 | 6–8 |
| Extended aeration waste activated | 1–3 | 30–45 | 200–500 | 11–22 | 8–14 |
| Heat-treated primary plus waste activated | 4–10 | 35–50 | 1000–1800 | 30–40 | 1–2 |

[a] 1.0 gpm/m = 0.225 m³/m·h.
[b] 1.0 lb/m/hr = 0.454 kg/m·h.
[c] 1.0 lb/ton = 0.500 kg/tonne.

design phase, a rented trailer unit can be used to determine the dewaterability of the sludge and to establish testing criteria for the performance specifications. After sizing and design of the press facility, selection of the press manufacturer can be based on both competitive bidding and qualification testing using trailer units, either individually, with the lower bidder's press first, or as a group of several proprietary machines operating in parallel. This procedure reduces the risk in design by demonstrating that the seleced manufacturer's press can achieve the results required by the performance specifications. This testing does not replace acceptance testing after construction, when the installed presses are evaluated to ensure compliance with the specifications.

## ■ EXAMPLE 13.14

An existing wastewater treatment plant is going to install belt filter presses to dewater anaerobically digested sludge prior to stockpiling and spreading on agricultural land. The current system of lagooning and disposal of liquid sludge is expensive and has created environmental concerns. Past sludge production records were studied to determine the following values for design. The average annual sludge quantity equals 80,000 gpd, with an average solids concentration of 5.0%. The design quantity during the peak month is 130,000 gpd at 4.0% solids, which contain 30% more dry solids than the annual average. The digesters have sufficient capacity to equalize the variations in raw-sludge feed during the peak month.

Field testing using a trailer-mounted press resulted in the following performance data. When the sludge feed had a solids concentration of 4.0%, the allowable hydraulic loading was 50 gpm/m (solids loading of 1000 lb/m/hr) producing a cake with a solids content of 22% with a polymer dosage of 9.6 lb/ton. At 5.0% sludge feed, acceptable operation was a hydraulic loading of 45 gpm/m (solids loading of 1100 lb/m/hr) producing a 24% cake with a polymer dosage of 8.1 lb/ton. Solids recoveries for all tests were between 94% and 96%. The polymer solution was 0.20% dry powder by weight, and wash-water usage was 32 gpm/m.

The design operating schedule is a maximum of 12 hr/day during the peak month. A minimum of two machines is desired so that sludge can be continuously dewatered if one unit is out of service. Based on these criteria, size the belt filter presses and determine the operating times under design conditions. For the average annual sludge quantity, calculate the polymer usage and weight of cake produced and estimate the wastewater generated.

*Solution.*    The belt width required based on peak month operation is

$$\frac{130{,}000 \text{ gal}}{50 \text{ gpm/m} \times 60 \text{ min/hr} \times 12 \text{ hr}} = 3.6 \text{ m}$$

or

$$\frac{130{,}000 \text{ gal} \times 8.34 \times 0.040}{1000 \text{ lb/m/hr} \times 12 \text{ hr}} = 3.6 \text{ m}$$

Specify two belt filter presses with effective belt widths of 2.0 m. For these units, the operating time at the peak monthly load is

$$\frac{130{,}000 \text{ gal}}{2 \times 2.0 \text{ m} \times 50 \text{ gpm/m} \times 60 \text{ min/hr}} = 10.8 \text{ hr/day}$$

with both presses operating; with one unit operating, the time required is 21.6 hr/day. The operating time at the average annual load is

$$\frac{80{,}000}{2 \times 2.0 \times 45 \times 60} = 7.4 \text{ hr/day}$$

with both presses operating. Polymer usage at the average daily annual load is

$$\frac{80{,}000 \text{ gal} \times 8.34 \times 0.05 \times 8.1 \text{ lb/ton}}{2000 \text{ lb/ton}} = 140 \text{ lb/day}$$

The weight of cake produced is

$$\frac{80{,}000 \times 8.34 \times 0.05}{0.24 \times 2000} = 70 \text{ tons/day}$$

The flow of wastewater generated equals the wash water plus filtrate. The filtrate can be estimated by subtracting the theoretical volume of sludge cake from the sludge feed plus polymer solution.

sludge feed = 45 gpm/m

polymer solution feed

$$= \frac{1100 \text{ lb/m/hr} \times 8.1 \text{ lb/ton}}{60 \text{ min/hr} \times 2000 \text{ lb/ton} \times 0.002 \text{ lb/lb} \times 8.34 \text{ lb/gal}} = 4 \text{ gpm/m}$$

theoretical flow of sludge cake

$$= \frac{1100 \text{ lb/m/hr}}{60 \text{ min/hr} \times 0.24 \text{ lb/lb} \times 8.34 \text{ lb/gal} \times 1.05} = 9 \text{ gpm/m}$$

   filtrate $= 45 - 4 - 9 = 32 \text{ gpm/m}$

wash water $= 32 \text{ gpm/m}$

Therefore, the wastewater flow is $32 + 32 = 64 \text{ gpm/m}$. The quantity of wastewater at the average daily annual load is

   $64 \text{ gpm/m} \times 4.0 \text{ m} \times 60 \text{ min/hr} \times 7.4 \text{ hr} = 110,000 \text{ gpd}$  ∎

# 13.22  DESCRIPTION OF FILTER PRESS DEWATERING

The two types of plate-and-frame filter presses are the fixed-volume press and the variable-volume diaphragm press. Removal of dewatered cake from a fixed-volume press is done by manually separating the press frames and loosening the layers of cake from the recessed plates with a wooden paddle if they do not drop by force of gravity. The diaphragm press is designed for automatic operation. After opening, the cakes are forcefully discharged and the filter cloths automatically washed before the press closes for another cycle. Although reducing the manual labor in operation, this modification increases the mechanical complexity of a press filter [7].

Compared to belt presses, filter presses are more expensive, have higher operating costs, and are substantially larger machines for the same sludge processing capacity. Dewatering of wastewater sludge requires lime and ferric chloride conditioning; polymer flocculation is not suitable. High cake dryness is the principal advantage of pressure filtration with cake solids content greater than 35% and up to 40%–50% possible.

A pressure filter consists of depressed plates held vertically in a frame for proper alignment and pressed together by a hydraulic cylinder (Fig. 13.20). Each plate is constructed with a drainage surface on the depressed portion of the face. Filter cloths are caulked onto the plate and peripheral gaskets seal the frames when the press is closed. Influent and filtrate ports are formed by openings that extend through the press. Sludge is pumped under pressure into the chambers between the plates of the assembly, and water passing through the media drains to the filtrate outlets. Solids retained form cakes between the cloth surfaces and ultimately fill the chambers. High pressure consolidates the cakes by applying air to the sludge inlet at about 225 psi (1550 kN/m$^2$). After this filter cycle, which requires from 2 to 3 hr, compressed air blows the feed sludge remaining in the influent ports back to a holding tank. The filter plates are separated and dewatered cakes drop out

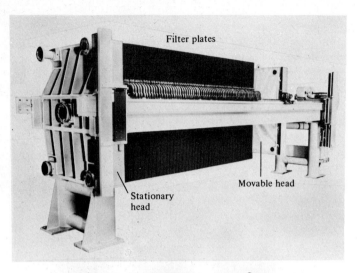

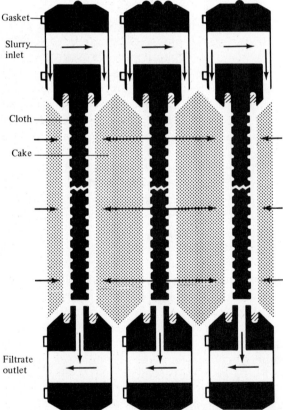

**Figure 13.20** Photograph of a pressure filter and diagram of the filter chamber assembly showing slurry channel and filtrate discharge. (Courtesy of Shriver-Eimco Process Equipment Co.)

of the chambers into a hopper equipped with a conveyor mechanism. Cake release is assisted by introducing compressed air behind the filter cloths and by manually prodding with a paddle if necessary.

Chemical conditioning improves sludge filterability by flocculating fine particles so that the cake remains reasonably porous, allowing passage of water under high pressure. Dosages for conditioning wastewater sludges, expressed as percentages of dry solids in the feed sludge, are commonly 10%–20% CaO and 5%–8% $FeCl_3$. Precoating the media with diatomaceous earth or fly ash helps to protect against blinding and ensures easy separation of the cake for discharge. Filter aid for the precoat is placed by feeding a water suspension through the filter before applying sludge. In some cases, the aid may be added to the conditioned sludge mixture to improve porosity of the solids as they collect. Solids capture in pressure filtration is very high, commonly measuring 98%–99%. The organic sludge solids content in a typical cake is 35%. If the application of conditioning chemicals were 20%, the cake would have a total solids concentration including chemicals of 40%.

The operation of a variable-volume diaphragm press is diagrammed in Fig. 13.21. Sludge is pumped into the recessed chambers at a pressure of 100 psi (690 kN/m²) for 10–20 min, allowing filtrate to drain out through the clothes on both sides of the chamber. Following this filling process, sludge feed is turned off; the water under high pressure is pumped into the space between the diaphragm and one side of the plate body. Compressing the partially dewatered cake for 15–30 min concentrates the cake to the desired final solids content. At the end of the compression stage, the sludge feed and filtrate lines are blown out with compressed air. The press then automatically opens and the cakes are released by pulling the filter cloths down around small-diameter rollers. The filter cloths are then washed on both sides as they are retracted up to their original position. The cycle time of the diaphragm press is 15%–25% of the cycle time of a fixed-volume machine, and the cakes are one-half or less in thickness. Therefore, the yield for a diaphragm press can be more than twice the yield of a fixed-volume press in terms of mass of solids dewatered per unit area of filter cloth per unit of time.

## 13.23  APPLICATION OF PRESSURE FILTRATION

Water treatment plant wastes are suited to pressure filtration since they are often difficult to dewater, particularly alum sludges and softening precipitates containing magnesium hydroxide. Figure 13.22 shows a schematic flow diagram for a pressure filtration process. Gravity-thickened alum wastes are conditioned by the addition of lime slurry. A precoat of diatomaceous earth or fly ash is applied prior to each cycle, and conditioned sludge is then fed continuously to the pressure filter until filtrate ceases and the cake is consolidated under high pressure. A power pack holds the chambers closed during filtration and transports the movable head for opening and closing. An equalization tank provides uniform pressure across the filter chambers as

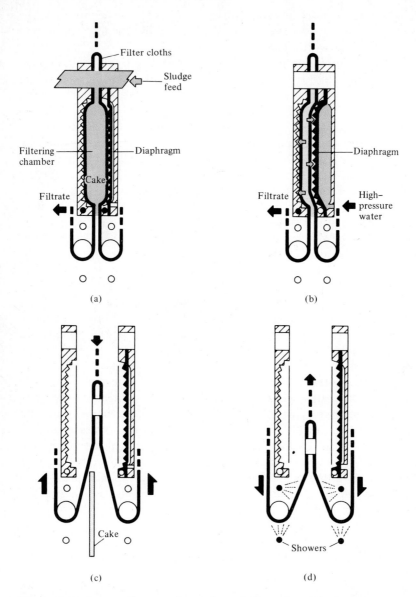

**Figure 13.21** Automatic operations of a variable-volume diaphragm pressure filter: (a) Filtering. (b) Compression. (c) Cake discharge. (d) Washing of filter cloths. (Courtesy of Ingersoll-Rand and Ishigaki Mechanical Industry Company Ltd., patented.)

the cycle begins. Prior to cake discharge, excess sludge in the inlet ports of the filter is removed by air pressure to a core separation tank. Filtrate is measured through a weir tank and recycled to the inlet of the water treatment plant. Cake is transported by truck to a disposal site.

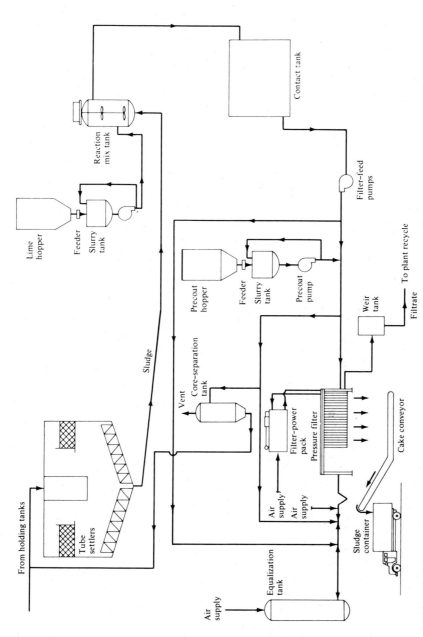

**Figure 13.22** Schematic flow diagram for conditioning and filter-press dewatering of alum sludge from surface-water treatment. [From G. P. Westerhoff and M. P. Daly, "Water-Treatment-Plant Waste Disposal," *J. Am. Water Works Assoc.* 66(7) (1974): 443. Copyright 1974 by the American Water Works Association, Inc.]

Alum sludges are conditioned using lime and/or fly ash. Lime dosage is in the range 10%–15% of the sludge solids. Ash from an incinerator, or fly ash from a power plant, is applied at a much higher dosage, approximately 100% of dry sludge solids. Polymers may also be added to aid coagulation. Fly ash and diatomaceous earth are used for precoating; the latter requires about 5 lb/100 ft$^2$ of filter area. Under normal operation, cake density is 40%–50% solids and has a dense, dry, textured appearance.

Wastewater sludges are amenable to dewatering by pressure filtration after conditioning with ferric chloride and lime or fly ash. Minimum ferric chloride and lime dosages are 5% and 10%, respectively, for a waste concentration of 5% solids or greater. Conditioning with fly ash requires 100%–150% additions. Either diatomaceous earth or fly ash is used for precoating the filter media.

# Centrifugation

Centrifuges are used for both dewatering sludges and thickening waste slurries for further processing. The most common type is the solid-bowl scroll centrifuge. Centrifugation is used in larger wastewater plants, generally with a capacity of 10 mgd or more, where solids recovery is adequate and centrifuged cake is dense enough for disposal. Applications in chemical treatment of water or wastewater are sludges difficult to dewater by gravity separation, such as alum coagulation residues, and lime-softening precipitates prior to recalcining.

## 13.24  DESCRIPTION OF CENTRIFUGATION

Solids are removed from the waste stream flowing through a centrifuge under the influence of a centrifugal field of 100–600 times the force of gravity. The separated solids are deposited against the spinning solid bowl, while the overflow is a clear supernatant. The fundamental difference in centrifuges is the manner of solids collection and discharge. For the variety of granular, fibrous, flocculent, and gelatinous solids in sanitary sludges, the solid-bowl scroll centrifuge illustrated in Fig. 13.23 is best adapted to provide satisfactory supernatant clarity and cake dryness. A scroll centrifuge is also referred to as a conveyor or decanter centrifuge. The two principal components are a rotating solid bowl in the shape of a cylinder with a cone section on one end and an interior rotating-screw conveyor. Feed slurry enters at the center and is spun against the bowl wall. Settled solids are moved by the conveyor to one end of the bowl and out of the liquid for drainage before discharge while clarified effluent discharges at the other end over a dam plate. This system is best suited for separating solids that compact to a firm cake and can be conveyed easily out of the water pool. If solids compact poorly, moving a soft cake causes redispersion, resulting in poor clarification and a

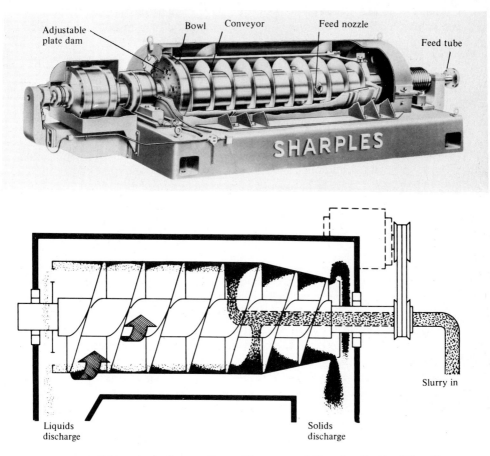

**Figure 13.23** Solid-bowl scroll centrifuge. (Courtesy of Sharples-Stokes Div., Pennwalt Corp.)

wet concentrate. Flocculent solids can generally be made scrollable by chemical conditioning of the sludge, and the redispersing action that occurs in a scroll machine is advantageous in classification of particles by centrifugation.

A major advantage of scroll dewatering is operational flexibility. Machine variables include pool volume, bowl speed, and conveyor speed. The depth of liquid in the bowl and the pool volume can be controlled by an adjustable plate dam. Pool volume adjustment varies the liquid retention time and changes the drainage deck surface area in the solids-discharge section. The bowl speed affects gravimetric forces on the settling particles, and conveyor rotation controls the solids retention time. The driest cake results when the bowl speed is increased, the pool depth is the minimum allowed, and the differential speed between the bowl and conveyor is the maximum possible. Flexibility of operation allows a range of densities in the solids discharge varying from a dry cake to a thickened liquid slurry. Feed rates, solids content, and prior chemical conditioning can also be varied to influence performance. Solids removal and cake consolidation can both be enhanced by adding polymers or other coagulants with the feed sludge.

The performance of centrifugal dewatering for given feed and machine-operating conditions depends on the dosage of chemical coagulants. Suspended solids removal and usually cake dryness increase with greater chemical additions, while carryover of solids in the centrate decreases. There is, however, a saturation point at which flocculent dosage does not significantly improve centrate clarity. Optimum chemical conditioning without overdosing can be determined most reliably by full-scale or pilot-plant tests. For some wastes, centrate recycling can improve overall suspended-solids removal, but for others it may cause upset, owing to an accumulation of fine particles.

## 13.25  APPLICATIONS OF CENTRIFUGATION

The characteristics of a wastewater sludge to be dewatered determines the centrifugation capacity, chemical conditioning, cake dryness, and solids recovery. The ratio of primary to secondary solids in a sludge has a significant effect. For illustration, consider the following typical performance data in dewatering different kinds of sludges by using an adequate polymer dosage for a minimum of 90% solids recovery: raw primary, cake solids of 28%–34% and polymer dosage of 2–4 lb/ton of dry solids; raw primary plus waste activated sludge, cake solids of 18%–25% and polymer dosage of 6–14 lb/ton; and raw waste activated sludge, cake solids of 14%–18% and polymer dosage of 12–20 lb/ton [2]. Usually, the cake solids can be increased a few percentage points by applying an excessive polymer dosage. Machine loadings have notable influence on performance. Underloading allows reduced polymer dosage and produces a drier cake. As the loading increases toward the capacity of the centrifuge, the required polymer dosage increases and the cake becomes wetter. Therefore, to be cost effective, dewatering must take into account machine loading relative to hours of operation per day and the operation of standby centrifuge capacity.

The number of centrifuges and operating hours are designated in design. An installation requires at least one standby machine. More may be required depending on availability of maintenance service and alternative methods of disposal. For example, wastewater plants under 20 mgd should have two full-sized machines (one + one standby); between 20–100 mgd, a minimum of three machines (two + one); and over 250 mgd, a minimum of six machines (four + two) [2]. The designer's performance specification designates the sludge characteristics (raw or digested, solids concentrations, ratio of primary to waste activated, and temperature) and performance requirements (sludge flowrates, minimum cake solids, minimum solids recovery, and polymer dosage). The design engineer does not normally select the actual model and size of the centrifuges unless on-site testing has been conducted. Field tests of two or more machines are run concurrently so that all are dewatering sludge with the same characteristics. The testing range of flow and solids loadings should be adequate to fully evaluate plant operations. The test machines must represent the full-scale centrifuges that will be provided by the manufacturer for installation.

Water treatment plant wastes are also amenable to centrifuge dewatering. For alum sludges from surface-water treatment, centrifugation performance must be verified by testing at each plant since sludge characteristics vary considerably. In general, aluminum hydroxide slurries from coagulation settling and gravity-thickened backwash waters can be concentrated to a truckable pasty sludge of about 20% solids. The removal efficiency in a scroll centrifuge ranges from 50% to 95% based on operating conditions and polymer dosage, and the centrate is correspondingly turbid or clear. Lime-softening precipitates compact more readily than alum floc. A gravity-thickened sludge of 15%–25% solids can be dewatered to a solidified cake of 65%. Solids recovery is often 85%–90% with polymer flocculation.

## 13.26  SUSPENDED-SOLIDS-REMOVAL EFFICIENCY

Centrate from dewatering sludge is returned to the head of a treatment plant contributing suspended solids and BOD to the raw-wastewater influent. In addition to centrate from centrifuges, other recycled flows from sludge processes are supernatant from anaerobic or aerobic digesters, overflow from gravity thickeners, and filtrate from belt, pressure, and vacuum filters. Consequently, poor solids capture in sludge thickening and dewatering contribute to the load on the plant and cycling of solids within the system. Being colloidal in nature, many of the finer solids pass through primary sedimentation for capture in biological aeration and return in the waste secondary sludge for thickening and dewatering again. Cycling solids can lead to overloading and upset of all treatment processes. However, their presence is normally first noticed by a thinner primary sludge and an increase in the oxygen demand of biological aeration.

A relatively easy method of estimating solids capture by a sludge-thickening or dewatering unit is to measure solids concentrations in the process flows. The relationship is developed as follows. The solids mass balance is given in Eq. 13.15, and the liquid volumetric balance in Eq. 13.16:

$$M_S = M_R + M_C \qquad (13.15)$$

$$Q_S = Q_R + Q_C \qquad (13.16)$$

where  $M_S, Q_S$ = mass of solids and quantity of flow of feed sludge
$M_R, Q_R$ = mass of solids and quantity of return flow
$M_C, Q_C$ = mass of solids and quantity of flow of thickened sludge or cake

Without introducing significant error, the specific gravity of all flows can be assumed to be 1.00. Hence, the mass of solids $M$ in a process flow stream is the rate of flow $Q$ times the solids concentration $S$.

$$M = Q \times S \qquad (13.17)$$

Combining these equations results in the following relationships:

$$\frac{M_C}{M_S} = \frac{S_C(S_S - S_R)}{S_S(S_C - S_R)} \qquad (13.18)$$

where   $M_C/M_S$ = fraction of solids removal (solids capture)
$S_C$ = solids concentration in thickened sludge or cake
$S_S$ = solids concentration in feed sludge
$S_R$ = solids concentration in return flow

$$\frac{Q_R}{Q_S} = \frac{S_C - S_S}{S_C - S_R} \qquad (13.19)$$

where   $Q_R/Q_S$ = fraction of feed sludge appearing as return flow

The solids concentrations $S$ can be either total solids (residue upon evaporation) or suspended solids (nonfilterable residue). Testing for suspended solids by the standard laboratory technique of filtration through a glass filter is feasible for dilute suspensions where dissolved solids are a major portion of the total solids. In contrast, wastes with high solids contents are difficult to test accurately by laboratory filtration. Therefore, suspended-solids analysis of a sludge sample is performed by a total-solids test that is corrected by subtracting an estimated dissolved-solids concentration. The procedure does not create significant error since the filterable solid content usually amounts to less than 5% of the total solids in the sludge samples.

■ **EXAMPLE 13.15**
A scroll centrifuge dewaters an alum–lime sludge containing 8.0% solids at a feed rate of 20 gpm. The cake produced has a solids concentration of 55% and the centrate contains 9000 mg/l. Calculate the solids-removal efficiency and the centrate flow.

*Solution.*   From Eq. 13.18,

$$\text{solids removal} = \frac{M_C}{M_S} = \frac{55(8.0 - 0.9)}{8.0(55 - 0.9)} = 0.90 = 90\%$$

Substituting into Eq. 13.19 yields

$$Q_R = 20 \text{ gpm} \times \frac{55 - 8.0}{55 - 0.9} = 17.4 \text{ gpm} \qquad ■$$

■ **EXAMPLE 13.16**
The operation of a vacuum filter dewatering wastewater sludge was analyzed by sampling and testing the feed sludge and cake for total-solids content and the filtrate for both total-solids and suspended-solids concentrations. The presence of conditioning chemicals was ignored since the polymer addition amounted to only 3% of the solids

content in the sludge. Based on the following laboratory results, estimate the suspended-solids capture and the fraction of flow of feed sludge appearing as filtrate.

|  | Total solids (mg/l) | Suspended solids (mg/l) |
|---|---|---|
| Sludge | 36,600 | |
| Cake | 158,000 | |
| Filtrate | 10,800 | 8700 |

*Solution.*  Calculate the dissolved solids in the filtrate:

$$DS = TS - SS = 10,800 - 8700 = 2100 \text{ mg/l}$$

Applying this value to the sludge and cake, one sees that suspended-solids concentrations for all three flows are

$$S_C = 158,000 - 2100 = 156,000 \text{ mg/l}$$

$$S_S = 36,600 - 2100 = 34,500 \text{ mg/l}$$

$$S_R = 8700 \text{ mg/l}$$

Substituting into Eqs. 13.18 and 13.19 yields

$$\text{solids removal} = \frac{M_C}{M_S} = \frac{156,000 \times (34,500 - 8700)}{34,500 \times (156,000 - 8700)} = 0.79 \quad (79\%)$$

$$\frac{\text{filtrate flow}}{\text{sludge feed}} = \frac{Q_R}{Q_S} = \frac{156,000 - 34,500}{156,000 - 8700} = 0.82 \quad (82\%) \qquad ■$$

# Recovery of Chemicals

## 13.27  LIME RECOVERY

Calcination for recovery of lime involves burning calcium carbonate precipitate at temperatures of 600°–2000°F to yield calcium oxide and carbon dioxide:

$$CaCO_3 \rightarrow CaO \text{ (lime)} + CO_2 \uparrow \qquad (13.20)$$

Water-softening sludges can be pretreated to remove magnesium by applying carbon dioxide to preferentially dissolve magnesium hydroxide precipitate without affecting the calcium carbonate solids. Carbonation also improves mechanical thickening by providing better calcium recovery and a denser cake. Water-softening precipitates include both calcium hardness removed from the raw water and calcium from the added lime; therefore, recalcination should theoretically produce more recycled lime than needed. Impurities in the sludge and calcium losses during dewatering and dust recovery in the recalciner, however, reduce the actual yield such that practical recovery is usually only about 25% greater than the original lime applied.

A lime recovery process generally consists of the following steps: gravity thickening to 25% solids or greater; sludge carbonation using stack gases from the recalcining furnace to dissolve magnesium hydroxide; centrifugation to increase solids content to greater than 60%; flash drying of dewatered sludge using hot off-gases from the recalciner; and conversion of calcium carbonate to calcium oxide in a rotary kiln or fluidized-bed calciner. In the rotary kiln process, dewatered paste is conveyed to a sloping, cylindrical enclosure for drying and firing. The kiln shell has a refractory lining, rotates at approximately 1 rpm, and is sloped so that dried sludge travels toward the firing end. Solids retention time is approximately 1.5 hr, and the unit is heated by burning oil or natural gas. Burned lime is cooled and discharged to a conveyor for transfer to a storage bin, exhaust gases are scrubbed to remove dust, and a portion is diverted back to the sludge-processing step for use in carbonation.

The fluidized-bed calciner requires pretreatment of the wet cake with soda ash and previously dried calcium carbonate to form lumps that move and roll freely. Particles, now 80% solids, are applied to a cage mill, where they are dried and beaten into a fine powdery dust that becomes suspended in the exhaust gases. Calcium carbonate dust, removed from the gases by a cyclone separator, is fed to a tall cylindrical furnace, where it is rapidly burned, forming hard spherical pellets of calcium oxide that are held suspended in the vertical air stream passing through the furnace. After growing to sufficient size, they drop out of suspension and are removed from the reactor. Recalcining requires a fuel rate of about 10 million Btu per ton of lime produced.

Lime recovery from wastewater coagulation requires greater pretreatment since the precipitate contains organics, phosphorus, and other undesirable constituents in addition to magnesium. Calcium recovery is less efficient and makeup lime is required. A multiple-hearth furnace is employed for calcining the dewatered cake.

Adoption of recalcination depends on the relative expense and feasibility of alternative sludge-disposal methods. Lime recovery may not be feasible for many treatment plants, owing to the presence of undesirable impurities. Appreciable quantities of magnesium, aluminum, iron, or clay may hinder dewatering operations and limit the purity and value of the product lime. Economical operation is generally limited to large treatment plants that can process more than 20 tons/day, although in certain locations smaller plants may be feasible. The best method for disposal of lime sludges is related to the plant size, the nature of the sludge, and the disposal options available.

## 13.28  ALUM RECOVERY

Aluminum sulfate applied in coagulation of water precipitates as aluminum hydroxide. The alum can be recovered as a dilute liquid alum by reacting sulfuric acid with the settled sludge as follows:

$$2Al(OH)_3 + 3H_2SO_4 \rightarrow Al_2(SO_4)_3(alum) + 6H_2O \tag{13.21}$$

A complete recovery process includes concentrating the alum sludge, reacting with sulfuric acid, and removing impurities. Pretreatment by gravity thickening is generally sufficient to produce the desired minimum solids concentration of 2%, thereby assuring adequate strength of the recycled aluminum sulfate solution (a solution with less than 1.0% alum is not considered suitable for reuse). A recovery reaction with sulfuric acid is carried out at a pH of about 2.0. Acid in excess of the chemical reaction is needed to establish this lower pH and allow for chemical decomposition of organic matter present in the sludge. The final step of removing undesirable suspended matter is accomplished by either gravity clarification or filtration. Recovered alum solution is held in tanks and metered to points of application in the treatment plant. Waste residue from processing is treated with lime, dewatered, and disposed of by land burial.

The most critical problem in alum regeneration is the elimination of impurities that cannot be separated by gravity. Recovered alum solution may carry over resolubilized iron and manganese, inert solids such as clay, carbon added for odor control, and colloidal organics charred by the sulfuric acid. One proposed scheme is pressure filtration of the reclaimed alum solution using fly ash as a filter aid. Near the end of each filter cycle, lime is applied to neutralize the cake. Gravity clarification recovers only about 70% of the original alum fed, while pressure filtration allows nearly 100% recovery. Treatment plants should be equipped to supply new alum in case of failure in the alum recovery process. For example, high turbidity in the raw water during spring runoff may result in contaminated sludge that cannot be processed to recover the alum.

# Ultimate Disposal

The majority of wastewater sludges is disposed of on land, with approximately three quarters being used as soil conditioner and the remainder buried in landfill. Other principal methods of disposal are incineration and discharging in the ocean. Land application is increasing because of the rising cost of energy required to burn sludge and the regulatory restrictions on ocean disposal.

## 13.29 AGRICULTURAL LAND APPLICATION

The main environmental concerns with regard to disposal of wastewater sludges on agricultural land are the degree of stabilization of the putrescible organic matter, reduction of pathogenic organisms, and presence of toxic chemicals.

Anaerobic and aerobic digestion are the common processes for reducing the organic content in sludges. Besides eliminating unaesthetic conditions, particularly offensive odors, stabilization reduces the attraction of vectors like flies and rodents. Although disease transmission in an agricultural setting is unlikely, vectors can be a nuisance and do have the potential of carrying pathogens from sludge. Digestion processes also substantially reduce the numbers of enteric bacteria and viruses. The adverse environments and passage of time during storage and spreading on the land contributes significantly to continued die-off of pathogens. The criteria for adequate stabilization for anaerobic digestion are a volatile solids reduction of 38%; for aerobic digestion, an oxygen uptake rate of the digested sludge of 1.5 g/hr per gram of volatile suspended solids or less [8].

Haul distance, climate, and availability of storage facilities are key factors in land application of sludge. At large plants, mechanical dewatering is used to reduce the sludge mass and cost of hauling. The cake can be distributed by a spreader with backbeaters that throw the solids outward from the rear of a wagon or truck-mounted box. Disk cultivators are used to incorporate the sludge solids into the surface soils. Biologically stabilized cake can be stockpiled between planting seasons and during winter months. Storage sites are prepared to prevent pollution of air, groundwater, and surface waters. At small plants, liquid sludge is often transported in a tank truck or tractor-drawn wagon for spreading on fields. Storage can be provided at the plant in digesters or holding tanks. Also, drying beds can be used for dewatering and storage of sludge when spreading is not feasible. When digested sludge is applied to farmland, the cultivation of fodder, fiber, and seed crops is unrestricted. Food crops with harvested parts that contact the soil–sludge mixture cannot be grown for 18 months after sludge application, and the proposed restriction for crops with underground harvested parts is 5 yr [8]. Animals should be prevented from grazing for 30–60 days after application, and public access should be prevented for 1 yr.

The variety and numbers of bacteria, viruses, and parasitic organisms pathogenic to humans and animals found in wastewater relate to the state of health of the contributing community. Although treatment processes reduce their numbers, often considerably, the effluent and sludge still contain the species present in the wastewater. The certainty of pathogens being present in raw sludge is the reason for spreading only stabilized sludges on agricultural land. The effectiveness of reducing pathogenic populations during sludge stabilization is a subject of considerable controversy. In general, anaerobic or aerobic digestion of sludge is effective in reducing the numbers of viruses and bacteria, but not in reducing roundworm and tapeworm ova or other resistant parasites. Being the most fragile microorganisms, bacteria and viruses are inactivated by sunlight, drying, and competition in the soil environment. In contrast, parasitic ova are more resistant and may persist in soils or on vegetation for several weeks or months. Even though the risk of infecting livestock is not great, farmers are usually advised to allow 4 weeks or more after sludge application before grazing animals or harvesting a fodder crop. Regarding human health, despite the possibility of communicable disease transmission, the lack of epidemiological evidence suggests that the current practice of sludge disposal to land is safe [9].

Liquid sludge may be applied by a vehicle equipped with a rear splash plate for surface spreading or by chisel plows for subsurface incorporation. The flexibility of vehicular hauling allows application at a variety of locations, often privately owned farmlands. Spraying from fixed or portable irrigation nozzles can be practiced where odor and insects are not problems. Subsurface injection is the most environmentally acceptable method since the sludge is incorporated directly with the soil, eliminating exposure to the atmosphere. The system illustrated in Fig. 13.24 has seven injectors on spring-loaded chisel plow shanks with wide cultivator sweeps. As a sweep passes through the soil, a cavity is formed approximately 6 in. deep for injecting the sludge. Sludge is supplied to the injection unit directly from a remote holding tank through underground piping and a flexible delivery hose. Another means of incorporating sludge, which is less effective in covering the liquid, is disk plowing with hoses discharging sludge ahead of each disk. Valves and sludge pumps located near the holding tank are operated by radio control from the tractor.

The environmental concerns of chemicals in sludge spread on agricultural land are surface-water and groundwater pollution and contamination of soil and crops with toxic substances. Laboratory analyses of a sludge normally include solids content; nitrate, ammonia, and organic nitrogen; soluble and organic phosphate; potassium; heavy metals of cadmium, copper, nickel, lead, and zinc; and selected organic compounds such as chlorinated hydrocarbons and polychlorinated biphenyls (PCBs). To eliminate the possibility of transmitting toxic contaminants to humans, non-food-chain crops are preferred. Grasses and feed grains for animal consumption are considered acceptable. At the disposal site, the sludge, soil, groundwater, and any water runoff are monitored on a regular basis for fecal coliforms, nutrients, and contaminants.

Cadmium is the heavy metal of greatest concern to human health when applying municipal wastewater sludges to land, since it can be taken up by plants to enter the human food chain [9]. The movement of cadmium to groundwater is very unlikely to occur at the pH values of greater than 6.0 commonly associated with agricultural soils. The primary chronic health effect of excessive dietary intake is damage to the kidney.

The cadmium content of agricultural soils ranges from near zero to several hundred milligrams per kilogram of soil contaminated by industrial wastes. In addition to sludge, many phosphate fertilizers contain cadmium. Predicting plant uptake from soils with accumulated cadmium is difficult because of interactive controlling conditions. Soil pH is an important chemical factor, but the kind of plant species is just as important. Within plants, the highest amounts of cadmium occur in the fibrous roots followed by the leaves, and the lowest concentrations are in fruits, seeds, and storage organs. Potential high-risk food plants for humans are leafy vegetables (e.g., lettuce and spinach). Fodder crops and cereal grains appear to present the least risk. Davis and Coker [10] have written a comprehensive literature review and discussion of the distribution of cadmium, plant uptake, and hazards associated with sludge used in agriculture.

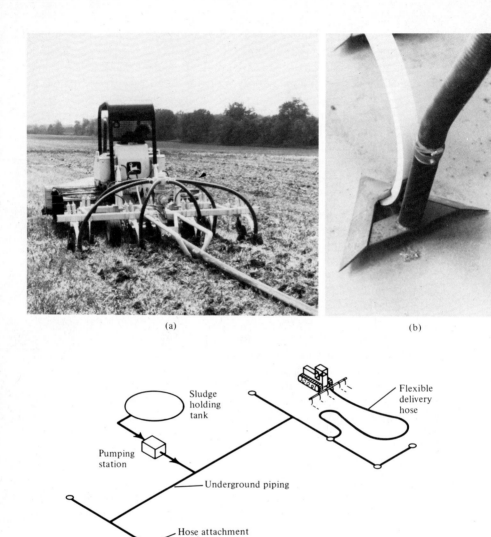

(a)

(b)

Sludge holding tank

Pumping station

Underground piping

Hose attachment

Flexible delivery hose

(c)

**Figure 13.24** Subsurface injection of liquid sludge. (a) Tractor pulling the injection unit that is supplied sludge through the trailing hose. (b) Individual injector with sludge supply hose and cultivator sweep attached to the shank of a chisel plow. (c) Schematic layout of a sludge supply system for an injector site. (Courtesy of Briscoe Maphis.)

Guidelines for the addition of cadmium in sludge to agricultural land suggest a wide range of allowable application rates. The maximum permissible level recommended by the Environmental Protection Agency is 0.5 kg/ha·yr on agricultural land for food-chain crops. The maximum cumulative application is specified in the range of 5–20 kg/ha based on characteristics of the soil. The concentrations of cadmium relative to dry solids in municipal sludges are usually 20 mg/kg or less, although uncontrolled discharge of industrial wastewaters to the treatment plant can dramatically increase this amount.

Nitrogen is another important factor in determining the allowable rate of applying sludge to land. The concern is pollution of groundwater with nitrate. On porous soils, the recommended rate of nitrogen application is an amount equivalent to the nutrient need of the vegetation.

■ **EXAMPLE 13.17**

A municipal wastewater contains 200 mg/l of BOD, 220 mg/l of suspended solids, 35 mg/l of nitrogen, and 0.016 mg/l of cadmium. Wastewater processing is primary sedimentation and secondary activated sludge with the waste sludge anaerobically digested prior to spreading on agricultural land. The plant effluent is discharged to a river. Based on operational data, 25% of the influent cadmium appears in the 110 mg of digested sludge solids produced per liter of wastewater processed. The fertilizer value of the dried sludge solids is 4.0% available nitrogen.

a. Calculate the concentration of cadmium in the plant effluent in milligrams per liter. If the concentration limit of cadmium in the river water is 0.0013 mg/l to protect aquatic life, what dilution ratio of plant effluent to cadmium-free river water is needed?

b. Calculate the concentration of cadmium in the dried sludge solids in units of kilograms per tonne. The digested sludge is being applied to land cultivated for growing soybeans. The recommended annual nitrogen application is 220 kg/ha, and the maximum allowable cadmium loading is 0.5 kg/ha·yr. Based on these criteria, compute the maximum sludge application in units of tonnes of dry solids per hectare.

c. Ferric chloride (waste pickle liquor) added to the wastewater for increasing phosphorus extraction in treatment also increases the cadmium removal to 70%. Calculate the data requested in questions 1 through 4 for this degree of cadmium removal.

*Solution*

a. The cadmium in effluent amounts to $0.75 \times 0.016 = 0.012$ mg/l. Assume the dilution ratio is $X$ (i.e., when the effluent flow is 1, the river flow is $X$). Then

$$X \times 0 + 1 \times 0.012 = (1 + X) \times 0.0013$$

$$X = \text{flow of river/flow of wastewater discharge} = 8.2$$

b. The concentration of cadmium in sludge solids equals

$$\frac{0.25 \times 0.016 \, \text{mg/l} \times 1000 \, \text{kg/t}}{110 \, \text{mg/l}} = 0.036 \, \text{kg/t}$$

The nitrogen content of the sludge solids equals

$$\frac{110 \text{ mg/l} \times 0.04 \times 1000 \text{ kg/t}}{110 \text{ mg/l}} = 40 \text{ kg/t}$$

The maximum allowable sludge application rate is

$$\text{based on cadmium} = \frac{0.5}{0.036} = 14 \text{ t/ha}$$

$$\text{based on nitrogen} = \frac{220}{40} = 5.5 \text{ t/ha} \quad (\text{Use this loading.})$$

**c.** The cadmium in effluent is $0.30 \times 0.016 = 0.0048$ mg/l.

$$\text{dilution ratio required} = 2.7$$

concentration of cadmium in sludge $= 0.10$ kg/t

The maximum allowable sludge application rate is

based on nitrogen $= 5.5$ t/ha

$$\text{based on cadmium} = \frac{0.5}{0.10} = 5.0 \text{ t/ha} \quad (\text{Use.})$$

These computations illustrate that treatment for removal of phosphate by chemical precipitation also increases the removal of cadmium. Therefore, while the effluent quality improves, the contamination of the sludge is greater. ∎

## 13.30  LAND BURIAL OF DEWATERED SLUDGE

Raw or digested wastewater sludges and chemical residues from water treatment may be buried if a suitable site is available [11]. Except in highly urbanized areas, land and transportation costs are less expensive than incineration or chemical recovery. Sludges are often buried at municipal sanitary landfills along with other solid wastes, requiring systematic depositing, compacting, and covering the wastes. Usually 6–12 in. of earth are placed over each 2 ft of compacted fill. The top earth cover should have a minimum depth of 2 ft and be grassed to prevent erosion.

Site selection considers soil conditions, groundwater levels, location relative to populated areas, and future land-use planning. Conditions must be such that gas leachate, water seepage, and runoff do not cause pollution, nuisance, or health hazards. A monitoring program should be established to ensure that an adequate environment is maintained at the site. Projected land use may be a park with recreational facilities that are not affected by gradual subsidence of the ground surface.

## 13.31  COMBUSTION OF ORGANIC SLUDGES

Incineration involves drying sludge cake to evaporate the water, followed by burning for complete oxidation of the volatile matter. Drying occurs at a temperature of approximately 700°F, and burning is sustained in the range

1200°–1400°F. A minimum temperature of 1350°F is needed to deodorize exhaust gases. Excess air is required to ensure complete combustion of organics and minimize the escape of odor-producing compounds in stack gases. The amount needed is 25%–100% over the stoichiometric air requirement and varies with the nature of the sludge and type of incineration equipment. Supplying excess air has the adverse effects of reducing the burning temperature and increasing heat losses from the furnace. Heat emitted from burning volatile solids, minus losses, is available for drying the incoming sludge and heating the air supply. Self-sustaining combustion of dewatered raw sludge, after the incinerator temperature has been raised to the ignition point by burning an auxiliary fuel, is possible if the organic solids content is greater than approximately 35%. The exact solids content necessary for autogenous burning depends on several factors related to the characteristics of the sludge and the design and operation of the furnace (e.g., preheating of incoming air and cooling exhaust gases).

Heat yield from sludge combustion is related directly to moisture content and volatile solids concentration. Several theoretical equations have been proposed for calculating calorific values based on elemental composition, principally the carbon, hydrogen, oxygen, and sulfur contents. However, experience has shown that calculated results often are inaccurate. A laboratory calorimeter test is the only reliable method for determining the heat value of a waste. Equation 13.22 [12] is an empirical formula for fuel values of different types of vacuum-filtered sludges, taking into account the amount of coagulant added before filtration:

$$Q = a\left(\frac{100P_v}{100 - P_c} - b\right)\left(\frac{100 - P_c}{100}\right) \tag{13.22}$$

where  $Q$ = fuel value, Btu/lb of dry solids

$a$ = coefficient (131 for raw primary sludge, 107 for waste-activated sludge)

$b$ = coefficient (10 for raw primary sludge, 5 for activated sludge)

$P_v$ = volatile solids in sludge, %

$P_c$ = inorganic conditioning chemicals applied for dewatering the sludge, % ($P_c$ = 0 for organic polymers)

Heat-balance calculations for an incinerator include the heat evolved from burning of volatile solids and auxiliary fuel, the quantity absorbed for evaporation of moisture in the wet cake, losses in the stack gases and ash, and radiation through the furnace walls [3]. The fuel value of raw sludge ranges from 6500 to 9500 Btu/lb of dry solids, while the calorific value for digested dry solids is 2500–5500 Btu, depending on the fraction of inerts. Evaporation of moisture from wet sludge varies from 1800 to 2500 Btu/lb of water. Information on heat losses from a furnace is available from the equipment manufacturer. The amount of excess air used during operation defines the heat loss in exhaust gases. Natural gas, or fuel oil, is frequently used as a supplemental fuel for complete combustion.

■ **EXAMPLE 13.18**

Raw primary sludge is dewatered by vacuum filtration prior to incineration. The cake contains a total of 24% dry solids, of which 65% are volatile. The combined ferric chloride and lime dosage in sludge conditioning was 8.0% of the dry solids. With allowance for heat losses in the stack gases and ash plus radiation from the multiple-hearth furnace, the heat consumed in evaporating water from the sludge is assumed to be 3500 Btu/lb. Compute the fuel values for the dry solids and the wet sludge and the heat required to dry the sludge.

*Solution.*    The fuel value of dry solids, from Eq. 13.22, is

$$Q = 131\left(\frac{100 \times 65}{100 - 8} - 10\right)\left(\frac{100 - 8}{100}\right) = 7310 \text{ Btu/lb of dry solids}$$

The fuel value of wet sludge is

$$Q = 0.24 \times 7310 = 1750 \text{ Btu/lb of wet sludge}$$

The heat required to dry 1 lb of wet sludge is

$$Q = 0.76 \times 3500 = 2660 \text{ Btu/lb}$$

From these calculations, the heat from combustion of the dry solids is less than the heat required to evaporate moisture from the wet cake. Therefore, auxiliary fuel will be needed to maintain combustion.    ■

## 13.32  SLUDGE INCINERATORS

Two major incineration systems employed in the United States are the multiple-hearth furnace and the fluidized-bed reactor. The multiple-hearth unit has received widest adoption because of its simplicity and operational flexibility. Its furnace consists of a lined circular shell containing several hearths arranged in a vertical stack and a central rotating shaft with rabble arms (Fig. 13.25). Its operating capacity is related to the total hearth area (diameter and number of stages) and varies from 200 to 8000 lb/hr of dry sludge. Sludge cake is fed onto the top hearth and is raked slowly in a spiral path to the center. Here, it falls to the second level, is pushed to the periphery, and drops in turn to the third hearth, where it is again raked to the center. The upper levels allow for evaporation of water, the middle hearths burn the solids, and the bottom zone cools the ash prior to discharge.

The hollow central shaft is cooled by forced air vented out the top. A portion of this preheated air from the shaft is piped to the lowest hearth and is further heated by the hot ash and combustion as it passes up through the furnace. The gases are then cooled as heat is absorbed by the incoming sludge. The countercurrent flow pattern of air and sludge solids reduces heat losses and increases incineration efficiency. Stack gases are discharged through a wet scrubber to remove fly ash and other air pollutants. Emission control requirements specify that the exhaust cannot contain particulate matter in excess of 70 mg/m$^3$ or exhibit 10% or greater opacity (shadiness), except for 2 min in any 1-hr period. The presence of uncombined water is the only reason for failure to meet the latter requirement.

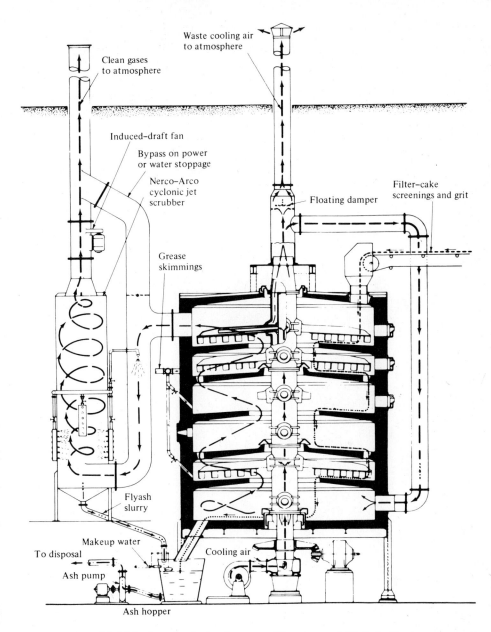

**Figure 13.25** Multiple-hearth sludge furnace. (Courtesy of Nichols Engineering and Research Corp., a Wheelabrator-Frye Co.)

Multiple-hearth furnaces can also be designed to dry organic sludge or recalcine calcium carbonate precipitate. The solids and airflow system for calcination is similar to the pattern used for sludge incineration, with heat being supplied by burning natural gas or oil. For sludge drying, hot gases from an external furnace flow concurrent with the sludge down through the multiple hearths in order to dry the organics without scorching.

Sludge can also be dried in a flash-dryer system. First, wet sludge cake is mixed with recycled dried solids and pulverized in a cage mill. Hot gases from a separate incinerating furnace suspend the dispersed sludge particles up into a pipe duct, where they are dried, and a cyclone separator removes the dried solids from the moisture-laden hot gas. Part of the dried sludge returns to the mixer for blending with incoming wet cake, while the remainder is either withdrawn for a soil conditioner or burned in the furnace for fuel.

A fluidized-bed incinerator uses a sand bed as a heat reservoir to promote uniform combustion of sludge solids (Fig. 13.26). The bed is expanded by

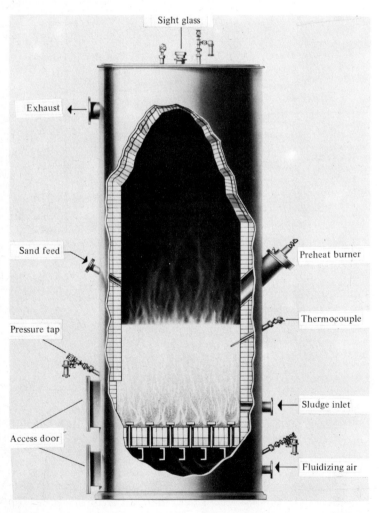

**Figure 13.26** Fluidized-bed incinerator. (Courtesy of Dorr-Oliver, Inc.)

upflow of air through the sand. Dewatered sludge is injected into the fluidized sand above the grid. Violent mixing of the solids and gases in the hot sand promotes rapid drying and burning at a temperature of 1400°–1500°F. The quantity of excess air needed is about 25%. The sand bed acts as a heat reservoir, enabling reduced startup time when the unit is operated only 4–8 hr/day. When necessary, the bed is preheated by using an auxiliary fuel. Water vapor and ash are carried out of the bed by combustion gases. A cyclonic wet scrubber removes ash from the exhaust and finally separates it from the scrubber water in a cyclone separator.

## 13.33  OCEAN DISPOSAL OF SLUDGE

Coastal cities have discharged digested sludge into the ocean for decades. Dewatered wastes may be transported to offshore sites in barges and dumped, or sludge slurry may be pumped to deep water through a submarine outfall.

In recent years, this practice has been questioned by regulatory agencies, particularly in the United States. The principal environmental concerns are degradation of recreational waters, buildup of solids on the sea bottom, and toxicity to marine life. The contaminants involved are the same as those related to disposal on land—heavy metals, pathogens, and organic pollutants.

Ocean disposal sites should have adequate current velocities for initial dilution and waste dispersion to prevent concentration of pollutants. Some waste slurries have a lower specific gravity than seawater and rise to the surface, while the solids of others accumulate on the bottom creating sludge deposits that decompose. The content of heavy metals in a municipal sludge can be limited by instituting and enforcing a sewer ordinance controlling industrial wastewaters. As with land disposal, the sludge should be biologically stabilized to reduce pathogens.

## PROBLEMS

**13.1**  The settled sludge from coagulation of a surface water is 1.5% solids, of which 30% are volatile. Compute the specific gravity of the dry solids and specific gravity of the wet sludge. Assume specific gravity values of 2.5 for the fixed matter and 1.0 for the volatile. What is the volume of waste in cubic meters per 1000 kg of dry solids?

**13.2**  A primary wastewater sludge contains 6.0% dry solids that are 65% volatile. Calculate the specific gravities of the solid matter and the wet sludge. If this residue is thickened to a cake of 22.0% solids, what is the specific gravity of the moist cake?

**13.3**  Compute the volume of a waste sludge with 96% water content containing 1000 lb of dry solids. If the moisture content is reduced to 92%, what is the sludge volume?

**13.4**  An activated-sludge wastewater plant with primary clarification treats 10 mgd of wastewater with a BOD of 240 mg/l and suspended solids of 200 mg/l.

(a) Calculate the daily primary and waste-activated sludge yields in pounds of dry solids and gallons, assuming the following: 60% SS removal and 35% BOD reduction in primary settling; a primary sludge concentration of 6.0% solids; an operating food/microorganism ratio of 1 : 3 in the aeration basin; and a solids concentration of 15,000 mg/l (70% volatile) in the waste-activated sludge. (b) What would be the solids content of the sludge mixture if the two waste volumes were blended together?    (c) If the waste activated is thickened separately to 4.5% solids before blending, what would be the combined sludge volume and solids content?

**13.5** Estimate the excess solids production of an extended aeration process treating unsettled wastewater with a BOD of 200 mg/l and suspended solids concentration of 240 mg/l assuming $k = 0.35$. Express the answer in milligrams per liter of wastewater treated.

**13.6** A municipal wastewater with 220 mg/l of BOD and 260 mg/l of suspended solids is processed in a two-stage trickling-filter plant. Calculate the following per $1.0 \text{ m}^3$ of wastewater treated:    (a) the dry solids production in primary and secondary treatment assuming 50% SS removal and 35% BOD reduction in primary settling and a $k$ value of 0.40 applicable for the trickling-filter secondary;    (b) the daily sludge volume, both primary and secondary solids, assuming 5.0% solids content. (A specific gravity of 1.0 can be assumed for the wet sludge.)

**13.7** What is the estimated waste-activated sludge generated from a conventional aeration process treating 29,000 $\text{m}^3/\text{d}$ with 173 mg/l of BOD after primary settling. The operating F/M is 0.24 g BOD/day per gram MLSS. Assume a suspended solids concentration of 9800 mg/l in the waste sludge.

**13.8** A river-water treatment plant coagulates raw water having a turbidity of 12 units with an alum dosage of 40 mg/l.    (a) Estimate the total sludge solids production in lb/mil gal of water processed.    (b) Compute the volumes of waste from the settling basins and filter backwash water assuming that 70% of the total solids are removed in sedimentation and 30% in filtration, a settled sludge solids concentration of 1.2%, and a solids content of 600 mg/l in the filter wash water.    (c) Calculate the composite sludge volume after the two wastes are gravity thickened to 3.5% solids.

**13.9** A water plant treats a surface water with a mean turbidity of 18 NTU by applying an average of 19.8 mg/l of alum. From plant records, the sludge yield from the sedimentation basins averages 1400 $\text{m}^3$ per $\text{Mm}^3$ of water treated with a solids content of 1.4%. The filter wash water, at a volume of 3.2% of the water treated, contains an average of 160 mg/l of solids. Determine the total dry solids produced per 1.0 $\text{Mm}^3$ of water treated and compare this value with the total sludge solids calculated by using Eq. 13.8.

**13.10** Example 11.4 describes softening of a water by excess lime treatment in a two-stage system. Based on the appropriate chemical reactions, both precipitation softening and recarbonation removal of excess lime, calculate the total residue produced per million gallons of water processed.

**13.11** Example 11.6 describes softening a water by split treatment in a two-stage system. Based on the appropriate chemical reactions, calculate the precipitate produced per million gallons of water processed.

**13.12** The superintendent of a water treatment plant requests your assistance in calculating the dissolved mineral solids removed in the lime softening of a well water and the sludge solids produced. He wants to present these data to the city council to emphasize the improvement in water quality resulting from treatment. The plant processes 0.45 mgd by the addition of 300 mg/l of hydrated lime, which is 74% CaO by weight, followed by recarbonation. The lime treatment also removes iron and manganese from the water. Analyses of the well water and treated water are listed below.   (a) Tabulate the ionic concentrations, calculate the milliequivalents per liter, and sketch milliequivalent-per-liter bar graphs. (Refer to Example 11.1.)   (b) Calculate the dissolved minerals (calcium, magnesium, iron, manganese, and bicarbonate) removed in pounds per day for the 0.45 mil gal of water treated.   (c) Calculate the dry sludge solids produced per day and the weight of the wet sludge, assuming a solids concentration of 5%.

| Well water | | | | Treated water | | | |
|---|---|---|---|---|---|---|---|
| | | | | | | | |

(All values are given in milligrams per liter except pH.)

| | | | | | | | |
|---|---|---|---|---|---|---|---|
| Ca | 94 | $HCO_3$ | 390 | Ca | 21 | $HCO_3$ | 107 |
| Mg | 24 | $SO_4$ | 73 | Mg | 14 | $SO_4$ | 64 |
| Na | 27 | Cl | 2 | Na | 23 | Cl | 6 |
| Fe | 0.7 | F | 0.3 | Fe | 0 | F | 0.2 |
| Mn | 0.4 | | | Mn | 0 | | |

Total hardness = 332                 Total hardness = 112

pH = 7.2                                     pH = 7.7

**13.13** The proposed sludge-processing scheme for a conventional step-aeration activated-sludge plant is as follows: return of waste-activated sludge to the head of the plant, withdrawal of the combined sludge (raw and waste activated) from the primary clarifier, concentration in a gravity thickener, application of plant effluent to the thickener for dilution water, return of the thickener overflow to the plant inlet, and vacuum filtration of the thickened underflow. Briefly comment on the operating problems you would anticipate with this system. What type of sludge handling and thickening would you recommend to replace the proposed scheme?

**13.14** Outline the alternative methods for disposal of wastes from a precipitation softening plant processing groundwater.

**13.15** A gravity thickener handles 33,000 gpd of wastewater sludge, increasing the solids content from 3.0% to 7.0% with 90% solids recovery. Calculate the quantity of thickened sludge.

**13.16** A waste sludge flow of 40 $m^3$/day is gravity thickened in a circular tank with a diameter of 6.8 m. The solids concentration is increased from 4.5% to 7.5% with 85% suspended-solids capture. Calculate the solids loading in $kg/m^2 \cdot d$ and the quantity of thickened sludge in $m^3/d$.

**13.17** Size a gravity thickener based on 10 $lb/ft^2$/day for a waste sludge flow of 25,000 gpd with 5.0% solids. Assume a side water depth of 10.0 ft. After installation, operation at design flow yields an underflow of 8.0% with 90% solids removal. What is the flow of dilution water needed to maintain an overflow rate of 400 gpd/ft$^2$? If the consolidated sludge blanket in the tank is 4.0 ft thick, compute the estimated solids retention time in the thickener.

**13.18** Settled sludge and filter wash water from water treatment are thickened in the clarifier–thickeners prior to mechanical dewatering. The settled sludge volume is 1150 gpd with 1.0% solids, and the backwash is 9800 gpd containing 500 mg/l of suspended solids. Calculate the daily quantity of thickened sludge, assuming a concentration of 3.0% solids.

**13.19** Lagoon disposal of alum sludge is being considered for a water treatment plant processing 10,000 m$^3$/d. The proposed site can accommodate four lagoon cells each with a surface area of 800 m$^2$ and depth of 3.0 m. Sludge from the sedimentation basins averages 15 m$^3$/d and contains 1.4% solids; the quantity of filter wash water is 400 m$^3$/d and contains 160 mg/l of suspended solids. The wash water is to be gravity-thickened to greater than 1% with polymer addition before discharging to the lagoons. The plan of operation is to discharge from the sedimentation basins and thickeners to fill one cell at a time, withdrawing supernatant and allowing consolidation and air drying for at least one year after filling before cleaning. This aging period is expected to enhance dewatering by subjecting the sludge to freeze–thaw cycles during the winter months. Based on field studies, this procedure will shrink the sludge layer to a depth of less than one meter with an average solids content of 10%. Is this proposed plan likely to be successful?

**13.20** A dissolved-air flotation thickener with a surface area of 400 ft$^2$ processes 230,000 gal of waste-activated sludge containing 1.0% solids in a 24-hr period. Calculate the solids loading in pounds per square foot per day. If the solids content of the float is 4.0%, what is the volume of the thickened waste assuming 100% solids capture? Compute the hydraulic loading.

**13.21** A flotation thickener processes 250 m$^3$ of waste-activated sludge in a 16-hr operating period. The solids content is increased from 10,000 mg/l to 40,000 mg/l with 92% solids capture. Calculate the quantity of float in cubic meters per 16-hr period. If the solids loading is 12 kg/m$^2$·hr, what is the hydraulic loading in m$^3$/m$^2$·d?

**13.22** Waste-activated sludge processed by dissolved-air flotation is concentrated from 9800 mg/l to 4.7% with 95% suspended-solids capture. During a 24-hr operating period, 50,000 gal of sludge was applied with a polymer dosage of 32 mg/l. The thickener surface area is 75 ft$^2$. Calculate the solids loading, volume of float produced in 24 hr, and polymer addition in lb/ton of dry solids.

**13.23** A single-stage anaerobic digester has a capacity of 13,800 ft$^3$, of which 10,600 ft$^3$ is below the landing brackets. The average raw-waste sludge solids fed to the digester are 580 lb of solids/day.  (a) Calculate the digester loading in lb of volatile solids fed/ft$^3$ of capacity below the landing brackets/day. Assume that 70% of the solids are volatile.  (b) Determine the digester capacity required based on Eq. 13.11, using the following data:

$$\text{average daily raw-sludge solids} = 580 \text{ lb}$$

$$\text{raw-sludge moisture content} = 96\%$$

$$\text{digestion period} = 30 \text{ days}$$

$$\text{solids reduction during digestion} = 45\%$$

digested-sludge moisture content = 94%

digested-sludge storage required = 90 days

**13.24** Calculate the digester capacity in cubic meters required for conventional single-stage anaerobic digestion based on the following parameters:

daily raw-sludge solids production = 630 kg

volatile solids in raw sludge = 70%

moisture content of raw sludge = 95%

digestion period = 30 days

volatile solids reduction during digestion = 50%

moisture content of digested sludge = 93%

storage volume required for digested sludge = 90 days

**13.25** The design wastewater loading for a proposed two-stage trickling-filter plant is 1200 m³/d containing 260 mg/l BOD and 310 mg/l suspended solids.    (a) Calculate the required capacity of a conventional digester based on a raw sludge concentration of 4.5%, digested sludge solids concentration of 8.0%, total solids reduction during digestion of 40%, an $f$ value of 0.50 for Eq. 13.5, a $k$ value of 0.40 for Eq. 13.6, and a storage period for digested sludge of 90 days.    (b) Compute the digester capacity per population equivalent load on the treatment plant.

**13.26** The average daily quantity of thickened raw waste sludge produced in a municipal treatment plant is 15,000 gal containing 10,000 lb of solids. The solids are 70% volatile.    (a) Calculate the percentage of moisture in the thickened sludge.    (b) Determine the volume required for a first-stage high-rate digester based on the following criteria: a maximum loading of 100 lb of volatile solids per 1000 ft³/day and a minimum detention time of 15 days.    (c) If the volatile solids reduction in the completely mixed digester is 60%, what is the percentage of moisture in the digested sludge?

**13.27** The waste sludge production from a trickling-filter plant is 12.5 m³/d containing 620 kg of solids (70% volatile). The proposed anaerobic digester design to stabilize this waste consists of two floating-cover heated tanks, each with a volume of 480 m³ when the covers are fully raised and a volume of 310 m³ with the covers resting on the landing brackets.    (a) Calculate the digester volume per equivalent population with the covers fully raised. Assume 100 g of solids production per capita.    (b) Calculate the digester loading in terms of kg volatile solids fed/m³ of tank volume below the landing brackets/day.    (c) Calculate the digested-sludge storage time available between lowered and raised cover positions. Assume a volatile solids reduction of 60% during digestion and a digested-sludge moisture content of 93%.

**13.28** A domestic wastewater plant using rotating biological contactors for secondary treatment (Fig. 12.30) has a two-stage anaerobic digestion system (Fig. 13.11). The first stage is a heated, completely mixed fixed-cover digester with a liquid volume of 15,500 ft³. The second stage is an unheated and unmixed digester with a liquid volume of 15,500 ft³ equipped with a gas-dome cover and supernatant draw-offs. Alum is added to the RBC effluent ahead of the final clarifiers to precipitate phosphate. (The plant was retrofitted with alum feeders and new final clarifiers after original construction to meet a revised effluent standard for phosphorus.) The average wastewater flow is 0.63 mgd. Influent characteristics are 166 mg/l of BOD, 128 mg/l of suspended solids, and 7.1 mg/l of inorganic phosphorus. Effluent characteristics are 7 mg/l of BOD, 16 mg/l of suspended solids, and 1.3 mg/l of inorganic phosphorus. Alum addition is 405 lb/day. The quantity of raw sludge produced averages 2400 gpd with 4.4% solids, which are 67% volatile. The quantity of digested sludge accumulated in the second-stage tank averages 1500 gpd with 4.3% solids, which are 45% volatile. The total gas production from digestion is 4200 ft³/day.    (a) Calculate the volatile solids loading and liquid detention time of the first-stage digester. How do these values compare with those in Table 13.7?    (b) Calculate the digestion gas production per pound of volatile solids added and per pound of volatile solids destroyed. Compare these values with those in Table 13.6.    (c) Estimate the total dry solids sludge yield as the sum of raw primary solids (Eq. 13.5), biological sludge solids (Eq. 13.6), and alum precipitate as aluminum phosphate (Eq. 14.5). How does this compare with actual sludge solids yield?

**13.29** The two-stage anaerobic digesters at a trickling-filter plant are a fixed-cover tank for first-stage high-rate sludge stabilization with a liquid volume of 42,000 ft³ and a floating-cover second-stage storage and thickening tank with a volume of 70,000 ft³ between the lowered and fully raised cover positions. The quantity of raw sludge applied is 2800 ft³/day containing 5.0% solids that are 70% volatile.    (a) If 60% of the volatile solids in the raw sludge is reduced by digestion, calculate the solids concentration in the digested sludge leaving the first-stage digester.    (b) Estimate the daily methane production in first-stage digestion.    (c) If the digested, thickened sludge contains 7.0% solids, compute the maximum number of days of digested-sludge storage available.    (d) Assume the first-stage digester has been operated at a digesting sludge temperature of 35°C for several weeks. How would the gas production change if the operating temperature were decreased slowly to 25°C? How would gas production be affected by a slow increase in operating temperature from 35° to 45°C?

**13.30** A two-stage stone-media trickling-filter plant was evaluated for the town of Nancy in Prob. 12.34. Determine the anaerobic digester capacity for stabilizing the sludge from this plant in two identical single-stage floating-cover tanks, and the required area for open-air drying beds for a northern state in the Great Lakes region. Using Eq. 13.11, calculate the volume required based on the following:    (a) The design quantity of raw sludge solids during the maximum month is the amount calculated by Eq. 13.4 using Eq. 12.37 and Eq. 12.38.    (b) A raw sludge solids content of 4.0%, digested sludge solids content of 6.0%, total solids reduction of 35%, and required storage period of 120 days. How does the calculated volume compare with the suggested capacity of 5 ft³/PE?    (c) Calculate the required area of open-air drying beds.

**13.31** A wastewater sludge is stabilized and thickened in anaerobic digestion. The daily raw-sludge feed is 100,000 lb containing 5500 lb of dry solids. Forty percent of the matter applied is converted to gases during digestion, and the digested residue is increased to 10% solids by supernatant withdrawal. Calculate the volumetric reduction of wet sludge achieved in the digestion process.

**13.32** The raw sludge pumped to an anaerobic digester contains 5.0% solids that are 65% volatile. The digested sludge withdrawn from the tank has 6.5% solids that are 40% volatile. Based on these data, estimate by calculations (a) the percentage of volatile solids converted to gas during digestion and (b) the quantity of supernatant withdrawn for every 1000 gal of raw sludge fed to the digester.

**13.33** The solutions to Prob. 12.50 include flow diagrams for trickling-filter and RBC plants to treat a wastewater flow of 2.0 mgd. Expand one of these diagrams to include anaerobic digestion of the waste sludge and open-air drying of the digested sludge. Show the arrangement for two single-stage digesters, drying beds, sludge and supernatant-return piping, and utilization of the digestion gas. List numerical design guidelines for sizing the digesters and drying beds.

**13.34** The solution to Prob. 12.50 is a flow diagram for a treatment plant with primary sedimentation and secondary step aeration. Expand this diagram to include anaerobic digestion of the waste sludge and dewatering of the digested sludge by belt-filter presses. Show the arrangement for handling both the primary sludge and waste-activated sludge, a two-stage digestion system, belt filter presses, sludge piping, and piping for return of digester supernatant and press wastewater. List numerical design guidelines for sizing the digesters, waste sludge thickeners, and belt presses.

**13.35** Aerobic digesters with a total capacity of 50,000 $ft^3$ stabilize waste-activated sludge from an extended-aeration treatment plant without primary clarification. The average daily sludge flow pumped to the digestion tanks is 32,000 gal with 1.5% solids, of which 65% are volatile. The estimated solids production per capita is 0.17 lb of VS/person/day. Calculate the sludge detention time in the digesters, VS loading lb/$ft^3$/day and volume provided per capita.

**13.36** The oxidation-ditch treatment plant described in Prob. 12.47 has two aerobic digesters each with a liquid volume of 8000 $ft^3$ and 200 cfm aeration capacity. The digesters are operated as a semibatch process. Each morning the air is turned off, and the digesting sludge solids are allowed to thicken. After several hours of quiescent settling, supernatant is withdrawn and returned to the plant inlet so waste sludge can be pumped into the digesters. Once a week, settled sludge is withdrawn from the hopper bottom of the digesters and hauled away for disposal by spreading on agricultural land. In order to maintain the MLSS concentration at 2500 mg/l in the oxidation ditches, an average of 2000 $ft^3$/day of sludge containing 7600 mg/l of suspended solids (65% volatile) is pumped to the digesters, after withdrawal of supernatant. The average suspended-solids concentration in the digesting sludge is 11,000 mg/l, and the lowest temperatures measured in the aerating sludge are 8–10°C. The withdrawal of digested gravity-thickened sludge averages 700 $ft^3$/day containing 14,000 mg/l of suspended solids (55% volatile). Calculate the volatile-solids loading and air supply and compare the results with the recommended values given in Section 13.14. Calculate the sludge aeration period in degree-days. (The sludge age can be calculated by dividing the product of the volume of the digesters and the suspended solids concentration in the digesting sludge by the average daily sludge withdrawal times its suspended-solids concentration.) How does the calculated aeration period in degree-days compare to the recommended value for adequate sludge stabilization?

**13.37** A waste-activated sludge with a total-solids concentration of 12,500 mg/l and volatile-solids content of 8800 mg/l is applied to an aerobic digester with a

detention time of 25 days. The volatile-solids destruction is 50% during digestion. Calculate the volatile-solids loading in units of $kg/m^3 \cdot d$ and the solids
content of the digested sludge, assuming no supernatant is withdrawn. If the
solids content at the bottom of the digester can be increased to 1.5% by
quiescent settling, what percentage of the waste sludge can be decanted as
supernatant?

**13.38** The extended aeration system for the town of Nancy is described in Prob.
12.49. Determine the aerobic digestion capacity for stabilizing the sludge from
this plant in two diffused-air digesters equipped with supernatant draw-offs.
Although sludge dewatering on open-air drying beds is being considered, the
feasibility of dewatering by belt filter presses is also under consideration because of the long cold winter. The aerating sludge is expected to be as low as
8°C, and drying beds probably could not be cleaned for up to 3 months.   (a)
Calculate the digester volume required based on the following: (1) The daily
waste-solids yield for design can be calculated from average flow and BOD,
the waste sludge has a total solids content of 1.0%, and the solids are 70%
volatile. (2) The volatile solids loading of the digester should not exceed 0.04
$lb/ft^3/day$, and the aeration period should not be less than 200 degree-days,
computed by multiplying the estimated sludge age by 8°C. (3) A thickened
digested sludge with a solids concentration of 1.5%. (4) Volatile solids reduction
during digestion of 50%. (5) An average operating suspended-solids concentration in the digesters of 1.0%.   (b) For the drying beds, assume a maximum
allowable solids loading of 10 $lb/ft^2/yr$.   (c) For sizing belt filter presses,
assume a maximum hydraulic sludge loading of 30 gpm per meter of belt width
(Table 13.9) and a maximum operating period of 35 hr per week, 7 hr/day
for 5 days. Two presses are required for reliability by the design reviewing
authority.

**13.39** The design wastewater flow for a small community is 400 $m^3/d$, containing 200
mg/l BOD and 240 mg/l suspended solids. The proposed treatment is extended
aeration without primary sedimentation. After aerobic digestion, the waste
sludge can be taken directly from the aerobic digesters for spreading on agricultural land or pumped into a large asphalt-lined storage basin for winter storage.
(a) Calculate the required capacity for the aerobic digesters assuming the
F/M loading on the aeration tank is 0.15 g BOD/d per gram MLSS, the waste-
activated sludge solids content is 1.5% (70% volatile), and the criteria for sizing
the aerobic digesters are a maximum VS loading of 0.60 $kg/m^3 \cdot d$ and a minimum
digestion period of 15 days.   (b) How large must the storage basin be to hold
a 90-day volume of digested sludge assuming that the total solids reduction
during aerobic digestion is 25% of the total solids in the applied sludge and the
solids concentration in the storage basin can be increased to 2.0% by withdrawing supernatant.

**13.40** What are the reasons for mixing either an organic amendment or a recoverable
bulking agent with dewatered sludge cake before composting?

**13.41** A primary sludge feed of 10,000 gpd containing 6.0% dry solids that are 65%
volatile is dewatered by vacuum filtration, producing a cake with 22.0% solids.
Compute the specific gravities of the wet sludge and filter cake. Assuming 100%
solids capture in filtration and neglecting the addition of conditioning chemicals,
calculate the volume and weight reductions achieved in the dewatering process.

**13.42** A treatment plant produces an average of 75,000 gpd of primary sludge with a solids concentration of 6.0% and 230,000 gpd of waste-activated sludge having a water content of 99.0%.   (a) If the waste-activated sludge is thickened to 4.0% solids by flotation and then combined with the primary sludge prior to dewatering, what is the quantity and solids content of the blended sludge? (Assume 100% solids capture in flotation thickening.)   (b) If vacuum filtration dewaters the sludge to 23% solids, how many 8-ton truckloads of cake must be hauled to land burial each day? (Assume 100% solids capture in filtration and negligible weight added by chemical conditioning.)

**13.43** A vacuum filter with an area of 23 $m^2$ processes 60 $m^3$ of sludge containing 65,000 mg/l of solids in 6 hr of operation. Calculate the filter yield and estimate the volume of cake with 28% solids, assuming a specific gravity of 1.3 for the dry solids. The weight of polymer added to the raw sludge for conditioning may be neglected.

**13.44** The raw sludge produced by a trickling-filter plant is 6600 gpd containing 2700 lb of dry solids. What area of vacuum-filter surface would you recommend assuming a 7-hr operating period each day? Calculate the weight of the cake produced, assuming 95% capture of the waste solids, chemical additions of 10%, and a moisture content of 75%.

**13.45** A vacuum filter with a surface area of 25 $m^2$ dewatered 100 $m^3$ of digested sludge containing 5.5% solids in a 7-hr period. The dosages of conditioning chemicals are 4.0% ferric chloride and 18% lime. The total mass of cake produced was 29,000 kg with an average solids content of 22%.   (a) Calculate the percentage of solids capture.   (b) Calculate the filter yield in kilograms of organic solids per square meter per hour.

**13.46** The smallest vacuum filter produced commercially by a leading manufacturer has a 7-ft-diameter drum that is 2 ft 10 in. in width; the nominal surface area is 60 $ft^2$. Estimate the equivalent population that can be served by this unit assuming 0.20 lb of dry solids/capita/day of sludge production, a filter yield of 5.0 $lb/ft^2/hr$, and a 7-hr operating period each day.

**13.47** The smallest belt filter press manufactured for sludge dewatering has an effective belt width of 1 m. Estimate the equivalent population that can be served by this unit assuming 0.20 lb of dry solids/capita/day of sludge production, a solids loading of 100 lb/m/hr, and a 7-hr operating period each day.

**13.48** A belt filter press with an effective belt width of 2.0 m dewaters 100 gpm of anaerobically digested sludge with a solids content of 6.5%. The polymer dosage is 6.4 gpm, containing 0.20% powdered polymer by weight. The wash-water consumption for belt cleaning is 60 gpm. The cake solids content is 30% and the suspended-solids concentration in the wash water measures 1800 mg/l. Calculate the hydraulic loading rate, solids loading rate, and polymer dosage and estimate the solids recovery.

**13.49** A belt filter press housed in a mobile trailer was used to test the dewaterability of an anaerobically digested sludge. The effective width of the belt was 0.50 m. The operating data were a sludge feed rate of 4.5 $m^3/hr$, polymer dosage of 0.45 $m^3/hr$ containing 0.20% powdered polymer by weight, and wash-water flow of 1.07 l/s. Based on laboratory analyses, total solids in the feed sludge equaled 3.5%, total solids in the cake equaled 32%, wastewater from belt washing contained 2600 mg/l suspended solids, and the filtrate was 1.20 l/s with 500 mg/l of suspended solids. Calculate the hydraulic loading, solids loading, polymer dosage, and solids recovery.

**13.50** At an existing treatment plant, the vacuum filters are being replaced by belt filter presses to dewater anaerobically digested sludge prior to stockpiling and spreading on agricultural land. The design sludge production, which is the average quantity during the maximum month of the year plus 25% for future plant expansion, equals 90,000 gpd with an average solids concentration of 5.0%. Using a trailer-mounted press, the performance data for dewatering the 5.0% sludge were a hydraulic loading of 45 gpm/m, cake solids content of 24%, polymer dosage of 4.0 gpm/m with a concentration of 0.20% powdered polymer, wash-water usage of 32 gpm/m, and wastewater production of 64 gpm/m containing 2200 mg/l of suspended solids. Calculate the following: solids loading, polymer dosage, wash-water usage and wastewater production per 1000 gal of sludge dewatered, solids recovery, and daily cake production. For a design operating schedule of 12 hr/day with two presses, recommend the size of the presses for installation.

**13.51** An alum sludge is dewatered by pressure filtration. A daily volume of 40 m³ of slurry is pressed from 2.0% solids concentration to a cake of 40% solids, after conditioning with 10% lime. Calculate the volume of filtrate and weight of the cake produced per day. (The 40% concentration of cake solids includes the alum precipitate and the lime addition.)

**13.52** A diaphragm plate-and-frame filter press is used to dewater alum sludge from a water treatment plant. Each chamber of the press has a filtering area of 36 ft². During normal operation, the total cycle time (feed, compression, discharge, and wash) is 20 min. The volume of alum sludge applied per chamber per cycle is 300 gal, and the lime dosage as CaO is 10% of the alum solids. The alum sludge feed averages 3.6% solids, and the sludge cake is 38.5% solids including the lime. Solids capture can be assumed to be 100% since the filtrate contained less than 200 mg/l of suspended solids. Calculate the cake production per gallon of alum sludge in pounds, assuming a specific gravity of 1.0, and the filter yield in pounds per square foot per hour based on applied alum solids.

**13.53** The solution to Prob. 11.19 is a process flow diagram for a surface-water treatment plant, and the solution to Prob. 11.27 is a process flow diagram for a water-softening plant. Expand each diagram to include a scheme to process the sludge from sedimentation and the wash water from filtration. The wastes must be thickened and dewatered for hauling to land disposal, since sufficient land area at the plant sites is not available for lagoons or drying beds. (Refer to Sections 13.5 and 13.3.)

**13.54** A solid-bowl centrifuge dewaters an anaerobically digested blend of primary and waste-activated sludges at a flowrate of 8.0 l/s with a polymer addition of 10 kg per 1000 kg of dry solids. The sludge feed contains 2.4% solids, the cake averages 19%, and the suspended solids concentration in the centrate is 3200 mg/l. Calculate the solids recovery and centrate flowrate.

**13.55** A lime precipitate from a water-softening plant is concentrated in a scroll centrifuge from 10% to 60%. The suspended-solids capture is 90%. Calculate the suspended-solids concentration in the centrate and the volumetric reduction of the waste slurry as a percentage of the feed.

**13.56** A combined raw-primary and waste-activated sludge is dewatered by a vacuum filter with a coil-spring media. Typical operating data are as follows:

daily sludge feed = 55,000 gal/day with 4.0% solids

filter cake = 78,000 lb/day at 16.6% solids

filter yield = 1400 lb/hr (4.0 lb/ft$^2$/hr from 350 ft$^2$ area)

filtrate = 13,000 mg/l suspended solids

(a) The filtrate from sludge dewatering is returned to the plant influent. Assuming all of the returned solids reappear in the sludge fed to the filter, how many pounds of solids are applied to the filter for each pound of solids that appears in the cake? (Actually, a portion of the solids returned to the influent probably passes through the primary clarifiers and is partially decomposed in the aeration basin before being recycled to the filter in the waste-activated sludge.) (b) Compute the daily operating time, in hours, of the filter. Remember that filter yield is related to the solids removed by the filter, not those applied. (c) Assume that the solids removal could be improved to 95% at the same hydraulic loading applied to the filter (5900 gal/hr). Calculate the pounds of solids in the filtrate, the volume of daily sludge feed, the daily operating time, and the improved filter yield.

**13.57** What are the two principal criteria used in determining the maximum rate of sludge application on agricultural land?

# REFERENCES

1. *Water Treatment Plant Waste Management*, (Denver: AWWA Research Foundation, 1987).
2. "Design Manual, Dewatering Municipal Wastewater Sludges," U.S. Environmental Protection Agency, Office of Research and Development, EPA/625/1-87/014 (September 1987).
3. "Process Design Manual for Sludge Treatment and Disposal," U.S. Environmental Protection Agency, Municipal Environmental Research Laboratory, EPA 625/1-79-011 (September 1979).
4. G. Hoyland, A. Dee, and M. Day, "Optimum Design of Sewage Sludge Consolidation Tanks," *J. Inst. Water Env. Manage.* 3 (1989): 505–516.
5. *Recommended Standards for Sewage Works, Great Lakes–Upper Mississippi River Board of State Public Health & Environmental Managers* (Albany, NY: Health Research Inc., Health Education Services Division, 1978).
6. "Composting of Municipal Wastewater Sludges," U.S. Environmental Protection Agency, Technology Transfer, EPA 625/4-85-014 (August 1985).
7. K. A. Pietila and P. J. Joubert, "Examination of Process Parameters Affecting Sludge Dewatering with a Diaphragm Filter Press," *J. Water Poll. Control Fed.* 53(12) (1981): 1708–1716.
8. "Proposed Standards for the Disposal of Sewage Sludge," 40 CFR Parts 257 and 503, U.S. Environmental Protection Agency, Federal Register, Vol. 54, No. 23 (February 1989).
9. H. R. Pahren, J. B. Lucas, J. A. Ryan, and G. K. Dotson, "Health Risks Associated with Land Application of Municipal Sludge," *J. Water Poll. Control Fed.* 51(11) (1979): 2588–2599.
10. R. D. Davis and E. G. Coker, *Cadmium in Agriculture With Special Reference*

*to the Utilization of Sewage Sludge on Land,* Water Research Centre, Stevenage Laboratory, England, TR 139 (June 1980).

11. "Process Design Manual, Municipal Sludge Landfills," U.S. Environmental Protection Agency, Office of Solid Waste, EPA 625/1-78-010 SW-705 (October 1978).

12. G. M. Fair, J. C. Geyer, and D. A. Okun, *Water and Wastewater Engineering,* Vol. 2 (New York: Wiley, 1968), p. 36–20.

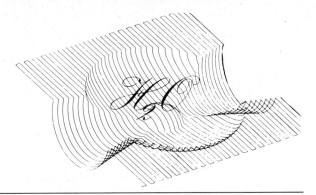

# Chapter 14

# Advanced Wastewater Treatment Processes

*Advanced wastewater treatment* refers to methods and processes that remove more contaminants from wastewater than are usually taken out by conventional techniques. The term *tertiary treatment* is often used as a synonym for advanced waste treatment, but the two are not precisely the same. Tertiary suggests an additional step applied only after conventional primary and secondary waste processing, for example, chemical coagulation of a secondary effluent for removal of phosphorus. Advanced treatment actually means any process or system that is used after conventional treatment, or to modify or replace one or more steps, to remove refractory contaminants. Several of these pollutants not taken out by secondary biological methods can adversely affect aquatic life in streams, accelerate eutrophication of lakes, hinder use of surface waters for municipal supplies, and restrict direct reuse of wastewater for irrigation, groundwater recharge, or other beneficial applications.

The term *water reclamation* implies that the combination of conventional and advanced treatment processes employed returns the wastewater to nearly original quality (i.e., reclaims the water). Reclamation is an introduction to renovating wastewater by improving the quality to such an extent that it may be directly reused. Renovation and reuse of wastewater are becoming more important as water needs cannot be met by the relatively fixed natural supply available. The envisioned reuse of wastewater ranges from agricultural and industrial process water to potable supplies, with the degree of purification required varying according to the specific use.

## Limitations of Secondary Treatment

Contamination of municipal water results from human excreta, food preparation wastes, and a wide variety of organic and inorganic industrial wastes. Conventional treatment employs physical–biological processes, and possibly chlorination, to reduce biochemical oxygen demand, suspended solids, and pathogens. Water-quality deterioration from municipal use results in the following approximate increases of pollutants after ordinary secondary treatment: 300 mg/l of total solids, 200 mg/l of volatile solids, 30 mg/l of suspended solids, 30 mg/l of BOD, 20 mg/l of nitrogen, and 7 mg/l of phosphorus. Treated wastewater may also contain traces of organic chemicals, heavy metals, excreted pathogens, and other contaminants.

## 14.1 EFFLUENT STANDARDS

The maximum acceptable level of organic matter in a wastewater effluent after biological treatment is defined in terms of BOD and suspended-solids concentrations. For each of these parameters, the arithmetic mean of daily measurements in a period of 30 consecutive days is limited to 30 mg/l, and that in a period of 7 consecutive days is limited to 45 mg/l. Conventional wastewater processing can include chemical disinfection, usually chlorination. Where public health is a concern, the effluent standard is often a maximum geometric mean concentration of fecal coliform bacteria in a period of 30 consecutive days of 200 per 100 ml and in a period of 7 consecutive days 400 per 100 ml. In some cases, these standards are not adequate to protect the receiving water body to ensure adequate quality for uses involving human contact, that is, recreation and water supply. Advanced wastewater processes can further reduce BOD and suspended-solids concentrations, increase removals of pathogens, precipitate phosphorus, oxidize ammonia–nitrogen, and diminish the content of trace organics and heavy metals.

Phosphorus is the key nutrient for growth of algae and aquatic plants associated with eutrophication. Where effluents are discharged to lakes and estuaries, either directly or via rivers, standards have been adopted by most states to limit phosphorus input. A commonly established maximum effluent concentration is 1.0 mg/l. This criterion requires a reduction of about 90% in domestic wastewater treatment.

Ammonia nitrogen, the most common form of nitrogen in an effluent from biological treatment, is toxic to fish at relatively low concentrations and can exert a significant oxygen demand. The maximum allowable ammonia–nitrogen concentration to protect warm-water fish is about 3 mg/l at a pH of 8; for cold-water fish, the limit is about 1 mg/l. With regard to oxygen demand, 4.6 lb of oxygen are used theoretically to bacterially nitrify 1.0 lb of ammonia–nitrogen. For disposal in a receiving stream with limited dilutional flow, oxidation of a major portion of the ammonia is sometimes needed along with more stringent BOD and suspended-solids removals. For example, effluent limits may be established at 10 mg/l of BOD, 10 mg/l of SS, and 5 mg/l of $NH_3$–N. Partial denitrification is occasionally specified where the receiving watercourse is used for a public water supply. The limit for nitrate concentration in drinking water is 10 mg/l $NO_3$–N.

Wastewater effluent can be chlorinated in an attempt to reduce the number of pathogens. Usually the chlorine is applied at a dosage up to 15 mg/l with a minimum contact time of 30 min. Even though fecal coliforms in conventional wastewater treatment plus chlorination are removed to a level of 200 per 100 ml, a 99.99 + reduction, the effluent may not be free of excreted pathogens. Helminth eggs are unharmed by chlorination and protozoal cysts are extremely resistant, so they must be physically removed, and viruses are more resistant to chlorination than fecal bacteria. Furthermore, the presence of suspended solids inhibits disinfection by harboring bacteria and viruses within biological floc, shielding them from the action of the chlorine.

The ability of gravity sedimentation to remove suspended solids following biological aeration is a major limitation of conventional treatment. Plain granular-media filtration can remove suspended solids to a concentration of about 10 mg/l to further reduce BOD. With chemical coagulation, the suspended solids can be reduced to less than 5 mg/l, and in a well-operating coagulation and filtration system an effluent turbidity of less than 10 NTU can be achieved. A pathogen-free effluent is produced by tertiary coagulation and filtration, with or without sedimentation, and chlorination with an extended contact time. Helminth eggs, protozoal cysts, and suspended solids that interfere with the disinfecting action of chlorine are removed in filtration. Chlorination of the clear filtered effluent with a high chlorine residual-contact time inactivates enteroviruses and kills pathogenic bacteria. The microbiological effluent standard for reuse of tertiary effluent in unrestricted irrigation and groundwater recharge is commonly 2.2 fecal coliforms per 100 ml. Sometimes allowable limits are also specified for enteric viruses and helminth eggs, such as *Ascaris*.

Effluent chlorination results in the formation of chloramines from the reaction between free chlorine and ammonia. Chloramines are toxic to fish at very low levels. The recommended maximum concentrations of residual chlorine are 0.01 mg/l for warm-water fish and 0.002 mg/l for cold-water fish. In advanced wastewater treatment, dechlorination is performed by the addition of sulfur dioxide to the effluent.

Many organic and inorganic chemicals are refractory to conventional treatment, and local conditions often dictate discharge limits for specific compounds. Some cause esthetic problems, e.g., foam or color in the receiving watercourse or taste and odor in downstream water supplies. Toxic organic compounds or metal ions are injurious to aquatic life and, in sufficiently high concentrations, prevent safe reuse of the receiving water as a public supply source. The main problems are detecting, identifying, and establishing discharge limits for toxicants. The traditional approach has been to compare instream concentrations of known pollutants with surface-water-quality standards. However, sometimes the toxicants cannot be identified by chemical testing, standards may not exist, and the impact of toxicity may depend on interacting factors such as pH, alkalinity, and hardness. As a result, biomonitoring has been developed for toxicity assessment.

Biomonitoring to evaluate acute and chronic toxicities is performed by short- and long-term tests on aquatic organisms like minnows, crustaceans, and invertebrates. The tests are conducted on full-strength wastewater and several dilutions, using water from the receiving stream to determine the NOEL (no-observed-effect-level) values for acute and chronic exposures. Based on these data and the dilution provided by the receiving stream, exposure criteria can be calculated and discharge permit limits established. The toxicants and their sources are determined by additional bioassays in combination with chemical testing on samples from the wastewater collection system.

## 14.2 FLOW EQUALIZATION

Wastewater flows have a diurnal variation ranging from less than one half to more than 200% of the average flowrate. In addition, daily volumes are increased by inflow and infiltration into the sewer collection system during wet weather. The strength of a municipal waste also has a pronounced diurnal variation resulting from nonuniform discharge of domestic and industrial wastes. Treatment plants traditionally have been sized to handle these deviations without significant loss of efficiency. For conventional primary and secondary treatment, designing units to accommodate peak hourly flow is more economical than installing flow equalization basins to eliminate the diurnal influent pattern. Industrial wastes entering a municipal system can cause excessively large flows and peak organic loads, so it is better to install facilities at the industrial site for flow smoothing prior to discharge.

Many advanced waste treatment operations, such as filtration and chemical clarification, are adversely affected by flow variation and sudden changes in solids loading. Maintaining a relatively uniform influent allows improved chemical feed control and process reliability. Costs saved by installing smaller units for chemical precipitation and filtration, together with reduced operating expenses, may more than compensate for the added costs of flow equalization facilities.

Equalization basins to level diurnal flow variations may be designed for either side-line or in-line operation. A side-line basin may receive, temporarily store, and return either raw wastewater after screening and degritting or clarified wastewater after primary sedimentation. A raw-wastewater basin requires aeration mixing to reduce odors and maintain organic solids in suspension. The tank is often equipped with a sludge scraper to prevent consolidation of settled solids. If the side-line basin follows primary sedimentation, aeration mixing and sludge scrapers may not be necessary. In a warm climate, the basin may be a paved pond that is flushed clean with chlorinated effluent after each use by an automatic sprinkling system. All the wastewater passes through an in-line basin. If placed between grit removal and primary sedimentation, mechanical or diffused aeration is required to keep solids in suspension and to prevent septicity. Holding tanks may also be placed at other locations in the treatment scheme. For example, a basin serving as a pump suction pit can be located just ahead of filters to dampen hydraulic surges without providing complete flow equalization.

Basin volume required for flow equalization is determined from mass diagrams based on average diurnal flow patterns. The capacity needed is usually equivalent to 10%–20% of the average daily dry-weather flow. Extra volume should be provided below the low water level, since both mechanical and diffused aeration systems must have a minimum depth to maintain mixing. Example 14.1 includes sample calculations for determining equalization basin volume.

■ **EXAMPLE 14.1**
Determine the basin volume needed to equalize the diurnal wastewater flow pattern diagrammed in Fig. 12.16b.

*Solution.*    Hourly flowrates measured from Fig. 12.16b and calculated cumulative volumes are listed in Table 14.1; the mass diagram of wastewater flow from Table 14.1 is drawn in Fig. 14.1. The slope of the line connecting the origin and final point on the mass curve equals the average 24-hr rate of 18.2 mgd. To find the required volume for equalization, construct lines parallel to the average flowrate and tangent to the mass curve at the high and low points. The vertical distance between these two parallels is the required basin capacity; in this case, the value is 2.5 mil gal or 13.7% of the daily influent of 18.2 mil gal.    ■

# Selection of Advanced Wastewater Treatment Processes

Table 14.2 lists popular advanced treatment methods. Upgrading treatment to increase BOD and suspended-solids removal is usually accomplished by tertiary operations, such as filtration. Emphasis is also being placed on unit

**TABLE 14.1**    WASTEWATER FLOWS FROM
FIG. 12.16b

| Time | Flowrate (mgd) | Cumulative volume (mil gal) |
|---|---|---|
| Midnight | — | 0 |
| 2 A.M. | 13.1 | 1.1 |
| 4 | 10.7 | 2.0 |
| 6 | 9.7 | 2.8 |
| 8 | 14.8 | 4.0 |
| 10 | 22.5 | 5.9 |
| Noon | 26.6 | 8.1 |
| 2 P.M. | 25.6 | 10.3 |
| 4 | 23.5 | 12.2 |
| 6 | 21.8 | 14.0 |
| 8 | 18.9 | 15.6 |
| 10 | 16.7 | 17.0 |
| Midnight | 14.4 | 18.2 |
| Average | 18.2 | |

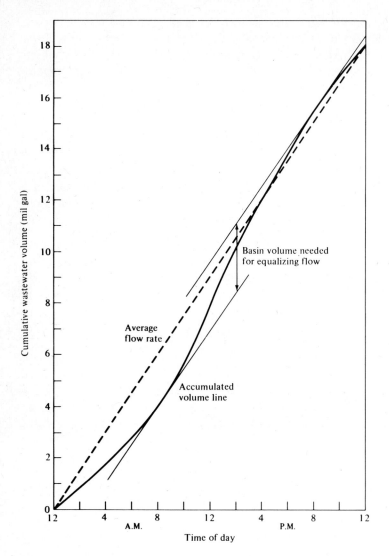

**Figure 14.1** Mass diagram of wastewater flow, from Table 14.1 for Example 14.1, to determine the basin volume needed to equalize flow.

operations for nutrient removal, particularly precipitation of phosphates by either separate chemical coagulation or biological–chemical aeration and by nitrogen extraction by air stripping or biological nitrification–denitrification.

## 14.3 SELECTING AND COMBINING UNIT PROCESSES

Culp and Culp [1] present the following major factors affecting unit process selection: influent wastewater characteristics, the effluent quality required, reliability, sludge handling, process compatibility, and costs. The degrees of

**TABLE 14.2**  SELECTED ADVANCED
WASTEWATER TREATMENT
PROCESSES

Suspended-solids removal
  Filtration through granular media
  Chemical coagulation and clarification
Organic matter removal
  Adsorption on granular activated carbon
  Extended biological oxidation
Phosphorus removal
  Biological–chemical precipitation and clarification
  Chemical coagulation and clarification
  Irrigation of cropland
Nitrogen removal
  Biological nitrification–denitrification
  Ammonia reduction by air stripping
  Irrigation of cropland
Heavy metals removal
  Lime precipitation
Dissolved-solids removal
  Reserve osmosis

treatment needed at present and in the future are the primary considerations, because neither raw-wastewater characteristics nor effluent quality specifications can be safely considered as permanent. When considering alternative schemes, the best choice for the original plant allows for expansion and change to meet future needs. Maximum flexibility for modifying operations to improve performance is of major importance.

The reliability of a process is directly related to the experience gained from operating plant-scale systems. Only limited data are available on actual operating results and the costs of new treatment techniques. Pilot-plant studies can provide satisfactory information on liquid processing, but they do not often adequately evaluate sludge problems or compatibility with previous and subsequent unit operations. Problems involved in going from pilot to full scale are not easily recognized; thus, for maximum reliability, future plant designers should take advantage of prior art and practice developed from existing full-scale installations.

Sludge handling and disposal dictate to a considerable extent the selection of processes that are most feasible for separating contaminants from wastewater. A unit operation, even though successful in extracting pollutants from water, can be unacceptable if the waste sludge produced is difficult and costly to dewater; hence, sludge disposal must always be considered an integral part of any treatment technique. The elimination of residue by spreading on land, burial, or incineration necessitates concentration of waste slurries by biological or chemical stabilization and mechanical dewatering methods. Advanced treatment processes require greater study because they often include chemical precipitates as well as organic sludges. A critical question is whether organic and inorganic wastes should be kept separate or combined for thickening and disposal. The answer often determines whether chemical treatment is applied as a tertiary step or combined with conventional operations of primary sedimentation and biological aeration.

A fourth consideration involves compatibility of unit processes applied in the overall treatment scheme. Optimum pH is often important and can influence the sequence of operations. For instance, ammonia stripping is most efficient from alkaline water, while disinfection by chlorine is more effective at low pH values. Possible effects of recycled waste streams from individual unit operations in the overall treatment process must be considered. Segregation or blending of centrate, filtrate, backwash water, and other return flows and their point of return demand evaluation in design.

Both capital and operating costs influence process selection and often dictate design decisions. Common factors include electric power consumption; recovery and reuse of chemicals; carbon regeneration; the choice of methods for sludge disposal; and separation or combination of biological and chemical treatment. Regarding the latter, phosphorus-removal schemes incorporated in present-day conventional processes may appear to be cheaper than tertiary treatment until the cost of handling the biological–chemical sludge mixture is included in the evaluation, making separate disposal of the two sludges more economical except in locations where the blended sludge can be disposed of by land burial.

# Suspended-Solids Removal

Removal of suspended solids from the effluent of a conventional treatment plant may be in order to reduce the organic content or to pretreat the wastewater for subsequent processing. Effective disinfection requires removal of suspended solids that can harbor and protect pathogenic bacteria and viruses from the oxidizing action by chlorine or ozone. Carbon adsorption columns are preceded by filtration to prevent fouling, and reverse osmosis for removal of dissolved salts requires an influent of extremely low turbidity.

## 14.4 GRANULAR-MEDIA FILTRATION

Design criteria for wastewater filters cannot be derived directly from experience in potable water systems. Waterworks filters are generally operated at constant rates under relatively steady suspended-solids loading. Unless equalization is provided, a wastewater plant must handle a varying rate of flow with peak hydraulic and solids loadings occurring simultaneously. Particulate matter found in typical wastewaters is less predictable and much more "sticky" than water-plant solids, thus making filter backwashing more difficult. Also, the nature of suspended matter is not consistent and varies with the preceding treatment processes. Microbial flocs are the dominant suspended solids following secondary biological treatment, while carryover from biological–chemical and physical–chemical methods contains a significant quantity of coagulant residue.

Dual-media anthracite–sand filters with downward flow by either gravity or pressure dominate in American practice. This design permits production of a high-quality effluent with reasonable filter runs between bed cleanings. Typically, the layer of coarse coal with a specific gravity of 1.35–1.75 is 12–18 in. (0.3–0.5 m) deep, and the underlying sand layer with a specific gravity of 2.65 is 8–12 in. (0.2–0.25 m) [2]. After backwashing, the media are arranged with the coarse lighter coal on top and the finer heavier sand on the bottom. The actual distribution and degree of intermixing depend on both relative particle-size gradations and specific gravities of the media. A filter bed with coarser coal over finer sand allows both surface straining and ''in-depth'' filtration, reducing premature surface plugging and delaying solids breakthrough. Other types of filters are trimedia with crushed garnet or ilmenite beneath the silica sand layer, and deep-bed single-medium filters of sand or anthracite ranging in size from 1 to 3 mm with bed depths of 3–6 ft (1–2 m).

Cleaning solids from granular media requires air or mechanical scouring in addition to fluidizing the bed by upflow of wash water. Fluidization alone does not provide the necessary abrasive action among media grains. Common sequences are surface wash by a rotating arm with high-pressure orifices followed by hydraulic backwash; air scour by vigorous aeration up through the underdrain followed by backwash; and air scour with simultaneous sub-fluidization before initiating the full flow of backwash water. The source of backwash water is effluent from the tertiary filters. If the filtrate is disinfected, the chlorine contact tank serves as the source of supply. Otherwise, a storage tank is needed with sufficient capacity to clean the filters during their peak usage. Dirty wash water should be collected in an equalizing tank and returned to the plant influent at a nearly constant rate for treatment. The quantity of backwash water returned is 2–5% of the filtered water.

Design considerations for granular-media filters [2, 3, 4] include the filter configuration; types, size, gradation, and depths of filter media; method of flow control; backwash requirements; filtration rate; terminal head loss; quantity and characteristics of the water applied; and desired effluent quality. Figure 14.2 is a schematic diagram of a biological treatment plant with tertiary filtration. Normally, at least two and usually four filter cells are provided for operational flexibility. The bed area should be sufficient to allow peak design flows with one unit out of service for backwashing or repair.

Gravity filters in wastewater processing are usually designed with flow control by either influent flow splitting or declining-rate filtration. These methods are described in Section 10.20.

Pressure filters (Fig. 10.41) are normally operated at a constant rate of flow, although some units use a constant applied pressure without effluent control, resulting in declining-rate filtration. Two advantages of enclosed filters are that high pressure can be used to overcome a greater head loss and negative head conditions are not created in the filter.

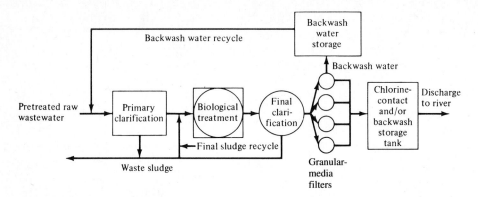

**Figure 14.2** Typical layout of a biological treatment plant with tertiary granular-media filters. [From *Wastewater Filtration,* Environmental Protection Agency, Technology Transfer (July 1974): 6.]

The filtration rate, terminal head loss, length of filter run, and solids loading are interrelated parameters. Production from gravity-flow beds is normally 2–6 gpm/ft$^2$ (120–350 m$^3$/m$^2$·d), and terminal head loss 8–10 ft (2.5–3.0 m) for gravity filters and 20–30 ft when using pressure filters. The length of filter runs should be a minimum of 6 hr to avoid excessive use of wash water and shorter than 40 hr to reduce bacterial decomposition of organics trapped in the media. These design limits provide an average run length of about 24 hr. The highest solids loading on tertiary filters coincides with maximum hydraulic flow, since clarification following secondary treatment is least efficient at peak overflow. Even in well-operated plants, the suspended-solids content during high flows can range from 30–50 mg/l (15–25 turbidity units). Therefore, the critical design condition is often based on the maximum 4-hr flowrate at the worst expected suspended-solids concentration.

Experience indicates that filtration following secondary biological treatment can reduce suspended solids to a level of 4–9 mg/l, so the expected performance of a well-designed and properly operated system is an effluent with SS and BOD concentrations of less than 10 mg/l. However, chemical treatment prior to filtration is required to consistently produce an effluent of 5 mg/l of SS and 5 mg/l of BOD.

### ■ EXAMPLE 14.2
Compute the area of granular-media filters required for suspended-solids removal from a trickling-filter plant effluent. The average daily flow is 18.2 mgd (68,900 m$^3$/d), and the maximum wet-weather flow for a 4-hr period is 31.0 mgd (117,000 m$^3$/d). Data from a pilot-plant study are plotted in Fig. 14.3. Filters are dual-media, gravity-flow beds, downtime for backwashing is 30 min, and water usage equals 150 gal/ft$^2$ (6.1 m$^3$/m$^2$). Assume a nominal filtration rate of 3 gpm/ft$^2$ (54 m$^3$/m$^2$·d). Check the peak rate with one filter cell out of service.

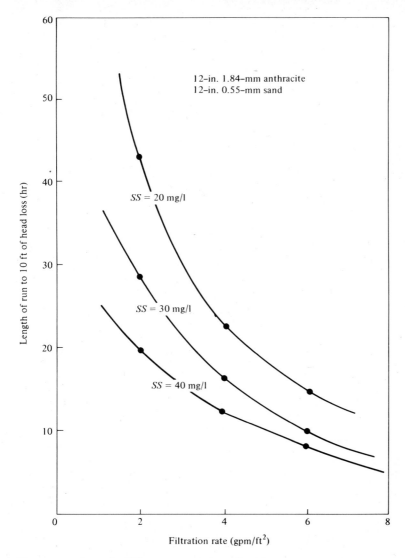

**Figure 14.3** Length of filter run at various filtration rates and suspended-solids concentrations for a bed of anthracite coal and sand media, and an applied wastewater of trickling-filter plant effluent. [From E. R. Baumann and J. Y. C. Huang, "Granular Filters for Tertiary Wastewater Treatment," *J. Water Poll. Control Fed.* 46(8) (August 1974).]

*Solution.* For the average daily wastewater flow and nominal filtration rate of 3.0 gpm/ft$^2$,

$$\text{filter area} = \frac{18.2 \text{ mgd} \times 694 \text{ gpm/mgd}}{3.0 \text{ gpm/ft}^2} = 4200 \text{ ft}^2$$

Use four filters, each 1050 ft$^2$ (97.5 m$^2$).

From Fig. 14.3, the length of filter run is 21 hr for 3.0 gpm/ft$^2$ and 30 mg/l of SS.

$$\text{filter cycle time} = \text{run length} + \text{backwash time}$$

$$= 21.0 + 0.5 = 21.5 \text{ hr}$$

$$\text{volume of filtrate/cycle} = 21 \times 3.0 \times 60 = 3800 \text{ gal/ft}^2$$

$$\text{backwash volume/cycle} = 150 \text{ gal/ft}^2$$

$$\text{backwash as a percentage of filtrate} = \frac{150}{3800} = 4.0\%$$

$$\left(\begin{array}{c}\text{needed filtration rate to account for}\\\text{downtime and return of backwash}\\\text{water}\end{array}\right) = 3.0 \times 1.04 \times \frac{21.5}{21}$$

$$= 3.2 \text{ gpm/ft}^2 \ (57 \text{ m}^3/\text{m}^2\cdot\text{d})$$

The quantity of suspended solids removed assuming 30 mg/l applied and 5 mg/l remaining in the effluent is

$$(30 - 5) \times 18.2 \times 8.34 = 3800 \text{ lb of dry SS/day } (1700 \text{ kg/d})$$

For the maximum 4-hr flow of 31.0 mgd, estimate the required filtration rate with three filter cells operating and the fourth out of service if the return backwash water is 10% of the wastewater flow.

$$\text{filtration rate} = \frac{31.0 \times 1.1 \times 694}{3 \times 1050} = 7.5 \text{ gpm/ft}^2 \ (130 \text{ m}^3/\text{m}^2\cdot\text{d})$$

The length of filter run from Fig. 14.3 at 40 mg/l is 6 hr, which should be sufficient to allow operation when one cell is out of operation for cleaning or emergency repair. ∎

## 14.5 FILTRATION AND CHLORINATION FOR VIRUS REMOVAL

In Southern California, inland wastewater treatment plants discharge to flood control channels or river beds during periods of little or no dilutional flow. Impoundments in these watercourses are open to the public for recreation, including swimming. They also serve as recharge basins to replenish groundwater withdrawn by municipal wells. As a result of these activities in water reuse, effluent standards are needed to protect public health, with emphasis on virus removal.

The State of California Health Department established both quality and treatment criteria for wastewaters discharged into recreational impoundments. The quality criterion was a 7-day median coliform concentration less than or equal to 2.2 per 100 ml. Because of the difficulties in monitoring for the presence of viruses, no specific limit on virus concentration was proposed. The specified tertiary treatment included coagulation, sedimentation, filtration, and disinfection. Alternative tertiary treatment may be substituted for this criterion provided the same degree of virus removal is achieved by the proposed processing. The prospect of being able to apply advanced wastewater treatment other than conventional water treatment resulted in researching less expensive alternatives.

The Pomona Virus study [5] investigated virus-removal efficiency of several tertiary treatment systems on a pilot-plant scale. Flow diagrams for three of the systems are given in Fig. 14.4. The first scheme, serving as the baseline for alternative processing, includes coagulation with 150 mg/l of alum, sedimentation, dual-media filtration at 5 gpm/ft$^2$, and chlorine contact for 2 hr with minimum residuals of 5 and 10 mg/l. Figure 14.4b is direct filtration (no flocculation or sedimentation) with a low alum dosage of 5 mg/l. The rate of filtration was 5 gpm/ft$^2$ and the time and dosage of chlorine was 2-hr contact with residuals of 5 and 10 mg/l, which are the same as the baseline system. Figure 14.4c illustrates two-stage carbon adsorption with intermediate disinfection. The first granular-carbon bed with 10-min contact also provided gravity filtration at 3.5 gpm/ft$^2$, intermediate chlorination was

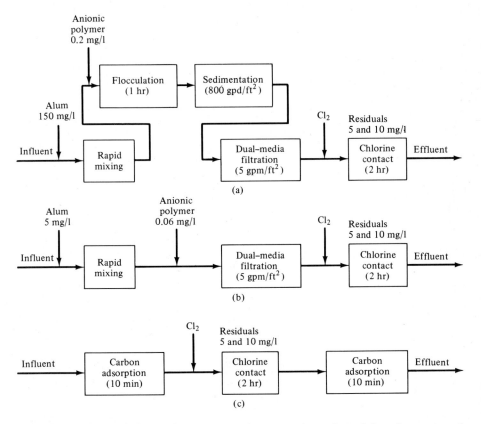

**Figure 14.4** Three of the tertiary treatment systems investigated for virus-removal efficiency in the Pomona, CA, pilot-plant studies.   (a) Coagulation, sedimentation, filtration, and disinfection.   (b) Coagulation, filtration, and disinfection.   (c) Two-stage carbon adsorption and disinfection.

operated the same as the other systems, and second-stage carbon adsorption had a gravity flowrate of 4 gpm/ft$^2$ with 10-min contact. Other processes tested, but not discussed here, were ozonation in place of chlorination for disinfection and processing of a nitrified wastewater effluent. The major objectives of this study were to rank the treatment systems based on their efficiency and reliabilty of virus removal and to estimate treatment costs for those systems that performed equivalent to the baseline system.

Naturally occurring viruses were present in relatively low concentrations in the wastewater and were rarely detected in the tertiary effluents. Therefore, seeding experiments had to be conducted with sufficient virus concentrations to produce measurable effluent virus counts. With polio I as a seed, viruses were added to the influent in the concentration range of $10^5$–$10^6$ PFU/gal. Viruses and detection procedures are discussed in Section 12.4.

The key conclusions of the study were that the majority of virus inactivation occurred during disinfection, and the main function of the preceding treatment was to remove substances that interfere with effluent disinfection. Also, the magnitude of the chlorine residual directly affected the effluent virus concentration. The systems equivalent in efficiency and reliability of virus removal were the baseline system (Fig. 14.4a) with 5 mg/l of combined chlorine residual, direct filtration (Fig. 14.4b) with 10 mg/l of chlorine residual, two-stage carbon adsorption (Fig. 14.4c) with a 5-mg/l residual, the baseline system with a 10-mg/l ozone dosage, and direct filtration with a 50-mg/l ozone dosage. The least-cost system that performed equivalent to the baseline system was direct filtration and disinfection at the higher chlorine residual of 10 mg/l.

As a result of the Pomona Virus Study, tertiary treatment has been implemented at five plants of the Sanitation Districts of Los Angeles County (Long Beach, Los Coyotes, Pomona, San Jose Creek, and Whittier Narrows) treating approximately 140 mgd [6]. The water reclamation at all of these plants is conventional biological treatment followed by direct tertiary filtration, with prior chemical coagulation using alum and anionic polymer, and plug-flow chlorination with an extended contact time (Fig. 14.4b). The effluent quality is microbiologically and chemically safe for unrestricted reuse, except for direct drinking water [6]. After years of effluent monitoring, no enteric viruses have been detected, and the median coliform count has been less than 1 per 100 ml. The concentrations of heavy metals, nitrate nitrogen, and trihalomethanes meet drinking-water standards. The excellent chemical quality is controlled by both limiting the industrial wastes entering the sewer system and efficient treatment plant operations. In addition to groundwater recharge and recreational impoundments, the reclaimed water is reused for landscape irrigation by communities near the reclamation plants.

# Carbon Adsorption

Activated carbon effectively removes dissolved organics at low concentrations in water and wastewater. Powdered carbon, commonly used in water treatment, has not received widespread application in wastewater processing, owing to the difficulty of regeneration. The current trend is installation of granular-carbon columns as tertiary conditioning following chemical precipitation and granular-media filtration. When the adsorptive capacity is exhausted, the spent carbon can be regenerated for reuse.

## 14.6 GRANULAR-CARBON COLUMNS

Dissolved organics not extracted by conventional biological treatment can be removed to a considerable degree by adsorption and biodegradation on activated carbon. The large surface area of activated carbon assimilates organics while microbial degradation reopens pores in the granules. Because of this biological contribution, toxic substances in the applied wastewater can reduce the removal capacity. Some readily biodegradable substances are difficult to adsorb on carbon, thereby making it difficult to predict the quality of effluent for a given wastewater. The ability to remove soluble organics should be demonstrated by pilot-plant tests as well as experience from existing full-scale plants.

The configuration and operation of a carbon contactor, in addition to the nature of waste organics and character of the granular carbon, influence the effectiveness of adsorption. Contact times are generally less than 30 min since longer periods do not substantially enhance removals but do produce a favorable environment for generating hydrogen sulfide. Anaerobiosis can be controlled by decreased contact time, more frequent backwashing, oxygen addition, or prechlorination.

The two alternative carbon-contacting systems are downward passage through the bed, either under pressure or gravity flow, and upflow through a packed or expanded column. Granular carbon in a downflow unit adsorbs organics and filters out suspended materials. However, this dual-purpose approach can result in an unsatisfactory effluent owing to losses of efficiency in both filtration and adsorption. Downflow beds are periodically backwashed to remove accumulated solids, and therefore usually have an underdrain system and water piping similar to granular-media filters.

Figure 14.5 diagrams the upflow, countercurrent, packed-bed carbon columns installed in Water Factory 21, Orange County Water District, California. After chemical clarification and granular-media filtration, water enters the bottom of the column through a screen manifold and overflows via outlet screens at the top. Fresh carbon slurry is added by gravity, and the spent packing is withdrawn from the bottom. With the column in service, replacement can be performed continuously at a slow rate, or intermittently by replacing 5%–10% of the column contents at a time. About 10% of the contactor volume is void space at the top of the tank. This permits upward flow to be increased for bed expansion and flushing particulate matter to waste to reduce head loss through the bed. The upflow-to-waste cycle can also be used to clean the medium of excess carbon fines, if necessary. Countercurrent operation allows carbon near the inlet to become fully saturated with impurities prior to withdrawal for regeneration.

Expanded-bed upflow contactors can be either pressurized or con-

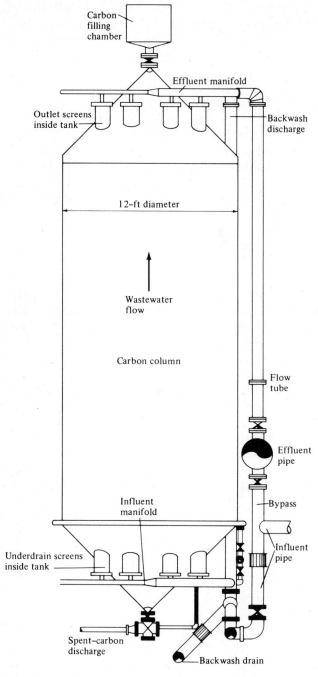

**Figure 14.5** Upflow countercurrent carbon column, Orange County Water District, CA.

structed with open tops. Tank volume includes a space of about 50% for bed enlargement during operation. Expanded-bed units have the advantage of being able to treat wastewaters relatively high in suspended solids without the excessive head losses that occur in a downflow unit.

The general ranges of design criteria for both tertiary and physical–chemical carbon columns are listed in Table 14.3 [7]. Granular-media filters are generally placed ahead of carbon units to reduce the suspended-solids loading on the columns. Upsets in pretreatment, which produce a rapid rise in turbidity of the influent, can suddenly increase the head loss through downflow and packed-upflow beds. Installation of upflow expanded-bed contactors employs downstream filtration to remove bacterial floc and other suspended matter flushed from the carbon columns.

## 14.7  CARBON REGENERATION

Exhausted activated carbon must be regenerated and reused to make adsorption economically feasible. Restoration is accomplished by heating in a multiple-hearth furnace with a low-oxygen steam atmosphere (Fig. 13.25). Adsorbed organics are volatilized and released in gaseous form at a temperature of about 1700°F. With proper control, granular carbon can be restored to near virgin adsorptive capacity with only 5%–10% weight loss. Thermal regeneration involves drying, baking (pyrolysis of adsorbates), and activating by oxidation of the remaining residue. The time required is about 30 min—15 min for drying, 5 min for gasifying the volatiles, and 10 min for reactivation. Furnace controls include the temperature, rate of steam feed, and rotational speed of the rabble arms. The latter two determine the carbon depth on the hearths and residence time in the furnace. Temperature is the most critical factor since insufficient heat will not volatilize the organic matter, while too high a temperature burns the carbon.

**TABLE 14.3**  GENERAL DESIGN PARAMETERS FOR
GRANULAR-CARBON COLUMNS

| | |
|---|---|
| Carbon dosage (regeneration requirement) | |
|    Tertiary treatment | 200–400 lb/mil gal |
|    Physical–chemical process | 500–1800 lb/mil gal |
| Contact time (empty bed basis) | 10–50 min |
| Hydraulic loading | 2–10 gpm/ft$^2$ |
| Backwash rate | 15–20 gpm/ft$^2$ |
| Flow configuration | Upflow or downflow; one-stage or multistage |
| Contactor configuration | Gravity or pressure vessels; steel or concrete construction |

*Source: Process Design Manual for Carbon Adsorption*, Environmental Protection Agency, Technology Transfer (October 1973): 6–1.

Figure 14.6 shows a process flow diagram for carbon contacting and regeneration. Spent carbon withdrawn from the bottom of the columns is transported in water slurry to a dewatering tank. After draining, the wet carbon feeds by a screw conveyor into the top of a multihearth furnace. Regenerated carbon, discharging from the bottom of the furnace, is quenched in water. The slurry is then washed to remove carbon fines and hydraulically transported to filling chambers of the carbon columns or to storage. Makeup carbon is also transported in water suspension to the contactors. Regeneration furnaces should be sized to allow for a substantial maintenance downtime. The fuel requirement for regeneration, including both furnace heating and steam generation, approximates 4000–4500 Btu/lb of carbon.

## Phosphorus Removal

Most phosphorus entering surface waters is from human-generated wastes and land runoff. Contributions from nonpoint sources in surface drainage vary from 0 to 15 lb of phosphorus/acre/year, depending on land use, agricultural practice, fertilizer additions, topography, soil conservation practices, and other factors. Domestic waste contains approximately 3.5 lb (1.6 kg) of

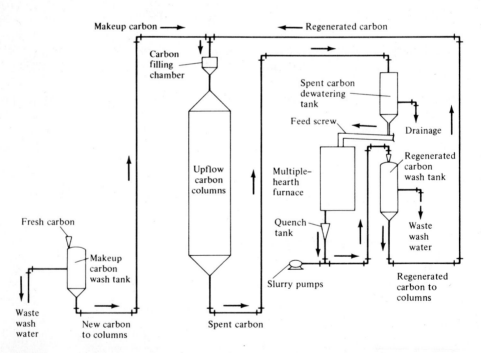

**Figure 14.6** Carbon regeneration diagram for upflow columns. [From *Process Design Manual for Carbon Adsorption,* Environmental Protection Agency, Technology Transfer (October 1973): 3–66.]

phosphorus/capita/year, of which about 60% is from phosphate builders used in synthetic detergents.

The most common forms of phosphorus are organic phosphorus, orthophosphates ($H_2PO_4^-$, $HPO_4^{2-}$, $PO_4^{3-}$), and polyphosphates. Typical polyphosphates are sodium hexametaphosphate, $Na_3(PO_3)_6$; sodium tripoly-phosphate, $Na_5P_3O_{10}$; and tetrasodium pyrophosphate, $Na_4P_2O_7$. All polyphosphates gradually hydrolyze in aqueous solution and revert to the ortho form. Domestic wastewater contains approximately 10 mg/l of total phosphorus, of which about 70% is soluble.

Reactions involving phosphorus are

$$PO_4 + NH_3 + CO_2 \xrightarrow{\text{sunlight}} \text{green plants} \tag{14.1}$$

$$\text{organic P} \xrightarrow[\text{decomposition}]{\text{bacterial}} PO_4 \tag{14.2}$$

$$\text{polyphosphates} \xrightarrow[\text{in water}]{\text{hydrolysis}} PO_4 \tag{14.3}$$

$$PO_4 + \text{multivalent metal ions} \xrightarrow[\text{coagulant}]{\text{excess}} \text{insoluble precipitates} \tag{14.4}$$

Reaction 14.1 is photosynthesis. Equations 14.2 and 14.3 are decomposition and hydrolysis reactions in which complex phosphates are converted to the stable orthophosphate form. Equation 14.4 is the precipitation of orthophos-phate by chemical coagulation. Substantial excess concentrations of hydro-lyzing coagulants (aluminum or iron) or lime are required for effective phos-phate precipitation.

## 14.8  BIOLOGICAL PHOSPHORUS REMOVAL

Primary sedimentation in conventional treatment settles only a small percent-age of the phosphorus in wastewater, since the majority is in solution. Sec-ondary biological processing involves removal of soluble phosphate taken up by the microbial floc. The amount synthesized into growth is related primarily to the concentration of phosphates in the wastewater relative to the BOD content. Treating waste with a high BOD/P ratio eliminates a large percentage of the phosphorus, whereas processing with phosphorus in excess of biologi-cal needs results in lower removal efficiency. Domestic wastewater has a surplus of phosphorus relative to the quantities of nitrogen and carbon neces-sary for synthesis. In general, the amount of P embodied in the biological floc of a conventional activated-sludge process is equal to about 1% of the BOD applied. Anticipated removal in treatment of a typical wastewater with 200 mg/l of BOD is 2 mg/l of P, or a 20% phosphorus reduction.

A study of Menar and Jenkins [8] concluded that conventionally designed primary and activated-sludge secondary treatment can remove a maximum of 20%–30% of an influent 10 mg/l of phosphorus by biological means. These and other authors [9] attribute greater phosphate reduction, reported in some activated-sludge plants, to hard-water wastes where phosphates are complexed by calcium followed by enmeshing the precipitate into the biological floc. Another conclusion was that operational parameters such as organic loading, mixed-liquor suspended solids, and dissolved oxygen have no effect on enhanced removal of phosphate by activated sludge. Others contend that environmental control of the biological process can result in improved phosphorus removal in wastewater aeration. Carberry and Tenney [10] concluded, based on laboratory investigations, that luxury uptake of phosphate by activated sludge occurs by a biological mechanism.

The method of processing and disposal of sludge withdrawn from primary and secondary settling tanks is an important consideration in nutrient removal. The only phosphorus considered extracted is that portion which does not end up in surface waters, namely, the amount in solids disposed of or hauled away from the treatment plant site. An extended aeration system operating without sludge wasting extracts no phosphorus. Dewatering of raw-waste sludge followed by land burial of solids results in maximum phosphorus removal. Conventional sludge stabilization by anaerobic or aerobic digestion returns to the influent of the treatment plant a supernatant liquid containing nutrients.

## 14.9 BIOLOGICAL–CHEMICAL PHOSPHORUS REMOVAL

Chemical precipitation, using aluminum and iron coagulants or lime, is effective in phosphate removal [11]. Three alternatives employed with conventional biological treatment are illustrated in Fig. 14.7. The process most frequently employed is that in Fig. 14.7a, chemical precipitation with biological treatment.

### Chemical Precipitation and Biological Treatment

Coagulation with biological aeration is used both to upgrade existing plants and in new design. For this process (Fig. 14.7a), chemicals are added to the activated-sludge tank or to the effluent of the aeration basin before final settling. In a trickling-filter or rotating biological contactor system, the coagulant is usually applied only to the effluent of the biological process. Proper mixing at the point of addition and a few minutes of flocculation prior to clarification are essential for maximum effectiveness. The location for best floc formation and subsequent settling is determined experimentally in the field by varying the position of chemical application and monitoring settleability of the solids.

Both alum and ferric chloride are used in combined chemical–biological

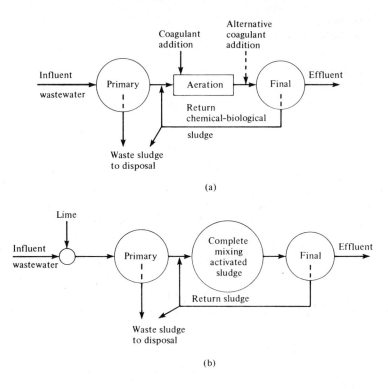

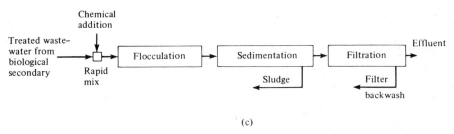

**Figure 14.7** Phosphorus removal schemes.    (a) Chemical precipitation with biological treatment.    (b) Lime precipitation in primary sedimentation followed by secondary completely mixed activated sludge.    (c) Tertiary treatment by chemical precipitation.

flocculation, with lime and polymers occasionally applied as coagulation aids. The theoretical chemical reaction between alum and phosphate is

$$Al_2(SO_4)_3 \cdot 14.3H_2O + 2PO_4^{3-}$$
$$= 2AlPO_4 \downarrow + 3SO_4^{2-} + 14.3H_2O \qquad (14.5)$$

The molar ratio of aluminum to phosphorus in this equation is 1:1, which is equivalent to a weight ratio of 0.87:1.00. Since alum contains 9.0% Al, 9.7 lb of coagulant is theoretically required to precipitate 1.0 lb of P. The actual coagulation reaction in wastewater is only partially understood and more complex than Eq. 14.5 because of secondary reactions with colloidal solids and alkalinity (Eq. 11.34).

Alum demand is also a function of the degree of phosphorus removal, as diagrammed in Fig. 14.8 [12]. The data are from operation of a bench-scale activated-sludge unit consisting of an aeration chamber and separate settling tank with a sludge return line. The wastewaters applied were settled domestic and a glucose–glutamic acid medium, both with a BOD of 150–190 mg/l and average phosphorus level of 10 mg/l. Without alum added to the aeration basin, phosphorus removal averaged 23% (Fig. 14.8). Chemical–biological processing improved phosphate extraction with increased Al to P dosages (aluminum applied to total influent phosphorus), indicating a drop in coagulation efficiency as the amount of phosphorus remaining decreases. The aluminum to phosphorus ratios for 90% P removal from the municipal and synthetic wastes were 1.5:1 and 2.0:1 Al to P, respectively, equivalent to 170 and 220 mg/l of alum. In full-scale activated-sludge and trickling-filter plants, alum applications vary from 50 to 200 mg/l for 80%–95% phosphorus removal [11]. Diffused aeration provides more efficient flocculation and, for the same degree of phosphorus removal, uses less alum than trickling-filter installations.

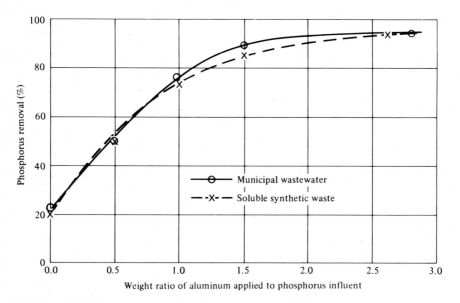

**Figure 14.8** Biological–chemical phosphorus removal in a bench-scale, activated-sludge unit with alum added to the aeration chamber. The substrates applied were settled municipal wastewater and a synthetic medium of glucose and glutamic acid, with a phosphorus content of approximately 10 mg/l of P. [From D. T. Anderson and M. J. Hammer, "Effects of Alum Addition on Activated Sludge Biota," *Water and Sewage Works* 120(1) (January 1973).]

A large addition of aluminum coagulant has a marked influence on the biota of activated sludge. Anderson and Hammer [12] reported that free-swimming and stalked protozoans are adversely affected to the extent that higher life forms are virtually absent with alum dosages in excess of 150 mg/l. Under these conditions it appears that chemical flocculation replaces the role of protozoans in clarifying settled effluent. BOD and suspended-solids removals are enhanced by coagulant addition to aeration basins and trickling filters and generally result in effluent concentrations in the range of 10 to 20 mg/l.

Ferric chloride can also be applied with biological aeration to complex the phosphate ion. The hypothetical chemical reaction is

$$FeCl_3 + PO_4^{3-} = FePO_4\downarrow\ + 3Cl^- \tag{14.6}$$

The molar and weight ratios of Fe to P are 1:1 and 1.8:1, respectively. Theoretically, 5.2 lb of $FeCl_3$ is needed to precipitate 1 lb of P since ferric chloride is 34% iron. The actual dosage is usually greater than this equation predicts for 85%–95% phosphate removal, and lime is commonly applied to maintain optimum pH and aid coagulation. The ferric iron reacts with both natural alkalinity and lime to precipitate as ferric hydroxide (Eqs. 11.40 and 11.41).

The least expensive and most common source of iron coagulants is waste pickle liquor from the steel industry [13]. Pickle liquor is variable in composition depending on the metal treatment process. Ferrous sulfate resulting from pickling with sulfuric acid and ferrous chloride from pickling with hydrochloric acid are the two common waste liquors from metal finishing. Waste pickle liquors have an iron content of 5%–10% and free acid in the range of 0.5%–15%. The addition of lime or sodium hydroxide is necessary for neutralization and pH adjustment.

Aluminum and iron coagulants can be mixed with raw wastewaters to precipitate phosphates in primary clarification. This point of chemical addition, however, is not as common as secondary chemical–biological processing. The advantage of first-stage chemical settling is increased suspended-solids removal, thus reducing the organic load to the biological secondary. The major disadvantage is higher coagulant usage for the same degree of phosphorus removal. In many plants, chemical feeders permit coagulation at either location to allow more flexibility in operation.

Characteristics of waste chemical–biological sludge are influenced by the coagulant dosage, the nature of the wastewater solids, the system design, and operating conditions. The quantity of chemical precipitate produced can be estimated using the theoretical equations for a given dosage, and the organic residue calculated as described in Section 13.2. Addition of metal salts improves settleability of microbial floc resulting in a denser waste slurry. Still, it is difficult to predict a specific value for the sludge-volume index because of many other influencing factors. Chemical residues do not hamper sludge thickening and stabilization by any conventional methods; in fact, mechanical dewatering is improved. Studies indicate that coagulation precipitates do not release phosphate back into solution during biological digestion [14]. Of course, the greater inert content of inorganic–organic sludge will influence ultimate disposal by incineration.

## Lime Precipitation of Raw Wastewater

Lime applied prior to primary clarification (Fig. 14.7b) precipitates phosphates and hardness cations along with organic matter. Reaction with alkalinity (Eq. 14.7) consumes most of the lime and produces calcium carbonate residue that aids in settling suspended solids. Calcium ion also combines with orthophosphate in an alkaline solution to form gelatinous calcium hydroxyapatite (Eq. 14.8). Treating domestic wastewater requires a dosage of 100–200 mg/l as calcium hydroxide to remove 80% of the phosphate. The actual amount applied depends primarily on phosphorus concentration and hardness of the wastewater.

$$Ca(HCO_3)_2 + Ca(OH)_2 = 2CaCO_3\downarrow\ + 2H_2O \tag{14.7}$$

$$5Ca^{2+} + 4OH^- + 3HPO_4^{2-} = Ca_5(OH)(PO_4)_3\downarrow\ + 3H_2O \tag{14.8}$$

The system sequence of lime precipitation followed by activated-sludge treatment, rather than reversing their order, uses less lime when an effluent of low-phosphorus concentration is desired. The principal reason is that a biological system can readily extract low concentrations of phosphorus that would require excessive lime addition to precipitate. Secondary treatment by complete-mixing activated sludge is not adversely affected by chemical pretreatment. With proper control exercised, microbial production of carbon dioxide in the activated-sludge unit is sufficient to maintain a pH near neutral in the aeration compartment. A chemical–biological system of phosphate extraction can remove 90%–95% of the total phosphorus from a domestic wastewater containing lime dosages generally less than 150 mg/l as calcium hydroxide.

Use of excess lime in chemical treatment has two potential problems: scale formation on tanks, pipes, and other equipment, and disposal of the large quantity of lime sludge produced. Only operation of a full-scale installation will reveal the significance of these possible troubles. The quantity of sludge produced is about 1.5–2 times that obtained by conventional treatment.

## Tertiary Treatment by Chemical Precipitation

This process, delineated in Fig. 14.7c, is biological secondary treatment followed by chemical treatment with a flow diagram similar to that used in processing surface-water supplies. The mixing and sedimentation system can consist of either separate rapid mix, flocculation, and sedimentation units in series or a flocculator–clarifier with these three operations in a single-compartmented tank. Filters are usually multimedia beds operated by pressure or gravity flow. Possible chemical additives are lime, alum, ferric chloride, and ferric sulfate, with polymers as flocculation aids. A major design consideration of tertiary treatment is processing and disposal of settled sludge and filter backwash water.


## 14.10  TRACING PHOSPHORUS THROUGH TREATMENT PROCESSES

Tracing phosphorus through a hypothetical treatment plant assists in understanding the mechanisms of phosphate removal. In order to do this, the phosphorus content of primary and biological sludges must be estimated. The phosphorus in dry primary-sludge solids from settling domestic wastewater is in the range of 0.4%–1.3% with a typical value of 0.9%, which is used in subsequent calculations. The phosphorus content of waste-activated sludge is more variable; nevertheless, recorded values are often in the 1.5%–2.5% range. In the following calculations, biological sludge solids are assumed to contain 2.0% phosphorus on a dry-weight basis.


### Biological Phosphorus Removal

Consider a conventional activated-sludge plant with primary clarification. Assume an influent with the characteristics of an average domestic wastewater as listed in Table 14.1. A flow diagram for the plant is shown in Fig. 14.9 with the influent concentrations of BOD, suspended solids (SS), insoluble organic phosphorus (oP), soluble inorganic phosphorus (iP), and total phosphorus (tP). Sedimentation reduces the wastewater BOD to 130 mg/l (35% removal), SS to 120 mg/l (50% removal), and tP to 8.9 mg/l (11% removal). From the sedimentation of 1 liter of wastewater, the quantity of sludge solids is 120 mg of SS containing $0.009 \times 120 = 1.1$ mg of oP.

Next, consider biological phosphorus removal by the activated-sludge process. Based on the method presented in Section 13.2, the waste biological solids produced per liter of wastewater treated using $k = 0.5$ (Fig. 13.1) and an applied BOD of 130 mg/l is $0.5 \times 130 = 65$ mg of SS. Assuming a phosphorus content of 2.0%, the oP in the waste sludge is $0.02 \times 65 = 1.3$ mg.

The tP in the plant effluent equals the influent (tP = 10.0 mg/l) minus the oP removals in the primary and waste-activated sludges (1.1 and 1.3 mg/l, respectively) for a remainder of 7.6 mg/l. The oP concentration in the effluent is 2.0% of the 30 mg/l of SS for 0.6 mg/l, which leaves an iP of 7.0 mg/l. Overall phosphorus reduction, assuming none of the phosphorus in the sludge is recycled to the plant in return flows, is from 10 to 7.6 mg/l, for a removal efficiency of 24%.

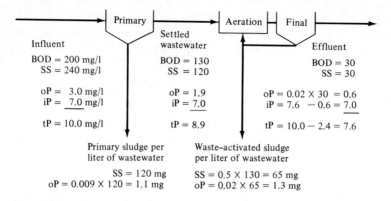

**Figure 14.9** Tracing phosphorus through a conventional biological treatment plant.

## Biological–Chemical Phosphorus Removal

Figure 14.10 is a flow diagram for a hypothetical biological–chemical treatment plant using alum addition in the activated-sludge process to precipitate the inorganic phosphate. (Phosphorus removal in primary sedimentation is the same as in Fig. 14.9.) The required alum dosage was established by gradually increasing the amount applied while monitoring the total phosphorus in the plant effluent. For this hypothetical plant, the minimum alum dosage to reduce the effluent phosphorus concentration from 7.6 mg/l without coagulation to the maximum allowable tP of 1.0 mg/l is assumed to be 110 mg/l alum (molecular weight, 600). Addition of this coagulant also enhances the SS removal, reducing the effluent SS from 30 to 15 mg/l. The calculated

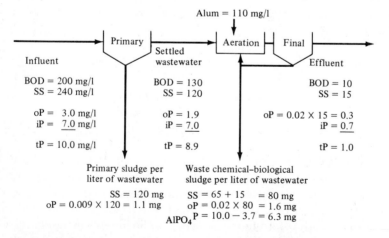

**Figure 14.10** Tracing phosphorus through a biological–chemical treatment plant.

oP in the effluent is therefore $0.02 \times 15 = 0.3$ mg/l, and the iP concentration is $1.0 - 0.3 = 0.7$ mg/l.

Now consider phosphorus removal by the biological–chemical process. The waste biological solids include 65 mg of SS resulting from the applied BOD (Fig. 14.9) plus the 15 mg/l from improved SS removal, for a total of 80 mg. The calculated oP in this organic sludge is $0.02 \times 80 = 1.6$ mg. The quantity of phosphorus precipitated by the alum as $AlPO_4$ (Eq. 14.5) is the influent tP (10.0 mg/l) minus the effluent tP (1.0 mg/l) and the oP removal in the organic sludge solids $(1.1 + 1.6$ mg) for a remainder of 6.3 mg. The overall phosphorus removal efficiency is 90%. For an alum dosage of 110 mg/l and phosphorus feed concentration of 8.9 mg/l, the weight ratio of aluminum to phosphorus applied to the activated-sludge process is

$$\frac{Al}{P} = \frac{110(2 \times 27/600)}{8.9} = \frac{110 \times 0.09}{8.9} = \frac{1.1}{1.0}$$

Example 14.3 illustrates the procedure for calculating the theoretical sludge production for this biological–chemical treatment plant.

### ■ EXAMPLE 14.3
Calculate the theoretical sludge solids production for the hypothetical biological–chemical plant diagrammed in Fig. 14.10. Assume the influent wastewater characteristics listed in Fig. 14.10, primary removals of 50% SS and 35% BOD, an operating F/M in the activated-sludge process of 0.40, an alum dosage of 110 mg/l, 15 mg/l of effluent SS, and an effluent phosphorus concentration of 1.0 mg/l.

*Solution.*  The following calculations express sludge solids production in terms of milligrams per liter of wastewater treated:

$$\text{SS removal in primary sedimentation} = 0.50 \times 240 = 120 \text{ mg}$$

$$\text{SS removal in the activated-sludge process} = k \times \text{BOD} \quad \text{(Eq. 13.6)}$$

From Fig. 13.1, $k = 0.50$ for an F/M of 0.40, and therefore

$$\text{SS removal} = 0.50 \times 0.65(200) = 65 \text{ mg}$$

Since the $k$ from Fig. 13.1 assumes an effluent SS of 30 mg/l and the actual effluent concentration is 15 mg/l,

$$\text{total SS removal} = 65 + (30 - 15) = 80 \text{ mg}$$

From Eq. 14.5,

$$AlPO_4 \text{ precipitate} = \frac{(\text{P precipitated})(\text{MW of } AlPO_4)}{(\text{MW of P})}$$

Based on the 6.3 mg of phosphorus precipitate calculated in Fig. 14.10,

$$AlPO_4 \text{ precipitate} = \frac{(6.3)(122)}{31} = 25 \text{ mg}$$

From Eq. 14.5,

$$\text{unused alum} = \text{alum dosage} - \frac{(\text{P precipitated})(\text{MW of alum})}{(2 \times \text{MW of P})}$$

$$= 110 - \frac{6.3 \times 600}{2 \times 31} = 49 \text{ mg/l}$$

The alum not used in phosphorus precipitation reacts with the natural alkalinity to precipitate as $Al(OH)_3$. From Eq. 11.34 in Section 11.9,

$$Al(OH)_3 \text{ precipitate} = \frac{(\text{unused alum})[2 \times \text{MW of } Al(OH)_3]}{(\text{MW of alum})}$$

$$= \frac{(49)(2 \times 78)}{600} = 13 \text{ mg}$$

The total sludge solids production equals the sum of the primary SS removal, activated-sludge SS removal, effluent SS reduction below 30 mg/l, $AlPO_4$ precipitate, and $Al(OH)_3$ precipitate. For this hypothetical plant, solids production per liter of wastewater treated equals $120 + 65 + 15 + 25 + 13 = 238$ mg. ∎

# Nitrogen Removal

Most nitrogen found in surface waters is derived from land drainage (3–24 lb of N/acre/year) and dilution of wastewater effluents. Feces, urine, and food-processing discharges are the primary sources of nitrogen in domestic waste with a per capita contribution in the range of 8–12 lb of N/yr. About 40% is in the form of ammonia and 60% bound in organic matter. Conventional primary and secondary processing extracts approximately 40% of the total nitrogen, leaving most of the remainder as ammonia in the effluent.

The nitrogen forms of interest are organic, inorganic, and gaseous nitrogen. Bacterial decomposition releases ammonia by deamination of nitrogenous organic compounds (Eq. 14.9), and continued aerobic oxidation results in nitrification (Eq. 14.10). Equation 14.11 is biochemical denitrification that occurs with heterotrophic metabolism in an anaerobic environment. These three reactions in sequence define the biological nitrification–denitrification process. Water-soluble inorganic nitrogens ($NH_3$, $NO_2^-$, $NO_3^-$) serve as plant nutrients in photosynthesis (Eq. 14.12). Finally, ammonia can be air stripped from solution at high pH.

$$\text{Organic N} \xrightarrow[\text{decomposition}]{\text{bacterial}} NH_4^+ \tag{14.9}$$

$$NH_4^+ + O_2 \xrightarrow[\text{bacteria}]{\text{nitrifying}} NO_2^- \tag{14.10a}$$

$$NO_2^- \xrightarrow[\text{bacteria}]{\text{nitrifying}} NO_3^- \tag{14.10b}$$

$$NO_3^- \xrightarrow[\text{denitrification}]{\text{bacterial}} N_2 \uparrow \tag{14.11}$$

$$\text{Inorganic N} + CO_2 \xrightarrow{\text{sunlight}} \text{green plants} \qquad (14.12)$$

$$NH_4OH \xrightarrow[\text{at basic pH}]{\text{air stripping}} NH_3 \uparrow \qquad (14.13)$$

## 14.11 TRACING NITROGEN THROUGH TREATMENT PROCESSES

Tracing nitrogen through a hypothetical treatment plant assists in understanding the transformations of the various forms of nitrogen. For this purpose, nitrogen contents must be assumed for the raw wastewater and waste sludges. Wastewater characteristics are assumed to be those of an average domestic wastewater as listed in Table 12.1. The nitrogen content of dry sludge solids in primary sludge is in the range 2%–4%, that in waste-activated sludge is 2%–6%, and that in anaerobically digested solids is 2%–6%. For the subsequent calculations, the values are assumed to be 4.0% for primary solids, 6.0% for activated sludge, and 4.0% for digested solids.

Consider a conventional activated-sludge plant with primary clarification. Figure 14.11 is a flow diagram listing the concentrations of BOD, suspended solids (SS), insoluble organic nitrogen (org-N), soluble ammonia–nitrogen ($NH_3$–N), soluble nitrate–nitrogen ($NO_3$–N), and total nitrogen (tN). Sedimentation reduces the wastewater BOD to 130 mg/l (35% removal), SS to 120 mg/l (50% removal), and tN to 30 mg/l (14% removal). From the sedimentation of 1 liter of wastewater, the quantity of sludge solids is 120 mg of SS containing $0.04 \times 120 = 5$ mg/l of org-N.

Next consider biological nitrogen removal by the activated-sludge secondary. In the method presented in Section 13.2, the waste biological solids produced per liter of wastewater treated, assuming $k = 0.5$ (Fig. 13.1) and an applied BOD of 130 mg/l, is $0.5 \times 130 = 65$ mg/l of SS. The nitrogen content at 6.0% is org-N $= 0.06 \times 65 = 4$ mg/l. The org-N in the effluent is 6.0% of the SS, which is $0.06 \times 30 = 2$ mg/l. During biological metabolism in the activated-sludge process, nitrogen in the waste organic matter is released to solution in the form of ammonia (deamination). Therefore, the remainder of the 26 mg/l of tN in the effluent is 24 mg/l of $NH_3$–N.

Aeration in the activated-sludge process can induce nitrification converting a portion of the $NH_3$–N to $NO_3$–N. The occurrence of nitrification, and the degree to which it proceeds, depends on the environmental and operating conditions including temperature, dissolved oxygen concentration, and sludge age. In Fig. 14.11 the diagram with substantial nitrification assumes that the effluent inorganic nitrogen concentrations are 5 mg/l of $NH_3$–N and 15 mg/l of $NO_3$–N. When the wastewater in a final clarifier becomes anaerobic, microorganisms can use the oxygen in the nitrate for respiration, releasing nitrogen gas. In this example, denitrification is assumed to release 5 mg/l of $N_2$–N, reducing the effluent tN to 22 mg/l.

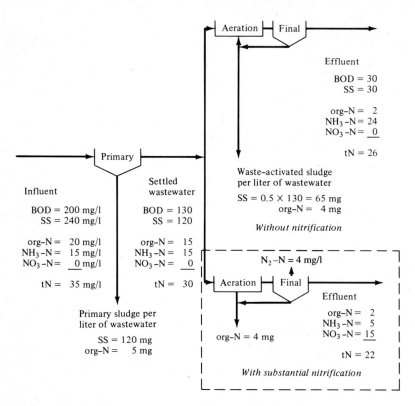

**Figure 14.11** Tracing nitrogen through a conventional biological treatment plant without nitrification and with substantial nitrification in activated-sludge aeration.

Figure 14.11 is based on the assumption that none of the nitrogen withdrawn in the waste sludges is recycled to the plant in return flows. While this is possible by mechanical dewatering of the raw sludges, biological stabilization by either anaerobic or aerobic digestion releases nitrogen to supernatant returned to the plant influent. Anaerobic digestion is likely to return 40%–50% of the org-N in the sludge solids as ammonia, 4 mg/l of the 9 mg/l removed in the wastewater processing. Depending on operation, aerobic digestion can result in nitrogen loss through nitrification–denitrification.

Nitrogen removal is affected by several factors: The forms and concentrations in the raw wastewater, synthesis in aerobic treatment, nitrification–denitrification, and methods of sludge processing. From the scheme in Fig. 14.11, without nitrification and assuming a return of 4 mg from anaerobic digestion, the effluent tN would be 30 mg/l for a removal of 5/35 = 14%. With nitrification and no nitrogen return from sludge processing, the removal would be (35 − 22)/35 = 37%.

## 14.12 BIOLOGICAL NITRIFICATION

Nitrification does not remove ammonia but converts it to the nitrate form, thereby eliminating problems of toxicity to fish and reducing the nitrogen oxygen demand (NOD) of the effluent. Ammonia oxidation to nitrate is a diphasic process performed by autotrophic bacteria with nitrate as an intermediate product (Eq. 14.10). These aerobic reactions yield energy for metabolic functions such as synthesis of carbon dioxide into new cell growth. Conversion of ammonia to nitrite is the rate-limiting step that controls the overall reaction; therefore, nitrite concentrations normally do not build up to significant levels. The rate of nitrification in wastewater, being essentially linear, is a function of time and independent of ammonia–nitrogen concentration (zero-order kinetics):

$$NH_4^+ + 1.5O_2 \xrightarrow{Nitrosomonas} NO_2^- + 2H^+ + H_2O + energy \qquad (14.14)$$

$$NO_2^- + 0.5O_2 \xrightarrow{Nitrobacter} NO_3^- + energy$$

Temperature, pH, and dissolved oxygen concentration are important parameters in nitrification kinetics studied by Wild, Sawyer, and McMahon [15]. The nitrification rate decreases by about one half for every 10°–12°C temperature drop above 10°C and then decreases more rapidly in cold wastewater such that lowering the temperature from 10° to 5°C halves the rate of ammonia oxidation. Based on temperature alone, a winter aeration period would have to be several times longer than in the summer; however, this seasonal effect can be overcome to a considerable degree by increasing mixed-liquor suspended solids (MLSS) and adjusting pH to a more favorable level. The optimum pH for nitrification is 8.2–8.6, with 90% of the maximum occurring at 7.8 and 8.9, and less than 50% of optimum below 7.0 and above 9.8 (Fig. 14.12). The laboratory studies also showed no detectable inhibition of nitrification at dissolved-oxygen levels exceeding 1.0 mg/l, or at ammonia–nitrogen concentrations up to 60 mg/l.

Sludge age and temperature are interrelated factors in establishing and maintaining healthy nitrifier populations essential to efficient ammonia oxidation. In continuous-flow aeration systems, a long sludge age (retention time) is required to prevent excessive loss of viable bacteria (i.e., the growth rate must be rapid enough to replace microbes lost through sludge wasting and washout in the plant effluent). The supply of organic matter controls the growth of heterotrophic organisms, while the quantity of ammonia applied governs synthesis of nitrifiers. Increased sludge wasting, as a result of organic loading, reduces the sludge retention time and removes both heterotrophs and nitrifiers from the system. On a substrate of domestic waste having a BOD/total nitrogen ratio of approximately 200 mg/l to 35 mg/l, growth rates of nitrifying bacteria are substantially lower than those of decomposers. Therefore, in activated-sludge processes under normal operating conditions, nitrification is limited because of loss of the autotrophic populations, and

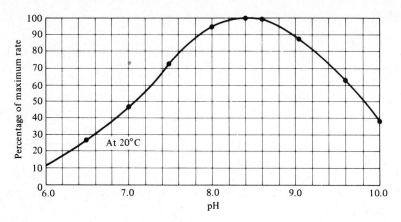

**Figure 14.12** Rate of nitrification versus pH at constant temperature. [From H. E. Wild, Jr., C. N. Sawyer, and T. C. McMahon, "Factors Affecting Nitrification Kinetics," *J. Water Poll. Control Fed.* 43(9) (1971): 1852.]

current design for biological nitrification calls for two-step treatment. The first stage reduces BOD without oxidation of the ammonia–nitrogen to produce an effluent having a lower BOD/ammonia ratio—about 40 mg/l to 25 mg/l. Applying this flow to a second-stage nitrification unit provides an adequate growth potential for nitrifiers relative to heterotrophs, since the system can be operated at an increased sludge age.

The relative reproduction rates of heterotrophic and nitrifying bacteria are also influenced to a measurable extent by temperature. In southern climates, nitrification may be possible in a single-stage extended aeration unit treating domestic waste if pH and sludge wasting are carefully controlled to compensate for the relatively high BOD/ammonia–N feed. However, winter wastewater temperature in northern states is often 10°–15°C, requiring a two-stage system with the best combination of aeration tank capacity, MLSS concentration, and pH control. Operation is possible at a less favorable pH level and lower mixed-liquor solids at a warm temperature provided the first stage is properly controlled. If nitrification is allowed to occur in carbonaceous aeration, the reduced ammonia supply to the secondary leads to starving the nitrifier populations. The sludge age, rate of recirculation, and air supply can be adjusted to minimize ammonia oxidation in the first stage; nevertheless, provision for chlorinating the effluent of the aeration tank prior to clarification is recommended in design. The addition of 2–8 mg/l of chlorine is effective in inhibiting nitrifying bacteria and can also help control sludge bulking caused by denitrification in the carbonaceous-phase clarifier.

## Nitrification by Suspended-Growth Systems

This discussion is based on tentative design criteria for ammonia oxidation [16]. The optimum aeration tank is either a long, narrow, spiral-flow basin with diffused aeration or a shorter tank divided into a series of at least three compartments for diffused or mechanical aeration equipment, with intervening ports (Fig. 14.13). This tank configuration simulates plug flow compatible with the zero-order kinetics of ammonia oxidation.

Biological nitrification destroys alkalinity, which can result in a drop of pH when processing wastewaters of moderate hardness, or where alum precipitation has been used for phosphate removal in the preceding activated-sludge phase. Theoretically, 7.2 lb of alkalinity is destroyed per pound of ammonia–nitrogen oxidized to nitrate, as follows:

$$2NH_4HCO_3 + 4O_2 + Ca(HCO_3)_2 = Ca(NO_3)_2 + 4CO_2 + 6H_2O$$

$$(14.15)$$

Whether pH should be controlled by chemical addition depends on the rate of nitrification desired, as limited by other environmental conditions. For example, in operation at low temperature, lime can be applied to maintain oxidation efficiency for the aeration tank capacity available. New plant design should provide for installation of chemical feeders and instrumentation for monitoring pH in the aeration basin.

The recommended design mixed-liquor concentration for a nitrification process receiving normal secondary effluent is in the range 1500–2000 mg/l of volatile suspended solids. Mean cell residence time must be greater than in a carbonaceous activated-sludge process, and quantitative values available indicate that a sludge age up to 20 days is needed.

Loading on nitrification basins is expressed in units of lb ammonia–N/ 1000 ft³/day of aeration tank volume. Figure 14.14 shows recommended loadings for various temperatures and volatile mixed-liquor concentrations (MLVSS) at pH 8.4, based on studies at Marlboro, Massachusetts. Corrections for permissible loadings at pH values other than 8.4 can be taken from Fig. 14.12. The value selected for design loading should include a factor to allow for reasonable peaking of the influent ammonia content. A commonly adopted peak loading is 1.5 times the average daily nitrogen load under low-temperature conditions.

Stoichiometrically, nitrification of 1.0 lb of ammonia–N in the form of

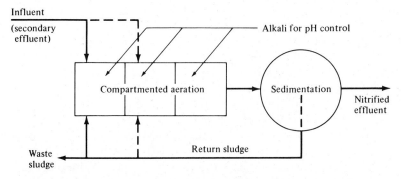

**Figure 14.13** Flow diagram for nitrification by diffused or mechanical aeration of wastewater following conventional biological treatment.

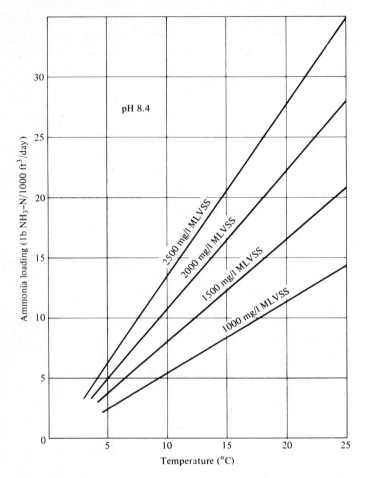

**Figure 14.14** Permissible nitrification-tank loadings at an optimum pH of 8.4. [From *Nitrification and Denitrification Facilities,* Environmental Protection Agency, Technology Transfer (August 1973): 22.]

ammonium bicarbonate requires 4.6 lb of oxygen (Eq. 14.15); however, additional oxygen allowance must be made for carbonaceous BOD carried over to the nitrification stage. A dissolved oxygen concentration of 3.0 mg/l in the mixed liquor is suggested under average loading conditions with a lower concentration permitted during peak loads, but not below 1.0 mg/l.

Suggested design criteria for final clarifiers following nitrification are an overflow rate of 400–500 gpd/ft$^2$ (16–20 m$^3$/m$^2$·d) based on average daily discharge, with a maximum permissible value of 1000 gpd/ft$^2$ (41 m$^3$/m$^2$·d) at peak hourly flow, and a side-water depth of at least 10 ft (3.0 m). Because of the relatively slow settling velocities of nitrifying sludges, more than two clarifiers are desirable to ensure satisfactory operation when one tank is out of service for maintenance. Hydraulic-type collector arms are recommended for rapid sludge return since denitrification can occur in settled sludge, creating problems of floating solids. When present, the float should be collected by skimmers and returned to the aeration basins. A desirable capacity for sludge recirculation pumps is 100% of the raw-wastewater influent, although normal operation will probably be at a rate of only 50% return. Accumulation of biological floc is limited by the low organic loading and slow growth of nitrifying bacteria. Consequently, the volume of excess sludge produced is small, with a reasonable estimate being less than 1% of the quantity of wastewater processed.

## Nitrification by Fixed-Growth Systems

Although suspended-growth aeration is used extensively, fixed-growth systems have been employed for nitrification. Rotating biological contactors for nitrification are operated in multistage at low hydraulic loadings that have been reduced to compensate for low-temperature conditions. Nitrifying bacteria adhering to the media must be in sufficient numbers to provide ammonia oxidation in a single passage of wastewater flow since the RBC process does not include sludge recirculation. Furthermore, to prevent interference by heterotrophic growths covering the disk surfaces, the BOD of the applied wastewater should be relatively low, preferably not exceeding 20–30 mg/l. Biological towers with plastic packing or redwood media rely on recirculation to provide adequate contact time for nitrification. Underflow from clarification of the tower effluent is recirculated for uniform hydraulic loading on the packing and for return of settled floc to increase the mass of nitrifying bacteria in contact with the wastewater.

■ **EXAMPLE 14.4**

Calculate the aeration basin volume for suspended-growth nitrification following conventional secondary treatment. The wastewater characteristics are the following:

$$\text{average daily design flow} = 10 \text{ mgd}$$
$$\text{ammonia–nitrogen} = 20 \text{ mg/l}$$
$$\text{BOD} = 40 \text{ mg/l}$$
$$\text{minimum operating temperature} = 10°C$$
$$\text{operating pH} = 7.8$$
$$\text{design volatile MLSS} = 1500 \text{ mg/l}$$

*Solution*

$$\text{average ammonia load} = 10 \times 20 \times 8.34 = 1670 \text{ lb/day of N}$$

$$\text{maximum ammonia load} = 1.5 \times 1670 = 2500 \text{ lb/day of N}$$

Permissible nitrification tank loading from Fig. 14.14 for a temperature of 10°C and 1500 mg/l of MLVSS is 8.1 lb of $NH_3$–N/1000 $ft^3$/day. Correcting this to a pH of 7.8 using Fig. 14.12, the allowable loading is 8.1 × 0.88 = 7.1 lb/1000 $ft^3$/day.

$$\text{aeration basin volume} = 2500 \frac{1000}{7.1} = 350{,}000 \text{ ft}^3$$

$$\text{resulting aeration period} = \frac{350{,}000 \times 7.48 \times 24}{10{,}000{,}000} = 6.3 \text{ hr}$$

$$\text{BOD load on aeration basin} = 10 \times 40 \times 8.34 = 2300 \text{ lb/day}$$

The oxygen uptake using 4.6 lb of $O_2$/lb of $NH_3$–N and 1.0 lb of $O_2$/lb of BOD is 2500 × 4.6 + 2300 × 1.0 = 13,800 lb/day.                                   ■

## 14.13  BIOLOGICAL DENITRIFICATION

Nitrite and nitrate are bacterially reduced to gaseous nitrogen by a variety of facultative heterotrophs in an anaerobic environment. An organic carbon source, such as acetic acid, acetone, ethanol, methanol, or sugar, is needed to act as a hydrogen donor (oxygen acceptor) and to supply carbon for biological synthesis. Certain autotrophic bacteria are also capable of denitrification by oxidizing an inorganic compound for energy and using carbon dioxide for synthesis. While denitrification is considered an anoxic process because it occurs in the absence of dissolved oxygen, strict anaerobiosis characterized by hydrogen sulfide and methane production is not necessary.

Methanol is the preferred carbon source because it is the least expensive synthetic compound available that can be applied without leaving a residual BOD in the process effluent—but this does not imply that methanol treatment is cheap. Introduction of methanol first reduces the dissolved oxygen present by Eq. 14.16; then biological reduction of nitrate and nitrite occurs (Eqs. 14.17 and 14.18).

$$3O_2 + 2CH_3OH = 2CO_2\uparrow + 4H_2O \tag{14.16}$$

$$6NO_3^- + 5CH_3OH = 3N_2\uparrow + 5CO_2\uparrow + 7H_2O + 6OH^- \tag{14.17}$$

$$2NO_2^- + CH_3OH = N_2\uparrow + CO_2\uparrow + H_2O + 2OH^- \tag{14.18}$$

From these reactions, the amount of methanol required as a hydrogen donor for complete denitrification is

$$CH_3OH = 0.7DO + 1.1NO_2\text{–}N + 2.0NO_3\text{–}N \tag{14.19}$$

where   $CH_3OH$ = methanol, mg/l
          DO = dissolved oxygen, mg/l
     $NO_2$–N = nitrite–nitrogen, mg/l
     $NO_3$–N = nitrate–nitrogen, mg/l

Approximately 30% excess methanol feed is needed for synthesis; hence chemical consumption to satisfy both energy and synthesis can be estimated from the relationship

$$CH_3OH = 0.9DO + 1.5NO_2\text{–}N + 2.5NO_3\text{–}N \tag{14.20}$$

Limited information is available at present on the kinetics of the denitrification reaction. The optimum pH falls in the same range as for most heterotrophic bacteria (between 6.5 and 7.5), with the rate reducing to about 80% of maximum when the pH is lowered to 6.1 or raised to 7.9. Nitrified wastewaters, which tend to be basic, are naturally controlled from excessively high pH by carbon dioxide generated in a denitrification unit; thus there appears to be no need for addition of chemicals to control pH in actual systems. The effect of temperature on the rate of denitrification is sketched in Fig. 14.15 [17].

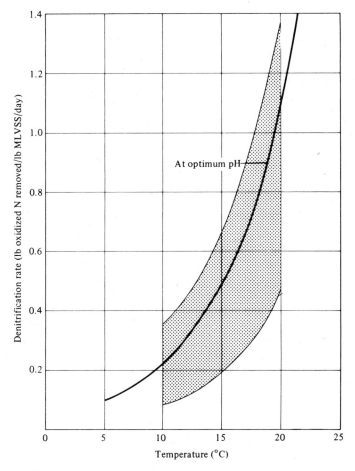

**Figure 14.15** Effect of temperature on the rate of denitrification. [From *Nitrification and Denitrification Facilities,* Environmental Protection Agency, Technology Transfer (August 1973): 28.]

## Denitrification by Suspended-Growth Systems

The process studied most extensively consists of a completely mixed basin followed by a clarifier for sludge separation and return (Fig. 14.16). Although a single, mixed chamber is common in laboratory studies, plug flow minimizes short-circuiting and becomes more suitable for the relatively short detention periods required. Underwater stirrers, comparable to those used in water-works flocculation tanks, mix the contents sufficiently to keep microbial floc in suspension without producing undue aeration. The power supply in the range of 1 hp for each 2000–4000 ft$^3$ of tank volume appears to be adequate. Whether basins should be covered to minimize absorption of oxygen is a matter of conjecture, but certainly airtight covers should be avoided.

Denitrification reactions form carbon dioxide and nitrogen gas bubbles that inhibit gravity settling by adhering to the biological floc. Supersaturation of the mixed liquor with gases can be relieved by short-term aeration in an open channel, or tank, between the denitrification basin and final clarifier. The settleability of sludge solids following this air stripping appears to be similar to that of other biological sludges. Recommended clarifier depths and overflow rates are the same as those suggested for final settling tanks following the nitrification process. Basins should be equipped with rapid-sludge-return collector arms and skimming devices. A sludge recirculation capacity equal to the average wastewater flow is recommended, and scum may be returned to the denitrification tank or routed to disposal. Withdrawal of excess microbial solids, to keep the biological system in balance, ranges from 0.2 to 0.3 lb/lb of methanol applied.

The volumetric capacity needed for denitrification can be estimated using Fig. 14.17, which is based on pilot-plant studies. Denitrifying sludges after degasification have good settling characteristics, allowing design mixed-liquor solids of 2000–3000 mg/l that are approximately 65% volatile.

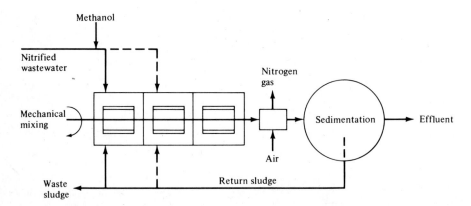

**Figure 14.16** Flow diagram for a completely mixed, compartmented denitrification basin and clarifier.

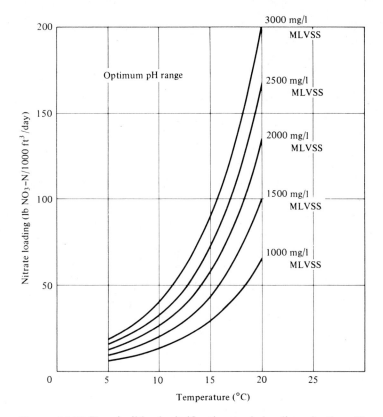

**Figure 14.17** Permissible denitrification-tank loadings in the pH range 6.5–7.5. [From *Nitrification and Denitrification Facilities,* Environmental Protection Agency, Technology Transfer (August 1973): 30.]

## Nitrification–Denitrification

A two-stage system composed of the units illustrated in Figs. 14.13 and 14.16, following secondary biological treatment, can achieve about 90% inorganic nitrogen reduction and 80%–85% total nitrogen removal under normal operating conditions. The biological cultures performing ammonia oxidation are more sensitive to heavy metals and organic toxins than conventional activated sludge. Therefore, industrial wastes discharged to municipal sewers must be carefully monitored and necessary controls established to ensure that the nitrifying microorganisms are not inhibited.

### ■ EXAMPLE 14.5
Based on the following data, calculate the volume needed for suspended-growth denitrification.

$$\text{average daily design flow} = 10 \text{ mgd}$$

$$\text{nitrate–nitrogen} = 20 \text{ mg/l of N}$$

$$\text{dissolved oxygen} = 8 \text{ mg/l}$$
$$\text{minimum operating temperature} = 8°C$$
$$\text{operating pH} = 7.8$$
$$\text{design volatile MLSS} = 2000 \text{ mg/l}$$

*Solution*

$$\text{average nitrate load} = 10 \times 20 \times 8.34 = 1670 \text{ lb/day of N}$$

$$\text{maximum nitrate load} = 1.5 \times 1670 = 2500 \text{ lb/day of N}$$

Permissible denitrification tank loading from Fig. 14.17 for a temperature of 8°C and 2000 mg/l is 20 lb of $NO_3$–N/1000 ft³/day. Correcting this to a pH of 7.8, the allowable loading is $20 \times 0.9 = 18$ lb/1000 ft³/day.

$$\text{denitrification basin volume} = 2500 \, \frac{1000}{18} = 140{,}000 \text{ ft}^3$$

$$\text{resulting detention time} = \frac{140{,}000 \times 7.48 \times 24}{10{,}000{,}000} = 2.5 \text{ hr}$$

The average methanol dosage for 8 mg/l of DO and 20 mg/l of $NO_3$–N, from Eq. 14.20, is

$$CH_3OH = 0.9 \times 8 + 2.5 \times 20 = 57 \text{ mg/l} \qquad \blacksquare$$

## 14.14 BIOLOGICAL NITRIFICATION–DENITRIFICATION

Rather than supplying a chemical like methanol as a carbon compound for denitrification, one can use the organic matter in wastewater as an oxygen acceptor for nitrate reduction. Several processes have been proposed for biological nitrification–denitrification. All of them rely on an aerobic zone for nitrification and an anoxic zone (a zone lacking in oxygen) for denitrification.

The activated-sludge process, shown in Fig. 14.18, blends nitrified recir-

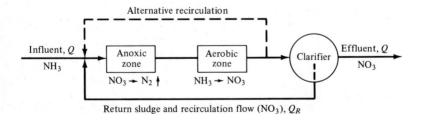

**Figure 14.18** Biological nitrification–denitrification by recirculation of return sludge for blending with raw wastewater before flow through an anoxic zone and subsequent aerobic zone.

culation with raw wastewater for denitrification in the anoxic zone and nitrification in the aerobic zone. Both zones reduce the BOD content of the wastewater. The anoxic zone provides the oxygen in nitrate for biological respiration, releasing nitrogen gas. In the aerobic zone, the BOD is reduced by microbial uptake of dissolved oxygen, and the majority of the nitrogen is converted to nitrate. Since only that nitrate returned to the process influent is denitrified, the degree of nitrogen removal is limited by the rate of recirculation and the nitrate respiration achievable in the anoxic zone. Ideally, with high recirculation and efficient denitrification in the anoxic zone, nitrogen removal could be 70%–80% in processing domestic wastewater [18]. In reality, the removal is more likely to be about 50% for operational reasons. For instance, the recirculation flow cannot reduce excessively the hydraulic detention time in the nitrification–denitrification process. The relative concentrations of BOD and nitrogen also influence the efficiency of nitrogen removal since the oxygen demand of the wastewater controls the rate and extent of microbial nitrate respiration. A reduction of 1 mg/l of $NO_3$–N theoretically satisfies 3.4 mg/l of BOD, which is the oxygen/nitrogen ratio in nitrate.

Nitrification–denitrification by modifying a plug-flow activated-sludge process was studied by the Water Research Centre, Great Britain [19]. The laboratory units, illustrated schematically in Fig. 14.19, had a series of four completely mixed reactors to simulate plug flow in an aeration tank and to allow isolation of an anoxic zone. The influent was a settled municipal wastewater with an average BOD of 240 mg/l and total nitrogen of approximately 75 mg/l. Operating parameters are listed on the diagrams. The nitrogen concentrations are expressed as percentages of the total influent nitrogen.

Figure 14.19a is the flow scheme and nitrogen data for operation as a normal activated-sludge process. A mass balance based on measurements of nitrogen in the influent, effluent, and waste sludge showed an unaccountable loss of 14%. This was assumed to result from nitrate respiration in the final clarifier resulting in denitrification. Total nitrogen removal was 33% (100 to 67), with 93% of the effluent nitrogen in the form of nitrate.

Figure 14.19b is the flow scheme and nitrogen data for operation as an activated-sludge process with denitrification. The first reactor was converted to an anoxic zone by replacing the air supply with a stirrer to blend the wastewater influent with the return activated sludge. The total nitrogen removal averaged 59% (100 to 41), and the nitrate content in the effluent was 40% less than during normal operation.$([(62 - 37)/62] \times 100 = 40\%)$.

Following these laboratory studies, a full-scale research project was undertaken at the Rye Meads wastewater plant, Thames Water Authority, Great Britain [19, 20]. The plug-flow aeration tanks were modified to provide anoxic zones for denitrification. The arrangements of anoxic zones and wastewater feed are diagrammed in Fig. 14.20; each aeration tank received a settled wastewater flow of approximately 9000 $m^3$/d. Unit 1 was modified to create a mechanically stirred (unaerated) anoxic zone in the first one-fourth of the tank; this was the same process tested in the laboratory (Fig. 14.19). Unit 2

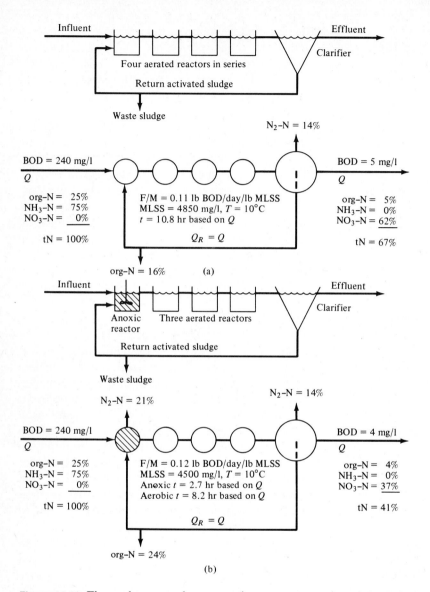

**Figure 14.19** Flow schemes and average nitrogen amounts as percentages of influent nitrogen from laboratory testing of (a) normal activated-sludge process with nitrification and (b) modified activated-sludge process with nitrification–denitrification [19].

was the same arrangement except that mixing in the anoxic zone was obtained by reduced aeration rather than by mechanical stirring. Two-thirds of the air diffusers had been removed from this section of the tank. Nitrate reductions in these units relative to normal activated-sludge processing were both in the range 47%–49%.

Units 3 and 4, depicted in Fig. 14.20, were placed into operation with

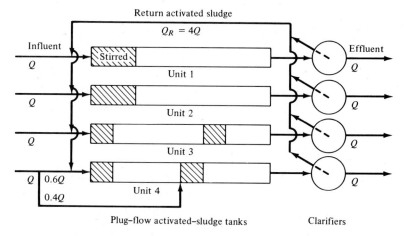

**Figure 14.20** Modified activated-sludge arrangements tested for nitrification–denitrification efficiency at the Rye Meads wastewater plant, Thames Water Authority, Great Britain. Shaded areas: anoxic zones; unshaded rectangles: aerobic zones.

two anoxic zones, each one-eighth of the tank length, separated by an aerobic zone. The anoxic zones were mixed by reduced aeration with three-quarters of the diffusers removed. As in previous schemes, the first anoxic zone was to reduce nitrate in the return activated sludge. The purpose of the second anoxic zone was to increase denitrification efficiency by nitrate respiration of the influent ammonia oxidized in the immediately preceding aerobic zone. Unit 3, when operated with all of the wastewater entering at the inlet of the tank, did not achieve any greater denitrification than units 1 and 2. This was attributed to difficulty in achieving an anoxic condition in the second zone. In the operation of unit 4, anoxia was created in the second zone by applying 40% of the influent wastewater directly to the head of this zone. Under these operating conditions, the nitrate removal was 54% greater than normal activated-sludge treatment, which is approximately 6% greater than the removals in units 1 and 2. Denitrification was expected to be greater than the observed 54%. Based on subsequent laboratory testing, the process with dual anoxic zones—which were mechanically stirred—achieved up to 70% nitrate reduction using a similar mode of operation. Apparently, mixing the anoxic zone with diffused air limited nitrate respiration.

After the field studies at Rye Meads, the Water Research Centre, Great Britain, conducted a laboratory investigation to evaluate the use of dual anoxic zones in a modified activated-sludge process [20]. As shown in Fig. 14.21, the laboratory system consisted of a series of eight equal-sized compartments followed by a clarifier. Underflow from the final clarifier was returned to the first reactor, and the first and fifth compartments were stirred anoxic reactors receiving influent wastewater. A second laboratory system with eight aerated compartments (no anoxic zones) was operated as a control

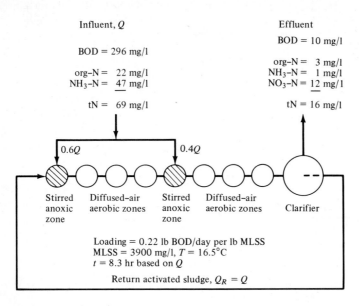

Influent, $Q$

BOD = 296 mg/l

org–N = 22 mg/l
NH$_3$–N = 47 mg/l

tN = 69 mg/l

Effluent

BOD = 10 mg/l

org–N = 3 mg/l
NH$_3$–N = 1 mg/l
NO$_3$–N = 12 mg/l

tN = 16 mg/l

0.6$Q$          0.4$Q$

Stirred      Diffused–air      Stirred      Diffused–air
anoxic       aerobic zones     anoxic       aerobic zones      Clarifier
zone                           zone

Loading = 0.22 lb BOD/day per lb MLSS
MLSS = 3900 mg/l, $T$ = 16.5°C
$t$ = 8.3 hr based on $Q$
Return activated sludge, $Q_R = Q$

**Figure 14.21** Flow scheme of the laboratory-scale modified activated-sludge system tested for nitrification–denitrification efficiency using two anoxic zones. The division of flow between anoxic zones and nitrogen concentrations in the effluent is given for optimum denitrification based on the listed operating conditions.

unit to determine the effluent nitrate concentrations processed by a normal activated-sludge process. In a series of test runs the proportions of wastewater applied to the fifth compartment were 0%, 30%, 35%, 40%, and 50% of the influent. At zero flow to the second anoxic zone, only the nitrate in the return activated sludge was denitrified for a reduction of 41% of the oxidized nitrogen in the effluent. Anoxia in the second zone was inadequate to produce significant denitrification. At a 30% feed rate, the percentage of nitrate removal increased to 52%. When 35%–40% of the influent was applied to the second anoxic zone, the degree of denitrification increased to 68%–70% since part of the nitrate produced from the wastewater fed to the first anoxic zone was removed in the second. The operating conditions and wastewater data for a 40% portion applied to the second zone are given in Fig. 14.21. (The normal activated-sludge effluent from the control unit contained 40 mg/l of NO$_3$–N as compared to 12 mg/l from the modified process.) At a 50% feed rate to the second zone, the performance deteriorated to 40% nitrate reduction since insufficient BOD was applied to the first anoxic zone to satisfy the nitrate respiration demand.

Two other modes of operation were tested using this laboratory apparatus. First, with a single initial anoxic zone consisting of compartments 1 and 2, the sludge recycle was doubled, so $Q_R = 2Q$. The measured nitrate removal was 66% because more effluent nitrate was being returned for denitrification. The second test was conducted with a 60%–40% division of influent between the first and second anoxic zones and a sludge recycle of 2$Q$. Compared with the normally operated process, the nitrate reduction was 77% with an effluent nitrate level of 9 mg/l; however, this was really a false value because the effluent contained 4 mg/l of $NH_3$–N, which is an increase of 3 mg/l from previous tests. Thus, the nitrate reduction relative to the other tests should be considered to be approximately 70%. The additional ammonia–nitrogen appeared in the effluent since the detention times in the aerobic zones were too short to achieve complete nitrification.

## 14.15 AMMONIA STRIPPING

Equilibrium of the ammonium ion and dissolved ammonia gas in water is controlled by both pH and temperature, as shown in Fig. 14.22. Only $NH_4^+$ ions are present in neutral solution at ambient temperatures, while at a pH

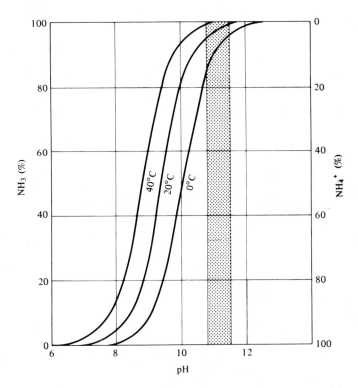

**Figure 14.22** Distribution of ammonia and ammonium ion in water relative to pH and temperature. The normal pH range of 10.8–11.5 for ammonia stripping is shown shaded.

of 11, essentially all the ammonia appears as $NH_3$ gas. The percentage of dissolved ammonia relative to ammonium ion decreases with the cooling of the wastewater.

The physical–chemical process of ammonia stripping consists of (1) raising the wastewater pH to a value in the range 10.8–11.5 with lime applied for preceding phosphate precipitation, (2) formation and reformation of water droplets in a stripping tower, and (3) providing air–water contact and droplet agitation by circulation of large quantities of air through the tower. The rate of gas transfer from liquid to air is influenced by pH, temperature, relative ammonia concentrations, and agitation at the air–water interface. The latter two factors depend primarily on the hydraulic loading, the quantity of air flow, and the configuration of the column packing. Common hydraulic loadings are in the range of 1–2 gpm/ft$^2$ (0.7–1.4 l/m$^2$·s), and air flow during warm weather is usually 300–500 ft$^3$/gal of wastewater (2200–3700 m$^3$/m$^3$) [16]. Countercurrent towers, where air enters at the bottom and exhausts from the top while the water drips and splatters down through the medium, have been found to be most efficient. In cross-flow units, air is pulled in through louvered sides, drawn through the packing, and discharged from the top.

Figure 14.23 is a schematic diagram of a countercurrent stripping tower. Towers of this kind at Water Factory 21, Orange County, California, are

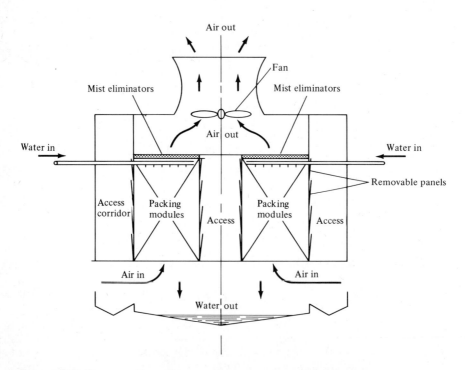

**Figure 14.23** Cross section of a countercurrent ammonia-stripping tower.

designed for a hydraulic loading of 1.0 gpm/ft$^2$ and an air flow of 400 ft$^3$/gal. The tower packing is constructed of 0.5-in. diameter plastic pipe laid in a crisscross pattern on 3-in. centers horizontally. Prefabricated modules of this fill, about 6 × 6 × 4 ft high, are positioned behind the removable air baffle panels. Corridors around the stripping columns in the tower allow access for replacing modules, if necessary, and removing excessive calcium carbonate scale by spraying with water.

Advantages of air stripping for nitrogen removal are simplicity of operation, ease of control, and low cost relative to other extraction processes. Disadvantages are the inability to operate at an ambient air temperature below 32°F, and deposition of calcium carbonate scale on the packing.

Towers cannot be employed in northern climates where cold-weather nitrogen removal is needed without providing an alternative method for winter operation. Experience with full-scale stripping towers demonstrated both the excellent reduction possible during warm weather and the serious problems that occur during low-temperature operation [16]. The efficiency decreases from 90%–95% removal at 20°C to a maximum of 75% at 10°C. When air temperature drops below freezing, the process has to be shut down owing to ice formation in the tower.

Another serious operating problem is accumulation of calcium carbonate scale on the tower packing. Scale can result in a decrease in efficiency caused by interference with both droplet formation and air flow. Recently designed countercurrent towers have smooth plastic fill, which is less susceptible to scaling, and removable panels placed along corridors within the tower for access to clean or remove packing modules.

The fate of ammonia discharged to the atmosphere is not normally considered to be a problem. Concentration in the stack discharge, before dilution in the atmosphere, is about 6 mg/m$^3$, which is significantly lower than the odor threshold of 35 mg/m$^3$ for ammonia. The United States has no emission standard, or air-quality criterion, for ammonia, since it is not considered an air pollutant. The ammonia content in rainfall is directly related to the atmospheric concentration; thus washout can occur where tower discharge increases in natural background concentration in the air. Ammonia in rainfall decreases to about the natural level at a downwind distance of 3 mi from a tower processing 15 mgd of wastewater [16].

Recarbonation of lime-treated water is necessary to convert hydroxide ions to bicarbonates for control of scaling in subsequent treatment units. Carbon dioxide neutralizes excess lime, precipitating calcium carbonate at a pH of about 10 (Eq. 14.21); additional recarbonation reduces the pH further and stabilizes the CaCO$_3$ to soluble calcium bicarbonate (Eq. 14.22). A two-stage system with intermediate clarification can be used to separate these two reactions and extract a major portion of the calcium carbonate precipitate from solution. Advantages are a finished water lower in hardness and total dissolved solids, and a sludge of almost pure calcium carbonate suitable for recalcining. Single-stage recarbonation, which performs Eqs. 14.21 and 14.22 in a single basin, stabilizes water, leaving all the calcium in solution.

$$Ca(OH)_2 + CO_2 = CaCO_3 \downarrow + H_2O \qquad (14.21)$$

$$CaCO_3 + CO_2 + H_2O = Ca(HCO_3)_2 \qquad (14.22)$$

# Wastewater Reclamation

The major reuse of wastewater is for agricultural irrigation, amounting to about 60% of the total. The second largest reuse, accounting for approximately 30%, is for industrial cooling and process waters. Since the public is generally not exposed to these reused waters, high-quality biological processing with or without chemical disinfection is often considered to be satisfactory treatment. The remainder of wastewater reuse, less than 10% of the total, is for fish and wildlife, recreation, and groundwater recharge, mostly through percolation. For these applications, control of industrial wastewaters entering the sewer system and tertiary treatment for a higher degree of organic matter removal and disinfection for virus inactivation are usually the prescribed pretreatment. Dilution with surface water and dispersion of seepage in groundwater are relied on for reducing the concentrations of refractory contaminants.

A small fraction of groundwater recharge is performed by direct injection through wells in coastal areas as a barrier against underground intrusion of saltwater from the ocean. Because a portion of this injected water seeps toward inland potable groundwater supplies, the wastewater is reclaimed to meet drinking-water standards. This is the highest degree of wastewater renovation required in the United States, since direct potable reuse is not recommended.

## 14.16  WATER QUALITY AND REUSE APPLICATIONS

Protection of public health is the primary concern in establishing water-quality standards for reuse. Important environmental considerations are protection of groundwater, soils, and crops. In establishing standards, consideration must be given to the economics of integrating wastewater treatment with the reuse application. For example, a properly designed agricultural irrigation system restricted to fodder, fiber, and seed crops can reuse conventionally treated wastewater without health or environmental risks. In contrast, surface infiltration of wastewater, whether planned or unintentional, can result in significant deterioration of groundwater quality unless the groundwater in storage provides abundant dilution or the infiltration water is treated to drinking-water quality.

The water reclamation regulations for the state of California [21] specify both treatment processes and quality standards as diagrammed in Fig. 14.24. Conventional biological treatment of the wastewater is the first step in reclamation. For irrigation of fodder, fiber and seed crops, stabilization ponds may be used for biological treatment; however, for other reuse applications, the coliform and turbidity standards dictate high-quality processing in a conventional plant, with the preferred process being activated sludge. Al-

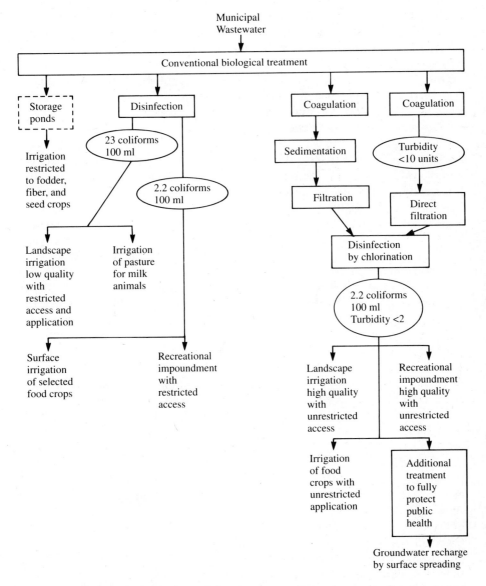

**Figure 14.24** General scheme for reclaimed water-quality and reuse applications based on Wastewater Reclamation Criteria, Environmental Health, State of California [21].

though no quality standards are applied to irrigation of fodder, fiber, and seed crops, a biologically stabilized effluent is required to prevent offensive odors, and a coliform standard is usually established for the health protection of the agricultural workers. The two basic degrees of treatment in Fig. 14.24 are (1) for lower quality reclaimed water, a coliform concentration not to exceed 23 per 100 ml, as determined from results of the last seven tests performed, and a maximum concentration not to exceed 240 per 100 ml in any two consecutive samples and no turbidity limit, and (2) for higher-quality reclaimed water, a median coliform concentration not to exceed 2.2 per 100 ml, as determined from the last seven tests, and a maximum concentration not to exceed 23 per 100 ml in any sample and a turbidity limit not to exceed either 2 units for the average operating value or 5 units more than 5% of the time during a 24-hr period. The wastewater processing necessary to reliably produce the higher quality of reclaimed water is biological treatment, chemical coagulation, granular-media filtration, and disinfection by chlorination with an extended contact time. (Refer to Section 14.5.)

The regulations for water reuse established by the state of Arizona [22] are listed in Table 14.4. The allowable limits for various reuses are specified without reference to any required treatment processes, and the water quality for several water reuse applications differ from the California standards. For irrigation of fodder, fiber, and seed crops, the fecal coliform allowable limit is 1000 per 100 ml as the geometric mean of a minimum of five samples and 4000 per 100 ml as the maximum of a single sample.

The Arizona standards for landscape irrigation are less restrictive than the California standards. Landscape irrigation with restricted access has a fecal coliform allowable limit of 200 per 100 ml and a maximum of 1000 per 100 ml, compared with the California total coliform standard of 23 per 100 ml and 240 per 100 ml, respectively. Landscape irrigation with open access has a fecal coliform allowable limit of 25 per 100 ml and a maximum of 75 per 100 ml, compared with the California total coliform standard of 2.2 per 100 ml and 25 per 100 ml, respectively. Nevertheless, the coliform and turbidity standards for irrigation of food consumed uncooked are similar: Arizona specifies a fecal coliform geometric mean of 2.2 per 100 ml and maximum of 25 per 100 ml, which are numerically the same as the California total coliform limits, and an allowable turbidity of 1 NTU, which is similar to the California turbidity standard of less than 2 units.

For higher-quality reuse applications, allowable limits are also specified for enteric viruses, protozoa, and helminths. The operators of reuse systems are not required to monitor routinely for these pathogens; however, if the Arizona Department of Environmental Control requires corrective action to improve the microbiological quality of the reclaimed water, monitoring for pathogens will be necessary for verification.

Both of these reuse regulations specify reliability in the design to ensure that the plant can meet the designated water-quality standards, an operational plan for safety of public health, and a monitoring program for quality control. Reliability in performance requires provision for safe disposal or storage of

**TABLE 14.4** ALLOWABLE PERMIT LIMITS FOR SPECIFIC REUSES OF RECLAIMED WATER, STATE OF ARIZONA [22]

| Parameter | A Orchards | B Fiber, seed, & forage | C Pastures | D Livestock watering | E Processed food | Landscaped areas F Restricted access | Landscaped areas G Open access | H Food consumed raw | I Incidental human contact | J Full-body contact |
|---|---|---|---|---|---|---|---|---|---|---|
| pH | 4.5–9 | 4.5–9 | 4.5–9 | 6.5–9 | 4.5–9 | 4.5–9 | 4.5–9 | 4.5–9 | 6.5–9 | 6.5–9 |
| Fecal coliform (CFU/100 ml)[a] geometric mean (5 sample minimum) | 1000 | 1000 | 1000 | 1000 | 1000 | 200 | 25 | 2.2 | 1000 | 200 |
| single sample not to exceed | 4000 | 4000 | 4000 | 4000 | 2500 | 1000 | 75 | 25 | 4000 | 800 |
| Turbidity (NTU)[b] | — | — | — | — | — | — | 5 | 1 | 5 | 1 |
| Enteric virus[c] | — | — | — | — | — | — | 125 per 40 l | 1 per 40 l | 125 per 40 l | 1 per 40 l |
| Entamoeba Histolytica | — | — | — | — | — | — | — | None detectable | — | None detectable |
| Giardia lamblia | — | — | — | — | — | — | None detectable | None detectable | None detectable | None detectable |
| Ascaris lumbricoides | — | — | None detectable | — | — | — | — | — | None detectable | — |
| Common large tapeworm | — | — | — | None detectable | — | — | — | — | — | — |

[a] CFU = colony-forming units.

[b] NTU = nephelometric turbidity units.

[c] Expressed as PFU, plaque-forming units; MPN, most probable numbers; or immunofluorescent foci per liter.

"None detectable" means no pathogenic microorganisms observed during examination.

wastewater in case of emergency, installation of multiple treatment units for flexibility in operational control and shutdown for maintenance, standby chemical feeders and chlorinators, alarm devices installed to warn of process malfunction, alternative electric power supply, and adequate funding for operation and maintenance. Protection of public health includes isolation of restricted agricultural irrigation areas by buffer zones and fencing, signs identifying the water supply as reclaimed water, isolation and identification of reclaimed water piping and appurtenances from potable water supply, landscape irrigation scheduling, and other management practices to reduce the contact of people with reclaimed water. Management of a reuse system includes scheduled laboratory testing of reclaimed water quality and recording of plant operations and maintenance.

The wastewater reclamation regulations established in the United States have been replicated or used as the basis for similar regulations in many countries. Unfortunately, the construction and operation of complex wastewater treatment systems dictated by these restrictive quality standards have inhibited beneficial water reuse in developing areas of the world. As a result, wastewaters are often not officially approved for agricultural irrigation awaiting some unresolved solution in order to achieve compliance with standards that cannot be attained within local resources. This results in the practice of totally uncontrolled unsafe irrigation of salad and other food crops with inadequately treated wastewater.

A group of environmental experts and epidemiologists meeting in Engelberg, Switzerland, under the auspices of several international organizations, formulated new tentative microbiological guidelines for treated wastewater reuse in agricultural irrigation [23]. As shown in Table 14.5, the guidelines require effective wastewater treatment to remove helminths to a level of one egg per liter, since helminthic disease transmission has been identified as the top-priority health problem in developing countries. For unrestricted irrigation, the guideline of 1000 fecal coliforms per 100 ml is recommended. In conventional wastewater treatment, tertiary coagulation and filtration are needed to ensure removal of helminth eggs, but in stabilization ponds removal of eggs can be achieved by simply impounding the water for a sufficiently long time to allow settlement of the eggs. Other pathogens are reduced by natural die-off. Therefore, with the availability of adequate land area and tolerance for a reduced aesthetic quality of the reclaimed water, a low-cost easy-to-operate stabilization pond system in a warm climate can provide a reclaimed water for irrigation.

Investigations on the effectiveness of stabilization ponds to remove helminth eggs and protozoal cysts showed 100% removal in all cases where the total retention time in multicelled ponds was greater than 20 days. Hookworm larvae may survive in aerobic ponds with an overall retention time of less than 10 days but not if the retention time is greater than 20 days [24]. Inactivation of enteric viruses is rapid in warm waters and is in the range of 1 to 2 log units per 5 days retention in ponds at a temperature greater than 25°C. Thus, a pond system with an overall retention time of 20 days in a

**TABLE 14.5**  TENTATIVE MICROBIOLOGICAL QUALITY GUIDELINES FOR TREATED WASTEWATER REUSE IN AGRICULTURAL IRRIGATION,[a] *THE ENGELBERG REPORT* [23]

| Reuse process | Intestinal nematodes[b] (eggs/liter) | Fecal coliforms (no./100 ml) |
|---|---|---|
| Restricted irrigation[c] Irrigation of trees, industrial crops, fodder crops, fruit trees,[d] and pasture[e] | Less than 1 | Not applicable[c] |
| Unrestricted irrigation Irrigation of edible crops, sports fields, and public parks[f] | Less than 1 | Less than 1000[g] |

[a] In specific cases, local epidemiologic, sociocultural, and hydrogeologic factors should be taken into account, and these guidelines modified accordingly.

[b] *Ascaris, Trichuris* and hookworms.

[c] A minimum degree of treatment equivalent to at least a 1-day anaerobic pond followed by a 5-day facultative pond or its equivalent is required in all cases.

[d] Irrigation should cease 2 weeks before fruit is picked, and no fruit should be picked off the ground.

[e] Irrigation should cease 2 weeks before animals are allowed to graze.

[f] Local epidemiologic factors may require a more stringent standard for public lawns, especially hotel lawns in tourist areas.

[g] When edible crops are always consumed well cooked, this recommendation may be less stringent.


warm climate would be expected to achieve a reduction of excreted viruses of 4 to 6 log units (99.99% to 99.9999%) [24]. Fecal bacteria are also significantly reduced in stabilization ponds with warm water and long retention times. A summary of reported bacterial removal efficiencies in multicelled ponds with retention time greater than 25 days reduced fecal coliforms to 100 per 100 ml or less [25]. (Raw domestic wastewater contains approximately 3,000,000 per 100 ml, so the reduction was between 99.99% and 99.999%.)

## 14.17 AGRICULTURAL IRRIGATION

Reuse of wastewater for agricultural irrigation is often considered land treatment and/or disposal. Regardless of designated primary objective, irrigation systems are being used extensively in regions where available sites have suitable soil conditions and groundwater hydrology and a climate favorable to grow grasses, crops, or trees. Land irrigation is also used for advanced wastewater treatment to recycle nutrients to land instead of polluting surface waters. Reuse of reclaimed water has a strong ecological appeal; however, the requirements for large tracts of land and reclaimed water storage are major disadvantages in humid climates, northern states, and metropolitan areas.

Agricultural irrigation has the following features [26]: Water loading rate of 0.5–4 in./wk (13–100 mm/wk); annual application of 2–20 ft/yr (0.6–6 m/yr); field area for 1 mgd (3785 m$^3$/d) flow of 56–560 acres (23–230 ha); minimum depth to groundwater of 5 ft (1.5 m); soils: moderately permeable with good productivity when irrigated; disposition of wastewater: evapo-transpiration and percolation; and climatic restrictions: storage needed for cold weather and runoff from irrigation and precipitation.

Water distribution is by fixed or moving sprinkling systems, or surface spreading. Fixed nozzles are attached to risers from either surface or buried pipe networks. The most popular moving sprinkling system is a center-pivot spray boom that rotates around a central tower with the distribution piping suspended between wheel supports riding on circumferential tracks. On flat land, having less than 1% slope, surface irrigation is possible by the ridge-and-furrow method. Water applied to furrows, spaced about 3 ft apart, flows down slope by gravity and seeps into the ground. Border-strip irrigation, the second method for surface spreading, uses parallel soil ridges constructed in the direction of slope. Water introduced between the ridges at the upper end flows down the 20–100-ft wide strips several hundred feet long.

## Restricted Irrigation

Restricted irrigation refers to the use of low-quality reclaimed water in specific areas where only fodder, fiber, and seed crops are grown, for example, alfalfa, cotton, and wheat. Public access is controlled by fencing and warning signs around the perimeter. If watering is by spray irrigation, buffer zones are established along the boundaries of the site to prevent aerosols from drifting into adjacent public access areas. Thus, the only health risk is to the agricultural workers.

The objectives of wastewater treatment for restricted agricultural irrigation are (1) maximum removal of helminth eggs and protozoal cysts, (2) effective removal of pathogenic bacteria, (3) reduction of enteric viruses, and (4) substantial removal of organic matter to clarify the water and eliminate offensive odors. This treatment can be accomplished best by conventional biological processing followed by detention in storage reservoirs or by completely mixed aerated lagoons followed by stabilization ponds in series prior to detention in storage reservoirs. Storage is required to equalize the demand for irrigation water with the supply of reclaimed water. The crop selections, growing seasons, soil conditions, and types of irrigation systems affect the volume of storage needed. In warm semiarid or arid regions, reservoirs have water depths of 5 to 10 m with a storage volume equivalent to about 90 days of reclaimed water production. Thus, the detention time of the reclaimed water in the reservoirs varies from two to several months. In cold humid regions, the required storage volume may be significantly greater to accumulate wastewater flow for 4 or more months when irrigation is not possible.

Besides equalizing storage, reservoirs provide supplementary treatment for removal of helminth eggs, protozoal cysts, pathogenic bacteria, enteric viruses, and organic matter. The potential unaesthetic conditions that can occur in storage are the growth of algae, which increase the turbidity and suspended solids, and the potential generation of odors if prior biological treatment is not adequate.

The management of an agricultural irrigation system is an interdisciplinary function of agronomy and environmental engineering. The engineering aspect is to ensure adequate operation and maintenance of the biological treatment system to produce a reclaimed water suitable for storage in the reservoirs and to monitor water-quality by conducting routine tests, such as biochemical oxygen demand, suspended solids, and coliform bacteria and maybe occasional tests for helminth eggs and enteric viruses. The agronomic aspect involves selection of crops, fertilizer and pesticides applications, and scheduling of water flow through the reservoirs to the fields. The cultivation of only approved fodder, fiber, and seed crops must be monitored and strictly enforced; food crops for human consumption, particularly those eaten uncooked, cannot be irrigated with the low-quality reclaimed water designated for restricted irrigation.

## Unrestricted Irrigation

Unrestricted irrigation refers to the application of high-quality reclaimed water for irrigation of food crops for human consumption, even those eaten uncooked. Public access to the irrigation site is not controlled, but persons must be warned not to use the water for drinking or domestic purposes. Signs reading "Irrigated with Reclaimed Wastewater," or a similar warning, should be posted at the boundaries of the site, and faucets discharging reclaimed water should be posted with signs reading "Reclaimed Water, Do Not Drink," or a similar warning, or be secured to prevent public use. To preclude inadvertent connection as a potable water source, the irrigation piping in areas accessible to the public should be color-coded, buried with colored tape, or otherwise suitably marked to indicate nonpotable water.

The wastewater for unrestricted irrigation must be treated for the removal of helminth eggs and protozoal cysts, elimination of pathogenic bacteria, and inactivation of enteric viruses. Of greatest importance is the removal of helminth eggs, since they are extremely persistent, surviving in harsh environmental conditions, and because of their latency period are transmitted primarily through salad crops eaten uncooked. Recommended microbiological standards published for reclaimed water applied to unrestricted irrigation of food crops are extremely restrictive, with the quality near that of drinking water (Fig. 14.24 and Table 14.4). An extensive study in California using reclaimed water of this quality to irrigate food crops eaten uncooked demonstrated no health or environmental risks [27]. To achieve this high quality, the required wastewater processing is conventional biological treatment, chemical coagulation, granular-media filtration, and disinfection by chlorination. The coliform standard in Table 14.5 is applicable in developing countries only with warm climates and abundant sunshine where stabilization ponds are effective for both wastewater treatment and disinfection.

## Chemical Quality

The chemical characteristics of irrigation water are important for public health and agronomy. Reuse standards applied to edible crops include heavy metals and organic compounds detrimental to the health of consumers. The agronomic standards include salinity, sodium absorption ratio, and specific ion toxicity of sodium, chloride, boron, and trace elements affecting sensitive crops. Normally, sanitary wastewater from domestic and commercial sources does not have chemical contaminants in excess of the allowable limits for agricultural cultivation. Table 14.6 gives the guidelines for interpretation of

**TABLE 14.6** GUIDELINES FOR INTERPRETATION OF WATER QUALITY FOR IRRIGATION

| Potential irrigation problem | Units | Degree of restriction on use | | |
| --- | --- | --- | --- | --- |
| | | None | Slight to moderate | Severe |
| Salinity (affects crop water availability)[a] | | | | |
| $EC_w$ | dS/m | <0.7 | 0.7–3.0 | >3.0 |
| TDS | mg/l | <450 | 450–2000 | >2000 |
| Infiltration (affects infiltration rate of water into the soil. Evaluation using $EC_w$ and SAR together)[b] | | | | |
| SAR = 0–3 and $EC_w$ = | | >0.7 | 0.7–0.2 | <0.2 |
| = 3–6 = | | >1.2 | 1.2–0.3 | <0.3 |
| = 6–12 = | | >1.9 | 1.9–0.5 | <0.5 |
| = 12–20 = | | >2.9 | 2.9–1.3 | <1.3 |
| = 20–40 = | | >5.0 | 5.0–2.9 | <2.9 |
| Specific ion toxicity (affects sensitive crops) | | | | |
| Sodium (Na)[b] | | | | |
| surface irrigation | SAR | <3 | 3–9 | >9 |
| sprinkler irrigation | me/l | <3 | <3 | |
| Chloride (Cl)[c] | | | | |
| surface irrigation | me/l | <4 | 4–10 | >10 |
| sprinkler irrigation | me/l | <3 | >3 | |
| Boron (B) | mg/l | <0.7 | 0.7–3.0 | >3.0 |

*Source:* Adapted from the Food and Agricultural Organization of the United Nations (FAO) (1985) [28].

[a] $EC_w$ means electrical conductivity, a measure of the water salinity, reported in deciSiemens per meter at 25°C (dSm) or in units millimhos per centimeter (mmho/cm). Both are equivalent. TDS means total dissolved solids, reported in milligrams per liter (mg/l).

[b] SAR means sodium adsorption ratio, and is sometimes reported by the symbol RNa. At a given SAR, infiltration rate increases as water salinity increases. Evaluate the potential infiltration problem by SAR as modified by $EC_w$.

[c] For surface irrigation, most tree crops and woody plants are sensitive to sodium and chloride; use the values shown. Most annual crops are not sensitive. With overhead sprinkler irrigation and low humidity (<30%), sodium and chloride may be absorbed through the leaves of sensitive crops.

water quality for irrigation related to salinity from the Food and Agricultural Organization of the United Nations (FAO) [28]. The nitrogen content of irrigation water is restricted where percolation to groundwater is possible or where the crops are sensitive to high nitrate. Trace elements, including several heavy metals, are toxic to selected plants. Of the 20 elements listed by FAO [28], only a few are likely to be present in treated municipal wastewater in concentrations exceeding the recommended guidelines for irrigation water. An example of a possible trace element in wastewater is cadmium, with a guideline limit of 0.01 mg/l. This conservative value is recommended due to its potential for accumulation in plants and soils in amounts that may be harmful to humans. Also, cadmium is toxic to beans, beets, and turnips at concentrations as low as 0.1 mg/l in nutrient solutions.

## Design Process for Land Treatment

The recommended design for an irrigation system is the iterative process diagrammed in Fig. 14.25 [26]. The initial step in design is to define the characteristics of the wastewater, effluent quality, and the site. Regulatory limits on effluent quality are established to protect both the groundwater and surface water. As outlined in Table 14.7, investigations of the site encompass climate, geology, soils, plant cover, and topography. The iterative procedure for determining the field area needed for irrigation involves the interdependence of the hydraulic loading rate, nitrogen loading rate, water storage volume, and crop selection. System monitoring, the method of wastewater distibution, discharge control, and agricultural management are final considerations.

The water balance for a land application system can be calculated by the relationship

$$\text{Precipitation} + \text{wastewater loading} = \text{evapotranspiration} + \text{percolation} + \text{runoff} \qquad (14.23)$$

Runoff is zero for most irrigation and infiltration–percolation installations, while precipitation and evapotranspiration are of little significance compared to wastewater loading and percolation for rapid infiltration systems.

The storage required can be calculated using monthly, weekly, or daily water-balance determinations. If the sum of evapotranspiration and percolation is less than precipitation plus available effluent, the balance must be stored. Conversely, when losses exceed the quantity available, water can be drawn from storage to supplement irrigation. These relationships are summarized in the following statement:

$$\begin{pmatrix} \text{Precipitation} + \\ \text{effluent available} \end{pmatrix} \pm \begin{pmatrix} \text{change in} \\ \text{storage} \end{pmatrix} = \begin{pmatrix} \text{evapotranspiration} \\ + \text{percolation} \end{pmatrix} \qquad (14.24)$$

The nitrogen balance for an irrigation system can be calculated from the relationship

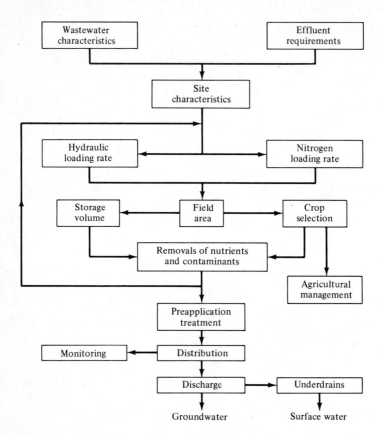

**Figure 14.25** Iterative design process for an irrigation system. [From *Process Design Manual for Land Treatment of Municipal Wastewater,* U.S. Environmental Protection Agency, EPA 625/1-77-008 (October 1977): 5-5.]

$$\text{Applied nitrogen} = \text{crop uptake} + \text{loss by denitrification}$$
$$+ \text{loss by percolation} \qquad (14.25)$$

The greatest nitrogen uptake is by perennial grasses, often in the range of 200–500 lb/acre/yr (220–560 kg/ha·y). Field crops (e.g., corn, soybeans, and cotton) assimilate considerably less nitrogen in one growing season, 70–170 lb/acre/yr (80–190 kg/ha·y). Loss by denitrification, commonly assumed to be 15%–25% of the applied nitrogen, is difficult to accurately predict. Based on the maximum contaminant level for drinking water, the allowable nitrogen concentration in percolate is usually taken as 10 mg/l.

### ■ EXAMPLE 14.6

These data are from sample calculations by Pound and Crites [29]. Compute the storage requirements using monthly water balances for a 1-mgd irrigation system based on the following: (1) Design evapotranspiration and precipitation data, listed

**TABLE 14.7** GENERAL DESIGN CONSIDERATIONS FOR LAND TREATMENT SYSTEMS

| Wastewater characteristics | Climate | Geology | Soils | Plant cover | Topography | Application |
|---|---|---|---|---|---|---|
| Flow | Precipitation | Groundwater | Type | Indigenous | Slope | Method |
| Constituent | Evapotranspiration | Seasonal | Gradation | to region | Aspect of | Type of |
| load | Temperature | depth | Infiltration/ | Nutrient- | slope | equipment |
| | Growing | Quality | permeability | removal | Erosion | Application |
| | season | Points of | Type and | capability | hazard | rate |
| | Occurrence | discharge | quantity of | Toxicity | Crop and farm | Types of |
| | and depth of | Bedrock | clay | levels | management | drainage |
| | frozen ground | Type | Cation exchange | Moisture | | |
| | Storage | Depth | capacity | and shade | | |
| | requirements | Permeability | Phosphorus | tolerance | | |
| | Wind velocity | | adsorption | Marketeability | | |
| | and direction | | potential | | | |
| | | | Heavy metal | | | |
| | | | adsorption | | | |
| | | | potential | | | |
| | | | pH | | | |
| | | | Organic matter | | | |

*Source:* R. D. Johnson, "Land Treatment of Wastewater," *The Military Engineer* 65(428) (1973): 375.

in columns 2 and 5 of Table 14.8, are with average monthly distibutions for the wettest year in 25. (2) A perennial grass is grown and irrigated year-round. (3) Runoff is contained and reapplied. (4) The design year begins in October with the storage reservoir empty. (5) Nitrogen is the limiting factor for a land area of 120 field acres to balance the quantity applied with the amount of nitrogen in harvested grass. (6) The design allowable percolation rate is 10 in./month from March through November and 5 in./month for the remaining months (column 3, Table 14.8).

*Solution.*   The effluent available per month is

$$\frac{10 \text{ mil gal/day} \times 30.4 \text{ days/month} \times 36.8 \text{ acre·in./mil gal}}{120 \text{ acres}} = 9.3 \text{ in.}$$

Water losses (column 4 of Table 14.8) are found by summing evapotranspiration and percolation. The effluent applied (column 6) is the difference between water losses and precipitation. The total water available (column 8) equals the sum of the effluent available and precipitation.

The monthly change in storage (column 9) from Eq. 14.24 is the difference between the total water available (column 8) and water losses (column 4). The total accumulated storage (column 10), computed by summing the monthly changes, reaches a maximum of 22.7 in. in March.

$$\text{basin capacity equivalent to this value} = \frac{22.7 \text{ in.} \times 120 \text{ acre}}{12 \text{ in./ft}}$$

$$= 227 \text{ acre·ft} = 74 \text{ mil gal}$$

The wastewater effluent applied is 111.7 in. (9.3 ft) on an annual basis, the maximum rate, in July, being about 3.5 in./wk.   ∎

## ■ EXAMPLE 14.7

Calculate the crop uptake of nitrogen required for the irrigation design outlined in Example 14.6. Base calculations on an allowable concentration of nitrogen in the percolate of 10 mg/l and an assumed denitrification loss of 20%. The average nitrogen content of the applied wastewater is 25 mg/l with a hydraulic loading of 1.0 mgd on 120 acres (112 in./yr). From Table 14.8, the percolation is 105 in./yr.

*Solution*

$$\text{applied nitrogen} = \frac{1.0 \text{ mil gal/day} \times 365 \text{ days/yr} \times 25 \text{ mg/l} \times 8.34}{120 \text{ acres}}$$

$$= 634 \text{ lb/acre/yr}$$

$$\text{loss by denitrification} = 0.20 \times 634 = 127 \text{ lb/acre/yr}$$

$$\text{loss by percolation} = \frac{105 \text{ in./yr} \times 10 \text{ mg/l} \times 8.34}{36.8 \text{ acre·in./mil gal}} = 238 \text{ lb/acre/yr}$$

From Eq. 14.25,

$$\text{crop uptake required} = 634 - 127 - 238 = 269 \text{ lb/acre/yr}$$   ∎

**TABLE 14.8** WATER BALANCE AND STORAGE CALCULATIONS FOR EXAMPLE 14.6 (all values given in inches)

| (1) Month | (2) Evapotranspiration | (3) Percolation | (4) Water losses: (2) + (3) | (5) Precipitation | (6) Effluent applied | (7) Effluent available | (8) Total water available: (5) + (7) | (9) Change in storage: (8) − (4) | (10) Total storage |
|---|---|---|---|---|---|---|---|---|---|
| Oct. | 2.3 | 10.0 | 12.3 | 1.6 | 10.7 | 9.3 | 10.9 | −1.4 | 0 |
| Nov. | 1.0 | 10.0 | 11.0 | 2.4 | 8.6 | 9.3 | 11.7 | 0.7 | 0.7 |
| Dec. | 0.5 | 5.0 | 5.5 | 2.7 | 2.8 | 9.3 | 12.0 | 6.5 | 7.2 |
| Jan. | 0.2 | 5.0 | 5.2 | 3.0 | 2.2 | 9.3 | 12.3 | 7.1 | 14.3 |
| Feb. | 0.3 | 5.0 | 5.3 | 2.8 | 2.5 | 9.3 | 12.1 | 6.8 | 21.1 |
| March | 1.1 | 10.0 | 11.1 | 3.4 | 7.7 | 9.3 | 12.7 | 1.6 | 22.7 |
| April | 3.0 | 10.0 | 13.0 | 3.0 | 10.0 | 9.3 | 12.3 | −0.7 | 22.0 |
| May | 3.5 | 10.0 | 13.5 | 2.1 | 11.4 | 9.3 | 11.4 | −2.1 | 19.9 |
| June | 4.8 | 10.0 | 14.8 | 1.0 | 13.8 | 9.3 | 10.3 | −4.5 | 15.4 |
| July | 6.0 | 10.0 | 16.0 | 0.5 | 15.5 | 9.3 | 9.8 | −6.2 | 9.2 |
| Aug. | 5.7 | 10.0 | 15.7 | 1.1 | 14.6 | 9.3 | 10.4 | −5.3 | 3.9 |
| Sept. | 3.9 | 10.0 | 13.9 | 2.0 | 11.9 | 9.3 | 11.3 | −2.6 | 1.3 |
| Total | 32.3 | 105.0 | 137.3 | 25.6 | 111.7 | 111.8 | 137.2 | | |

*Source:* C. E. Pound and R. W. Crites, *Land Treatment of Municipal Wastewater Effluents*, Environmental Protection Agency, Technology Transfer (August 1975).

## 14.18  LANDSCAPE IRRIGATION

Restricted landscape irrigation can be used in areas where public access is limited and/or where the water application is controlled to prevent direct contact with people. Examples are areas landscaped for beautification only, highway median strips, and golf courses. Where transient human activities take place in the area, the water is applied only during night hours without airborne drift or surface runoff into public areas. The vegetation is allowed to dry, and excess water is allowed to soak into the ground before start of use. Warning signs should be prominently displayed reading "Irrigation with Reclaimed Wastewater," or a similar warning, and faucets should be posted with signs reading "Reclaimed Water, Do Not Drink," or a similar warning, or be secured to prevent public use. The irrigation piping should be color-coded, buried with colored tape, or otherwise suitably marked to indicate nonpotable water.

Unrestricted landscape irrigation with a high-quality reclaimed water can be used on public parks, lawns, playgrounds, gardens, and other places where people might contact the reclaimed water. This quality of water can be used to spray irrigate areas where airborne drift or surface runoff into public areas may occur; however, the timing of watering should minimize public contact. This reclaimed water is not intended for irrigation of private lawns or gardens. Warning signs, secure faucets, and marking of piping as nonpotable water are required.

The allowable limits for reclaimed water quality for reuse in restricted and unrestricted landscape irrigation are given for California in Fig. 14.24 and for Arizona in Table 14.4. For restricted irrigation, high-quality biological treatment and chlorination can meet the coliform standards. The water applied for unrestricted irrigation of parks, playgrounds, lawns, and gardens must be free of pathogens; thus, the standards are very restrictive, requiring tertiary filtration and disinfection.

## 14.19  GROUNDWATER RECHARGE AND POTABLE SUPPLY

Groundwater recharge from surface infiltration of wastewater may be intentional for aquifer recharge or inadvertent, resulting from disposal of wastewater. Deterioration of groundwater quality because of infiltration depends on the characteristics of the recharge water, removal of contaminants by the soils, and dilution by the groundwater in storage. Before undertaking aquifer recharge with the intention of recovering the reclaimed water as a drinking-water supply, a demonstration project is recommended to evaluate the feasibility of storage and controlled subsurface movement of the recharged water, improvement of recharge water quality by filtration through the soil profile, and treatment of the reclaimed water necessary to maintain drinking-water quality in the recovered groundwater withdrawn from wells. Numerous monitoring wells are necessary for tracking water movement and sampling for water-quality testing.

The minimum quality of reclaimed water for surface infiltration, from basins or river channels, is equivalent to the quality for unrestricted landscape irrigation with processing by biological treatment, chemical coagulation, granular-media filtration, and chlorine disinfection. Where the soils do not provide significant removal of contaminants, and dilution of recharged water in the aquifer is limited, more extensive treatment is necessary to ensure that the groundwater remains in compliance with drinking-water standards. Injection of reclaimed water into an aquifer through recharge wells requires that the water meet drinking-water standards. Water reclamation is required to remove microorganisms, heavy metals, organic chemicals, and inorganic salts. The sequence of biological, chemical, and physical processes often includes activated sludge, lime precipitation, denitrification, granular-media filtration, carbon adsorption, disinfection, and reverse osmosis.

None of the water-reuse regulations recommend reclaiming wastewater directly to drinking water. The highest level of reclamation is for groundwater recharge after advanced treatment in an attempt to comply with drinking-water standards. Although a surface-water supply is polluted, usually only a small percentage of the raw water is actually derived from treated wastewater. Thus, the practice of indirect potable reuse through polluted water sources is only marginally related to direct water reuse. The revision of drinking-water regulations, currently being undertaken nationally and internationally, is resulting in much more restrictive maximum contaminant levels for more chemicals, and for many of the trace organic compounds the proposed maximum contaminant level goal is zero. Although monitoring and advanced treatment technologies are available with respect to pathogens, inorganic chemicals, and radionuclides, the large number and low concentrations of organic chemicals pose uncertainties. Most have not been identified or quantified. The health effects over a lifetime of consumption of these trace organic chemicals are unknown. Therefore, direct reuse of reclaimed wastewater for a domestic water supply, even when diluted with a natural water, is not considered safe. Until the effectiveness of fail-safe advanced treatment can be clearly demonstrated, the risk of wastewater contaminants being present in the reclaimed water is too great to justify domestic reuse, particularly when natural water sources are available.

## 14.20 WATER FACTORY 21, ORANGE COUNTY, CALIFORNIA

Orange County in southern California borders the Pacific Ocean in a location where a geological discontinuity allows access between the ocean and the inland groundwater basin. Lowering the water table below sea level, by extensive pumping of the groundwater basin, can cause encroachment of seawater. To prevent saltwater intrusion, a hydraulic barrier system consisting of two separate well fields has been established. Extraction wells about 2 mi from the coast intercept and return salty water to the ocean, while injection wells further inland pump fresh water to the underground aquifers to form a water mound blocking passage of seawater. The injection water is reclaimed wastewater from Water Factory 21 processed to meet drinking-water standards. The reclaimed water can be blended with high-quality deep-well water, if necessary, for dilution to meet the regulatory limit for dissolved salts if the reverse osmosis process is shut down for maintenance.

Water Factory 21 is designed to reclaim unchlorinated effluent from biological treatment of municipal wastewater containing a low proportion of industrial wastes at a uniform flowrate of 15 mgd (0.66 $m^3/s$). The reverse osmosis plant is limited to a design flowrate of 5 mgd (0.22 $m^3/s$), since demineralization of one-third of the water is adequate to provide a blended effluent of suitable quality. As diagrammed in Fig. 14.26, the processing consists of lime clarification with sludge recalcination, air stripping, recarbon-

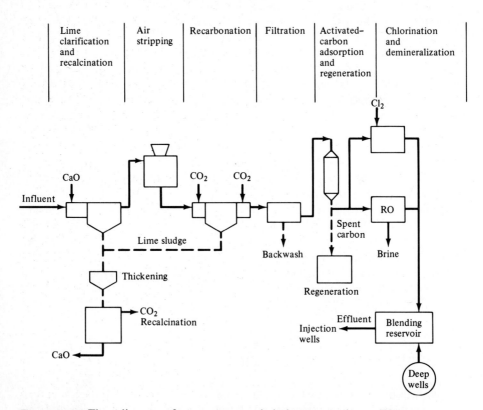

**Figure 14.26** Flow diagram of wastewater and sludge processing at Water Factory 21.

ation, filtration, activated-carbon adsorption and carbon regeneration, disinfection by chlorination, and demineralization by reverse osmosis. Figure 14.27 is an aerial view of the plant, showing the orientation of the treatment units.

Lime clarification is performed in separate rapid-mix, flocculation, and sedimentation basins at a controlled pH of 11.0 by application of 350–400 ng/l of CaO. An anionic polymer dose of approximately 0.1 mg/l is applied to aid in coagulation. The principal objectives in lime precipitation are removal of heavy metals and reduction of organic solids. Concentrations of heavy metals with the exception of mercury were significantly reduced. Because of the low amount of mercury present in the influent wastewater, the concentration in the plant effluent did not exceed the standard for drinking water. In contrast, the regulatory concentration limits of cadmium and chromium were exceeded occasionally when 70%–80% removal efficiencies could not reduce adequately the periodic high influent concentrations [30]. The

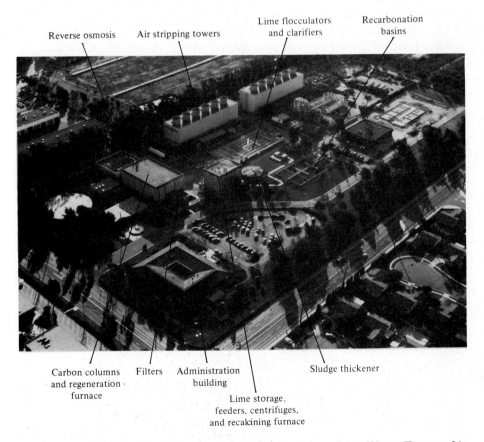

**Figure 14.27** Aerial view of the wastewater reclamation plant, Water Factory 21. (Courtesy of Orange County Water District.)

difficulty in reducing peak concentrations of cadmium and chromium emphasized the importance of adequate pretreatment at industrial sources. Reduction in suspended organic solids resulted in a COD removal averaging 50%
and turbidity reduction of over 90%. A secondary benefit of high-lime treatment is disinfection; Coliform reduction was greater than 99.999% and virus
removal was 98%–99.9% [30].

Lime sludge from wastewater clarification and calcium carbonate precipitate from first-stage recarbonation are concentrated in a gravity thickener to
8%–15% solids. Centrifuges dewater the thickened sludge for both recalcination and disposal. Approximately 50% of the sludge solids are recalcined in
a multiple-hearth furnace for lime recovery and reuse. By wasting one-half
of the precipitated solids, the buildup of impurities in the recalcined lime is
controlled to prevent recycling of contaminants to the wastewater. The lime
processing building contains two storage bins, one for recalcined lime and the
other for purchased lime. Carbon dioxide in the stack gas from recalcination is
used for recarbonation. Figure 14.28 pictures the unit processes related to
lime treatment.

Air stripping is accomplished in countercurrent induced-draft towers
with a design air/water ratio of 400 ft$^3$ per gal of wastewater (3000 m$^3$/m$^3$).
When Water Factory 21 was processing a trickling-filter effluent, ammonia–nitrogen was reduced from an influent concentration in the range of
20–30 mg/l to an effluent average of 5 mg/l. In 1978 biological treatment of
the wastewater entering Water Factory 21 was upgraded to activated-sludge
processing. As a result, the ammonia–nitrogen concentration was reduced
to less than 5 mg/l and continuous forced air is no longer needed. Normal
operation is to allow the wastewater to cascade over the tower packing with
only natural ventilation. This process is effective for removing a wide variety
of volatile organic compounds, such as chlorinated benzenes and halogenated
methanes [30]. After air stripping, the wastewater is stabilized by two-stage
recarbonation and filtered through gravity multimedia beds for suspended-
solids removal before activated-carbon adsorption.

The granular activated-carbon columns have been operated in both upflow and downflow directions. In upflow operation, as originally intended,
carbon fines smaller than 25 $\mu$m were suspended in the effluent and accumulated in the reverse osmosis modules, fouling the membranes. Therefore, the
columns have been modified for downflow operation at an empty-bed contact
time of 34 min [31]; COD removal efficiency by carbon adsorption averages
60% with a mean influent of 42 mg/l and 49% at a mean influent of 24 mg/l.
In general, chlorinated benzenes, some phthalates and aromatic hydrocarbons, and brominated trihalomethanes have been removed at least as efficiently as COD. On the other hand, chlorinated compounds with one or two
carbons were not absorbed. Chromium, copper, and lead were partially
removed by activated carbon [31].

A flow diagram of the reverse osmosis plant at Water Factory 21 is shown
in Fig. 14.29. Effluent from the carbon columns is treated with sodium
hexametaphosphate as a scale inhibitor and chlorine to control biological

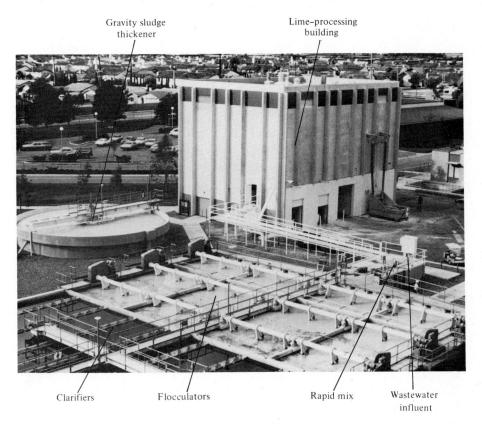

Gravity sludge
thickener

Lime–processing
building

Clarifiers          Flocculators                    Rapid mix        Wastewater
                                                                      influent

**Figure 14.28** Picture of Water Factory 21 showing the inlet end of lime clarification
(rapid mixing, flocculation, and sedimentation), the gravity sludge-thickening tank,
and the lime processing building, which contains centrifuges for sludge dewatering,
a lime recalcining furnace, lime feeders and slakers, and carbon dioxide compressors.

growth on the membranes. Particulates are removed by filtration through
replaceable polypropylene elements with an effective size of 25 $\mu$m. The
water is pressurized by feed pumps to 550 psi (3800 kPa). Sulfuric acid is
injected into the discharge from the pumps to adjust the pH to approximately
5.5 to minimize membrane hydrolysis and scale formation, primarily calcium
carbonate and calcium sulfate.

Each reverse osmosis module consists of six spiral-wound, cellulose
acetate elements, for a total of 1920 ft$^2$ (178 m$^2$) of active membrane area
enclosed in an 8-in.- (20.3-cm-) diameter fiberglass cylinder 23 ft (7.0 m) long.
A total of 252 modules are arranged into six independent sections of 42
modules each. In order to obtain the specified 85% permeate recovery with
90% salt removal, the water is passed through three stages in each section.
The feedwater flows in parallel to the first 24 modules, and the concentrate
from the first pass flows to 12 modules; the second pass concentrate flows

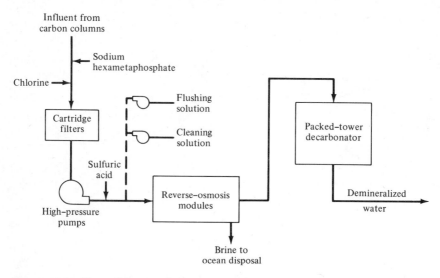

**Figure 14.29** Flow diagram of the reverse-osmosis plant at Water Factory 21.

through the 6 remaining units. (In the original design, each section consisted of 35 modules, as pictured in Fig. 14.30, arranged in three stages of 20, 10, and 5. Seven additional modules were added to each section to compensate for reduced productivity.) The concentrated brine, approximately 15% of the feed, is discharged to an outfall sewer for ocean disposal. The demineralized water is air stripped in packed-tower decarbonators to remove carbon diox-

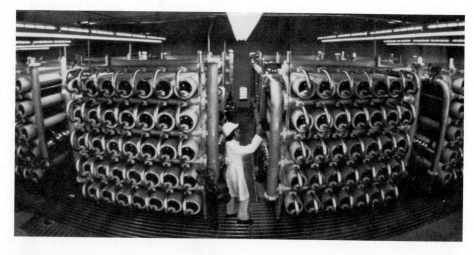

**Figure 14.30** View of the reverse-osmosis modules at Water Factory 21. (Courtesy of Orange County Water District.)

ide, which resulted from the prior acidic pH adjustment. Volatile trace organics are also removed during this aeration.

Membranes require periodic cleaning to restore the water flux rate. The loss of productivity is a major problem of membrane separation in wastewater reclamation. The fouling layer consists primarily of bacteria that attach to the surfaces of the cellulose acetate membranes, even though a combined chlorine residual is maintained in the feedwater. Of the cleaning solutions available, combinations of detergents and enzymes appear to be most effective. Flushing the system with permeate displaces feedwater from the modules to inhibit bacterial growth when the plant is to be out of operation for several days. Flushing is also used after cleaning to remove chemical solutions.

The primary purpose of reverse osmosis is removal of dissolved inorganic solids. This is routinely accomplished with 90% reduction to a level of about 100 mg/l total dissolved solids and recovery of approximately 85% of the applied water. In addition, membranes reject the high-molecular-weight polymeric substances that represent the majority of organics as measured by the COD test. The COD reductions are about 90% resulting in concentrations in the permeate of less than 2 mg/l and often less than 1 mg/l [32]. Nonionic organic compounds of low molecular weight are poorly removed by reverse osmosis; however, concentrations of these trace organics are reduced in air stripping of the permeate.

Water Factory 21 has a high reliability for producing water with contaminant concentrations less than the maximum levels dictated by drinking-water standards, particularly since the wastewater pretreatment was upgraded to activated sludge and the proportion of industrial wastes in the municipal wastewater has been reduced. Of the heavy metals, only chromium exceeded the maximum contaminant level in the carbon effluent, but that occurred only 2% of the time. Although trace organic compounds were detectable in the effluent, no standards exist for these substances for reclaimed waters. Many of these appear to be the result of chlorination for disinfection. Of the trace organic chemicals suspected from industrial origin, most were reduced in concentrations to the low nanogram per liter level [30]. All the blended injection-water samples tested for enteric viruses were negative. The safety of reusing the reclaimed water for groundwater recharge is increased since any peak concentration of a contaminant will be reduced by dilution and evened out by alternate adsorption and desorption as the water moves through the aquifer. Also, the reclaimed water can be diluted with high-quality deep-well water prior to injection.

# PROBLEMS

**14.1**  List the common characteristics of a wastewater discharged from conventional biological processing. Which contaminants are most likely to cause pollution if the effluent is discharged to a stream with limited dilutional flow? Which constituents adversely affect a lake or reservoir receiving the wastewater discharge?

**14.2**  What are the potential benefits of flow equalization in municipal wastewater processing?

**14.3**  Compute the basin volume required to equalize the diurnal wastewater flow pattern diagrammed in Fig. 12.16a. Follow the procedure described in Example 14.1.

**14.4**  The feasibility of an in-line equalization chamber is being considered to reduce diurnal variations in flow through a conventional activated-sludge plant. The design wastewater flow is 10 mgd. The treatment units are sized based on the following parameters:

$$\text{equalization chamber volume} = 15\% \text{ of the design flow}$$
$$\text{overflow rate and depth of primary clarifiers} = 600 \text{ gpd/ft}^2, 8 \text{ ft}$$
$$\text{aeration period and loading of aeration tanks} = 5 \text{ hr}, 40 \text{ lb BOD/} 1000 \text{ ft}^3/\text{day}$$
$$\text{overflow rate and detention time of final clarifiers} = 800 \text{ gpd/ft}^2, 2.2 \text{ hr}$$

(a) Compare the volume of the equalization chamber to the total liquid volume of the clarifiers and aeration tanks.  (b) The typical design values used in sizing treatment tanks compensate for diurnal flow variation; therefore, the parameters listed above are too conservative for design based on equalized flow. What criteria do you suggest for sizing the primary clarifiers, aeration tanks, and final clarifiers with influent flow equalization? If these values are used in tank sizing, is the equalization chamber justified?

**14.5**  Discuss how sludge handling and disposal influence selection of unit processes for removing pollutants from wastewater.

**14.6**  What are basic differences between the characteristics of water and wastewater that influence the design of granular-media filters?

**14.7**  The solution to Prob. 12.50 is a flow diagram for a treatment plant with primary sedimentation and secondary step aeration. Expand this diagram to include direct tertiary granular-media filtration and effluent chlorination. Show the arrangement of filters and backwash system. Recommend the kind of filter media, underdrain system, method of backwashing, and flow control. (Refer to Sections 14.4 and 11.18–11.20.) List numerical design guidelines for sizing the processing units.

**14.8**  Assume that flow equalization is provided prior to filtration in Example 14.2, thus allowing a nominal application rate of 5.0 gpm/ft² for design. Compute the filter area required, cycle time, backwash as a percentage of filtrate, and rate of filtration with three of four cells operating.

**14.9**  Calculate the area of granular-media filters needed for suspended-solids removal from a wastewater effluent with an average daily flow of 1500 m³/d and peak 4-hr discharge of 3300 m³/d. The nominal filtration rate should exceed neither 180 m³/m²·d with all beds operating nor 350 m³/m²·d at maximum discharge with only three of four beds in service.

**14.10**  Sketch a schematic flow diagram of the least-cost system for virus removal tested in the Pomona Virus Study. Label the treatment units and the chemicals applied.

**14.11** What effluent quality in terms of BOD, SS, chlorine residual, dissolved solids, nitrogen, and coliforms would you anticipate for treatment including primary sedimentation, conventional activated sludge, dual-media filtration, chlorination, and dechlorination?

**14.12** What kinds of pollutants are removed by granular-carbon columns?

**14.13** Discuss the advantages of upflow, countercurrent, packed-bed carbon columns relative to downflow gravity contactors.

**14.14** How is granular activated-carbon regenerated?

**14.15** For Eq. 14.5, verify by appropriate calculations the molar ratio for Al to P of 1 to 1 and the weight ratio of commercial alum to phosphorus of 9.7 to 1.0.

**14.16** A laboratory activated-sludge system was used to evaluate chemical–biological phosphorus removal by applying alum to the aeration tank. The laboratory apparatus was a diffused-air complete-mixing aeration tank, separate gravity clarifier, and airlift return sludge pump. Twelve liters of settled municipal wastewater were applied daily at a constant rate to the aeration tank, which had a volume of 3.6 l. The aeration tank MLSS concentration was held near 2000 mg/l by wasting 200–250 ml of mixed liquor each day. The temperature was 22°–24°C, pH 7.3–7.7, and SVI varied from 90–130 ml/g. The alum solution feed to the aeration tank had a strength of 10.0 mg of commercial alum per milliliter. The following data were collected at various alum feed rates after arriving at steady-state conditions.

| Wastewater feed (l/d) | Alum applied (ml/d) | Influent BOD (mg/l) | SS (mg/l) | P (mg/l) | Alk (mg/l) |
|---|---|---|---|---|---|
| 12.0 | 0 | 150 | 94 | 10.3 | 350 |
| 12.0 | 70 | 158 | 104 | 10.9 | 350 |
| 12.0 | 140 | 169 | 114 | 10.6 | 340 |
| 12.0 | 210 | 173 | 135 | 10.4 | 360 |
| 12.0 | 390 | 173 | 123 | 9.3 | 320 |

| Wastewater feed (l/d) | Alum applied (ml/d) | Effluent BOD (mg/l) | SS (mg/l) | P (mg/l) | Alk (mg/l) |
|---|---|---|---|---|---|
| 12.0 | 0 | 6 | 7 | 7.9 | 230 |
| 12.0 | 70 | 6 | 8 | 5.4 | 200 |
| 12.0 | 140 | 5 | 11 | 2.4 | 160 |
| 12.0 | 210 | 10 | 10 | 1.0 | 140 |
| 12.0 | 390 | 8 | 17 | 0.5 | 80 |

(a) Calculate the phosphorus removal and the weight ratio of alum applied to total phosphorus in the influent wastewater for each run. Plot a graph of percentage of phosphorus removal versus the weight ratio of alum to phosphorus.    (b) Calculate the average BOD and SS (suspended solids) removal efficiencies. Suggest reasons why the effluent BOD and SS values are lower and resulting efficiencies higher than are normally achieved by a full-scale treatment plant.    (c) Why did the concentration of alkalinity in the effluent decrease with increasing alum dosage?

**14.17** A domestic wastewater, with 200 mg/l of BOD, 240 mg/l of SS, and 10 mg/l of P, is treated by extended aeration without primary sedimentation at a loading of 12.5 lb of BOD/1000 ft$^3$/day and an aeration period of 24 hr. Alum dosing for phosphate precipitation is graphed in Fig. 14.8. What is the alum addition in milligrams per liter required for 80% phosphorus removal? Estimate the biological–chemical solids produced per million gallons of wastewater processed using the technique presented in Sec. 13.2 for biological waste and Eqs. 14.5 and 11.34 to approximate the amount of chemical precipitate. Compute the volume of sludge in gal/mil gal of wastewater treated assuming a solids content of 1.5%.

**14.18** The characteristics of a settled wastewater (primary effluent) are as follows:

| | | | |
|---|---|---|---|
| calcium = | 34 mg/l | chloride = | 43 mg/l |
| magnesium = | 12 mg/l | BOD = | 130 mg/l |
| sodium = | 64 mg/l | suspended solids = | 120 mg/l |
| potassium = | 12 mg/l | volatile SS = | 100 mg/l |
| alkalinity = | 185 mg/l | nitrogen = | 30 mg/l |
| sulfate = | 44 mg/l | phosphorus = | 9 mg/l |

This wastewater is processed in a diffused-air activated-sludge system at a F/M loading of 1.0 g of BOD applied per day for each 3.0 g of MLSS in the aeration basin. The final clarifiers are appropriately sized.    (a) During conventional operation, what are the estimated effluent BOD, SS, and P concentrations in milligrams per liter? Calculate the average waste organic solids production in milligrams per liter of wastewater treated.    (b) Alum is then added to the aeration tanks at a dosage equivalent to an Al/P ratio (aluminum added to phosphorus in the applied wastewater) of 2.0 : 1.0. With this chemical addition, the effluent BOD, SS, and P decrease to 10, 15, and 1.0 mg/l, respectively. No aluminum appears in the effluent. Calculate the average amounts of chemical precipitates and waste organic solids produced per liter of wastewater processed. How do these reactions affect the ionic character of the wastewater?    (c) Assume that granular-media filtration is added as a tertiary step following the biological–chemical phosphorus removal. What are the estimated BOD, SS, and P concentrations in the effluent? Estimate the additional waste solids removed by tertiary filtration.

**14.19** Problem 13.28 describes the operation of an RBC plant that adds alum to the biodisk effluent to precipitate phosphate in the final clarifiers. Assume the total phosphorus in the raw unsettled wastewater is 10.1 mg/l consisting of 3.0 mg/l organic phosphorus and 7.1 mg/l inorganic phosphorus.    (a) Trace phosphorus through the plant from influent to effluent as illustrated in Fig. 14.9, and write the appropriate chemical and biological reactions resulting in phosphorus removal.    (b) Calculate the alum dosage in milligrams per liter as alum and as the weight ratio of Al/P. How do these values compare with values given in Section 14.9 under the heading "Chemical Precipitation and Biological Treatment"?

**14.20**  A conventional activated-sludge plant treats a domestic wastewater with 220 mg/l of suspended solids, 180 mg/l of BOD, and 9.0 mg/l of total phosphorus. After primary settling, the characteristics are 110 mg/l of suspended solids, 120 mg/l of BOD, and 7.5 mg/l of total phosphorus. Based on a pilot-plant study, a dosage of 116 mg/l of alum is required to reduce the effluent total phosphorus to 1.0 mg/l with the aeration tank operating at an F/M of 0.20 g BOD/day/g MLSS. The effluent suspended-solids concentration averages 10 mg/l. Calculate the estimated production of sludge solids per cubic meter of wastewater treated. If the primary sludge has a solids content of 5.0% and the waste sludge from the chemical–biological aeration system has a concentration of 1.5%, calculate the liters of combined sludge produced per cubic meter of wastewater treated.

**14.21**  A domestic wastewater with 220 mg/l of suspended solids, 180 mg/l of BOD, and 9.0 mg/l of phosphorus is treated by lime precipitation in primary settling followed by conventional activated-sludge processing. A dosage of 200 mg/l of CaO removes 80% of the influent phosphorus. Of the remaining 1.8 mg/l of phosphorus, 0.8 mg/l is removed by biological uptake during aeration and 1.0 mg/l appears in the plant effluent. Lime precipitation in the primary also removes 80% of the suspended solids and 60% of the BOD. The plant effluent suspended-solids concentration averages 20 mg/l. Calculate the estimated production of sludge solids per cubic meter of wastewater treated based on the following data: The chemical precipitate produced is 2.0 mg of calcium hydroxyapatite and calcium carbonate per 1.0 mg of CaO applied. The aeration process operates at an F/M of 0.2  g BOD/day/g MLSS. How does this sludge production compare to the sludge solids produced by the biological–chemical treatment in Prob. 14.20?

**14.22**  Advanced treatment of the wastewater from a small city consists of aeration tanks without primary sedimentation with the addition of alum for phosphorus precipitation, final clarifiers with sludge recirculation back to the aeration tanks, tertiary granular-media filters with a backwash storage tank that drains back to the head of the plant, effluent chlorine-contact chambers for disinfection, aerobic digesters with supernatant return to the head of the plant, and a belt filter press for dewatering digested sludge withdrawn from the digesters with filtrate returned to the head of the plant. The influent wastewater is 2000 m$^3$/d with 200 mg/l BOD, 240 mg/l SS, and 10.0 mg/l P. The effluent characteristics are 3 mg/l BOD, 3 mg/l SS, and 0.5 mg/l P. The fecal coliform count is less than 200 per 100 ml. The activated-sludge process is operated at F/M = 0.15, and the alum addition is 190 mg/l. Aerobic digestion reduces the organic content of the waste-activated sludge by 25% and thickens the digested sludge to 2.0% organic-chemical solids. The dewatered filter cake has 18% solids.    (a) Draw a flow diagram of the treatment system.    (b) Estimate the weight of filter cake produced per day by calculating the following: the organic solids produced by aeration, organic solids removed by tertiary filtration, organic solids remaining after digestion, chemical precipitates, and weight of total solids at 18% concentration. (c) What is the principal form of chlorine residual formed in chlorination of the wastewater? What chemical can be used for dechlorination?

**14.23** Phosphorus is precipitated during biological–chemical treatment by the addition of waste pickle liquor (ferrous sulfate) to the aeration tanks of an activated-sludge process. The influent wastewater flow is 17.7 mgd containing 210 mg/l BOD, 190 mg/l SS, and 7.1 mg/l P. The effluent characteristics are 20 mg/l BOD, 26 mg/l SS, 1.0 mg/l P, and 1.4 mg/l Fe. The waste pickle liquor feed is 2300 gpd with an iron content of 8.8% by weight. The dry weight of sludge solids produced per day is 28,000 lb. (a) Calculate the weight ratios of iron applied to influent phosphorus and iron precipitated to phosphorus removed. (b) Compute the percentage of iron in the dry sludge solids.

**14.24** Chemical coagulation and sedimentation of a raw municipal wastewater using 60 mg/l of $FeCl_3$ and 50 mg/l of CaO precipitates 230 mg/l of suspended solids and 8 mg/l of phosphorus. What is the weight ratio of iron applied to phosphorus removed? Calculate the weight of organic–chemical sludge solids per million gallons of wastewater treated using the method in Section 13.2 for organic matter and Eqs. 14.6 and 14.7 for chemical precipitates. If the sludge has an 8.0% solids concentration, compute the volume of waste in gal/mil gal of wastewater processed.

**14.25** Why is it impractical to perform biological nitrification concurrently with BOD reduction in an activated-sludge aeration tank operating at the wastewater temperatures common in most regions of the United States?

**14.26** Convert the concentration data for phosphorus given in milligrams in Fig. 14.9 to percentages with 10 mg/l equivalent to 100%. Convert the concentration data for nitrogen given in milligrams in Fig. 14.11 to percentages with 35 mg/l equivalent to 100%. How do these percentages compare with percentages that would be expected for BOD removal?

**14.27** How would anaerobic digestion of the waste sludge affect the nitrogen data given in Fig. 14.11?

**14.28** What is the reduction in rate of biological nitrification (a) if the wastewater temperature decreases from 20° to 10°C and (b) if the operating pH is 7.0 rather than 8.5? How can reduced temperature and pH be compensated for in operating a suspended-growth nitrification system?

**14.29** Calculate the aeration basin volume for suspended-growth nitrification of an average daily design flow equal to 1500 $m^3/d$ containing 25 mg/l of ammonia–nitrogen. Assume a minimum operating temperature of 16°C, operating pH 7.4, and design MLVSS of 1500 mg/l. If flow equilization is provided for the plant influent, what basin volume would you recomend?

**14.30** For the town of Nancy, the solution to Prob. 12.32 gives design data, the solution to Prob. 12.49 sizes an extended aeration system, and the solution to Prob. 13.38 sizes aerobic digesters and belt filter presses. Assume this plant has been constructed and is in operation. Proposed new effluent standards specified for this plant are a maximum total phosphorus concentration of 2.0 mg/l and a maximum ammonia nitrogen concentration of 5.0 mg/l. The city council has employed you to study the implications of these new effluent limitations on the operation and modification of the treatment plant. You are to submit a preliminary report within 12 months. (a) Outline a plan to evaluate performance of the existing treatment system by wastewater sampling, laboratory testing, and modifying or changing the operation of unit processes of the treatment system if necessary. These data are also to be used for future design if the new effluent standards are implemented.   (b) Propose a treatment system with processing schemes for phosphorus removal and nitrification. Based on assumed data, calculate preliminary sizes of treatment units and estimate chemical usage.   (c) Propose full-scale and/or pilot-plant studies to evaluate modified or additional treatment processes.

**14.31** (a) Why is methanol, or some other carbon source, needed in biological denitrification?   (b) Calculate the methanol demand for denitrifying 1 $m^3$ of nitrified effluent with a nitrate nitrogen concentration of 25 mg/l and dissolved oxygen concentration of 8 mg/l. Express answers in grams of methanol and liters of 95% methanol with a specific gravity of 0.82.

**14.32** Tertiary nitrification–denitrification processing is being considered for a 30-mgd secondary-treatment plant. The wastewater characteristics selected for nitrification process design are 30 mg/l of $NH_3$–N, 30 mg/l of BOD, a temperature of 18°C, pH 7.4, and 1500 mg/l of MLVSS. Using these parameters, compute the aeration basin volume required, recommended clarifier surface area, and estimated oxygen utilization. Assume the following characteristics for subsequent denitrification design: 27 mg/l of $NO_3$–N, 5 mg/l of DO, a temperature of 16°C, pH 7.0, and 1500 mg/l of MLVSS. Calculate the denitrification basin volume, recommended clarifier surface area, and methanol dosage.

**14.33** The overall nitrogen removal in Fig. 14.21 is $(69 - 16)100/69 = 77\%$, and in Fig. 14.19b the removal is $100 - 41 = 59\%$. Why is the flow scheme in Fig. 14.21 more effective in nitrogen removal than the one in Fig. 14.19b?

**14.34** How many milligrams of oxygen are in 1.0 mg of nitrate nitrogen? Using this value, estimate the milligrams per liter of BOD satisfied in the anoxic zones of the scheme in Fig. 14.21.

**14.35** The Citizens Committee on Environmental Issues has proposed to the state environmental protection agency the adoption of an effluent standard for all the wastewater treatment plants in the state to include a maximum allowable inorganic nitrogen limit of 10 mg/l, which is based on the maximum contaminant level (MCL) for drinking water of 10 mg/l of nitrate nitrogen. The committee's main arguments are that treatment plants are not properly designed to remove nitrogen and that nitrate nitrogen in surface waters is an environmental risk and health hazard. The state agency has data showing that no surface waters used as drinking-water sources exceed 10 mg/l of nitrate nitrogen, and analyses of groundwaters near flowing waters downstream from wastewater discharges show little or no nitrate contamination. Nevertheless, these data have not deterred the committee from insisting on reducing the concentration of contaminants listed in the Safe Drinking Water Act to less than their MCLs in wastewater discharges. You have been asked to present a general assessment of nitrogen removal by conventional wastewater treatment and the feasibility of available technology to retrofit existing plants and design new plants to meet an effluent limitation of 10 mg/l of inorganic nitrogen. (Assume other speakers will address the topics of economic feasibility and other sources of nitrogen contamination, such as agricultural land drainage to surface waters and percolation of nitrate fertilizers to groundwaters under cropland.)

**14.36** Discuss the major factors limiting the use of air-stripping towers for nitrogen removal.

**14.37** Compare the reclaimed water quality required by California and Arizona for irrigation of food crops consumed uncooked.

**14.38** Define the concept of reliability in wastewater processing and discuss some ways to increase reliability in tertiary treatment.

**14.39** A city in a semiarid region plans to reuse wastewater for agricultural irrigation. The soils at the proposed irrigation site are composed of poorly drained alkaline loams underlain by dense clayey strata. The groundwater is unsuitable for potable use because of naturally occurring high salt and nitrate contents. The proposed crops include alfalfa, barley, corn, cotton, grain sorghums, and grasses. What kind of wastewater treatment is necessary, and what quality standards must be met for this reuse application? (Assume the wastewater is primarily from domestic water use and the concentrations of heavy metals, boron, and toxic chemicals are negligible.) What controls are necessary for operation of the irrigation site?

**14.40** Using Eq. 14.23, estimate the land area in acres required for irrigation of 1.0 mgd based on precipitation of 20 in./yr, evapotranspiration of 40 in./yr, percolation equal to 10 in./month, and zero runoff.

**14.41** The total field area at an agricultural irrigation site in a semiarid region is 4800 acres. The proposed total surface area of the storage reservoirs is 300 acres. By water-balance calculations, determine the storage requirement based on the data listed below. The total water available each month is calculated in acre-inches by summing the effluent available plus precipitation minus evaporation loss from the storage reservoirs. The monthly irrigation requirements listed in the table were determined by estimating the consumptive use by the crops most likely to be grown at the site taking into account optimum economic farming operation, crop rotation, and planting and harvesting times. Assume the storage reservoirs are empty on October 1.

| Month | Effluent available (acre-in.) | Precipitation (in.) | Evaporation (in.) | Irrigation requirement (acre-in.) |
|---|---|---|---|---|
| Oct. | 20,100 | 0.3 | 4.6 | 0 |
| Nov. | 19,400 | 0.5 | 2.3 | 0 |
| Dec. | 19,800 | 1.0 | 1.0 | 0 |
| Jan. | 19,000 | 1.2 | 1.1 | 0 |
| Feb. | 16,600 | 1.1 | 1.6 | 0 |
| Mar. | 17,800 | 1.1 | 4.0 | 22,000 |
| Apr. | 17,900 | 0.7 | 5.0 | 22,000 |
| May | 19,400 | 0.2 | 7.4 | 36,500 |
| Jun. | 20,400 | 0.1 | 8.9 | 36,500 |
| Jul. | 22,300 | 0 | 9.5 | 25,500 |
| Aug. | 22,300 | 0 | 8.6 | 44,000 |
| Sep. | 20,700 | 0.1 | 6.2 | 33,000 |
| | 235,700 | 6.3 | 60.2 | 219,500 |

**14.42** What quality standards and regulations apply to irrigation of a golf course (tee boxes, fairways, and greens), excluding the landscaped area around the clubhouse? The play area for a golf course is considered to have restricted access if fenced along the boundary with public property.

**14.43** What are the main purposes for each of the following processes in water reclamation: lime precipitation, granular activated-carbon adsorption, and reverse osmosis?

**14.44** What are the functions of air stripping at Water Factory 21? Why are the granular activated-carbon columns operated in a downflow direction? What pretreatment is performed to prepare water for reverse-osmosis treatment?

**14.45** What is the primary operational problem of the reverse-osmosis process at Water Factory 21? What maintenance procedures are used to reduce the problem?

**14.46** A 4.0-mgd tertiary treatment plant processing a domestic wastewater had the following treatment processes: preliminary screening and grit removal, primary sedimentation, flow equalization to eliminate diurnal variation, conventional diffused-air activated-sludge processing at a BOD loading of 30–35 lb/1000 ft³/ day, anaerobic sludge digestion with land disposal of digested solids, alum application to the secondary clarifier overflow of approximately 5 mg/l, dual-media gravity filtration at a rate of 3–4 gpd/ft², chlorination at a dosage up to 25 mg/l for a detention time of 1.0 hr in a plug-flow tank, sulfur dioxide addition for dechlorination, and sodium hydroxide for increasing pH if necessary. The effluent is discharged to an irrigation channel. The plant performance specified by the plant designer was an effluent BOD of 5 mg/l and suspended solids of 10 mg/l. After the plant operated for several years, the state revised the effluent limitations in the NPDES permit as listed below. Which of these effluent standards do you think this plant can meet? Which ones do you think this plant cannot meet and why? Which ones are in doubt?

| Parameter | Influent wastewater | Effluent limitations | | |
| --- | --- | --- | --- | --- |
| | | 90-Day average | 30-Day average | Daily maximum |
| BOD, mg/l | 290 | — | 5.0 | 10.0 |
| Total dissolved solids, mg/l | 900 | 250 | — | 500 |
| Suspended solids, mg/l | 270 | — | 5.0 | 10.0 |
| Total nitrogen, mg/l | 50 | — | — | — |
| Ammonia nitrogen, mg/l | 34 | — | — | 5.0 |
| Chlorine residual, mg/l | | — | — | 0 |
| Coliform bacteria | — | 2.2/100 ml for 7-day average | | |
| Dissolved oxygen | — | 5.0 mg/l or more 80% of time | | |
| Heavy metals | — | MCLs for drinking water | | |
| pH | 7 | In the range of 6.5–8.5 | | |

## REFERENCES

1. R. L. Culp and G. L. Culp, *Advanced Wastewater Treatment* (New York: Van Nostrand Reinhold, 1978).
2. "Tertiary Filtration of Wastewaters," by the Task Committee on Design of Wastewater Filtration Facilities, *Proc. Am. Soc. Civil Engrs., Env. Eng. Div.* 112 (EE6) (December 1986): 1008–1025.
3. "Wastewater Filtration," EPA Technology Transfer Seminar Publication, Environmental Protection Agency (July 1974).
4. E. R. Baumann and J. Y. C. Huang, "Granular Filters for Tertiary Wastewater Treatment," *J. Water Poll. Control Fed.* 46(8) (August 1974): 1958–1972.
5. *Pomona Virus Study,* Sanitation Districts of Los Angeles County (Sacramento, CA: California State Water Resources Control Board, 1977).
6. E. C. Hartling, "Expanding the Use of Reclaimed Water in Los Angeles County," in *Implementing Water Reuse* (Denver: AWWA Research Foundation, 1987), pp. 385–399.
7. "Process Design Manual for Carbon Adsorption," U.S. Environmental Protection Agency, Technology Transfer (October 1973).
8. A. B. Menar and D. Jenkins, "Fate of Phosphorus in Waste Treatment Processes: Enhanced Removal of Phosphate by Activated Sludge," *Environ. Sci. Technol.* 4(12) (December 1970): 1115–1121; *Proceedings, 24th Industrial Waste Conference* (Lafayette, IN: Purdue University, 1969), pp. 655–673.
9. J. T. Riding, W. R. Elliott, and J. H. Sherrard, "Activated Sludge Phosphorus Removal Mechanisms," *J. Water Poll. Control Fed.* 51(5) (May 1979): 1040–1053.
10. J. B. Carberry and M. W. Tenney, "Luxury Uptake of Phosphate by Activated Sludge," *J. Water Poll. Control Fed.* 45(12) (December 1973): 2444–2462.
11. "Process Design Manual for Phosphorus Removal," U.S. Environmental Protection Agency, Technology Transfer, EPA 625/1-76-001a (April 1976).
12. D. T. Anderson and M. J. Hammer, "Effects of Alum Addition on Activated Sludge Biota," *Water and Sewage Works* 120(1) (January 1973): 63–67.
13. R. D. Leary, L. A. Ernest, R. M. Manthe, and M. Johnson, "Phosphorus Removal Using Waste Pickle Liquor," *J. Water Poll. Control Fed.* 46(2) (February 1974): 284–300.
14. J. C. O'Shaughnessy, J. B. Nesbitt, D. A. Long, and R. R. Kountz, "Digestion and Dewatering of Phosphorus-Enriched Sludges," *J. Water Poll. Control Fed.* 46(8) (August 1974): 1914–1926.
15. H. E. Wild, Jr., C. N. Sawyer, and T. C. McMahon, "Factors Affecting Nitrification Kinetics," *J. Water Poll. Control Fed.* 43(9) (September 1971): 1845–1854.
16. "Process Design Manual for Nitrogen Control," U.S. Environmental Protection Agency, Technology Transfer (October 1975).
17. "Nitrification and Denitrification Facilities," EPA Technology Transfer Seminar Publication, U.S. Environmental Protection Agency (August 1973).
18. W. E. Wruble, C. C. Plummer, and M. J. Hammer, "Biological Denitrification and Phosphate Removal," *Proc. Am. Soc. Civil Engrs., J. Env. Eng. Div.* 100 (EE5) (October 1974): 1199–1203.
19. P. F. Cooper, E. A. Drew, D. A. Bailey, and E. V. Thomas, "Recent Advances in Sewage Effluent Denitrification: Part I," *Wat. Pollut. Control* 76(3) (1977): 287–300.

20. P. F. Cooper, B. Collinson, and M. K. Green, "Recent Advances in Sewage Effluent Denitrification: Part II," *Water Pollut. Control* 76(4) (1977): 389–401.
21. *Wastewater Reclamation Criteria, California Administrative Code, Title 22, Division 4, Environmental Health,* Department of Health Services, State of California (Berkeley, CA: 1978).
22. *Environmental Quality, Title 18, Chapter 9, Water Pollution Control, Article 7, Regulations for the Reuse of Wastewater,* Department of Environmental Quality, State of Arizona (Phoenix, AZ: 1985).
23. *Health Aspects of Wastewater and Excreta Use of Agriculture and Aquaculture,* The Engelberg Report (Dubendorf, Switzerland: International Reference Centre for Wastes Disposal, 1985).
24. R. G. Feachem, D. J. Bradley, H. Garelick, and D. D. Mara, *Sanitation and Disease, Health Aspects of Excreta and Wastewater Management; World Bank Studies in Water Supply and Sanitation 3* (Chichester: Wiley, 1983).
25. C. R. Bartone, "Development of Health Guidelines for Water Reuse in Agriculture: Management and Institutional Aspects," in *Implementing Water Reuse* (Denver: AWWA Research Foundation, 1987), pp. 489–504.
26. "Process Design Manual for Land Treatment of Municipal Wastewater," U.S. Environmental Protection Agency, EPA 625/1-77-008; U.S. Army Corps of Engineers, COE EM 1110-1-501 (October 1977).
27. R. Cort, B. Sheikh, R. Jaques, and R. Cooper, "Safety, Feasibility, and Cost of Reuse of Wastewater Irrigation of Raw-Eaten Vegetables," in *Implementing Water Reuse* (Denver: AWWA Research Foundation, 1987), pp. 445–474.
28. R. S. Ayers and D. W. Westcot, *Water Quality for Agriculture 8,* FAO Irrigation and Drainage Paper No. 29, Rev. 1 (Rome: Food and Agriculture Organization of the United Nations, 1985).
29. C. E. Pound and R. W. Crites, *Wastewater Treatment and Reuse by Land Application,* Vols. I and II, Office of Research and Development, Environmental Protection Agency (August 1973).
30. "Wastewater Contaminant Removal for Groundwater Recharge at Water Factory 21," U.S. Environmental Protection Agency, Municipal Environmental Research Laboratory, EPA 600/2-80-114 (August 1980).
31. P. L. McCarty, D. Argo, and M. Reinhard, "Operational Experiences with Activated Carbon Adsorbers at Water Factor 21," *J. Am. Water Works Assoc.* 71(11) (1979): 683–689.
32. P. L. McCarty and M. Reinhard, "Trace Organics Removal by Advanced Wastewater Treatment," *J. Water Poll. Control Fed.* 52(7) (1980): 1907–1922.

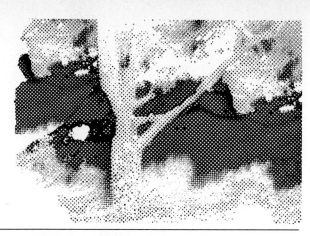

# Chapter 15
# Water-Quality Models

This chapter is intended as a brief introduction to the subject of water-quality modeling. It is included because this subject is so relevant to the design of water and wastewater treatment processes and to other aspects of water supply and pollution control. Water-quality models facilitate analysis of the impact of treatment levels on receiving waters and tracing effects of contaminants introduced into surface and groundwater systems. The references at the end of the chapter provide a beginning point for a more in-depth review of the development and application of water-quality models.

Water-quality models may be physical, analog, or digital, although our focus here will be on digital simulation. This form of modeling requires that there be mathematical expressions that equate water quality at a location of interest with factors that determine it. Such models vary in their complexity. Their nature depends on the application to be made of the model, the availability of data, and the level of understanding of the hydrochemical and hydrobiological processes involved. Unfortunately, the complexities of these processes, which are great, make the difficulties associated with hydrologic modeling seem inconsequential in comparison.

In general, water-quality models should permit acceptance of inputs in terms of pollutant (constituent) concentration versus time at points of entry to the system, description of the mixing and reaction kinetics in the stream element or groundwater element of concern, and synthesis of a time-distributed output indicating pollutant concentration at the outlet of the element (segment) being modeled. An analogy may be drawn to the streamflow routing performed in a downstream sequence from one stream channel segment to another. In the case of water-quality modeling, the common representation is the calculation of change in constituent concentration as it passes through successive elements of the water body being modeled.

As in the case of other water-resources-modeling processes, the approach may be deterministic or stochastic. In the case of water-quality models, the stochastic approach is often ruled out because actual records of water-quality parameters are unavailable for long enough periods to permit frequency methods to be used. Of course, generated sequences can be used for this purpose if adequate mathematical statements representing the kinetics of the system can be developed and their parameters determined [1, 2].

The deterministic approach to water-quality modeling requires that relationships between water-quality loading and the flow or hydraulic features of the system be established and that the appropriate chemical and biological reactions be tractable for solutions. Where theory-based relationships cannot be employed, empirical relationships are often used. The optimum model to use would naturally be the one best defining the actual water-quality response of the system. Many models have been developed; some of them are described in the references [3–25].

Pollutants may be classified as conservative or nonconservative (nonconservative pollutants decay with time). Somewhat more specifically, they may be organic, inorganic, radiological, thermal, or biological. Finally, they may be categorized by specific forms, such as BOD, phosphorus, nitrogen, bacteria, viruses, or specific toxic substances. These pollutants may be loaded into a watercourse or groundwater system from either point or nonpoint sources.

The time rate of delivery of a pollutant must be determined if its characteristics are to be modified by management practices or its impact on some element of the system evaluated. For example, the consequences of some quantity of silt delivered to a lake would not be the same if it were introduced over a period of 5 days as opposed to 2 hr. Thus, monitoring of water quality must generally be on a continuous basis if the data are to be of value for water resources planning and for developing continuous modeling processes.

Unstable pollutants such as radioactive materials, heat, BOD, and living organisms all have time-dependent decays and are thus nonconservative in nature. To deal with such constituents, it is necessary that both the mixing properties and the reaction kinetics of the system be known or approximated. On the other hand, many inorganic pollutants are conservative in nature, and their handling depends mainly on an ability to model the mixing mechanics of the receiving body of water.

The problems associated with modeling chemical and biological changes in a water body are many and complex. The field conditions encountered in natural water systems are highly varied and often negate the validity of reaction rate and other mechanisms determined under laboratory conditions. Furthermore, pollutants derived from nonpoint sources are subjected to many alterations in their travels over and through the ground before they reach a watercourse. The highly varied chemical, biological, and hydraulic characteristics of the land must be dealt with in estimating pollutant loadings from these sources. Although qualitative descriptions of hydrochemical and hydrobiological processes are easy to come by, their quantification is something else again. Fortunately, in some cases, empirical relationships between pollutant concentration and stream flow or other hydrologic variables can be used to describe water-quality loading mechanisms that cannot be obtained on a more theoretical basis. One such relationship is displayed in Fig. 15.1. It should be stressed, however, that a model is only as good as the data and theory on which it is based.

## 15.1 TYPES OF WATER-QUALITY MODELS

Most water-quality models in use today are designed to trace the movement of pollutants through streams, rivers, lakes, estuaries, and other open bodies of water. Groundwater-quality models have also been developed, but these generally lag the surface-water models in their state-of-the-art status. Water-quality models may be equipped to accept pollutant loadings at specific locations (points) and to accept pollutant inflows along a watercourse (nonpoint) as well.

Point-source water-quality models generally deal only with confined bodies (channels, groundwater systems) of water while nonpoint models must also take into consideration most other phases of the hydrologic cycle (Fig. 15.2). Since nonpoint pollutants are moved to streams, estuaries, etc., in overland flow, interflow, and groundwater flow processes, the representative models must include these phases of the hydrologic cycle in addition to the

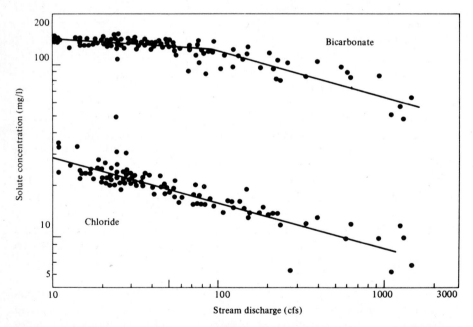

**Figure 15.1** Concentration–discharge relations for selected solutes—San Lorenzo River at Big Trees, California (From T. D. Steele, "Seasonal Variations in Chemical Quality of Surface Water in the Pescadero Creek Watershed, San Mateo County, California," Ph.D. dissertation, Stanford University, May 1968.)

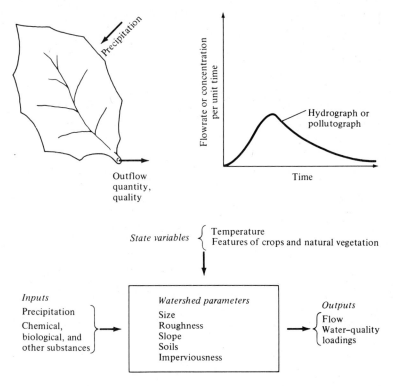

**Figure 15.2** Black-box concept of watershed modeling.

channel phase. As a result, nonpoint models are often thought of as *loading models* that act to trace the movement of pollutants from their originating locations to watercourses. Once a watercourse is reached, these loads (pollutants' inputs) can be handled by stream-quality models or groundwater-quality models, as the case may be. Since nonpoint models represent the introduction of pollutants to watercourses from land surfaces, they are strongly associated with the occurrence of precipitation events. Point-source models, on the other hand, usually represent continuous inputs of pollutants, primarily from the discharge of waste treatment works.

Many water-quality models are extensions of equations developed in 1925 by Streeter and Phelps for predicting BOD and DO in flowing streams [1, 4]. These equations are often supplemented by various simple first-order reaction equations and mixing and sedimentation models. The trend, however, is toward more complex multiconstituent water-quality models that can handle interactions involving numerous chemical constituents and biological organisms. Although such models require considerable data and are often demanding of computer time, they offer the promise of a better understanding of water-quality mechanics and have the potential for bringing about more efficient policies for water-quality management.

Water-quality models may be steady state or time varying in design. The former classification can be used where the principal variables are not time dependent or can be assumed to be so within a given stream reach or segment. These models are simpler and less expensive to operate than time-varying models, but they cannot handle rapidly changing situations, such as the introduction of nonpoint pollutants during a storm. In general, steady-state models are more suited to long-range planning, whereas time-varying models fit the need for setting policies for event management such as that associated with an intense storm.

Models may be deterministic or stochastic. Deterministic models deal mainly with projections of mean values of pollutant concentrations, while stochastic models incorporate the randomness of the physical, chemical, and biological processes being studied. While all real systems are three dimensional, the models employed to represent these systems may be one, two, or three dimensional in character. The choice depends on the use of the model's results and on the nature of the system being modeled. Where one-dimensional models are used, complete mixing is assumed in vertical and lateral directions. For two-dimensional models, either vertical or lateral mixing may be assumed; the choice depends on the nature of the system.

Water-quality models can be structured for solution by hand, desk calculator, or digital computer. If the situation to be modeled requires little computational effort, the results may be achieved quickly by manual procedures. For more elaborate cases, digital models may be required. The simplest model that can be relied on to produce results of the desired level of reliability should be chosen.

As in the case of water-quantity modeling, both simulation and optimization models may be employed. Each has its own place. Simulation models calculate the values of water-quality variables for given hydrologic, waste treatment, boundary, and initial conditions. Optimization models are used to identify management options that best fit preset management conditions.

Finally, both lumped- and distributed-parameter approaches may be taken. Lumped-parameter models are especially suited to large-scale systems analysis, while distributed-parameter models can provide a greater level of detail where localized decisions must be made. In either case, the models may be operated continuously or tailored to the simulation of specific events. Continuous simulation can produce water-quality histories that can be further analyzed by frequency methods so that inferences can be drawn about the risk associated with possible happenings. Event simulation models can be used to gather rather detailed information about the policies needed to cope with extreme occurrences.

## 15.2  AN ELEMENTARY WATER-QUALITY MODEL

Water-quality modeling efforts have been on the increase since the early 1960s. In the beginning, the most common parameters included in such models were temperature, BOD, and DO. The list has steadily increased until today many models incorporate salinity; carbonaceous, nitrogenous, and benthic BOD; temperature; total organic carbon; refractory organic carbon; sedimentary, soluble, and organic phosphorous; ammonia, nitrite, nitrate, and organic nitrogen; dissolved oxygen; toxic compounds; phytoplankton; and zooplankton [5]. The precision with which all of these constituents can be modeled is not uniform, however, and in some cases more research is needed. Modeling natural systems is difficult due to the many interactions and influences involved.

To provide some understanding of the way in which a water-quality model might be formulated, we shall consider a simple problem involving the discharge of waste to a receiving stream and the impact of this on the DO level at a downstream location. The nature of the problem is illustrated in Fig. 15.3. The waste flow from a city $q$ is found to have a DO level of zero and an ultimate BOD of $BOD_{UW}$. This waste flow is discharged to a receiving

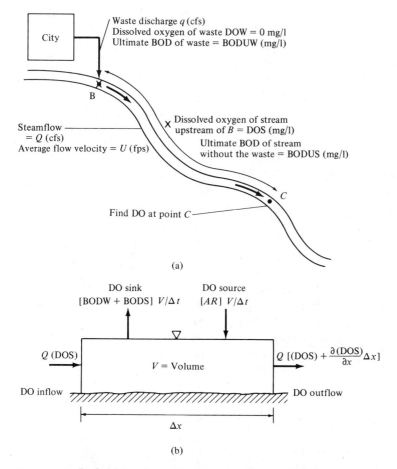

Figure 15.3 Definition sketch for DO model. (a) Problem setting. (b) DO, accounting in stream setting.

stream having a flow of $Q$, a DO level of DOS, and an ultimate BOD of $BOD_{US}$. Le us assume that the BOD rate constant $K_1$ and the stream reaeration coefficient $K_2$ are known. The problem is to formulate a model that can predict the DO level at point $C$ in the sketch, given the information above. Since the concern here is with the model formulation, it will not be necessary to assign numerical values to the variables and parameters specified above.

The principal components of the model will be streamflow $Q$, waste flow $q$, average flow velocity $u$, the DO upstream of waste discharge DOS, the DO of the waste $DO_W = 0$, the BOD of the waste flow and the stream $BOD_{UW}$ and $BOD_{US}$, respectively, coefficients $K_1$ and $K_2$, and any sources or sinks of oxygen in the region from $B$ to $C$. The sources of oxygen supply could include photosynthesis, reaeration, and the supply available in the stream. Oxygen sinks would be BOD in the flows and could also include sludge deposits on the stream bottom and algal respiration.

For illustrative purposes, we shall make the following assumptions:

1. There is steady flow (no variation with time).
2. The wastes are distributed uniformly across the stream cross section.
3. There is no dispersion along the stream path (no mixing in the downstream direction).
4. The decay rate for the waste may be represented by a first-order reaction.
5. There is only one point of waste entry in the stream segment of concern.
6. The effects of algae and bottom sludge deposits may be neglected.

It should be recognized that the foregoing assumptions may not be valid under many of the actual conditions encountered in the field. They are made here to simplify the model development and to emphasize that all models include assumptions of various severity and that these must be understood by the model user before deciding on its suitability.

Now that we have described the problem, made some simplifying assumptions, and made some statements about what is known (DOS, for example), we can begin to formulate the model. Refer again to Fig. 15.3 and note that the stream segment of interest is of length $\Delta x$, that its cross-sectional area will be defined as $A$, and that the volume of water contained in the stream from $B$ to $C$ will be designated $V$. With this additional definition, we can begin the model formulation by considering that between sections $B$ and $C$ of the stream, a continuity equation involving DO can be written. This takes the form

DO inflow + sources of DO − (DO outflow + DO sinks)
= change in DO storage in segment

or

$$Q(\text{DOS}) + g(\text{AR})V - \left\{ Q\left[ (\text{DOS}) + \frac{\partial(\text{DOS})}{\partial x} \Delta x \right] \right.$$

$$\left. + f(\text{BOD}_W + \text{BOD}_S)V \right\} = \frac{V\Delta(\text{DOS})}{\Delta t} \qquad (15.1)$$

where $g(\mathrm{AR})$ is a function of reaeration and $f(\mathrm{BOD_W} + \mathrm{BOD_S})$ is a function of the BOD exerted by the waste in the stream segment per unit of time. We now insert for $Q$ its equivalent $u(A)$ (the product of velocity and cross-sectional area), collect terms, rearrange, and divide through by $V$. Equation 15.1 becomes

$$\frac{\Delta(\mathrm{DOS})}{\Delta t} = -u\frac{\partial(\mathrm{DOS})}{\partial x} - f(\mathrm{BOD_W} + \mathrm{BOD_S}) + g(\mathrm{AR}) \qquad (15.2)$$

Now remembering that we have made the assumption that the flow in the stream segment is steady (stream flow and waste flow are constant, so eventually the DO concentration in the stream will also be unchanging), the left-hand term in Eq. 15.2 becomes zero, and we can write

$$u\frac{\partial(\mathrm{DOS})}{\partial x} = -f(\mathrm{BOD_W} + \mathrm{BOD_S}) + g(\mathrm{AR}) \qquad (15.3)$$

This equation may be written in terms of the dissolved oxygen deficit at a given location (Fig. 15.4) [5]. Thus, we may deal with the difference between DO under saturated conditions and the level of DO that actually exists at the location. If the equation is written in this form and integrated, the result is the classical Streeter-Phelps equation for dissolved-oxygen deficit [1, 4, 5]:

$$D = \frac{KL_0}{K_2 - K_1}(e^{-K_1 x/u} - e^{-K_2 x/u}) + D_0 e^{-K_2 x/u} \qquad (15.4)$$

where $L$ is the initial BOD load exerted on the stream by the waste ($\mathrm{BOD_W}$) and organic material in the stream itself ($\mathrm{BOD_S}$), $D_0$ is the initial level of DO in the stream, and $D$ is the dissolved-oxygen deficit at the downstream location a distance of $x$ units from the initial point (point $C$ in Fig. 15.3).

This equation and variations of it find wide use in water-quality modeling

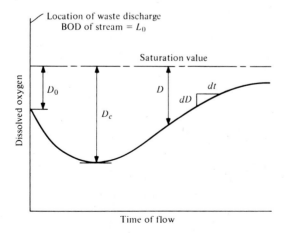

**Figure 15.4** Dissolved-oxygen sag curve.

of DO. It is the standard deterministic form and can be used successfully for its purpose if the assumptions made are valid for the situation, and satisfactory values of the constants $K_1$ and $K_2$ can be obtained. Note that these two parameters are actually functions of pollution loading and rate of streamflow. Thus, the assignment of fixed values is only an approximation that must be evaluated for its reasonableness at the conditions under which the model is to be used.

Equation 15.4 often appears in the form

$$D = \frac{k_1' L_0}{k_2' - k_1'} (e^{-k_1' t} - e^{-k_2' t}) + D_0 e^{-k_2' t} \tag{15.5a}$$

and since $e^{-k' t} = 10^{-kt}$, where $k = 0.434 k'$, the equation may also be written in terms of common logarithms as

$$D = \frac{k_1 L_0}{k_2 - k_1} (10^{-k_1 t} - 10^{-k_2 t}) + D_0 (10^{-k_2 t}) \tag{15.5b}$$

The proportionality factor $k_1$ is a temperature function. The proportionality factor $k_2$ is also a temperature function but, more important, it is a function of the turbulence of the stream.

A general approximate formula for the reaeration coefficient of natural rivers is given by O'Connor and Dobbins:

$$k_2' = \frac{(D_L U)^{1/2}}{H^{3/2}} \tag{15.6}$$

where    $k_2'$ = reaeration coefficient (base $e$) per hour
$\phantom{where}$    $D_L$ = diffusivity of oxygen in water = 0.000081 ft²/hr at 20°C
$\phantom{where}$    $U$ = velocity of flow, ft/hr
$\phantom{where}$    $H$ = depth of flow, ft

The effect of temperature on the reaeration coefficient $k_2$ is as follows [20]:

$$k_{2T} = k_{2-20} \times 1.047^{T-20} \tag{15.7}$$

where    $k_{2T}$ = reaeration coefficient at temperature $T$
$\phantom{where}$    $k_{2-20}$ = reaeration coefficient at 20°C

The value of $k_2'$ ranges from 0.20 to 10.0 per day, the lower values representing deep slow-moving rivers, the higher values shallow streams with steep slopes.

From an engineering design viewpoint, the dissolved-oxygen sag curve indicates the point of minimum DO. This critical point is the place in the stream where the rate of change of the deficit is zero and the demand rate equals the reaeration rate:

$$k_2' D_c = k_1' L = k_1' L_0 e^{-k_1' t_c} \tag{15.8}$$

Solving for the critical time $t_c$, one obtains

$$t_c = \frac{1}{k_2' - k_1'} \ln\left[\frac{k_2'}{k_1'}\left(1 - D_0\frac{k_2' - k_1'}{k_1'L_0}\right)\right] \tag{15.9}$$

These equations have constants that must be carefully evaluated. The $k_1'$ term reflects the rate at which bacteria demand oxygen and is calculated from the BOD test by running BOD determinations. The $k_2$ term is the reaeration characteristic of the stream and varies from reach to reach in most streams. Constant $k_1$ can be evaluated in the laboratory; $k_2$ must be determined from field studies. In the development of these equations, it is assumed that $k_1$ and $k_2$ are constant and that only one source of pollution exists and that the only oxygen demand is the BOD. Variations from these assumptions may be taken into account in any practical case. Some of the following processes, in addition, may be taking place in any given river stretch:

1. removal of BOD by adsorption or sedimentation;
2. addition of BOD along the river stretch by tributary inflow;
3. addition of BOD or removal of oxygen from the water by the benthal layer;
4. addition of oxygen by the photosynthetic action of plankton;
5. removal of oxygen by plankton respiration.

■ **EXAMPLE 15.1**

A city of 200,000 population produces sewage at the rate of 120 gpcd, and the sewage plant effluent has a BOD of 28 mg/l. The temperature of the sewage is 25.5°C, and there is 1.8 mg/l of DO in the plant effluent. The stream flow is 250 cfs at 1.2 fps and the average depth is 8 ft. The temperature of the water is 24°C before the sewage is mixed with the stream. The stream is 90% saturated with oxygen and has a BOD of 3.6 mg/l. The deoxygenation coefficient $k_1'$ is equal to 0.50 at 20°C. Determine the following:

a. the sewage flow, cfs;
b. DO of the mixture of water and sewage plant effluent;
c. the temperature of the mixture of water and sewage plant effluent;
d. the value of the initial oxygen deficit for the river just below the plant discharge;
e. the distance downstream to the point of minimum DO;
f. the minimum DO in the stream below the sewage plant.

*Solution*

a.  The sewage flow, cfs:

$$(120 \text{ gpcd})(2 \times 10^5) = 240 \times 10^5 \text{ gpd} = 24 \text{ mgpd}$$

$$1 \text{ mgpd} = 1.547 \text{ cfs}$$

$$\text{sewage flow} = (24)(1.547) \text{ cfs} = 37.1 \text{ cfs}$$

b.  The DO of the mixture of water and sewage plant effluent: Assuming a pressure of 1 atm and a chloride concentration of zero, the solubility of oxygen in water at 24°C is 8.5 mg/l. The DO of the river water is $(8.5$ mg/l$)(0.90) = 7.65$ mg/l. The DO of the mixture is

$$DO_m = \frac{Q_r(DO_r) + Q_s(DO_s)}{Q_r + Q_s}$$

$$= \frac{(250 \text{ cfs})(7.65 \text{ mg/l}) + (37.1 \text{ cfs})(1.8 \text{ mg/l})}{287.1 \text{ cfs}}$$

$$= 6.89 \text{ mg/l}$$

c. The temperature of the mixture of water and sewage plant effluent: As in b,

$$T_m = \frac{(250 \text{ cfs})(24°C) + (37.1 \text{ cfs})(25.5°C)}{287.1 \text{ cfs}}$$

$$= 24.2°C$$

d. The value of the initial oxygen deficit: Assuming pure water at 24.2°C. The saturation value of $O_2$ is 8.48 mg/l, but we have 6.89 mg/l.

deficit $= 8.48 - 6.89 = 1.59$ mg/l

e–f. The distance downstream to the minimum point of DO and the value of the minimum DO. The constant $k_2'$ may be computed from Eq. 15.6.

$$k_{2-20}' = \frac{(D_L U)^{1/2}}{H^{3/2}}$$

$$= \frac{(81 \times 10^{-6} \text{ ft/hr} \times 1.2 \text{ ft/sec} \times 3600 \text{ sec/hr})^{1/2} \times 24 \text{ hr/day}}{(8 \text{ ft})^{3/2}}$$

$$= 0.627$$

The values of $k_2'$ and $k_1'$ at 24.2°C are found by Eq. 15.7.

$$k_{2T} = k_{2-20} \times 1.047^{T-20}$$

$$k_{2-24.2}' = 0.627 \times 1.047^{24.2-20}$$

$$= 0.76$$

$$k_{1-24.2}' = 0.50 \times 1.047^{4.2}$$

$$= 0.607$$

The value of the initial BOD, $L_0$, of the mixture of river water and plant effluent is

$$L_0 = \frac{(250 \text{ cfs})(3.6 \text{ mg/l of BOD}) + (37.1 \text{ cfs})(28 \text{ mg/l of BOD})}{287.1 \text{ cfs}}$$

$$= 6.75 \text{ mg/l of BOD}$$

The time $t_c$ to the point of minimum DO is found by Eq. 15.9.

$$t_c = \frac{1}{k_2' - k_1'} \ln\left\{\frac{k_2'}{k_1'}\left[1 - \frac{D_0(k_2' - k_1')}{k_1' L_0}\right]\right\}$$

$$= 1.08 \text{ days}$$

The value of the minimum DO is found by using Eq. 15.5.

$$D_c = \frac{k_1' L_0}{k_2' - k_1'}(e^{-k_1' t_c} - e^{-k_2' t_c}) + D_0 e^{-k_2' t_c}$$

$$= 2.80 \text{ mg/l}$$

The distance downstream at which the critical DO occurs is calculated from the value of $t_c$ and the velocity of flow.

distance = 1.2 ft/sec × 3600 sec/hr × 24 hr/day × 1.08 days
= 112,000 ft

Thus, under the given conditions the oxygen sag curve takes the form shown in Fig. 15.5.  ■

The foregoing discussion should convey to the reader that the formulation of water-quality models follows the same general pattern as that for hydrologic models. Both include continuity considerations combined with appropriate equations of motion and reaction (BOD decay, for example). The beginning of the modeling process is thus an understanding of the mechanics of the system of concern and an ability to represent the mechanics in adequate

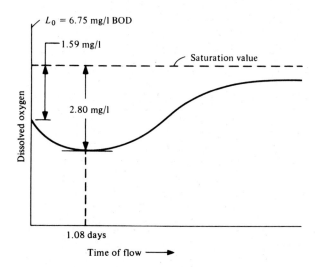

**Figure 15.5** Dissolved-oxygen sag curve for Example 15.1.

mathematical terms. The following descriptions of several types of water-quality models will bring these ideas into sharper focus.

## 15.3 EPA STORMWATER MANAGEMENT MODEL

A widely used stormwater runoff model with capability for modeling the movement of certain water-quality constituents in urban areas is the EPA stormwater management model (SWMM) [2, 3, 6, 7]. The model can simulate the runoff from an area for any prescribed rainfall pattern. In using the model, the drainage area is broken up into a number of subareas having approximately uniform properties. The flow diagram for the runoff portion of the model is given in Fig. 15.6.

The model has the capability of determining stormwater flows and water quality at locations in a stormwater system and receiving body of water. It is really a nonpoint pollution model since it deals with the movement of pollutants from the land surface of the area to combined sewers or storm drainage outfalls. The model is of the event simulation type and is not designed to develop long continuous histories.

The computer program for SWMM consists of a control segment and five computational blocks: (1) executive, (2) runoff, (3) transport, (4) storage, and (5) receiving water.

The *executive* block is the control for computational blocks, with all access and transfers between the other blocks routed through its subroutine MAIN. The *runoff* block accepts rainfall data and a description of the drainage system as inputs and provides both hydrographs and time-dependent pollutional graphs (pollutographs). The *transport* block routes through the distribution system flows, which are corrected for both dry-weather flow and infiltration. It is also capable of producing hydrographs and pollutographs at specified points. In the *storage* segment, flows from the transport block are received, and the effect of any treatment provided is considered. The effects of the discharge on a receiving stream are determined in the *receiving water* block. Segments may be run separately to facilitate adjustments.

The drainage area is divided into subcatchments, gutters, and pipes. Subcatchments are divided into three parts: pervious, impervious with surface detention, and impervious without surface detention. Subcatchments are defined by the area, width, slope, and ground cover, while gutters and pipes are described by the slope, length, and Manning roughness coefficient. This information and rainfall data in the form of hyetographs are read into the runoff block. A step-by-step computation of runoff volume is then initiated [2].

The input to the water-quality portion of the SWMM consists of hydrographs developed in the hydrologic phase of the model. The output takes the form of pollutographs for each of the pollutants modeled. The hydrographs and pollutographs that are calculated are then introduced into the transport block, which combines them with the dry-weather and infiltrated flow compo-

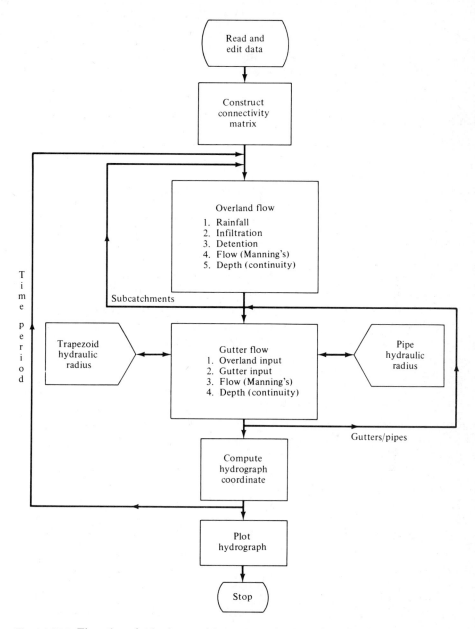

**Figure 15.6** Flowchart for hydrographic computation. (After Metcalf and Eddy, Inc., University of Florida, and Water Resources Engineers, Inc., [3].)

nents to produce the actual outfall graphs for water quality and quantity. The SWMM is capable of predicting the concentrations of suspended solids, BOD, total coliform, COD, settleable solids, nitrates, phosphates, and grease in stormwater runoff [7].

The SWMM makes use of the assumption that the amount of a pollutant that can be removed from a drainage area during a storm event is a function of the storm duration and initial quantity of the pollutant. This can be represented by a first-order differential equation of the form

$$-\frac{dP}{dt} = kP \tag{15.10}$$

which integrates to

$$P_0 - P = P_0(1 - e^{-kt}) \tag{15.11}$$

where    $P_0$ = initial amount of pollutant per unit area
$P$ = remaining amount of pollutant per unit area at time $t$
$k$ = decay rate

The value of $k$ is assumed to be directly proportional to the rate of runoff. In the model, this is represented as $k = br$, where $b$ is a constant and $r$ is the runoff rate. A value of 4.6 has been used for $k$ based on analyses of storm event data from urban areas. This implies identical rainfall intensities and wash-off rates for all storms, a condition not met with in reality.

For the prediction of suspended solids and BOD, it has been determined that a modification of Eq. 15.11 is needed. This change incorporates an availability factor $A_0$, which represents the percentage of pollutant $P$ that is available for capture by the storm [7]. Thus, Eq. 15.11 becomes

$$P_0 - P = A_0 P_0(1 - e^{-kt}) \tag{15.12}$$

Coliform densities are predicted as the product of the suspended-solids concentration and an appropriate conversion factor. The EPA provides complete information on methods used to establish the required model parameters [3].

For each time step in the modeling process, the rate of runoff is calculated using the hydrologic model. A value of $P$ is also determined (Eq. 15.11 or 15.12) and then becomes the value of $P$ for the next time step. Then, during the time interval, the change in value of $P$ can be related to the quantity of flow from the area to produce the pollutograph of the constituent of interest. Calculations proceed from one time step to the next until the storm event has ended.

Calibration of the model centers around a trial-and-error procedure to determine the ideal combination of loading rate and removal coefficient that would result in a satisfactory match of the observed and computed pollutographs. The parameters derived in this manner are valid only for the particular storm used in the calibration and should not be used for other storms unless their features are considered to be quite similar. The SWMM has the flexibility to determine pollutant loading from a variety of urban land use characterizations. It can also be used to generate input data for use in stream quality models such as the one discussed in the preceding section. The EPA summarizes the model's application to a number of urban areas and storm events [7].

## 15.4  EPA QUAL II MODEL

Qual II is a comprehensive stream water-quality model that can simulate up to 13 water-quality constituents in any desired combination. The model can accommodate DO; BOD; temperature; algae as chlorophyll *a*; ammonia, nitrite, and nitrate nitrogen; dissolved orthophosphate as P; coliforms; an arbitrary nonconservative constituent; and three conservative constituents [8]. The model may be applied to branching stream systems that are well mixed. The assumption is made that advection and dispersion are significant only in the principal direction of flow. Multiple waste discharges, withdrawals, tributary flows, and incremental (lateral) inflows can be incorporated in this model.

The model is limited to the simulation of time periods during which stream flows are approximately constant and waste inputs are constant as well. It is possible, however, to operate the model in both steady-state and dynamic modes. In the latter case, diurnal variations in water quality can be studied. The model is an outgrowth of an earlier version, Qual I, which was developed in 1970 [9]. It has seen wide use in water-quality planning and management programs. Qual II permits input or output of data in metric units [8, 10].

Qual II makes use of several different types of computational elements to represent the prototype stream system: the headwater element, standard element, element just upstream from a junction, junction element, last element in system, input element, and withdrawal element. Figure 15.7 shows how the prototype system is represented and also how stream branches may be subdivided into computational elements. River reaches are the basis for entering most data. They are aggregates of computational elements. Within a specified reach, hydraulic data, reaction rate coefficients, initial conditions, and incremental runoff data are considered to be constant. The definition of a reach is also illustrated in Fig. 15.7. The dimensional limitations of the model are summarized at the bottom of the figure.

The following paragraphs describe the conceptual form of Qual II, its underlying functional representation, its hydraulic features, the longitudinal dispersion model, and constituent reactions and interactions [8, 10].

### The Qual II Conceptual Form

Figure 15.8 illustrates a stream reach that has been separated into several subreaches or computational elements. For each computational element of length $\Delta x$, a hydrologic balance can be written in terms of the flow into the upstream face of the element, external sources of withdrawals, and the outflow across the downstream face of the element. The nomenclature for these flow components is given in Fig. 15.8. Note that this accounting follows that for the elementary DO model discussed earlier. In a similar fashion, a materials balance for any water-quality constituent of interest can also be struck. In the materials balance, we consider both advection (transport) and

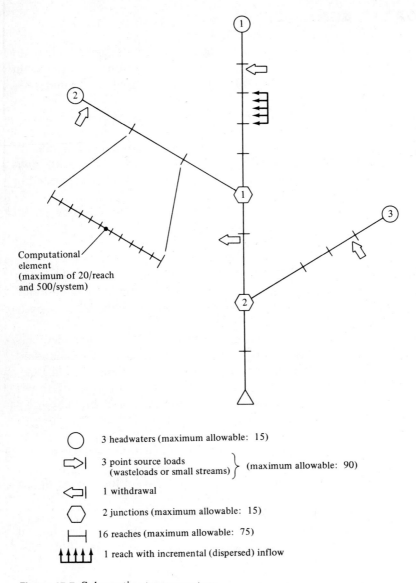

Computational
element
(maximum of 20/reach
and 500/system)

○    3 headwaters (maximum allowable: 15)

⇨⌐    3 point source loads
         (wasteloads or small streams) ⎱ (maximum allowable: 90)

⧀⌐    1 withdrawal

⬡    2 junctions (maximum allowable: 15)

⊢⊣    16 reaches (maximum allowable: 75)

↑↑↑↑↑    1 reach with incremental (dispersed) inflow

**Figure 15.7** Schematic stream system.

dispersion as mechanisms for moving the constituent mass along the stream
axis. Mass can be added by waste streams and internal sources such as
bottom deposits (benthic sources). It can also be removed by internal sinks
of the same nature. The model assumes that there is complete mixing within
each computational element.

Refer again to Fig. 15.8; it will be seen that the stream can be likened to
a series of completely mixed reactors (computational elements) connected
by the mechanisms of advection and dispersion. Furthermore, groups of such

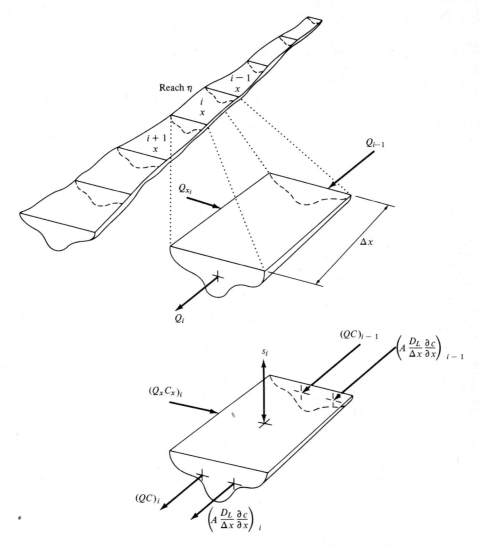

**Figure 15.8** Discretized stream system. (After Water Resources Engineers Inc. [15].)

elements in sequence can be seen to constitute reaches as defined in Fig. 15.7. These reaches will be assumed to have the same properties relative to hydraulic character, biological and chemical rate constants, and other factors. The groupings of reaches ultimately define the entire stream system.

## Functional Representation

The basic equation of the Qual II model is the advection–dispersion mass transport equation conceptualized in Fig. 15.8 [8]. This equation is numerically integrated over time for each water-quality constituent of concern.

Embodied in the equation are the effects of advection, dispersion, dilution, the reactions and interactions of constituents, and any identified sources or sinks. For a particular constituent $C$ the equation may be written

$$\frac{\partial M}{\partial t} = \frac{\partial(A_x D_L \partial C/\partial x)}{\partial x} dx - \frac{\partial(A_x \bar{u} C)}{\partial x} dx + (A_x \, dx)\frac{dC}{dt} + s \qquad (15.13)$$

where   $M$ = mass (M)
       $x$ = distance (L)
       $t$ = time (T)
      $C$ = concentration (M/L$^3$)
     $A_x$ = cross-sectional area (L$^2$)
      $D$ = dispersion coefficient (L$^2$/T)
      $\bar{u}$ = mean velocity (L/T)
      $s$ = external source or sinks (M/T)

Since $M = VC$, we can write

$$\frac{\partial M}{\partial t} = \frac{\partial(VC)}{\partial t} = V\frac{\partial C}{\partial t} + C\frac{\partial V}{\partial t} \qquad (15.14)$$

where $V$ is the incremental volume and is the product of the cross-sectional area $A_x$ and the length of the element $\Delta x$. Now, if we assume that steady-flow conditions prevail, the volume contained in the computational element must remain unchanged with time. Thus, $\partial V/\partial t = 0$ and Eq. 15.14 reduces to

$$\frac{\partial M}{\partial t} = V\frac{\partial C}{\partial t} \qquad (15.15)$$

Inserting $V(\partial C/\partial t)$ for $\partial M/\partial t$ in Eq. 15.13 and rearranging gives

$$\frac{\partial C}{\partial t} = \frac{\partial(A_x D_L \partial C/\partial x)}{A_x \partial x} - \frac{\partial(A_x \bar{u} C)}{A_x \partial x} + \frac{dC}{dt} + \frac{s}{V} \qquad (15.16)$$

The terms on the right-hand side represent dispersion, advection, constituent changes, external sources and sinks, and dilution, in that order. The term $dC/dt$ reflects constituent changes such as decay or growth. Examples of such changes are reaeration, coliform die-off, and algal respiration and photosynthesis.

## Hydraulic Features

It has been noted that an assumption of Qual II is that stream flow is steady state. Thus, $\partial Q/\partial t = 0$. From Fig. 15.8 it can be seen that the flow balance for a computational element can be written

$$\left(\frac{\partial Q}{\partial x}\right)_i = (Q_x)_i \qquad (15.17)$$

where $(Q_x)_i$ is the sum of external withdrawals and/or inflows to the element. After the flow is determined using Eq. 15.17, it is a simple matter to determine other hydraulic properties such as the average velocity $\bar{u}$ and the depth of flow and cross-sectional area.

## Longitudinal Dispersion

*Dispersion* is defined as being associated with spatially averaged velocity variation [8]. It is a convective transport mechanism. Various investigators have studied the dispersion process. The Qual II model recognizes an expression for the dispersion coefficient $D_L$ developed by Elder for relatively narrow channels where one-dimensional flow can be assumed:

$$D_L = 5.93du^* \tag{15.18}$$

where $d$ is the mean depth in feet in the stream and $u^*$ is the average shear velocity [8]. For very wide channels, an expression derived by Fisher is used:

$$D_L = 22.6n\bar{u}d^{0.833} \tag{15.19}$$

where   $D_L$ = longitudinal dispersion coefficient, ft$^2$/sec
$\quad\quad\quad n$ = Manning's roughness coefficient
$\quad\quad\quad \bar{u}$ = mean velocity, fps
$\quad\quad\quad d$ = mean depth, ft

## Constituent Reaction

A number of rate parameters as well as parameters describing various chemical and biological processes are required for use in the Qual II model. Reactions affecting the increase or decrease of constituent concentrations are often expressed in first-order kinetic terms. Thus, it is assumed that reaction rates are proportional to constituent concentrations. Although the actual processes modeled are sometimes characterized by higher-order kinetics, the assumption that they can be approximated by first-order relationships often proves adequate for many natural river systems [11]. Equation 15.4, representing the dissolved-oxygen deficit in a river reach, is illustrative of the types of reaction equations used in many water-quality models.

In the Qual II model, major interactions of the nutrient cycles, atmospheric aeration, algal production, benthic oxygen uptake, carbonaceous oxygen demand, and their effect on dissolved oxygen in a stream are considered. These interactions are illustrated in Fig. 15.9. The arrows show the direction of normal system progression in a moderately polluted situation [8]. Under some circumstances, as when oxygen supersaturation occurs, oxygen can be driven from solution and the direction shown in the figure will be in error.

The coliforms and the arbitrary nonconservative constituent are considered as decaying constituents. They are not treated as having interactions with other constituents. By definition, the conservative constituents do not

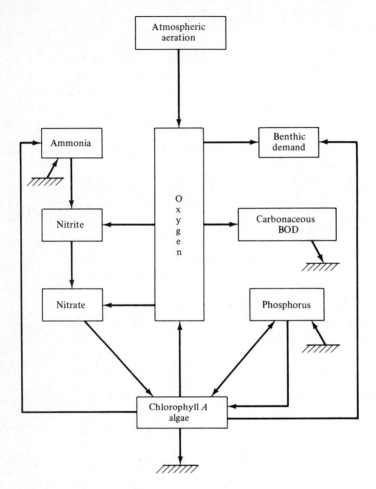

**Figure 15.9** Major constituent interactions.

decay or interact. They are subject only to mixing and thus dilution. The mathematical relationships describing individual constituent reactions and interactions are described in the EPA computer program documentation manual for Qual II [8]. A more detailed treatment of these is beyond the scope of this text. For further information, the EPA manual and other references should be consulted [1, 6, 9, 10, 12].

## 15.5 LAKE AND RESERVOIR MODELING

The modeling of water quality in lakes and reservoirs has common features with the stream system models discussed. That is, they both involve the water budget and the reactions and interactions of water-quality constituents.

Lakes and reservoirs have another feature, however, that makes their modeling different and more complex than that for flowing streams. The reason for this is that lakes have very different dimensions from flowing streams and their flow characteristics are also different. The factors that must be reckoned with in modeling large open bodies of water include depth, length, width, surface area, the nature of bottom materials, surrounding ground cover, prevailing winds, climate, and surface inflows and outflows. In general, each lake has its own special features that require individual treatment. Thus, lake models tend to vary considerably from the generalized conceptual model of an idealized lake.

In developing a lake model, as for a stream system, it is necessary to construct a water budget. The rate of inflow and outflow and the flow pattern determine, in part, the manner in which constituents are introduced and the degree of flushing from the system. The mixing processes of the lake are also important determinants of the fate of pollutants. The rate of removal of a particular constituent from a lake is highly dependent on the degree of mixing occurring in the system. The principal mixing mechanisms for lakes are wind actions, temperature changes, and atmospheric pressure. In small lakes with shallow depths, even power boat activities may be influential.

The thermal properties of lakes play an important role in the mixing of their waters. In the temperate zone, for example, most lakes have two thermal circulation periods, one in the spring and the other in the fall. During these times, there occurs a condition of temperature homogeneity associated with intense vertical circulation.

Most lakes and reservoirs have inlets and outlets; the flow generated by these establishes what is known as a gradient or slope current. Such a current is usually small for large lakes and thus its importance in mixing is usually insignificant compared to that of the wind. Currents may also be generated by oscillations in the lake surface resulting from differences in barometric pressure on opposite shores and by the reaction that takes place, after wind subsidence, with the water level at the windward shore rising and that at the leeward shore falling. Of all of these influences, the wind is usually the most important in generating currents.

Exacting analyses of circulation in stratified lakes are difficult to obtain. This is due to the complexity of the circulation processes that must be dealt with. There have been many studies of these processes, and various models are reported [1, 13–15]. In general, numerous assumptions are incorporated and the models are highly site specific.

A water-quality model for use in lake systems analysis was developed by Chen and Orlob [13]. This model is based on a multilayered hypothesis in which horizontal sections are assumed to be completely mixed and therefore unstratified. The lake's hydrodynamic behavior is considered to be density dependent, which means that temperature is the principal determining variable.

Inflows from tributaries are considered to enter the lake at a level where the surrounding fluid density is their equal. Because of incompressibility considerations, the inflows are assumed to generate advective flows between the elements above their level of entry. Use of the model requires various input data. These include hydrologic and water-quality parameters of tributary inflows, the quantity and nature of waste discharges, the character of the outflows, and meteorologic conditions. Computations can be performed over time periods ranging up to 24 hr. To facilitate computations, each layer in the lake is considered as a discrete continuously stirred tank reactor. The inflow and outflow from each element is determined by the use of hydrodynamic relationships. Changes in concentration of water-quality constituents are tied to conditions in effect in the assumed reactors. Mass balance equations describing the physical, chemical, and biological processes are used to estimate the rates of change. Numerical solution techniques are employed to calculate constituent concentrations as a function of time.

## 15.6 GROUNDWATER-QUALITY MODELS

Groundwater quality has become a major source of concern in recent years. This has come about from the realization that many groundwater sources that were at one time considered almost pristine have now been degraded in quality by seepages from dumps, leakage from industrial waste holding ponds, and by other waste disposal and industrial and agricultural practices. To deal with such problems, there has been an expanding movement to develop quantitative techniques to understand the mechanics of groundwater quality. These models, although not as advanced as their surface-water counterparts, are now beginning to play an important role in water-quality management [16–18].

In 1974 Gelhar and Wilson developed a lumped parameter model for dealing with water quality in a stream-aquifer system. The nomenclature and conceptualization of their model are shown in Fig. 15.10 [18]. The rationale for using a lumped-parameter approach was that when dealing with changes in groundwater quality over long periods of time, temporal rather than spatial variations are most important.

Changes in the water table in the Gelhar-Wilson (GW) model are represented by the following equation:

$$p\frac{dh}{dt} = -q + \varepsilon + q_r - q_p \tag{15.20}$$

where $h$ = average thickness of the saturated zone
$p$ = average effective porosity
$\varepsilon$ = natural recharge rate
$q$ = natural outflow from the aquifer
$q_r$ = artificial recharge per unit area
$q_p$ = pumping rate per unit area
$t$ = time

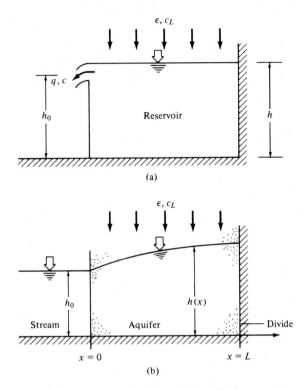

**Figure 15.10** Schematic of the Gelhar-Wilson model.
(After Novotny and Chesters [6].)

This is just another form of the continuity equation relating inflow, outflow, and change in storage (the left-hand term in Eq. 15.20). The change in concentration of a constituent is given by

$$ph\frac{dc}{dt} + (\varepsilon + q_r + \alpha ph)c = \varepsilon c_L + q_r c_r \qquad (15.21)$$

where  $c$ = concentration
  $c_L$ = concentration of the natural recharge
  $c_r$ = concentration of the artificial recharge
  $\alpha$ = first-order rate constant for degradation of the contaminant

The GW model assumes that dispersion is negligible. This assumption may be made on the basis that the objective of the model is to estimate regional average concentrations [6]. The model also provides for the determination of hydraulic and solute response times for the system. These are measures of the lag that occurs in the movement of both water and constituent inputs to the system. Gelhar and Wilson assume that the response of an aquifer to a specific input can be likened to that of a well-mixed linear reservoir. Their studies showed that the model's determination of the concentration of constituents leaving an aquifer is representative of the average concentration of the constituent in the aquifer. On this basis, it appears that the model is well suited to estimating the quality of groundwater discharging to a surface stream, provided the aquifer is narrow relative to the length along which discharge occurs.

## 15.7  RELIABILITY OF WATER-QUALITY MODELS

Although water-quality models have become recognized tools to aid planners and managers, it must be understood that their usefulness, reliability, and acceptance are quite variable. Even when the theory underlying a given situation to be modeled is well understood, the complexity of most issues requires that simplifying assumptions be made. The introduction of such assumptions creates an element of uncertainty in model output. The degree of uncertainty depends on the nature of the model and the conditions specified.

The blind acceptance by planners and managers of the sophisticated-looking outputs of many models is cause for concern. Most models are only rough approximations and their use must incorporate this understanding. On the other hand, the wise application of modeling techniques to planning and management issues adds a dimension of power that should be made the most of. If one understands the model and its limitations, interpretation of its results can be enlightening. Models are often most useful for comparing alternatives rather than being depended on to produce single outcomes.

According to Novotny and Chesters, the most accurate models are hydrologic models relating to small impervious areas. The least reliable are models of water quality for large watersheds [6].

## PROBLEMS

**15.1**  The 5-day, 20°C BOD of a waste is 250 mg/l, and its temperature is 85°F. The 5-day 20°C BOD of the water in a stream is 5 mg/l, and its temperature is 55°F. What will be the 5-day, 20°C BOD of the stream below the point of waste discharge when the total flow is 150 cfs and the waste flow is 0.5 mgpd? Assume that the BOD rate is constant and the $K$ at 20°C is 0.35 for both the stream and the waste.

**15.2**  The sewage flow from a city is $25 \times 10^6$ gpd. If the average 5-day 20°C BOD is 265 mg/l, compute the total daily oxygen demand in pounds and the population equivalent of the sewage.

**15.3**  A city of 150,000 population produces sewage at the rate of 150 gpcd, and the waste treatment plant effluent has a BOD of 19 mg/l. The temperature of the waste is 26.1°C, and there is 1.4 mg/l of DO in the plant effluent. The streamflow is 161 cfs at 1.3 fps and an average depth of 5 ft. The temperature of the stream before the waste flow is introduced is 21.2°C. The stream is 85% saturated with oxygen and has a BOD of 1.8 mg/l. The deoxygenation coefficient $k_1$ is equal to 0.45 at 20°C. Graph the oxygen sag curve.

**15.4** If all 7 mg/l of DO in a clean turbulent stream can be counted on for the oxidation of organic matter in a domestic waste (the reaeration is a factor of safety against nuisance), what should be the dilution factor in cfs per 1000 population if the waste flow is 150 gpd and its BOD is 230 mg/l?

**15.5** A city discharges 4 mgpd of 15°C sewage with a 5-day, 20°C BOD of 250 mg/l into a stream whose discharge is 78 cfs; the water temperature is 17°C. If $k_2$ is equal to 0.28 at 20°C, what is the critical oxygen deficit and the time at which it occurs? Assume that the stream is 100% saturated with oxygen before the sewage is added.

**15.6** A stream with $k_2$ equal to 0.31, temperature of 28°C, and minimum flow of 800 cfs receives 10 mgd of sewage from a city. The river water is saturated with DO above the city. What is the maximum permissible BOD of the sewage if the dissolved-oxygen content of the stream is never to be below 4.0 mg/l?

# REFERENCES

1. D. P. Loucks, J. R. Stedinger, and D. A. Haith, *Water Resources Systems Planning and Analysis* (Englewood Cliffs, NJ: Prentice-Hall, 1981).
2. W. Viessman, Jr., G. L. Lewis, and J. W. Knapp, *Introduction to Hydrology* (New York: Harper and Row, 1989).
3. W. C. Huber, J. P. Heaney, S. J. Nix, R. E. Dickinson, and D. J. Polmann, *Storm Water Management Model User's Manual, Version III,* EPA-600/2-84-109a (NTIS PB84-198423). Cincinnati, OH: Environmental Protection Agency, Nov. 1981.
4. N. W. Streeter and E. B. Phelps, *U.S. Public Health Service Bull.* No. 146 (1925).
5. R. H. French, "Fundamentals of Modeling: Concepts, Components, How to Formulate a Model, Applicability to Water Quality Problems," in Lecture Notes on Water Quality Modeling, Water Resources Center, Desert Research Institute, University of Nevada, Reno, 1981.
6. V. Novotny and G. Chesters, *Handbook of Nonpoint Pollution, Sources and Management* (New York: Van Nostrand Reinhold, 1981).
7. U.S. Environmental Protection Agency, *Maximum Utilization of Water Resources in a Planned Community, Application of the Storm Water Management Model,* Vol. I, Cincinnati, OH: Municipal Environmental Research Laboratory, 1979.
8. U.S. Environmental Protection Agency, *Computer Program Documentation for Stream Quality Model (Qual II),* EPA-600/9-81-014. Athens, GA: Environmental Research Laboratory, 1981.
9. Texas Water Development Board, "Qual-I-Simulation of Water Quality in Streams and Lands," *Program Documentation and Users' Manual,* PB 202973. Springfield, VA: National Technical Information Service, 1970.
10. U.S. Environmental Protection Agency, *User's Manual for Stream Quality Model (Qual II),* EPA-600/9-81-015. Athens, GA: Environmental Research Laboratory, 1981.
11. A. K. Biswas (ed.), *Models for Water Quality Management* (New York: McGraw-Hill, 1981).

12. Nebraska Natural Resources Commission, "Lower Platte River Basin Water Quality Management Plan," Lincoln, NE, June 1974.

13. C. W. Chen and G. T. Orlob, "Ecologic Simulation for Aquatic Environments, Final Report," OWRR Project No. C-2044, Office of Water Resources Research, Washington, DC: Department of the Interior, Dec. 1972.

14. P. S. Lombardo and D. D. Franz, *Mathematical Model of Water Quality in Rivers and Impoundments* (Palo Alto, CA: Hydrocomp, Inc., 1971).

15. Water Resources Engineers, Inc., *Prediction of Thermal Energy Destruction in Streams and Reservoirs*. Prepared for the California Department of Fish and Game, 1967.

16. R. A. Freeze and J. A. Cherry, *Groundwater* (Englewood Cliffs, NJ: Prentice-Hall, 1979).

17. E. F. Wood, R. A. Ferrara, W. G. Gray, and G. F. Pinder, *Groundwater Contamination from Hazardous Wastes* (Englewood Cliffs, NJ: Prentice-Hall, 1984).

18. L. W. Gelhar and J. L. Wilson, "Ground Water Quality Modeling," in *Proceedings of the 2nd National Ground Water Quality Symposium*. Washington, DC: U.S. Environmental Protection Agency, 1974.

19. R. H. French, "A Statistical Approach to the Modeling of Conservative and Nonconservative Substances," in Lecture Notes on Water Quality Modeling, Water Resources Center, Desert Research Institute, University of Nevada, Reno, 1981.

20. P. S. Lombardo, *Critical Review of Currently Available Water Quality Models* (Palo Alto, CA: Hydrocomp, Inc., 1973).

21. Texas Water Development Board, "DOSAG-I Simulation of Water Quality in Streams and Canals," *Program Documentation and Users' Manual*. Springfield, VA: National Technical Information Service, 1970.

22. W. Viessman, Jr. and C. Welty, *Water Management: Technology and Institutions* (New York: Harper & Row, 1984).

23. J. D. O'Connor and W. E. Dobbins, "Mechanism of Reaeration in Natural Streams," *Trans. Am. Soc. Civil Eng.* **123,** 655 (1958).

24. R. V. Thomann and J. A. Mueller, *Principles of Surface Water Quality Modeling and Control* (New York: Harper & Row, 1987).

25. R. C. Johanson, J. C. Imhoff, and H. Davis, Jr., *Users Manual for Hydrological Simulation Program—Fortran (HSPF)*, EPA/9-80-015. Athens, GA: Environmental Protection Agency, Apr. 1980.

# APPENDIX

**TABLE A.1** WEIGHTS AND MEASURES

### Length

| cm | m | in. | ft | yd | mi |
|---|---|---|---|---|---|
| 1 cm = 1 | 0.01 | 0.3937008 | 0.03280840 | 0.01093613 | $6.213712 \times 10^{-6}$ |
| 1 m = 100 | 1 | 39.37008 | 3.280840 | 1.093613 | $6.213712 \times 10^{-4}$ |
| 1 in. = 2.54 | 0.0254 | 1 | 0.08333333... | 0.02777777... | $1.578283 \times 10^{-5}$ |
| 1 ft = 30.48 | 0.3048 | 12 | 1 | 0.3333333... | $1.893939... \times 10^{-4}$ |
| 1 yd = 91.44 | 0.9144 | 36 | 3 | 1 | $5.681818... \times 10^{-4}$ |
| 1 mi = $1.609344 \times 10^5$ | $1.609344 \times 10^3$ | $6.336 \times 10^4$ | 5280 | 1760 | 1 |

### Area

| cm² | m² | in.² | ft² | acre | mi² |
|---|---|---|---|---|---|
| 1 cm² = 1 | $10^{-4}$ | 0.155003 | $1.076391 \times 10^{-3}$ | $2.5 \times 10^{-8}$ | $3.861022 \times 10^{-11}$ |
| 1 m² = $10^4$ | 1 | 1550.003 | 10.76391 | $2.5 \times 10^{-4}$ | $3.861022 \times 10^{-7}$ |
| 1 in.² = 6.4516 | $6.4516 \times 10^{-4}$ | 1 | $6.944444 \times 10^{-3}$ | $1.59 \times 10^{-7}$ | $2.490977 \times 10^{-10}$ |
| 1 ft² = 929.0304 | 0.09290304 | 144 | 1 | $2.3 \times 10^{-5}$ | $3.587007 \times 10^{-8}$ |
| 1 acre = $40.47 \times 10^6$ | 4047 | $6.27 \times 10^6$ | 43,560 | 1 | $1.56 \times 10^{-3}$ |
| 1 mi² = $2.589988 \times 10^{10}$ | $2.589988 \times 10^6$ | $4.014490 \times 10^9$ | $2.78784 \times 10$ | 640 | 1 |

## Volume

| | ml | l | in.³ | ft³ | gal | acre·ft |
|---|---|---|---|---|---|---|
| 1 cm³ = 1 | 1 | $10^{-3}$ | 0.06102374 | $3.531467 \times 10^{-5}$ | $2.641721 \times 10^{-4}$ | $8.1 \times 10^{-10}$ |
| 1 l = 1000 | 1000 | 1 | 61.02374 | 0.03531467 | 0.2641721 | $8.1 \times 10^{-7}$ |
| 1 in.³ = 16.38706 | 16.38706 | 0.01638706 | 1 | $5.787037 \times 10^{-4}$ | $4.329004 \times 10^{-3}$ | $1.33 \times 10^{-8}$ |
| 1 ft³ = 28,316.85 | 28,316.85 | 28.31685 | 1728 | 1 | 7.480520 | $2.3 \times 10^{-5}$ |
| 1 gal (U.S.) = 3875.412 | 3875.412 | 3.785412 | 231 | 0.1336806 | 1 | $3.07 \times 10^{-6}$ |
| 1 acre·ft = $1.23 \times 10^{9}$ | $1.23 \times 10^{6}$ | $1.23 \times 10^{6}$ | $75.3 \times 10^{6}$ | 43,560 | 325,851 | 1 |

## Mass

| | g | kg | oz | lb | ton |
|---|---|---|---|---|---|
| 1 g = 1 | 1 | $10^{-3}$ | 0.03527396 | $2.204623 \times 10^{-3}$ | $1.102311 \times 10^{-6}$ |
| 1 kg = 1000 | 1000 | 1 | 35.27396 | 2.204623 | $1.102311 \times 10^{-3}$ |
| 1 oz (avdp) = 28.34952 | 28.34952 | 0.02834952 | 1 | 0.0625 | $5 \times 10^{-4}$ |
| 1 lb (avdp) = 453.5924 | 453.5924 | 0.4535924 | 16 | 1 | 0.0005 |
| 1 ton = 907,184.7 | 907,184.7 | 907.1847 | 32,000 | 2000 | 1 |

# SYSTEME INTERNATIONAL D'UNITÉS (SI)

The SI metric system is based on meter-kilogram-second units. The principal units applicable to water and wastewater engineering are listed in Tables A.2 and A.3. Derived SI units are consistent, since the conversion factor among various units is unity (e.g., 1 joule = 1 newton × 1 meter). Prefixes given in Table A.4 may be added to write large or small numbers and thus avoid the use of exponential values of 10. Groups of three digits, on either side of the decimal point, are separated by spaces. Tables A.5 and A.6 list common conversion factors from customary units to SI metric. [Reference: *Units of Expression for Wastewater Management,* Manual of Practice No. 6 (Washington, DC: Water Pollution Control Federation, 1982).]

**TABLE A.2**    BASIC SI UNITS

| Quantity | Unit | Symbol |
|----------|------|--------|
| Length | meter | m |
| Mass | kilogram | kg |
| Time | second | s |
| Thermodynamic temperature | Kelvin | K |
| Molecular weight | mole | mol |
| Plane angle | radian | rad |

**TABLE A.3**    DERIVED SI UNITS

| Quantity | Unit | Symbol | Formula |
|----------|------|--------|---------|
| Energy | joule | J | $N \cdot m$ |
| Force | newton | N | $kg \cdot m/s^2$ |
| Power | watt | W | $J/s$ |
| Pressure | pascal | Pa | $N/m^2$ |

**TABLE A.4**    MULTIPLES AND SUBMULTIPLES OF SI UNITS

| Multiplier | | Prefix | Symbol |
|-----------|---|--------|--------|
| 1 000 000 | $= 10^6$ | mega | M |
| 1 000 | $= 10^3$ | kilo | k |
| 0.001 | $= 10^{-3}$ | milli | m |
| 0.000 001 | $= 10^{-6}$ | micro | $\mu$ |

**TABLE A.5**  CONVERSION FACTORS FROM ENGLISH UNITS TO SI METRIC UNITS

| Customary units | | | Metric units | |
| --- | --- | --- | --- | --- |
| Description | Symbol<br>Multiply ... | Multiplier<br>by ... | Symbol<br>to obtain ... | Reciprocal |
| Acre | acre | 0.404 7 | ha | 2.471 |
| British thermal unit | Btu | 1.055 | kJ | 0.947 0 |
| British thermal units<br>  per cubic foot | Btu/ft$^3$ | 37.30 | J/l | 0.026 81 |
| British thermal units<br>  per pound | Btu/lb | 2.328 | kJ/kg | 0.429 5 |
| British thermal units<br>  per square foot<br>  per hour | Btu/ft$^2$/hr | 3.158 | J/m$^2$·s | 0.316 7 |
| Cubic foot | ft$^3$ | 0.028 32 | m$^3$ | 35.31 |
| Cubic foot | ft$^3$ | 28.32 | l | 0.035 31 |
| Cubic feet per minute | cfm | 0.471 9 | l/s | 2.119 |
| Cubic feet per minute<br>  per thousand<br>  cubic feet | cfm/1000 ft$^3$ | 0.016 67 | l/m$^3$·s | 60.00 |
| Cubic feet per second | cfs | 0.028 32 | m$^3$/s | 35.31 |
| Cubic feet per second<br>  per acre | cfs/acre | 0.069 98 | m$^3$/s·ha | 14.29 |
| Cubic inch | in.$^3$ | 0.016 39 | l | 61.01 |
| Cubic yard | yd$^3$ | 0.764 6 | m$^3$ | 1.308 |
| Fathom | f | 1.839 | m | 0.546 7 |
| Foot | ft | 0.304 8 | m | 3.281 |
| Feet per hour | ft/hr | 0.084 67 | mm/s | 11.81 |
| Feet per minute | fpm | 0.005 08 | m/s | 196.8 |
| Foot-pound | ft·lb | 1.356 | J | 0.737 5 |
| Gallon, U.S. | gal | 3.785 | l | 0.264 2 |
| Gallons per acre | gal/acre | 0.009 35 | m$^3$/ha | 106.9 |
| Gallons per day<br>  per linear foot | gpd/lin ft | 0.012 42 | m$^3$/m·d | 80.53 |
| Gallons per day<br>  per square foot | gpd/ft$^2$ | 0.040 74 | m$^3$/m$^2$·d | 24.54 |
| Gallons per minute | gpm | 0.063 08 | l/s | 15.85 |
| Grain | gr | 0.064 80 | g | 15.43 |
| Grains per gallon | gr/gal | 17.12 | mg/l | 0.058 41 |
| Horsepower | hp | 0.745 7 | kW | 1.341 |
| Horsepower-hour | hp·hr | 2.684 | MJ | 0.372 5 |
| Inch | in. | 25.4 | mm | 0.039 37 |
| Knot | knot | 1.852 | km/h | 0.540 0 |
| Knot | knot | 0.514 4 | m/s | 1.944 |
| Mile | mi | 1.609 | km | 0.621 5 |
| Miles per hour | mph | 1.609 | km/h | 0.621 5 |
| Million gallons | mil gal | 3 785.0 | m$^3$ | 0.000 264 2 |
| Million gallons per day | mgd | 43.81 | l/s | 0.022 82 |
| Million gallons per day | mgd | 0.043 81 | m$^3$/s | 22.82 |
| Ounce | oz | 28.35 | g | 0.035 27 |
| Pound (force) | lbf | 4.448 | N | 0.224 8 |
| Pound (mass) | lb | 0.453 6 | kg | 2.205 |
| Pounds per acre | lb/ac | 1.121 | kg/ha | 0.892 1 |

*(continued)*

**TABLE A.5** (*continued*)

| Description | Customary units Symbol Multiply ... | Multiplier by ... | Metric units Symbol to obtain ... | Reciprocal |
|---|---|---|---|---|
| Pounds per cubic foot | pcf | 16.02 | $kg/m^3$ | 0.062 42 |
| Pounds per foot | lb/ft | 1.488 | kg/m | 0.672 0 |
| Pounds per horsepower-hour | lb/hp·hr | 0.169 0 | mg/J | 5.918 |
| Pounds per square foot | $lb/ft^2$ | 0.047 88 | $kN/m^2$ | 20.89 |
| Pounds per square inch | psi | 6.895 | $kN/m^2$ | 0.145 0 |
| Pounds per thousand cubic feet per day | $lb/1000 \ ft^3/day$ | 0.016 02 | $kg/m^3 \cdot d$ | 62.43 |
| Square foot | $ft^2$ | 0.092 90 | $m^2$ | 10.76 |
| Square inch | $in.^2$ | 645.2 | $mm^2$ | 0.001 550 |
| Square mile | $mi^2$ | 2.590 | $km^2$ | 0.386 1 |
| Square yard | $yd^2$ | 0.836 1 | $m^2$ | 1.196 |
| Ton, short | ton | 0.907 2 | t | 1.102 |
| Yard | yd | 0.914 4 | m | 1.094 |

Acceleration of gravity $g = 32.174 \ ft/s^2 = 9.806 \ 65 \ m/s^2$.

**TABLE A.6** SELECTED ENGLISH–METRIC CONVERSION FACTORS

| | English unit | Multiplier | Metric unit |
|---|---|---|---|
| Mass | lb | 0.453 6 | kg |
| Length | ft | 0.304 8 | m |
| Area | ft$^2$ | 0.092 90 | m$^2$ |
| | acre | 4 047 | m$^2$ |
| | acre | 0.404 7 | ha |
| Volume | gal | 0.003 785 | m$^3$ |
| | gal | 3.785 | l |
| | ft$^3$ | 0.028 32 | m$^3$ |
| | ft$^3$ | 28.32 | l |
| Velocity | fpm | 0.005 08 | m/s |
| Flow | mgd | 3 785 | m$^3$/d |
| | gpm | 5.450 | m$^3$/d |
| | gpm | 0.063 09 | l/s |
| | cfs | 0.028 32 | m$^3$/s |
| BOD loading | lb/1000 ft$^3$/day | 16.02 | g/m$^3$·d |
| | lb/acre/day | 1.121 | kg/ha·d |
| Solids loading | lb/ft$^2$/day | 4.883 | kg/m$^2$·d |
| | lb/ft$^3$/day | 16.02 | kg/m$^3$·d |
| Hydraulic loading | gpd/ft$^2$ | 0.040 75 | m$^3$/m$^2$·d |
| | gpm/ft$^2$ | 0.679 0 | l/m$^2$·s |
| Concentration | lb/mil gal | 0.119 8 | mg/l |

| | Metric unit | Multiplier | English unit |
|---|---|---|---|
| Mass | kg | 2.205 | lb |
| Length | m | 3.281 | ft |
| Area | m$^2$ | 10.76 | ft$^2$ |
| | m$^2$ | 0.000 247 | acre |
| | ha | 2.471 | acre |
| Volume | m$^3$ | 264.2 | gal |
| | m$^3$ | 35.31 | ft$^3$ |
| | l | 0.264 2 | gal |
| | l | 0.035 31 | ft$^3$ |
| Velocity | m/s | 196.8 | fpm |
| Flow | m$^3$/d | 0.000 264 2 | mgd |
| | m$^3$/d | 0.183 5 | gpm |
| | m$^3$/s | 35.31 | cfs |
| | l/s | 15.85 | gpm |
| BOD loading | g/m$^3$·d | 0.062 43 | lb/1000 ft$^3$/day |
| | kg/ha·d | 0.892 1 | lb/acre/day |
| Solids loading | kg/m$^2$·d | 0.204 8 | lb/ft$^2$/day |
| | kg/m$^3$·d | 0.062 42 | lb/ft$^3$/day |
| Hydraulic loading | m$^3$/m$^2$·d | 24.54 | gal/ft$^2$/day |
| | l/m$^2$·s | 1.473 | gpm/ft$^2$ |
| Concentration | mg/l | 8.345 | lb/mil gal |

**TABLE A.7** RELATIVE ATOMIC WEIGHTS

| Name | Symbol | Atomic number | Atomic weight | Name | Symbol | Atomic number | Atomic weight |
|---|---|---|---|---|---|---|---|
| Actinium | Ac | 89 | — | Mercury | Hg | 80 | 200.59 |
| Aluminum | Al | 13 | 26.9815 | Molybdenum | Mo | 42 | 95.94 |
| Americium | Am | 95 | — | Neodymium | Nd | 60 | 144.24 |
| Antimony | Sb | 51 | 121.75 | Neon | Ne | 10 | 20.183 |
| Argon | Ar | 18 | 39.948 | Neptunium | Np | 93 | — |
| Arsenic | As | 33 | 74.9216 | Nickel | Ni | 28 | 58.71 |
| Astatine | At | 85 | — | Niobium | Nb | 41 | 92.906 |
| Barium | Ba | 56 | 137.34 | Nitrogen | N | 7 | 14.0067 |
| Berkelium | Bk | 97 | — | Nobelium | No | 102 | — |
| Beryllium | Be | 4 | 9.0122 | Osmium | Os | 76 | 190.2 |
| Bismuth | Bi | 83 | 208.980 | Oxygen | O | 8 | 15.9994 |
| Boron | B | 5 | 10.811 | Palladium | Pd | 46 | 106.4 |
| Bromine | Br | 35 | 79.904 | Phosphorus | P | 15 | 30.9738 |
| Cadmium | Cd | 48 | 112.40 | Platinum | Pt | 78 | 195.09 |
| Calcium | Ca | 20 | 40.08 | Plutonium | Pu | 94 | — |
| Californium | Cf | 98 | — | Polonium | Po | 84 | — |
| Carbon | C | 6 | 12.01115 | Potassium | K | 19 | 39.102 |
| Cerium | Ce | 58 | 140.12 | Praseodymium | Pr | 59 | 140.907 |
| Cesium | Cs | 55 | 132.905 | Promethium | Pm | 61 | — |
| Chlorine | Cl | 17 | 35.453 | Protactinium | Pa | 91 | — |
| Chromium | Cr | 24 | 51.996 | Radium | Ra | 88 | — |
| Cobalt | Co | 27 | 58.9332 | Radon | Rn | 86 | — |
| Copper | Cu | 29 | 63.546 | Rhenium | Re | 75 | 186.2 |
| Curium | Cm | 96 | — | Rhodium | Rh | 45 | 102.905 |

| Element | Symbol | Atomic Number | Atomic Weight | Element | Symbol | Atomic Number | Atomic Weight |
|---|---|---|---|---|---|---|---|
| Dysprosium | Dy | 66 | 162.50 | Rubidium | Rb | 37 | 85.47 |
| Einsteinium | Es | 99 | — | Ruthenium | Ru | 44 | 101.07 |
| Erbium | Er | 68 | 167.26 | Samarium | Sm | 62 | 150.35 |
| Europium | Eu | 63 | 151.96 | Scandium | Sc | 21 | 44.956 |
| Fermium | Fm | 100 | — | Selenium | Se | 34 | 78.96 |
| Fluorine | F | 9 | 18.9984 | Silicon | Si | 14 | 28.086 |
| Francium | Fr | 87 | — | Silver | Ag | 47 | 107.868 |
| Gadolinium | Gd | 64 | 157.25 | Sodium | Na | 11 | 22.9898 |
| Gallium | Ga | 31 | 69.72 | Strontium | Sr | 38 | 87.62 |
| Germanium | Ge | 32 | 72.59 | Sulfur | S | 16 | 32.064 |
| Gold | Au | 79 | 196.967 | Tantalum | Ta | 73 | 189.948 |
| Hafnium | Hf | 72 | 178.49 | Technetium | Tc | 43 | — |
| Helium | He | 2 | 4.0026 | Tellurium | Te | 52 | 127.60 |
| Holmium | Ho | 67 | 164.930 | Terbium | Tb | 65 | 158.924 |
| Hydrogen | H | 1 | 1.00797 | Thallium | Tl | 81 | 204.37 |
| Indium | In | 49 | 114.82 | Thorium | Th | 90 | 232.038 |
| Iodine | I | 53 | 126.9044 | Thulium | Tm | 69 | 168.934 |
| Iridium | Ir | 77 | 192.2 | Tin | Sn | 50 | 118.69 |
| Iron | Fe | 26 | 55.847 | Titanium | Ti | 22 | 47.90 |
| Krypton | Kr | 36 | 83.80 | Tungsten | W | 74 | 183.85 |
| Lanthanum | La | 57 | 138.91 | Uranium | U | 92 | 238.03 |
| Lead | Pb | 82 | 207.19 | Vanadium | V | 23 | 50.942 |
| Lithium | Li | 3 | 6.939 | Xenon | Xe | 54 | 131.30 |
| Lutetium | Lu | 71 | 174.97 | Ytterbium | Yb | 70 | 173.04 |
| Magnesium | Mg | 12 | 24.312 | Yttrium | Y | 39 | 88.905 |
| Manganese | Mn | 25 | 54.9380 | Zinc | Zn | 30 | 65.37 |
| Mendelevium | Md | 101 | — | Zirconium | Zr | 40 | 91.22 |

Source: "Report of the International Commission on Atomic Weights—1961." *J. Am. Chem. Soc.* 84 (1962): 4192.

**TABLE A.8** PROPERTIES OF WATER IN ENGLISH UNITS

| Temperature (°F) | Specific weight $\gamma$ (lb/ft³) | Mass density $\rho$ (lb·sec²/ft⁴) | Absolute viscosity $\mu$ ($\times 10^{-5}$ lb·sec/ft²) | Kinematic viscosity $\nu$ ($\times 10^{-5}$ ft²/sec) | Vapor pressure $p_v$ (psi) |
|---|---|---|---|---|---|
| 32 | 62.42 | 1.940 | 3.746 | 1.931 | 0.09 |
| 40 | 62.43 | 1.938 | 3.229 | 1.664 | 0.12 |
| 50 | 62.41 | 1.936 | 2.735 | 1.410 | 0.18 |
| 60 | 62.37 | 1.934 | 2.359 | 1.217 | 0.26 |
| 70 | 62.30 | 1.931 | 2.050 | 1.059 | 0.36 |
| 80 | 62.22 | 1.927 | 1.799 | 0.930 | 0.51 |
| 90 | 62.11 | 1.923 | 1.595 | 0.826 | 0.70 |
| 100 | 62.00 | 1.918 | 1.424 | 0.739 | 0.95 |
| 110 | 61.86 | 1.913 | 1.284 | 0.667 | 1.24 |
| 120 | 61.71 | 1.908 | 1.168 | 0.609 | 1.69 |
| 130 | 61.55 | 1.902 | 1.069 | 0.558 | 2.22 |
| 140 | 61.38 | 1.896 | 0.981 | 0.514 | 2.89 |
| 150 | 61.20 | 1.890 | 0.905 | 0.476 | 3.72 |
| 160 | 61.00 | 1.896 | 0.838 | 0.442 | 4.74 |
| 170 | 60.80 | 1.890 | 0.780 | 0.413 | 5.99 |
| 180 | 60.58 | 1.883 | 0.726 | 0.385 | 7.51 |
| 190 | 60.36 | 1.876 | 0.678 | 0.362 | 9.34 |
| 200 | 60.12 | 1.868 | 0.637 | 0.341 | 11.52 |
| 212 | 59.83 | 1.860 | 0.593 | 0.319 | 14.70 |

**TABLE A.9** PROPERTIES OF WATER IN SI METRIC UNITS

| Temperature (°C) | Specific weight $\gamma$ (kN/m³) | Mass density $\rho$ (kg/m³) | Absolute viscosity $\mu$ ($\times 10^{-3}$ kg/m·s) | Kinematic viscosity $\nu$ ($\times 10^{-6}$ m²/s) | Vapor pressure $p_v$ (kPa) |
|---|---|---|---|---|---|
| 0 | 9.805 | 999.8 | 1.781 | 1.785 | 0.61 |
| 5 | 9.807 | 1000.0 | 1.518 | 1.518 | 0.87 |
| 10 | 9.804 | 999.7 | 1.307 | 1.306 | 1.23 |
| 15 | 9.798 | 999.1 | 1.139 | 1.139 | 1.70 |
| 20 | 9.789 | 998.2 | 1.002 | 1.003 | 2.34 |
| 25 | 9.777 | 997.0 | 0.890 | 0.893 | 3.17 |
| 30 | 9.764 | 995.7 | 0.798 | 0.800 | 4.24 |
| 40 | 9.730 | 992.2 | 0.653 | 0.658 | 7.38 |
| 50 | 9.689 | 988.0 | 0.547 | 0.553 | 12.33 |
| 60 | 9.642 | 983.2 | 0.466 | 0.474 | 19.92 |
| 70 | 9.589 | 977.8 | 0.404 | 0.413 | 31.16 |
| 80 | 9.530 | 971.8 | 0.354 | 0.364 | 47.34 |
| 90 | 9.466 | 965.3 | 0.315 | 0.326 | 70.10 |
| 100 | 9.399 | 958.4 | 0.282 | 0.294 | 101.33 |

**TABLE A.10** SATURATION VALUES OF DISSOLVED OXYGEN IN WATER EXPOSED TO WATER-SATURATED AIR CONTAINING 20.90% OXYGEN UNDER A PRESSURE OF 760 mm Hg[a]

| Temperature (°C) | Chloride concentration in water (mg/l) | | | Difference per 100 mg chloride | Temperature (°C) | Vapor pressure (mm) |
|---|---|---|---|---|---|---|
| | 0 | 5000 | 10,000 | | | |
| | Dissolved oxygen (mg/l) | | | | | |
| 0 | 14.6 | 13.8 | 13.0 | 0.017 | 0 | 5 |
| 1 | 14.2 | 13.4 | 12.6 | 0.016 | 1 | 5 |
| 2 | 13.8 | 13.1 | 12.3 | 0.015 | 2 | 5 |
| 3 | 13.5 | 12.7 | 12.0 | 0.015 | 3 | 6 |
| 4 | 13.1 | 12.4 | 11.7 | 0.014 | 4 | 6 |
| 5 | 12.8 | 12.1 | 11.4 | 0.014 | 5 | 7 |
| 6 | 12.5 | 11.8 | 11.1 | 0.014 | 6 | 7 |
| 7 | 12.2 | 11.5 | 10.9 | 0.013 | 7 | 8 |
| 8 | 11.9 | 11.2 | 10.6 | 0.013 | 8 | 8 |
| 9 | 11.6 | 11.0 | 10.4 | 0.012 | 9 | 9 |
| 10 | 11.3 | 10.7 | 10.1 | 0.012 | 10 | 9 |
| 11 | 11.1 | 10.5 | 9.9 | 0.011 | 11 | 10 |
| 12 | 10.8 | 10.3 | 9.7 | 0.011 | 12 | 11 |
| 13 | 10.6 | 10.1 | 9.5 | 0.011 | 13 | 11 |
| 14 | 10.4 | 9.9 | 9.3 | 0.010 | 14 | 12 |
| 15 | 10.2 | 9.7 | 9.1 | 0.010 | 15 | 13 |
| 16 | 10.0 | 9.5 | 9.0 | 0.010 | 16 | 14 |
| 17 | 9.7 | 9.3 | 8.8 | 0.010 | 17 | 15 |
| 18 | 9.5 | 9.1 | 8.6 | 0.009 | 18 | 16 |
| 19 | 9.4 | 8.9 | 8.5 | 0.009 | 19 | 17 |
| 20 | 9.2 | 8.7 | 8.3 | 0.009 | 20 | 18 |
| 21 | 9.0 | 8.6 | 8.1 | 0.009 | 21 | 19 |
| 22 | 8.8 | 8.4 | 8.0 | 0.008 | 22 | 20 |
| 23 | 8.7 | 8.3 | 7.9 | 0.008 | 23 | 21 |
| 24 | 8.5 | 8.1 | 7.7 | 0.008 | 24 | 22 |
| 25 | 8.4 | 8.0 | 7.6 | 0.008 | 25 | 24 |
| 26 | 8.2 | 7.8 | 7.4 | 0.008 | 26 | 25 |
| 27 | 8.1 | 7.7 | 7.3 | 0.008 | 27 | 27 |
| 28 | 7.9 | 7.5 | 7.1 | 0.008 | 28 | 28 |
| 29 | 7.8 | 7.4 | 7.0 | 0.008 | 29 | 30 |
| 30 | 7.6 | 7.3 | 6.9 | 0.008 | 30 | 32 |

[a] Saturation at barometric pressures other than 760 mm (29.92 in), $C'_\sigma$ is related to the corresponding tabulated values $C_S$ by the equation

$$C'_\sigma = C_S \frac{P - p}{760 - p}$$

where  $C'_\sigma$ = solubility at barometric pressure $P$ and given temperature, mg/l
 $C_S$ = saturation at given temperature from table, mg/l
 $P$ = barometric pressure, mm
 $p$ = pressure of saturated water vapor at temperature of the water selected from table, mm

# Index